A Manual of
Organic Materia Medica
and Pharmacognosy

An Introduction to
the Study of the Vegetable Kingdom and
the Vegetable and Animal Drugs

MAVEN BOOKS

A Manual of
Organic Materia Medica and Pharmacognosy

An Introduction to
the Study of the Vegetable Kingdom and
the Vegetable and Animal Drugs

LUCIUS E. SAYRE, B.S., Ph.M.

Dean of the School of Pharmacy;
Professor of Materia Medica in the University of KANSAS;
Member of the Committee of Revision of
the United States Pharmacopoeia;
Director of Drug Laboratory for State of Kansas

MAVEN BOOKS

Chennai   Trichy   Tirunelveli   New Delhi

MAVEN BOOKS

An Imprint of **MJP Publishers**

ISBN 978-93-87867-73-4   **MAVEN Books**

All rights reserved   No. 44, Nallathambi Street,
Printed and bound in India   Triplicane, Chennai 600 005

MJP   778   © Publishers, 2020

Publisher :   **C. Janarthanan**

Publisher's Note

The legacy of a country is in its varied cultural heritage, historical literature, developments in the field of economy and science. The top nations in the world are competing in the field of science, economy and literature. This vast legacy has to be conserved and documented so that it can be bestowed to the future generation. The knowledge of this legacy is slowly getting perished in the present generation due to lack of documentation.

Keeping this in mind, the concern with retrospective acquiring of rare books has been accented recently by the burgeoning reprint industry. MAVEN Books is gratified to retrieve the rare collections with a view to bring back those books that were landmarks in their time.

In this effort, a series of rare books would be republished under the banner, "MAVEN Books". The books in the reprint series have been carefully selected for their contemporary usefulness as well as their historical importance within the intellectual. We reconstruct the book with slight enhancements made for better presentation, without affecting the contents of the original edition.

Most of the works selected for republishing covers a huge range of subjects, from history to anthropology. We believe this reprint edition will be a service to the numerous researchers and practitioners active in this fascinating field. We allow readers to experience the wonder of peering into a scholarly work of the highest order and seminal significance.

MAVEN Books

PREFACE TO THE FOURTH EDITION

The Ninth Revision of the United States Pharmacopœia, as in no previous edition, makes it important, and even necessary, that all works of a pharmaceutical character be revised.

The last revision of the Pharmacopœia has required, on the part of the revisers, very exceptional work directed toward the subject of standards; and inasmuch as the United States Pharmacopœia, as well as the National Formulary, is mentioned in the statute, known as the Food and Drugs Law, this revision has become of greatest importance.

Recognizing this, great pains have been taken in the revision of the present edition, that the standards, whenever mentioned, shall conform to the legal standard above referred to.

Many changes have been made necessary by the fact that the U.S.P. IX has deleted fifty-three vegetable drugs and has added, or raised to official recognition, but four of well known drugs.

Among the conspicuous changes in U.S.P. IX, is the adoption of "Mil" (singular), "Mils" (plural), for cubic centimeter (cc.). This coined word, Mil—for Milliliter, is more accurate than cubic centimeter, (cc.) for the thousandth of a liter, which the cubic centimeter was intended to express. Throughout this present edition "mil" and "mils" have been used, replacing the less accurate "cc."

The Families of plants yielding organic drugs have been rearranged in the present volume. The order of arrangement adopted is that which is followed by all botanists of any note at the present time, commencing with the Algæ, Fungi, and other cryptogamous growths, the order and sequence of such authors as Engler and Prantl have been practically followed. This has required an entire transposition of the natural orders of the former edition.

The Chapter on Inorganic Chemicals has been enlarged to meet the demand of many students. Added to this is a brief Chapter on Therapeutic Action, which is intended as a suggestion to students of how to expand their knowledge in this direction by reference to other works.

The chapters relating to histological study of plant tissues have been entirely omitted in the present edition in order to economize space for new material, and, secondly, because Professor Stevens, formerly associated in this work, has published an entirely satisfactory volume for class-room work and covered the ground more completely in his "Plant Anatomy."

A Chapter on Serotherapy has been incorporated which, in treatment, while it is concise, it is hoped will meet the present demand of students of

Materia Medica, who first must have studied the elements of this very extensive subject.

The author desires to make special mention of valuable service rendered by his associate, Mr. Chas. M. Sterling, who has revised that portion of the work included in the various chapters of Part IV. The author regrets that he has been obliged to reduce rather than lengthen many articles in Materia Medica in order that the present volume should not be unduly expanded.　　　　　　　　　　　　　　　　　　L. E. S.

PREFACE TO THE FIRST EDITION

The present volume is, in a slight degree, a revision of a work written by the author in 1879, entitled "Organic Materia Medica and Pharmacal Botany." This work has been out of print a number of years, and until recently the author has had no time to rewrite it in such a manner as seemed necessary to bring it up to the present standard; it has also been deemed advisable to change completely the model of the former work. The task now accomplished presents not so much a revision, as a new treatise.

Two methods of classification of drugs are here brought into use—a classification according to physical characteristics, and a classification according to botanical relationships—both of which are, though occupying separate divisions of the book, so brought together by a system of numbering that the place of the drug in each of the classes is at once apparent. The author would here suggest that those who make use of the work in connection with a cabinet of specimens, should have the containers in the cabinet numbered to accord with numbers in the book, in order that students may readily find specimens for identification and study.

It is perhaps needless to state that the nomenclature and general character of the text is made to conform with the present standard—The United States Pharmacopœia; but the capitalization of specific names derived from proper nouns has been discarded, in accordance with present botanical practice. The descriptive heading of each of the official drugs has been in most cases given in the pharmacopœial language. The unofficial drugs are distinguished in the text by the use of a different type and by a different setting of the article from that which treats of the official drugs. In this connection the author desires to give credit to Mr. George S. Davis, who has aided in the work by placing at the author's disposal most excellent material regarding rare unofficial drugs, and the use of material from his publication, credited under Bibliography.

The scope of the work, it will be seen, embraces not only the official drugs of the vegetable and animal kingdoms, but a vast variety of unofficial drugs, some of which are of rare occurrence in the market. These have been included because of the greater field this inclusion gives for pharmacal and botanical study; the greater variety of forms presented to the student of pharmacognosy, the wider will be his range of observation. It is hoped that in the 624 drugs mentioned, the student or instructor will be able to make a selection which will be ample to supply material to illustrate the principles of the subject under consideration. In a work of this size an exhaustive treatment of this number of drugs could not be given, but by a brief mention of them material for study is indicated. It may be mentioned in this connection that wherever metric measurements are given, these are stated in millimeters; this has been deemed advisable for the purpose of comparison.

The illustrations included in Part I are taken mainly from Bentley's "Manual of Botany," to the author of which our thanks are due. An exception, however, is found in the drawings of the starches, which were prepared from original specimens. The remaining illustrations, with the exception of those in the Chapter on Animal Drugs, have been prepared under the direction of C. E. McClung, Ph. G., a graduate of the Kansas State University School of Pharmacy, class of '92. All the drawings of the cross-sections are drawn directly from sections prepared by him, the cell contents being first removed by the method described in Appendix C. It has been our aim to present the elements of each drug in their true proportions. As often as possible the cells in their exact shape and relative size have been drawn, and in no case has meaningless shading been employed. For some of the drawings of the medicinal plants credit is given below in the Bibliography. The illustrator has kindly furnished a Chapter on Pharmacal Microscopy, which will be found in Appendix C.

The author is much indebted to Professor Vernon Kellogg for information concerning animal drugs used in pharmacy; also for Appendix B, in which he treats of insects attacking drugs. The drawings to illustrate the material furnished by Professor Kellogg are hereby credited to Miss Mary Wellman, artist.

For aid in the preparation of the text in Part I our thanks are due to Mr. A. O. Garrett, who, in his university course, has made botany a special study.

Appendix B, upon the synthetic remedies, is the work of Mr. F. B. Dains, who has made a specialty of organic chemistry and was instructor in this subject in the University of Kansas during the year 1894. In this section the new spelling of chemicals has been adopted only in a few cases.

To Dr. S. W. Williston, Professor of Physiology and Anatomy, who has aided in the condensed description of therapeutic action; to Mr. O. H. Parker and Mr. William Clark, members of the Senior Class of '94, who assisted in the study of characteristics from crude specimens of drugs in the open market; to Mr. W. O. Strother, of the same class, who supplied a few drawings of cross-sections; and to Mr. W. F. Newton, of the Junior Class, who materially aided not only in the study of drug characteristics, but also in arranging the material, our thanks are due.

L. E. S.

BIBLIOGRAPHY

The following works have been consulted, and credit is due to the authors of the same:

United States Pharmacopœia.

United States Dispensatory.

National Formulary.

National Dispensatory.

"Reports of the American Pharmaceutical Association."

Friederich A. Flückiger and Daniel Hanbury: "Pharmacographia," London, 1874.

John Harley: "Royle's Materia Medica," sixth edition, London, 1876.

Parke, Davis & Co.: "Organic Materia Medica," Detroit, 1888.

Robert Bentley: "A Manual of Botany," London, 1887.

Samuel O. L. Potter: "Materia Medica, Pharmacy, and Therapeutics," eleventh edition, 1909, Philadelphia.

Asa Gray: "New Manual of the Botany of the Northern United States," seventh edition, N. Y.

William Dymock, C. J. H. Warden, David Hooper: "Pharmacographia Indica," London and Bombay, 1893.

Bernard Fischer: "Die Neueren Arzneimittel" (sixth edition), Berlin.

A. Brestowski: "Die Neueren und Neuesten Arzneimittel," Leipzig.

David M. R. Culbreth: "Materia Medica and Pharmacology," Philadelphia.

Walter A. Bastedo: "Materia Medica, Pharmacology, Therapeutics, Prescription Writing," 1913, Philadelphia.

William Chase Stevens: "Plant Anatomy," Philadelphia, 1907.

Walter E. Dixon: "Manual of Pharmacology," fourth edition, 1915, London.

Henry H. Rusby: "Structural Botany," 1911, Philadelphia and New York.

D. S. Rabow: "Die Neuesten Arzneimittel und Spezialitäten," 1910, Strassburg, Ger.

New and Non-official Remedies, Amer. Med. Asso., 1916, Chicago.

Burt E. Nelson: "Analysis of Drugs and Medicines."

Tschirch und Oesterle: "Anatomischer Atlas der Pharmakognosie und Nahrungs-mittlekunde."

Greenish and Collin: "Anatomical Atlas of Vegetable Powders."

Henry Kraemer: "Applied and Scientific Pharmacognosy."

A. E. Vogl: "Die Wichtigsten Vegetabilischen Nahrungs und Genussmittel."

Ludwig Koch: "Einfuhrung in der Mikroskopische Analyse der Drogenpulver."

C. A. Winslow: "Elements of Applied Microscopy."

Henry George Greenish: "Food and Drugs."

Smith Ely Jelliffe: "Introduction to Pharmacognosy."

Karsten and Altmans: "Lehenbuch der Pharmakognosie."

A. L. Winton: "Microscopy of Vegetable Foods."

Ludwig Koch: "Mikroskopische Analyse der Drogen pulver."

W. Griffiths: "The Principal Starches Used as Food."

Also, the following works, from which, in addition, illustrations have been taken:

Kohler's "Medizinal-Pflanzen," Hrs'g von G. Pabst (2 vols.), 1887–1890.

Robert Bentley and Henry Trimen: "Medicinal Plants" (4 vols.), London, 1880.

A. B. Frank: "Lehrbuch der Botanik," Leipzig, 1892.

Emm. Le Maut and Th. Decaisne: "Traité Général de Botanique," Paris, 1876.

Anton Kerner von Marilau: "Pflanzenleben" (2 vols.), Leipzig, 1887.

J. Godfrin, De Nancy, et Ch. Noël: "Atlas manuel de l'histologie de drogues simples," Paris, 1887.

L. Trabut: "Précis de Botanique Médicinale," 1891.

William Woodville: "Medical Botany" (second edition), London, 1810.

Eduard Strasburger: "Das Botanische Practicum," Jena, 1897.

Naegeli and Schwendener: "The Microscope," Swan, Sonneschein & Co., London, 1892.

Eduard Strasburger: "Bau und Verrichtungen der Leitungsbahnen in den Pflanzen," Jena, 1891.

Sachs: "Text Book of Botany."

De Bary: "Comparative Anatomy of Phanerogams and Ferns," Oxford, 1884.

Haberlandt: "Physiologische Pflanzenanatomie," Leipzig, 1896.

Pfeffer: "Pflanzenphysiologie," Leipzig, 1897.

Meyer: "Untersuchungen über die Stärkekörner," Jena, 1895.

Thomas and Dudley: "Laboratory Manual of Plant Histology," Crawfordsville, Ind., 1894.

Vines: "Student's Text Book of Botany," London and New York, 1895.

Strasburger: "Cytologische Studien," Berlin, 1897.

Beilstein: "Handbuch der Organischen Chemie," Hamburg und Leipzig, 1898.

Strasburger, Noll, Schenck, Schimper: "Lehrbuch der Botanik," Jena, 1894.

Zimmerman, A.: "Botanical Microtechnique," New York, 1893.

Tschirch and Oesterle: "Anatomischer Atlas der Pharmakognosie und Nahrungs-mittelkunde," Leipzig, 1900.

Dr. A. E. Vogl: "Die Wichtigsten Vegetabilishen Nahrungs und Genussmittel," Berlin and Wien, 1899

Dr. Albert Schneider: "Powdered Vegetable Drugs," San Francisco, Cal.

Dr. Henry Kraemer: "Botany and Pharmacognosy," Philadelphia.

Henry George Greenish, F. I. C., F. L. S.: "Foods and Drugs," Philadelphia, 1903.

Dr. George Karsten: "Lehrbuch der Pharmakognosie des Pflanzenreiches," Jena, 1903.

Dr. Arthur Meyer: "Die Grundlagen und die Methoden für die Mikroskopische Unterendung von Pflanzenpulver," Jena, 1901.

Dr. Ludwig Koch: "Die Mikroskopische Analyse der Droguenpulver," Berlin, 1904.

Greenish and Collin: "An Anatomical Atlas of Vegetable Powders," London, 1904.

White and Humphrey: "Pharmacopedia," London, 1904.

TABLE OF CONTENTS

PART I

A STUDY OF DRUGS

PART II

DRUG DESCRIPTION

PART III

INSECTS INJURIOUS TO DRUGS.

PART IV

POWDERED DRUGS

CHAPTER I

CHAPTER II

CHAPTER III

LIST OF ILLUSTRATIONS

ORGANIC MATERIA MEDICA

AND

PHARMACOGNOSY

PART I
A STUDY OF DRUGS

CLASSIFICATION

Drugs may be arranged in several different ways, to suit the aim and convenience of the student. The prominent systems of classification in common use are as follows:

I. **Therapeutical.**—This system of classification is especially valuable to the student of medicine. Here the physiological action and therapeutical application are made most prominent.

II. **Chemical.**—Classification of organic drugs is not infrequently based upon the character of the constituents. In this way alkaloidal drugs, glucosidal drugs, drugs containing volatile oil, etc., form the subgroups. Other subgroups of chemical classification are:

Inorganic Chemicals.—To the pharmacist the chemical action, the crystalline form, the solubility, and other physical properties are of especial value. For mineral substances, therefore, he adopts the classification of the chemist. Some therapeutists, seeing a certain relation between therapeutical action and chemical constitution, adopt the same method of grouping also for these mineral substances.

Synthetical Remedies.—This class of remedial agents is most difficult to classify in a manner consistent with science, partly because our materia medica is becoming overloaded with proprietary combinations and mixtures of synthetic medicinal products with various adjuvants to modify their action. These latter have oftentimes certain unscientific names, which give little or no idea of their composition.

III. **Physical.**—According to this method, drugs having allied physical properties are brought together. Roots, leaves, flowers, fruits, and seeds form the principal divisions. Under this head two different arrangements are present in this volume: (*a*) Classification into subgroups based upon such prominent features as odor, taste, etc. By this means the aromatic, bitter, acrid, sweet, and mucilaginous drugs are brought together. (*b*) Classification into subgroups based upon structural characteristics. Here drugs having similar structure are found associated. In the table having this arrangement the official drugs only are found. Appended to each there is a physical description in the fewest possible terms—such prominent terms as are used in describing the physical and structural characteristics.

Each drug has a number, so that a ready reference to the same drug in the body of the work is made easy. Here a fuller description is found.

Instructors in pharmacognosy who use this book are recommended to employ this conspectus and to have the students use these numbers. When labeling the drug (or its container) for class work, these numbers should be employed. The experience of the author in teaching the subject under consideration has been most favorable to this method. By the use of the numbers at first, the student quickly grows to learn, not only the drug, but the place in the system to which it belongs. The subject grows in interest until he is able to recognize the drug and to properly classify it.

IV. Botanical.—By this arrangement drugs belonging to the same natural order are brought together. In subdividing these orders botanical relationship is emphasized to as large an extent as is practicable in dealing with drugs from a pharmaceutical standpoint. From the point of view of the scientist this is the ideal system. This method has been adopted in the body of this work.

Geographical.—Drugs are rarely classified according to the locality of their occurrence. It is, however, instructive to the student to refer individual, or classes of drugs, to their locality. Drugs of ancient times were obtained chiefly from Asia. Many of these have survived, and are official to-day; notably aloes, myrrh, etc. With the discovery of the new world many important drugs were made accessible. Geographical classification is therefore of interest from many points of view. The presentation of this subject is facilitated by outline maps with the drugs indicated in their natural localities. As an example of such a map, see Cinchona.

Alphabetical Arrangement.—In all the standard books of reference, such as the "Pharmacopœia" and the "Dispensatories," a strictly alphabetical arrangement is followed, no attention being paid to systems of classification. The arrangement is made wholly subservient to quick and ready reference.

In the following order four classifications will be presented: 1. A synopsis of therapeutical agents. 2. Chemical agents. 3. Classification of organic drugs, as indicated under (*a*) and (*b*). 4. Botanical arrangement, where drugs will be treated at some length.

TABLE OF THERAPEUTICAL AGENTS

I. INTERNAL REMEDIES

A. *Affecting Nutrition:*

              Hæmatics (Blood Tonics).
              Alkalies.
              Acids.
              Digestants.
              Antipyretics.
              Alteratives.

B. *Affecting the Nervous and Muscular Systems:*
        (*a*) The Brain—
                Cerebral Excitants.
                Cerebral Depressants.
                Narcotics.
                Hypnotics.
                Analgesics.
                Anæsthetics.
        (*b*) The Spinal Cord—
                Motor Excitants.
                Motor Depressants.
        (*c*) Nerve Centers and Ganglionic System—
                Antispasmodics.
                Tonics.
                Antiperiodics.
        (*d*) Heart and Circulatory System
                Cardiac Stimulants.
                Cardiac Sedatives.
                Vascular Stimulants.
                Vascular Sedatives
                    Vasoconstrictors.
                    Vasodilators.
        (*e*) Excretories—
                Diuretics.
                Renal Depressants.
                Vesical Tonics and Sedatives.
                Urinary Sedatives.
                Diaphoretics and Sudorifics.
                Anhidrotics.
                Antilithics.

C. *Affecting Special Organs—Partly through the Nervous System:*
        (*a*) Organs of Respiration—
                Expectorants.
                Pulmonary Sedatives.
                Errhines.
                Sternutatories.
        (*b*) Alimentary Canal—
                Sialagogues.
                Emetics.
                Purgatives.
                Astringents.
                Stomachics.
        (*c*) The Liver—
                Hepatic Stimulants.
                Cholagogues. ·
                Hepatic Depressants.
        (*d*) Generative System—
                Ecbolics or Oxytocics.
                Emmenagogues.
                Aphrodisiacs.
                Anaphrodisiacs.
        (*e*) Eyes (Ciliary Muscle)—
                Mydriatics.
                Myotics.

II. EXTERNAL REMEDIES

A. *Irritants:*
                Rubefacients.
                Epispastics.
                Pustulants.
                Escharotics.

B. *Local Sedatives:*
                · Demulcents.
                Emollients.

III. AGENTS WHICH ACT UPON ORGANISMS WHICH INFEST THE HUMAN BODY

> Antiseptics.
> Disinfectants.
> Antizymotics.
> Anthelmintics.
> Antiparasitics.
> Antiperiodics.

THERAPEUTICAL AGENTS DEFINED

HÆMATICS restore the quality of the blood to normal condition. They exert a direct influence on the composition of the blood: *e.g.*, preparations of iron, of manganese, cod-liver oil, etc.

ALKALIES act, in the concentrated form, as caustics (escharotics), but when diluted, as antacids. Dilute alkalies, if given before meals, however, will stimulate the production of the acid gastric juice. The carbonates of potassa and soda and the bicarbonates, also preparations of the alkaline earths, such as lime-water and mixtures of magnesium carbonate, are good examples. Some of the salts of the alkalies have a remote antacid effect, becoming decomposed in the blood and excreted in the urine, which they render less acid.

ACIDS.—These have an action opposite to that of the alkalies. When much diluted, they are administered for the purpose of checking hyper-acidity of the stomach, by stimulating the production of the alkaline pancreatic juice and checking the acid gastric juice. Examples: Dilute hydrochloric acid, phosphoric acid.

DIGESTANTS.—Agents which effect solution (digestion) of food in the alimentary canal. Examples: Pepsin, pancreatin, trypsin, papain, etc.

ANTIPYRETICS.—Agents which reduce the temperature of the body, either by reducing the circulation or diminishing tissue change, or metabolism, or favoring the loss of heat through radiation, conduction, etc. Examples: Quinine, aconite, antipyrine, antimony, etc.

ALTERATIVES.—A term used to designate a class of agents which alter the course of morbid conditions, modifying the nutritive processes while promoting waste, by stimulating secretion, absorption, and the elimination of morbid deposits; especially used in the chronic diseases of the skin. Employed in the treatment of phthisis, syphilis, gout, neuralgia, asthma, etc. Examples: Arsenious acid, mercury, iodine and the iodides, sarsaparilla, guaiac, colchicum, stillingia, etc.

CEREBRAL EXCITANTS.—Agents which increase the functional activity of the cerebrum, without causing any subsequent depression of brain function. Examples: Camphor, valerian, caffeine, cannabis (in small doses), etc.

CEREBRAL DEPRESSANTS have an opposite effect to the preceding, lessening brain activity. Some of the drugs of this class are employed as hypnotics or as analgesics.

NARCOTICS.—Agents which lessen the sensibility to pain and cause sleep. A narcotic will abolish pain, while an anodyne will frequently merely overcome wakefulness. Examples: Opium, cannabis indica, belladonna, humulus, etc.

HYPNOTICS.—Agents which induce sleep and will often abolish pain and cause neither deliriant nor narcotic effects. Examples: Chloral, sulphonal, trional, the bromides, etc.

ANALGESICS.—Agents which relieve pain by their effect upon the sensory centers; the term is synonymous with anodynes. The general anodynes, which taken internally, affect the whole organism; local anodynes affect the part to which they are applied. Examples: Opium, belladonna, hyoscyamus, aconite, antipyrine, acetanilid, aspirin, chloral hydrate, etc.

ANÆSTHETICS.—Agents which suspend consciousness and temporarily destroy sensation. The local anæsthetics affect only the part to which they are applied. Examples: Ether, chloroform, nitrous oxide, etc. Local anæsthetics: Cocaine, carbolic acid, ether spray, etc.

MOTOR EXCITANTS.—Agents which increase the functional activity of the spinal cord and the motor apparatus, invigorating the action of the heart and lungs. Examples: Nux vomica, strychnine, etc.

MOTOR DEPRESSANTS have an opposite effect to the motor excitants, lowering the functional activity of the spinal cord and motor apparatus. Examples: Alcohol, opium, aconite, conium, belladonna, etc.

ANTISPASMODICS.—Agents acting on the nervous system in various ways. They prevent or allay irregular action or spasm of voluntary and involuntary muscles. This is accomplished frequently by a sedative influence upon the nerve centers, while a few others exert their influence by stimulating the nerve centers employed to relieve spasms. Examples: Alcohol, ether, valerian, camphor, asafœtida, musk, the bromides, hydrocyanic acid, etc.

TONICS.—Agents which increase the vigor and tone of the system by improving the appetite, favoring digestion and assimilation, and adding strength to the circulatory system. Examples: Gentian, columbo, quinine, etc.

ANTIPERIODICS.—Agents which prevent or check the return of diseases which recur periodically, possibly by a toxic action upon the microbes in the blood, which are supposed to cause the disease; but little is known of their mode of action. The typical antiperiodic, quinine, has, however, a decided effect upon the heart and brain, as well as other parts of the nervous system.

CARDIAC STIMULANTS, as the name implies, are agents which increase the heart's action, the force and frequency of the pulse. Examples: Ether, alcohol, atropine, sparteine, nitroglycerine, etc.

CARDIAC SEDATIVES allay and control palpitation and overaction of the heart. Examples: Aconite, veratrum viride, digitalis, antimony, etc.

VASCULAR STIMULANTS.—Agents which dilate the peripheral vessels and increase the peripheral circulation.　Members of this class also strengthen the heart's action, and are advantageously employed in debilitated conditions of the central organs of the circulation.　Examples: Alcohol, preparations of ammonia, caffeine, digitalis, strophanthus, epinefrin, etc.

VASCULAR SEDATIVES.—Agents which lessen the capillary circulation and raise the blood pressure by stimulating the vasomotor center or its mechanism and the walls of the vessels.　Examples: Ergot, digitalis, opium, salts of iron, etc.

DIURETICS.—Agents which increase the secretion of urine, acting either directly upon the secreting cells of the kidneys or by raising the general or local arterial tension.　Employed in acute congestion and inflammation of the kidneys and in dropsies.　Examples: Squill, scoparius, triticum, and organic salts of the alkalies.

RENAL DEPRESSANTS.—Agents which lower the activity of the renal cells, thereby lessening the urinary secretion.　Examples: Morphine, quinine, ergot, etc.

VESICAL TONICS AND SEDATIVES.—Agents acting upon the bladder, in the one case increasing the tone of the muscular fibers and in the other lessening the irritability of that organ.　Examples: Tonics—strychnine, cantharis, belladonna, etc.; sedatives—opium, buchu, uva ursi, pareira, etc.

RENAL SEDATIVES.—Agents which exert a sedative action upon the whole urinary tract.　Examples: Copaiba, cubebs, etc.

DIAPHORETICS AND SUDORIFICS.—Agents which increase the action of the skin and promote perspiration.　Examples: Dover's powder, jaborandi, camphor, sweet spirits of niter, etc.

ANHIDROTICS.—Agents which check perspiration.　Examples: Acid camphoric, atropine, zinc salts, acids, alum, etc.

ANTILITHICS.—Agents used to prevent the formation of insoluble concretions or to dissolve concretions when formed in the ducts.　Examples: Salts of lithia, potassium, benzoic acid, etc.

EXPECTORANTS.—Agents which are employed to facilitate the expulsion of bronchial secretions and to modify the character of these when abnormal.　Examples: Ammonium chloride, the aromatic balsams, squill, licorice, senega, etc.

PULMONARY SEDATIVES.—Agents which allay the irritability of the respiratory center and the nerves of the lungs and bronchial tubes.　Examples: Belladonna, opium, hyoscyamus, hydrocyanic acid, etc.

ERRHINES AND STERNUTATORIES.—The latter are agents which affect locally the nasal mucous membrane, producing sneezing; the former produce an increase of nasal secretion and discharge.　They also indirectly stimulate the vasomotor centers and at the same time excite the respiratory centers.　Examples: Ipecacuanha, sanguinaria, veratrine, etc.

SIALAGOGUES.—Agents which promote the secretion and flow of saliva from the salivary glands. Examples: Pyrethrum, mezereum, the mercurials and antimonials, etc.

EMETICS.—Agents which cause vomiting, acting directly upon the nerves of the stomach or acting through the blood upon the vomiting center, or by reflex irritation of the vomiting center. Examples: Mustard, zinc sulphate, apomorphine, ipecacuanha, tartar emetic, etc.

PURGATIVES produce evacuation of the contents of the intestinal canal by increasing secretion along the tract, by exciting peristaltic action, etc. Examples: Podophyllum, colocynth, jalap, croton oil, magnesium sulphate, etc.

ASTRINGENTS.—Agents which produce contraction of muscular fiber, which coagulate albumen and lessen secretion from mucous membranes, arresting discharges. Examples: Tannic and gallic acids, alum, lead acetate, persulphate of iron, etc.

STOMACHICS.—Agents which increase the appetite and promote gastric digestion. They also check fermentation and dispel accumulation of flatus. Examples: Peppermint, cardamom, ginger, capsicum, etc.

HEPATIC STIMULANTS (Cholagogues).—Agents which excite the liver and increase the functional activity of that organ so that the amount of bile is augmented, etc. Hepatic stimulants increase the activity of the liver-cells, while cholagogues remove the bile from the duodenum. Examples: Podophyllum, aloes, jalap, colocynth, mercurous chloride, etc.

HEPATIC DEPRESSANTS.—Agents which reduce the functional activity of the liver, having the opposite effect of the foregoing, that of diminishing the formation of the bile, urea, and glycogen. Examples: Opium, quinine, arsenic, antimony, etc.

ECBOLICS, OR OXYTOCICS.—Agents which stimulate the pregnant uterus and produce contraction of that organ, either by direct irritation of the muscles of the womb, or indirectly by affecting the uterine center of the cord. Examples: Ergot, cotton-root bark, savin, cimicifuga, etc.

EMMENAGOGUES.—Agents which stimulate the uterine muscular fibers and restore the normal menstrual function. Examples: Ergot, apiol, iron, etc.

APHRODISIACS.—Agents used to excite the function of the genital organs when they are morbidly depressed. Examples: Phosphorus, zinc phosphide, salts of iron, gold, or arsenic, etc.

ANAPHRODISIACS.—Agents which diminish the sexual desire. Examples: The bromides, camphor, etc.

MYDRIATICS.—Agents which cause dilatation of the pupil; used to temporarily destroy accommodation by causing paralysis of the ciliary muscle. Examples: Atropine and homatropine.

MYOTICS.—Agents acting in a manner contrary to that of the above, producing contraction of the pupil by stimulating the circular muscular fibers of the iris and at the same time contracting the ciliary muscle. Examples: Pilocarpine, eserine, etc.

IRRITANTS.—Agents which are applied locally to the skin to produce certain effects, as rubefacients (simply reddening the skin); epispastics (blistering); pustulants (causing blebs in which is found pus); escharotics, or caustics (actually destroying the tissue). Examples: Mustard (rubefacient); cantharides (epispastic); croton oil (pustulant); caustic potassa, carbolic acid, and strong mineral acids (escharotics). .

LOCAL SEDATIVES.—Agents which diminish irritation in the part to which applied, relieving local inflammation. Examples: Acetate of lead, opium, belladonna, etc.

DEMULCENTS.—Bland remedies used to allay and mechanically protect inflamed surfaces. They are used also internally for this purpose, as in acute inflammation of the alimentary canal. Examples: Mucilages of acacia, flaxseed, Iceland and Irish moss, elm, etc.

EMOLLIENTS resemble the above; are used externally to soften and soothe the irritated and abraded skin. Examples: Lard, olive oil, cacao-butter, etc.

ANTISEPTICS.—Agents which arrest putrefaction, either by preventing the growth of micro-organisms causing putrefactive decomposition or by destroying these micro-organisms. Examples: Carbolic acid, corrosive sublimate, etc.

DISINFECTANTS.—Some authorities limit the use of this term to those agents which destroy the micro-organisms. The terms antiseptic and disinfectant are frequently used interchangeably. Examples: Corrosive sublimate, carbolic acid, iodoform, zinc chloride, eucalyptol, etc.

ANTIZYMOTICS.—A term applied to agents which arrest fermentation. Examples: See above.

ANTHELMINTICS.—Agents which destroy such parasitic worms as infest the alimentary canal. Tæniafuges destroy tape-worms; vermifuges expel these intestinal parasites. Examples: Santonin, spigelia, chenopodium, etc. Tæniafuges: Filix mas, pelletierin, cusso, etc.

ANTIPARASITICS.—Agents which destroy those parasites which infest the human body externally. Examples: Mercurial preparations, chrysarobin, carbolic acid, cocculus, etc.

ANTIPERIODICS.—See above.

Thus far we have only very briefly called attention to therapeutical and physiological action of drugs, giving but a few examples. We will temporarily leave the further consideration of this, and for the time refer to the therapeutical agents themselves.

INORGANIC THERAPEUTICAL AGENTS

(Titles printed in capitals, U.S.P.; all others without asterisk are unofficial; N.F. Salts indicated by *.)

GROUP I.—THE ALKALINE METAL.—POTASSIUM, SODIUM AMMONIUM AND LITHIUM

THE OFFICIAL U.S.P., N.F. AND UNOFFICIAL POTASSIUM SALTS

POTASSII ACETAS.—Potassium Acetate, $KC_2H_3O_2$. White crystalline powder, odorless, warm saline taste, deliquescent, soluble in water.
Actions: Alterative, diuretic, refrigerant.
Dose: 2 Gm. (30 gr.) (in solution).
Preparations: Elixir Potassii Acetatis, N.F.
Elixir Potassii Acetatis et Juniperi, N.F.

POTASSII BICARBONAS.—Potassium Bicarbonate, $KHCO_3$. Transparent prisms, odorless or granular powder, having a saline and slightly alkaline taste, soluble in 3 parts water.
Action: Antacid, diuretic.
Dose: 2 Gm. (30 gr.).
Preparations: Liquor Potassii Citratis, U.S.P.
Liquor Antisepticus Alkalinus, N.F.

POTASSII BITARTRAS.—Potassium Bitartrate, Cream of Tartar, $KHC_4H_4O_6$. White crystalline powder, odorless, having a pleasant acidulous taste, soluble in 200 parts water.
Action: Remote antacid.
Dose: 2 Gm. (30 gr.).
Preparations: Pulvis Jalapæ Compositus, U.S.P.

POTASSII BROMIDUM.—Potassium Bromide, KBr. White cubical crystals or granular powder, odorless, having a saline taste, soluble in 1.5 parts water.
Action: Sedative, hypnotic.
Dose: 1 Gm. (15 gr.).
Preparations: Elixir Potassii Bromidi, N.F.
Syrupus Bromidorum, N.F.
Mistura Chlorali et Potassii Bromidi Composita, N.F.
Pulvis Potassii Bromidi Effervescens, N.F.
Pulvis Potassii Bromidi Effervescens cum Caffeina, N.F.

POTASSII CARBONAS.—Potassium Carbonate, Salts Tartar, K_2CO_3. White granular powder, very deliquescent, odorless, having a strongly alkaline taste, soluble in water.
Action: cholagogue, rarely internally.
Dose: 1 Gm. (15 gr.) (largely diluted).
Preparations: Syrupus Rhei, U.S.P.
Syrupus Rhei Aromaticus, U.S.P.

Potassii Chloridum.—Potassium Chloride, KCl. Colorless elongated, prismatic or cubical crystals, or as a white granular powder, odorless, taste saline, soluble in 2.8 parts of water, insoluble in alcohol.
Action: Similar to Sodium Chloride.
Dose: Same as NaCl.

POTASSII CHLORAS.—Potassium Chlorate, $KClO_3$. Tabular plates, or white granular powder, odorless, having a cooling characteristic taste, soluble in 16 parts water, in 1.7 parts boiling water, insoluble in alcohol, etc.
Caution: Explosives with carbon compounds.
Action: Alterative, antiseptic, astringent.
Dose: 250 mg. (4 gr.).
Preparations: Trochisci Potassii Chloratis.

POTASSII CITRAS.—Potassium Citrate, $K_3C_6H_5O_7$. Prismatic crystals, or white granular powder, odorless, having a cooling, saline taste, soluble in water.
Action: Diaphoretic, refrigerant.
Dose: 1 Gm. (15 gr.).
Preparations: Potassii Citras Effervescens, U.S.P.

POTASSII CITRAS EFFERVESCENS.—Effervescent Potassium Citrate. Effervescent mixture containing 20 per cent. Potassii Citras.
Action: Refrigerant.
Dose: 4 Gm. (60 gr.).

Potassii Cyanidum.—Potassium Cyanide, KCN. White, opaque pieces, or white granular powder, deliquescent, odorless when dry, soluble in 2 parts water. *Very Poisonous.*

Action: Antispasmodic, cough sedative. (*With Caution*).
Dose: 10 mg. (⅙ gr.) (largely diluted).
Poison Antidote = Ferri Hydroxidum cum Magnesio Oxido.
Potassii Dichromas.—Potassium Dichromate, $K_2C_2O_7$. Orange-red prisms, or tabular crystals, odorless, having an acidulous, metallic taste, soluble in 9 parts water.
Action: Alterative, astringent, caustic.
Dose: 10 mg. (⅙ gr.).
Potassii Ferrocyanidum.—Potassium Ferrocyanide, $K_4Fe(CN)_6 + H_2O$. Yellow tabular crystals, odorless, having a mild saline taste, soluble in 4 parts water.
Action: Sedative, antihydrotic (in night sweats).
Dose: 5 Gm. (7½ gr.) (rarely).
POTASSII HYDROXIDUM.—Potassium Hydroxide, Caustic Potash, KOH. Fused masses or in pencils, odor of lye, taste very acrid, caustic, soluble in water.
Action: Antacid in 5 per cent. solution, escharotic.
Dose: 1 mil.
Preparations: Liquor Potassii Hydroxidi, U.S.P.
POTASSII HYPOPHOSPHIS.—Potassium Hypophosphite, KH_2PO_2. White opaque plates, odorless, taste pungent, saline, soluble in 0.5 parts water, 7 parts alcohol.
Action: Tonic in wasting diseases.
Dose: 0.5 Gm. (7½ gr.).
Preparations: Syrupus Hypophosphitum, U.S.P.
POTASSII IODIDUM.—Potassium Iodide, KI. Opaque, white, cubical crystals or white granular powder, odor of iodine; pungent, saline, afterward bitter taste, soluble in 0.7 parts water, 12 parts alcohol, 2.5 parts glycerin.
Action: Alterative, resolvent (only in solution).
Dose: 0.5 Gm. (7½ gr.).
Preparations: Liquor Iodi Compositus, U.S.P.
　　Tinctura Iodi, U.S.P.
　　Unguentum Iodi, U.S.P., Ung. Pot. Iod., N.F.
　　Liquor Hydrargyri et Potassii Iodidi, N.F.
POTASSII NITRAS.—Potassium Nitrate, Saltpeter, KNO_3. Prisms or white crystalline powder, odorless having a pungent, cooling, saline taste, soluble in 3.6 parts water.
Action: Antiseptic, diuretic, refrigerant.
Dose: 0.5 Gm. (7½ gr.) largely diluted.
POTASSII PERMANGANAS.—Potassium Permanganate, $KMnO_4$. Dark purple colored prisms, odorless and having a sweet but afterward astringent taste, soluble in 15 parts of water.
Action: Antiseptic, disinfectant.
Dose: 1 grain (pill form), 1 Gm. (15 gr.) (aqueous solution).
Antidote for morphine poisoning.
*POTASSII SULPHAS.**—Potassium Sulphate, K_2SO_4.—Prisms or white powder, odorless, having a saline, bitter taste, soluble in 9 parts water.
Action: Cathartic, diuretic.
Dose: 2 Gm. (30 gr.).
POTASSII ET SODII TARTRAS.—Potassium and Sodium Tartrate, Rochelle Salt. Transparent prisms, or white powder, odorless having a cool saline taste, soluble in 1.2 parts water.
Action: Refrigerant, cathartic.
Dose: 8 Gm. (120 gr.).
Preparations: Pulvis Effervescens Compositus (Seidlitz Powder).
POTASSA **SULPHURATA.**—Sulphurated Potassa, Liver of Sulphur. Irregular pieces, liver-brown when freshly made changing to greenish-yellow and finally grey, with age odorless, when dry, fetid odor when moist, having a nauseous bitter taste, very soluble in water.
Action: Antacid, alterative.
Dose: Not given in the U.S.P.
Potassii Tartras.—Potassium Tartrate, $K_2C_4H_4O_6 \cdot H_2O$. Small transparent or white monoclinic crystals or white powder, somewhat deliquescent, odorless, having a saline slightly bitter taste, soluble in water, almost insoluble in alcohol.

Action: Mild purgative.
Dose: 1 drachm to 1 oz.

THE OFFICIAL U.S.P. AND UNOFFICIAL SODIUM SALTS

SODII ACETAS.—Sodium Acetate, $NaC_2H_3O_2 + 3H_2O$. Transparent prisms or granular crystalline powder, odorless, having a cooling saline taste, soluble in water, in 23 parts alcohol.
Action: Remote antacid.
Dose: 1 Gm. (15 gr.).

SODII ARSENAS.—Sodium Arsenate, $Na_2HAsO_4 \cdot 7H_2O$. Colorless, transparent, monoclinic prisms, odorless, having a mild alkaline taste, soluble in 1.2 parts of water, very sparingly soluble in hot water, nearly insoluble in alcohol.
Action: Alterative, tonic.
Dose: 0.0005 Gm. ($\frac{1}{10}$ gr.).
Preparations: Sodii Arsenas Exsiccatus, U.S.P

SODII ARSENAS EXSICCATUS.—Exsiccated Sodium Arsenate, Na_2HAsO_4. An amorphous white powder, soluble in 3 parts of water, nearly insoluble in boiling alcohol.
Action: Alterative.
Dose: 3 mg. ($\frac{1}{20}$ gr.).
Preparations: Liquor Sodii Arsenatis, U.S.P.
Liquor Sodii Arsenatis N.F. (Pearsons).

SODII BENZOAS.—Sodium Benzoate, $NaC_7H_5O_2$. White amorphous, granular or crystalline powder, odorless, having a sweetish astringent taste, soluble in 1.6 parts of water, 43 parts of alcohol.
Action: Antiseptic, antipyretic.
Dose: 1 Gm. (15 gr.).
Preparations: Liquor Antisepticus Alkalinus, N.F.

SODII BICARBONAS.—Acid Sodium Carbonate, Baking Soda, $NaHCO_3$. White powder odorless, having a cooling, mildly alkaline taste, soluble in 12 parts water, decomposes in boiling water.
Action: Antacid, alterative.
Dose: 1 Gm. (15 gr.).
Preparations: Pulvis Effervescens Compositus, U.S.P.
Liquor Sodæ et Menthæ N.F.
Trochisci Sodii Bicarbonatis, U.S.P.

SODII BISULPHIS.—Acid Sodium Sulphite, $NaHSO_3$. Opaque, prismatic crystals or granular powder, odor of sulphur dioxide, disagreeable taste, soluble in 3.5 parts of water, 70 parts alcohol.
Action: Rarely used in medicine.
Dose: 0.5 Gm. ($7\frac{1}{2}$ gr.).

SODII BORAS.—Sodium Borate, Borax, $Na_2B_4O_7 + 10H_2O$. Transparent prisms or white powder, inòdorous, sweetish alkaline taste, soluble in 20 parts water, in 1 part glycerin.
Action: Antiseptic, astringent, detergent.
Dose: 0.5 Gm. ($7\frac{1}{2}$ gr.).
Preparations: Liquor Sodii Boratis Compositus, N.F.
Liquor Antisepticus Alkalinus, N.F.
Sodii Boro-Benzoas, N.F.

SODII BROMIDUM.—Sodium Bromide, $NaBr$. Cubical crystals or white granular powder, odorless, having a saline slightly bitter taste, soluble in 1.7 parts water, 12.5 parts alcohol.
Action: Nerve sedative, hypnotic.
Dose: 1 Gm. (15 gr.).
Preparations: Elixir Sodii Bromidi, N.F.
Syrupus Bromidorum, N.F.

SODII CACODYLAS.—Sodium Cacodylate, $Na(CH_3)AsO_2$. White, deliquescent, prisms or powder, odorless, when dry, emitting a garlic odor when burned, soluble in water and in 2.5 parts alcohol.
Action: Alterative.
Dose: 0.06 Gm. (1 gr.).

SODII CARBONAS MONOHYDRATUS.—Monohydrated Sodium Carbonate, $Na_2CO_3 + H_2O$. White crystalline, granular powder, odorless, having a strong alkaline taste, soluble in 3 parts water, 7 parts glycerin, insoluble in alcohol.

Action: Antacid.

Dose: 0.25 Gm. (4 gr.).

Sodii Chloras.—Sodium Chlorate, $NaClO_3$. Crystals or crystalline powder, odorless, having a cooling saline taste, soluble in water.

Caution: Explosive with organic compounds.

Action: Antiseptic, similar to potassium chlorate.

Dose: 0.25 Gm. (4 gr.).

SODII CHLORIDUM.—Sodium Chloride, Salt, $NaCl$. White crystalline, granular powder, odorless, having a purely saline taste, soluble in 2.8 parts water, 10 parts glycerin; slightly soluble in alcohol.

Action: Emetic.

Dose: 15 Gm. (240 gr.).

Preparation: Liquor Sodii Chloridi Physiologicus U.S.P.

SODII CITRAS.—Sodium Citrate, $2Na_3C_6H_5O_7 + 11H_2O$. White granular powder, odorless, having a cooling saline taste, soluble in 1.1 parts water, insoluble in alcohol.

Action: Diuretic, refrigerant.

Dose: 1 gm. (15 gr.).

Preparations: Liquor Sodii Citratis, N.F.

Liquor Sodii Citro-Tartratis Effervescens, N.F.

SODII CYANIDUM.—Sodium Cyanide, $NaCN$. White, opaque, amorphous pieces, or granular powder, deliquescent, freely soluble in cold water.

Action: Sedative, no dose given in U.S.P., Poison.

SODII GLYCEROPHOSPHAS.—Sodium Glycerophosphate. White, mono-clinic, plates or scales, or white powder, odorless, having a saline taste. Soluble in cold or hot water, insoluble in alcohol.

Action: Nutritive Tonic.

Dose: 0.25 Gm. (4 gr.).

Preparations: Liquor Sodii Glycerophosphatis, U.S.

Elixir Glycerophosphatum Co. N.F.

Elixir Calcii et Sodii Glycerophosphatum, N.F.

SODII HYDROXIDUM.—Sodium Hydroxide, Caustic Soda. Dry, white, or nearly white flakes, fused masses or sticks, deliquescent, odorless, having a caustic taste, soluble in water or alcohol. Caution is necessary in handling as it destroys organic tissue. (See Potassium Hydroxide.)

Preparations: Liquor Sodii Hydroxidi, U.S.P.

Soda cum Calce (London Paste) N.F.

SODII HYPOPHOSPHIS.—Sodium Hypophosphite. Pearly plates or granular powder, deliquescent, odorless, having a bitter, sweet, saline taste, soluble in water, alcohol or glycerin.

Action: Tonic, nutrient.

Dose: 1 Gm. (15 gr.).

Preparations: Syrupus Hypophosphitum, U.S.P.

Syrupus Calcii et Sodii Hypophosphitum, N.F.

Syrupus Sodii Hypophosphitis, N.F.

Liquor Hypophosphitum, N.F.

Liquor Hypophosphitum Compositus, N.F.

Elixir Hypophosphitis, N.F.

SODII INDIGOTIN DISULPHONAS.—Sodium Indigotin Disulphonate, Indigo Carmine, $C_{16}H_8O_2N (SO_3Na)_2$. Blue powder or dark purple paste, sparingly soluble in water, almost·insoluble in alcohol.

Dose: Not given in U.S.P.

SODII IODIDUM.—Sodium Iodide, NaI. Colorless, cubical crystals or white crystalline powder, deliquescent, odorless, having a saline bitter taste, soluble in water or 2 parts alcohol; also in glycerin.

Action: Alterative.

Dose: 0.3 Gm. (5 gr.).

SODII NITRAS.—Sodium Nitrate, Chili Saltpeter, $NaNO_3$. Colorless, transparent, rhombohedral crystals, hygroscopic, odorless, having a cooling saline and slighty bitter taste, soluble in 1.1 parts water, 100 parts alcohol.

Action: Rarely used in medicine.

SODII NITRIS.—Sodium Nitrite, $NaNO_2$. White, opaque-fused masses or sticks, or colorless, hexagonal crystals, or granular powder, deliquescent, odorless, having a mild saline taste, soluble in 1.5 parts water sparingly soluble in alcohol.

Action: Seldom used in medicine.
Dose: 0.06 Gm. (1 gr.).
Preparations: Spiritus Aetheris Nitrosi, U.S.P.

SODII PERBORAS.—Sodium Perborate, $NaBO_3 + 4H_2O$. White, crystalline granules or powder, odorless, having a saline taste, soluble in water.
Action: Antiseptic.
Dose: 0.06 Gm. (1 gr.).

SODII PHENOLSULPHONAS.—Sodium Phenolsulphonate, $C_6H_5O \cdot SO_3Na + 2H_2O$. Transparent rhombic prisms, or crystalline granules. Colorless, odorless, having a cooling saline bitter taste, efflorescent, soluble in 4.2 parts water, or 140 parts alcohol, or 5 parts glycerin.
Action: Antiseptic, disinfectant.
Dose: 0.25 Gm. (4 gr.).

SODII PHOSPHAS.—Sodium Phosphate, $Na_2HPO_4 + 12H_2O$. Large, color less, monoclinic, crystals or granular salt, odorless, having a cooling saline taste, soluble in 2.7 parts water insoluble in alcohol.
Action: Purgative.
Dose: 4 Gm. (60 gr.).
Preparations: Sodii Phosphas Exsiccatus, Liq. Sodii Phos. Co., N.F.

SODII PHOSPHAS EFFERVESCENS.—Effervescent Sodium Phosphate.
Action: Purgative.
Dose: 10 Gm. (2½ drachms).

SODII PHOSPHAS EXSICCATUS.—Exsiccated Sodium Phosphate, Na_2HPO_4.
Action: Purgative.
Dose: 2 Gm. (30 gr.).
Preparation: Sodii Phosphas Effervescens.

SODII SALICYLAS.—Sodium Salicylate, $NaC_7H_5O_3$. White, micro-crystalline-powder, or scales, or as an amorphous powder, nearly odorless, having a sweet saline taste, soluble in water and in 9.2 parts of alcohol; soluble in glycerin. Contained in Liq. Antisepticus, N.F.
Action: Antirheumatic.
Dose: 1 Gm. (15 gr.).
Preparations: Elixir Sodii Salicylatis, N.F.

SODII SULPHAS.—Sodium Sulphate, Glauber's Salt. $Na_2SO_4 + 10H_2O$. Large, colorless transparent monoclinic prisms, or granular crystals, odorless, having a bitter saline taste, soluble in water, insoluble in alcohol, soluble in glycerin.
Action: Cathartic.
Dose: 15 Gm. (4 drachms).

SODII SULPHIS EXSICCATUS.—Exsiccated Sodium Sulphite, Na_2SO_3. White powder, odorless, having a cooling saline sulphurous taste, soluble in 3.2 parts water, almost insoluble in alcohol.
Action: Antiferment.
Dose: 1 Gm. (15 gr.).

SODII THIOSULPHAS.—Sodium Thiosulphate, Hypo, $Na_2S_2O_3 + 5H_2O$. Colorless, transparent, monoclinic prisms, efflorescent, odorless, having a cooling saline, bitter taste, soluble in water, insoluble in alcohol.
Action: Used internally instead of sulphur.
Dose: 1 Gm. (15 gr.).

THE OFFICIAL U.S.P., N.F. AND UNOFFICIAL LITHIUM SALTS

Lithii Benzoas.—Lithium Benzoate, $LiC_7H_5O_2$. Light, white powder or small shining crystalline scales; odorless or slight benzoin-like odor, cooling sweetish taste, soluble in 3 parts of water, 13 parts alcohol.
Action: Antirheumatic, antiseptic.
Dose: 1 Gm. (15 gr.).

LITHII BROMIDUM.—Lithium Bromide, LiBr. White granular salt, odorless, sharp bitter taste, deliquescent, very soluble in water; freely soluble in alcohol or ether.
Action: Hypnotic.
Dose: 1 Gm. (15 gr.).
Preparations: Elixir Lithii Bromidi, N.F.

LITHII CARBONAS.—Lithium Carbonate, Li_2CO_3. Light, white powder, odorless, alkaline taste, soluble in 78 parts water, 140 parts boiling water; almost insoluble in alcohol; soluble in dilute acids.

Action: Antirheumatic, antacid.
Dose: 0.5 Gm. (8 gr.).

LITHII CITRAS.—Lithium Citrate, $Li_3C_6H_5O_7 + 4H_2O$. White powder, of granular form, odorless, cooling slightly alkaline taste, deliquescent, soluble in 14 parts water, slightly soluble in alcohol.
Action: Antirheumatic, for gout.
Dose: 0.5 Gm. (8 gr.).
Preparations: Elixir Lithii Citratis, N.F.
Lithii Citras Effervescens.

Lithii Salicylas.—Lithium Salicylate, $LiC_7H_5O_3$. White or greyish-white powder, deliquescent, odorless, taste sweetish. Very soluble in water and in alcohol.
Action: Antirheumatic.
Dose: 1 Gm. (15 gr.).
Preparations: Elixir Lithii Salicylatis, N.F.

THE OFFICIAL U.S.P., N.F. AMMONIUM SALTS

AMMONII BENZOAS.—Ammonium Benzoate $C_6H_5COONH_4$. Thin, white, laminar crystals or crystalline powder; gradually losing ammonia upon exposure, odorless, or having a slight odor of benzoic acid, a saline, bitter, afterward slightly acrid taste; soluble in 10 parts of water, 35 parts alcohol and about 8 parts glycerin.
Action: Antirheumatic, diuretic.
Dose: 1 Gm. (15 gr.).

AMMONII BROMIDUM.—Ammonium Bromide, NH_4Br. Colorless, transparent, prismatic crystals, or as a white crystalline or granular powder; odorless, of a somewhat pungent saline taste; somewhat hygroscopic; soluble in 1.3 parts water, 12 parts alcohol, readily soluble in boiling water or boiling alcohol.
Action: Hypnotic.
Dose: 1 Gm. (15 gr.).
Preparation: Elixir Ammonii Bromidi N.F.

AMMONII CARBONAS.—Ammonium Carbonate, NH_4HCO_3. White, internally translucent masses, having a strong odor of ammonia, and a sharp ammoniacal taste, soluble in 4 parts water; decomposed by heat.
Action: Pulmonary stimulant.
Dose: 0.3 Gm. (5 gr.).
Preparations: Liquor Ammonii Acetatis, U.S.P.
Spiritus Ammonii Aromaticus, U.S.P.
Mistura Pectoralis (Stokes) N.F.

AMMONII CHLORIDUM.—Ammonium Chloride (Muriate of Ammonia) NH_4Cl. White crystalline or granular powder, without odor, having a cooling saline taste, somewhat hygroscopic, soluble in 2.6 parts water, 100 parts alcohol, 8 parts glycerin.
Action: Stimulant, expectorant.
Dose: 0.3 Gm. (5 gr.).
Preparations: Trochisci Ammonii Chloridi, U.S.P.
Mistura Ammonii Chloridi, N.F.

Ammonii Hypophosphis.—Ammonium Hypophosphite, $NH_4PH_2O_2$. Colorless, hexagonal plates, or granular powder, deliquescent, odorless, having a saline bitter taste; soluble in water, in 15 parts alcohol.
Action: Stimulant, tonic. Contained in Syr. Ammon. Hypophos, N.F
Dose: 0.2 Gm. (3 gr.).

AMMONII IODIDUM.—Ammonium Iodide, NH_4I. Minute, colorless, cubical crystals, or as a white, granular powder, odorless, having a sharp, saline taste, soluble, in water, in 3.7 parts alcohol, in 1.5 parts glycerin.
Action: Alterative (rarely internally).
Dose: 0.3 Gm. (5 gr.).
Preparations: Tinctura Iodi Decolorata, N.F.

Ammonii Phosphas.—Ammonium Phosphate $(NH_4)_2HPO_4$. Colorless crystals, or white crystalline powder, odorless, having a saline taste, soluble in 4 parts water, insoluble in alcohol.
Action: For gout.
Dose: 0.3 Gm. (5 gr.).

AMMONII SALICYLAS.—Ammonium Salicylate, $C_6H_4(OH)COONH_4$. Color-less, lustrous, monoclinic, prisms or plates or white crystalline powder; odorless, having at first a slightly saline bitter taste, after-taste sweetish, soluble in water, 3 parts alcohol.
Action: Antirheumatic.
Dose: 0.5 Gm. (8 gr.).

AMMONII VALERAS.—Ammonium Valerate. A compound of ammonium and valeric acid having somewhat varying composition. It occurs in color-less, or white quadrangular plates, odor of valeric acid, deliquescent, having a sharp sweetish taste, very soluble in water or alcohol, soluble in ether.
Action: Antispasmodic.
Dose: 0.5 Gm. (8 gr.).
Preparations: Elixir Ammonii Valeratis, N.F.

GROUP II.—ALKALINE EARTHS AND ALLIES.—CALCIUM, BARIUM, STRONTIUM AND MAGNESIUM

THE U.S.P. AND N.F. AND UNOFFICIAL CALCIUM SALTS

CALCII BROMIDUM.—Calcium Bromide, $CaBr_2$. White granular salt; odorless of a sharp, saline taste and very deliquescent. Soluble in water and in alcohol, insoluble in chloroform or ether.
Action: Hypnotic.
Dose: 1 Gm. (15 gr.).
Preparations: Elixir Calcii Bromidi, N.F.

CALCII CARBONAS PRÆCIPATUS.—Precipitated Calcium Carbonate, Precipi-tated Chalk, $CaCO_3$. Fine, white, micro-crystalline powder, without odor or taste, and permanent in air, nearly insoluble in water.
Action: In diarrheal conditions.
Dose: 1 Gm. (15 gr.).
Preparations: Unguentum Sulphuris Compositum, N.F.

CALCII CHLORIDUM.—Calcium Chloride, $CaCl_2$. White, slightly translucent, hard fragments, granules, or sticks; odorless, having a sharp saline taste. Is very deliquescent, soluble in water, in 10 parts alcohol, boiling increases solubility in either.
Uses: In internal bleeding. (Drying agent.)
Dose: 0.5 Gm. (8 gr.).

CALCII GLYCEROPHOSPHAS.—Calcium Glycerophosphate, $C_3H_5(OH)_2PO_4Ca$. Fine, white powder, odorless and almost tasteless, somewhat hygroscopic, soluble in 50 parts water, souble in less water at a lower temperature, citric acid increases its solubility; insoluble in alcohol.
Action: Nutritive, Tonic.
Dose: 0.25 Gm. (4 gr.).

CALCII HYPOPHOSPHIS.—Calcium Hypophosphite, $Ca(PH_2O_2)_2$. Colorless, transparent, monoclinic prisms, as small lustrous scales, or as a white crystalline powder; odorless, having a nauseous bitter taste permanent in air. Soluble in 6.5 parts water, insoluble in alcohol.
Action: Nutritive tonic in phthisis.
Dose: 0.5 Gm. (8 gr.).
Preparations: Syrupus Hypophosphitum, U.S.P.
Syrupus Hypophosphitum Compositus, U.S.P. VIII.
Emulsum Olei Morrhuæ cum Hypophosphitibus, U.S.P. VIII
Elixir Calcii Hypophosphitis, N.F.
Liquor Hypophosphitum, N.F.
Liquor Hypophosphitum Compositus, N.F.
Syrupus Calcii Hypophosphitis, N.F.
Syrupus Calcii et Sodii Hypophosphitum, N.F.

CALCII LACTAS.—Calcium Lactate (Ca $(C_3H_5O_3)_2 + 5H_2O$). White, granular, masses or powder; odorless and nearly tasteless, efflorescent, soluble in 20 parts water, insoluble in alcohol.
Dose: 0.5 Gm. (8 gr.).

Calcii Lactophosphas.—Calcium Lactophosphate. White granules or as a white powder, without odor, soluble in water, almost insoluble in alcohol.
Dose: 0.5 Gm. (8 gr.). Syr. Calc. Lactophos. et Ferr., N.F.

*_Calcii Phosphas Præcipitatus._—Precipitated Calcium Phosphate, U.S.P. VIII, $Ca_3(PO_4)_4$. White, amorphous or micro-crystalline, bulky powder, odorless and tasteless, permanent in air. Almost insoluble in cold water, decomposed by boiling water, insoluble in alcohol.
Used in preparations.
Preparations: Syrupus Calcii Chlorhydrophosphatis, N.F.
Emulsum Morrhuæ cum Calcii Phosphate, N.F.

Calcii Sulphas Exsiccatus.—Exsiccated Calcium Sulphate, Plaster Paris, U.S.P. VIII, $CaSO_4 + 2H_2O$. Fine white powder, forming with an equal weight of water a smooth paste which rapidly hardens. Used in making plaster casts.

CALCII SULPHIDUM CRUDUM.—Crude Calcium Sulphide, Calx Sulphurata, U.S.P. VIII. Pale grey or yellowish powder, having a faint odor of hydrogen sulphide and a nauseous and alkaline taste, decomposed in moist air. It is very slightly soluble in cold water, more so in hot water, readily soluble in solution of ammonium salts, insoluble in alcohol.
Action: Antiseptic, alterative.
Dose: 0.06 Gm. (1 gr.).
Preparations: Liquor Calcis Sulphuratæ, N.F.

CALX.—Calcium Oxide, Lime, CaO. Hard, white or greyish-white masses or granules, or as a white powder; odorless, caustic taste. Soluble in 840 parts water, in 1,740 parts boiling water; insoluble in alcohol or glycerin or syrup. When moistened it is converted into calcium hydroxide.
Preparations: Liquor Calcis, U.S.P.
Syrupus Calcis, U.S.P. VIII.

CALX CHLORINATA.—Chlorinated Lime, Chloride of Lime. White, or greyish white, granular powder, having the odor of chlorine. It is partially soluble in water or alcohol.
Action: Disinfectant, deodoramt.
Dose: Not given.
Preparations: Liquor Sodæ Chlorinata, U.S.P.
Liquor Potassæ Chlorinata, N.F.

THE OFFICIAL U.S.P. AND N.F. MAGNESIUM SALTS

MAGNESII CARBONAS.—Magnesium Carbonate (approximately) ($MgCO_3$). $Mg(OH)_2 + 5H_2O$. Light, white, friable masses, or in a bulky, white powder, without odor, and having a slight earthy taste; permanent in air; practically insoluble in water, insoluble in alcohol.
Action: Antacid, laxative.
Dose: 3 Gm. (45 gr.).
Preparations: Magma Magnesia, U.S.P.
Liquor Magnesii Citratis, U.S.P.

*_Magnesii Chloridum._—Magnesium Chloridi. Colorless, transparent crystals, or as white translucent pieces; deliquescent in moist air, soluble in water or alcohol.
Action: Purgative in solution.
Dose: 30 Gm. (450 gr.).

MAGNESII OXIDUM.—Magnesium Oxide. Light Magnesia, MgO. White, very bulky and very fine powder, without odor, and having an earthy, but not a saline taste; is almost insoluble in water; insoluble in alcohol, soluble in dilute acids.
Action: Antacid. Cathartic.
Dose: 2 Gm. (30 gr.).
Preparation: Pulvis Rhei Compositus.

MAGNESII OXIDUM PONDEROSUM.—Heavy Magnesium Oxide, Heavy Magnesia, MgO. White, dense and very fine powder, which conforms to the reactions and tests for Magnesii Oxidum.
Preparation: Pulvis Rhei et Magnesiæ Anisatus, N.F.

MAGNESII SULPHAS.—Magnesium Sulphate, Epsom Salt, $MgSO_4 + 7H_2O$. Small colorless, prismatic needles or rhombic prisms, without odor, and having a cooling saline taste; slowly efflorescent, soluble in water and almost insoluble in alcohol.

Action: Cathartic.
Dose: 16 Gm. (240 gr.).
Preparations: Infusum Sennæ Compositum, U.S.P.
    Liquor Magnesii Sulphatis Effervescens, N.F.
Magnesii Sulphas Effervescens.—Effervescent Magnesium Sulphate, U.S.P. VIII. This unofficial article is a reliable cathartic and an agreeable method of administrating Epsom salt. The dose is the same as that of Magnesii Sulphas.

THE OFFICIAL U.S.P. AND N.F. STRONTIUM SALTS

STRONTII BROMIDUM.—Strontium Bromide, $SrBr_2 + 6H_2O$. Colorless, transparent, hexagonal crystals; odorless and having a bitter, saline taste; deliquescent soluble in water or alcohol, insoluble in ether.
Action: Hypnotic.
Dose: 1 Gm. (15 gr.).
**Strontii Carbonas.*—Strontium Carbonate, $(SrCO_3 + H_2O_n)$. White powder, odorless and tasteless, insoluble in water, soluble in dilute hydrochloric, nitric, or acetic acid, but not sulphuric.
Seldom used in medicine.
STRONTII IODIDUM.—Strontium Iodide, $SrI_2 + 6H_2O$. Colorless, transparent, hexagonal plates, or white granular powder, or in crystalline crusts; odorless having a bitter, saline taste; deliquescent, soluble in water or alcohol, slightly soluble in ether.
Action: Alterative.
Dose: 0.3 Gm. (5 gr.).
STRONTII SALICYLAS.—Strontium Salicylate, $Sr(C_7H_5O_3)_2 + 2H_2O$. White, crystalline powder; odorless and somewhat sweet, saline taste, soluble in 19 parts water, 61 parts alcohol. Solubility increasing with boiling water or alcohol.
Action: Antirheumatic.
Dose: 1 Gm. (15 gr.).

THE BARIUM SALTS

Seldom used medicinally, see p. 28.

THE OFFICIAL CERIUM SALT U.S.P.

CERII OXALAS.—Cerium Oxalate. Fine, white or pinkish-white powder, without odor or taste, insoluble in water, alcohol, or ether.
Valuable in vomiting of pregnancy.
Dose: 0.2 Gm. (3 gr.).

THE CADMIUM SALTS

Salts obsolete as therapeutic agents.

GROUP III.

THE OFFICIAL U.S.P. AND UNOFFICIAL ZINC SALTS

ZINCI ACETAS.—Zinc Acetate, $Zn(C_2H_3O_2)_2 + 2H_2O$. Soft, white plates, faintly acetous odor, astringent metallic taste, soluble in 2.5 parts water, 36 parts alcohol. Solubility increasing with heat.
Action: Mild, astringent.
Dose: 0.125 Gm. (2 gr.).
Preparations: Oleatum Zinci, N.F.
Zinci Bromidum.—Zinc Bromide, $ZnBr_2$, U.S.P. VIII. White, granular powder, odorless, having in solution a sharp, saline and metallic taste. Very deliquescent. Readily soluble in water or alcohol. Rarely used.
Dose: 0.125 Gm. (2 gr.).

2

ZINCI CARBONAS PRÆCIPITATUS.—Precipitated Zinc Carbonate. Impalpable, white powder without odor or taste, permanent, insoluble in water or alcohol.
Action: Astringent, externally.
ZINCI CHLORIDUM.—Zinc Chloride, $ZnCl_2$. White granular powder, or fused masses, very deliquescent, odorless. In solution, the taste is astringent and metallic.
Action: Caustic, externally.
Preparation: Liquor Zinci Chloridi, U.S.P.
Zinc Iodidum.—Zinc Iodide, ZnI_2. White granular powder, odorless, taste, saline, sharp metallic, readily soluble in neutral liquids.
Action: Alterative, astringent, antiseptic.
Dose: 0.065 Gm. (1 gr.).
ZINCI OXIDUM.—Zinc Oxide, ZnO. White, fine, amorphous powder, odorless, tasteless, insoluble in water or alcohol.
Action: Antiseptic, astringent (rarely internally).
Dose: 0.250 Gm. (4 gr.).
Preparations: Unguentum Zinci Oxidi, U.S.P.
Unguentum Extensa, N.F.
Pasta Zinci, Pasta Zinci Mollis, Pasta Zinci Sulphurata, N.F.
ZINCI PHENOLSULPHONAS.—Zinc Phenolsulphonate, Zinc Sulphocarbolate, $Zn(C_6H_5O_4S)_2 - 8H_2O$. Colorless, transparent, rhombic prisms or tabular crystals, odorless; astringent, metallic taste, soluble in 1.7 parts water or alcohol.
Action: Antiseptic, astringent.
Dose: 0.125 Gm. (2 gr.).
ZINCI STEARAS.—Zinc Stearate. Very fine white powder, tasteless, having a very faint odor resembling fat, insoluble in water, alcohol or ether, soluble in 3 parts glycerin.
Action: Used as dusting powder and in ointments.
Preparations: Unguentum Zinci Stearatis, U.S.P. VIII.
ZINCI SULPHAS.—Zinc Sulphate, White Vitriol, $ZnSO_4 + 7H_2O$. Colorless, transparent, rhombic crystals or a granular, crystalline powder, without odor, and having an astringent, metallic taste, efflorescent, soluble in water or in 3 parts glycerin, insoluble in alcohol.
Action: Astringent, externally.
Emetic, internally.
Dose: 1 Gm. (15 gr.).
Preparations: Liquor Zinci et Ferri Compositus, N.F.
Liquor Zinci et Alumini Compositus, N.F.
ZINCI VALERAS.—Zinc Valerate, $Zn(C_5H_9O_2)_2 + 2H_2O$. White, pearly scales, having the odor of valeric acid, and a sweetish, astringent, metallic taste, soluble in about 50 parts of water, and in about 35 parts of alcohol.
Action: Antispasmodic.
Dose: 0.125 Gm. (2 gr.).
Preparation: Elixir Zinci Valeratis, N.F.
ZINCUM.—Zinc, Zn. Bluish-white metal, having a peculiar taste and a perceptible odor or when rubbed, soluble in dilute sulphuric acid, etc.
Action: Used to prepare $ZnSO_4$ and $ZnCl_2$.

THE OFFICIAL U.S.P. AND N.F. ALUMINUM SALTS

ALUMEN.—Alum, $AlNH_4(SO_4)_2 + 12H_2O$ or $AlK(SO_4)_2 + 12H_2O$. Both occur in large colorless crystals, or as white powders. Alum is odorless and has a sweetish and strongly astringent taste. Potassium alum is soluble in 7.2 parts of water, insoluble in alcohol, soluble in glycerin. Ammonium alum is less soluble than potassium alum.
Action: Astringent.
Dose: 0.5 Gm. (8 gr.).
Preparation: Alumen Exsiccatum.
**Alumni Chloridum.*—Aluminum Chloride, $AlCl_3 + 6H_2O$. White, or yellowish-white deliquescent crystalline powder, nearly odorless; taste, sweet and very astringent; soluble in water and glycerin, also in 3 parts alcohol.
Dose: 0.3 Gm. (5 gr.).
ALUMEN EXSICCATUM.—Exsiccated Alum, Dried Alum, Burnt Alum, $AlNH_4(SO_4)_2$ or $AlK(SO_4)_2$. White, granular powder, without odor, having a

sweetish astringent taste, and attracting moisture upon exposure to the air; soluble in 14 parts of water; also in 1.4 parts of boiling water; insoluble in alcohol.

Used externally as an escharotic.

ALUMINI HYDROXIDUM.—Aluminum Hydroxide (Principally) $Al(OH)_3$. White, bulky, amorphous powder, odorless and tasteless; permanent in air, insoluble in water or alcohol, soluble in hydrochloric or sulphuric acid, or alkaline hydroxides.

Use: Used externally as an astringent.

Aluminum Sulphas.—Aluminum Sulphate, $Al_2(SO_4)_3 + 16H_2O$. U.S.P. VIII. White, crystalline powder, shining plates, or fragments, without odor, having a sweetish astringent taste, soluble in water, insoluble in alcohol.

Uses: Used as caustic and preservative of cadavers.

GROUP IV.—Lead, Silver, Copper and Bismuth

THE OFFICIAL U.S.P. AND N.F. LEAD SALTS

PLUMBI ACETAS.—Lead Acetate, Sugar of Lead $(Pb(C_2H_3O_2) + 3H_2O)$. Colorless, shining, transparent, monoclinic prisms or plates or as heavy white, crystalline masses, or granular crystals, having a faintly acetous odor and a sweetish astringent, afterward metallic, taste; efflorescent, and absorbing carbon dioxide on exposure to the air. Soluble in 1.4 parts of water, 38 parts of alcohol; soluble in boiling water and in glycerin.

Action: Astringent.

Dose: 0.065 Gm. (1 gr.).

Preparations: Liquor Plumbi Subacetatis, U.S.P. (Contained in Lead water). Ceratum Plumbi Subacetatis, U.S.P. VIII.
Pilula Opii et Plumbi, N.F.
Lotio Plumbi et Opii, N.F.

Plumbi Carbonas.—Lead Carbonate, White Lead, approximately $(PbCo_3)_2Pb(OH)_2$. Heavy, white, opaque powder, or as a pulverulent mass, without odor or taste; permanent in air, insoluble in water or alcohol, but soluble in acetic acid or dilute nitric acid, with effervescence.

Seldom used in medicine.

Plumbi Iodidum.—Lead Iodide, U.S.P. VIII, PbI_2. Heavy, bright yellow powder, without odor or taste; permanent in air, soluble in about 1,300 parts of water, 200 parts of boiling water, very slightly soluble in alcohol, soluble in fixed alkalies.

Action: Alterative, discutient, externally.
Preparation: Unguentum Plumbi Iodidi, N.F.

PLUMBI OXIDUM.—Lead Oxide, Litharge, PbO. Heavy, yellowish or reddish-yellow powder, or in minute scales, without odor or taste, almost insoluble in water, insoluble in alcohol.

Preparations: Emplastrum Plumbi, U.S.P.
Liquor Plumbi Subacetatis, U.S.P.

Plumbi Oxidum Rubrum.—Red Oxide of Lead, Red Lead, Pb_2PbO_4. Heavy, orange-red powder without odor or taste, almost insoluble in water, insoluble in alcohol.

THE OFFICIAL U.S.P. AND UNOFFICIAL SILVER SALTS

Argenti Cyanidum.—Silver Cyanide, AgCN, U.S.P. VIII. A white powder, without odor or taste, insoluble in water, alcohol or cold nitric acid; soluble in boiling nitric acid.

Formerly used in preparing acidum hydrocyanicum dilutum.

ARGENTI NITRAS.—Silver Nitrate, $AgNO_3$. Colorless, transparent tabular, rhombic crystals becoming grey or greyish black on exposure to the light in the presence of organic matter; without odor, but having a bitter caustic and strongly metallic taste, soluble in water and in 30 parts alcohol.

Action: Externally, caustic, antiseptic; internally, in gastritis and diarrhea.

Dose: 0.01 Gm. (⅙ gr.) (pills).

Preparation: Argenti Nitras Fusus.

ARGENTI NITRAS FUSUS.—Moulded Silver Nitrate, Fused Silver Nitrate, Lunar Caustic. Moulded silver nitrate is a white, hard solid, generally in the form of pencils and resembles the crystals in physical properties.
Uses: Caustic (externally).

Argenti Nitras Mitigatus.—Mitigated Caustic 33 per cent. $AgNO_3$. Silver nitrate fused with twice its weight of potassium nitrate and moulded into pencils.
Uses: Used when a modified caustic effect is desired.

ARGENTI OXIDUM.—Silver Oxide, Ag_2O. Heavy, dark brownish-black powder, odorless and having a metallic taste, slightly soluble in water, insoluble in alcohol but soluble in nitric acid.
Uses: Astringent, internally.
Dose: 0.06 Gm. (1 gr.).

THE OFFICIAL COPPER SALT

CUPRI SULPHAS.—Copper Sulphate, Blue Vitriol, $CuSO_4 + 5H_2O$. Deep blue, triclinic crystals or as a blue, granular powder; odorless, of a nauseous metallic taste; slowly efflorescent in dry air; soluble in 2.5 parts of water, 500 parts of alcohol and in 2.8 parts glycerin, very soluble in boiling water.
Uses: Astringent, styptic, tonic, emetic.
Dose: (astringent) 0.01 Gm. (⅕ gr.).
        (emetic) 0.25 Gm. (4 gr.).

THE OFFICIAL U.S.P. AND UNOFFICIAL BISMUTH SALTS

BISMUTHI BETANAPHTHOLAS.—Bismuth Betanaphthol. Buff colored to greyish-brown, amorphous powder of a varying composition of Bi_2O_3 and $C_{10}H_7OH$; odor, faint; tasteless; insoluble in water, alcohol, chloroform or ether.
Action: Antiseptic, astringent.
Dose: 0.5 Gm. (8 gr.).

Bismuthi Citras.—Bismuth Citrate, $BiC_6H_5O_7$. White, amorphous or micro-crystalline powder, odorless and tasteless, permanent in air, insoluble in water or alcohol, soluble in ammonia water.
Preparation: Bismuthi et Ammonii Citras, U.S.P.

BISMUTHI ET AMMONII CITRAS.—Bismuth and Ammonium Citrate. Shining, pearly or translucent scales; odorless, having a metallic taste, very soluble in water but sparingly soluble in alcohol.
Action: Astringent.
Dose: 0.125 Gm. (2 gr.).
Preparation: Liquor Bismuthi, N.F.

Bismuthi Oxidum.—Bismuth Oxide, Bi_2O_3. Lemon-yellow powder, resembling bismuth subnitrate in its medical properties.
Dose: 0.65 Gm. (10 gr.).

BISMUTHI SUBCARBONAS.—Bismuth Subcarbonate. White or pale yellowish-white powder; odorless and tasteless permanent in air, insoluble in water or alcohol.
Action: Astringent.
Dose: 0.5 Gm. (8 gr.).

BISMUTHI SUBGALLAS.—Bismuth Subgallate, Dermatol. Amorphous, bright yellow powder, odorless and tasteless, permanent in air, insoluble in water, alcohol, chloroform, or ether.
Action: Astringent.
Dose: 0.5 Gm. (8 gr.).

BISMUTHI SUBNITRAS.—Bismuth Subnitrate. White powder, odorless, almost tasteless, and slightly hygroscopic, insoluble in water or alcohol.
Action: Astringent, contained in Glyc. Bismuthi, N.F., Magma Bismuthi, U.S.P.
Dose: 0.5 Gm. (8 gr.).

BISMUTHI SUBSALICYLAS.—Bismuth Subsalicylate. White or nearly white amorphous or crystalline powder; odorless, tasteless, and permanent in air, almost insoluble, in cold water, and in alcohol.
Action: Intestinal, antiseptic.
Dose: 0.5 Gm. (8 gr.).

GROUP V.—MANGANESE AND IRON

THE OFFICIAL U.S.P. AND N.F. MANGANESE SALTS

Mangani Citras Solubilis.—Soluble Manganese Citrate, Manganese and Sodium Citrate. Yellowish or pinkish-white powder, or as translucent scales; odorless and having a slightly bitter and astringent taste. It is permanent in air, soluble in 4 parts of water, nearly insoluble in alcohol.
Action: Tonic.
Dose: 0.2 Gm. (3 gr.).

MANGANI DIOXIDUM PRÆCIPITATUM.—Precipitated Manganese Dioxide, MnO_2. Heavy, black, fine, odorless, tasteless powder insoluble in water or alcohol.
Action: Alterative, emmenagogue.
Dose: 0.25 Gm. (4 gr.) (pill form).

Mangani Glycerophosphas Solubilis.—Soluble Manganese Glycerophosphate, Manganous Glycerinophosphate. ($MnC_3H_7O_6P$ and citric acid.) Yellowish or pinkish-white powder, odorless, and having an acid taste, soluble in 4 parts of water. Alcohol dissolves out the citric acid, leaving an insoluble residue.
Action: Nutritive, tonic.
Dose: 0.2 Gm. (3 gr.).

Mangani Hypophosphis.—Manganese Hypophosphite, $Mn(PH_2O_2)_2$ + H_2O (U.S.P. VIII). Pink, granular, or crystalline powder; odorless and nearly tasteless; permanent in air, soluble in water, insoluble in alcohol.
Action: Nerve tonic.
Dose: 0.2 Gm. (3 gr.).
Preparation: Syrupus Hypophosphitum Compositus, U.S.P. VIII.

Mangani Sulphas.—Manganese Sulphate, $MnSo_4$ + $4H_2O$ (U.S.P. VIII). Translucent, pale rose-colored prisms, soluble in water, insoluble in alcohol.
Action: Alterative, cholagogue.
Dose: 0.2 Gm. (3 gr.) (pill form).

THE OFFICIAL U.S.P. AND N.F. AND UNOFFICIAL IRON SALTS

FERRI CARBONAS SACCHARATUS.—Saccharated Ferrous Carbonate. Greenish-brown powder, gradually becoming oxidized by contact with the air; odorless, and having at first a sweetish, afterward slightly ferruginous taste. It is only partially soluble in water, but dissolves readily in hydrochloric acid.
Action: Chalybeate, tonic.
Dose: 0.25 Gm. (4 gr.).

FERRI CHLORIDUM.—Ferric Chloride, Iron Perchloride, $FeCl_3$. Orange yellow, crystalline pieces, odorless or having a faint odor of hydrochloric acid and having a strongly styptic taste, deliquescent, soluble in water, alcohol, glycerin, or ether. (In Liq. Ferri Chloridi, U.S.P.)
Action: Externally styptic, internally tonic (rarely).
Dose: 0.065 Gm. (1 gr.).

Ferri Citras.—Ferric Citrate. Thin, transparent, garnet red scales, without odor and having a slight ferruginous taste, slowly soluble in water, insoluble in alcohol.
Action: Chalybeate, tonic.
Dose: 0.2 Gm. (3 gr.).

FERRI ET AMMONII CITRAS.—Iron and Ammonium Citrate, Soluble Ferric Citrate, Ammonia Ferric Citrate. Thin, transparent, garnet-red scales, odorless and having a saline, mildly ferrunginous taste, deliquescent, readily soluble in water, insoluble in alcohol.
Action: Chalybeate, tonic.
Dose: 0.25 Gm. (4 gr.).

Ferri et Ammonii Sulphas.—Ferric Ammonium Sulphate, Iron Ammonium Alum, $FeNH_4(SO_4)_2 + 12H_2O$, U.S.P. VIII. Pale violet, octahedral crystals, without odor and having an acrid, styptic taste, efflorescent, soluble in 2.7 parts of water, insoluble in alcohol.
Action: Astringent, styptic.
Dose: 0.5 Gm. (7½ gr.).

Ferri et Ammonii Tartras.—Iron and Ammonium Tartrate, U.S.P. VIII. Thin, transparent scales, varying from garnet red to reddish-brown, without odor and having a sweetish, slightly ferruginous taste, slightly deliquescent, very soluble in water, insoluble in alcohol.
Action: Non-astringent, chalybeate, tonic.
Dose: 0.25 Gm. (4 gr.).
Ferri et Potassii Tartras.—Iron and Potassium Tartrate, U.S.P. VIII. Thin, transparent scales, varying in color from a garnet red to reddish brown, without odor and having a sweetish, ferruginous taste, slightly deliquescent in air, soluble in water, insoluble in alcohol.
Action: Non-astringent, hematic.
Dose: 0.5 Gm. (8 gr.).
FERRI ET QUININÆ CITRAS.—Iron and Quinine Citrate, Ferri et Quininæ Citras Solubilis, U.S.P. VIII. Thin, transparent scales, of a greenish or golden yellow color, odorless, and having a bitter, mildly ferruginous taste, deliquescent, soluble in water, slightly soluble in alcohol.
Action: Bitter, chalybeate, tonic.
Dose: 0.25 Gm. (4 gr.).
Ferri et Strychninæ Citras.—Iron and Strychnine Citrate, U.S.P. VIII. Thin, transparent scales, varying in color from a garnet red to yellowish brown, without odor and having a bitter, slightly ferruginous taste, deliquescent, readily soluble in water, partially soluble in alcohol.
Action: Chalybeate, tonic.
Dose: 0.125 Gm. (2 gr.).
Ferri Glycerophosphas.—Ferric Glycerophosphate (variable quantity of $Fe_2(C_3H_7O_6P)_3$). Yellowish-green, transparent, amorphous scales, or as a yellowish-green powder, odorless, tasteless, slowly soluble in 2 parts water, insoluble in alcohol.
Action: Nutritive, tonic.
Dose: 0.2 Gm. (3 gr.).
Ferri Hydroxidum.—Ferric Hydroxide, $Fe(OH)_3$, U.S.P. VIII. Brownish-red magma or reddish-brown, tasteless, insoluble powder.
Action: Used in arsenical antidote, Ferric Hydroxidum cum Magnesii Oxide, U.S.P.
Ferri Hypophosphis.—Ferric Hypophosphate, U.S.P. VIII. White or greyish-white powder, odorless and nearly tasteless, permanent in air, soluble in 2,300 parts of water, 1,200 parts of boiling water; more readily soluble in the presence of hypophosphorous acid, or in warm concentrated solutions of alkali citrates.
Action: Nutrient, tonic.
Dose: 0.2 Gm. (3 gr.).
Preparation: Elixir (Liquor and Syrup.) Ferri Hypophosphitis, N.F.
Ferri Lactas.—Ferrous Lactate, $Fe(C_3H_5O_3)_23H_2O$. Greenish-white crystalline masses having a slight, characteristic odor and a mild, sweet, ferruginous taste, soluble in 40 parts of water, and in 12 parts of boiling water; freely soluble in alkali citrates, almost insoluble in alcohol.
Action: Used in diarrhea.
Dose: 0.3 Gm. (5 gr.).
FERRI PHOSPHAS.—Ferric Phosphate (Ferric Phosphas Solubilis, U.S.P. VIII). Thin, bright green, transparent scales, without odor, and having an acidulous, slightly saline taste, permanent in dry air, freely soluble in water, insoluble in alcohol.
Action: Tonic.
Dose: 0.25 Gm. (4 gr.).
Preparations: Elixir Ferri, Quininæ et Strychninæ Phosphatum, U.S.P. VIII. Elixir Ferri Phosphatis, N.F.
Ferri Pyrophosphas.—Ferric Pyrophosphate, U.S.P. VIII, Soluble Ferric Pyrophosphate. Thin, apple-green, transparent scales, without odor and having an acidulous, slightly saline taste, permanent in dry air, freely soluble in water, insoluble in alcohol.
Action: Chalybeate, tonic.
Dose: 0.25 Gm. (4 gr.).
Preparation: Elixir Ferri Pyrophosphates, N.F.
Elixir Ferri Pyrophosphatis Quininæ et Strychninæ, N.F.
FERRI SULPHAS.—Ferrons Sulphate, Iron Protosulphate, Green Vitriol, $FeSO_4 + 7H_2O$. Pale, bluish-green, monoclinic prisms, without odor, and

having a saline, styptic taste, efflorescent, soluble in 1.4 parts of water, 0.4 part of boiling water; insoluble in alcohol.
Action: Disinfectant, deodorant.
Dose: 0.1 Gm. (1½ gr.).
Preparations: Ferri Sulphas Exsiccatus, U.S.P.
  Ferri Sulphas Granulatus, U.S.P.
  Mistura Ferri Compositus, U.S.P.
  Liquor Zinci et Ferri Compositus, N.F.
  Ferri Carbonas Saccharatus, U.S.P.
  Liquor Ferri Subsulphatis, U.S.P.
  Liquor Ferri Tersulphatis, U.S.P.

FERRI SULPHAS EXSICCATUS.—Exsiccated Ferrous Sulphate, Dried Ferrous Sulphate, $FeSO_4$. Greyish-white powder, slowly soluble in water. It is used in making pills the crystallized sulphate not being adapted to that purpose.
Dose: 0.06 Gm. (1 gr.).

FERRI SULPHAS GRANULATUS.—Granulated Ferrous Sulphate, Precipitated Ferrous Sulphate, $FeSO_4 + 7H_2O$. Very pale, bluish-green, crystalline powder, which conforms in every respect to the reactions and tests of Ferri Sulphas.
Dose: 0.1 Gm. (1½ gr.).

FERRUM.—Iron, Fe. Fine, bright and non-elastic wire, used in the making of iron preparations such as: Syr. Ferr. Iod. and Syr. Ferr. et Mangan. Iod., N.F.

FERRUM REDUCTUM.—Reduced Iron, Ferrum Redactum, Iron by Hydrogen, Quevenne's Iron. Very fine, greyish-black, lusterless powder, without odor or taste, permanent, insoluble in water or alcohol.
Action: Tonic.
Dose: 0.06 Gm. (1 gr.).
Preparations: Pilulæ Ferr. Quin. Stryc. et Arsen. Fort., N.F.
  Pilulæ Ferri Iodidid, U.S.P.

GROUP VI.—Gold and Mercury

THE OFFICIAL U.S.P. GOLD SALT (NONE N.F.)

AURI ET SODII CHLORIDUM.—Gold and Sodium Chloride. Orange-yellow powder, odorless, having a saline metallic taste, deliquescent, very soluble in water; alcohol or ether dissolving the gold chloride and leaving the sodium chloride.
Action: Alterative, tonic.
Dose: 0.005 Gm. (1⁄12 gr.).

THE OFFICIAL U.S.P. MERCURY SALTS (NONE N.F.)

HYDRARGYRI CHLORIDUM CORROSIVUM.—Corrosive Mercuric Chloride, Bichloride of Mercury, Corrosive Sublimate, Mercuric Chloride, Perchloride of Mercury, $HgCl_2$. Heavy, colorless, rhombic crystals, or crystalline masses, or as a white powder, odorless, and having a characteristic and persistent metallic taste; permanent in air, soluble in 13.5 parts of water, 3.8 parts of alcohol, about 12 parts of glycerin. The solubility increases in each case with heat.
Action: Antiseptic.
Dose: 0.003 Gm. (1⁄20 gr.).
Preparations: Toxitabellæ Hydrargyri Corrosivi, U.S.P.
  Lotio Flava, N.F.
  Hydrargyri Iodidum Rubrum, U.S.P.
  Hydrargyrum Ammoniatum, U.S.P.

HYDRARGYRI CHLORIDUM MITE.—Mild Mercurous Chloride, Mercurous Chloride, Calomel, Protochloride of Mercury, Subchloride of Mercury, HgCl. White, impalpable powder, becoming yellowish white on being triturated with strong pressure, odorless, tasteless, insoluble in water or neutral liquids.
Action: Laxative, alterative.
Dose: 0.15 Gm. (2½ gr.) (laxative).
  0.015 Gm. (¼ gr.) (alterative).
Preparations: Pilulæ Catharticæ Compositæ, U.S.P., Lotio Nigra, N.F.
  Pulv. Hydrarg. Clor. Mit. et Jalap, N.F.

HYDRARGYRI IODIDUM FLAVUM.—Yellow Mercurous Iodide, Mercurous Iodide, Protoiodide of Mercury, Yellow Iodide of Mercury, HgI. Bright yellow, amorphous powder, odorless and tasteless, almost insoluble in water and wholly insoluble in alcohol or ether.

Action: Alterative, antisyphilitic (in pill).

Dose: 0.01 Gm. ($\frac{1}{6}$ gr.).

HYDRARGYRI IODIDUM RUBRUM.—Red Mercuric Iodide, Mercuric Iodide, Biniodide of Mercury, Red Iodide of Mercury, HgI_2. Scarlet-red, amorphous powder, odorless and tasteless, permanent in air, almost insoluble in water, soluble in 115 parts of alcohol, 120 parts of ether.

Action: Alterative, antisyphilitic.

Dose: 0.003 Gm. ($\frac{1}{20}$ gr.)

Preparations: Liquor Arseni et Hydrargyri Iodidi, U.S.P.

HYDRARGYRI OXIDUM FLAVUM.—Yellow Mercuric Oxide, HgO. Light, orange-yellow, amorphous, heavy, impalpable powder, odorless, and having a somewhat metallic taste, permanent in air, but turning dark when exposed to the light, almost insoluble in water, insoluble in alcohol; soluble in dilute hydrochloric or nitric acid, forming colorless solutions.

Action: Used externally.

Dose: Not given in U.S.P.

Preparations: Oleatum Hydrargyri, U.S.P.

Unguentum Hydrargyri Oxidi Flavi, U.S.P.

HYDRARGYRI OXIDUM RUBRUM.—Red Mercuric Oxide, Red Precipitate, HgO. Heavy, orange-red, crystalline scales, or as a crystalline powder, acquiring a yellow color when finely divided, odorless and having a somewhat metallic taste, permanent in air, almost insoluble in water, insoluble in alcohol, soluble in dilute hydrochloric or nitric acids.

Action: Used externally.

Dose: Not given in U.S.P.

Preparation: Unguentum Hydrargyri Oxidi Rubri, U.S.P. VIII.

HYDRARGYRI SALICYLAS.—Mercuric Salicylate, Mercuric Subsalicylate. White, slightly yellowish or slightly pinkish powder, odorless, tasteless insoluble in water or alcohol.

Action: Antirheumatic.

Dose: 0.004 Gm. ($\frac{1}{15}$ gr.).

HYDRARGYRUM.—Mercury, Quicksilver, Hg. Shining, silver-white metal; liquid at ordinary temperatures, odorless, tasteless, insoluble in ordinary solvents, but soluble in nitric acid. Used in making the preparations of mercury.

Preparations: Hydrargyrum cum Creta, U.S.P.

Massa Hydrargyri, U.S.P.

Unguentum Hydrargyri, U.S.P.

Unguentum Hydrargyri Dilutum, U.S.P.

HYDRARGYRUM AMMONIATUM.—Ammoniated Mercury, White Precipitate, $HgNH_2Cl$. White, pulverized pieces, or as a white amorphous powder, without odor, and having an earthy, afterward styptic and metallic taste, permanent in air, insoluble in water or alcohol, soluble in warm hydrochloric or nitric acid, etc.

Action: Used externally.

Preparations: Unguentum Hydrargyri Ammoniati, U.S.P.

HYDRARGYRUM CUM CRETA.—Mercury with Chalk, Grey Powder. Light grey, rather damp powder, free from grittiness, without odor, and having a slightly sweet taste, insoluble in neutral liquids.

Action: Cholagogue, alterative.

Dose: 0.25 Gm. (4 gr.).

GROUP VII.—ANTIMONY AND ARSENIC

THE OFFICIAL U.S.P. AND N.F. ANTIMONY SALTS

ANTIMONII ET POTASSII TARTRAS.—Antimony and Potassium Tartrate, Tartar Emetic, $2K(SbO)C_4H_4O_6 + H_2O$. Crystals, or white granular powder, odorless, having a sweet, afterward metallic taste, soluble in 15.5 parts water, insoluble in alcohol.

Action: Expectorant, emetic.
Dose: (expectorant) 0.005 Gm. ($\frac{1}{10}$ gr.).
　　.　(emetic) 0.03 Gm. ($\frac{1}{2}$ gr.).
Preparations: Syrupus Scillæ Compositus, U.S.P.
　　Vinum Antimonii, U.S.P. VIII.
* *Antimonii Oxidum.*—Antimony Oxide, Antimony Trioxide, Sb_2O_3. Heavy, greyish-white powder, permanent in air, odorless and tasteless, almost insoluble in water or alcohol, readily soluble in hydrochloric or tartaric acids. (Contained in Pulvis Antimonialis, N.F.),
Dose: 0.06 Gm. (1 gr.).

THE OFFICIAL U.S.P. ARSENIC SALTS

Arseni Iodidum.—Arsenous Iodide, AsI_3. Orange-red, inodorous, crystalline powder, soluble in 12 parts water, 28 parts alcohol, completely soluble in chloroform or ether.
Action: Alterative.
Dose: 0.006 Gm. ($\frac{1}{10}$ gr.).
Preparations: Liquor Arseni et Hydrargyri Iodidi, U.S.P.
ARSENI TRIOXIDUM.—Arsenic Trioxide (Acidum Arsenosum, 1890). White powder, or in heavy masses, odorless, tasteless, sparingly soluble in 100 parts water, sparingly in alcohol, soluble in 5 parts of glycerin and readily soluble in acids and alkalies.
Action: Irritant, alterative.
Dose: 0.002 Gm. ($\frac{1}{30}$ gr.).
Antidote: Ferri Hydroxidum cum Magnesio Oxido.
Preparations: Liquor Acidi Arsenosi, U.S.P.
　　Liquor Arseni et Hydrargyri Iodidi, U.S.P.
　　Liquor Potassii Arsenitis, U.S.P.
　　Liquor Sodii Arsenatis, U.S.P.
　　　(All official U.S.P. arsenical solutions 1 per cent. strength.)
　　Liquor Auri et Arseni Bromidi, N.F.
　　Liquor Arsenicalis, Clemens, N.F.
　　Pilulæ Metallorum, N.F.

GROUP VIII.—Inorganic Acids

Acidum Hydrochloricum.—Hydrochloric Acid, strength 31 to 33 per cent.
Acidum Hydrochloricum Dilutum.—Dilute Hydrochloric Acid, strength, $9\frac{1}{2}$ to $10\frac{1}{2}$ per cent.
Acidum Hydrobromicum Dilutum.—Dilute Hydrobromic Acid, $9\frac{1}{2}$ to $10\frac{1}{2}$ per cent.
Acidum Hydriodicum Dilutum.—Dilute Hydrochloric Acid, $9\frac{1}{2}$ to $10\frac{1}{2}$ per cent.
Acidum Hypophosphorosum.—Hypophosphorous Acid, 30 to 32 per cent.
Acidum Hypophosphorosum Dilutum.—Diluted Hypophosphorous Acid, $9\frac{1}{2}$ to $10\frac{1}{2}$ per cent.
Acidum Nitricum.—(Aqua Fortis), Nitric Acid, 67 to 69 per cent.
Acidum Nitricum Dilutum.—Dilute Nitric Acid, $9\frac{1}{2}$ to $10\frac{1}{2}$ per cent.
Acidum Nitrohydrochloricum.—(Aqua Regia) Nitrohydrochloric Acid.
Acidum Nitrohydrochloricum Dilutum.—Dilute Nitrohydrochloric Acid, $9\frac{1}{2}$ to $10\frac{1}{2}$ per cent.
Acidum Sulphuricum.—(Oil Vitriol) Sulphuric Acid, 93 to 95 per cent.
Acidum Sulphuricum Dilutum.—Dilute Sulphuric Acid, $9\frac{1}{2}$ to $10\frac{1}{2}$ per cent.
Acidum Sulphuricum Aromaticum.—(Elixor Vitriol) Aromatic Sulphuric Acid, 19 to 21 per cent. by weight of H_2SO_4.
Acidum Phosphoricum.—Phosphoric Acid, 85 to 88 per cent.
Acidum Phosphoricum Dilutum.—Dilute Phosphoric Acid, $9\frac{1}{2}$ to $10\frac{1}{2}$ per cent.
　Almost all of the inorganic acids can be classed as anhydrotics and remote alkalyzers.
　The dose of dilute acids is from 0.3 to 1 mil (5 to 15 minims).

GROUP IX.—Halogens

Chlorine is unofficial.
Bromine was official in U.S.P. VIII and is official in the present N.F.
Iodine is official in U.S.P. IX and is found in the following preparations:

Tinctura Iodi, U.S.P.
Liquor Iodi Compositus, U.S.P.
Unguentum Iodi, U.S.P.
Liquor Iodi Phenolatus, N.F.
Syrupus Iodotannicus, N.F.
Tinctura Iodi Churchill, N.F.
Tinctura Iodi Decolorata, N.F.
Linimentum Ammon. Iod., N.F.
Collodium Iodi-(Iodi), N.F.
Phenol-(Phenol Iodatum) Iodatum, N.F.
Amylum Iodatum, U.S.P. VIII.

GROUP X.—Oxygen and Nitrogen Monoxide

Since the addition of both these gases to the U.S.P. IX it has become necessary to add them to the list.

OXYGENIUM.—Oxygen, O.
NITROGENII MONOXIDUM.—Nitrogen Monoxide, N_2O (Laughing Gas).

GROUP XI.—Sulphur and Phosphorus

Sulphur is official in three forms:

Sulphur Lotum.—Washed Sulphur. (In Troches Sulph. et Pot. Bitart, N.F.).
Sulphur Præcipitatum.—Precipitated Sulphur.
Sulphur Sublimatum.—Sublimed Sulphur, contained in 2 N.F. ointments.

Phosphorus is official as Phosphorous, and in the following preparations:

Pilulæ Phosphori, U.S.P.
Oleum Phosphoratum, N.F.
Elixir Phosphori, N.F.
Elixir Phosphori et Nucis Vomicæ, N.F.
Liquor Phosphori, N.F.

GROUP XII.—Carbon, Boron, Silicon

ACTION OF INORGANIC SALTS

Brief Synopsis.—The following synopsis is introduced merely to illustrate how inorganic remedial agents may be grouped for therapeutical study. The memorizing of therapeutic action of individual drugs is too laborious and unprofitable, especially for beginners. The student is advised to consult works on Therapeutics, construct such groups for himself and expand the treatment of them with possibly the help of an instructor.

Alkalies (Sodium, Potassium, Lithium, and Ammonium).—The alkalies as a class exhibit an extremely interesting set of reactions, the results of which may be attributed to the whole molecule (salt action) on one hand and to the positive and negative ions on the other. The chief effects are probably due to the negative ions—as in bromides, iodides, etc. (where salt action does not play a part), and in a large measure the salts of the alkalies are administered for the effects of this part of the molecule. Dixon gives as typical actions for the alkalies as a class that they neutralize acids, saponify fats, dissolve proteins, and act as disinfectants. The hydroxides in a concentrated form are extremely caustic, and by virtue of their solvent action on protein material (formation of alkali albuminates) destroy any tissue with which they come in contact in a remarkably short time. Their action in this respect is even more severe than that of concentrated acids, since the healing process seems to go on more rapidly with the latter class of compounds. They also exercise an inhibiting action on the gastric glands and indirectly in this way diminish the flow of the pancreatic juice, through a lessening of the amount of secretin produced. The alkalies, especially lithium, are said to prevent or arrest the precipitation of uric acid, although it does not dissolve any acid previously precipitated. Diuresis is produced by alkalies by virtue of their salt action rather than by the action of the individual ions.

Individually the potassium ion acts as a universal depressant, affecting especially the circulation and the central nervous system. When injected, potassium lowers the blood pressure markedly. The lithium ion acts much like the potassium ion, only less powerfully. The sodium ion exerts the least effect of all the alkalies.

Forscheimer gives the following disorders in which the alkalies as a class are indicated: acute endocarditis, gastric hypersecretion, acute peritonsillitis, acute urethral gonorrhœa, carcinoma of stomach, to change reaction of urine in tuberculous kidney, gastric hypersecretion, diabetes before surgical operation, diabetic acidoses, diabetic coma, expectorants in chronic bronchitis, gastric hyperacidity, with iodides in syphilitic gastritis, in non-bleeding gastric ulcer, gastric neuroses.

ALKALINE EARTH SALTS (Calcium, Barium, Strontium, Magnesium).—*Calcium*, also Cerium and Aluminum.—Calcium and its salts play a very important part in our daily metabolism and as a result the calcium balance must in general remain positive or nervous and other complications will arise. Administered by mouth, the calcium ion has very little specific effect, probably largely because of its slow absorption. When injected into the circulation, however, it exhibits a very definite and decided action on muscular tissue particularly. It increases and prolongs normal contraction and retards relaxation, and it constricts the walls of blood vessels producing a rise in blood pressure. It is particularly useful in retarding inflammatory processes. The alkaline salts of calcium are used in diarrhea and the hydroxide is often given with the milk to infants to render it more digestible. Again lime water mixed with linseed oil is used externally in burns. Lime salts are also given to increase the coagulability of the blood.

Ten different lime salts enter into the Materia Medica of this element. The salts include the oxide, hydroxide, carbonate, phosphate, chloride, hypophosphite, and lactate. The carbonate enters into the composition of two important preparations, *e.g.*, mercury and chalk, and chalk mixture (Mistura Cretæ). Lime salts are indicated in the following disorders (Forscheimer): CaCl₂ used with normal saline in asiatic cholera; in addition to salt bath in chronic bronchitis; to control gastric hemorrhage; in epidemic dropsy, in hematuria complicating scarlet fever, before operation in catarrhal jaundice, to obviate quinine intolerance.

Barium.—The barium ion, like calcium, has special affinity for muscle. The action in general is that of a tonic constriction more forceful than that given by calcium. When the chloride is given by mouth, it produces violent pains, nausea, and vomiting, due to its action on the plain muscles of the stomach. Like calcium it is absorbed very slowly. Barium chloride is sometimes given internally in cardiac disease. Forscheimer says it is indicated to prevent the anaphylactic reaction.

Strontium.—The bromide, iodide, salicylate, and lactate are used at times. The specific effects produced are, however, attributable to the negative ions. The specific action of the metal ion has not been fully determined.

Magnesium.—When magnesium salts are injected, they affect primarily the medullary centers, producing unconsciousness, and a fall of blood pressure. All tissues are depressed in fact. Calcium relieves this condition immediately, due probably to the formation of slightly ionized triple phosphate. When applied to a nerve trunk, $MgSO_4$ blocks the nerve impulse. These results are not obtained, however, from its oral administration, because of the slight extent to which it is absorbed. All salts of magnesium are said to change to the acid carbonate in the intestine. $MgSO_4$ is probably the best of our saline laxatives, its action being due to osmotic changes due to salt concentration. The sulphate, carbonate, and oxide, the latter in two forms each light and heavy, are the chief compounds used. The oxide and the carbonate are antacids as well as laxatives. The salts of magnesium are indicated in the following disorders: *Magnesia*, gastric hyperacidity, acute gastritis, chronic constipation, milk of Mg, following calomel in membranous laryngitis, mouth wash in diabetes mellitus, in nephritis. *Mg. citrate*, to expel tape-worms, to empty gastro-intestinal tract in pneumonia. *Mg. salicylate*, in peristaltic unrest. $MgSO_4$, in acute alcoholism, fibrinous pleurisy, acute gastritis, acute nephritis, in bacillary dysentery, in biliousness, chronic alcoholism, in chronic catarrhal jaundice, constipation, wood alcohol poisoning, modification of the severity of pleurisy, prevention of lead poison.

Cerium.—Only one salt of cerium, the oxalate, is important. It is insoluble in water so that ion action seems improbable. Just how efficient this compound is medicinally is hard to say, since absorption seems to be nil. It has been used as a remedy in vomiting in pregnancy, and also for relief of gastric crises in tabes (Forscheimer). Its beneficial effects are attributed to local protective action similar to that of bismuth.

Aluminum.—The salts of aluminum are typical mineral astringents and antiseptics. The most used salt is the aluminum potassium sulphate or common alum and the hydroxide exsiccated alum is also used; besides these we have kaolin, aluminum-aceto tartrate, and aluminol or aluminum naphthol sulphonate. Taken by mouth aluminum salts have until recently been held to be very slightly absorbed. There seems to be some evidence, however, that it is absorbed into the blood stream. When injected, the aluminum ion acts as a poison to the liver and kidney, producing fatty degeneration. Nervous disturbances, tremors, paralysis, and local anesthesia are also reported as being due to its injection. The sulphate is used in acute and chronic lead poisoning. Locally, aluminum salts act as astringents and antiseptics. They are also used for hardening the skin, checking local hemorrhage, and as a styptic. They are indicated in the following disorders: *Alum*, external use in smallpox, external use in varicella, as gargle in nasopharyngitis, as spray in laryngitis. *Alumnol*, in nasopharyngitis. *Kaolin*, used in gastric hyperacidity, chiefly in making of cataplasma kaolin.

SILVER, BISMUTH, COPPER, LEAD.—*Silver.*—The medicinal preparations of silver include the cyanide, nitrate (also fused and mitigated), and oxide. We have in addition numerous compounds in which silver is in organic combination, such as protargol, argyrol, and argonin. Silver salts were at one time used as alternatives, in diseases such as epilepsy and locomotor ataxia. Their use, however, has been abandoned since evidence for the slight absorption of silver compounds has accumulated. It is, however, of value in inflammatory conditions of the stomach, such as subacute gastritis and gastric ulcer.

As a local remedy, silver salts are very important. The nitrate is very effectively used as caustic, germicide, and astringent. It is thus used in pharyngitis, laryngitis, urethritis, and colitis. In tonsillitis it is also of value. It is indicated and used in cases of gonorrheal conjuncti vitis, also. Forscheimer gives the following as cases where silver salts are indicated: *Organic salts*, acute gonorrhea, chronic gonorrhea urethritis. $AgNO_3$, acute tonsillitis, uvulitis, antiseptic wash, cauterization of bite, chronic laryngitis, rhinitis, nasopharyngitis, bacillary dysentery, gastric hypersecretion, foot and mouth disease, irritable stricture, irrigation of bladder, mercurial stomatitis, tabes, etc.

Bismuth.—Soluble salts of bismuth are toxic and not administered. They produce in large doses a primary stimulation of the nerve centers which is shortly followed by depression, lowering of blood pressure, and irritation of the organs of excretion. The insoluble salts find their chief use in treatment of inflammations of the mucous membrane and in catarrhal conditions. The most commonly used salt is the subnitrate. The citrate, subcarbonate, subgallate, and subsalicylate are also employed. Locally the subnitrate is used as a dusting powder. The subgallate finds much use in dermatology. Bismuth salts are indicated in the following conditions: mucous colitis, cirrhosis of liver, hypersecretion, control of diarrhea, to control gastric hemorrhage, non-bleeding gastric ulcer, enteritis; *bismuth subcarbonate*, in typhoid fever; *bismuth subgallate*, in enterocolitis; $BiONO_3$ in bacillary dyspepsia, to control diarrhea, dusting powder.

Copper.—Copper Salts are used medicinally for local effects. Here they act as astringent, caustic and antiseptic. Copper is very destructive of certain forms of life. The typhoid bacillus is very sensitive toward it. As an astringent, it is often used in chronic enteritis and in treating small ulcers of the conjunctiva it acts as a caustic. Only the sulphate is official. Copper is indicated in chronic gonorrheal urethritis, membranous rhinitis, oriental sore, scurvy, blastomycosis.

Lead.—The following preparations are employed: acetate, iodide, oxide, plaster, and cerate, also liquor subacetates and its diluted solution. Internally lead is at times used as an astringent and in the treatment of diarrhea, especially in combination with opium. The solution of lead subacetate is used in various acute inflammatory conditions, such as sprains and bruises. Lead has a peculiarity of being absorbed by the tissues of the body where it is gradually stored up and produces so-called chronic lead poisoning. The first signs of this condition often appears as an anemia, or as a colic. Lead has much the same action on muscle fibers that barium has and there is thus observed a contraction in all plain muscular tissue throughout the body. This action is also visible on the vascular system and the arteries. Other characteristic signs of lead poisoning are the "blue line" at the margin of gum and teeth, and the "wrist drop" produced by paralysis of the extensors of wrist and fingers. Lead salts are indicated in the following conditions: lead acetate, abscess and gangrene of lung, hematuria, relief of edema.

IRON, ARSENIC, MERCURY, ZINC, ANTIMONY, PHOSPHORUS.—Iron and its salts are drugs which act largely by affecting metabolic processes. There are, of course, other actions of iron salts as will be shown later. In surveying the medicinal preparations of this element and its salts, one meets with no less than thirty-five separate compounds. Fourteen of these are soluble solid preparations; nine, insoluble solid preparations; and twelve, liquid preparations. There are eight scale salts of iron, besides the reduced iron itself. The iodide, chloride, carbonate, and sulphate are the most commonly used salts. Since iron is a normal constituent of body tissues and since we know that there is a continual breaking down of the hemoglobin of the blood to form many body pigments which subsequently are eliminated from the body. The iron thus lost must be replaced. This is normally done by the iron contained in the ingested food stuffs.

The physiological action of iron has been a subject much debated, particularly the question of the availability of so-called inorganic iron. That inorganic iron can be absorbed seems now to be quite certain, but what its drug action is is not always so clear.

Iron has been called "a general tonic" and given in a host of conditions without any particular reason except that it would "build up" the system. It does, to be sure, have some action on the blood-making organs and as far as hemoglobin construction is concerned, where iron is lacking, it is beneficial. We have it thus indicated in the several types of anemias. In the form of Basham's mixture it has been given in cases of chronic nephritis, although pharmacologists agree that it has no direct action on the kidneys. Ferric hydroxide enters into the official arsenic antidote and is very beneficial in this connection. Iron salts are also used as powerful astringents and styptics. The subsulphate is most commonly employed. Forscheimer mentions the use of iron in more than thirty-five disorders, of which the following are a few: Addison's disease, amyloid kidney, anemia, bronchial asthma, cerebrospinal meningitis, convalescence from pneumonia, chlorosis, hay-fever, hookworm disease, nephritis, occupation paresis, rachitis, chronic rhinitis, acute tonsillitis.

Arsenic.—Arsenic and its preparations are well-known in the inorganic materia medica. Salts of arsenic have been used for centuries as curative agents. Metallic arsenic itself is not used, but in its stead the trioxide As_2O_3, arsenate, iodide, and several liquors containing these salts.

.As to physiological action, arsenic belongs to drugs that affect general metabolic processes. Arsenic is a protoplasmic poison capable of destroying all forms of life. Its extreme toxicity renders it unfit for use as a bactericide. On the blood-making organs, arsenic has a pronounced but as yet not understood effect. In small doses hyperemia of the bone marrow is produced, which brings about a stimulation of the leucoblastic cells concerned in the formation of white corpuscles. Small doses of arsenic also diminish the catabolic processes, thus bringing about an increase in body weight. Given repeatedly it shows a marked action on the skin, producing a thickening of the epithelium and a keratosis not unlike that seen in the regular callous. In many cases of malnutrition it is indicated.

In organic combination, arsenic is used as an antizymotic. It has been used successfully in certain types of malarial infection. In trypanosomiasis and syphilis it has been particularly useful. Forscheimer mentions arsenic as being indicated in the following diseases: Addison's disease, bronchial asthma, chronic malaria, epilepsy, goiter, hay-fever, neuralgia, leukemia, neurasthenia, neuralgic headache, pernicious anemia, syphilis, diabetes mellitus, Hodgkin's disease, scarlet fever, tetanus, pellagra, arthritis.

Mercury.—Metallic mercury in a fine state of subdivision, as well as many of its salts, forms one of the most important series of inorganic remedial agents we know anything about. The United States Pharmacopœia recognizes thirteen official preparations to be used both internally and externally.   The physiological and therapeutic action of mercury and its salts is extremely varied.   Locally they may be entirely inert or they may exert an escharotic influence on the skin.   The soluble salts are particularly toxic to lower forms of life and form therefore very fine antiseptics. We have mercury salts used as antisyphilitics, as cathartics, as antiseptics, as antiphlogistics, and as diuretics.   As an antiphlogistic mercury acts particularly on the so-called endothelial membranes.   It has been used for inflammation of mucous surfaces, but is not nearly as beneficial in these conditions.   For iritis mercury forms a valuable remedy.   Mercury is a specific for syphilis, and although many preparations of arsenic, such as salvarsan and neosalvarsan, the cacodyllates, etc., are used a great deal, mercurial inunctions and injections of the so-called "grey oil" are very common in modern clinical practice.

Mercury may be used as a cathartic and is so used particularly in the form of calomel.   Mercury has been used with success as a diuretic, being particularly valuable in chronic parenchymatous nephritis.   As an antiseptic mercury is used a great deal in the form of its bichloride. This salt combines with proteins, forming compounds which inhibit regular functioning of life processes.

Forscheimer mentions the following uses for mercury and its salts: syphilis, absorption in pleurisy, tuberculous peritonitis, chancre, gummata, acute myelitis, optic neuritis, throat affections, secondary anemias.

Zinc.—Zinc and its salts come under the general head of mineral astringents.   The Pharmacopœia recognizes eleven salts and preparations.   Most of the zinc salts are either astringents or antiseptics.   The astringent action in all probability comes from the fact that a typical insoluble salt is formed when coming in contact with protein.

The oxide is probably more used than any other salt of zinc.   It is especially indicated in conditions demanding sedative and astringent action. The stearate of zinc also finds some use along this line.   Zinc has been used in epilepsy, although with doubtful success.   The sulphate of zinc is a powerful and rapid emetic and is so used at times, especially in cases of poisoning when apomorphine is not available or indicated.   Zinc salts are not absorbed from the alimentary tract to any large extent.   We do have, however, cases of chronic zinc poisoning.   The symptoms here . are much like those of lead poisoning, producing colic, and other derangements of the alimentary canal.   Forschheimer mentions its use in the following disorders: chronic laryngitis, oriental sore, bronchopneumonia, stomatitis, vaginitis, gonorrheal urethritis, herpes of lips, gastritis, acute opium poisoning, nasopharyngitis, peristaltic unrest.

Antimony.—Antimony to-day finds comparatively little use.   One . salt and one preparation are official in the Pharmacopœia.   In the middle ages about the time of the iatrochemists, we find the virtues of antimony lauded to the skies.   And we find one of the foremost iatrochemists,

Basil Valentine, writing a separate treatise on this element, whose powers, were so wonderful that he entitled it "The Triumphal Chariot of Antimony." Locally tartar emetic, the most important salt of antimony, is an irritant, producing pustules that resemble those of smallpox. Internally tartar emetic is an emetic. It is, however, little employed to-day in this way. In very small doses it is an expectorant, being used in cough mixtures with considerable success. Its action here is in all probability due to an increase in the bronchial secretions. Antimony has also been used for its quieting action of the heart, it unlike aconite or veratrum acting directly on the heart muscle. As a destroyer of trypanosomes it has been used in sleeping sickness. The trichloride of antimony, so-called butter of antimony, is used as a caustic. Forscheimer mentions the use of antimony in only three conditions, viz.; expectorant in bronchitis, relapsing fever, and laryngitis.

Phosphorus.—Phosphorus is a very important element in the human economy. It enters into the making of the constituents of cytoplasm and nucleoplasm, and the formation of bones. It is thus constantly ingested and needed as a food stuff in the form of phosphoproteins, nucleoproteins, phosphorized fats, and inorganic phosphates. We find it used in medicine, for its influence on the development of bone, for building up the nervous tissue in nervous exhaustion and degeneration of nerve center. The pharmacopoeia recognizes elementary phosphorus and the pill of phosphorus; besides this we have seven preparations of the salts of phosphorus, mostly hypophosphites. Forscheimer mentions its use in the following conditions: beri-beri, atropic laryngitis, chronic myelitis, laryngeal tuberculosis, neurasthenia, neuralgia due to anemia, rachitis, mediastinla lymph glands.

ORGANIC CHEMICALS

The organic agents of a medicinal character are extremely numerous. They are sometimes referred to as hydrocarbon derivatives. They are, in fact, of extremely varied composition. Some of them are purely synthetic products, so-called derivatives of the paraffins and benzins, for example. Among these are phenols, amines, ketones, pyrazolones, etc., with their innumeral derivatives. Classified with the organic agents there are also products, characterized as alkaloids, glucosides and neutral principles. Many combinations of metallic bases (silver, mercury, bismuth, copper, etc.) with organic radicles constitute a very important group in the general class of organic agents often called "New Remedies."

For a most complete and satisfactory reference to these remedial agents, see "New and Non-Official Remedies," published by the American Medical Association.

SYNOPSIS OF HYDROCARBONS AND THEIR DERIVATIVES

CLASS I.—METHANE DERIVATIVES:

1. Hydrocarbons—
   Saturated.
   Unsaturated.
2. Halogen Substitution Products of the Hydrocarbons.
3. Alcohols—
   Primary, Secondary, and Tertiary.
   Monatomic, Diatomic, Polyatomic (or monacid, etc.).
4. Derivatives of Alcohols—
   Esters.
   Sulphur Derivatives of Alcohols and Ethers.
   Inorganic Esters.
   Nitriles.
   Amines or Ammonium Bases, Hydroxylamines, and
     Hydrazines.
   Metalloid Compounds, Phosphorus, Arsenic, etc.
   Metallic Compounds.
5. Aldehydes and Ketones.
6. Acids (monobasic, dibasic, tribasic, etc.).
7. Derivatives of Acids—
   Esters (ethereal salts).
   Halogen Substitution Products and Haloids of the Acid
     Radicals.
   Acid Anhydrides.
   Thio-acids and Anhydrides.
   Acids Amides and Amido Acids.
8. Cyanogen Compounds and Their Derivatives.
9. Carbonic Acid Derivatives.
10. Carbohydrates

   Transition to the Aromatic Compounds

1. Polymethylenes.
2. Furfurane, Thiophene, and Pyrrol.
3. Azoles, etc.

CLASS II.—BENZENE OR AROMATIC DERIVATIVES:

1. Hydrocarbons.
2. Halogen Derivatives.
3. Sulphur Derivatives.
4. Nitro Derivatives.
5. Amido Derivatives.
6. Other Nitrogen Derivatives.
   Diazo Compounds.
   Azo Compounds.
   Hydrazines.
. Phenols.
. Alcohols, Aldehydes, and Ketones.
7. Acids.
19. Combinations of the above classes.

DOUBLE RING COMPOUNDS

11. Indigo Group.
12. Diphenyl and its Derivatives.
13. Diphenyl-methane and similar compounds.
14. Triphenyl-methane and Derivatives, including certain dyes—
   1. Malachite Green Group.
   2. Rosaniline Group.
   3. Aurin Group.
   4. Eosin Group.

CONDENSED RING COMPOUNDS

15. Naphthalene and its Derivatives.
16. Anthracene and its Derivatives.
17. Phenanthrene and its Derivatives.
18. Pyridine Group.
19. Chinoline Group.
20. Alkaloids of Complicated Composition.
21. Hydrated Benzenes (Terpenes and Camphors).
22. Tars and Glucosides.
23. Albumins and Albuminoids.

The organic compounds of the various classes may be briefly defined as follows:

CLASS I.—METHANE OR FATTY ACID SERIES

1. HYDROCARBONS of this series are the compounds of carbon and hydrogen, having the carbon atoms connected in a chain—thus, methane, CH_4; ethane, CH_3–CH_3; propane, CH_3–CH_3–CH_2.

   These compounds are the first of a series of compounds varying by the increment CH_2. They may be taken as illustrative of many such series of organic compounds, called homologous series.

   When there are four or more atoms of carbon in the molecule, the carbon atoms may form branching chains, as in isobutane.

   This compound has the same percentage composition, but has different properties from butane, CH_3–CH_2–CH_2–CH_3. This is called isomerism.

   Unsaturated hydrocarbons or derivatives have atoms of carbon united to one another by two or three bonds of affinity —thus, ethylene, $CH_2 = CH_2$; acetylene, $CH \equiv CH$. These compounds will unite with halogens or halogen acids without an equivalent loss of hydrogen.

2. HALOGEN SUBSTITUTION PRODUCTS are hydrocarbons in which one or more atoms of hydrogen are replaced by a corresponding number of atoms of a halogen—thus, chloroform, $CHCl_3$; iodoform, CHI_3.

3. ALCOHOLS are formed by the replacement of one or more hydrogen atoms of a hydrocarbon by a corresponding number of hydroxyl (OH) groups. They are of neutral reaction, but analogous to metallic hydroxides. They combine with acids, losing water, forming compounds analogous to salts, termed esters. They may also be defined as hydroxyl combined with an alkyl radical* —thus, alcohol (ethyl hydroxide), C_2H_5OH.

* *Radicals.*—It is usual to designate as radicals those groups of atoms which are found repeating themselves in a comparatively large number of compounds derived from one another, and in which these combinations play the part of simple elements: *e.g.*, CH_3 is called the methyl radical, CH_3–Cl is methyl chloride, CH_3OH is methyl alcohol, etc. CH_3–CO– is termed the acetyl radical and C_2H_3O–Cl is acetyl chloride, etc.

Alcohols with one (OH) group are termed monatomic—thus, alcohol, ethyl alcohol, CH_3CH_2OH; with two (OH) groups are termed diatomic—thus, glycol, $CH_2OH–CH_2OH$; with three (OH) groups are termed triatomic—thus, glycerin, $CH_2OH–CHOH–CH_2OH$, etc. Those alcohols with two or more groups are called polyatomic. Alcohols are also divided into three classes. If the hydrogen substituted is in the methyl radical (CH_3), making the group CH_2OH, *primary* alcohols are formed; or, if the substitution is in the methylene radical (CH_2), making the group $CHOH$, *secondary* alcohols are formed; or, if the substitution is in the methine radical $(\equiv CH)$, making the group $\equiv COH$, *tertiary* alcohols are formed.†

The principal groups of organic chemistry are the following:

Primary alcohols,	$—CH_2OH.$	Aldehydes,	$—COH.$
Secondary alcohols,	$=CHOH.$	Ketones,	$=CO.$
Tertiary alcohols,	$\equiv COH.$	Acids,	$—COOH.$
Ethers,	$\equiv C–O–C\equiv$	Sulphonic acids,	$—HSO_3.$
Nitriles,	$—C\equiv N.$	Nitro compounds,	$—NO_2.$
Amido compounds,	$—NH_2.$	Oximes,	$=NOH.$
Imido compounds,	$=NH.$		

4. **DERIVATIVES OF ALCOHOLS.**—(*a*) Ethers are compounds of neutral reaction, derived from alcohols by the elimination of one molecule of water from two molecules of alcohol. They are analogous to the metallic oxides—thus, ether (or di-ethyl-oxide), $(C_2H_5)_2O$; ethyl-propyl-ether, $C_2H_5–O–C_3H_7$.

(*b*) Sulphur derivatives of alcohols and ethers are formed by replacing one or more atoms of oxygen by sulphur—thus, mercaptan, C_2H_5SH; ethyl-sulphid, $(C_2H_5)_2S$.

(*c*) Inorganic esters are compounds derived from the inorganic acids by the exchange of the replaceable hydrogen by an alcohol radical—thus, ethyl-nitrate, $C_2H_5–O–NO_2$. They are analogous to inorganic salts.

(*d*) Nitriles are compounds of hydrocyanic acid (HCN) in which the hydrogen is replaced by an alcohol radical—thus, aceto-nitrile or methyl-cyanide, CH_3CN. Iso-nitriles differ in properties from the nitriles by having the radical joined to the nitrogen—thus, CH_3NC.

(*e*) Nitrogen bases: Amines and ammonium bases are compounds formed by the introduction of one or more alcohol radicals in place of the hydrogen in ammonia or ammonium salts—thus, methylamine, CH_3NH_2; trimethylamine, $(CH_3)_3N$.

Amines are designated as primary, secondary (imines), tertiary (nitrile bases), or quarternary (ammonium bases), as one, two, or three atoms of hydrogen are replaced in ammonia, or as the four atoms of hydrogen are replaced in ammonium.

† Groups of elements like the above are always found to have constant properties, and are said to be the *characteristic groups* in the classes in which they are found.

Hydroxylamines and hydrazines are compounds derived respectively from hydroxylamine and hydrazine as the amines are derived from ammonia—thus, methyl-hydroxylamine, NH_2OCH_3; methyl-hydrazine, CH_3–$NHNH_2$.

(*f*) Metalloid compounds: Phosphorus, arsenic, etc.

Phosphines are compounds derived from phosphine as the amines are derived from ammonia—thus, methyl phosphine, CH_3PH_2.

Arsines are compounds derived from arsine in the same manner, trimethyl arsine, $(CH_3)_3As$. Among the derivatives in this class are the cacodyles.

(*g*) Metallic compounds are combinations of the alcohol radicals with the metals—thus, zinc methyl, $Zn(CH_3)_2$; zinc ethyl, $Zn(C_2H_5)_2$.

5. ALDEHYDES AND KETONES are substances which result from the oxidation of primary and secondary alcohols respectively, with the separation of two atoms of hydrogen. Thus, aldehyde, (CH_3CHO), characterized by the group –COH; acetone dimethyl-ketone, CH_3–CO–CH_3, characterized by the group $= CO$.

Oximes are compounds derived from aldehydes and ketones by replacing the oxygen with the group $= NOH$. Thus, aldoxime, CH_3–$CH = NOH$; ketoxime, $(CH_3)C_2 = NOH$.

6. ACIDS are oxidation products of the primary alcohols and the corresponding aldehydes, and contain the characteristic group, –COOH, the hydrogen of which is replaceable by a metal to form a salt. Acids may be monobasic, dibasic, tribasic, etc., as they contain one or more of these groups—thus, acetic acid, CH_3–COOH; oxalic acid, COOH–COOH, etc.

7. DERIVATIVES OF ACIDS.—(*a*) Esters are compounds formed by replacing the typical hydrogen of an acid by an alcohol radical— thus, acetic ether (ethyl-acetic-ether, ethyl-acetate), CH_3–$COOC_2H_5$.

(*b*) Halogen Derivatives.—1. Substitution products in which the halogen replaces the hydrogen of the alcohol radical—thus, monochlor-acetic acid, $CH_2ClCOOH$.

2. Chlorides of the Acid Radicals.—The halogen replaces the hydroxyl (OH) of the acid group—thus, acetyl chloride, CH_3COCl.

(*c*) Acid Anhydrides.—Two molecules of an acid combined with the loss of water—thus, acetic acid anhydride, $(CH_3CO)_2O$.

(*d*) Thio-acids and Anhydrides.—Oxygen of the acids substituted by sulphur—thus, thiacetic acid, CH_3COSH.

(*e*) Amido Acids.—Compounds formed (1) by the replacement of the hydrogen of ammonia by acid radicals—thus, glycocoll (amido acetic acid), $CH_2(NH_2)COOH$.

(2) Acid Amides.—Compounds formed by replacing the OH of the acid by the amido group, NH_2—thus, acetamide, CH_3-CONH_2.

8. CYANOGEN COMPOUNDS AND THEIR DERIVATIVES.—Those compounds derivable from cyanogen, C_2N_2; hydrocyanic or prussic acid, HCN; potassium ferrocyanide, $K_4Fe(CN)_6$; ethylthiocyanate, C_2H_5SCN.

9. CARBONIC ACID DERIVATIVES.—Compounds derivable by substitution from carbonic acid (H_2CO_3)—thus, ethyl carbonate, $CO(OC_2H_5)_2$; carbon oxychloride (phosgene gas), $COCl_2$; carbamide (urea), $CO(NH_2)_3$; guanidine, $CNH(NH_2)_2$; uric acid, xanthine, etc.

10. CARBOHYDRATES.—Compounds of carbon, hydrogen, and oxygen containing two atoms less of hydrogen than the corresponding polyatomic alcohol. Chemically, they are aldehyde alcohols or ketone alcohols.

The principal groups are:

1. Grape sugar group, $C_6H_{12}O_6$.
2. Cane sugar group, $C_{12}H_{22}O_{11}$.
3. Cellulose group, $(C_6H_{11}O_5)_x$.

TRANSITION TO THE AROMATIC COMPOUNDS

1. Polymethylenes are compounds containing three or more methylene (CH_2) groups joined in a ring—thus, tri-methylene,

$$\begin{array}{c} CH_2 \\ \diagup \diagdown \\ H_2C\!-\!CH_2 \end{array}$$

2. Furfurane, thiophene, and pyrrol are compounds in which four carbon atoms with one atom of either oxygen, sulphur, or nitrogen are joined in a ring—thus, furfurane,

$$\begin{array}{c} HCCH \\ \| \ \| \\ HCCH \\ \diagdown \diagup \\ O \end{array}$$

thiophene,

$$\begin{array}{c} HC\!-\!CH \\ \| \ \ \| \\ HC \ \ CH \\ \diagdown \diagup \\ S \end{array}$$

pyrrol,

$$\begin{array}{c} HC\!-\!CH \\ \| \ \ \| \\ HC \ \ CH \\ \diagdown \diagup \\ NH \end{array}$$

3. Azoles contain two or more atoms other than carbon in a ring; and may be considered as derived from furfurane, thiophene, and pyrrol by replacing $=CH-$ by $=N-$, — thus, pyrazole,

$$\begin{array}{c} C\!-\!C \\ \| \ \ \| \\ C \ \ N \\ \diagdown \diagup \\ NH \end{array}$$

antipyrine (phenyldimethylpyrazolon), C_6H_5N

$$\begin{array}{c} O\!=\!C\!-\!CH \\ | \ \ \| \\ C\!CH_3 \\ \diagdown \diagup \\ N.CH_3 \end{array}$$

CLASS II.—BENZENE OR AROMATIC SERIES

1. Hydrocarbons of this series are compounds containing carbon and hydrogen, having the carbon atoms connected in a ring—thus,

benzene, C_6H_6,

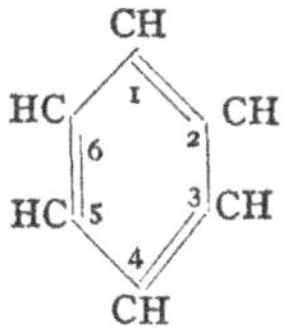

2. Halogen Substitution Products have an atom of hydrogen replaced by a halogen atom—thus, mono-chlor-benzene, C_6H_5Cl; dibrom-benzene, $C_6H_4Br_2$. C_6H_5 is called the phenyl radical; C_6H_4, the phenylene radical.

The di-substitution products may form three isomers according as the two are adjacent in the ring—thus,

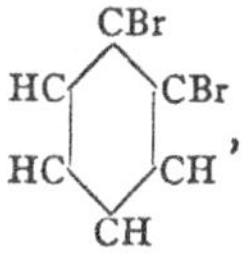

called ortho-di-brom-benzene; or as they have an atom of hydrogen between—thus, called meta-di-brom-benzene; or as the atoms are opposite—thus, called para-di-brom-benzene.

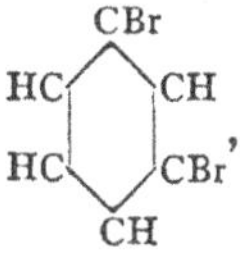

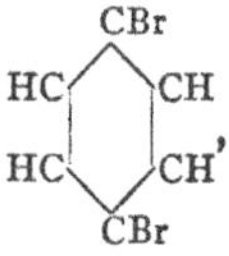

When three atoms are substituted, we have symmetrical 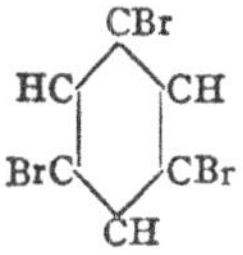 tri-brom-benzene; asymmetrical 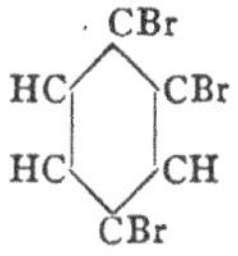 tri-brom-benzene; and adjacent 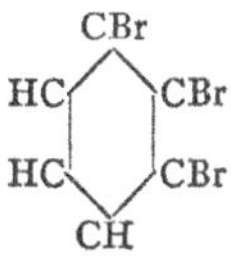 tri-brom-benzene.

This method of nomenclature is used whenever any element or group takes the place of hydrogen.

3. Sulphur Derivatives of the aromatic series are analogous to those of the fatty acid series—thus, benzene sulphonic acid, $C_6H_5SO_3H$.

4. **NITRO DERIVATIVES** are analogous to those of the methane series, but are more stable, and can be made by the direct treatment of the hydrocarbon with nitric acid—thus, nitro-benzene, $C_6H_5NO_2$; tri-nitro-toluene, $C_6H_2CH_3(NO_2)_3$.

5. **AMIDO DERIVATIVES.**—(1) Compounds formed by replacing one or more atoms of hydrogen in benzene or derivative hydrocarbons by one or more amido groups—thus, aniline (amido-benzene), $C_6H_5NH_2$; phenylene-diamine, $C_6H_4(NH_2)_2$.

(2) Compounds formed by replacing one or more atoms of hydrogen in ammonia by the aromatic hydrocarbon radicals—thus, diphenyl-amine, $(C_6H_5)_2NH$.

6. **OTHER NITROGEN DERIVATIVES.**—Diazo, azo-compounds, and hydrazines: (a) Diazo-compounds are intermediate products in the conversion of amido compounds to alcohols by means of nitrous acid. They contain the characteristic group $-N=N--$—thus, diazo-benzene-chloride, $C_6H_5-N=N-Cl$.

(b) Azo-compounds contain the same group as the diazo-compounds, but joined on each side to an alkyl radical—thus, azo-benzene (benzene-azo-benzene), $C_6H_5-N=N-C_6H_5$

(c) Hydrazines are compounds derived by the replacement of the hydrogen of hydrazine (N_2H_4) by one or more aromatic hydrocarbon radicals—thus, phenylhydrazine, $C_6H_5NH-NH_2$. They contain the charactistic group $=N-N=$.

7. **PHENOLS** are oxygenated derivatives of the benzenes. Chemically, they are midway between the alcohols and acids, and are formed by the replacement of H of the benzene nucleus by hydroxyl—thus, phenol (carbolic acid), C_6H_5OH; creosol, $C_6H_4(CH_3)OH$.

When two or more of the hydrogen atoms are replaced by the hydroxyl group, the polyacid phenols are obtained—thus, pyrocatechin (o-dioxy-benzene), $C_6H_4(OH)_2$; resorcin (m-di oxybenzene; and hydroquinone (p-dioxy-benzene). Tri-acid phenols: $C_6H_3(OH)_3$, pyrogallic acid = o-trioxy-benzene; phloroglucin (s-trioxybenzene), oxyhydroquinone (a-trioxybenzene).

8. **ALCOHOLS, ALDEHYDES, AND KETONES.**—Analogous to the same compounds of the methane series, containing the same groups, replacing the hydrogen of the side-chains—thus, $C_6H_5CH_2OH$, benzyl alcohol; C_6H_5CHO, benzaldehyde; $C_6H_5C=OC_2H_5$, acetophenone.

9. **ACIDS.**—Compounds analogous to the acids of the methane series, capable of forming the same kinds of derivatives—thus, benzoic acid, C_6H_5COOH; toluic acid, $C_6H_4CH_3COOH$; phthalic acid, $C_6H_4(COOH)_2$.

10. **COMBINATIONS OF THE ABOVE CLASSES.**

DOUBLE RING COMPOUNDS

11. **INDIGO GROUP.**—Compounds containing double rings similar to those of indigo—thus, indigo, $C_6H_4<^{NH}_{CO}>C=C<^{NH}_{CO}>C_6H_4$; isatin, $C_6H_4\diagup^N_{CO}\diagdown C(OH)$; indol, $C_6H_4\diagup^{CH}_{NH}\diagdown CH$.

12. **DIPHENYL AND ITS DERIVATIVES.**—Compounds containing two phenyl groups joined directly to each other—thus, diphenyl, $C_6H_5–C_6H_5$; benzidine (p-diamidodiphenyl), $C_6H_4NH_2–C_6H_4NH_2$.

13. **DIPHENYL-METHANE AND SIMILAR COMPOUNDS.**—Compounds in which two H atoms of methane are replaced by two phenyl groups, $(C_6H_5)_2$,—thus, diphenyl-methane, $CH_2(C_6H_5)_2$; benzophenone, $CO(C_6H_5)_2$.

14. **TRIPHENYL-METHANE GROUP.**—Compounds in which three H atoms of methane are replaced by the phenyl radical—thus, triphenyl-methane, $CH(C_6H_5)_3$; triphenyl-methane-carbinol, $C(OH)(C_6H_5)_3$.

These compounds are of especial interest, including extensive series of dyes. The following groups of dyes are distinguished:

(1) Diamido-triphenyl-methane group (the bitter-almond-oil green group).
(2) Triamido-triphenyl-methane (the rosaniline group).
(3) Trioxy-triphenyl-methane (the aurin group).
(4) Triphenyl-methane-carboxylic acid (the eosin group).

For a more complete description of these dyes the student is referred to works on organic chemistry.

CONDENSED RING COMPOUNDS

15. **NAPHTHALENE AND ITS DERIVATIVES.**—Naphthalene contains two condensed rings and has the composition $C_{10}H_8$; or, graphically,

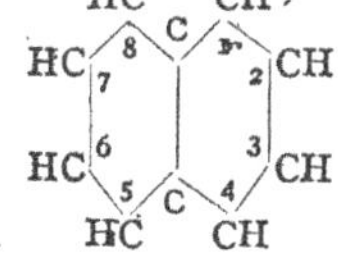

This is an increment of C_2H_4 over benzene. The hydrogen can be replaced as in benzene, forming derivatives—thus, $–C_{10}H_7OH$, naphthol; $C_{10}H_7NH_2$, naphthylamine, etc. When an atom of hydrogen connected to a carbon atom adjacent to either of the atoms of carbon common to both rings is replaced, alpha (α) derivatives of naphthalene are made. If those not adjacent are replaced, we have beta (β) derivatives.

16. ANTHRACENE AND ITS DERIVATIVES.—Anthracene contains three condensed rings, $C_{14}H_{10}$, or, graphically,

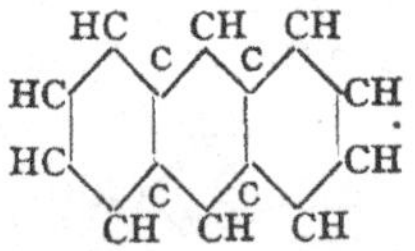

is an increment of C_4H_2 over naphthalene. The hydrogen can be replaced as in benzene, forming derivatives—thus, $C_{14}H_8O_2$, anthraquinone; $C_{14}H_8O_4$, alizarine.

17. PHENANTHRENE AND ITS DERIVATIVES.—Phenanthrene is an isomer of anthracene, containing three condensed rings—thus,

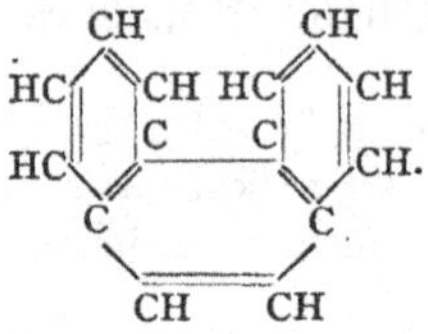

It can form derivatives in the same way.

Pyridine derivatives, alkaloids, and compounds related to them. These are compounds that may be considered as derived from benzene, naphthalene, anthracene, by the exchange of $-N=$ for $-CH=$ in the rings. All may be considered as derived from benzene on the one hand and from pyridine on the other.

18. PYRIDINE GROUP.—Pyridine may be considered as benzene in which $=CH-$ is exchanged for $=N-$. The hydrogen of pyridine is replaceable, forming derivatives such as picoline (methyl-pyridine), $C_5H_4NCH_3$.

Hydrated pyridine or piperidine, $C_5H_{11}N$; conine, α-normal-propyl-piperidine, $C_5H_{10}N(C_3H_7)$.

19. QUINOLINE AND ACRIDINE GROUPS bear the same relation to pyridine that naphthalene and anthracene bear to benzene.

20. ALKALOIDS OF COMPLEX OR UNKNOWN COMPOSITION.—Included in this class are the tropine, opium, narcotine, cinchona, strychnine, and solanine bases.

21. HYDRATED DERIVATIVES OF BENZENE.—Terpenes are hydrocarbons of the general formula $(C_5H_8)_x$, or, most commonly, $C_{10}H_{16}$. Camphors are oxygen derivatives of the terpenes: $C_{10}H_{16}O$, camphor.

22. GLUCOSIDES are vegetable substances that, when treated with alkalies, acids, or enzymes, are so broken up that one of the products of the decomposition is a glucose. They are ethereal derivatives of these sugars.

23. RESINS.—The resins are closely related to the terpenes and are formed from them. Their composition is as yet unknown.

24. ALBUMIN AND ALBUMINOIDS make up the greater part of the animal

organism, and are also found in plants, especially in the seeds. The composition is as yet in doubt.

As before stated, many of these organic compounds are mere mixtures of synthetical chemicals. They have the alluring titles of "New Remedies," for which special merit is claimed. Many of them have certain euphonic titles, which give no information as regards their constituents; others have proper scientific names, which tell at once their composition. Virgil Coblentz, referring to their nomenclature, divides them into two classes, as follows:

CLASS I. TITLES OF ORGANIC CHEMICALS

(*a*) Titles of this class express concisely the composition (chemical) of the compound—as, for example, acet-anilid, benz-anilid, ethylene-diamine, ethoxyl-caffein, acety-ethyl-phenyl-hydrazine.

(*b*) Such titles as embrace euphonic combinations of different syllables of names of the bodies entering into the composition of the remedy—for example, tann-albin (compound of tannin and albumin); amyl-form (a combination of starch and formaldehyde); salipyrine (a compound of salicylic acid and antipyrine); lactophenin (lactic acid derivative of phenetidin); gall-al (aluminum gallate); gall-anol (gallic acid and anilid), etc.

CLASS II.—DESCRIPTIVE TITLES

These are especially coined euphonic titles, which are generally of Greek and Latin origin, and partake of a descriptive character. These describe, in a way, either the uses, properties, or physical characters of the compound—as, for example, pyoktanin is made up of the Greek words πύον, meaning pus, and χτείνο, to kill; thalline, from the Greek, θαλλός, meaning a green twig, referring to the bright green color produced by the action of the oxidizing agents.

Other titles are of arbitrary character, such as loretin, an adaptation from *laura*, or *lorenit*, in which the last three letters of loretin have been reversed.

Owing to the entire absence of any data upon the nomenclature of these remedies, the derivation of many of these titles is entirely a matter of conjecture.

TITLES OF NEW REMEDIES

The following synopsis of new remedies aims to include such agents as have become established and those which have some promise of becoming permanent additions to the Materia Medica, giving merely the name, chemical formula, brief statement as to physical properties, use and dose; the idea being to give simply a general survey of the newer remedies admitted, or seeking admission, into the list of recognized therapeutical agents. Some are recognized as modern synthetic medicinal products. See "New and Non-official Remedies," 1916, American Medical Association.

ACETANILID (**Acetanalidum** U. S.).—$C_6H_5NH.CCH_3$. Analgesic, nerve sedative, and germicide. Dose, 0.2 to 0.5 GM. (3 to 8 gr.).

ACETONE (**Acetonum** U. S.).—$CH_3.CO.CH_3$. Nervine. Dose, 5 to 15 ♏ (0.3 to 1 Gm.).

ACETOPYRIN.—Compound of antipyrin and acetyl salicylic acid, sparingly soluble in alcohol. Antiseptic. Prompt and energetic in migraine, acute articular rheumatism, etc., in doses of 0.5 to 1 Gm. (7½ to 15 gr.) in cachets.

Acetphenetedinum.—See Phenacetin.

ACETOZONE.—(Benzoyl-acetyl-peroxide.) $CH_3COOOCOC_6H_5$. An exceedingly hygroscopic powder, therefore diluted with 50 per cent. inert substance. Decomposed by water contact into its respective hydrogen peroxides, a most powerful germicide, without toxicity. Intestinal antiseptic, especially valuable in typhoid fever. One Gm. (15 gr.) in one liter (1 qt.) water; 100 mils (4 fluid oz.) to be taken every four hours.

Acid Camphoric (U. S.).—$C_8H_{14}(COOH)_2$. Anhidrotic. Dose, 15 gr. (1 Gm.).

Acid Phenylcinchonicum.—$C_6H_5.C_9H_5N(COOH.)$ Atophan, Phenyl-Quinolin Carboxylic Acid.) Colorless needles, or yellowish white crystalline powder; odor, faintly suggesting Benzoic acid; taste, bitter. Gout remedy. Dose, 0.5 Gm. (8 gr.).

Acidium Trichloraceticum.—See Trichloracetic Acid.

ACTOL.—(Silver Lactate.) $AgC_3H_5O_3 + H_2O$. Grayish white powder, soluble in 15 parts water. Without caustic action on wounds. Solutions must be protected from light.

ADRENALIN.—Solution of adrenalin chloride. Active principle of the adrenal gland. The most powerful astringent and hemostatic known; one drop of the solution 1 in 10,000 will blanch the mucous membrane of the eyelid in one minute. Valuable in coryza, hay-fever, hemorrhage, iritis, laryngitis, surgical operations, etc.

Æthyl Carbamas.—See Urethane.

Æthylis Chloridum.—See Ethyl chloride.

ÆTHYMORPHINÆ HYDROCHLORIDUM.—See Dionin.

ALUMNOL.—$[(C_{10}H_5OH(SO_3)_2]_3Al_2$. (Aluminum b-naphthol-di-sulfonate.) Antiseptic astringent. In 1 to 2 per cent. solutions, principally in gonorrhea, also as a gargle.

Antipyrina.—$C_2N_2HO(CH_3)_2C_6H_5$. Antipyretic, antirheumatic, and antineuralgic. Dose, 15 to 30 gr. (1 to 2 Gm.).

ARGENTAMINE.—Ethylene-Diamine-Silver Nitrate. Astringent, Antiseptic.

ARGONIN.—(Argentum-caseinicum.) Compound of silver with casein, representing about 7 per cent. of its weight of silver nitrate. Soluble in water, non-irritant, and non-precipitated by soluble chlorides. Antiseptic, chiefly in gonorrhea, as a 2 per cent. solution.

ARGYROL.—(Silver Vitellin.) Compound of nuclein and silver, 30 per cent. Closely allied to argonin. Therapeutically, used locally.

ARISTOL.—$[C_6H_2(CH_3)(C_3H_7)OI]_2$. (**Thymolis Iodidum** U. S.) A red-brown sticky powder, soluble in absolute alcohol, ether, chloroform, fixed oils, and camphor carbolate, insoluble in water and glycerin. Employed in most skin affection, etc. Dose, 0.125 Gm. (2 gr.).

ASPIRIN.—(Acetyl-salicylic-acid.) White powder, sparingly soluble in water, freely in alcohol. Antirheumatic in doses 0.1 to 0.3 Gm. in capsule; 0.5 to 1 Gm. per diem.

ATOPHAN.—See **Acid Phenylcinchonicum.**

ATOXYL.—Sodium arsanilate (Proprietary). Dose, ⅓ gr. (0.02 Gm.).

Benzaldehydum.—$C_6H_5.COH$. Contained in bitter almond oil.

Benzinum Purificatum.—Petroleum Ether. An immiscible solvent. Should not be confounded with Benzene, or Benzol (C_6H_6).

BENZOSAL.—(Guaiacol-Benzoate.) $C_{14}H_{12}O_3$. Colorless, crystalline powder, nearly tasteless and odorless. Intestinal antiseptic. Dose, 0.2 to 0.6 Gm. (1 to 3 gr.). per day.

Benzosulphinidum.—Saccharin, Glucidum. $C_6H_4SO_2CONH$. Dose, 0.2 Gm. (3 gr.).

Betaeucaineæ Hydrochloridum.—Eucaine (C_6H_5COO)HCl. White, crystalline powder, odorless. Local anæsthetic. Used in 2 or 3 per cent. solution; stronger for nose and throat.

Betanaphthol.—Naphthol, $C_{10}H_7OH$. Antiseptic and disinfectant. Dose, 0.25 Gm. (4 gr.).

Bromoform.—$CHBr_3$. Anæsthetic, a remedy in whooping-cough. Dose, 2 to 5 drops (0.1 to 0.3 Gm.).

Caffeinæ Sodio-Benzoas.—A mixture of caffeine and sodium benzoate. Diuretic and antirheumatic. Average dose, by mouth, 0.3 Gm., or 5 gr.; hypodermic dose, 0.2 Gm., or 3 gr.

Camphora Monobromata.—$C_{10}H_{15}OBr$. Colorless prismatic needles or scales, or as a powder having a mild but characteristic champhoraceous odor and taste; permanent in air. Heart depressant, vasoconstrictor and hypnotic. Dose, 0.125 Gm., or 2 gr.

Chinosol.—(Potassium oxyquinolin-sulphate.) Bright yellow crystalline powder. Powerful antiseptic in the treatment of catarrh, ulcers, etc.; of great value in dentistry as an antiseptic mouth-wash (1 : 1,000), not affecting injuriously the gums or teeth.

Chloral Hydrate (Chloralum Hydratum U. S.).—$C_2HCl_3O + H_2O$. Hypnotic. Dose, 10 to 20 ♏ (0.6 to 1.25 Gm.).

Chloralamide.—(Chloralformamide U. S.) $CCl_3CH.OH.CONH_2$. Soluble in nine parts of water. Hypnotic. Dose, 10 to 30 gr. (0.65 to 2.0 Gm.).

Chloretone.—(Chloroform Acetone.) $HO.C(CH_3)_2CCl_3$. White crystals. The saturated solution is used as a local anæsthetic. Internally hypnotic. Dose, 1 to 4 Gm. (15 to 60 gr.).

Chrysarobinum.—A mixture of neutral principles, extracted from Goa Powder. A brownish to orange yellow, microcrystalline powder, tasteless, odorless and irritating to the mucous membrane. Antiseptic. Dose, 0.03 Gm. (½ gr.).

Collargol.—Colloidal Silver, Argentum Credé. 78 per cent. metallic silver, used locally in 15 per cent. ointment; internally in infectious gastric and intestinal diseases, in tablets, etc., containing 1 gr. (0.06 Gm.).

Cinnaldehyde (U. S.).—$C_6H_5.CH = CH.COH$. In cinnamon oil (or synthetic oil).

COTARNINÆ HYDROCHLORIDUM.—(Stypticin), (CH_3O) $(CH_2O_2).C_9H_6N$-$(CH_3)Cl$. Obtained by hydrolyzing narcotine. A yellow, crystalline powder, odorless, and deliquescent in moist air. Styptic. Dose, 0.06 Gm. (1 gr.).

Creasotal.—(Beechwood Creosote 90 per cent. and carbolic acid.) A viscid, amber-colored, nearly odorless and tasteless liquid, insoluble in water and glycerin. Preferred to creasote in the treatment of tuberculosis, also in typhoid fever. In capsules, in oil, or in emulsion. Dose, 1 to 16 Gm.

Cresol (U. S.).—$C_6H_4(CH_3)OH$. Antiseptic. Dose, 1 ♏ (0.05 mil.).

Creosoti Carbonas.—A mixture of the carbonate of various constituents of creosote, chiefly guaiacol and cresol. A clear, colorless or yellowish viscid liquid, odorless and tasteless, or having a slight odor and taste of creosote. Substitute for creosote for internal use. Dose, 1 Gm. (15 gr.).

Croton Chloral.—(Butyl Chloral Hydrate.) $C_4H_5Cl_3O + H_2O$. Action and dose same as chloral hydrate.

Dermatol or **Bismuth Subgallate (U. S.).**—A fine saffron yellow powder. A substitute for iodoform in the treatment of wounds, ulcers, etc.

Diabetin.—$C_6H_{12}O_6$. A variety of levulose used as a substitute for cane sugar in the regimen of diabetic patients. Only an inconsiderable portion of it is excreted with the urine.

DIACETYLMORPHINA.—$C_{17}H_{17}(O.C_2H_3O)_2NO$. Prepared from morphine by acetylization. White crystal powder without odor. Used the same as morphine. Dose, 0.003 Gm. (1/20 gr.).

DIACETYLMORPHINÆ HYDROCHLORIDUM.—Dose, 0.003 Gm. (1/20 gr.).

Di-iodoform.—$C_2H_2I_4$. Used as a substitute for iodoform.

Dionin.—(Ethyl-morphine-hydrochlorate.) $C_2H_5O(OH)C_{17}H_{17}NO.HCl + H_2O$. Local anæsthetic, sedative, analgesic, chiefly used in ophthalmic practice. Dose, 1/64 to 1/16 gr.

Diuretin.—(Theobromine Sodium Salicylate.) $C_7H_7N_4O_2Na + C_6H_4(OH)$-$COONa$. White amorphous powder. Diuretic. Acts directly upon the kidneys without producing insomnia and depression. Dose, 15 gr. (1 Gm.).

EMETINÆ HYDROCHLORIDUM.—$C_{30}H_{44}N_2O_4.2HCl$. Obtained from Ipecac, white or very slightly yellowish, crystalline powder without odor. A new remedy in pyorrhea. Hypodermic dose, 0.02 Gm. (1/3 gr.).

ERYTHROL TETRANITRATE.—Tetranitrol. C_4H_6. Vasodilator similar to nitroglycerin. Action slower and more lasting. Dose, $\frac{1}{2}$ to 1 gr. (0.03 to 0.06 Gm.).

ETHYL BROMIDE.—C_2H_5Br. Colorless, very volatile, non-inflammable liquid of a chloroformic taste and odor. Employed in minor surgery for general anæsthesia.

ETHYL CHLORIDE (**Æthylis Chloridum** U. S.).—C_2H_5Cl. Local anæsthetic, producing no shock, vomiting, or nausea.

ETHYLENE DIAMINE.—Said to be an albumen solvent, non-corrosive, for the solution of diphtheritic membrane, etc.

EUCAINE HYDROCHLORATE B.—(Benzoyl-vinyl-diaceton-alkamine.) $C_{15}H_{21}NO_2$·HCl. A white crystalline powder. Less toxic than cocaine, does not produce mydriasis or corneal disturbances. In ophthalmic practice used in 2 per cent., in genito-urinary diseases, 0.5 to 2 per cent., for infiltration anæsthetic 0.1 to 1 per cent., solutions.

Eucalyptol (U. S.).—(Cineol). $C_{10}H_{18}O$. In eucalyptus oil. Dose, 5℞ (0.3 Cc.).

Eugenol (U. S.).—$C_6H_3(OH)(OCH_3).C_3H_5$. Synthetic clove oil.

EUPHORIN.—Phenyl-urethane. $C_6H_5NH—CO—OC_2H_5$. Colorless crystalline powder. Antipyretic, analgesic, etc. Dose, 0.13 to 0.5 Gm. (2 to 8 gr.).

EUQUININE; EUCHININ.—$C_2H_5O.CO.OC_{20}H_{23}N_2O$. Carbonic acid ester derivative of quinine. White crystals, devoid of bitterness, tasteless quinine compound. Given to children in mucilaginous vehicle.

EUROPHEN.—$C_4H_9(OCH_3)C_6H_3.C_6H_3.C_4H_9(CH_3)OI$. A cresol derivative containing 22 per cent. of iodine. A yellow powder insoluble in water. Antiseptic as a 3 per cent. ointment for burns, scalds, and ulcers; as antisyphilitic, in solution in a fixed oil hypodermically. Dose, $\frac{1}{4}$ to 1 gr. ($\frac{1}{60}$ to $\frac{1}{15}$ Gm.). Must not be confounded with euphorin.

EXALGIN.—$C_6H_5N(CH_3)(CH_3CO)$. Analgesic, dose, 3 to 6 gr. (0.2 to 0.4 Gm.); antipyretic, $7\frac{1}{2}$ gr. (0.5 Gm.).

EXODIN.—(Diacetyl-rufigallic-acid-tetramethyl-ether.) Purgative. Dose, $7\frac{1}{2}$ to 12 gr. (0.5 to 0.8 Gm.).

Formaldehyde.—HCOH. A 37 per cent. solution (U. S. P.) of formic aldehyde (HCOH). Antiseptic and disinfectant. A powerful bactericide even when largely diluted. A spray of 2 per cent. solution completely disinfects fabrics; 0.5 to 1 per cent. solution to disinfect rooms, walls, furniture, etc.

Guaiacol (U. S.).—$C_6H_4—OHOCH_3$. The chief constituent of creasote, which contains it in varying proportions of 60 to 90 per cent. A colorless liquid of a strong aromatic odor. Dose, 1 to 2 drops for children, 3 to 5 drops for adults, dissolved in water with cognac and wine. Must be administered continuously for months.

Guaiacol Carbonate (U. S.).—$CO_3(C_6H_4OCH_3)_2$. An odorless, tasteless, crystalline powder, insoluble in water. Dose, 0.2 to 0.5 Gm., increased to 6 Gm. per day. More readily borne by the stomach than guaiacol itself. (*Synonym*, Duotal.)

HEROIN.—(Morphine di-acetic-ester.) $C_{17}H_{17}NO_3\begin{Bmatrix} CH_3CO \\ CH_3CO \end{Bmatrix}$. White crystals. The hydrochlorate in small doses, from 0.005 to 0.03 Mgm. ($\frac{1}{12}$ to $\frac{1}{2}$ gr.), in laryngeal cough, bronchitis, pulmonary tuberculosis; usually associated with other agents—terpin hydrate, etc.

HOLOCAIN.—$CH_3C\begin{Bmatrix} N—C_6H_4OC_2H_5 \\ N—C_6H_4OC_2H_5 \end{Bmatrix}$. Muscular anæsthetic, germicidal (poison).

HYPNONE.—$C_6H_5COCH_3$. Soporific, hypnotic. Dose, 1 to 3 ℞.

ICHTHYOL.—$C_{28}H_{36}S_3O_6(NH_4)_2$. Ammonium sulphichthyolate and sodium sulphichthyolate are both employed under the name ichthyol. The latter, owing to its density, being dispensed when pills are prescribed, the former in ointments. Dark-brown semi-liquids of a fetid odor. Employed in a host of maladies, including eczema, bruises, burns, rheumatism, migraine, chilblains, etc.; also used in form of impregnated cotton, gauze, or soap.

IODOL (**Iodolum** U. S.).—C_4I_4NH. Antiseptic and alterative. Dose, 5 to 10 gr.

LACTOPHENIN.—$C_6H_4\begin{Bmatrix} OC_2H_5 \\ NH.CO.CH(OH)CH_3 \end{Bmatrix}$. Chemically resembling phenacetine. Antipyretic and antirheumatic. Soluble in 40 parts of water. The daily dose is 10 to 40 gr. in divided quantities.

LUTEIN.—Dispensed in tablets representing about 20 gr. of fully developed corpora lutea of pigs. Used in cases of gastric tetany, etc. Dose, 1 tablet three times a day.

LYSIDIN.—(Ethylene-ethenyl-diamine.) $(CH_2N)(CH_2NH)CCH_3$. (Methyl-gly-oxalidin.) A very hygroscopic, pinkish-white, alkaline mass. Comes in the form of 50 per cent. aqueous solution only. Said to possess five times the uric-acid solvent power of piperazine. Thirty to 150 minims in one pint or more of aerated water.

MEDINAL.—A sodium salt of veronal, *q.v.*

MERCUROL.—(Mercurous Nuclein.) Brownish-white powder; contains 10 per cent. of mercury. It does not precipitate albumen and is employed dissolved in physiologic salt solution.

Methylene Blue.—(Methylthioninæ Hydrochloridum U. S.). $C_{16}H_{18}N_3SCl$. A blue aniline similar to the agent following. Anodyne, antiperiodic, analgesic, bactericide. Employed especially in carcinomatous growths. Dose, 0.1 to 0.5 Gm. (hypodermically). The crystalline powder has a dark-green color.

METHYLENE VIOLET.—(Pyoktanin Cœruleum.) Also called methyl-blue. Violet crystalline powder. Excellent bactericide and deodorant. Deeply penetrates the tissues and colors the urine. Applied to purulent wounds, malignant tumors, chancroids, etc.

MIGRANIN.—A combination of 89.4 per cent. of antipyrin, 8.2 per cent. of caffein, 0.56 per cent. of citric acid. For migraine and as a general analgesic.

Naphthalenum.—$C_{10}H_8$. Antiseptic.

NAPHTOL (Betanaphthol U. S.).—$C_{10}H_7OH$. Antiseptic. Dose, 0.25 Gm. (4 gr.).

NARGOL.—(Silver Nuclein.) Contains 10 per cent. of silver. (Allied to Argyrol.)

NEO-SALVARSAN.—A modified (sodium combination) of salvarsan. Three parts of this salt is approximately equal to that of 2 parts of Salvarsan. See "New and Non-Official Remedies."

NOSOPHEN.—(Tetra-iodo-phenolphthalein.) $(C_6H_2I_2OH)_2C\diagup^{C_6H_4CO}_{\diagdown O}$. Also known as iodophen; containing 60 per cent. of iodine. Pale yellow, odorless, and tasteless powder. Its sodium salt is known as antinosin, the bismuth salt as eudoxin. A harmless yet efficient substitute for iodoform.

NOVOCAINE.—Cocaine substitute. Para-amino-benzoyl, diethyl-amino-ethanol hydrochloride. $CH_2(C_6H_4NH_2COO)[N(C_2H_5)_2]HCl$. Use, same as cocaine.

OREXIN.—(Phenyl-dihydro-chinazoline.) $C_6H_4CH_2NC_6H_5NCH$. Colorless, lustrous, odorless crystals of bitter, pungent taste. Dose, 0.13 to 0.4 Gm. (2 to 6 gr.).

OUABAIN, CRYSTALLIZED.—Crystallized Strophanthin. G-Strophanthin Thoms. $C_{30}H_{46}O_{12}+9H_2O$. A glucoside, obtained from Acocanthera ouabaio by Arnaud, or, as now commonly prepared, from Strophanthus gratus, in which case it is also called crystallized strophanthin, or g-strophanthin Thoms. (The official strophanthin is methyl ouabain, $C_{31}H_{48}O_{12}$.)

PARAFORMALDEHYDUM.—(Paraform) (Trioxymethylene). It contains not less than 95 per cent. of $(HCHO)_3$ (90.05), a polymeric form of formaldehyde. White friable masses or as a powder having a slight odor of formaldehyde. Antiseptic and escharotic. Dose, 0.5 Gm. (8 gr.).

Paraldehyde (U. S.).—$C_6H_{12}O_3$. Hypnotic. Dose, ♏xv to ℥j (1 to 4 mils.).

PERONIN.—$(C_6H_5CH_2O(OH)C_{17}H_{17}NO.HCl$. Feeble narcotic. Dose, ⅓ to 1 gr. (1⁄45 to 1⁄15 Gm.).

PHENACETIN.—$C_6H_4\diagdown^{NH-CH_3CO}_{\diagup OC_2H_5}$ (Acetphenetidinum U. S.).—Antipyretic and analgesic. Dose, 7.5 gr. (0.5 Gm.)

PHENOCOLL HYDROCHLORIDE.—(Amido-acetparaphenetidin Chloride.) $C_6H_4(OC_2H_5)NHCOCH_2NH_2HCl$. A white powder of a slightly saline taste. Antipyretic, in typhoid fever and pneumonia and in various forms of rheumatism and neuralgia. Dose, 8 to 15 gr. (0.5 to 1.0 Gm.).

PHENOLPHTHALEINUM.—$(C_6H_4OH)_2CO.C_6H_4CO$. White or faintly yellowish-white crystalline powder, odorless and tasteless. Permanent in the air. Purgative. Dose, 0.15 Gm. (2½ gr.).

PHENYLIS SALICYLAS.—See Salol.

PIPERAZINE.—(Piperazidine, Diethylen-diamine, Dispermine.) $NH(CH_2)_4NH$. A colorless crystalline body of alkaline reaction, hygroscopic, taste not unpleasant. A powerful solvent of uric acid; recommended for gravel, renal and vesical calculi, for gout and diabetes. Dose, 1 Gm. (15 gr.)

PROTARGOL.—Protein combination with silver, 8.3 per cent. Yellowish hygroscopic powder. Used in ophthalmic practice, etc.; also in suppositories and as dusting powder on chancre. (See also *Nargol*.)

PYROGALLOL.—$(C_6H_3(OH)_3$. Antiseptic for external use.

RESORCIN (**Resorcinol** U. S.).—(Meta-dioxy-benzol.) $C_6H_4(OH)_2$. Antiseptic.

RADIUM AND RADIUM SALTS.—See New and Non-Official Remedies, 1916, page 264.

SAFROL.—$C_6H_3(C_3H_5) \genfrac{}{}{0pt}{}{O}{O}{>}CH_2$. (**Safrolum** U. S.). In sassafras oil.

SALIPYRIN.—$C_{11}H_{12}N_2OC_7H_6O_3$. Compound of antipyrine and salicylic acid. Dose, 0.1 to 0.5 Gm.

SALOL.—(**Phenyl Salicylas** U. S.). $C_6H_4{<}\genfrac{}{}{0pt}{}{OH}{COOC_6H_5}$ Antipyretic. Dose, 0.3 to 0.6 Gm. (5 to 10 gr.). .

SALOPHEN.—$C_6H_1{<}\genfrac{}{}{0pt}{}{OH}{CO_2C_6H_4NHCOCH_3}$. Derivative of phenol and salicylic acid, resembling salol. White leaflets, odorless and tasteless. Antirheumatic, in doses of 0.2 to 0.4 Gm.

SALVARSAN "606."—Diamino-dihydroxy-arsenobenzene-hydrochloride. $(C_6H_3As.-OHNH_2HCl)_2$. Bright yellow powder of acid reaction, slowly soluble in 10 parts of water. Dose, 10 gr. (0.6 Gm.) which for intravenous use is dissolved in 300 mils of normal salt solution, neutralized with NaOH.

Serum Antitoxins.—The blood-serums of immunized animals. A class of preparations employed hypodermically for the treatment of diseases of germ origin, such as diphtheria, etc. They have the power of neutralizing the toxin produced by the micro-organism or germs. **Serum Antidiphthericum** U. S.

SODIUM CACODYLATE.—$O = As{-}\genfrac{}{}{0pt}{}{CH_3}{CH_3}{\diagdown}{ONa}$ Dose, 0.25 Gm. (4 gr.) per os.

SODIUM ETHYLATE.—CH_3CH_2ONa. Whitish powder, decomposed in the presence of water into alcohol and caustic soda. Depilatory.

SOZOIODOL.—$C_6H_2I_2(OH)SO_3H + 3H_2O$. Usually supplied similar to potassium sozoiodal (which see), but more soluble.

SULPHONETHYLMETHANUM.—See Trional.

SULPHONMETHANUM.—See Sulphonal.

SULPHONAL (**Sulphonmethanum** U. S.)—(Diethyl-sulphon-dimethyl-methane.) A whitish crystalline substance, devoid of odor or taste. Dose, 1 Gm. (15 gr.).

TANNALBIN.—Compound of tannin and albumen, tasteless powder, containing 50 per cent. of tannin. Astringent. Dose, 1 to 2 Gm. (8 to 15 gr.).

TANNIGEN.—(Diacetyl Tannin.) $C_{14}H_8(COCH_3)_2O_9$. Derivative of tannin, greyish-yellow, odorless, tasteless powder. An intestinal antiseptic, capable of passing the stomach unaltered. Dose, 1 Gm.

TANNOPIN.—(Tannon.) $2(CH_2)_6N_4(C_{14}H_{10}O_9)_3$. Compound of urotropin and tannin. Brown, odorless, tasteless, powder. Dose, 0.5 to 1.0 Gm.

TEREBENUM.—A liquid consisting of dipentene and other hydrocarbons, obtained by the action of concentrated sulphuric acid on oil of turpentine. Colorless, thin liquid, having a rather agreeable thyme-like odor and an aromatic, somewhat terebinthinate taste. Antiseptic. Dose, 0.25 mil (4 minims).

TERPINI HYDRAS.—$C_{10}H_{18}(OH)_2 + H_2O$. Colorless, lustrous, rhombic prisms, nearly odorless, and having a slightly aromatic and somewhat bitter taste. Antiseptic in bronchitis. Dose, 0.25 Gm. (4 gr.).

TETRONAL.—$(C_2H_5)_2C(SO_2C_2H_5)_2$.

THYMOLIS IODIDUM.—See Aristol.

THIOCOL.—$CH_3{-}\genfrac{}{}{0pt}{}{OCH}{}{\diagup}{OH}{\diagdown}{SO_3K}$ (Potassium Guaiacol Sulphonic Acid.) Combines the full power of creasote and guaiacol.

THIOL.—A water soluble mixture of sulphurated and sulphonated petroleum oils, said to have a drying astringent, antiphlogistic and disinfecting action. In skin diseases said to be useful; also used internally. Dose, 1½ to 3 gr. (0.1 to 0.2 Gm.).

TRICHLORACETIC ACID.—CCl_3COOH. Used as a caustic.

TRICRESOL.—(Ortho, meta and para cresol). Disinfectant and germicide.

TRINITROPHENOL.—(Picric Acid.) $C_6H_2(OH)(NO_2)_3$. Pale yellow, rhombic prisms or scales, odorless and having an intensely bitter taste. It explodes when heated rapidly and when subjection to percussion. Used mostly externally for burns, etc. Dose, 0.03 Gm. (½ gr.).

TRIONAL (**Sulphonethylmethanum** U. S.).—$C_2H_5CH_3C(SO_2C_2H_5)_2$. Derivative of sulphonal. Lustrous scales. Nerve sedative and hypnotic. Dose, 0.2 to 0.3 Gm.

URETHANE.—$CO\diagup^{NH_2}_{\diagdown OC_2H_5}$ (**Æthylis Carbamas**, U. S.). Hypnotic. Dose, 1 to 2 Gm. (15 to 30 gr.).

URICEDIN.—Produced by action of sulphuric and hydrochloric acids on lemon juice and neutralizing the product with sodium bicarbonate. Slightly yellow granular substance. In the treatment of the uric acid diathesis. Dose, ½ to 1 teaspoonful in hot water, two or three times a day, up to 300 gr. per day.

UROTROPIN.—(**Hexamethylenamina** U. S.) $(CH_2)_6N_4$. Formed by the union of formaldehyde and ammonia. Diuretic and uric acid solvent. For uric acid calculi, cystitis. Dose, 0.25 to 1 Gm. (4 to 15 gr.).

Vanillin.—$C_6H_3(OH)(OCH_3)COH$. From vanilla bean.

VERONAL.—Diethyl-barbaturic acid. $CO\diagup^{NHCO}_{\diagdown NHCO}\diagdown C\diagup^{C_2H_5}_{\diagdown C_2H_5}$ Hypnotic and somnifacient. Dose, 5 to 15 gr. (0.3 to 1 Gm.) in hot liquid.

XEROFORM.—$(C_6H_2Br_3O)_2BiOH$—Bi_4O_3. Deodorant and astringent and antiseptic.

SYNONYMS

Much confusion exists in dispensing remedies of this class because an agent is recognized by so many different names, as for example:

Argentum Colloidale; *syn.*, ...	Argentum Crede, Collargol, Colloidal silver.
Beta-naphthol Benzoate; *syn.*,..	Benzo-naphthol, Benzoyl-beta-naphthol.
Beta-naphthol Salicylate; *syn.*, ..	Betol, Naphtalol, Naphthosalol, Salinaphthol
Bromacetanilid; *syn.*,..........	Antisepsin, Asepsin.
Bismuth-iodo-subgallate; *syn.*,..........	Airol, Airogen, Airoform.
Calcium Beta-naphthol Sulphonate; *syn.*,	Abrastol, Asaprol.
Creasote Tannate; *syn.*, ...	Creosal, Tannosal.
Dimethyl-ethyl-carbinol Chloral; *syn.*,...:	Dormiol, Amylene-chloral.
Dithymol Diiodid; *syn.*, ...	Aristol, Annidalin, Di Thymol Iodid, Di Iodo Dithymol, (And several other similar names).
Epinephrin; *syn.*,.	Adnephrin, Adrenalin, Adrenamine, Adrenol, Adrin, Caprenalin, Hemisine, Hemostatin, Suprarenalin.

Ethyl Chlorid; *syn.*,................... { Antidolorin, Ethylol, Kelene, Mono-chlor-ethane.

Hexamethylene-tetramine; *syn.*, .   . { Aminoform, Ammonio-formaldehyde, Crystamine, Cystogen, Formin, Saliformin, Urotropin.

Hexamethylene Anhydromethylen Citrate; *syn.*, ... { Helmitol.

Levulose; *syn.*,........................ { Diabetin, Fructose, Fruit sugar.

Ortho-ethoxy-ana-mono-benzoyl-amido-chinolin; *syn.*,.................... { Benzanalgene, Analgen, Quinalgen.

Paraphenetin Carbamid; *syn.*,............ { Dulcin, Sucrol.

Phenyl-dimethyl-parazolon; *syn.*,......... (Pharm. Germ.) { Analgesin, Anodynin, Antipyrin, Dimethyloxy-quinizin, Methozan, Phenazon (Br. Ph.), Phenylon, Pyrazin, Pyrazolin, Parodyn, Salozolon, Sedatin.

Phenylacetamide; *syn.*,................. { Acetanilid, Antifebrin, (And several hundreds of trade names for headache powders, etc.).

Phenylmethyl-ketone; *syn.*,.............. { Acetophenone, Hypnone.

Plant Pepsin; *syn*...................... { Papain, Papoid, Papayotin, Caroid.

Salicylic Acid Ester of Quinine; *syn.*,...... { Salochinin, Saloquinin.

Trioxymethylen; *syn.*,................... { Paraformaldehyde, Paraform, Triformol.

Abrin, *syn.*,..............................Jequiritin.
Acetyl-salicylic Acid; *syn.*,...................Aspirin.
Aluminum Aceto-tartrate; *syn.*,.............Alsol.
Australian Oil Eucalyptus; *syn.*,............Flucol.
Bismuth Chryssophanate; *syn.*,.............Dermol.
Bismuth Subgallate; *syn.*,..................Dermatol.
Bismuth Beta-naphtholate; *syn.*,.............Orphol.
Cotarnine Hydrochlorid; *syn.*,..............Stypticin.
Creosote Carbonate; *syn.*,.................Creosotal.
Diethylen-diamin; *syn*,....................Piperazin.
Dimethyl-xanthine; *syn.*,...................Theobromine.
Guaiacol Carbonate; *syn.*,.................Duotal.
Laricinic Acid; *syn.*,......................Agaricin.
Magnesium Dioxid; *syn.*,...................Biogen.

Oxyquinaseptol; *syn.*,.......................Diaphtherin.
Phenyl-ethyl-urethan; *syn.*,..................Euphorin.
Saccharin; *syn.*,.............................Garanotose.
Subgallate of Bismuth; *syn.*,.................Dermatol.
Sodium chlorate; *syn.*,.......................Oxychlorin.
Sodium beta-naphtholate; *syn.*,..............Microcidin.
Tang-Ki, Fl'ext.; *syn.*,.......................Eumenol.
Trichloracetic Acid (50 per cent. solution); *syn.*, Acetocaustic, etc.

CONSPECTUS A.—OFFICIAL DRUGS ARRANGED ACCORDING TO STRUCTURAL CHARACTERISTICS

SYNOPSIS OF CLASSIFICATION

VEGETABLE

I. PHANEROGAMS.
    A. *Subterranean or Underground Organs—*
        1. Roots.
        2. Rhizomes.
        3. Tubers.
        4. Bulbs.
        5. Corms.

    B. *Overground Stems—*
        a. Herbaceous.
        6. Herbs.
        β. Woody.
        7. Barks.
        8. Twigs.
        9. Woods.
        10. Piths.

    C. *Outgrowths from Overground Stems—*
        11. Leaves.
        12. Leafy tops.
        13. Plant hairs and glandular outgrowths.
        14. Flowers and parts of flowers.
        15. Fruits and parts of fruits.
        16. Seeds and seed coverings.

II. CRYPTOGAMS.
    A. *Equisetaceæ.*
    B. *Filices.*
    C. *Lycopodiaceæ.*
    D. *Lichenes.*
    E. *Fungi.*
    F. *Algæ.*

III. ABNORMAL GROWTHS CAUSED BY PARASITES.
    A. *Excrescences.*

IV. NON-CELLULAR DRUGS DERIVED FROM CELL-CONTENTS AND SECRETIONS.
    A. *Farinaceous.*
    B. *Extractives.*
    C. *Concrete juices.*
    D. *Sugars.*
    E. *Gums.*
    F. *Gum resins.*
    G. *Resins.*
    H. *Oleoresins.*
    I. *Balsams.*
    J. *Stearoptens.*
    K. *Fatty substances.*

ANIMAL

I. INSECTS.
II. TISSUES AND SECRETIONS.
III. SERUMS AND BACTERIAL PRODUCTS.

DERIVED FROM THE VEGETABLE KINGDOM

(With Brief Description)

NOTE.—Drugs marked N.F. were transferred from U.S.P. viii to National
ormulary.

PHANEROGAMS.

. **Subterranean or Underground Organs.**

1. ROOTS

(*a*) MONOCOTYLEDONOUS ROOTS:
   Orange, brown, thick, mealy or horny cortical layer, separated
   from the wood-bundles by the nucleus sheath; broad central
   pith, ...**Sarsaparilla, 58**

(*b*) DICOTYLEDONOUS ROOTS:
   a. Fleshy, with thin bark.
   In transverse slices,* externally grayish-brown; twisted and
   irregularly matted wood fibers; light and spongy meditullium,
                                             Sumbul, 400
   Irregular pieces; reddish-yellow; narrow medullary rays, producing
   mottled appearance (rhizome, U.S.P. 1900),.......**Rheum, 120**
   Externally grayish; fibrous; small, narrow, radiating wood-bundles
   in concentric cirlces,....................**Phytollacca, 126, N.F.**
   With thick bark.
   Subcylindrical, grayish-brown, narrow wood-wedges and medullary
   rays; porous meditullium,......................**Stillingia, 304**
   White; wood-bundles small, scattered; narrow medullary rays,
                                             Althea, 341
   Externally dull gray; wood-bundles near the center small and
   scattered; distinct cambium,......   ......**Belladonna, 503**
   Yellowish Brown. See *Rhizomes, Scopola,*...................**506**
   In sections, externally brown; small, yellow wood-bundles, radiate
   near the bark; dark cambium,..................**Calumba, 156**
   Gray-brown; white bark, containing numerous concentric circles of
   lacitiferous vessels; yellow, woody center,........**Taraxacum, 553**
   Sometimes sliced longitudinally; gray-brown, internally lighter;
   meditullium spongy; radiate, with broad medullary rays,
                                             Lappa, 558, N.F.
   Hard, somewhat fusiform, brownish; wood-wedges and medullary
   rays narrow, radiate; resin ducts in circles,.....**Pyrethrum, 555**
   Sometimes in longitudinal slices; yellowish-brown; meditullium
   spongy; medullary rays, indistinct; distinct cambium (root and
   rhizome),.......................................**Gentian, 441**
   Yellowish-gray, keeled when dry; wood porous; fine medullary rays;
   whitish inner bark excessively developed on one side, **Senega, 302**
   Grayish **to** reddish-brown, usually twisted, deeply longitudinally furrowed, marked by distinct root scars, otherwise nearly smooth,
                                             Scammonii Radix, 462

   b. Woody, with thin bark.
   Generally in sections; brownish-yellow, with purplish longitudinal
   lines; thin cork; porous wood-wedges; broad medullary rays
   (root and rhizome),.........................**Gelsemium, 438**
   Rust-brown; thick cork; very narrow medullary rays (very
   astringent), ...**Krameria, 301, N.F.**
   Tortuous, subcylindrical pieces; wood-wedges arranged in concentric
   circles, separated by compressed stone cells,.....**Pareira, 157, N.F.**
   With thick bark.
   Brownish color, internally tawny yellow, thin cork; narrow wood-
   wedges; distinct medullary rays,............**Glycyrrhiza, 230**
   Light brown; wood porous; thin cork; numerous narrow medullary
   rays (rhizome, U.S.P. 1900),..............**Apocynum, 446, N.F.**

* When not otherwise stated, roots in this list are cylindrical in form.

Blackish color; thick bark; meditullium radiate, distinct medullary
rays,..**Ipecac, 530**

2. RHIZOMES

(a) Monocotyledonous:
   a. With rootlets.
   Obconical, but usually in slices; internally whitish, with dark dots;
   wood bundles short, curved, inclosed by wavy nucleus sheath;
   benumbing, acrid, bitter taste,
                                    Veratrum Viride 60, and V. Album, 60 a
   Whitish, irregular pieces; parenchyma thin-walled; small number of
   vascular bundles within a thick-celled nucleus sheath,
                                    Flores and Radix, Convallaria, 59, N.F.
   Brownish color; deep stem scars on upper side; rather thick bark;
   wood-bundles scattered, distinct toward the center; nucleus
   sheath indistinct,.......................**Cypripedium, 84, N.F.**
   b. Without rootlets.
   Buff color; flattish, lobed on one side; resin cells scattered through
   the parenchyma; wood-bundles both sides the nucleus sheath,
                                    Zingiber, 78
   Light yellow, straw-like, hollow; cortical zone thick, composed of
   large-celled parenchyma, and containing about six wood-bundles
   near the outer surface,..........................**Triticum, 37**

(b) Dicotyledonous Rhizomes:
   a. With rootlets.
   Brown; thin bark covered with a thin cork; vascular bundles small,
   inclosing a thick pith; odor disagreeable,.......**Valerian, 543**
   Yellow-brown; bark thin; wood-wedges largest on the lower side;
   large-celled pith near top of rhizome,..........**Serpentaria, 118**
   Internally bright reddish-yellow; thick bark; narrow wood-wedges;
   broad yellow medullary rays, inclosing large pith,..**Hydrastis, 134**
   Brownish black, internally whitish; thick bark; narrow wood-wedges
   and medullary rays; large pith; wood-wedges of the rootlets form
   "Maltese cross,"..............................**Cimicifuga, 133**
   Blackish-brown; bark thin; wood in one or two circles, inclosing
   · a large, angular pith, usually about six-rayed,.**Leptandra, 521, N.F.**
   Purplisk-brown; bark thin; wood whitish, and arranged in a circle,
   thickest on lower side; pith dark-colored or decayed,..**Spigelia, 439**
   Yellowish-brown; tough (root and rhizome),.....**Berberis, 165, N.F.**
   b. Without rootlets.
   Orange-brown; bark thick; wood-wedges short, forming a circle and
   inclosing a large pith,.....................**Podophyllum, 161**
   Yellowish-brown to dark brownish gray, indistinctly radiate wood,
   horny center,......................................**Scopola, 506**
   Reddish-brown; internally whitish, usually with small red dots;
   bark thin; small vascular bundles in loose circle; large pith;
   numerous resin cells in parenchyma,.........**Sanguinaria, 185**
   Brown, internally brownish-red; thin bark; wood-wedges small,
   forming distinct circles near cambium; medullary rays broad;
   large central pith,........................**Geranium, 266, N.F.**
   Aspididium, Male Fern Rhizome. (See *Cryptogams, II*), 12.

3. TUBERS

(a) Dicotyledonous:
   Dark brown, with lighter colored warts; bark thin, with a dense
   zone of resin cells on the inner surface; vascular bundles small
   and indistinct, concentric circles of resin cells,........**Jalapa, 460**
   Dark brown, wrinkled; thick bark; cambium about seven-rayed;
   small vascular bundles at termination and base of cambium
   rays; large-celled pith (tuberous root),..........**Aconitum, 146**

4. BULBS

(a) Monocotyledonous:
Pear-shaped, yellow-white; usually in scales; parenchyma thin-walled, with numerous raphides, and traversed by parallel vascular bundles and lactiferous ducts,......................**Scilla, 67**

5. CORMS

(a) Monocotyledonous:
Ovoid, with groove on one side, often in transverse slices; uniform in shape; mostly parenchymatous tissue, containing occasional raphides; vascular bundles numerous, scattered,

Colchici Cormus, 68

. Overground Stems.
     a. HERBACEOUS.

6. HERBS

(a) Dicotyledonous:
     a. Petals distinct.
Leaves small, trifoliate; stems pentangular, nearly smooth; as found in market, drug consists mostly of the greenish-brown stems, free from leaves (tops of the plant),.....**Scoparius, 246 N.F.**
     b. Petals united.
Leaves opposite, lanceolate, united at base; the drug consists of broken fragments of dark green leaves, with downy, resin-dotted, lower surface,........................**Eupatorium, 574 N.F.**
Leaves smooth, light green; florets yellow; the drug consist of fragments of leaves, stems, and flower-heads with stiff, resinous bracts,

Grindelia, 576

Leaves ovate, alternate, pale green; stems furrowed, hairy; flowers small, pale blue; capsules thin and papery,......**Lobelia, 552**
Leaves petiolate, ovate-lanceolate, sharply serrate; flowers purplish, in terminal, conical spikes,............**Mentha Piperita,.. 473**
Leaves nearly sessile, lance-ovate; flowers in terminal acute spikes, aromatic,................................**Mentha Viridis, 474**
Leaves opposite, oblong-ovate; flowers small; axillary cymules, aromatic,................. ....**Hedeoma, 475**
Leaves opposite, lance-ovate, serrate; flows axillary, one-sided racemes; stem quadrangular, smooth,

Scutellaria, 478 N.F.

     c. Petals absent.
Leaves opposite, sessile, ovate; flowers numerous, small; whole plant smooth, pale brown, inodorous,.......**Chirata, 443 N.F.**
Leaves digitate, with lancelinear leaflets; the drug in market consists of leafy tops and flowers, often contains nearly ripe fruit, all forming brownish-green, resinous mass,.... **Cannabis Indica, 112**

β. WOODY.

7. BARKS

(a) Bast with Isolated Bast Cells:
Bast fibers short, in radial lines or small groups; inner bark reddish-brown, finely striate; cork cells thin-walled,... **Cinchona, 532**
(b) Bast Radially Striate:
Long, closely rolled quills composed of many papery, yellowish-brown layers; smooth outer surface composed of stone cells; wood-bundles in wavy lines; no outer bark,

Cinnamomum Zeylanicum, 167

Corky layer present, gray-brown with numerous white patches on the surface; an almost uninterrupted line of white striæ near cork,......................**Cinnamomum Saigonicum, 169**
Irregular fragments, deprived of gray corky layer; bright rust brown; bast fibers few; light medullary rays in inner layer,

Sassafras, 170 N.F.

Irregular pieces; periderm greenish-brown, glossy; inner surface lighter, finely striate; bast wavy, irregular,
Prunus Virginiana, 203

(*c*) BAST TANGENTIALLY STRIATE:
Long, thin bands rolled into disks; periderm greenish-orange; inner surface white, silky; bast fibers long; bark flexible,
Mezereum, 365
Thin, tough bands; periderm blackish; inner surface lighter, smooth; bast fibers in elongated wedge-shaped groups,
Rubus, 217 N.F.
Thin, flexible quills or bands; periderm thin, yellow-brown; inner surface white, finely striate; bast fibers long, silky,
Gossypii Cortex, 344a, N.F.
Quilled or curved pieces; periderm ash-gray, covered with blackish patches; scaly; inner surface smooth, tawny, **Euonymus, 326, N.F.**
In strips or quills; periderm brown-gray, with purplish-brown stripes, covered with small brownish lenticels, inner surface white or brownish,..................**Viburnum Opulus, 540, N.F.**

(*d*) BAST CHECKERED:
Brownish-white pieces, nearly deprived of corky layer; bast fibers in rows, crossed by numerous medullary rays; mucilaginous,
Ulmus, 109
Pale brown pieces, deprived of corky layer; inner surface ridged; astringent,................................**Quercus, 102, N.F.**
Pale brown-white pieces; smooth on both sides or outer surface with fragments of red-brown bark; bast fibers pale brown, imbedded in white wood and crossed by white medullary rays,
Quillaja, 212, N.F.

(*e*) BAST WITHOUT STRIATION:
Thin quills or fragments; outer surface dark brown, somewhat warty; inner surface yellowish, finely striate,......**Granatum, 366**
Irregular chips or longitudinal pieces with outer corky layer, brownish-gray or reddish-brown and deeply furrowed, **Aspidosperma, 447**
Thin pieces; periderm glossy brown, dotted; inner surface whitish, smooth; numerous groups of stone cells,
Viburnum Prunifolium, 541
Thin quills; periderm dark brown with white dots; inner surface smooth, brownish-yellow; rows of crystal cells,....**Frangula, 333**
Curved or quilled pieces, outer surface brown-gray; inner surface yellowish, darkening with age; finely striate; bast bundles in groups; medullary rays narrow,....**Rhamnus Purshiana, 334**
Curved fragments; periderm brownish gray, with black dots; inner surface whitish, smooth; medullary rays narrow,
Xanthoxylum, 270

8. TWIGS

Greenish-gray; bark rather thick; wood in one or two circles (U.S.P., 1890),. ....**Dulcamara, 514, N.F.**

9. WOODS

Yellowish-white billets; free from bark; consists mostly of prosenchyma, with large ducts, radially striate; narrow medullary rays; very bitter,...........................**Quassia, 287**
Wood dark red; parenchyma in about four rows, forming irregular circles; medullary rays single-rowed; ducts large,
Santalum Rubrum, 239
Brownish-red; wood parenchyma in broad, wavy circle; medullary rays composed of about two rows of cells; ducts fine,
Hæmatoxylon, 238, N.F.

10. PITHS

Cylindrical, white, very light and spongy, **Sassafras Medulla, 172**
C. Outgrowths from Overground Stems.

11. LEAVES

(a) SIMPLE:

a. Margin entire.

Grey-green, lanceolately scythe-shaped, distinct marginal veins, coriaceous, feather-veined,........ **Eucalyptus, 368**

Dark green, obovate; midrib prominent, with a curved line on each side,..**Coca, 265**

Brownish-green above, pale below, ovate, tapering at the apex; midrib prominent, smooth with occasional perforations, **Belladonnæ Folia, 504**

Gray-green below, rather thick, obovate; smooth, dark-green, glossy above; lighter and reticulate below,............**Uva Ursi, 411**

b. Margin toothed, or crenate.

Ovate-oblong, gray-green; veining prominent; soft, hairy, aromatic, **Salvia, 492**

Lanceolate; upper surface dull green, tesselated; under surface rough, prominently veined, meshed,...........**Matico, 92, N.F.**

Obovate, dull yellowish-green, coriaceous, pellucid-punctate, sharply serrate, with gland at the base of each tooth,.........**Buchu, 274**

Oblong-lanceolate; upper surface vanished, brownish-green; under surface, white, hairy, distinctly reticulate,....**Eriodictyon, 464**

Ovate-oblong; upper surface green, wrinkled, velvety; under surface hairy, prominent whitish meshes,...............**Digitalis, 518**

Ovate-oblong, with large triangular crenations; gray-green, hairy; midrib prominent, lighter colored; odor heavy, narcotic, **Hyoscyamus, 509**

Ovate, with sharp-pointed teeth; smooth, brownish-green; midrib hairy, ... **Stramonium, 507**

Oblanceolate, coriaceous, sharply serrate; upper surface dark green, ...**Chimaphila, 414, N.F.**

Ovate, nearly smooth, veins parallel from the midrib, **Hamamelidis Folia, 199, N.F.**

(b) COMPOUND:

a. Margin entire.

Ovate, coriaceous, dull green; veins forming one or two wavy lines parallel to the margin,... ....**Pilocarpus, 275**

Lance-oval, gray-green, somewhat pubescent, odor peculiar, nauseous, .. . **Senna (Alexandria), 240**

12. LEAFY TOPS

Branches quadrangular; leaves in four rows, scale-like, imbricated, with shallow groove on the back,.................... **Sabina, 19**

13. PLANT HAIRS AND GLANDULAR OUTGROWTHS

Subglobular, minute, yellowish-brown granules; resinous; under the microscope, hood-shaped,.............. **Lupulinum, 111, N.F.**

Fine powder, pale yellow, not wetted by water; under the microscope, tetrahedral,..................... . **Lycopodium, 18**

Hairs of seed long, white (see seed covering),...... **Gossypium, 344 b**

Styles and stigmas, fine silky hairs (see also seed coverings), **Zea, 40, N.F.**

14. FLOWERS AND PARTS OF FLOWERS

(a) FLOWERS AND BUDS:

a. Racemose or cymose inflorescence.

Sepals five, reddish, veined; petals small, hairy, **Brayera (cusso), 222, N.F.**

b. Compound flower-heads.

Oblong-ovoid, unexpanded, small; smooth, somewhat glossy, greenish-brown scales.......................**Santonica, 601**

Petals united; rays long, white; disk-flowers numerous, short, yellow, tubular; receptacle hollow, naked,...... **Matricaria, 599**

Petals united; rays white, in rows; flower-heads large; receptacle solid, densely chaffy; disk florets few, yellow, tubular,

Anthemis, 600

Petals united; rays yellow, strap-shaped, longitudinally veined; receptacle flat, naked; disk florets yellow, tubular,

Calendula, 602, N.F.

Petals united; rays yellow, three-toothed; flower-heads large; receptacle nearly flat, pitted, hairy,. .... **Arnica, 565**

c. Single flowers.

Unexpanded; long, dark brown; calyx four-cleft, solid, glandular; petals four, formed into a head, aromatic,...**Caryophyllus, 371**

(*b*) PETALS.

Petals unexpanded; in form of cone; deep-red, yellow at base; fragrant,.................................. **Rosa Gallica, 213**

(*c*) STIGMAS:

Thread-like, long, silky, longitudinally veined, yellowish; with styles,

Zea, 40, N.F.

15. FRUITS AND PARTS OF FRUITS

(*a*) COLLECTIVE FRUITS:

Strobiles glandular; thin, greenish scales, with delicate veins; aromatic,....................................... **Humulus, 110**

Fleshy compressed; yellowish or brownish; efflorescence of sugar; numerous yellow akenes; sweet,...............**Ficus, 114, N.F.**

(*b*) FRUITS FROM SINGLE FLOWERS:

a. Cremocarps.

Ovate, compressed at the sides, grayish, hairy, mericarps usually separated, curved, five-ribbed; many thin oil tubes; aromatic, sweet,..**Anisum, 381**

Ovate, compressed, gray-green, smooth; mericarps usually separated, each five-ribbed; six oil tubes,.......................**Carum, 385**

Oblong, smooth, nearly cylindrical, greenish-brown; ribs prominent, obtuse, ten oil tubes,........................**Fœniculum, 382**

Globular, smooth; mericarps united, each with two oil tubes on face and five wavy ribs and four prominent ridges on back,

Coriandrum, 386

Ovoid crescent shaped, prominent ribs, alternating with the coarsely roughened furrows,........................ **Petroselinum, 391a**

b. Capsular fruit, superior.

Cylindrical, long, black-brown; numerous transverse divisions; seeds glossy brown; pulp sweet, odor prune-like,

Cassia Fistula, 247, N.F.

Inferior.

Ovate-triangular, three-celled; pale buff color; seeds brownish, angular, numerous, ...**Cardamomum, 82**

Cylindrical, long, wrinkled, single-celled, containing the numerous small black seeds; pulp blackish-brown; fragrant, Vanilla, 87, N.F.

c. Fleshy.

Globular, blackish-gray, reticulately wrinkled; internally whitish, hollow; pericarp formed into a stalk,.............. **Cubeba, 88**

Globular, brownish-black; internally lighter, hollow; reticulately wrinkled; without stalk, hot taste,...................**Piper, 89**

Globular, glandular; two-celled; each cell contains a single brownish seed; aromatic............................**Pimenta, 372, N.F.**

Subglobular, dark red, densely hairy, single-seeded,

Rhus Glabra, 317, N.F.

Blackish-brown, one seeded, ovoid-oblong,.............**Sabal, 47**

Oblong, wrinkled; pericarp red, shining; two-celled; with numerous yellowish seeds; intensely hot taste,..............**Capsicum, 516**

Globular, deprived of rind; light, spongy, breaking into thin, wedge-shaped pieces; contains many light-colored seeds; intensely bitter,

Colocynthis, 544

Oblong, wrinkled, black-blue; pulp soft, brownish-yellow; single seed; sweet, acidulous,......................**Prunum, 205, N.F.**

(*c*) PARTS OF FRUITS:

Thin, curved sections; glandular; pericarp leathery, dark brownish-green; odor fragrant,........　....Aurantii Cortex, 278

Thin curved sections; glandular; epidermis deep lemon-yellow; odor fragrant,....................................Limonis Cortex, 283

Pulp fibrous, dark brown, sweet, acidulous, with glossy brown, flattish seeds,..............................Tamarindus, 249, N.F.

16. SEEDS AND SEED COVERINGS

a. SEEDS

(*a*) DICOTYLEDONOUS:

　a. Albuminous.

Orbicular disks, grayish-green, curved, with fine, silky hairs; internally whitish, hairy; bitter taste,....　. Nux Vomica, 435

Ovate; testa removed; hard, light brown, reticulately furrowed; internally lighter, with dark brown veins; strongly aromatic,
Myristica, 154

Ovate-lanceolate; flattish, hairy, grey-green; internally white, oily,
Strophanthus, 451

Triangular, flattish, brown, deeply pitted; albumen whitish, oily,
Staphisagria, 143

Ovate, flattish; tests brown, glossy; albumen thin, inclosing large cotyledons, ... Linum, 264

　b. Exalbuminous.

Ovate, flattish, curved, with a thin, brown, membranous testa; longitudinally veined; embryo white, oily, consists of two planoconvex cotyledons; bitter taste,............ Amygdala Amara, 209

Resembles above, but longer, more convex sides; a bland, sweetish taste, ... Amygdala Dulcis, 210

Globular, small; testa dark brown, hard, finely pitted; embryo greenish-yellow, oily; pungent taste, ...Sinapis Nigra, 189

Globular, larger than above; testa yellowish; taste pungent,
Sinapis Alba, 188

Ovate, broad, flat; tests whitish; shallow groove and flat ridge near margin,..................................... Pepo, 548

Oblong or reniform; testa deep brown, granular; broad black groove along convex edge; elliptic cavity between cotyledons,
Physostigma, 252

(*b*)" MONOCOTYLEDONOUS:

　a. Albuminous.

Subglobular; tests reddish-brown, thin, pitted, hard; albumen whitish, tough, horny,....................... Colchici Semen, 69

β. SEED COVERINGS.

(*a*) ARILLODE:

(*b*) SEED HAIRS:

Hairs of seeds long, white, curling,... Gossypium Purificatum, 344 b.

II. CRYPTOGAMS.

A. **Filices.**

Rhizome, with glossy brownish scales,.............Aspidium, 12

B. **Algæ.**

Yellow, horny, transparent, with numerous forks and branches; mucilaginous taste,..　.................Chondrus, 1

Yellowish-white or brownish-white, shiny, thin, translucent taste mucilaginous, ... Agar, 4

C. **Lichens.**

Brownish above, whitish beneath foliaceous (U.S.P., 1890),
Cetraria, 9

D. Fungi.
Oblong, narrow, curved; longitudinally grooved; black; peculiar heavy odor,..Ergota, 5

E. Club Mosses.
Fine powder, pale yellow, very mobile, floats on water; under the microscope, tetrahedral. (See also under 13),..Lycopodium, 18

III. ABNORMAL GROWTHS CAUSED BY PARASITES

A. Excrescences.
Globular, with a short stipe; externally dull blue or lead color, covered with prominent warts, interior whitish, central cavity, Galla, 105

IV. NON-CELLULAR DRUGS DERIVED FROM CELL-CONTENTS AND SECRETIONS

A. Farinaceous.
Fine white powder, sometimes in angular masses; odorless and tasteless; insoluble in cold water,...................... Amylum, 42

B. Extractives.
Irregular masses containing fragments of leaves; dark brown; fracture conchoidal, brittle, glossy (Gambir),..............Catechu, 257
Cylindrical cakes, hard, mottled, reddish-brown; fracture uneven, lighter colored,................................Guarana, 329
Cylindrical sticks, glossy brown-black; fracture conchoidal; taste sweet,........................Extractum Glycyrrhizæ, 230 a

C. Concrete Juices.
Irregular masses containing some fragments of leaves; chestnut-brown, plastic, coarsely granular; odor heavy; narcotic,.....Opium, 180
Angular pieces, red-brown; internally lighter; waxy lustre; bitter,
Lactucarium, 594
Yellowish to blackish masses, hard, brittle, somewhat glossy; taste bitter, ...Aloe, 70
Small angular pieces, brittle, dark brownish-red; thin layers, ruby-red and transparent, sweetish,....................... Kino, 258
Flat pieces, blackish-brown, internally lighter; very elastic; floats on water,... Elastica, 309

D. Sugars.
(*a*) SOLID:
Granular, white, crystalline, transparent, very sweet, very soluble in water, ...Saccharum, 39
Crystals or white crystalline powder, gritty; translucent, sandy; sweetish taste,.......... .....Saccharum Lactis, q.v.
Irregular fragments; light yellow; internally white; porous, crystalline, friable, sweet, ...Manna, 429
(*b*) LIQUID:
Translucent syrup, yellow to brown-yellow, ...Mel, 618

E. Gums.
Roundish tears, fissured, brittle, translucent; fracture glass-like,
Acacia, 255
Curved bands, marked with parallel wavy lines; white, translucent, tough, horny,................................Tragacantha, 256

F. Gum Resins.
Irregular pieces composed of whitish tears imbedded in a brown-gray sticky mass; odor sickening,....................Asafœtida, 397
Irregular masses or tears; red-brown, dusty; fracture waxy; taste bitter, acrid,..Myrrha, 294
Cylindrical pieces or lumps, sometimes hollow in the center; orange-yellow; fracture smooth, waxy, ...Cambogia, 348
Irregular pieces or circular cakes; dark gray internally; porous; fracture angular; odor peculiar, somewhat cheese-like,
Scammonium, 462

G. Resins.

Irregular lumps; reddish-brown, smooth, mottled, with milk-white tears; agreeable balsamic odor,.................Benzoinum, 428

Large lumps or masses, yellowish-brown, transparent, brittle,
Resina, 27 c

Small globular tears, transparent, yellowish, brittle, glossy,
Mastiche, 319, N.F.

Irregular masses, greenish or reddish-brown; internally of a glossy luster; brittle,..............................Guaiacum, 269

H. Oleoresins.

Brownish-yellow, viscid liquid; transparent; odor peculiar; taste bitter,...Copaiba, 259

Light yellow or faintly greenish, transparent or viscid liquid; agreeable odor,.....................Terebinthina Canadensis, 29

Irregular masses, tough, yellowish, opaque; fracture crumbly,
Terebinthina, 27

Thick, viscid semi-fluid, nearly black, opaque; odor empyreumatic,
Pix Liquida, 28

I. Balsams.

Thick, syrupy liquid; brownish-black, thin layers, transparent,
Balsamum Peruvianum, 262

Very thick, semi-liquid, brownish-yellow, solid in the cold, agreeable odor,........   ....Balsamum Tolutanum, 263

Viscid semi-liquid, opaque, brownish-gray, odor agreeably balsamic,
Styrax, 201

J. Stearoptens.

Translucent masses; tough, crystalline, granular; odor penetrating, peculiar,....................   ....Camphora, 178

Colorless prisms or small scales; thyme-like odor,......Thymol, 390

Fine, white, transparent needles or crystals; peppermint odor,
Menthol, 473 b

K. Fatty Substances.

(a) OF VEGETABLE ORIGIN:

Straw-colored liquid; clear, rather thin; nutty odor, bland taste,
Oleum Amygdalæ Expressum, 211

Yellow, limpid liquid; transparent,..............Oleum Lini, 264 a

Viscid, yellowish, nearly colorless, transparent; taste sickening,
Oleum Ricini, 312 a

Viscid, yellow to brownish, transparent, somewhat fluorescent; taste acrid, burning,.........................Oleum Tiglii, 313 a

Pale yellow liquid; thin transparent; odor and taste mild nutty,
Oleum Gossypii Seminis, 344 c

Yellow to greenish-yellow, thin clear,.............Oleum Olivæ, 430

Yellowish-white solid; rather hard, brittle, aromatic; taste chocolate-like, ...Oleum Theobromatis, 346 a

(b) OF ANIMAL ORIGIN:

Pale yellow liquid; thin, transparent; fishy odor,
Oleum Morrhuæ, 616

Yellowish, nearly colorless; thin, clear,.......Oleum Adipis, 628 a

Soft, white solid; unctuous,........   ....Adeps, 628

Light yellow or whitish solid, rather firm,
Adeps Lanæ Hydrosus, 631

White solid; smooth, unctuous,.....   Sevum Præparatum, 626

(c) WAXES:

White masses; translucent; fracture crystalline,.....Cetaceum, 617

Yellowish-white cakes; brittle; semi-translucent, Cera Alba, 618 b

Yellow to brownish, opaque; fracture granular, Cera Flava, 618 a

(d) MIXTURE OF HYDROCARBONS, 631 a

Petrolatum Album, Petrolatum Liquidum, Petrolatum = Petrolatum Molle, Petrolatum Spissum, Paraffinum.

DRUGS OF ANIMAL ORIGIN

A. INSECTS.

Long, bronze-green; body cylindrical; head triangular,
Cantharis, 605
Oval, gray or brownish, wrinkled, covered with a whitish down,
Coccus, 606

B. TISSUES AND SECRETIONS.

Granular, crumbly, various sizes; dark reddish-brown; peculiar
penetrating odor,............. ....**Moschus, 620**
Viscid liquid, of a brownish or dark green color,..**Fel Bovis, 621**
Yellowish-green solid, rather soft,......... **Fel Bovis Purificatum**
Yellowish powder or thin, yellow, translucent scales,..**Pepsinum, 629**
Yellowish-white powder or transparent, brittle scales,
Pancreatinum, 630
(Fatty substances of animal origin. See above under K (*b*), Adeps,
etc.)

BACTERIAL PRODUCTS

Antitoxic Serums. See chapter on.
Serum Antidiphtheriticum. See chapter on.

GLANDULAR PRODUCTS

Glandulæ Suprarenales Siccæ. See chapter on.
Glandulæ Thyreoideæ Siccæ. See chapter on.

CONSPECTUS B.—OFFICIAL AND UNOFFICIAL DRUGS, AR RANGED ACCORDING TO PROMINENT PHYSICAL PROPERTIES, AND SUBDIVIDED BY ODOR AND TASTE

(*With Natural Order or Family.*)

ROOTS

CLASS I.—AROMATIC

(*a*) Odor and Taste Pronounced.

SUMBUL.—Musk Root. 400*. Umbelliferæ.
ANGELICA.—ANGELICA. 396, N.F. Umbelliferæ.
ANGELICA ATROPURPUREA.—AMERICAN ANGELICA. 395, N.F. Umbelliferæ.
Armoracia.—Horse-radish. 191. Cruciferæ.
Imperatoria.—Masterwort. 401. Umbelliferæ.
Levisticum.—Lovage. 403. Umbelliferæ.
Pimpinella.—Pimpernel. 404, N.F. Umbelliferæ.
Vetiveria.—Vetivert. 38. Gramineæ.

(*b*) Feebly Aromatic.

GELSEMIUM.—Yellow Jasmine. 438. Loganiaceæ.
INULA.—Elecampane. 557. Compositæ.
Methysticum.—Kava Kava. 95, N.F. Piperaceæ.
PETROSELINUM.—Parsley. 391, N.F. Umbelliferæ.

The names in italics refer to unofficial drugs; some of these (marked thus, *) have been official in one or more of the former editions of the U.S.P. The numbers correspond to the numbers of the drugs in the body of this work.
*Bold-faced caps indicate drugs officials in U.S.P. ix. Those in small caps are contained in National Formulary.

CLASS II.—ODORLESS AND TASTELESS

Alkanna.—Alkanet. 465. Boraginaceæ.

CLASS III.—ACRID

(*a*) Acridity Pronounced.
PHYTOLACCÆ RADIX.—Poke Root. 126, N.F. Photolaccaceæ.
PYRETHRUM.—Pellitory. 555. Compositæ.
SARSAPARILLA.—58. Liliaceæ.
SENEGA.—Seneka. 302. Polygalaceæ.
STILLINGIA.—Queen's Delight. 304. Euphorbiaceæ.
Pyrethrum Germanicum.—555 a. Compositæ.

(*b*) Acridity Slight.
BELLADONNÆ RADIX.—Deadly Nightshade. 503. Solanaceæ Fl'ext.,
    liniment.
**Euphorbia Corollata.*—Large Flowering Spurge. 305 a. Euphorbiaceæ.
Euphorbia Ipecacuanha.—Ipecacuanha Spurge. 305 b. Euphorbiaceæ.
Hemidesmus.—Indian Sarsaparilla. 458. Asclepiadaceæ. EUPHORBIA
    PILULIFERÆ.—305 c. Euphorbiaceæ.
Scopola.—See *Rhizomes.* 506.

CLASS IV.—BITTER

APOCYNUM.—Canadian Hemp. 446, N.F. Apocynaceæ.
ASCLEPIAS.—Pleurisy Root. 454, N.F. Asclepiadaceæ.
BRYONIA.—Bryony. 545, N.F. Cucurbitaceæ. Tr.
CALUMBA.—Columbo. 156. Menispermaceæ.
GENTIANA.—441. Gentianacæ.
IPECACUANHA.—Ipecac. 530. Rubiaceæ.
PAREIRA.—Pareira Brava. 157, N.F. Menispermaceæ.
RHEUM.—Rhubarb. 120. Polygonaceæ.
RUMEX.—Yellow Dock. 121, N.F. Polygonaceæ.
TARAXACUM.—Dandelion. 553. Compositæ.
APOCYNUM.—Dog's Bane. 446 a, N.F. Apocynaceæ.
BAPTISIA.—Wild Indigo. 233, N.F. Leguminosæ.
BERBERIS.—Oregon Grape (root and rhizome). 165, N.F. Berberidaceæ.
**Berberis.*—Barberry. 163. Berberidaceæ.
Cichorium.—Chicory. 554. Compositæ.
Frasera.—American Columbo. 442. Gentianaceæ.
Rhaponticum. Crimean Rhubarb. 120 a. Polygonaceæ.

CLASS V.—SWEETISH

GLYCYRRHIZA.—Licorice Root. 230. Leguminosæ.
Ipomæa Pandurata.—Wild Jalap. 461 a. Convolvulaceæ.
**Panax.*—Ginseng. 378. Araliaceæ.
Saponaria.—Soapwort. 129. Caryophyllaceæ.
Saponaria Levantica.—128. Caryophyllaceæ.

CLASS VI.—MUCILAGINOUS

ALTHÆA.—Marshmallow. 341. Malvaceæ.
LAPPA.—Burdock. 558, N.F. Compositæ.
Symphytum Comfrey.—466. Borraginaceæ.

CLASS VII.—ASTRINGENT

KRAMERIA.—Rhatany, 301, N.F. Krameriaceæ.
**Heuchera.*—Alum Root. 196. Saxifragaceæ.
**Statice.*—Marsh Rosemary. 422. Plumbaginaceæ.

UNCLASSIFIED.—Abri Radix, 231. Baycuru, 423. Carnauba, 48. Ceanothus,
    336. Cicuta, 406. ECHINACEA, *563.* Eryngium, 407. Helianthella, 562.
    HYDRANGEA, *197.* Jambu Assu, 94. Laciniaria, 561. Laserpitium, 402.

Manaca, 505. Maregamia, 298. Osmorrhiza, 408. Osmunda, 16. Pæonia, 148. Polemonium, 463. Polymnia, 560. Thapsia, 405. Triosteum, 542. VERBENA HASTATA, *470.* Verbena Urticæfolia, 471. Yerba Mansa, 93.

RHIZOMES

CLASS I.—AROMATIC

(*a*) Odor and Taste Pronounced.
CALAMUS.—Sweet Flag. 52. Araceæ.
SERPENTARIA.—Virginia Snakeroot. 118. Aristolochiaceæ.
VALERIANA.—Valerian. 543. Valerianaceæ
ZINGIBER.—Ginger. 78. Zingiberaceæ.
ASARUM.—Wild Ginger. 119, N.F. Aristolochiaceæ.
GALANGA.—Galangal. 79, N.F. Zingiberaceæ.
ZEDOARIA.—Zedoary. 80, N.F. Zingiberaceæ.

(*b*) Slightly Aromatic.
Arnicæ Radix.—564. Compositæ.
Aralia Nudicaulis.—False Sarsaparilla. 379. Araliaceæ.
ARALIA RACEMOSA.—American Spikenard. 379 a. Araliaceæ.
* *Curcuma.*—Turmeric. 81. Zingiberaceæ.
IRIS FLORENTINA.—Orris Root. 76, N.F. Irideæ.

CLASS II.—ACRID

IRIS.—Blue Flag. 75, N.F. Iridaceæ.
SANGUINARIA.—Bloodroot. 185. Papaveraceæ.
VERATRUM VIRIDE OR ALBUM.—American (green) or European (white) Veratrum. 60. Liliaceæ.
DIOSCOREA.—Wild Yam. 74, N.F. Dioscoreaceæ.
Symplocarpus.—Skunk Cabbage. 53. Araceæ.
TRILLIUM.—Birthwort. 64, N.F. Liliaceæ.

CLASS III.—SLIGHTLY ACRID AND BITTER

CIMICIFUGA.—Black Snakeroot. 133. Ranunculaceæ.
CONVALLARIA.—Lily of the Valley. 59, N.F. Liliaceæ. Also flowers.
CYPRIPEDIUM.—Lady's Slipper. 84, N.F. Orchidaceæ.
PODOPHYLLUM.—Mandrake. 161. Berberidaceæ.
SCOPOLA.—Solanaceæ. 506.
CAULOPHYLLUM.—Blue Cohosh. 160, N.F. Berberidaceæ.
Asclepias Cornuti.—Milk-weed. 455. Asclepiadaceæ.
Asclepias Incarnata.—Swamp Milk-weed. 456. Asclepiadaceæ.
Helleborus Niger.—Black Hellebore. 137. Ranunculaceæ.
Helleborus Viridis.—Green Hellebore. 138. Ranunculaceæ.
Polygonatum.—Solomon's Seal. 62. Liliaceæ.

CLASS IV.—BITTER

(*a*) With Rootlets.
HYDRASTIS.—Golden Seal. 134. Ranunculaceæ.
LEPTANDRA.—Culver's Physic. 521, N.F. Scrophulariaceæ.
Menispermum.—Yellow Parilla. 158. Menispermaceæ.
ALETRIS.—Colic Root. 73, N.F. Hæmodoraceæ.
Chamællirium.—Starwort. 63. Liliaceæ.
Collinsonia.—Stone-root. 502. Labiatæ.
Gillenia.—American Ipecac. 226. Rosaceæ.
Triosteum.—Bastard Ipecac. 542. Caprifoliaceæ.
Xanthorrhiza.—Yellow-root. 139. Ranunculaceæ.

CLASS V.—SWEETISH

ASPIDIUM.—Male Fern. 12. Filices.
SPIGELIA.—Pinkroot. 439. Loganiaceæ.
TRITICUM.—Couch Grass. 37. Gramineæ.

CLASS VI.—ASTRINGENT

(*a*) With **Rootlets.**
  * *Geum Rivale.*—Water Avens. 225. Rosaceæ.
  Geum Urbanum.—Avens. 224. Rosaceæ.

(*b*) With **Few or no Rootlets.**
  GERANIUM.—Cranesbill. 266, N.F. Geraniaceæ.
  Bistorta.—Bistort. 124. Polygonaceæ.
  Nymphæa.—Water Lily. 132. Nymphæaceæ.
  **Tormentilla.*—Tormentil. 223. Rosaceæ.

UNCLASSIFIED.—All unofficial: Actæa, 135. Adrue, 45. Aralia Hispida, *380.*
  Asparagus, 65. Carex, 44. Cnicus Arvensis, 566. Corallorrhiza, 85.
  Jeffersonia, 162. Rubia, 531. Sarracenia, 192.

TUBERS, BULBS, AND CORMS

(*Mostly Acrid*)

TUBERS

ACONITUM.—Aconite. 146. Ranunculaceæ.
JALAPA.—Jalap. 460. Convolvulaceæ.
Corydalis.—Turkey Corn. 187. Fumariaceæ.
Salep.—Salep. 86. Orchidaceæ.

BULBS

ALLIUM.—Garlic. 66, N.F. Liliaceæ.
SCILLA.—Squill. 67. Liliaceæ.

CORMS

COLCHICI CORMIS.—Colchicum-root. 68. Liliaceæ.
Arisæma Dracontium.—Green Dragon. 55. Araceæ.
**Arum.*—Indian Turnip. 54. Araceæ.

TWIGS AND BRANCHES

DULCAMARA.—Bittersweet. 514, N.F. Solanaceæ.
Gouania.—Chewstick. 337. Rhamnaceæ.
Pichi.—513. Solanaceæ.

WOODS

GUAIACI LIGNUM.—Guaiac Wood. 268, N.F. Zygophyllaceæ.
HÆMATOXYLON.—Logwood. 238, N.F. Leguminosæ.
SANTALUM RUBRUM.—Red Saunders. 239. Leguminosæ.
QUASSIA.—287. Simarubaceæ.
Juniperus Oxycedrus.—24. Pinaceæ.
Ostrya.—Ironwood. 108. Cupuliferæ.
Sassafras Lignum.—Sassafras Wood. 171. Lauraceæ.
SANTALUM ALBUM.—Sandalwood. 116, N.F. Santalaceæ.

BARKS

CLASS I.—AROMATIC

(a) **Deprived of Corky Layer.**
CINNAMOMUM ZEYLANICUM.—Ceylon Cinnamon. 167. Lauraceæ.
Cinnamomum Cassia.—Cassia Cinnamon. 168. Lauraceæ.
Sassafras.—170, N.F. Lauraceæ.
Canella.—White Cinnamon. 357, N.F. Canellaceæ.
Cinnamodendron.—False Winter's Bark. 358. Canellaceæ.
Coto.—174. Lauraceæ.

(b) **With Periderm.**
Cascarilla.—311. Euphoribiaceæ.
CINNAMOMUM SAIGONICUM.—Saigon Cinnamon. 169. Lauraceæ.
Angustura.—272. Rutaceæ.
Wintera.—Winter's Bark. 151. Magnoliaceæ.
Betula Lenta.—Sweet Birch. 107. Betulaceæ.

CLASS II.—ACRID

Gossypii Cortex.—Cotton-root Bark. 344 a, N.F. Malvaceæ.
MEZEREUM.—Mezereum. 365. Thymelæaceæ.
Quillaja.—Soap Bark. 212, N.F. Rosaceæ
XANTHOXYLUM.—Prickly Ash. 270. Rutaceæ
Erythrophlœum.—Sassy Bark. 234. Leguminosæ.
Pisicidia.—Jamaica Dogwood. 237. Leguminosæ.

CLASS III.—BITTER

ASPIDOSPERMA.—Quebracho. 447. Apocynaceæ.
Euonymus.—Wahoo. 326, N.F. Celastrineæ.
FRANGULA.—Buckthorn. 333. Rhamnaceæ.
Juglans.—Butternut. 101. Juglandaceæ.
RHAMNUS PURSHIANA.—Cascara Sagrada. 334. Rhamnaceæ.
    ext. aromatic.
Azedarach.—Margosa Bark. 300. Meliaceæ.
Quassiæ Cortex.—Quassia Bark. 288. Simarubaceæ.
Simaruba.—289. Simarubaceæ.

CLASS IV.—BITTER AND ASTRINGENT

CINCHONA.—532. Rubiaceæ. .
CINCHONA RUBRA.—532 a. Rubiaceæ.
PRUNUS VIRGINIANA.—Wild Cherry. 203. Rosaceæ.
Viburnum Opulus.—Opulus. 540, N.F. Caprifoliaceæ.
VIBURNUM PRUNIFOLIUM.—Black Haw. 541. Caprifoliaceæ.
Pinus Alba.—White Pine. 21, N.F. Pinaceæ.
Cornus.—Dogwood. 409, N.F. Cornaceæ.
Berberis.—Barberry Bark. 164. Berberidaceæ.
Liriodendron.—Tulip-tree Bark. 152. Magnoliaceæ.
Magnolia. 150. Magnoliaceæ.
Nectandra.—Beeberu. 173. Lauraceæ.
Prinos.—Black Alder. 325. Ilicineæ.
Chaparro Amargoso.—Amargosa. Simarubaceæ. 293.

CLASS V.—ASTRINGENT

GRANATUM.—Bark of Pomegranate Root. 366. Punicaceæ. Pelletierine Tannate.
Quercus.—White Oak. 102, N.F. Cupuliferæ.
Rubus.—Blackberry. 217. Rosaceæ.
HAMAMELIDIS CORTEX.—Witchhazel Bark. 200. Hamamelidaceæ.
Salix.—Willow. 96. Saliaceæ.

CLASS VI.—MUCILAGINOUS

ULMUS.—Slippery Elm. 109. Ulmaceæ.
UNCLASSIFIED.—Acer Rubrum, 328. Æsculus Glabra, 330. Ailanthus, 291.

Alnus, 103. Alstonia Constricta, 448. Alstonia Scholaris, 449. Ampelopsis, 339. Calycanthus, 153. Carya, *101*. Cascara Amarga, 292. Celastrus, 327. Cephalanthus, 534. Cercis, 235. CHIONANTHUS, *433*. Choke Cherry, 204. COCILLANA, *299*. CONDURANGO, *459*. Conessi, 450. Fagus, 104. FRAXINUS, *431*. Fraxinus Sambucifolia, *432*. Hippocastanum, 331. Hoang-Nan, 437. Juglans Nigra, 101a. Larix, 26. Lindera, 175. Malus, 207. Mistletoe, 117. MYRICA, 99. Newbouldia, 526. Populus, 97. Ptelia, 273. Rhus Aromatica, 318. Saraca, 236. Tonga, 472. Tsuga, 25.

N.F. drugs are indicated by small capital letters.

LEAVES

CLASS I—AROMATIC

BUCHU (Short).—274. Rutaceæ.
Buchu (Long).—274. Rutaceæ.
COCA.—265. Erythroxylaceæ.
ERIODICTYON.—Yerba Santa. 464. Hydrophyllaceæ.
EUCALYPTUS.—368. Myrtaceæ.
MATICO. 92, N.F. Piperaceæ.
PILOCARPUS.—Jaborandi. 275. Rutaceæ.
Rosmarinus.—Rosemary. 493. Labiatæ.
SALVIA.—Sage. 492. Labiatæ.
Tabacum.—Tobacco. 511. Solanaceæ.
Chekan.—Cheken. 370. Myrtaceæ.
Myrcia.—Bay Leaves. 369, N.F. Myrtaceæ.
Melaleuca.—Cajuput. 374. Myrtaceæ.
Conii Folia.—Hemlock Leaves. 384. Umbelliferæ.
THYMUS.—Garden Thyme. 494, N.F. Labiatæ.

CLASS II.—ACRID

Ruta.—Rue. 276. Rutaceæ.

CLASS III.—BITTER

BELLADONNÆ FOLIA.—Deadly Nightshade. 504. Solanaceæ.
DIGITALIS.—Foxglove. 518. Scrophulariaceæ.
HYOSCYAMUS.—Henbane. 509. Solanaceæ
SENNA (Alexandria).—240. Leguminosæ.
Senna (India).—240. Leguminosæ.
STRAMONII FOLIA.—Stramonium Leaves. 507. Solanaceæ.
Duboisia.—512. Solanaceæ.
Hepatica.—Liverwort. 147. Ranunculaceæ.

CLASS IV.—ASTRINGENT

CASTANEA.—Chestnut. 106, N. F. Cupuliferæ.
CHIMAPHILA.—Pipsissewa. 414, N.F. Ericaceæ.
HAMAMELIDIS FOLIA.—Witchhazel. 199, N. F. Hamamelideæ.
Rhus Toxicodendron.—Poison Ivy. 316. Anacardiaceæ.
UVA URSI.—Bearberry. 411. Ericaceæ.
Gaultheria.—Wintergreen. 413. Ericaceæ.

UNCLASSIFIED.—Ambrosia, 571. Arctostaphylos, 412. Aurantii Folia, 280. Betonica, 500. BOLDO, *166*. Borago, 467. Caroba, 527. Cassia Marilandica, 241. Comptonia, 100. Duboisia, 512. Epigæa, 415. Erechthites, 567. Erythronium, 72. Eupatorium Purpureum, 575. Euphrasia, 519. Fragaria, 221. Garrya, 410. Guaco, 570. Kalmia, 417. Ilex Opaca, 323. Ilex Paraguayensis, 324. Laurocerasus, 220. Laurus, 176. Ledum, 418. Ligustrum, 434. Lippia Mexicana, 469. Mitella Nuda, 198. Monarda Fistulosa, 487. Ocimum, 499. Oleander, 452. Orthosiphon, 495. Oxydendrum, 419. Persica, 206. Polypodium, 15. Pterocaulon, 569. Pulmonaria, 468. Pycnanthemum, 496. Rhododendron, 420. Satureia, 497. Sesamum, 528 a. Spinosum, 573. Strumarium, 572. Thea, 347. Trilisa, 568. Turnera, 360. Umbellularia, 177. Vaccinium, 416. VERBASCI FOLIA (also Verbasci Flores), 520. Yerba Buena, 498.

N.F. drugs indicated by small capital letters.

LEAFY TOPS

All Balsamic, Camphoraceous and Bitter.
SABINA.—Savine. 19. Coniferæ.
Juniperus Virginiana.—Red Cedar. 20. Coniferæ.
THUJA.—Arbor Vitæ. 22. Coniferæ.

HERBS AND WHOLE PLANTS

CLASS I.—AROMATIC

Labiatæ.
HEDEOMA.—Pennyroyal. 475.
MARRUBIUM.—Horehound. 476.
MENTHA PIPERITA.—Peppermint. 473.
MENTHA VIRIDIS.—Spearmint. 474.
Melissa.—Balm. 477.
CATARIA.—Catnip. 489, N.F.
Glechoma.—Ground Ivy. 481.
Hyssopus.—Hyssop. 488.
Lycopus.—Bugle Weed. 482.
Mojorana.—Sweet Marjoram. 483.
Monarda.—Horsemint. 486.
Origanum.—Wild Marjoram. 479.

Compositæ.
GRINDELIA.—576.
Tanacetum.—Tansy. 577.
Achillea.—Yarrow. 584.
Cotula.—May Weed. Wild Chamomile. 598.
Parthenium.—Feverfew. 597.
Solidago.—Golden Rod. 593.

Hypocreaceæ (Cryptogamous).
ERGOTA.—Ergot. 5.
Ustilago.—Corn Smut. 6.

CLASS II.—BITTER

Compositæ.
ABSINTHIUM.—Wormwood. 578.
EUPATORIUM.—Boneset. 574, N.F.
Carduus Benedictus.—Blessed Thistle. 586.
Erigeron.—Fleabane. 580.
Erigeron Canadense. 581.
Helenium.—Sneezewort. 583.

Gentianaceæ.
CHIRATA.—Chiretta. 443, N.F.
CENTAURIUM.—Centaury. 444, N.F.

Leguminosæ.
SCOPARIUS.—Broom. 246, N.F.
MELILOTUS.—Sweet Clover. 242, N.F.
GALEGA.—Goat's Rue. 245, N.F.

Scrophulariaceæ.
Chelone.—Balmony. 524.
Scrophularia.—Figwort. 523.

Polygalaceæ.
**Polygala.*—Bitter Polygala. 303.

Violarieæ.
Viola Tricolor.—Pansy. 359.

Ranunculaceæ.
COPTIS.—Goldthread. 136, N.F.

Fungi.
AGARICUS.—White Agaric. 7, N.F.
Fungus Chirurgorum.—Surgeon's Agaric. 7 a.
Torula.—Yeast. 8.

CLASS III.—ACRID

Ranunculaceæ.
*PULSATILLA.—140, N.F.
Ranunculus.—Crowfoot. 142.

Cruciferæ.
Bursa Pastoris.—Shepherd's Purse. 190.

Campanulaceæ.
LOBELIA.—Indian Tobacco. 552.

Moraceæ.
CANNABIS INDICA.—Indian Hemp. 112.

Papaveraceæ.
**Chelidonium.*—Celandine. 183.

Cacteæ.
CACTUS GRANDIFLORUS.—Night-blooming Cereus. 363, N.F.

Droseraceæ.
DROSERA.—Sundew. 193, N.F

Hypericineæ.
Hypericum.—St. John's-wort. 350.

CLASS IV.—ASTRINGENT

Rosaceæ.
Agrimonia.—Agrimony. 227.
Potentilla.—Cinquefoil. 228.

Rubiaceæ.
Galium.—Cleavers. 536.
Mitchella.—Squaw Vine. 535.

Onagrarieæ.
Œnothera.—Evening Primrose. 377.
Epilobium.—Willow Herb. 376.

Cistineæ.
HELIANTHEMUM.—Frostwort. 354, N.F.

Plantagineæ.
Plantago.—Plantain. 529.

Orobanchaceæ.
Epiphegus.—Beech-drop. 525.

CLASS V.—MUCILAGINOUS

All Cryptogamous.
Cetraria.—Iceland Moss. 9. Lichenes.
CHONDRUS.—Irish Moss. 1. Algæ.
Adiantum.—Maidenhair Fern. 13. Filices.
Fucus Nodosus. 2 a. Algæ.
Fucus.—Bladderwrack. 2, N.F. Algæ.
Laminaria.—Sea-girdles. 3. Algæ.

Unclassified.—All unofficial: Adonis, *141.* Anagallis, 424. Anhalonium, 364. Artemisia, 579. A. Abrotanum, 579 a. A. Vulgaris, 579 b. A. Frigida, 579 c. Asclepias Curassavica, 457. Bidens, 591. Elephantopus, 589. Ephedra, 36. Equisetum, 17. Eschscholtzia, 184. Euphorbia Piluliferâ, *305.* Frankenia, 353. Gnaphalium, 582. Impatiens, 267. Lactuca Sativa, 595. Lactuca Canadensis, 596. Lamium, 491. Leonurus, 485. Litmus, 10. Menyanthes, *445*, N.F. Mercurialis, 308. Mutisia, 588. Passiflora, 362. Penthorum, 195. Polygonum, 123. Portulaca, 131. Rudbeckia, 590. Sedum Acre, 194. Senecio, *592*, N.F. Silphium, 587. Solanum, *515.* Spiræa, 229. Stellaria, 130. Stylosanthes, 244. Teucrium, 490. Trifolium, 243. Trifolium Repens, *243* a, N.F. Urechites, 453. Urtica, 113. Veronica Officinalis, 522. Commelina, 56.
N.F. drugs indicated by small capital letters.

FLOWERS, BULBS, AND PETALS

CLASS I.—AROMATIC

ANTHEMIS.—Chamomile. 600. Compositæ.
Calendula.—Marigold. 602. Compositæ.
CARYOPHYLLUS.—Cloves. 371. Myrtaceæ.
MATRICARIA.—German Chamomile. 599. Compositæ
ROSA GALLICA.—Red Rose. 213. Rosaceæ.
Rose Centifolia.—Pale Rose. 214. Rosaceæ.
Sambucus.—Elder. Caprifoliaceæ. 539.
SANTONICA.—Levant Wormseed. 601. Compositæ.
Aurantii Flores.—Orange Flowers. 281. Aurantiaceæ.
Lavandula.—Lavender. 501. Labiatæ.
Populi Gemmæ.—Balm of Gilead Buds. 98. Salicaceæ.

CLASS II.—ACRID

ARNICA.—565. Compositæ.

CLASS III.—BITTER

Brayera.—222, N.F. Rosaceæ.
Carthamus.—Safflower. 603. Compositæ.
Rhœas.—Red Poppy. 186. Papaveraceæ.
Ambrosia Artemisiæfolia.—Ragweed. Compositæ. 571.

CLASS IV.—DEMULCENT

Althæa Rosea.—Althaeæ Folia.—Hollyhock. 342. Malvaceæ.
Malva.—Mallow. 343. Malvaceæ.
Tilia.—Linden Flowers. 340. Tiliaceæ.

FRUITS

CLASS I.—AROMATIC

(a) Cremocarps.—Umbelliferous fruits, consisting of two carpels or mericarps joined by their flat sides but easily separable.

ANISUM.—Anise. 381. Umbelliferæ.
CARUM.—Caraway. 385. Umbelliferæ.
CORIANDRUM.—Coriander. 386. Umbelliferæ.
FŒNICULUM.—Fennel. 382. Umbelliferæ.
Anethum. Dill. 387. Umbelliferæ.
APII FRUCTUS.—Celery Fruit. 388. Umbelliferæ.
Cuminum.—Cumin Fruit. 393. Umbelliferæ.
Agelica.—Angelica. 396. Umbelliferæ.

(*b*) Globular.
   Chenopodium.—American Wormseed. 125. Chenopodiaceæ.
   CUBEBA.—Cubebs. 88. Piperaceæ.
   PIMENTA.—Allspice. 372. Myrtaceæ.
   SABAL.—Saw Palmetto. Palmæ. 47.
   PIPER.—Black Pepper. 89. Piperaceæ
   *JUNIPERUS.—Juniper. 23. Pinaceæ.
   Piper Album.—White Pepper. 90. Piperaceæ.
   Piper Longum.—Long Pepper. 91. Piperaceæ.

(*c*) Of Various Forms.
   AURANTII AMARI CORTEX.—Orange Peel. 278. Rutaceæ.
   AURANTII DULCIS CORTEX.—278 a.
   CAPSICUM.—Cayenne Pepper. 516. Solanaceæ.
   CARDAMOMUM.—Cardamom. 82. Zingiberaceæ.
   HUMULUS.—Hops. 110. Moraceæ.
   Illicium.—Star Anise. 149. Magnoliaceæ.
   LIMONIS CORTEX.—Lemon Peel. 283. Aurantiac.
   VANILLA.—87. Orchidaceæ.
   Caryophylli Fructus.—Mother Clove. 317 a. Myrtaceæ.
   Citrus Bergamia.—Bergamotte. 285. Aurantiaceæ.
   Dipteryx.—Tonka Bean. 250. Leguminosæ.
         CLASS II.—ODORLESS AND TASTELESS
   CONIUM.—Hemlock Fruit. *383.* Umbelliferæ.

CLASS III.—BITTER

COLOCYNTHIS.—Colocynth. 544. Cucurbitaceæ.
COCCULUS INDICUS.—Cocculus Indicus. Fishberry. *159.* Menisper-
   maceæ. Picrotoxin.
Lappæ Fructus.—Burdock Fruit. 559. Compositæ.
*PAPAVERIS FRUCTUS.—Poppy. *181.* Papaveraceæ.

CLASS IV.—SWEET

CASSIA FISTULA.—Purging Cassia. 247. Leguminosæ.
FICUS.—Fig. *114.* Moraceæ.
Phytolaccæ Fructus.—Poke Berries. 127. Phytolaccaceæ.
PRUNUM.—Prune. *205.* Rosaceæ.
TAMARINDUS.—Tamarind. *249.* Leguminosæ.

CLASS V.—ACIDULOUS

RHUS GLABRA.—Sumach. *317.* Anacardiaceæ.
*RUBI IDAEI FRUCTUS.—Raspberry. *218.* Rosaceæ.
Rosa Canina.—Hips. 215. Rosaceæ.

CLASS VI.—ASTRINGENT

Bela.—Bael Fruit. 277. Rutaceæ.
Granati Fructus Cortex.—Pomegranate Rind. 367. Punicaceæ.

UNCLASSIFIED.—Ajowan, 389. Anacardium, 321. Ananassa, 57. Carota, 394.
   Ceratonia, 248. Diospyros, 427. Embelia, 421. Luffa, 546. Lycopersicum,
   517. Mangostana, 349. Momordica, 547. Morus, 115. Myrobalanus,

375. Phellandrium, 392. RHAMNUS CATHARTICA, 335. Semecarpus, 322. Serenoa, 47. Uva Passa, 338. XANTHOXYLI FRUCTUS, *271*. Cratægus, 219. N.F. drugs indicated by small capital letters.

SEEDS

CLASS I.—AROMATIC

MYRISTICA.—Nutmeg. 154. Myristicaceæ.
**Caffea.*—Coffee. 537. Rubiaceæ.
Cola.—Cola. 345. Sterculiaceæ.
Fœnum Græcum.—Fenugreek. 251. Leguminosæ.
Granum Paradisi.—Grain of Paradise. 83. Zingiberaceæ.
Areca.—Areca Nut. 46. Palmaceæ.

CLASS II.—BITTER

AMYGDALA AMARA.—Bitter Almond. 209. Rosaceæ.
NUX VOMICA.—Dog Button. 435. Loganiaceæ.
STAPHISAGRIA.—Stavesacre. 143. Ranunculaceæ.
STROPHANTHUS.—451. Apocynaceæ.
**DELPHINIUM.*—Larkspur. *144.* Ranunculaceæ.
**IGNATIA.*—St. Ignatius' Bean. *436.* Loganiaceæ.

CLASS III.—ACRID

COLCHICI SEMEN.—Colchicum Seed. 69.
SINAPIS ALBA.—White Mustard. 188. Cruciferæ.
SINAPIS NIGRA.—Black Mustard. 189. Cruciferæ.
Curcas.—Purging Nut. 314. Euphorbiaceæ.
**Hyoscyami Semen.*—Henbane Seed. 510. Solanaceæ.
Nigella.—145. Ranunculaceæ.
Ricinus.—Castor Bean. 312. Euphorbiaceæ.
**Sabadilla.*—Cevadilla. 61. Liliaceæ.
Tiglium.—Croton Seed. 313. Euphorbiaceæ.

CLASS IV.—MUCILAGINOUS OR OILY

AMYGDALA DULCIS.—Sweet Almond. 210. Rosaceæ.
LINUM.—Flax Seed. 264. Linaceæ.
PEPO.—Pumpkin Seed. 548. Cucurbitaceæ.
PHYSOSTIGMA.—Calabar Bean. 252. Leguminosæ.
**Stramonii Semen.*—Stramonium Seed. 508. Solanaceæ.
**Cydonium.*—Quince Seed. 208. Rosaceæ.
Theobroma.—Cacao. 346. Sterculiaceæ.
**Papaver.*—Poppy, or Maw Seed. 182. Papaveraceæ.

UNCLASSIFIED.—Abri Semen, 232. Cannabis Semen, 112 a. Cedron, 290 Gynocardia, 355. Helianthus, 604. Jambul, 373. Persea, 179. Cucumis, 550. Sesamum, 528. White Zapote, 286.
N.F. drugs indicated by small capital letters.

CELLULAR DRUGS AND PRODUCTS FROM FRAGMENTS OF THE PLANT

CLASS I.—AROMATIC

**CROCUS.*—SAFFRON. 77. Iridaceæ. Tr.
LIMONIS SUCCUS.—Lemon Juice. 282. Aurantiaceæ. SUCCUS CITRI. SUCCUS POMORUM. N.F.
LUPULINUM.—Lupulin. Glands from hops. 111. Moraceæ.
**MACIS.*—Mace. Arillode of Nutmeg. 155. Myristicaceæ.

CLASS II.—ODORLESS AND TASTELESS

LYCOPODIUM.—18. Lycopodiaceæ.
GOSSYPIUM PURIFICATUM.—Cotton. 344 b. Malvaceæ.
Kamala.—Rottlera. 315. Euphorbiaceæ.
Cibotium.—Penghawar. 14. Filices.

CLASS III.—BITTER

Araroba.—Goa Powder 254. Leguminosæ.

CLASS IV.—ASTRINGENT

GALLA.—Nutgalls. 105. Cupuliferæ.
Mucuna.—Cowage. 253. Leguminoseæ.

CLASS V.—MUCILAGINOUS

SASSAFRAS MEDULLA.—Sassafras Pith 172. Lauraceæ.

CLASS VI.—SWEETISH

ZEA.—Corn Silk 40. Gramineæ.

CLASS VII.—FARINACEOUS

AMYLUM.—Starch. 42. Gramineæ.
Avenæ Farina.—Oat Meal. Gramineæ.
Hordeum.—Pearl Barley 43 a. Gramineæ.
MALTUM. 43 b.
Sago.—Pearl Sago. Gramineæ.
Tapioca.—Euphorbiaceæ.
Taro, Triticum vulgare, Oryza, Solanum tuberosum, Canna, Maranta, and
   Curcuma leucorrhiza.

NON-CELLULAR DRUGS DERIVED FROM CELL-PRODUCTS INCLUDING SECRETIONS

FARINACEOUS

AMYLUM.—Starch (see above).

EXTRACTIVE SUBSTANCES

Extracts.
CATECHU (Gambir).—Cutch. 257. Leguminosæ.
EXTRACTUM GLYCYRRHIZÆ.—Licorice. 230 a. Leguminosæ.
GUARANA.—329. Sapindaceæ.
Annatto.—Anotta. 356. Bixineæ.
Catechu Pallidum.—Gambir 538. Rubiaceæ.
Curara.—Curare. 440. Loganiaceæ.
Monesia.—426. Sapotaceæ.

Concrete juices.
ALOE—Aloes. 70. Liliaceæ.
KINO.—258. Leguminosæ.
LACTUCARIUM.—Lettuce Opium. 594. Compositæ.
OPIUM.—180. Papaveraceæ.
ELASTICA.—India Rubber 309. Euphorbiaceæ.

Alveloz Milk.—307.
GUTTA PERCHA.—425. Sapotaceæ.
Papaya.—Papain. *361.* Passifloraceæ.
N.F. drugs indicated by small capital letters.

SACCHARINE SUBSTANCES

Solid.
MANNA.—429. Oleaceæ.
SACCHARUM.—Sugar. Cane Sugar. 39. Gramineæ.
SACCHARUM LACTIS.—Sugar of Milk.

Liquid.
MEL.—Honey.

MUCILAGINOUS SUBSTANCES

Gums.
ACACIA.—Gum Arabic. 255. Leguminosæ.
TRAGACANTHA.—Gum Tragacanth. 256. Leguminosæ.

RESINOUS SUBSTANCES

Gum Resins.
Aromatic.—Containing volatile oil.
ASAFŒTIDA.—397. Umbelliferæ.
Ammoniacum.—Ammoniac. 399. Umbelliferæ.
MYRRHA.—Myrrh. 294. Burseraceæ.
Bdellium.—296. Burseraceæ.
Galbanum.—398. Umbelliferæ.
Olibanum.—Frankincense. 295. Burseraceæ.

Acrid.—Free from volatile oil.
CAMBOGIA.—Gamboge. 348. Guttiferæ.
SCAMMONIUM.—Scammony. 462. Convolvulaceæ.
Euphorbium.—306. Euphorbiaceæ.

Pure Resins.—*(Benzoin sometimes classified as Solid Balsam.)*
Aromatic.
BENZOINUM.—428.—Styraceæ.
RESINA.—Rosin. 27 c. Pinaceæ.
Xanthorrhœa.—Acaroid Resin. 71. Liliaceæ.

Odorless and Tasteless.
Copal.—261. Leguminosæ.
Dammara.—Dammar. 33. Coniferæ.
Draconis Resina.—Dragon's Blood. 49. Palmæ.
Kauri Resin.—Kauri Gum. 34. Coniferæ.
Succinum.—Amber. 32. Coniferæ.

Bitter.
MASTICHE.—Mastic. *319.* Anacardaceæ.
Elaterium.—551. Cucurbitaceæ.
Lacca.—Lac. 310. Euphorbiaceæ.
Sandaracca.—35. Coniferæ.

Acrid.
GUAIACUM.—Guaiac Resin. 269. Zygophylaceæ.

Oleoresins.
COPAIBA.—Copaiva. 259. Leguminosæ.
Pix Burgundica.—Burgundy Pitch. 31. Coniferæ.
PIX LIQUIDA.—Tar. 28. Pinaceæ.
TEREBINTHINA.—*27.* Pinaceæ.
TEREBINTHINA CANADENSIS.—29. Pinaceæ.
Elemi.—297. Burseraceæ.

Gurjun.—Gurjun Balsam. 351. Dipterocarpeæ.
Pix Canadensis.—Canada Pitch. 30. Coniferæ.
Terebinthina Chia.—Chian Turpentine. 320. Anacardiaceæ.
Venice Turpentine.—27 a. Coniferæ.
TEREBINTHINA LARICIS. ·

Balsams.
BALSAMUM PERUVIANUM.—Balsam of Peru. 262. Leguminosæ.
BALSAMUM TOLUTANUM.—Balsam of Tolu. 263. Leguminosæ.
STYRAX.—Storax. 201. Hamamelidaceæ.
Liquidambar.—Sweet Gum. 202. Hamamelidaceæ.

STEAROPTENS OR CAMPHOR

CAMPHORA.—Camphor. 178. Lauraceæ.
MENTHOL.—Pipmenthol. 473 b. Labiatæ.
THYMOL.—Thymol. 390. Umbelliferæ and Labiatæ.
Borneo Camphor.—352. Dipterocarpeæ.

FATTY SUBSTANCES

LIQUID

Of Vegetable Origin.

From Seeds.
OLEUM AMYGDALÆ EXPRESSUM.—Expressed Oil of Almond. **211.**
Rosaceæ.
OLEUM GOSSYPII SEMINIS.—Cotton-seed Oil. 344 c. Malvaceæ.
OLEUM LINI.—Linseed Oil. 264 a. Lineæ.
OLEUM RICINI.—Castor Oil. 312 a. Euphorbiaceæ.
OLEUM TIGLII.—Croton Oil. 313 a. Euphorbiaceæ.
Oleum Juglandis.—Nut Oil. 101. Juglandaceæ.
Macassar Oil.—332. Sapindaceæ.
Pongama Oil.—260. Leguminosæ.
Oleum Papaveris.—Poppyseed Oil. 182 a. Papaveraceæ.
Oleum Sesami.—Benne-seed Oil. 528 c. Pedalinæ.
Oleum Sinapis Expressum.—Expressed Oil of Mustard. 189 b. Cruciferæ.

From Fruits.
OLEUM OLIVÆ.—Sweet Oil. 430. Oleaceæ.
Oleum Cannabis.—Oil of Hempseed. 112 b. Moraceæ.
Oleum Maydis.—Maize Oil. 41. Gramineæ.

Of Animal Origin.
OLEUM ADIPIS.—Lard Oil.
OLEUM MORRHUÆ.—Cod-liver Oil.
Oleum Bubulum.—Neats-foot Oil.
N.F. drugs indicated by small capital letters.

SOLID

Of Vegetable Origin.

From Seeds.
OLEUM THEOBROMATIS.—Cacao Butter. 346 a. Sterculiaceæ.
Oleum Cocois.—Cocoanut Oil. Palmæ.
Oleum Gynocardiæ.—Chaulmoogra Oil. Bixineæ.

From Fruits.
Oleum Lauri.—Oil of Bays. Laurel Oil. Lauraceæ.
Oleum Palmæ.—Palm Oil. Palmæ.

Of Animal Origin.
ADEPS.—Lard. 628.
ADEPS LANÆ HYDROSUS.—Lanolin. Wool-fat. 631.

SEVUM.—Suet. 626.
Butyrum. Butter.

Waxes.
CETACEUM.—Spermaceti. 617.
CERA ALBA.—White Wax. 618 b.
CERA FLAVA.—Yellow Wax. 618 a.

Hydrocarbon Oils and Fats. Petrolatum Products.
Petrolatum Album.—White Petrolatum.
Petrolatum Liquidum.—Liquid (oil).
Petrolatum Molle.—Soft Petrolatum.
Petrolatum Spissum.—Hard Petrolatum.

DRUGS OF ANIMAL ORIGIN

I.—ANIMALS

Insecta.
CANTHARIS.—Spanish Flies. 605.
COCCUS.—Cochineal. 606.
Blatta.—Cockroach.

Vermes.
Hirudo.—Blood-sucking Leech.

II.—ANIMAL TISSUES AND SECRETIONS

Poriphera.
Spongia.—Sponge.

Polypiphera.
Corallium.—Coral.

Acephala.
Testa.—Oyster Shell.

Cephalopoda.
Os Sepiæ.—Cuttle Fish Bone.
Crustacea.

Calculi Cancrorum.—Crabs' Stones.

Pisces.
**Ichthyocolla.*—Isinglass.
OLEUM MORRHUÆ.—Cod-liver Oil. (See Fatty Substances.) 616.

Insecta.
CERA ALBA.—White Wax. (See Fatty Substances.) 618 b.
CERA FLAVA.—Yellow Wax. (See Fatty Substances.) 618 a.
MEL.—Honey. (See Saccharine Substances.) 618.

Aves.
**Vitellus.*—Egg-yolk. Glycerite.
Ovi Vitellum Recens. N.F.
Ovum.—Egg.
Albumen Ovi.—White of Egg.
Ovi Albumin Recens. N.F.
Testa Ovi.—Egg-shell.
Ovum Gallinaceum. N.F.

Mammalia.
ADEPS.—Lard. (See Fatty Substances.) 628.
ADEPS LANÆ HYDROSUS.—Lanolin. (See Fatty Substances.) 631.
CETACEUM.—Spermaceti. (See Fatty Substances.) 617.
FEL BOVIS.—Ox-gall. 621.
FEL BOVIS PURIFICATUM.—Purified Ox-gall.

MOSCHUS.—Musk. 620.
OLEUM ADIPIS.—Lard Oil.   (See Fatty Substances.)   628 a.
PANCREATINUM.—Pancreatin.   630.
PEPSINUM.—Pepsin.   629.
RENNINUM. N.F.
SEVUM.—Suet.   (See Fatty Substances.)
Castoreum.—Castor.
Civetta.—Civet.
Gelatinum.—Gelatin.
Hydraceum.—Hyraceum.
Lac.—Milk.
LAC VACCINUM.
      (*a*) Butyrum.—Butter.   (See Fatty Substances.)
      (*b*) **ACIDUM LACTICUM.**—Lactic Acid.
      (*c*) **SACCHARUM LACTIS.**—Sugar of Milk. (See Saccharine Substances.)   Used chiefly as diluent.
Os.—Bone.
Oleum Bubulum.—Neats-foot Oil.   (See Fatty Substances.)
Sanguis.—Blood.

Batcerial Products.   See chapter on.
   Antitoxic Serums.
   Serum Antidiptheriticum.

Glandular Products.   See chapter on.
   Glandulæ Suprarenales Siccæ.
   Glandulæ Thyreoideæ Siccæ.

Serotherapy.   Remedial agents in;   See chapter on.

PART II

DRUG DESCRIPTION

SECTION I.—ORGANIC DRUGS FROM THE VEGETABLE KINGDOM, DESCRIBED AND ARRANGED ACCORDING TO FAMILIES.

CRYPTOGAMS

(Plants Producing Spores)

ALGÆ

Structure very various, growing for the most part in water, mostly in stagnant water in warm climates, but some on moist rocks or ground, etc. Entirely cellular, producing fronds.

1. CHONDRUS.—Irish Moss
CARRAGEEN

The dried plant **Chon'drus cris'pus** Lyngbye. (Fam. Gigartinaceæ.)

Botanical Characteristics.—*Thallus* fleshy, cartilaginous, compressed, dividing into short, moniliform filaments. *Antheridia* or *oogonia* in superficial spots. *Chondrus crispus* has four vessels or capsules imbedded in the frond. *Gigartina mamillosa* (*Chondrus mamillosa*) has an oval one raised upon a short stalk, and its frond is slightly channeled toward the base.

Source.—These plants inhabit the rocks on the American and European shores of the Atlantic Ocean. In the spring they are collected on the coast of New England and Ireland, the Massachusetts coast yielding about 15,000 barrels annually.

Description of Drug.—Yellowish or white, horny, translucent; many times forked; when softened in water, cartilaginous; shape of the segments varying from wedge-shaped to linear; at the apex emarginate or 2-lobed. It has a slight seaweed-like odor, and a mucilaginous, somewhat saline, taste.

Test.—When one part of chondrus is boiled for about ten minutes with thirty parts of water replacing water lost by evaporation, the solution should form a thick jelly upon cooling.

When softened in cold water chondrus should become gelatinous and transparent the thallus remaining nearly smooth and uniform and not swollen except at the tips.

A solution made by boiling 0.3 Gm. in 100 mils of water and filtering gives no precipitate on the addition of tannic acid T.S. (gelatin), and does not give a blue color when cold upon addition of iodine T.S. (starch).

CONSTITUENTS.—The principal constituent (90 per cent.) is **mucilage,** which is precipitated by lead acetate; traces of iodine and bromine have also been detected. There seems to be no starch present, but the cell-walls acquire a dark blue color in contact with iodine (Flück iger). Literature rather contradictory as to the nature of its various constituents.

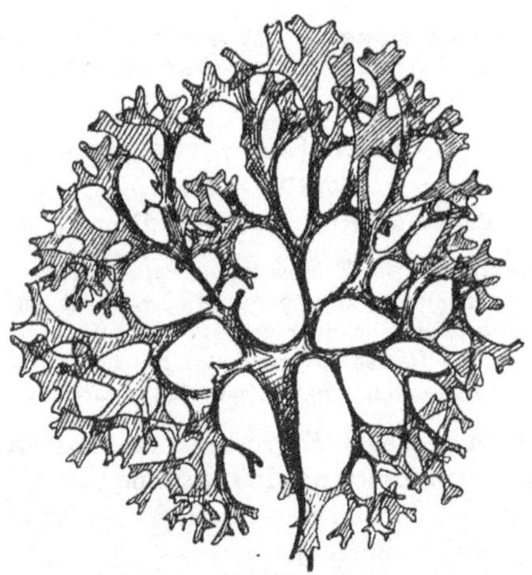

FIG. 1.—*Chondrus crispus.*

ACTION AND USES.—Demulcent and slightly nutritious. A dietetic is specially prepared from the powder, made in the form of jelly with water. Dose: 6 dr. (24 mils) in decoction.

2. **FUCUS VESICULOSUS, N.F.**—BLADDER-WRACK. The whole plant, **Fu'cus** vesiculo'sus Linné, growing on muddy rocks and floating to the shores of the North Atlantic and North Pacific Oceans, consists of long, flattened, branched fronds, upon which are dispersed blackish air-vessels (tubercles) in pairs, one on each side of the midrib. These cavities contain thin, gelatin. ous matter, and bear on their inner walls, when young, hair or transparent filiform cells. Odor marine-like; taste mucilaginous and saline. "Wracks" or rock weeds of other species are also collected, such as *Fucus nodosus.* 2a. The medicinal properties probably lie in the inorganic matter, the ash of the plant containing chlorides, bromides, iodides, phosphates, and sulphates; the organic matter is mainly mucilage. The medicinal value of the drug as an alterative has been questioned; it is used in obesity. "The fl'ext. and extract are irrational preparations, the only form in which to obtain the effects of the plant being the recent decoction (Shoemaker)." Fucus, N.F., constitutes the dried thallus of the above plant, yielding not more than 20 per cent. of ash.

3. **LAMINARIA.**—Sea-girdles or Tangles. From Lamina'ria digita'ta Lamouroux. A dark-spored seaweed having a ribless expansion resembling a leaf-blade. The stipitate portion has been used in gynecology as a substitute for sponge in making sponge tents for dilating the cervical canal. Contains salts, mucilage, and mannite; the latter principle is especially prominent in another species—*Laminaria saccharina*—like the above, abundant on the sea-coast.

4. **AGAR OR AGAR-AGAR** U.S.P. IX.—The dried mucilaginous substance extracted from Gracilaria (Sphoercocus) lichenoides.

Gracilaria and other marine Algæ growing along the eastern coast of Asia, particularly several species of Gelidium or Gloiopeltis (class Rhodophyceæ). Mostly in bundles 4 to 6 dm. in length, thin translucent, membranous, agglutinated pieces from 4 to 8 mm. in width; externally yellowish-white, shiny; tough when damp, brittle when dry; odor, slight; taste, mucilaginous. Tests show it to be insoluble in cold but slowly soluble in hot water. No gelatin or no starch, etc., Test.—Practically the same as that for chondrus. Ash, not more than 5 per cent. Average dose, 10 Gm. ($2\frac{1}{2}$ dr.).

Action and Uses.—Agar-agar is practically never used in medicine. It possesses demulcent and emulsifying properties in common with other species of Algæ. It is principally used at present in bacteriological laboratories as a culture-medium for micro-organisms.

Agar-agar in the dry state passes through the stomach undigested and on reaching the bowels takes up water and swells considerably, thereby increasing the volume of the evacuations; it is therefore considered a laxative.

FUNGI

Spore-bearing plants destitute of chlorophyll and reproduced by means of spores, not by true seeds.

5. ERGOTA.—Ergot

ERGOT.   (Ergot of Rye)

The carefully dried sclerotium of fungus **Clav'iceps purpu'rea** Tulasne (Fam. Hypocreaceæ), replacing the grains of rye, **Secale cereale** Linné (Gramineæ), with not more than 5 per cent. of harmless seeds, fruits and other foreign matter.

Development.—Sclerotium described: The early stage of the fungus consists of a profuse growth of *mycelium* in the tissues and upon the surface of the young ovary. In the "*sphacelia*" stage, as it is called, a multitude of *conidia* (non-sexual spores) are produced on the ends of the *hyphæ;* after the conidial stage the *mycelium* at the base of the ovary becomes greatly increased and assumes a hard and compact form. It grows with considerable rapidity, and carries upon its summit the old sphacelia and the remains of the now destroyed ovary. The compact, horn-shaped, dark-colored body which results (and is official) is called the *sclerotium*, which occupies the position of the displaced ovary. This sclerotium remains dormant.in winter, and in the spring produces spores, as follows: stalked receptacles (Fig. 3) grow up from the tissue of the ergot, in which are developed a number of *perithecia* (Fig. 4). These *perithecia* are somewhat flask-shaped cavities (Fig. 5) filled with *asci* (Fig. 5), the latter containing long, slender spores termed *ascospores* (Fig. 6), which again, by germinating on the rye and other grasses, give rise to a new growth, and to the development of *Claviceps*. Ergot, in short consists in its earliest stage of a mass of mycelium (threads or filaments of fungi) in and upon the

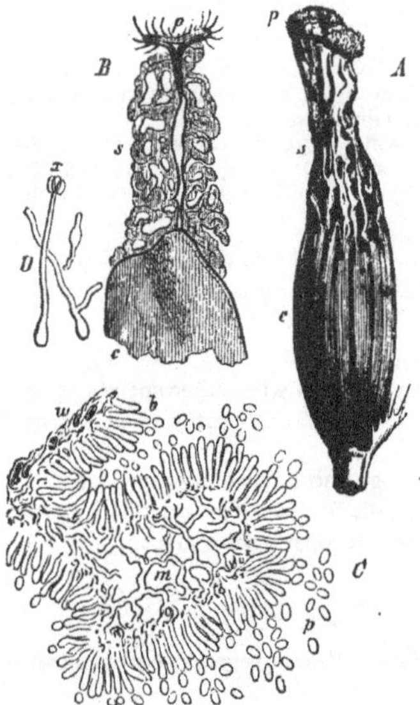

FIG. 2.—*Claviceps purpurea.* *A.* Young sclerotium, *c,* with old sphacelia, *s.* *p.* The apex of the dead ovary of rye. *B.* Upper part of *A,* in longitudinal section, showing sphacelia, *s.* *C.* Transverse section through the sphacelia, more highly magnified. *m.* The mycelium, surrounded with the hyphæ. *b.* Bearing conidia. *p.* Conidia fallen off. *w.* The wall of the ovary. *D.* Germinating conidia, forming sporidia, *x.*—(*Bachs.*)

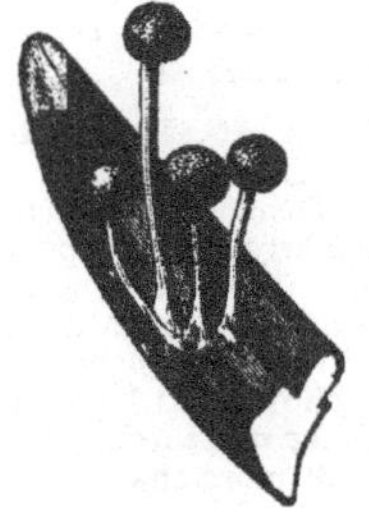

FIG. 3.—Portion of horn-shaped sclerotium of *Claviceps purpurea,* bearing four stalked receptacles.

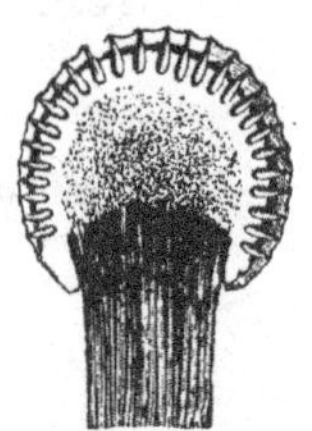

FIG. 4.—Longitudinal section of a receptacle, magnified, showing the perithecia.

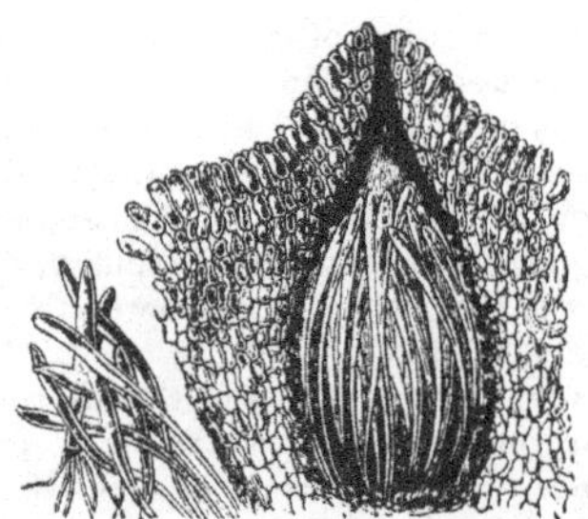

FIG. 5.—A single perithecium of *Claviceps purpurea,* magnified, showing the contained asci.

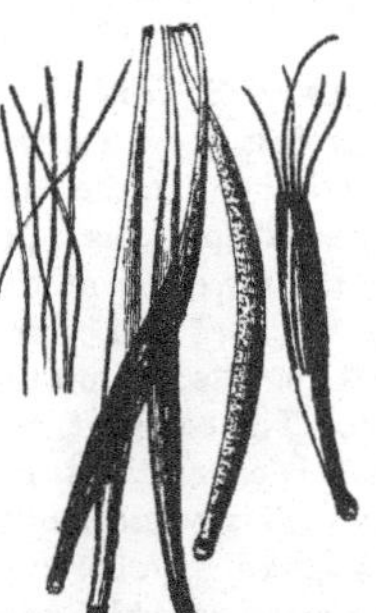

FIG. 6.—Asci containing the long, slender ascospores.

growing ovary. Conidia are produced (non-sexual spores) in great abundance which quickly germinate.

Following the conidial stage the mycelium at the base of the ovary assumes a hard and compact form, increases in size, becoming a horn-shaped and dark-colored body, the so-called ergot. Such a compact mass of hyphæ (the vegetative threads or filaments of the fungi) is called a sclerotium.

The official fungus grows on rye, but the ergot also grows on other grasses and some of these ergots from other grasses have been found to be stronger than that of rye. The different grades are *Russian*, *German*, *Austrian*, *Spanish* and *Swedish*.

PREPARATION AND PRESERVATION.—Ergot should be dried without artificial heat kept in tin or glass containers free from light. A few drops of chloroform or carbon tetrachloride should be added from time to time to prevent development of insects. The powdered drug should not be kept longer than one year. Suggestion for preservation—keep over slaked lime. Dip into ethereal solution of tolu and keep in stoppered bottles. Also by removing the oil from the drug.

DESCRIPTION OF DRUG.—The official ergot of rye is from 10 to 30 mm. ($\frac{2}{5}$ to $1\frac{1}{5}$ in.) long and from 2 to 6 mm. ($\frac{1}{12}$ to $\frac{1}{4}$ in.) in diameter. On other grasses it is usually of less size. Triangular, slightly curved, tapering toward, but obtuse at, the ends; externally purplish-black, internally whitish with pinkish lines; fracture short (not very brittle). If a portion be macerated in water containing hydrate of potassium or sodium, then carefully crushed under the blade of a spatula, the fragments of mycelium threads are plainly discernible under the microscope. Odor (especially in powder or when treated with an alkali) heavy and unpleasant; taste oily and disagreeable.

When more than one year old, it is unfit for use. Old ergot, which breaks with a sharp snap, is almost devoid of pinkish tinge upon the fracture, is hard and brittle between the teeth, and is comparatively odorless and tasteless, should be rejected.

CONSTITUENTS.—The active constituents of ergot are still somewhat in doubt due probably to the amorphous condition in which they exist. Barger and Carr have extracted a substance called ergotoxine (non-crystalline) to which the dangerously poisonous character of ergot is due including the power to produce gangrene. Barger and Dale have shown it to contain amines derived from amino acids. Two of especial physiological activity are:

1. p. Hydroxyphenylethylamine or (Tryamine) has action of same type as active constituents of suprarenal glands and substance chiefly concerned in standardization of ergot by rise in blood pressure.

2. b. Iminoazolethylamine (Ergamine) has an action of peculiar intensity on plain muscle especially on uterine muscle.

Ergotine an alkaloid thought by some to be identical with ergotonine. Ecboline same as cornutine. Others say ergotine and ecboline are

6

identical. Different samples of ergot may contain very different amounts of the three main constituents. The yield of ash should not exceed 5 per cent.

Assay of Ergot.—The physiological test for ergot, originated by E. M. Houghton, consists in feeding the preparation or drug to roosters, and noting the blackened and gangrenous appearance produced in the comb and wattles. The rapidity with which this change takes place and the depth of color produced denote the strength of the drug. An assay of the drug can be made by estimating the proportion of *cornutine* present, which, according to Beckurts, is as follows: 25 Gm. of the drug are freed from oil by percolation with petroleum spirits, then dried and well shaken with 100 Gm. of ether and 1 Gm. of magnesia, the latter having been suspended in 20 mils of water. After repeated agitation the mixture is allowed to stand for three or four hours. Then 60 Gm. of the clear ethereal solution (to 15 Gm. of ergot) are shaken four successive times with 25, 10, 10, and 10 mils of dilute HCl (0.5 per cent.), the united solutions rendered alkaline by NH_4OH, and the alkaloid shaken out with three successive portions of ether. On evaporation, drying, and weighing the somewhat crystalline yellowish-white cornutine the assay is completed. The results of such assay are unsatisfactory, but have proved of value as a check in qualitative estimations.

Preparation of Ergotin (Wiggers).—Treat ergot with ether to deprive it of fixed oil, then extract with hot alcohol, evaporate, and purify. It resembles cinchonic red, is soluble in alcohol, but insoluble in ether and water. Bonjeau's ergotin corresponds to a purified extract of ergot (aqueous extract, precipitated by alcohol, filtered, and evaporated); is soluble in alcohol and water.

ACTION AND USES.—Produces vascular contraction, especially of the arteries, all over the body. This property is said to be due to its action on the vasomotor centers in the cord. Because it contracts the arterioles it is hemostatic. The flow of urine is also diminished. It is ecbolic **and parturient,** powerfully exciting the pregnant uterus and expelling its contents. Recently it has been discovered to be of value in the treatment of insomnia, the sleep produced being more natural than that from other drugs.

Poisonous symptoms: dimness of vision, local anesthesia, and numbness are sometimes produced, even by medicinal doses. Antidotes: evacuants (stomach-pump, emetics, etc.), stimulants, nitrite of amyl, inhalations, friction, etc. Dose: 20 to 30 gr. (1.3 to 2 Gm.) in freshly prepared powder, wine, or fluidextract; ergotin solution, 1 to 3 gr. (0.65 to 0.2 Gm.).

OFFICIAL PREPARATIONS.
Extractum Ergotæ,......................Dose: 3 to 12 gr. (0.2 to 0.8 Gm.)
Fluidextractum Ergotæ,.........................½ to 2 fl. dr. (2 to 8 mils)

6. **USTILAGO.**—CORN SMUT. A fungous growth upon **Zea mays,** more particularly upon the inflorescence. Consists of blackish, irregular, roundish masses enveloping innumerable spores; of a disagreeable odor and taste. It contains probably sclerotic acid. Used as a parturient and emmenagogue. Dose: 15 to 30 gr. (1 to 2 Gm.).

7. **AGARICUS ALBUS, N.F.**—LARCH AGARIC. PURGING AGARIC. WHITE AGARIC. From **Polypo'rus officina'lis** Fries. The internal, decorticated portion of the fungus comes in light, colorless, spongy masses of irregular shape. Taste sweetish, acrid, and bitter. In large doses cathartic. In doses of 8 gr., gradually increased to 1 dr., it has been found useful in checking night-sweats of phthisis. Surgeon's agaric, from *Polyporus fomentarium* Fries, is used externally as a styptic in hemorrhage.

7 a. **FUNGUS CHIRURGORUM.**—Surgeon's Agaric. Same as Polyporus. See above.

8. **CEREVISIÆ** (*Saccharomyces*).—Fermentum Compressum (Compressed Yeast), N.F.—An organized ferment. Yeast is the name applied to the frothy scum that forms on the surface of saccharine liquids and rises from the bungholes of newly brewed beer. Under the microscope this froth is shown to consist of particles which multiply with extraordinary rapidity when placed in a moderately warm temperature. The globular forms are considered as the spores of a fungus belonging to the genus *Torula*, the cells of which are but slightly united, sometimes forming branching chains, the mycelium being almost absent. Yeast is employed in hastening the fermentation of worts and in leavening dough in bread-making. Bottom or sediment yeast is found on the bottom of fermenting vessels. Two quite distinct methods of brewing are produced, depending upon the employment of one or the other of these varieties of yeast. For the purpose of the bakery, yeast is dried and formed into cakes. Beer yeast is official in the B.P. Yeast, under the title of fermentum, was official in the U.S.P., 1820–'40, 1860–'80, used as a tonic, laxative, etc., but at present rarely employed. As a local remedy, as poultice, in treatment of eruptions of boils, it still finds some favor.

LICHENES

Consisting mainly of a thallus (often leaf-like), without stem and leaves, wholly cellular. Reproduced by spores.

9. **CETRARIA.**—Iceland Moss. The entire plant, **Cetra'ria islan'dica** Acharius. Off. U. S. P. 1890. The crisp, leaf-like lobes are cartilaginous, whitish

Fig. 7.—Section of thallus of *Cetraria islandica* through an apothecium. *as.* Asci, three of which contain ascospores. *gon.* Gonidia.

Fig. 8.—*Cetraria islandica.*

on the under surface, channeled and fringed at the margins. A strong decoction gelatinizes on cooling; taste mucilaginous and bitter. The Pharmacopœia calls attention to the fact that the drug is frequently mixed with pine leaves, moss, and other lichens; from these it should be freed. *Constituents:* It is largely composed (70 per cent.) of **lichen starch**, lichenin, and isolichenin, a

solution of the latter producing a blue color with iodine. Unlike the gum
of chondrus, it furnishes but a trace of mucic acid when treated with nitric
acid. Boiling with dilute acids converts the mucilage into sugar solution.
A solution of Iceland moss is precipitated by alcohol. The bitter principle,
cetraric acid (cetrarin, $C_{18}H_{16}O_8$), forms yellow salts, which are equal in
bitterness to quinine; this bitter principle may be removed by prolonged
maceration in water, or, still better, by treating the drug with twenty-four
times its weight of a weak solution of an alkaline carbonate. Demulcent,
nutritive, and, if the bitter principle be present, tonic; used in advanced
stages of phthisis when stronger remedies are unsuitable. Dose: 30 to 60
gr. (2 to 4 Gm.).
Preparation of Cetrarin: Boil drug with alcohol; express and add acidulated
(HCl) water to the filtrate; then allow cetrarin to deposit.

10. **LITMUS.**—A fermented coloring extract from various species of lichens
(*e.g.*, *Lecanora tartarea*), other varieties of which also yield the dyes orchil
and cudbear. *Habitat:* Northern Europe and African coast, and adjacent
islands. Litmus is in about ½ to 1 inch rectangular cakes, blue, light, friable,
finely granular. Unlike most vegetable dyes, it is not turned green by
alkalies. It is turned red by acids, for which it is used as a test in the form
of infusion (tincture), or litmus paper, made by dipping unsized paper in the
strong infusion.
    10 a. Orchil is a purplish-red, thickish liquid, with an ammoniacal odor.
    10 b. Cudbear (Persio, N.F.) is a purplish-red powder, sometimes used to
color preparations.

POLYTRICACEÆ

11. **POLYTRICHUM JUNIPERUM** Hedwig.—HAIR-CAP Moss. This common
moss is a powerful diuretic; in full doses given at very short intervals it has
proved very beneficial in dropsy. Dose: 1 to 2 dr. (4 to 8 Gm.), in infusion.

FILICES.—Ferns

Leafy plants with the *fronds* raised on a stipe (petiole) rising from a rhizome,
circinate in vernation. The *spore-cases* are found on the under side of the frond.
The life history of the fern is as follows:

When the minute spore from the sporangium on the frond drops to the ground,
it germinates into a more or less heart-shaped body called a prothallus. The
under surface of this body is provided with root-hairs and also female organs of
generation, archegonia, and male organs, antheridia; the frond-stage is a direct
outgrowth from the fertilized archeogonia.

Synopsis of Drugs 'from the Filices

A. *Rhizome.*	C. *Hairs.*	E. *Leaves.*
ASPIDIUM, 12.	Cibotium, 14.	Polypodium, 15.
B. *Herb.*	D. *Root.*	
Adiantum, 13.	Osmunda, 16.	

12. ASPIDIUM.—ASPIDIUM

MALE FERN.

The dried rhizome of **Dryop'teris fil'ix-mas** Schott, and of **Dryop'teris margina'lis**
Asa Gray (family Polypodiaceæ). Collected in autumn, freed from the
roots and dead portions of rhizome and stipes, and dried at a temperature
not exceeding 70° C.
BOTANICAL CHARACTERISTICS.—*Fruit-dots* round, borne at the back of the veins;
indusium covering the sporangia. Stipe continuous with the root-stock.

Frond lanceolate (*A. filix-mas*) or ovate-oblong (*A. marginalis*); fruit-dots in the former nearer the mid-vein than the margin, in the latter nearer the margin.

HABITAT.—North America.

DESCRIPTION OF DRUG.—As taken from the ground the rhizome consists of a caudex around which are arranged the dark brown, somewhat curved **leaf-stalk remnants or stipes**, about 25 to 50 mm. (1 to 2 in.) in length, imbricated like the shingles of a roof; at the base

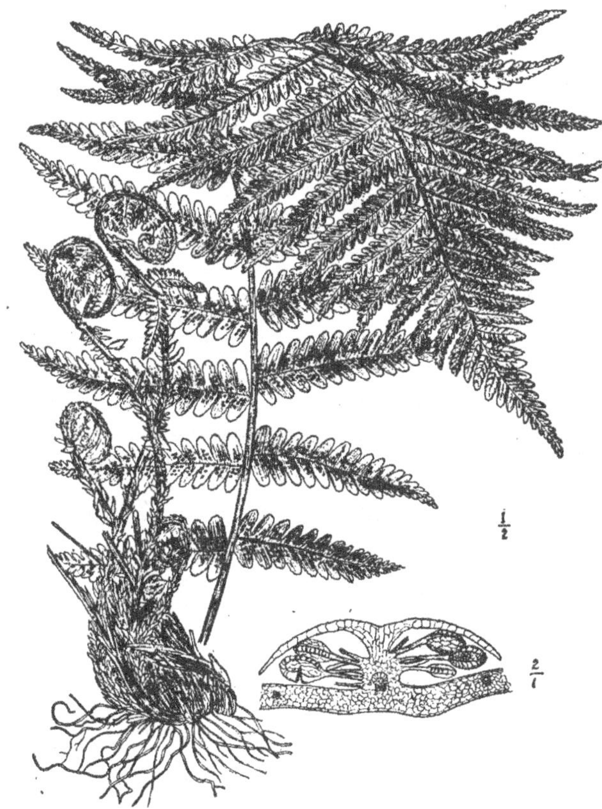

FIG. 9.—*Dryopteris filix-mas*—Plant and section through spore case.

they are densely surrounded by **thin, glossy, chaffy scales** of a lighter color and somewhat transparent. The entire rhizome is from 100 to 300 mm. (4 to 12 in.) long, and from 50 to 62 mm. (2 to 2½ in.) thick, flexible, tapering toward one end, usually split longitudinally, roughly scarred with remains of stipe bases or bearing several coarse longitudinal ridges or grooves, pale green when fresh and becoming pale brown and with occasional elongated areas of the still adhering brownish-black outer layers, fracture short, pale green in the inner half, the texture rather spongy and exhibiting from 6 to 12 fibrovascular bundles in a loose and interrupted circle; it generally comes into market broken into pieces of various lengths; internally pale green, spongy or corky; odor slight and disagreeable; taste sweetish, somewhat bitter and astringent, acrid and nauseous. Only such portions as are still green should be used in making prepara-tions. The deterioration of the root is rapid—loses its activity in one or two years.

MICROSCOPICAL STRUCTURE.—The prevailing tissue is parenchyma, the polyhedral, porous-walled cells of which contain starch, greenish or

brownish tannin-like substances, and drops of a greenish fixed oil. The thin subserous outer layer consists of smaller brown cells. Toward the center of the rhizome is an irregular circle of ten ($A.$ *filixmas*) or six ($A.$ *marginalis*) vascular bundles, outside of which are smaller scattered bundles. Distributed throughout the tissue are large air pores.

Powder.—Microscopical elements of: See Part iv, Chap. I, B.

CONSTITUENTS.—Filicic acid, $C_{35}H_{42}O_{13}$, filicin (filicic acid anhydrid, $C_{35}H_{40}O_{12}$), aspidin, $C_{23}H_{27}O_7$, the latter being poisonous, fixed oil, a trace of volatile oil, and chlorophyll. Ash 3 per cent.

Preparation of Filicic Acid.—This principle is deposited as a granular sediment when the oleoresin is allowed to stand.

FIG. 10.—Aspidium rhizome. (½ natural size.) A, Leaf-stalk. (Photograph.)

ACTION AND USES.—**Tæniafuge.** Dose: ½ to 2 dr. (2 to 8 Gm.). Theoleo resin is the most efficient preparation.

OFFICIAL PREPARATION.

Oleoresina Aspidii,   .Dose: ½ to 1 fl. dr. (2 to 4 mils).

13. **ADIANTUM.**—MAIDENHAIR. **Adian'tum peda'tum** Linné, an indigenous fern which has been used as a pectoral in chronic catarrh and other affections of the air-passages.

14. **CIBOTIUM.**—PENGHAWAR. PAKU-KIDANG. The chaffy hairs collected from the base of the fronds and stems of many varieties of ferns, especially of the genus Cibotium, growing in Sumatra and Java. Long, silky, yellowish or brownish, curling filaments (under the microscope flat and jointed), used to stop the flow of blood from capillaries by mechanical absorption of the serum.

15. **POLYPODIUM.**—POLYPODY. The leaves of **Polypo'dium vulga're** Linné, common in Europe and North America. Expectorant in chronic catarrh and asthma. Dose: 1 dr. (4 Gm.), in infusion.

16. **OSMUNDA REGALIS** Linné (order Osmundaceæ).—BUCKTHORN BRAKE.— A common fern in swamps, the root-stock of which is used as a demulcent, tonic, and styptic. Dose of fl'ext.: 1 to 3 fl. dr. (4 to 12 mils).

EQUISETACEÆ.—Horsetail Family

17. EQUISETUM.—Scouring Rush. The herb of **Equise′tum** hyema′le Linné. *Habitat:* Northern United States. Diuretic and astringent. Dose of fl′ext.: 15 to 60 ℥ (1 to 4 mils).

LYCOPODIACEÆ.—Club-moss Family

Low plants looking like very large mosses, more or less branching, and with the 1- to 3-celled *sporangia* (spore-cases) in the axils of the lanceolate, subulate, or rounded, persistent *leaves*. *Spores* homogeneous.

18. LYCOPODIUM.—Lycopodium

VEGETABLE SULPHUR

The spores of **Lycopo′dium** clava′tum Linné, and of other species of Lycopodium.

Botanical Characteristics.—*Stem* creeping extensively, with ascending very leafy branches. *Leaves* linear-awl-shaped, aristate. *Spikes* 1 to 4 on a slender peduncle 4 to 6 inches long.

Source and Collection.—Europe, Asia, and North America; collected mostly in Russia, Germany, and Switzerland, in July and August, by cutting off tops of the moss, shaking out spores, and sifting.

Description of Drug.—**A fine, pale-yellowish powder, very mobile,** free from odor and taste. It floats in water without being wet by it

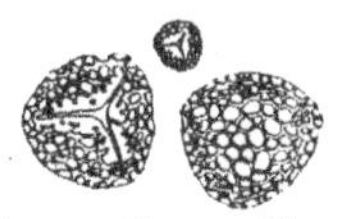

Fig. 11.—Spores of Lycopodium, as seen from the top and from bottom of the spore.

Fig. 12.—Lupulin; a gland viewed in profile and viewed obliquely. See Lupulinum, 111, p. 148.

Fig. 13.—Pollen of Pine: a grain seen in profile; another seen from the front side.

(due to the fixed oil), but sinks on being boiled. When slowly heated it burns quietly and should not leave more than 5 per cent. of ash, but when thrown into a flame it **flashes up.** Under the microscope the granules are seen to be tetrahedral, the basal side convex and the other three coming together to form a triangular pyramid. The surfaces are traversed in all directions by ridges which form regular, five- or six-sided meshes; at the points of intersection are small elevations, and along the edges short projections. Like lupulin, lycopodium is one of the interesting objects for microscopic study. Pollen of pine, an illustration of which is shown above, is sometimes used as an adulterant.

Adulterants.—These may be easily detected by the microscope or simple tests. Pine pollen consists of an elliptical cell with a globular cell attached to each end. Starch is detected with iodine; turmeric, by turning reddish-brown with alkalies; inorganic mixtures, by increasing the yield of ash over 5 per cent., and by sinking in carbon disulphide. Dextrin has been found in lycopodium to the extent of 50 per cent.

Powder.—Microscopical elements of: See Part iv, Chap. I, B.

CONSTITUENTS.—**Fixed oil 47 to 50 per cent.,** volatile bases in very small quantity, and ash containing alumina and phosphoric acid, not exceeding 3 per cent.

ACTION AND USES.—Absorbent and protective application to excoriated surfaces; in pharmacy, to facilitate the rolling of pill masses, and to prevent the adhesion of the pills.

PHANEROGAMS

(Plants producing true seed)

Pinaceæ.—Pine Family

Trees or shrubs with a resinous juice. The *wood* differs from that of dicotyledons in that it is destitute of ducts, but has instead bordered disks. The *leaves*

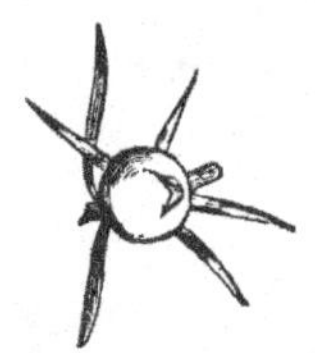

FIG. 14.—Berry (galbulus) and acicular leaves of Juniper.

FIG. 15.—Cone of the Larch.

are usually fascicled, and are mostly awl- or needle-shaped. *Fruit* a cone or galbulus.

Synopsis of Drugs from the Pinaceæ

A. *Tops.*
 SABINA, 19.
 Juniperus Virginiana, 20.
 Thuja, 22.
B. *Fruits.*
 *Juniperus, 23.
C. *Barks,*
 *Pinus Alba, 21.
 Tsuga, 25.
 Larix, 26.
D. *Oleoresins.*
 *Terebinthina, 27.

*Venice Turpentine, 27 a.
TEREBINTHINA CANADENSIS, 29.
Pix Canadensis, 30.
PIX LIQUIDA, 28.
Pix Burgundica, 31.
E. *Volatile Oils.*
 OLEUM SABINÆ, 19 a.
 OLEUM JUNIPERI, 23 a.

OLEUM CADINUM, 24.
OLEUM TEREBINTHINÆ, 27 b.
OLEUM PICIS LIQUIDÆ, 28 a.
Oleum Succini, 32 a.
F. *Resins.*
 RESINA, 27 c.
 Succinum, 32.
 Dammara, 33.
 Kauri, 34.
 Sandaracca, 35.

19. SABINA.—SABINA

SAVINE

The tops of Junip'erus sabi'na Linné. The young and tender green shoots are stripped off in the spring, coming into the market as short, thin, quadrangular

branchlets, clothed with alternate pairs of minute, opposite, scale-like leaves, appressed (more pointed and divergent in older twigs); each scale has a shallow groove and a conspicuous, depressed oil-gland in the back. The berry-like cone fruit is about the size of a pea, situated on a short, **recurved pedicel,** and covered with a bluish bloom; it is dry, but abounds in essential oil, and contains from 1 to 4 small, bony seeds. Odor strong, balsamic; taste bitter and acrid. Adulteration: Red cedar tops (20).

Powder.—Yellowish-brown. The microscopic elements consist of: Tracheids with bordered pits; parenchyma with numerous stomata; long bast fibers and starch grains.

CONSTITUENTS.—Tannin, resin, gum, etc., and **a volatile oil** (19 a) (2 per cent. in tops, 10 per cent. in berries) having the same composition as oil of turpentine.

ACTION AND USES.—Savine is an irritant, acting especially as a **uterine stimulant;** also diuretic, emmenagogue, and vermifuge. Dose: 5 to 15 gr. (0.3 to 1 Gm.). It is used externally in ointment as a stimulant dressing for bruises.

19 a. **OLEUM SABINÆ.**—OIL OF SAVINE. A nearly colorless, sometimes yellow, limpid, volatile oil, having a strong, terebinthinate odor, and a bitterish, intensely acrid taste. It has the same composition as oil of turpentine. Dose: 1 to 5 ℥ (0.065 to 0.3 mils).

20. **JUNIPERUS VIRGINIANA.**—The tops of the red cedar, or American saivne, are often used to adulterate sávine, from which they can scarcely be recognized except by difference in taste and smell. The galbulus of the false variety is borne on an **erect pedicel.**

21. **Pinus Alba N.F. Lin.**—WHITE PINE. The inner bark of **Pinus strobus** (Weymouth Pine), from eastern and central North America. In flat pieces about 6 inches long by 3 inches in width and $\frac{1}{24}$ inch in thickness. Bark brittle, fracture irregular, not fibrous, but showing several woody layers. Reddish-brown streaked with gray outside; inner, yellowish blotched with light brown; bland odor; mucilaginous, slightly bitter and astringent taste.
PROPERTIES.—Those of balsamic preparations generally.
USES.—An emollient and expectorant in chronic affections of air-passages. Dose of fluidextract: $\frac{1}{2}$ to 1 fluidrachm (2 to 4 mils).

21 a. **PINUS MONTANA MILLER.**—Pinus Pumilio Haenke—Dwarf Pine. From the fresh leaves of this dwarf pine a volatile oil is obtained which is official in the U.S.P. IX as **Oleum Pini Pumilionis,** Oil of Dwarf Pine Needles. It is employed as an inhalent in catarrh of the respiratory passages, chronic laryngitis and bronchitis; used locally in treatment of chronic rheumatic affections and when added to ether allays irritation and diminishes bronchial secretion.

22. **THUJA N.F.**—ARBOR VITÆ. The leafy tops of **Thu'ja** occidenta'lis Linné, a North American evergreen tree. Small flattened twigs having **a scalloped** appearance, due to the flat, lateral leaf-scales, each of which has an oil-gland near its apex; the other leaves folded lengthwise, boat-shaped, mostly glandless; odor balsamic, somewhat terebinthinate; taste pungently aromatic, camphoraceous, and bitter. The medicinal properties of Thuja depend mainly upon a volatile oil. It **resembles savine** in its general action. Dose: 15 to 60 gr. (1 to 4 Gm.), in infusion or fl'ext.

23. **JUNIPERUS, N.F.**—JUNIPER BERRIES. The fruit of **Junip'erus commu'nis** Linné, an evergreen shrub or small tree inhabiting the Northern Hemisphere, bearing small cones, the scales of which coalesce in threes, become fleshy, and ripen into the so-called berry. These berries or fruits are globular, about the size of a large pea, with a triangular depression at the top caused by a three-rayed furrow where the scales are united; at the base are a few small scales, remnants of undeveloped whorls; externally of a glossy, purplish-black color, covered with a grayish bloom; they contain a brownish-yellow pulp with oil-glands, in which are imbedded three small, bony, angular seeds, also covered with large oil-glands; odor disagreeably aromatic, balsamic;

taste sweetish, warm, and balsamic, slightly bitter.   The Smyrna berry from
J. *phœnicea* Linné, yields an oil of greater optical activity.

CONSTITUENTS.—Volatile oil, most abundant in the full-grown green
berries, being partially converted into resins on ripening, entirely so in the
dead-ripe, black berries; also juniperin, sugar (15 to 30 per cent.), wax, fat,
proteids, mucilage, etc.   Their virtues are extracted by water and alcohol.

ACTION AND USES.—Stimulant and diuretic, chiefly used as an adjuvant
to more powerful diuretics in dropsical complaints.   Dose: 15 to 60 gr.
(1 to 4 Gm.), in infusion, water spirit, etc., the volatile oil, however, obtained from the wood and branches, being principally used.   They are
largely used in the manufacture of gin, which owes its diuretic properties
to them.

23 a. **OLEUM JUNIPERI**, U.S.—OIL OF JUNIPER.   A colorless or greenish-yellow volatile oil, with a strong, terebinthinate odor and a hot,
acrid taste.   Specific gravity 0.850 to 0.865.   It consists of pinene,
$C_{10}H_{16}$, cadinene, and juniper camphor.

OFFICIAL PREPARATIONS.
Spiritus Juniperi (5 per cent.)   .Dose: 30 ♏ (2 mils).
Spiritus Juniperi Co. (0.4 per cent.).......Dose: 2 fldr. (8 mils).

24. OLEUM CADINUM.—OIL OF CADE

(Oleum Juniperi Empyreumaticum)

JUNIPER TAR OIL

An empyreumatic, oily liquid obtained from the heart-wood of **Junip'erus
oxyce'drus** Linné, by dry distillation in ovens.

BOTANICAL CHARACTERISTICS.—A tree 10 to 12 feet high, with spreading top and
drooping twigs.   *Leaves* awl-shaped.   *Fruit* globular, reddish-brown, about
the size of a filbert.

HABITAT.—Mediterranean Basin.

DESCRIPTION OF DRUG.—A brownish or dark brown, oily liquid, less
thick and more mobile than tar, having a tarry but characteristic
odor, and an aromatic, bitter, and acrid taste.

ACTION AND USES.—Used mostly externally in the treatment of **cutaneous
diseases** and as an insecticide in the form of liniments, ointments,
or soaps.   Dose: 3 ♏ (0.2 mil).

25. **TSUGA CANADENSIS** Carriere.—HEMLOCK SPRUCE.   (Bark.)   Tonic and
astringent.   Dose: 15 to 60 gr. (1 to 4 Gm.).

26. **LARIX AMERICANA** Michaux.—TAMARAC.   AMERICAN LARCH.   (Bark.)
Tonic and gently astringent, its chief action being upon mucous membranes.
Dose: ½ to 2 dr. (2 to 8 Gm.).

27. TEREBINTHINA, N.F.—TURPENTINE

TURPENTINE

A concrete oleoresin obtained from **Pi'nus palus'tris** Miller (Fam. Pinaceæ, U.S.P.
1900), and other species of Pinus.

BOTANICAL CHARACTERISTICS.—A large tree, 60 to 100 feet, with thin, scaled bark,
and hard, very resinous wood.   *Leaves* 10 to 15 inches long, in threes, from

long sheaths. *Sterile flowers* rose-purple. *Cones* large, cylindrical or conical-oblong.

SOURCE AND COLLECTION.—Southern United States, particularly North Carolina. The oleoresin is secreted in the sapwood; some of it flows spontaneously, but it is generally obtained by a process called "boxing," as follows: During the winter from one to four excavations, each holding from 4 to 8 pints, are cut into the tree through the sapwood. After a few days the bark above these cavities is removed for about a height of 3 feet, and some of the wood is hacked off, the hacks being in the shape of the letter L. The oleoresin begins to flow about the middle of March, and continues until September or October. The turpentine is removed by means of dippers constructed for the purpose, and then usually distilled. That which flows the first year is considered the best, being termed "virgin dip," and yields about 6 gallons of oil per barrel, and "window-glass rosin;" that of the next and subsequent years is known as "yellow dip," yielding about 4 gallons of oil per barrel, and medium grades of rosin. The turpentine which hardens on the tree is known as "scrapings," and yields about 2 gallons of oil per barrel, leaving a dark resin.

DESCRIPTION OF DRUG.—In yellowish, opaque, tough masses, brittle in the cold, crumbly-crystalline in the interior, of a terebinthinate odor and taste. In warm weather it is a yellowish, viscid semiliquid when fresh, but ultimately, through exposure to the air, becomes perfectly dry and hard.

CONSTITUENTS.—Volatile oil 20 to 30 per cent. (27 b), abietinic anhydride, $C_{44}H_{62}O_4$, in rosin (27 c), the acid of which, abietic acid, $C_{44}H_{64}O_5$, is crystalline, soluble in CS_2, benzol, alcohol, ether, chloroform, glacial acetic acid, and alkalies.

27 a. **Terebinthinæ Laricis, N. F.**—**Venice Turpentine.**—A yellowish or greenish liquid of honey-like consistence, collected in Switzerland and portions of France from *Larix europæa* De Candolle. Obtained by boring holes into the center of the wood and dipping the liquid out as it accumulates. It received its name from having formerly been almost entirely distributed from the Venetian port. Genuine Venice turpentine is comparatively scarce in the markets to-day, most of it being a factitious brown liquid made by dissolving rosin in oil of turpentine.

A number of other turpentines are obtained from various species of pine, larch, and fir, but hardly any of them enter our markets. The turpentines all agree in their medical properties, and differ only slightly in their physical characteristics, all of them being liquid at first, thickening through the evaporation and oxidation of their volatile oil, and ultimately solidifying They melt by heat, and at a high temperature ignite with a white flame attended with dense smoke.

CONSTITUENTS.—Volatile oil 20 to 30 per cent., resin (abietic anhydride, crystallizing out as abietic acid), a bitter principle, and traces of succinic and acetic acids.

ACTION AND USES.—The turpentines are rarely used internally, the volatile oil, to which the medicinal virtues are due, being used instead. Dose: 15 to 60 gr. (1 to 4 Gm.), in pills. Externally irritant and rubefacient, in ointments and plasters.

27 b. **OLEUM TEREBINTHINÆ,** U.S.—OIL OF TURPENTINE. SPIRITS OF TURPENTINE. A volatile oil distilled from turpentine, the markets of the United States being chiefly supplied by the North Carolina forests. A perfectly limpid, colorless liquid when pure, but generally somewhat colored from resin contained, or from oxidation; odor peculiar, strong, penetrating; taste hot, pungent, somewhat bitter. It is very volatile and inflammable. When purified by distilling with caustic soda, it constitutes the Oleum Terebinthinæ Rectificatum, U.S., which is officially directed to be dispensed when oil of turpentine is required for internal use.

CONSTITUENTS.—Oil of turpentine consists of several terpene hydrocarbons having the formula $C_{10}H_{16}$ (pinene), sp. gr. 0.855–0.870. When exposed to the air, it becomes thick from the oxidation of some of these hydrocarbons into resin. When the rectified oil is treated with nitric acid, large crystals of terpin hydrate (Terpini Hydras, U.S.) separate out, having properties similar to the oil of turpentine. Dose, 2 gr. (0.1 Gm.). The European turpentine oil contains pinene and sylvestrine; it forms with hydrochloric acid a crystalline compound, $C_{10}H_{16}HCl$ (artificial camphor). **Terebenum** is a liquid derived from the oil (consisting chiefly of pinene) by treatment with sulphuric acid, boiling point 156°–160°C. Dose: 8 ♏ (0.5 mil).

ACTION AND USES.—Stimulant, diuretic, hemostatic, occasionally diaphoretic; in large doses anthelmintic and cathartic; externally rubefacient, in rheumatism, etc. As a stimulant it is often beneficial in low forms of fever, and, when death is inevitable, to prolong life beyond the natural limit. Dose: 5 to 15 ♏ (0.3 to 1 mil) in emulsion.

OFFICIAL PREPARATIONS.

Linimentum Terebinthinæ (35 per cent.
  with resin cerate).

Oleum Terebinthinæ Rectificatum,....Dose, 5 to 15 ♏ (0.3 to 1 mil).

Ceratum Cantharidis. Emulsum Olei Terebinthinæ.

27 c. **RESINA,** U.S.—RESIN. ROSIN. COLOPHONY. The clarified residue left after distilling off the volatile oil from turpentine. It has been asserted that **Pinus palustris**, the official species, contains more resin than any other German or American pine. When pure, rosin is of a clear, pellucid, amber color, but the commercial rosin is yellowish-brown, more or less dark, sometimes almost black, the color depending upon its purity and the amount of heat used in its preparation; it breaks with a shining, shallow, conchoidal fracture; odor and taste faintly terebinthinate. White rosin is an opaque variety made by incorporating it with water.

CONSTITUENTS.—Rosin is the anhydride of abietic acid, $C_{44}H_{62}O_4$, into which acid it may be converted by warming with dilute alcohol. Ash, 0.05 per cent.

ACTION AND USES.—An important ingredient of ointments and plasters, and is said to have the property of preserving them from rancidity by preventing the oxidation of the fatty base.

OFFICIAL PREPARATION.

Emplastrum Resinæ.

28. PIX LIQUIDA.—TAR

TAR

SOURCE.—An empyreumatic oleoresin obtained by the destructive distillation of the wood of **Pinus palustris** Miller, and of other species of Pinus. The pine logs are cut into billets, and built up into a stack and covered with earth, as in making charcoal. Slow combustion is started through an opening in the top of the stack, and the resinous matter, as it melts out and collects in a cavity in the center, is drawn off into barrels.

DESCRIPTION.—A resinous, black semiliquid, of an empyreumatic, terebinthinate odor, and a sharp, bitterish, empyreumatic taste. Acid in reaction. Partly soluble in water.

  Birch tar, Dagget, or Oleum Rusci, from *Betula alba* Linné, has an odor similar to that of Russian leather.

CONSTITUENTS.—Tar is a very complex substance, varying with the kind of wood, amount of resins present therein, and the care exercised in its preparation, the chief constituents being an empyreumatic volatile oil, pyrocatechin, acetone, xylol, toluol, cresols (creosote), guaiacol, phenol, etc. The acid reaction which characterizes tar is due to acetic acid, obtained in an impure state as pyroligneous acid by distillation. In the retort is left behind the ordinary solid and fusible pitch of commerce.

ACTION AND USES.—Stimulant, irritant, insecticide, similar to, but less irritant than, the turpentines. Dose: 8 to 60 gr. (0.6 to 4 Gm.). The syrup is much used in **pulmonary affections.**

Official Preparations.

Syrupus Picis Liquidæ (0.5 per cent),....Dose: 1 to 4 fl. dr. (4 to 15 mils).
Unguentum Picis Liquidæ (50 per cent.).

28 a. OLEUM PICIS LIQUIDÆ RECTIFICATUM.—Oil of Tar. A volatile oil distilled from tar, the residue left being common pitch, pix nigra. A nearly colorless liquid when first distilled, but soon acquires a dark, reddish-brown color; it has the characteristic odor and taste of tar, which depends upon it for its medicinal properties. Dose: 1 to 5 ♏ (0.065 to 0.3 mil), in capsules or emulsion.

29. TEREBINTHINA CANADENSIS.—Canada Turpentine

CANADA BALSAM. BALSAM OF FIR

A liquid oleoresin obtained from A'bies **balsam'ea** Linné

Habitat.—Canada, Nova Scotia, Maine, and the mountainous regions further south.

Production.—The oleoresin is secreted in small vesicles in the bark, collected by puncturing and allowing the liquid to exude into a vessel having a broad and funnel-like lip. The vesicles contain only from a few minims to 1 fluid drachm.

Description of Drug.—A yellowish or faintly greenish, transparent liquid of honey-like consistence, becoming thicker and somewhat darker with age, but always retaining its transparency, and ultimately drying into a transparent mass; it has an agreeable, aromatic, terebinthinate odor, and a bitterish, feebly acrid, but not disagreeable taste, for which reason it is sometimes erroneously called balm of Gilead (98).

Action and Uses.—It has medical properties similar to the other turpentines and copaiba, but is rarely employed as a remedial agent. It is most valued for **mounting microscopic objects,** for which its beautiful and durable, uncrystalline transparency peculiarly fits it.

Official Preparation.

30. PIX CANADENSIS.—Canada Pitch or Hemlock Pitch. An oleoresin obtained from the North American hemlock spruce, A'bies **canaden'sis** Carriere. Resembles Pix Burgundica (31) in appearance, properties, and uses; it is somewhat darker red-brown in color and is much more fusible; odor weak, peculiar; taste very feeble. Rosin is a common adulteration.

31. PIX BURGUNDICA.—Burgundy Pitch. The resinous exudation prepared from Abies excelsa Poiret. A reddish-brown or yellowish-brown, opaque or translucent solid when pure, gradually taking the form of the vessel in which it is contained; brittle, breaking with a shining, conchoidal fracture; at body heat it becomes soft and adhesive; odor agreeable, somewhat aromatic, terebinthinate; taste aromatic and sweetish, not bitter. A mixture of common pitch, rosin, and turpentine melted together and agitated with water, is often substituted for Burgundy pitch, but may be detected by its insolubility in warm glacial acetic acid. Terebinthina cocta, a residue from the distillation of turpentine with water, and Resina pini (white turpentine), fused in hot water and strained, are allied products resembling the former, but these later become crystalline. *Constituents:* Volatile oil (smaller proportion than in turpentine), water, and resin. Gentle rubefacient and stimulant, in chronic rheumatism, etc., in plasters.

Emplastrum Picis Burgundicæ, U.S.P. 1890.
Emplastrum Picis **Cantharidatum** (92 per cent., with cerate of cantharides), U.S.P. 1890.

32. **SUCCINUM.**—AMBER. A fossil resin from extinct coniferous trees, found in greater or less quantities in every quarter of the globe; the largest deposits occur in the region surrounding the Baltic Sea, where it has been washed upon the shore. In small, irregular pieces, usually light or deep yellowish-brown, sometimes reddish-brown, generally translucent; tasteless and odorless, but emits an agreeable, aromatic odor when heated. It is almost insoluble in water, alcohol, ether, or oils, slightly soluble in chloroform. Used for fumigation, for the preparation of succinic acid and oil of amber, and in the arts.

32 a. **OLEUM SUCCINI.**—OIL OF AMBER. A light yellowish-brown or amber-colored liquid (colorless when pure), having a balsamic, empyreumatic odor, and a warm, acrid taste. On exposure to light and air it thickens and becomes darker, ultimately solidifying into a black mass. With fuming nitric acid it acquires a red color, changing after a time into a brown, resinous mass having a peculiar musk-like odor. It is often adulterated with oil of turpentine, which may be detected by its throwing down a solid camphor when hydrochloric acid gas is passed through the mixture. Stimulant, antispasmodic, and irritant. Dose: 5 to 15 ℥ (0.3 to 1 mil). Externally in liniments.

33. **DAMMARA.**—DAMMAR. GUM DAMMAR. A spontaneous, resinous exudation collected in the East Indies from A'gathis **dam'mara** Richard. Transparent, straw-colored, rounded masses, almost free from odor and taste, and breaking with a glossy, conchoidal fracture. Used mostly for varnishes.

34. **KAURI RESIN.**—KAURI GUM. A resin dug in large quantities from the soil in New Zealand, where it has exuded from **Dam'mara orienta'lis.** It is in large cream-colored or amber-colored masses. Used as a vulnerary in skin diseases; also used as a substitute for collodion, leaving an adherent, impervious, resinous varnish over the wound.

35. **SANDARACCA.**—SANDARAC. A resin exuding spontaneously from the bark of a North African evergreen tree, **Calli'tris quadrival'vis** Ventenat. Small rounded masses about the size of a pea, of a yellowish color; it resembles mastic somewhat, and is often substituted for it on account of its lower price, but a simple means of distinction is afforded in its becoming pulverulent (not adhesive) when chewed. It was formerly used as a mild stimulant in ointments and plasters, but is now mostly used for varnishes. Its powder is used as a pounce to prevent ink from spreading on paper or cloth.

GNETACEÆ

36. **EPHEDRA.**—The herb **Ephe'dra antisyphilit'ica** C. A. Meyer. This plant is a native of Arizona, where it is used in venereal diseases. Dose of fl'ext.: 1 to 2 fl. dr. (4 to 8 mils).

GRAMINEÆ.—Grass **Family**

A large order yielding the cereals (wheat, rye, etc.) and sugar cane, the source of most of the sugar of the market. The characteristics of the order are the *hollow stems* (culms), *flowers* in spikelets, and the *fruit*, a caryopsis.

Synopsis of Drugs from Gramineæ

A. *Rhizome.*
   TRITICUM, 37.
B. *Root.*
   Vetiveria, 38.
C. *Sugar.*
   SACCHARUM, 39.
D. *Styles and Stigmas.*
   *ZEA, 40.
E. *Fixed Oil.*
   Oleum Maydis, 41.

F. *Starches.*
   AMYLUM, 42.
a. Avenæ Farina.
b. Sago.
c. Tapioca.
d. Taro.
e. Triticum Vulgare.
f. Oryza.
g. Solanum Tuberosum.
h. Canna.
i. Maranta.
j. Curcuma Leucorrhiza.

G. *Fruit.*
   Hordei Fructus, 43.
H. *Decorticated Fruit.*
   Hordeum, 43 a.
I. *Germinated Seeds.*
   Maltum, 43 b.

*N.F.

37. TRITICUM.—Triticum

COUCH-GRASS

The dried rhizome of **Agropy'ron rep'ens** Beauvois.

Botanical Characteristics.—Creeping; *root-stocks* slender, numerous. *Spikelets* 4- to 8-flowered, glabrous; *glumes* 3- to 7-nerved; *rachis* glabrous; *leaves* flat.

Habitat.—Europe; naturalized and grows abundantly in North America.

Description of Drug.—Short, hollow sections from 3 to 6 mm. (⅛ to ¼ in.) long, and about the thickness and color of a straw; odorless; taste sweetish.

Powder.—Microscopical elements of: See Part iv, Chap. I, B.

Constituents.—No active constituent has been discovered in couch-grass; it contains glucose, mucilage, malates, triticin (a gummy substance resembling inulin), and inosit. Ash not to exceed 3 per cent.

Preparation of Triticin.—Obtained by exhausting powdered drug with water; neutralize with baryta; concentrate and precipitate with lead subacetate; remove lead; purify with charcoal; neutralize, concentrate, and precipitate with alcohol. It is an amorphous, white powder, inodorous, tasteless, deliquescent, and with HNO_3 is oxidized into oxalic acid.

Action and Uses.—Diuretic, demulcent. Dose: ½ to 3 dr. (2 to 12 Gm.).

Official Preparation.

Fluidextractum Tritici..............Dose: 1 to 4 fl. dr. (4 to 15 mils).

38. VETIVERIA.—Vetivert. The fibrous wiry roots of **Andropo'gon murica'tus** Retzius. *Habitat:* Eastern India. Tonic and stimulant, but mainly employed as a perfume in sachet powders, etc.

39. SACCHARUM.—Sugar

CANE-SUGAR

The refined sugar obtained from **Sac'charum officina'rum** Linné, and from various species or varieties of Sorghum, also from one or more varieties of **Be'ta vulga'ris** Linné (nat. ord. Chenopodiaceæ).

Source and Varieties.—The sugar cane is extensively cultivated in Africa, East and West Indies (especially Cuba), Brazil, and Southern United States, particularly Louisiana. The sugar beet is extensively cultivated in France and Spain, and has been introduced with varying success into some parts of the United States. Cane-sugar is also a constituent of the sugar maple; of the carrot and turnip; of cassia pulp, etc. The sugar in fresh fruit is mainly cane-sugar; by the action 'of the fruit acids, or a ferment, it is generally *inverted*, becomes uncrystalline, and influences polarized light in the opposite direction from that of cane-sugar, twisting the ray

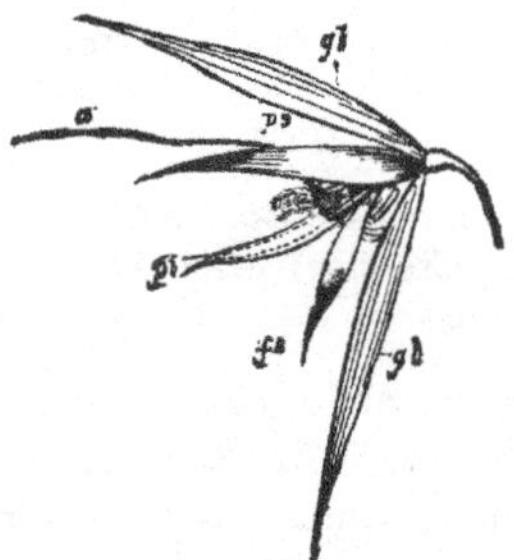

Fig. 16.—Spikelet of the Oat (*Avena sativa*).　*gl.* Glumes.　*ps. fs.* Paleæ or pales.　*a.* Awn. *fl.* An abortive flower.

Fig. 17.—*Triticum vulgare* (Wheat).　Plant and flowers (enlarged).

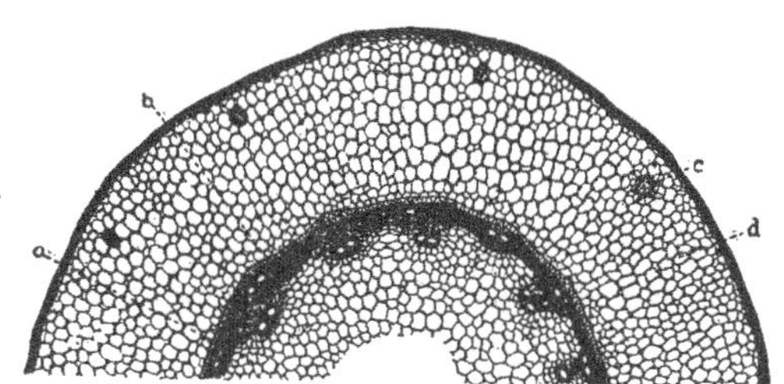

Fig. 18.—*Agropyron repens.*

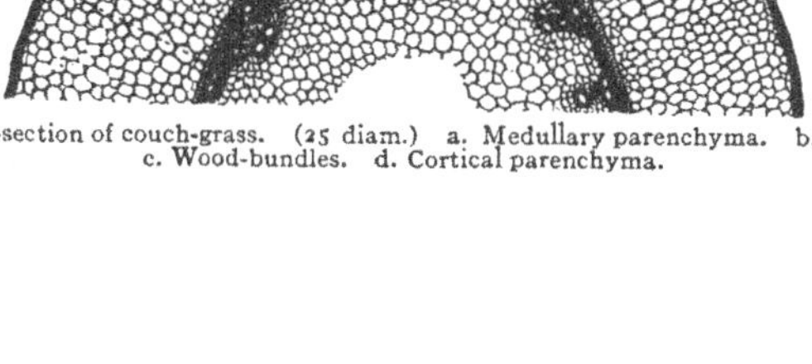

Fig. 19.—Cross-section of couch-grass. (25 diam.) a. Medullary parenchyma. b. Woody tissue.
c. Wood-bundles. d. Cortical parenchyma.

from right to left. Honey-sugar is probably a mixture of the two varieties—right- and left-handed. It is readily altered to a crystalline and granular mass of *grape-sugar* in dried fruit, as in the raisin, the prune, and solidified honey. This, the common form of grape-sugar, is right-handed, and is called dextrose (dextrogyrate), to distinguish it from lævulose. Barley-sugar is made by heating cane-sugar till it fuses, becoming thus, in a great measure, uncrystalline. Molasses (treacle)—*Syrupus fuscus* (official 1860–1870)—is the result from the evaporation of cane-sugar syrup; it is a mixture of cane-sugar with uncrystallizable sugar and coloring matter.

DESCRIPTION.—Sugar or sucrose, $C_{12}H_{21}O_{11}$, is in "white, dry, hard, distinctly crystalline granules, odorless, and having a purely sweet taste. Permanent in the air." The aqueous solution saturated at 15°C. (59°F.) has a sp. gr. of 1.345 and is miscible with water in all proportions, soluble in 175 parts of alcohol.

OTHER SUGARS. **Saccharum** Lactis.—Lactose obtained from the whey of cows' milk and purified by recrystallization.

SOURCE AND DESCRIPTION.—It is prepared from cows' milk by evaporating the whey after removing the curd. Cows' milk contains from 4.5 to 4.9 per cent. of sugar. It crystallizes in large hard prisms, has a feebly sweet taste and is soluble in six parts of cold water. It occurs in white, hard crystalline masses or as a white powder feeling gritty to the tongue, odorless, permanent in air. Like cane-sugar it forms compounds with metallic oxides, and reduces alkaline copper solutions. Practically insoluble in alcohol, ether, or chloroform. It is not effected directly by ferments. When heated with mineral acids it forms dextrose and galactose.

ACTION AND USES.—When injected into the blood-vessels it appears unaltered in the urine. When taken in the alimentary canal it is perfectly assimilated. When administered in large doses it acts as an active diuretic. Milk loses this diuretic effect on being boiled. Used in making tablet triturates.

MANNOSE (from mannite); maltose (from starch by the action of dilute acid or diastase); melitose (from eucalyptus).

CARAMEL, N.F. is a name applied to *burnt sugar* (*Saccharum ustum*), used in the liquid form as a coloring for spirits, vinegar, etc.

SACCHARUM UVEUM.—Grape-sugar. Glucose. Yellowish or whitish masses or granules much less sweet than cane-sugar. Composition $C_6H_{12}O_6H_2O$.

ACTION AND USES.—Demulcent and lenitive. Used in making the various syrups and compound syrups of the Pharmacopœia, etc.

OFFICIAL PREPARATION.—Syrupus.

40. ZEA, N.F.—ZEA.

CORN-SILK

The dried styles and stigmas of **Ze'a ma'ys** Linné (our common Indian corn)
Yellowish or greenish, soft, silky, hair-like threads, about 150 mm. (6 in.)
long; free from odor, with a sweetish taste. Constituents.—**Maizenic acid,**
fixed oil, resin, sugar, gum, albuminoids, phlobaphene, extractive, salt,
cellulose, and water.

Action and Uses.—Mild stimulant, diuretic. The infusion· may be taken *ad
libitum.*

Fluidextractum Zea (Unofficial)......Dose: ½ to 2 fl. dr. (2 to 8 mils).

41. **OLEUM MAYDIS.**—Maize Oil. A fixed oil expressed from the embryo
of the seed of **Zea mays** Linné. A yellow, viscid, transparent liquid, having
a peculiar odor like cornmeal, and a bland taste. This oil has become quite
valuable commercially, used as salad oil and by hydrogenation yields a valuable vegetable fat. In making of liniments and oleaginous preparations,
it is quite equal to olive oil. Demulcent.

AMYLUM.—Starch

STARCH

The starch grains obtained from the fruit of **Ze'a ma'ys** Linné.

Description.—Usually in opaque, angular or columnar masses, easily
pulverizable between the fingers, with a peculiar sound, into a fine,
white powder; odorless and tasteless. Under the microscope it is
seen to be composed of small granules striated concentrically or
excentrically around a nucleus or hilum. Insoluble in cold water,
but with boiling water it forms a glutinous paste on cooling. Iodine
is the test for starch, the characteristic blue color being produced
when only a minute quantity of the latter is present.

Other starches—chiefly distinguished by the size and shape of
the starch-granules as seen under the microscope:

(*a*) Avenæ Farinæ.—Oatmeal. From *Avena sativa* Linné, probably native to Western Asia, but now a common field crop. A grayish-white, not uniform meal, containing the gluten and fragments of
the integuments; bitterish. Demulcent and nutritive (due to the
gluten contained).

(*b*) Sago.—Pearl Sago. Globular, pearl-like grains, white or
brownish, prepared from *Metroxylon sagu, M. rumphii,* and other
species growing in the East India Islands.

(*c*) Tapioca.—Cassava Starch. Yielded by the rhizomes of
Brazilian plants, *Manihot utilissima* and *M. aipi,* nat. ord. Euphorbiaceæ. White and opaque, irregular lumps.

(*d*) Taro.—Taro Flour. A starch prepared from the corm of
Colocasia esculenta Schott, the food (poi) of the natives in Hawaii

and the West Indies. Recommended as a diet for dyspeptic and consumptive patients.

Starches from the underground parts of *Triticum vulgare* and *Oryza sativa*, Gramineæ; *Solanum tuberosum* (potato starch), Solanaceæ; *Canna edulis*, *Maranta arundinacea*, and *Curcuma leucorrhiza*, Scitamineæ.

Powder.—Microscopical elements of: See Part iv, Chap. I, B.

FIG. 20.—Maranta Starch. (× 250).

FIG. 21.—Curcuma Starch. (× 350).

FIG. 22.—Wheat Starch. (× 250).

FIG. 23.—Rice Starch. (× 250).

FIG. 24.—Potato Starch. (× 250).

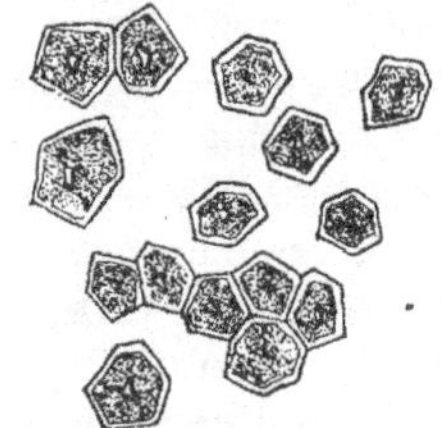

FIG. 25.—Corn Starch. (× 500).

See also Starches of Drugs, Part iv.

CHEMICAL COMPOSITION.—Starch is the basis of that class of organic compounds termed carbohydrates. Its composition is $C_6H_{10}O_5$. By hydrolysis it is converted into a gummy principle, dextrin, and glucose. Ferments convert it into alcohol and carbon dioxide—$C_6H_{10}O_5 = 2C_2H_5OH + 2CO_2$. Ash. Not more than 0.5 per cent.

ACTION AND USES.—Nutritive and demulcent.

OFFICIAL PREPARATION.

Glyceritum Amyli (10 per cent.).

Dextrinum Album, N.F. (White dextrine should not yield more than 0.5 per cent. of ash.)

43. **HORDEI FRUCTUS.**—BARLEY. The fruit of **Hor'deum dis'tichum** Linné, a common cultivated cereal indigenous to Western Asia. About 15 mm. (⅗ in.) long, tapering at the ends, on one side traversed by a longitudinal groove along which the grayish-yellow palea or husk is coalesced with the smooth, pale brown testa; underneath the testa is a layer of gluten surrounding the central starchy parenchyma. Nutritive.

43 a. **HORDEUM,** or pearl barley, is the fruit deprived of its brown integuments.

43 b. **MALTUM.**—MALT (U.S.P.IX). Prepared from the fruit of **Hordeum** distichum Linné by soaking, and then allowing fermentation to proceed until the young embryo is nearly the length of the fruit; the fruit is then dried in the sun and afterward kiln-dried in order to kill the germ. The object of this process is to develop the greatest possible amount of diastase, a peculiar ferment which has the property of converting starch into sugar. Malt occurs in yellowish or ambered-colored grains crisp when fractured with a whitish interior. Its odor is agreeable and characteristic. The taste is sweetish due to the conversion of some of the starch into maltose by the diastase present. Malt should float in cold water. Malt is demulcent and nutrient, given in the form of the extract.

ACTION AND USES.—Demulcent and nutritive given in conjunction with other substances chiefly.

PREPARATION.

Extractum Malti (liquid, of honey-like consistence).

CYPERACEÆ.—Sedge Family

44. **CAREX ARENARIA** Linné.—RED SEDGE. RADIX SARSAPARILLÆ GERMANICÆ. This sedge grows in the coast regions of Central and Northern Europe, where its rhizome is used as an alterative like sarsaparilla.

45. **ADRUE.**—GUINEA RUSH. The rhizome of **Cy'perus articula'tus** Linné, used in its native country to check vomiting and as a tonic. Dose of fl'ext.: 30♏ (2 mils).

PALMÆ.—Palm Family

Synopsis of Drugs from the Palmæ

A. *Seed.*
    Areca, 46.
B. *Fruit.*
    SABAL, 47.
C. *Root.*
    Carnauba, 48.

D. *Resin.*
    Draconis Resina, 49.
E. *Fixed Oils.*
    Oleum Palmæ, 50.
    Oleum Cocois, 51.

46. **ARECA.**—ARECA NUT. BETEL NUT. The seed of an East Indian tree, **Are'ca cat'echu** Linné. Roundish-conical, about 25 mm. (1 in.) long, flattened at the base; externally deep brown, varied with fawn-color, giving it a longitudinally-veined appearance; internally brownish-red with white veins. It abounds in tannin, and contains three alkaloids upon which its tæniafuge properties depend, arecoline, arecaine, and a trace of an undetermined alkaloid. Mixed with the leaves of *Piper betel* it forms the "betel" chewed so largely by the natives. It is strongly recommended as a tæniafuge and vermifuge. Dose: 2 to 3 dr. (8 to 12 Gm.).

47. SABAL

SABAL. (SAW PALMETTO)

The dried ripe fruit of Sereno'a serrula'ta (R. and S.) Hooker filius.

Irregularly spherical to oblong-ovoid; 10 to 25 mm. long, 10 to

15 mm. in diameter; externally blackish-brown, shrivelled, somewhat oily; epicarp thin, sarocarp about 1 mm. thick, greenish-yellow, soft, spongy, endocarp thin, friable; seed hard, chocolate-brown; odor aromatic; taste sweetish, acrid and oily. Tonic, diuretic, expectorant, and sedative, used in neuralgic affections to allay irritation of mucous membranes, and in pulmonary affections. Dose of fl'ext.: ½ to 2 fl. dr.

OFFICIAL PREPARATION.—Fluidextractum. Dose: 1 mil (15 ♏).

Powder.—Microscopical elements of: See Part iv, Chap. I, B.

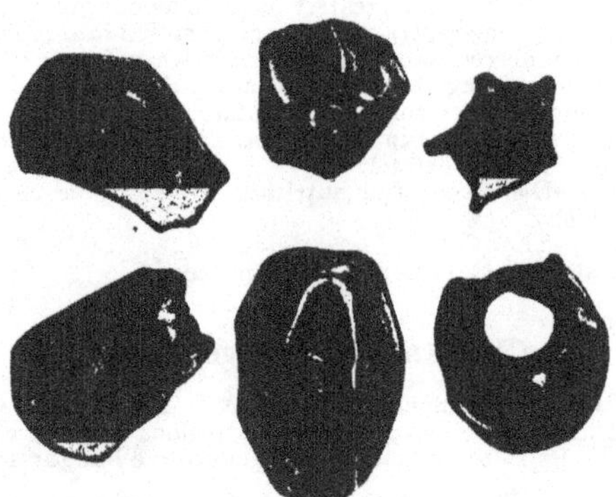

FIG. 26.—Sabal. Fruit. (Photograph.)

48. **CARNAUBA.**—The root of **Coper'nica cerif'era** Martius, used in Brazil, where the plant grows, as an alterative like sarsaparilla, stillingia, etc. Dose: 15 to 60 gr. (1 to 4 Gm.).

49. **DRACONIS RESINA.**—DRAGON'S BLOOD. A spontaneous resinous exudation from the ripening fruit of **Cal'amus dra'co** Willdenow. *Habitat:* East Indies, Siam, and the Molucca Islands. A dark brownish-red, internally brighter red resin, coming into market in various forms, small granules, oval pieces in bead-like strings, sticks, and the poorer varieties in cakes and disks; breaks with a dull, irregular fracture; tasteless and almost odorless, but when heated emits a benzoin-like odor, due to the benzoic acid which it contains. The red resin, constituting 90 per cent., has been termed draconin. The use of dragon's blood is almost entirely confined to the manufacture of paints and varnishes.

50. **OLEUM PALMÆ.**—PALM OIL. A fixed oil expressed from the fruit of **Elæ'is Guineen'sis** Jacquin, a West African palm cultivated in tropical America. A solid fat, harder than butter, of an orange-red color, bleaching upon exposure to light or heat. When fresh, it has a violet-like odor and a bland taste, but it rapidly becomes rancid and of an acrid taste. It is used principally in the manufacture of soaps and candles, occasionally in ointments.

51. **OLEUM COCOIS.**—COCOANUT OIL. A fixed oil expressed from the seeds of the tropical palm, **Co'cos nucif'era** Linné. A white solid, of the consistence of butter, and with a disagreeable odor. It is mostly used in soaps.

AROIDEÆ.—Arum Family

Herbs with an exceedingly acrid, colorless juice, and having a fleshy corm or rhizome. *Inflorescence* a spadix usually surrounded by a spathe. *Fruit* a berry.

Synopsis of drugs from the Aroideæ

A. *Rhizomes.*
    CALAMUS, 52
    Symplocarpus, 53.

B. *Corms.*
    Arum, 54.
    Arisæma Dracontium, 55.

52. CALAMUS.—Calamus

SWEET FLAG

The dried rhizome of **Acor'us cal'amus** Linné (Fam. Araceæ, U. S. P. 1900). Description of Drug.—Grows in swamps, and along the banks of streams and ponds. Subcylindrical sections of various lengths, about 20 mm. (⅘ in.)

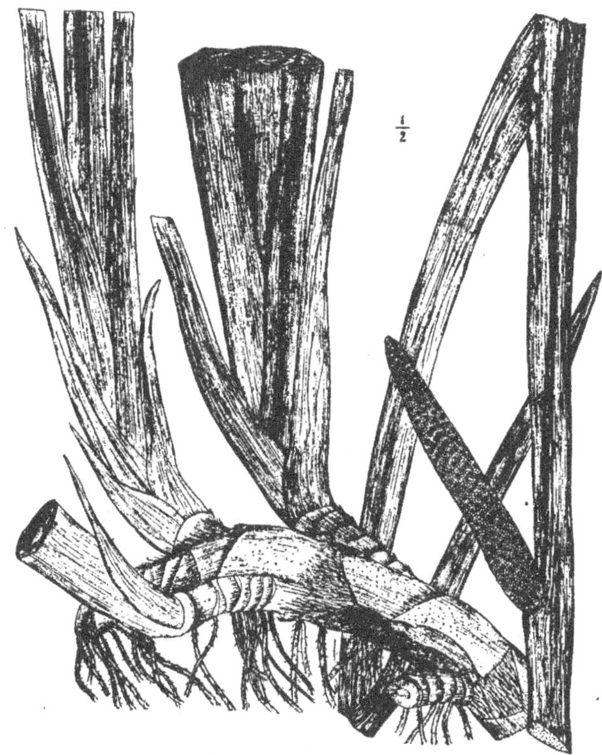

Fig. 27.—*Acorus calamus.*

thick; externally reddish-brown, deeply wrinkled, marked below with rootlet scars (little elongated dot-like rings) in wavy, longitudinal lines, above with leaf-scars; fracture short, corky, showing a pinkish or whitish interior dotted with yellowish or brownish dots, both in the thick cortical layer and in the spongy central column; odor aromatic; taste peculiar, very bitter. Although the unpeeled rhizome is directed, **the** pinkish-white **sections** deprived of the corky layer are often met with in market.

STRUCTURE.—The tissue is chiefly parenchyma, traversed by yellowish fibro-vascular bundles, most abundant just within and near the nucleus sheath. The cells of the parenchyma are filled with starch and volatile oil, the latter most abundant in the cortical layer. The spongy appearance of the central portion is due to large air-cells, as in all aquatic plants.

CONSTITUENTS.—Volatile oil 1 to 2 per cent., having the smell and taste of calamus, a bitter glucoside termed acorin (syrupy, yellow liquid), calamine, choline, resin, starch, and mucilage.

Isolation of Acorin.—A concentrated decoction of the drug is deprived of gum by precipitating with alcohol. The liquid is then treated with lead subacetate. The lead is removed by H_2S. The resulting liquid, after neutralization, is shaken with chloroform, which leaves on evaporation a thin, yellow, aromatic liquid, acorin. This splits into oil and sugar by hydration; by oxidation the resin and acoretin are obtained.

ACTION AND USES.—Tonic and carminative, and a feeble aromatic stimulant. Dose: 15 to 60 gr. (1 to 4 Gm.).

53. **SYMPLOCARPUS.**—SKUNK CABBAGE. The rhizome and roots of an indigenous herb, **Symplocar'pus fœ'tidus** Salisbury, so called from the disagreeable odor (depending upon a volatile oil) which is emitted by all parts of the fresh plant, and by the dried rhizome when triturated. It has an acrid taste, but the acrid principle has not yet been isolated. Stimulant, antispasmodic, and narcotic, causing nausea and vomiting, together with vertigo, headache, and dimness of vision. It has been used in asthma, whooping-cough, nervous and convulsive affections, and hysteria; also in chronic catarrh, chronic rheumatism, and bronchial and pulmonary affections. Dose: 10 to 20 gr. (0.6 to 1.3 Gm.).

54. **ARUM.**—INDIAN TURNIP. The corm of Arisæ'ma (Arum) triphyl'lum Torrey (Jack-in-the-pulpit or wake-robin). *Habitat:* North America, in rich woods. Depressed-globular, about 25 to 50 mm. (1 to 2 in.) in diameter, covered with a loose, wrinkled, brown epidermis; it often comes into market in white, starchy, transverse slices; inodorous; very acrid. This acrid principle is volatile, the fully dried corm being nearly inert. Arum has been used as a stimulant to the secretions in asthma, whooping-cough, chronic catarrh, and rheumatism. Dose: 8 to 15 gr. (0.5 to 1 Gm.).

55. **ARISÆMA DRACONTIUM** Schott.—GREEN DRAGON. *Habitat:* United States, west to Kansas. (Corm.) Diaphoretic and expectorant in dry, hacking coughs attended with irritation. Dose of fl'ext.: 1 to 10 ♏ (0.065 to 0.6 mil).

COMMELINACEÆ.—Spiderwort Family

56. **COMMELINA.**—ASIATIC DAY FLOWER. From **Com'melina com'munis.** This plant has recently been brought to notice as one of medicinal value. It is claimed to have peculiar hemostatic and healing properties. An account of the plant and a report of a chemical examination of it is found in the "Am. Jour. of Pharm.," July, 1898, p. 321.

BROMELIACEÆ.—Pineapple Family

57. **ANANASSA.**—PINEAPPLE. The fruit of **Ananas'sa sati'va** Schultz. The fresh juice contains the digestive ferment, bromelin, which is a powerful and rapid digestant of albumen, both animal and vegetable, acting in the presence of either acid or alkaline carbonates, but most energetically in neutral solutions. It is more nearly related to trypsin than to pepsin.

LILIACEÆ.—Lily Family

Herbs (rarely woody) with flowering stems springing from bulbs or corms with the *leaves* parallel-nerved, except in the tribe Smilaceæ, where they are netted-veined. The *perianth* consists of six divisions; *anthers* introrse; *ovary* superior, usually 3-celled.

Synopsis of Drugs from the Liliaceæ

A. *Root.*
    SARSAPARILLA,
      58.
B. *Rhizomes.*
    *Convallaria, 59.
    VERATRUM, 60.
    Veratrum Album,
      60 a.
    Polygonatum, 62.
    Chamælirium, 63.
    *Trillium, 64.
    Asparagus, 65.

C. *Bulbs.*
    *Allium, 66.
    SCILLA, 67.
D. *Corm.*
    COLCHICI COR-
      MUS, 68.
E. *Seeds.*
    COLCHICI SEMEN,
      69.
    Sabadilla, 61.

F. *Inspissated Juices.*
    ALOE, 70.
    Aloe Barbadensis,
      70 a.
    Aloe Capensis, 70 b.
    Aloe Socotrina, 70 c.
G. *Resin.*
    Xanthorrhœa, 71.
H. *Leaves.*
    Erythronium, 72.

58. SARSAPARILLA.—Sarsaparilla

SARSAPARILLA

The dried root of **Smi'lax officinalis** Kunth, **Smi'lax med'ica** Chamisso et Schlechtendal, **Smilax papyra'cea** Duhamel, **Smilax ornata** Hooker, and of other undetermined species of Smilax.

Botanical Characteristics.—Evergreen, climbing, shrubby plants. *Stem* prickly. *Leaves* alternate, netted-veined, coriaceous, ovate-oblong, with a cordate base, 1 foot long and 4 to 5 inches broad. *Flowers* in axillary clusters, dioecious; *stigmas* 3, sessile. *Fruit* a globular, 1- to 3-seeded berry.

Habitat.—Tropical America, in swampy forests.

Description of Drug.—The varieties used in medicine have a thick, knotty rhizome (which, if present, should be removed) from which grow in a horizontal direction the fleshy roots. These appear in the market several feet in length, cylindrical, about the thickness of a quill, very flexible; externally longitudinally wrinkled, of various colors, depending upon the variety, generally ash-colored, grayish-brown, or reddish-brown; internally whitish, horny, or occasionally mealy; nearly inodorous; taste mucilaginous, bitter, and acrid.

Structure.—A transverse section shows a thin, easily removed epidermis overlaying a thick cortical layer; this inner bark consists of loose parenchyma, the cells of which, when not devoid of solid contents, are filled with starch-granules or paste, and occasionally calcium oxalate raphides; a brownish ring (nucleus sheath) separates it from the woody center, which is made up of elongated woody cells. A small pith runs through the center of this woody zone.

Varieties.—There are four principal varieties of sarsaparilla, differing somewhat in appearance, and especially in the condition of the starch.

(*a*) Mealy—starch in granules (see Part iv).

The Honduras sarsaparilla is the kind most generally used in this country. It is grayish or grayish-brown from adhering dirt, beset with a few fibers, and comes in compact cylindrical bundles 2 or 3 feet long.

Brazilian sarsaparilla (Rio Negro, Para, or Lisbon sarsaparilla). Considered to be the finest variety. Dark brown or blackish-brown, with a thick cortical layer and pith, and a narrow, woody zone.

(*b*) Pasty—starch in a paste.

FIG. 28.—*Smilax officinalis*—Portion of vine and rhizome.

Jamaica or red sarsaparilla is of a reddish color externally; it is said to be the richest in extractive and to contain the best quality of starch. The name bearded sarsaparilla has been applied to it, from the numerous fibers attached.

Mexican sarsaparilla is deeply wrinkled, and brownish-gray from adhering earth. The woody zone and pith are about equal in thickness, each being about half as broad as the cortical layer.

Powder.—Characteristic elements: See Part iv, Chap. I, B.

CONSTITUENTS.—The activity of sarsaparilla depends upon an acrid glucoside, **parillin,** $C_{26}H_{44}O_{10} + 2\frac{1}{2}H_2O$ (variously termed smilacin, parillinic acid, pariglin, etc.), frothing with water and otherwise closely resembling saponin in action. Kobert states that two other glucosides are present, saponin (sarsaparilla saponin), $5(C_{20}H_{32}O_{10})$-$2\frac{1}{2}H_2O$, and sarsa-saponin, $12(C_{22}H_{36}O_{10}) + H_2O$. These two latter differ from parillin in their being soluble, while parillin is insoluble. The latter constituent is the most poisonous. Ash, not exceeding 10 per cent.

Preparation of Parillin.—Exhaust with warm alcohol and concentrate the liquid to a syrup; add $1\frac{1}{2}$ times its weight of water; macerate for several days, when a yellow precipitate will form; decant and mix with alcohol, and wash on a filter with 20 per cent. alcohol.

ACTION AND USES.—The efficiency of sarsaparilla as a remedial agent has been and is still much questioned, some declaring it almost inert, others ascribing to it valuable alterative and antisyphilitic properties. Preparations from good, well-preserved specimens are perhaps beneficial remedies in scrofulous affections, and as general blood-purifiers. Dose: 30 to 60 gr. (2 to 4 Gm.).

OFFICIAL PREPARATIONS.

Fluidextractum Sarsaparillæ,..............Dose: 30 to 60 ♏ (2 to 4 mils).
  Syrupus **Sarsaparillæ Compositus** (fl'ext.
  20 per cent., with the fluidextracts of
  glycyrrhiza and senna, and the oils of
  sassafras, anise, and gaultheria),.   2 to 4 fl. dr. (8 to 15 mils).

Fluidextractum Sarsaparillæ Compositum
  (75 per cent., with glycyrrhiza, sassafras,
  and mezereum),......... $\frac{1}{2}$ to $1\frac{1}{2}$ fl. dr. (2 to 6 mils)

59. CONVALLARIA.—(C. FLORES AND C. RADIX, N.F.)

LILY OF THE VALLEY

The dried rhizome and roots and dried inflorescence of **Convalla′ria majal′is** Linné.

BOTANICAL CHARACTERISTICS.—A low, perennial, glabrous herb with slender, running root-stocks. *Leaves* 2, oblong, bright green, and shining. Scape bearing a one-sided raceme of white, bell-shaped *flowers.* *Fruit* a few-seeded red berry.

HABITAT.—North America, Europe, and Northern Asia.

DESCRIPTION OF "ROOT."—In pieces from 50 to 75 mm. (2 to 3 in.) long, and about 3 mm. ($\frac{1}{8}$ in.) thick, the upper end gnarled and wrinkled, and with the remnants of the scape and petioles attached, tapering at the small end; annulate nodes beset with a circle of eight or ten long, branching, gray rootlets; externally white, fracture white, tough, and fibrous. Odor distinct; taste sweetish, somewhat bitter and acrid. C. Flores —see N.F.

CONSTITUENTS.—Two glucosides, **convallarin,** $C_{34}H_{62}O_{11}$ (the emetocathartic principle), acrid prisms, scarcely soluble in, but foaming when shaken with, water; and **convallamarin,** $C_{23}H_{44}O_{12}$, the cardiac acting principle, a sweetish, afterward bitter, crystalline powder.

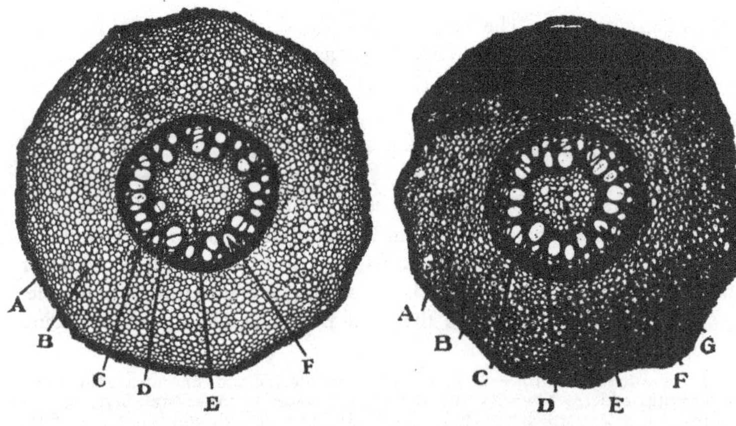

FIG. 29.—Sarsaparilla, Honduras—Cross-section of root. (18 diam.) A, Epidermis. B, Parenchyma of cortex. C, Endodermis. D, Wood parenchyma and fibers. E, Medulla. F, Water tube (Photomicrograph.)

FIG. 30.—Sarsaparilla, Jamaica—Cross-section of root. (21 diam.) A, Epidermis. B, Parenchyma of cortex. C, Endodermis. D, Wood parenchyma and fibers. E, Medulla. F, Water tube. G. Phloëm. (Photomicrograph.)

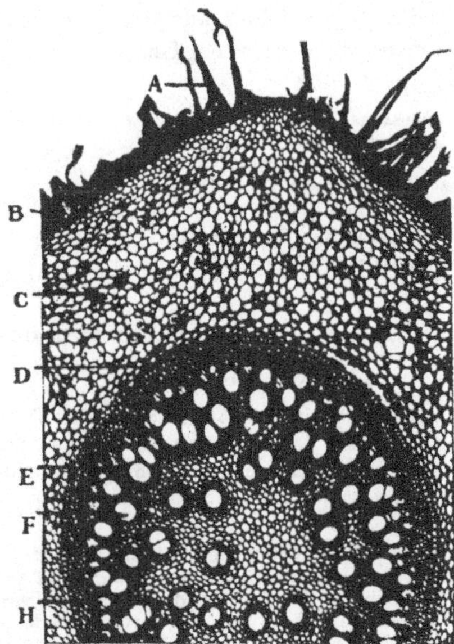

FIG. 31.—Sarsaparilla Mexican—Cross-section of root. (32 diam.) A, Root hairs. B, Cork. C, Parenchyma of cortex. D, Endodermis. E, Wood parenchyma and fibers. F, Water tube. H, Phloëm. (Photomicrograph.)

Preparation of Convallamarin.—The estimation of the value of the drug is based upon the separation of this constituent. The drug is extracted with alcohol, the tincture treated with subacetate of lead, and filtered; excess of lead removed by careful addition of H_2SO_4; filter, distil off alcohol, add water, neutralize carefully with Na_2CO_3, add solution of tannin. The precipitate of tannin compound is dissolved in 60 per cent. of alcohol, decolorized with animal charcoal, decomposed with zinc oxide. The filtrate is then evaporated to dryness.

ACTION AND USES.—Convallaria was introduced as a safer cardiac tonic than digitalis. Its absence of cumulative action was pointed out by therapeutists. "It does not disturb the stomach or cerebro-spinal functions if preparations free from convallarin are used." It is one of the most active diuretics, especially in cardiac dropsies. Dose: 5 to 30 gr. (0.3 to 2 Gm.); of convallamarin ½ to 2 gr. (0.0324 to 0.13 Gm.).

60. VERATRUM VIRIDE

AMERICAN HELLEBORE

The dried rhizome and roots of **Vera'trum vir'ide** Aiton (American).

BOTANICAL CHARACTERISTICS.—*Roots* fibrous; *stem* 2 to 7 feet high, stout and very leafy, somewhat pubescent. *Leaves* broadly oval, clasping. *Flowers* in dense panicles, yellowish-green. *Capsule* many-seeded.

HABITAT.—North America and Europe.

DESCRIPTION OF DRUG.—Usually in small pieces or large slices. When entire, obconical, from 50 to 75 mm. (2 to 3 in.) long, truncate at the base, tufted above with the inert stem-remnants and leaf-stalks, and beset on all sides with light yellowish-brown rootlets about the thickness of a knitting needle; externally blackish. A **transverse section** shows a dingy white surface dotted with darker colored dots and wavy lines within the nucleus sheath. The larger part of the tissue consists of parenchyma containing starch and calcium oxalate; nucleus sheath wavy, wood-bundles numerous. Rootlets have a thick, cortical parenchyma. Inodorous; taste bitter, very acrid, causing a tingling, benumbing sensation in the tongue. The powder is sternutatory. Starch grains of Veratrum, see Fig. 283.

Powder.—Characteristic elements: See Part iv, Chap. I, B.

CONSTITUENTS.—Veratrum viride contains the alkaloids **jervine,** $C_{26}H_{37}NO_3$ (to which the depressant action on the circulation is partly due) and **protoveratrine,** $C_{32}H_{51}NO_{11}$. This, the most important of the Veratrum alkaloids, occurs in colorless shining crystals, belonging to the monoclinic system, which are permanent in air and melt at 245° to 250°. Insoluble in water, benzene and petroleum benzin, and dissolves with difficulty in most other solvents. Chloroform and boiling 96 per cent. alcohol are its best solvents. Its alcoholic solution rapidly changes red litmus to blue. It forms a greenish colored solution with concentrated H_2SO_4 which gradually changes to blue and finally to violet.

If dissolved in diluted alcohol, it will usually be obtained in the form of a colored syrupy residue upon evaporation of the solvent, only a small portion crystallizing.

Jervine is a depressant to the respiratory center, to the vaso-motor center and to the heart muscles.

Rubijervine stimulates the cardio-inhibitory centers, but appears to depress the respiratory center.

There is no physiological relationship between protoveratrine and veratrine. The latter is the active principle of ASAGRÆA officinalis, (61).

FIG. 32.—*Veratrum viride*—Branch and rhizome.

ACTION AND USES.—The action of veratrum viride closely resembles that of aconite, being a powerful cardiac depressant and spinal paralyzant, but in addition it has a strong emetocathartic action, and consequently overdoses are less likely to prove fatal; death occurs by paralysis of the heart.   Dose: 1 to 5 gr. (0.065 to 0.3 Gm.).

OFFICIAL PREPARATIONS.

Tinctura Veratri Viridis (10 per cent.), Dose: 1 to 5 ♏ (0.065 to 0.3 mil).
Fluidextractum Veratri Viridis,   1 to 5 ♏ (0.065 to 0.3 mil).

61. SABADILLA.—Cevadilla. The seeds of **Vera'trum sabadil'la** Schlechtendal, and of **Asagræa officinalis** Lindley. *Habitat:* Mexico. They occur in commerce mixed with the fruit, which consists of three thin, papery, acuminate follicles, nearly erect, united at the base, opening by a ventral suture, and appearing like a single three-celled capsule. Each follicle contains one or two narrow, oblong or lance-linear seeds, about 6 mm. (¼ in.) long, dark brown or blackish, longitudinally shriveled, slightly winged, flat on one side, convex on the other, somewhat curved; apex pointed; the thin testa incloses a discolored, oily albumen, in the broader end of which is the small, linear embryo; inodorous; taste bitter, oily, strongly and persistently acrid.

CONSTITUENTS.—Sabadilla is the principal source of **veratrine**, $C_{37}H_{53}NO_{11}$ (Veratrina), a white powder, intensely acrid and sternutatory. The commercial veratrine is impure; it is a mixture of the alkaloid veratrine with other alkaloids extracted along with it, cevadine, $C_{32}H_{49}NO_9$, cevadilline, $C_{34}H_{53}NO_8$, sebadine, $C_{29}H_{51}NO_8$, and sabadinine.

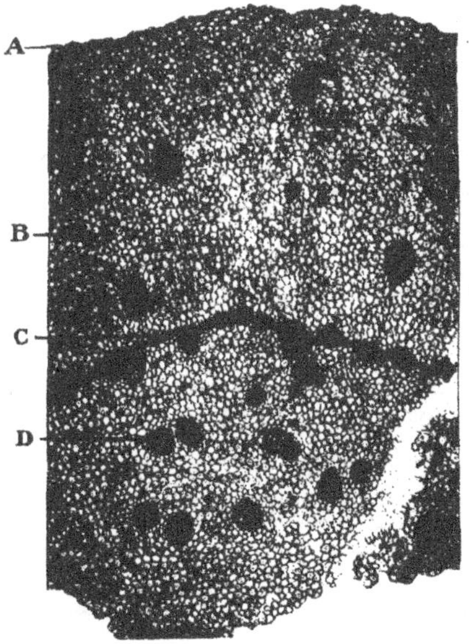

FIG. 33.—*Veratrum viride*—Cross-section of rhizome. (25 diam.) A, Cork. B, Parenchyma of ground tissue. C, Endodermis. D, Vascular bundle. (Photomicrograph.)

Preparation of Veratrine.—Remove resin and oil from alcoholic tincture by adding water *q.s.* Decompose native salt (veratrate of veratrine) in filtrate by means of KOH. Take up alkaloid with alcohol. Purify by converting into sulphate, decolorizing, and reprecipitating.

ACTION AND USES.—Sabadilla is rarely used except for the extraction of veratrine. It is a powerful irritant and is sometimes used to kill vermin in the hair.

62. POLYGONATUM.—Solomon's Seal. The rhizome of **Polygona'tum biflo'rum** Elliott, and of **P. gigante'um** Dietrich. *Habitat:* North America. A pale brownish-yellow or whitish root, annulate and jointed, each joint being surmounted by an obscurely seal-like stem-scar, which gives to the plant its name; internally whitish, spongy; inodorous; taste sweetish, mucilaginous, with an acrid, bitterish after-taste. Tonic, mucilaginous and mildly astringent; formerly much used in skin diseases and as a vulnerary, and has been recommended in gout and rheumatism. Dose: 1 to 2 dr. (4 to 8 Gm.), in fl'ext.

63. **CHAMÆLIRIUM LUTEUM** Gray. Helonias, N.F.—Helonias Dioica Pursh. False Unicorn. *Habitat:* United States. The rhizome, which is the part employed, is greenish-brown externally, closely annulate, about 25 mm. (1 in.) long, and 6 mm. (¼ in.) thick, beset on the lower side with numerous wiry rootlets; internally whitish, horny; bitter. Transverse surface is dirty white in hue and of a horny texture, and exhibits a well-defined central column occupying about one-third the diameter. It has been used as an adulterant for sanguinaria. Tonic, diuretic, anthelmintic. Dose: 15 to 60 gr. (1 to 4 Gm.).

64. **TRILLIUM,** N.F.—Birthroot. Wake-robin. The rhizome of **Trill'ium erec'tum** Linné, and other species of Trillium growing in the United States. Emmenagogue and emetic. Dose: 15 to 60 gr. (1 to 4 Gm.).

65. **ASPARAGUS.**—The rhizome of **Aspar'agus officina'lis** Linné. Cardiac sedative or palliative, diuretic, laxative. Dose: 30 to 60 gr. (2 to 4 Gm.).

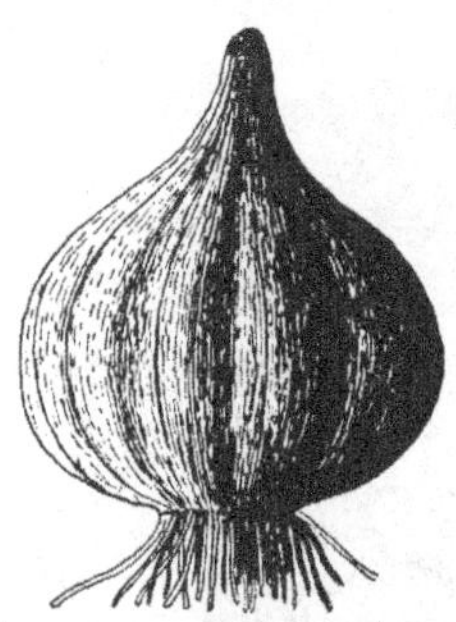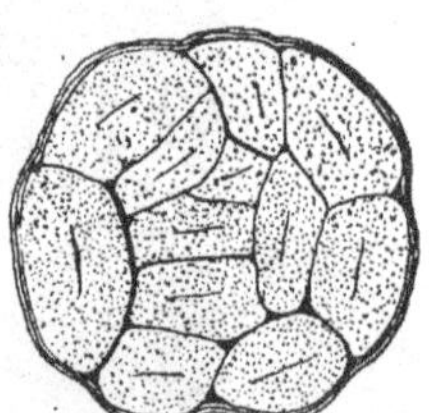

Fig. 34.—Allium.—Bulb and cross-section of same (twice natural size).

66. **ALLIUM,** N.F.—Garlic. The bulb of **Al'lium sati'vum** Linné. Official in U.S.P. 1890. A compound, subglobular bulb, flattened at the base, pointed at the apex, where several inches of the stem remains; it consists of five or six (in commercial garlic about eight) small, oblong, somewhat curved bulbs or "cloves" arranged around the central axis, each with a distinct coat, and internally whitish, moist, and fleshy; the whole bulb is inclosed by a dry, white, membranous coat, consisting of several delicate laminæ; odor pungent and disagreeable (alliaceous); taste warm, acrid. Used in the fresh state. Commercial garlic is a hybrid between *A. sativum* and *A. porrum* Linné. *Constituents:* Mucilage 35 per cent., albumen, fibrous matter, and moisture. The peculiar odor and taste are due to volatile oil, composed of the sulphide and oxide of allyl. Stimulant and expectorant, also diaphoretic and diuretic. Dose: 30 to 60 gr. (2 to 4 Gm.).

Syrupus Allii (20 per cent., with the addition of dilute acetic acid) (U.S.P. 1890)........ Dose: 1 to 2 fl. dr. (4 to 8 mils).

67. SCILLA.—Squill

SQUILLS

The inner freshly scaled bulb of the white variety **Urgin'ea maritima** (Linné) Baker, cut into slices and dried.

Botanical Characteristics.—*Bulb* semisuperficial. *Leaves* lanceolate, all radical, appearing after the flowers. *Scape* 2 to 4 feet high, terminated by a dense raceme of yellowish-green *flowers*, each one of which is accompanied by a long bract; *ovary* with 3 nectariferous glands at the apex.

Habits of Plant.—Grows in sandy places near the coast. The plant flowers in autumn, the leaves appear in the following spring. Bulb only half immersed in the soil.

Habitat.—Mediterranean shores, in dry, sandy places near the coast.

Description of Drug.—Squill comes into the market in **narrow horny** segments about 50 mm. (2 in.) long, often more or less contorted;

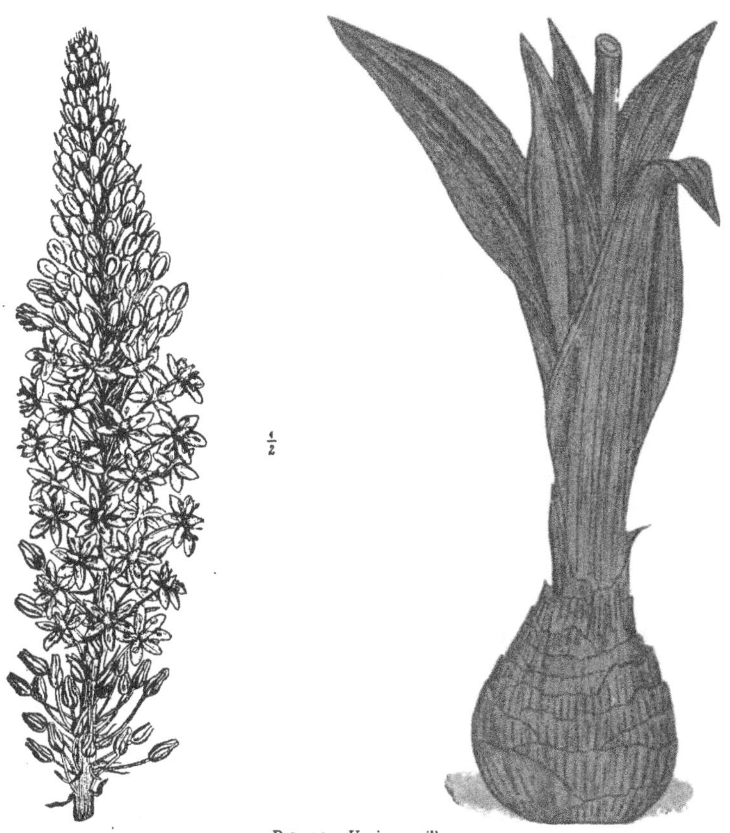

Fig. 35.—*Urginea scilla.*

color varying from **white or yellowish-white** to a reddish **tint**, slightly translucent; when dry, it is brittle and pulverizable, but by exposure to a moist atmosphere it becomes flexible. Occasionally vertical slices, sometimes adhering at the base, are met with. Odor slight; taste mucilaginous, bitter, nauseous, and acrid.

The fresh bulb is inversely pear-shaped, fleshy, varying in size from that of a man's fist to a child's head. There are two kinds,

8

differing only in color, one being entirely white, and the other reddish-brown externally, internally rose color, with white parenchyma. In preparing for market the outer scales are removed and the bulb is then sliced transversely, the central scales being also rejected as being too fleshy and mucilaginous; they lose about four-fifths of their weight in drying.

TEST.—If made into the official tincture and assayed biologically, the minimum lethal dose should not be greater than 0.006 mil of tincture or the equivalent in tincture of 0.0000005 Gm. of ouabain, for each gram of body weight of frog.

Powder.—Characteristic elements: See Part iv, Chap. I, B.

CONSTITUENTS.—Merck's analysis shows three active principles, **scillipicrin** (a bitter principle acting upon the heart), **scillitoxin**, glucoside (bitter, burning, also acting upon the heart), **scillin,** crystalline (producing numbness, vomiting, etc.), with mucilage, sugar, sinistrin, $C_6H_{10}O_5$, like dextrin, and calcium oxalate crystals. Later investigations point to the probability of the above principles being alkaloids, and they are named scillapicine, scillamarine, and scillamine respectively. Jamersted's scillain is a poisonous glucoside of a yellow color. Ash, not exceeding 8 per cent.

ACTION AND USES.—**Expectorant, diuretic,** in large doses emetic and cathartic. As an expectorant it is usually combined with tartar emetic or ipecac; as a diuretic, with stimulant expectorants. It is **very rarely given as an emetic** because of its uncertainty, having often proved fatal from its irritant action on the stomach and intestines, and by causing hypercatharsis, death occurring by arrest of the heart in systole. Dose: 1 to 3 gr. (0.065 to 0.2 Gm.).

OFFICIAL PREPARATIONS.

Acetum Scillæ (10 per cent.), ... Dose:	10 to 30 ♏ (0.6 to 2 mils).
Syrupus Scillæ (45 per cent. of the acetum), ...	30 to 60 ♏ (2 to 4 mils).
Fluidextractum Scillæ, ...	1 to 4 ♏ (0.065 to 0.25 mil).
Syrupus Scillæ Compositus (fl'ext. 8 per cent., with fl'ext. senega 8 per cent., and tartar emetic 2 per cent. or ⅛ gr. to the teaspoonful), ...	15 to 60 ♏ (1 to 4 mils); 1 to 2 fl. dr. (4 to 8 mils).

COLCHICUM.—MEADOW SAFFRON

The corm and the seed of **Colchicum Autumnale Linné.**

BOTANICAL CHARACTERISTICS.—*Col'chicum autumnal'e* Linné. *Corm* fibrous-rooted. *Leaves* about a foot long. *Flowers* several, lilac or purple, appearing in the autumn without the leaves.

HABITAT.—Europe and North Africa.

HABIT OF PLANT.—Flowers in autumn; the leaves appear in the spring.

In the latter part of spring a new corm begins to form at the expense of the old one. In September the upper portion of the flower emerges from the spathe just above ground unaccompanied with leaves. The rudimentary fruit at the base of the flower, below ground, in the following spring rises upon a stem above the surface, in the form of a 3-celled capsule. At the same time the leaves appear; so that, in fact, the leaves follow the flower, instead of preceding it. During the development of the fruit the new corm has been developing at the expense of the old parent one. It will be seen that the medicinal virtues depend upon the time of collection. Early in the spring it is too young, and late in the fall the parent corm has become exhausted by the nutriment furnished to the new plant. The proper period for collection, therefore, is said to be from June to the month of August, although April roots have been found to be of superior efficacy.

68. COLCHICI CORMUS

COLCHICUM CORM

The dried corm of **Col'chicum Autumnal'e** Linné, yielding by the official process not less than 0.35 per cent. of colchicine.

DESCRIPTION OF DRUG.—An ovoid corm about 25 to 40 mm. (1 to $1\frac{3}{5}$ in.), long flattened and deeply grooved on one side; when dried and deprived of its outer membranous covering it is wrinkled and of a brownish-gray color; internally whitish. It often comes into market in transverse starchy slices having a reniform outline, due to the lateral groove; inodorous; taste sweetish, bitter, and somewhat acrid. A very deep or large notch in the slices indicates that the corm has been partially exhausted by the offset which springs from the base.

Powder.—Microscopical elements: See Part iv, Chap. I, B.

CONSTITUENTS.—**Colchicine,** a methyl derivative of colchiceïn as will be seen from the following: Colchiceïn, $C_{15}H_9(NHCOCH_3)(OCH_3)_3$-COOH; colchicine, $C_{15}H_9(NHCOCH_3)(OCH_3)_3COOCH_3$. With mineral acids colchicine yields colchiceïne and methyl alcohol. Starch, gum, resin, fat, and sugar are also present.

Preparation of Colchicine.—Exhaust with alcohol, dilute with water, filter; add lead subacetate to precipitate coloring matter; add sodium phosphate to remove lead; precipitate solution with tannin, wash the precipitate and digest with lead oxide, dry, and dissolve out colchicine with alcohol. Occurs in whitish amorphous powder or crystals; odor saffron-like, taste bitter.

ACTION AND USES.—Colchicum is a gastro-intestinal irritant; the larger therapeutic doses sometimes cause nausea, vomiting and diarrhea. In poisoning there is intense gastro-intestinal irritation, bloody stools, irritation in the kidneys, sometimes an ascending paralysis.

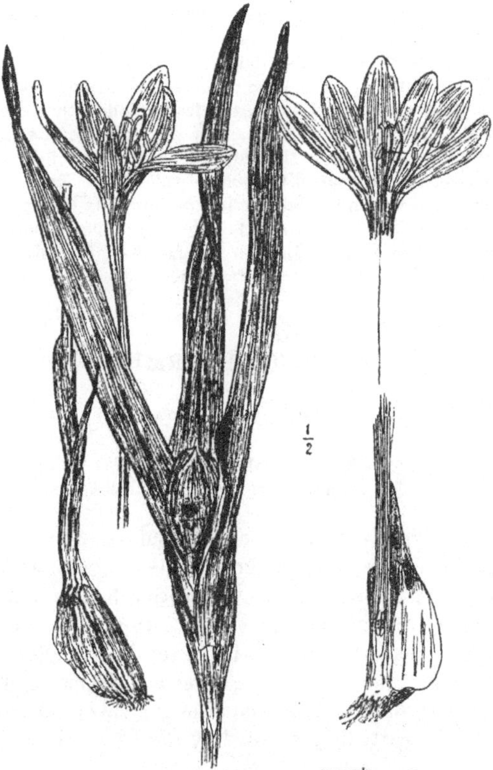

$\frac{1}{2}$

FIG. 36.—*Colchicum autumnale.*

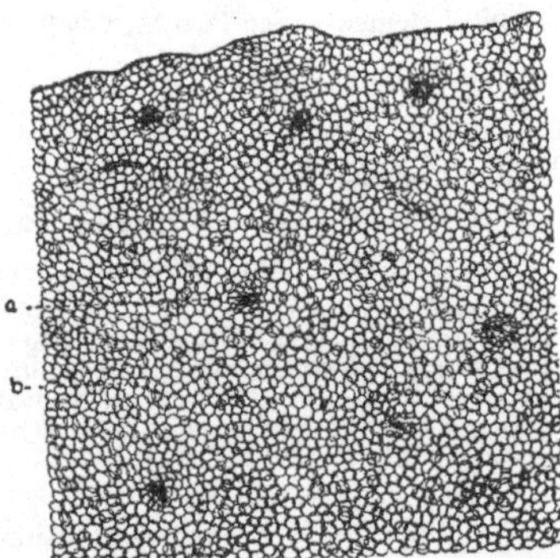

FIG. 37.—Cross-section of Colchicum root—outer portion. a, Vascular bundle. b, Parenchyma.

It is chiefly employed in gout and rheumatism, in which it is said to be very efficacious.   Dose: 2 to 8 gr. (0.13 to 0.5 Gm.).

OFFICIAL PREPARATION.

Extractum Colchici Cormi,................... Dose: 4 gr. (0.25 Gm.).

69. COLCHICI SEMEN

COLCHICUM SEED

The seed of **Col'chicum autumnal'e** Linné, yielding by the official process not less than 0.55 per cent. of colchicine.

DESCRIPTION OF DRUG.—These seeds have the same constituents and the same medicinal action as the roots, and are given in about the same doses.   They are hard, reddish-brown, subglobular, 3 mm. (⅛ in.) in diameter, somewhat pointed at the hilum and with a slight projection or caruncle on one side.   Testa thin, somewhat scurfy, closely adhering to the white albumen, which fills the entire seed and which is characterized by its **extreme hardness;** embryo small, nearly opposite the hilum; inodorous; taste oily, bitter, and somewhat acrid.   Dose: 3 gr. (0.2 Gm.).   Ash not exceeding 8 per cent.

Powder.—Characteristics: See Part iv, Chap. I, B.

OFFICIAL PREPARATIONS.

Tinctura Colchici Seminis (10 per cent.), Dose: 10 to 60 ♏ (0.6 to 4 mils).
Fluidextractum Colchici Seminis,   1 to 5 ♏ (0.065 to 0.3 mil).(

70. ALOE.—ALOES

Ger. ALOE

The inspissated juice of the leaves of **Aloe Perryi** Baker, yielding Socotrine Aloes; or **Aloe vera** Linné, yielding Curaçoa Aloes; or of **Aloe ferox** Miller, yielding Cape Aloes, U.S.P.

BOTANICAL CHARACTERISTICS.—Succulent plants with spicate inflorescence; *perianth* tubular; *style* equal in length to the stamens, or almost wanting. *Capsule* membranous, scarious; *seeds* in two rows, flattened or 3-cornered, winged.   Cape of Good Hope, etc.   The American aloe, or century plant (*Agave americana*), is a plant quite similar to the above.

COLLECTION.—The bitter, yellow, succulent portion of the leaf (which, when inspissated, constitutes the aloes of commerce) is found in thin-walled ducts near the surface.   The thick leaves are cut off near the base (March and April) and stood up in the sun to drain upon skins. Impurities are removed by skimming with a ladle, etc.   Artificial heat is sometimes used for evaporation. When of proper consistence, the evaporate is transferred to kegs, monkey skins, or boxes, and shipped by way of Bombay and Zanzibar.

Socotrine (Zanzibar) is a highly esteemed article, comes in yellowish-brown masses, sometimes soft, odor aromatic, saffron-like, never fetid or putrid and a nauseous and bitter taste, easily broken into pieces with conchoidal fracture and sharp edges, readily splinters. Does not possess any crystalline characteristics under the microscope.

Curaçoa, from the Dutch West Indies, is preferred by some to Socotrine. This variety comes in orange to blackish-brown, opaque masses, fracture surface, uneven, waxy, somewhat resinous; odor, characteristic but not aromatic as in the socotrine variety.

Cape Aloes, the South African Aloes, comes in reddish-brown or olive-black masses, usually covered with a yellowish dust, in thin fragments, transparent and of a reddish-brown color; fracture, smooth and glassy; odor, quite characteristic.

CONSTITUENTS.—The active principle of these different aloes is a bitter neutral principle having the general name of **aloin,** but slightly differing in each variety, forming possibly a homologous series; these aloes may be distinguished by their characteristic reactions with different reagents. It should be stated that the various processes of assay for aloes thus far proposed give discordant results. A small percentage of emodin is found in various varieties. Cape aloes contains 0.8 per cent. of this principle. Ash, not exceeding 4 per cent.

70 a. **ALOE BARBADENSIS.**—BARBADOES ALOES. Prepared from the leaves of **Aloe** chinensis, Steud and A. Vera, L. by boiling the juice or by making a decoction of the leaves; it is inferior to the other varieties. Its color varies, but it is usually dark **brown, approaching to** black, opaque even at the edges, and with a **dull fracture;** it is further distinguished by its **nauseous odor.** A solution of 1 part in 100,000 of distilled water produces a fine rose color on the addition of gold chloride or tincture of iodine, all the others, except Natal aloes, producing only a slow change, a feeble color, or no color whatever.

TESTS.—SOCOTRINE. The powder (dark brown) when mounted under the microscope in almond oil, shows yellowish- to reddish-brown, irregular or angular fragments; upon addition of nitric acid yields a yellowish- to reddish-brown solution.

CURAÇOA.—Powder (deep reddish brown) when treated as above shows numerous blackish brown more or less opaque and angular fragments; with nitric acid, yields a deep red liquid immediately.

CAPE.—Powder (greenish-yellow changing to light brown on aging). When treated as above and mounted under microscope it shows numerous distinctly angular bright yellow fragments. Nitric acid produces a reddish-brown liquid changing to purplish brown and finally greenish.

GENERAL TEST.—Intimately mix 1 Gm. of Aloes with 10 mils of hot water and dilute 1 mil of this mixture with 100 mils of water; a green fluorescence is produced upon the addition of an aqueous solution of sodium borate (1 in 20). Dilute 1 mil of the original aqueous mixture of Aloes with 100 mils of water, and shake it with 10 mils of benzene; upon separating the benzene solution and adding to it 5 mils of ammonia water a permanent deep rose color is produced in the lower layer, U.S.P. IX.

In the case of liquids it is best to evaporate about 10 mils, more or less, to a pasty consistency in a porcelain dish, acidulate, and extract from the dish with about 10 mils of ether by stirring with a glass rod and pouring off the ether into a test-tube. With pills or other solid material it is necessary only to powder, acidulate and extract as described. To this extract an equal volume of saturated borax solution is added, etc., U.S.D.A.

Preparation of Aloin.—From some varieties of aloes it is obtained by digesting in alcohol for twenty-four hours; then boil, filter, and set aside to crystallize. Can also be obtained by dissolving aloes (Barbadoes or Curaçoa) in acidulated boiling (HCl) water, and, when cold, resin will deposit; decant, evaporate, and set aside for two weeks, when aloin will crystallize. Shaking the crystals with acetic ether removes adhering resin. Dose: 2 to 5 gr. (0.12 to 0.32 Gm.). (See also 70 e.)

ACTION AND USES.—**Cathartic and emmenagogue.** As a cathartic aloes is slow in action but certain, having a peculiar affinity for the large intestine; it has produced beneficial effects as a cholagogue; as an emmenagogue it is extensively employed in amenorrhœa. Dose: 2 to 5 gr. (0.13 to 0.3 Gm.).

OFFICIAL PREPARATIONS.

Extractum Aloes,...................	Dose: 3 to 10 gr. (0.2 to 0.6 Gm.).
Tinctura Aloes (10 per cent. with glycyrrhiza 20 per cent.),	5 to 10 ♏ (0.3 to 0.6 mil); ½ to 4 fl. dr. (2 to 8 mils).
Tinctura Benzoini Composita (2 per cent. of aloes),..	10 to 40 ♏ (0.6 to 2.6 mils).
Extractum Colocynthidis Compositum (50 per cent.),.............	5 to 25 gr. (0.3 to 1.6 Gm.).
Pilulæ Aloes (about 2 gr. in each pill)	2 to 5 pills.
Pilulæ Rhei Compositæ (aloes 1½ gr. in each pill),...............	1 to 3 pills.

70 e. **ALOINUM.**—ALOIN (U.S.P. IX). A neutral principle from several varieties of aloes, chiefly Barbadoes aloes (yielding barbaloin), $C_{17}H_{20}O_7$, and Socotra or Zanzibar aloes (yielding socaloin), $C_{16}H_{16}O_7$, U.S.P. Nataloin, $C_{16}H_{18}O_7$, while not official, is a similar product. Minute acicular crystals, or a microcrystalline powder, yellow to yellowish-brown, of a slight odor and characteristic bitter taste. Barbaloin, soluble in 470 parts of ether; socaloin, soluble in 380 parts of ether. Both soluble in water and alcohol. It is rapidly decomposed in alkaline solution. Dose: 1 gr. (0.6 Gm.). Ash, not more than 0.5 per cent.

71. **XANTHORRHŒA.**—GUM ACAROIDES. BOTANY BAY RESIN. GRASS-TREE RESIN. A spontaneous resinous exudation from the stems of different shrubby Australian plants of the genus **Xanthorrhœa.** The yellow variety, from *X. hastilis* R. Brown, resembles gamboge in appearance; externally reddish yellow, internally a lighter yellow; odor agreeably balsamic, especially when heated, when it emits a tolu-like odor; taste balsamic, somewhat acrid. The red variety, from *X. australis* R. Brown, resembles dragon's blood in appearance, being externally deep brown-red; internally bright red; fracture glossy.

CONSTITUENTS.—Resin, benzoic and cinnamic acids, and a trace of volatile oil.

ACTION AND USES.—Resembles storax and tolu in medical properties. Dose: 8 to 30 gr. (0.5 to 2 Gm.). Chiefly used as a substitute for shellac, and for making colored varnishes.

72. **ERYTHRONIUM AMERICANUM** Smith.—ADDER'S TONGUE. DOG-TOOTH VIOLET. *Habitat:* United States. (Leaves.) Alterative. Sometimes applied as a poultice to scrofulous tumors.

HÆMODORACEÆ (Liliaceæ N.F.).—Bloodwort Family

73. **ALETRIS**, N.F.—COLIC ROOT. STARWORT. The rhizome of Alet'ris farino'sa Linné. *Habitat:* United States. Small, crooked, about the size of a quill, flattened and tufted above and beset with wiry, white rootlets below. Alcohol extracts its bitter principle. Bitter tonic, diuretic, and vermifuge; used extensively in the treatment of uterine diseases. Dòse: 10 to 30 gr. (0.6 to 2 Gm.).

DIOSCŒRACEÆ.—Yam Family

74. **DIOSCOREA,** N.F.—WILD YAM. COLIC ROOT. The rhizome of Diosco'rea villo'sa Linné. *Habitat:* United States. Expectorant, diaphoretic, antispasmodic, and a stimulant to the intestinal canal. It is a valuable remedy in bilious colic. Dose: 15 to 60 gr. (1 to 4 Gm.), in fluidextract.

IRIDEÆ.—Iris Family

Perennial herbs, with equitant, 2-ranked leaves, the flowering stem arising from a rhizome or corm.

75. **IRIS,** N.F.—IRIS VERSICOLOR. N.F. BLUE FLAG. (1890.) A horizontal, jointed rhizome, generally cut into longitudinal slices; externally brown, closely annulate from the leaf-sheath remnants, and near the broad flattened end crowded with long, simple rootlets. *Constituents:* Acrid resin 25 per cent., fixed oil, starch, gum, tannin, sugar, iridin, and indications of a brownish, viscid, amorphous alkaloid. *Preparation of Iridin:* Obtained by precipitating hot alkaline solution by an acid. The eclectic method of preparation is to precipitate concentrated alcoholic tincture with water; mix dried precipitate with equal quantity of licorice root. Cholagogue, cathartic and alterative. Dose: 10 to 30 gr. (0.6 to 2 Gm.).
Fluidextractum Iridis (U.S.P. 1890), . . . Dose: 10 to 30 ℥ (0.6 to 2 mils).
Extractum Iridis (U.S.P. 1890), ... 1 to 3 gr. (0.065 to 0.2 Gm.).

76. **IRIS FLORENTINA.**—ORRIS ROOT. The rhizome of I'ris florenti'na, Iris pallida, and Iris germanica Linné. *Habitat:* Northern Italy. In club-shaped pieces or joints, from 75 to 125 mm. (3 to 5 in.) in length, a broad depression or scar terminating the broad end. Externally white, peeled; fracture short, mealy, faintly yellowish-white; odor violet-like; taste mealy, bitterish, and somewhat acrid. It contains iridin, irone, $C_{13}H_{20}O$, a ketone of violet odor, acrid resin, starch, mucilage, bitter extractive, and orris camphor, consisting of a fat impregnated with volatile oil. Cathartic, diuretic. Dose: 5 to 15 gr. (0.3 to 1 Gm.). Chiefly used in tooth-powders and perfumes. (Highly magnified starch grains of Iris, see Fig. 286.)

77. **CROCUS,** N.F.—SAFFRON. The stigmas of Cro'cus sati'vus Linné. Asia Minor and Greece; cultivated for market in Spain, France, and other temperate countries of Europe; also cultivated in the southeastern counties of Pennsylvania. Commercial saffron is mostly of French or Spanish origin; a product of the Cape of Good Hope known as Cape saffron, resembling the genuine in odor, is a flower of a small plant belonging to the Scrophulariaceæ ("Pharm. Journal," VI, 462, 1865). "American saffron" consists usually of safflower. The commercial or "hay saffron" consists of orange-brown stigmas, separate, or united (three) to the top of the style, about 30 mm. (1⅕ in.) long, almost filiform, enlarging toward the top, which is toothed; their edges are rolled in, giving them a flattish-tubular appearance; crisp and

somewhat elastic; orange-brown; odor peculiar, aromatic; taste pungent, bitterish. In selecting saffron the above characteristics should be borne in mind; the drug should not emit an offensive smell when thrown upon live coals. If it has a musty flavor or a black, yellowish, or whitish color, it should be rejected. If the cake saffron be purchased, those should be selected which are close, tough, and firm in tearing. Owing to its high price, saffron offers a great field for adulteration, which is done in various ways. The commonest is to mix the stigmas with the styles, which may be distinguished by their

FIG. 38.—*Crocus sativus*—Plant, flower, and stigma.

lighter color. Old saffron and that deprived of its coloring matter leaves an oily stain when pressed between paper, due to the fixed oil with which they are covered to conceal their false nature. The florets of other flowers, as calendula, carthamus, and arnica, may be detected by dropping them into water, when their characteristic forms will come out. Mineral adulterants, which are sometimes found to the extent of 20 per cent., will subside to the bottom when the suspected drug is placed on water; carbonate of lime will effervesce when a drop of acid is placed on the suspected drug. *Constituents:* An orange-red coloring matter, which gives to saffron its chief value; a glucoside, usually called crocin, $C_{44}H_{70}O_{28}$, but formerly called polychroit, because of the many different colors it gives with acids; crocetin, $C_{34}H_{46}O_9$, and a volatile oil, $C_{10}H_{16}$, upon which its medicinal virtues depend. Saffron has fallen into almost complete disuse among practitioners of the United States and Great

Britain, but it is occasionally used in domestic practice in the form of a tea, to promote eruption in measles, scarlet fever, and other exanthematous diseases.   Dose: 5 to 30 gr. (0.3 to 2 Gm.).   Chiefly used for coloring preparations.

Tinctura Croci (10 per cent.).   (U.S.P. 1890),....Dose: 1 to 2 dr. (4 to 8 mils).

SCITAMINEÆ.—Banana Family

A tropical order, many species of which have a pungent principle in their rhizome or root; other species yield an abundance of starch and coloring matter.

Synopsis of Drugs from the Scitamineæ

A. *Rhizomes.*
   ZINGIBER, 78.
   *Galanga, 79
   *Zedoaria, 80.
   Curcuma, 81.

B. *Fruit.*
   CARDAMOMUM, 82.
C. *Seeds.*
   Granum Paradisi, 83.

78. ZINGIBER.—Ginger

GINGER

The dried rhizome of **Zin'giber officina'le** Roscoe (Fam. Zingiberaceæ, U.S.P. 1900), deprived of periderm.

BOTANICAL CHARACTERISTICS.—*Root-stock* biennial, creeping; *stem* 3 to 4 feet high; *leaves* linear-lanceolate, smooth. *Spikes* radical, each flower bracteate; *lip* 3-lobed; *stamens* 3, 2 abortive; *capsule* 3-celled, 3-valved.

HABITAT.—Africa, Hindustan; cultivated in the West Indies and tropics.

DESCRIPTION OF DRUG.—A flattened rhizome, from 25 to 100 mm. (1 to 4 in.) long, with large club-shaped lobes on one side; deprived of the corky layer by scraping, and bleached, leaving a pale buff-colored, striate surface, sometimes covered with a white powder of calcium carbonate from being steeped in milk of lime; fracture mealy and rather fibrous, showing a whitish interior dotted with numerous small, orange-colored oil and resin-cells. Transverse sections show a parenchymatous meditullium containing scattered resin-cells and numerous fibrovascular bundles, which latter are less abundant outside of the nuclear sheath. The central cylinder is quite broad as compared with the cortical layer; aromatic and spicy; pungent.

VARIETIES.—The above-described root, Jamaica ginger or white ginger, (deprived of corky layer), is the finest variety, yielding 5 per cent. oleoresin. African ginger is shorter, with broadly linear or oblong lobes, and is not deprived of its light brown, corky layer. Chinese ginger is also a coated rhizome, but has short stumpy lobes. East India ginger is scraped on the flat side, leaving the cork remaining on

the edges. It yields 8 per cent. of oleoresin. Green ginger consists
of the rhizome sent to market without drying; black ginger, of the
rhizome steeped in boiling water before drying, after which it has a
black, horny structure. The preserved ginger is an article on the

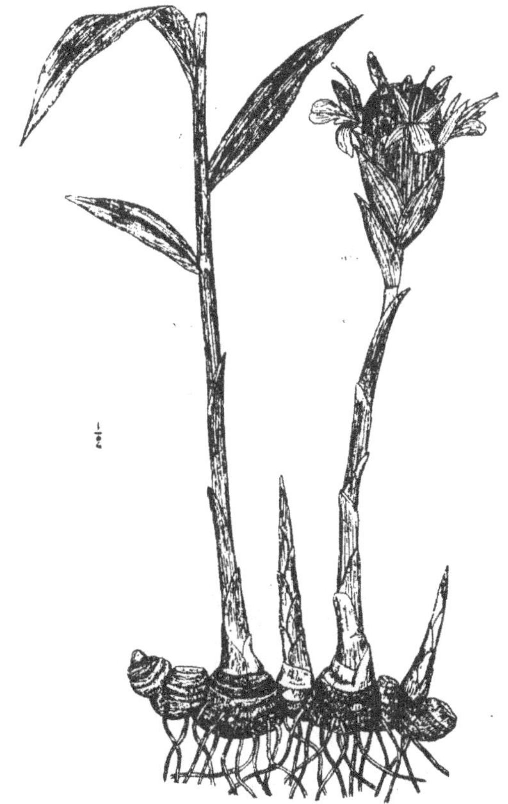

FIG. 39.—*Zingiber officinale.*

market which consists of soft, yellowish-brown pieces, obtained by
steeping the fresh ginger in hot syrup and carefully bottling.

Powder.—Characteristic elements: See Part iv, Chap. I, B.

CONSTITUENTS.—**Volatile oil,** 1 to 2 per cent. (consisting of camphene
and phellandrene), and gingerol, the former probably giving to it
its aromatic properties, and a resinous, viscid, inodorous extractive
its hot, pungent taste; also resin, starch (20 per cent.), and mucilage.
Jamaica ginger yields about 5 per cent. of oleoresin, the East India
ginger about 8 per cent.   Ash, not exceeding 8 per cent.

ACTION AND USES.—**Stimulant,** carminative, and stomachic, often used as an adjuvant to bitter, tonic preparations. When chewed it stimulates the secretion of the saliva and if snuffed into the nostrils in powder it occasions sneezing. It relieves abdominal cramp due to

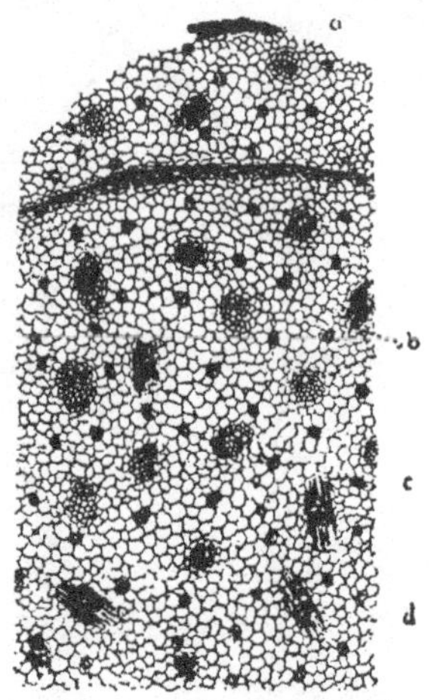

FIG. 40.—Cross-section of ginger, outer portion of peeled rhizome. *a.* Fragment of cork. *b.* Fibro-vascular bundles. *c.* Resin cell. *d.* Parenchyma. (15 diam.)

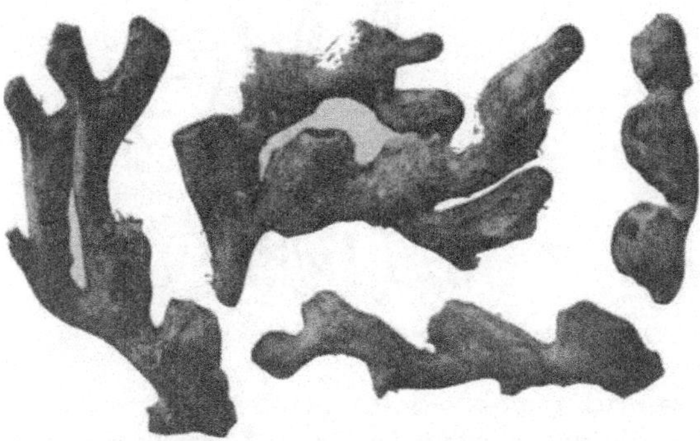

FIG. 41.—Zingiber—Jamaica, showing lobes of rhizome, peeled and unlimed. (½ natural size.) (Photograph.)

flatus and is useful to diarrhea mixtures, bitter tonics, and to preparations given to correct indigestion. As a rubifacient it is made into a cataplasm either alone or in combination with other species for the relief of colic, headache, myalgia, neuralgia, etc. Dose: 8 to 30 gr. (0.5 to 2 Gm.).

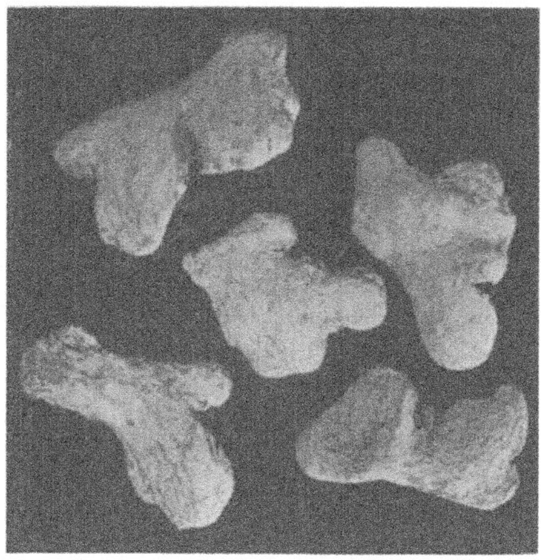

Fig. 42.—Zingiber—Chinese limed and peeled. (⅔ natural size.) (Photograph.)

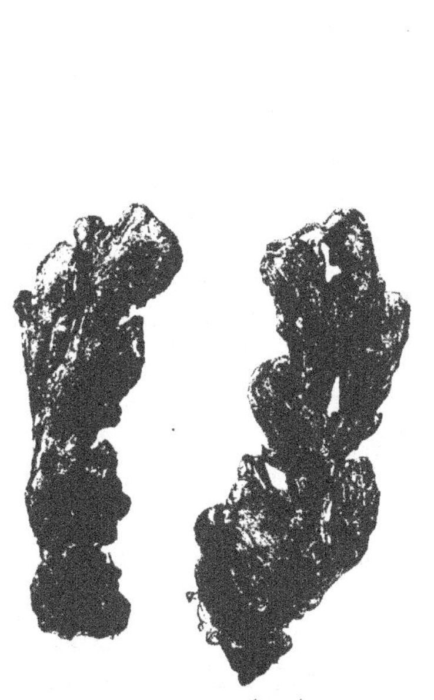

Fig. 43.—Zingiber—African, half peeled. (⅔ natural size.) (Photograph.)

Fig. 44.—*Curcuma longa.*

OFFICIAL PREPARATIONS.

Tinctura Zingiberis (20 per cent.),....Dose: 15 to 60 ♏ (1 to 4 mils).
Fluidextractum Zingiberis,.   8 to 30 ♏ (0.5 to 2 mils).
Fluidextractum Aromaticum,   8 to 30 ♏ (0.5 to 2 mils).
  Syrupus Zingiberis (3 per cent.),....   2 to 6 fl. dr. (8 to 24 mils).
Pulvis Aromaticum (35 per cent.),....   10 to 30 gr. (0.6 to 2 Gm.).
Pulvis Rhei Compositus (10 per cent.
  of ginger),............   1 to 3 dr. (4 to 12 Gm.).
Oleoresina Zingiberis,....   ½ to 2 ♏ (0.0324 to 0.13 mil).

79. **GALANGA.**—GALANGAL. N.F. The rhizome of **Alpi′nia officina′rum** Hance. *Habitat:* China. **Reddish-brown,** cylindrical, branched, about 100 mm. (4 in.) long, and about the thickness of the thumb, marked with circular or diagonally **annular, whitish rings,** the remains of former leaf-sheaths; internally orange-brown, dotted with numerous brownish-yellow resin-cells; **odor and taste ginger-like.** Small galangal, or galanga minor, does not exceed the little finger in size, is darker in color, and has a stronger taste and odor. Like ginger, their activity is due to a volatile oil and a resin, and they have the same medicinal action. (Highly magnified starch grains, see Part iv.)

80. **ZEDOARIA, N.F.**—ZEDOARY. The rhizome of **Cur′cuma zedoar′ia** Roxburgh. There are two kinds, the long and the round, both coming from the East Indies. Externally grayish-white, internally brown, hard, compact; odor aromatic; taste spicy, camphoraceous. The drug comes into market in slices and disks. It is used as an aromatic stimulant, and possesses properties similar to but inferior to those of ginger. Dose: 10 to 30 gr. (0.6 to 2 Gm.).

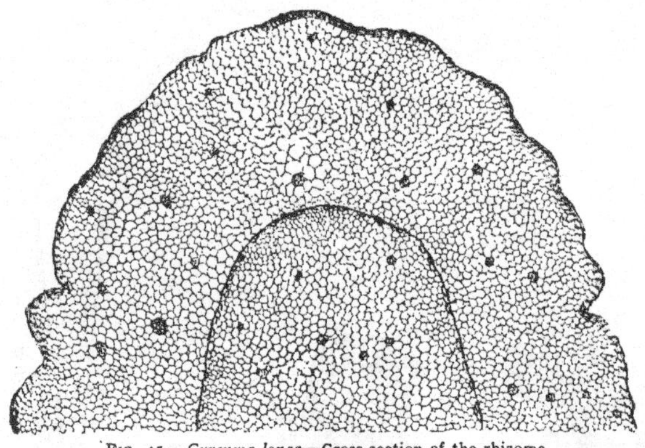

FIG. 45.—*Curcuma longa*—Cross-section of the rhizome.

81. **CURCUMA.**—TURMERIC. The rhizome of **Cur′cuma lon′ga** (Linné). *Habitat:* Southern Asia and East Indies, the best coming from China. Cylindrical pieces (*Curcuma longa*), about as thick, but not so long, as the finger, tuberculated and somewhat contorted; externally yellowish-gray, internally deep orange-yellow, with a darkish ring marking the circular nucleus sheath; hard, compact, breaking with a glossy, waxy fracture; odor feeble but peculiar; taste aromatic, pungent, bitter.

   Curcuma rotunda is round or oval, about the size of a pigeon's egg, or larger, marked externally with annular rings. Both forms of root are derived from the same plant, one being a modification of the other.

CONSTITUENTS.—Volatile oil, a viscid oil, a pungent resin, pasty starch, and a peculiar yellow coloring matter called curcumin, **turned brownish by alkalies,** becoming violet on drying; with boracic acid it produces an orange tint, changed to blue by alkaline solutions. Stimulant and tonic, but rarely used in that way, except in India, where it is used as a condiment, like ginger. It is used in pharmacy for **coloring ointments and tinctures,** and for preparing turmeric test-paper.

Preparation of Curcumin.—Obtained pure after removing the oil by exhausting the residual powder with ether, evaporating and recrystallizing from alcohol. Crystals yellow, with a vanilla-like odor.

82. CARDAMOMI SEMEN.—Cardamom Seed

CARDAMOM

The dried seed recently removed from capsules of **Eletta'ria Cardamomum** (White et Maton). (Fam. Zingiberaceæ.)

Botanical Characteristics.—Rhizome fleshy-fibrous. *Stem* 6 to 9 feet high. *Leaves* lanceolate, pubescent above, silky beneath. *Flowers* borne on scapes; *anthers* 2-lobed. *Capsules* 3-celled, 3-valved.

Habitat.—Malabar; cultivated in India.

Description of Fruit.—**Triangular-ovate,** from 12 to 37 mm. (½ to 1½ in.) long, with flat, ribbed sides, in the center of which are longitudinal furrows marking the positions of the cell-partitions; valves three, opening longitudinally at the rounded angles; central placenta. **The pericarp** is of a yellowish or buff color, leathery, and nearly tasteless. **Internally** 3-celled, each containing from 5 to 7 reddish-brown, irregularly angular, rugose seeds, having an aromatic odor and taste; these seeds form 75 per cent. of the fruit in the best varieties.

The inert pericarp is rejected in making preparations. The seeds are mostly agglutinated in groups of from 5 to 7, the individual seeds are oblong ovoid in shape 3- or irregularly 4-sided convex on the dorsal surface, longitudinally grooved on one side, about 3 to 4 mm. in length, externally reddish-gray to brown,

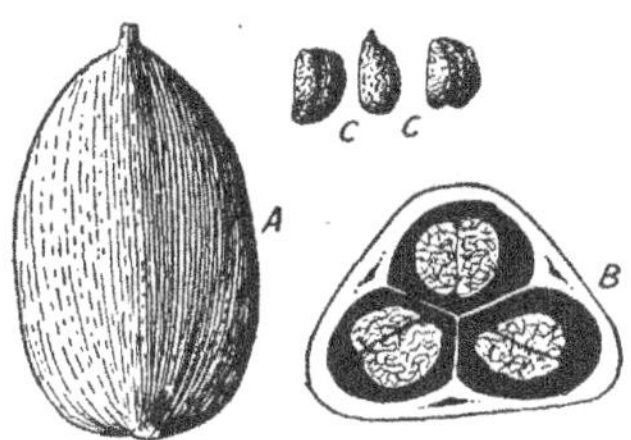

Fig. 46.—Cardamomum. *A.* Whole fruit enlarged. *B.* Cross-section. *C.* Seed.

coarsely tubercled, and often with adhering portions of the membraneous aril moderately hard but easily crushed, odor aromatic, taste aromatic and pungent. Seeds may be kept in the capsules until wanted for use.

Powder.—Greenish-brown, consisting chiefly of coarse angular fragments of cells of the reserve layers and seed coat.

Endosperm and perisperm of seeds filled with compound starch grains fragments of seed coat with dark brown stone cells, which are polygonal in surface view and about 0.020 in diameter.

Fragments of spiral tracheæ with accompanying bast fibers which are very slightly if at all lignified, relatively few or absent.

VARIETIES.—Malabar, the choicest, plump, light, and buff color; Aleppo, mostly short and greenish. These two kinds are mostly imported into the United States. Besides these, there are Madras cardamom, oblong, alternated above, pale in color; Ceylon, from *Elettaria major*, $1\frac{3}{5}$ in. (40 mm.) long, triangular, prolonged into a beak, dark gray and brown. This latter variety is of inferior flavor. Round cardamom, from *Amomum cardamomum* of Siam and Java, and *A. globosum* and *A. aromaticum* (Bengal cardamom) are known; also winged Java cardamom, from *A. maximum*. This latter variety has from 9 to 12 wings from the base of the apex, but the Bengal has 9 wings near the apex.

Powder.—Pale brownish-gray (of seed). Characteristic elements: (Powder of whole fruit.) Parenchyma of pericarp, thin-walled with prismatic calcium oxalate crystals; the pericarp valueless as an aromatic; parenchyma of endosperm with oil, proteid granules and starch, spherical or angular, simple or compound (1 to 4 μ in diam.); seed coat with dark brown stone cells (15 to 20 μ in diam.), inner wall thickened; pericarp has bast fibers very slightly lignified; outer epidermal cells elongated (20 to 30 μ in diam.), tangential walls thickened; oil cells with suberized walls; Ceylon differs from Malabar in containing trichomes and in the measurements of the elements.

CONSTITUENTS.—The pericarp is almost inert, consisting chiefly of lignin. The seeds abound in a fixed oil (10 per cent.) and a **volatile oil** (4.6 per cent.), consisting of terpene, diterpene, and terpineol, with rhombohedric masses of albuminous matter, gum. Ash, not exceeding 8 per cent.

Powder (of seed).—Characteristics: See Part iv, Chap. I, B.

ACTION AND USES.—Aromatic, stimulant, stomachic, and carminative, used principally in this country as an **adjuvant**. Dose: 5 to 15 gr. (0.3 to 1 Gm.).

OFFICIAL PREPARATIONS.

Tinctura Cardamomi (20 per cent.),...Dose: 1 to 2 fl. dr. (4 to 8 mils).
Tinctura Cardamomi Composita (2.5
  per cent., with cassia cinnamon, cara-
  way, and cochineal),..............   1 to 3 fl. dr. (4 to 12 mils).

83. **GRANUM PARADISI.**—GRAINS OF PARADISE. GUINEA GRAINS. The seeds of **Amo′mum gra′na paradi′si** and **Amo′mum melegue′ta.** Small, roundish, somewhat cuneiform; externally finely warty, reddish-brown; internally white. When rubbed, they emit a feebly aromatic odor; taste hot and peppery. Action somewhat resembles pepper.

ORCHIDACEÆ.—Orchis Family

Perennial herbs, sometimes parasitic, with perfect, irregular, and usually showy *flowers*, the *stigma* having a broad, glutinous surface (except in Cypripedium); the (usually single) *anther* is sessile on the style; it is 2-celled, each cell containing one or more waxy masses of pollen, pollinia (Fig. 115).

Synopsis of Drugs from the Orchidaceæ

A. *Rhizomes.*
    Cypripedium, 84.
    Corallorrhiza, 85.

B. *Tuber.*
    Salep, 86.
C. *Fruit.*
    *Vanilla, 87.

FIG. 47.—Cypripedium—Leaf and flower. (Photograph.)

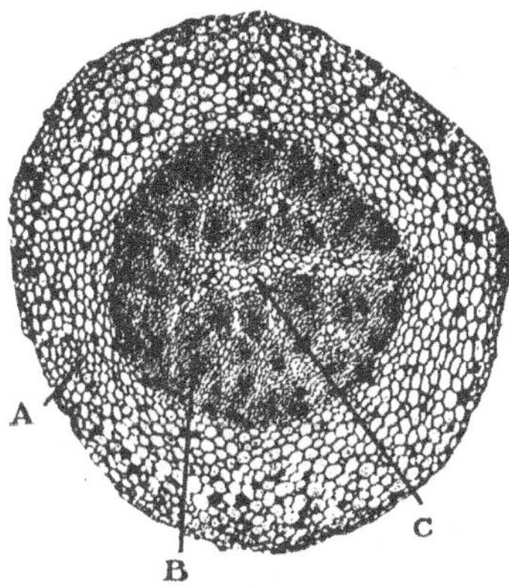

FIG. 48.—Cypripedium—Cross-section of rhizome. (20 diam.) A, Cortex. B, Vascular bundle. C, Ground tissue. (Photomicrograph.)

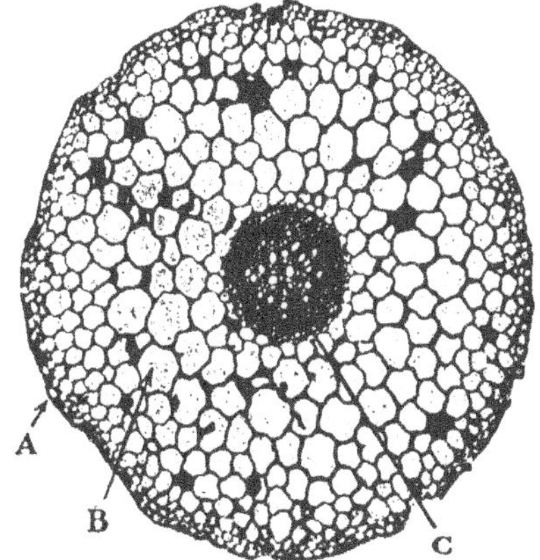

FIG. 49.—Cypripedium—Cross-section of rootlet. (30 diam.) A, Epidermis. B, Parenchyma of cortex. C, Xylem. (Photomicrograph.)

84. CYPRIPEDIUM.—CYPRIPEDIUM, N.F.

LADIES' SLIPPER ROOT

The dried rhizome and roots of **Cypripe′dium hirsu′tum** Miller, and **Cypripedium pubes′cens** Wildenow and **Cypripedium parviflo′rum** Salisbury.

HABITAT.—North America, in swampy regions.

DESCRIPTION OF DRUG.—A horizontal, somewhat curved rhizome, about the thickness of a quill and 100 mm. (4 in.) or less in length, of a dark brown or

9

light orange-brown color; on the upper side it is closely covered with **deeply concave stem-scars** about the width of the rhizome, and on the lower side with **smooth, simple, wavy rootlets,** abruptly descending, varying in length from 100 to 500 mm. (4 to 20 in.); cortical parenchyma thick, wood-bundles and nucleus sheath indistinct; fracture of rhizome short, of roots fibrous; odor somewhat valerian-like, diminishing with age; taste sweetish, bitter, somewhat pungent at the last.

Cypripedium parviflorum has the rhizome bent two or three times, almost at right angles, and is of a brighter orange-brown color; the rootlets are shorter and less wavy.

CONSTITUENTS.—Volatile oil (a trace), a volatile acid, resins, tannin, sugar, starch, and fixed oil. The active principle has not yet been isolated, but the virtues of the drug are supposed to reside in the volatile oil and a bitter principle (probably a glucoside). Ash, not more than 12 per cent.

ACTION AND USES.—Diaphoretic, **nerve stimulant, and antispasmodic,** less powerful than valerian. It is valuable as a substitute for opium in the treatment of children. Dose: 8 to 30 gr. (0.5 to 2 Gm.).

85. **CORALLORRHIZA ODONTORRHIZA** Nuttall.—CRAWLEY. CORAL ROOT. The rhizome of a parasitic, leafless herb growing throughout the United States east of the Mississippi. "A prompt and powerful diaphoretic, with sedative properties. A combination with blue cohosh is a good emmenagogue." Dose: 15 to 30 gr. (1 to 2 Gm.).

86. **SALEP.**—SALEP. The tubers of **Or'chis mas'cula** and **Orchis morio** Linné. *Habitat:* Europe. Frequently comes in powder. It is a farinaceous, gummy substance, somewhat analogous to tragacanth in composition. Demulcent and nutritive.

87. VANILLA, N.F.

VANILLA

The full-grown but immature fruit of **Vanil'la planifo'lia** Andrews, cured in the customary manner. Yielding to dilute alcohol 12 per cent. of extractive. Ash, 6 per cent.

BOTANICAL CHARACTERISTICS.—A fleshy, climbing orchid with long, smooth, dark green stem sending out at the nodes aërial rootlets which fasten it to the tree or other support. *Leaves* sessile, fleshy, tough, veinless. *Flowers* pale yellowish, in loose axillary racemes. *Fruit* a pod.

SOURCE AND VARIETIES.—Of the genus Vanilla there are some twenty-three species recognized, a few only of which are used and cured as the commercial vanilla, a product of cultivation mainly. The fruit is chiefly cultivated in Mexico and Bourbon, and to a greater or less extent in the West Indies, Java, Mauritius, Ceylon, the Fijis, and Straits Settlements.

COLLECTION AND CURING.—The fruits are collected before they are ripe, just as they begin to turn yellow, then placed between woolen blankets in a sweating-box and left there for thirty-six hours, being afterward exposed to the noonday sun just long enough to dry off the perspiration which was thus produced. This process is repeated until the fruit has a uniform blackish chocolate color, until the curer determines the process finished and the fruit ready for packing.

ARTIFICIAL POLLENIZATION OR FECUNDATION.—In Mexico and Guinea fertilization is left to natural influences, as by insects and by the wind; but in Reunion (Bourbon) artificial fecundation is resorted to because there is a total lack of the necessary insect life. Pollenization consists in holding the flower with the thumb and finger of the left hand,

FIG. 50.—*Vanilla planifolia*—Branch showing leaf and flowers.

and, with a splinter of wood or bamboo held in the right hand, raising up the labellum between the pollen and the stigma, then with the forefinger of the left hand pressing the former down upon the latter. Transversely are seen several rib-like processes extending inward. These are the placentæ which support the numerous minute seeds. Projecting into the central cavity and borne on the inner cell-wall are unicellular papillose hairs; these secrete oil and resin, which elaborate vanillin.

DESCRIPTION OF DRUG.—Linear, somewhat triangularly compressed pods from 150 to 250 mm. (6 to 10 in.) long, 8 mm. (⅓ in.) thick, attenuated at the base, where they are curved more or less into a hook; flexible; externally finely furrowed longitudinally, dark brown, shining, unctuous, often covered with an incrustation of fine, acicular crystals of vanillin;* they split lengthwise into two unequal valves, showing numerous minute, lenticular, glossy black seeds imbedded in a black, oily pulp, which also contains shining, acicular crystals. The peculiar, strong, aromatic odor resides chiefly in the pulp; taste warm, aromatic, sweetish.

CONSTITUENTS.—The aroma of vanilla, chiefly depends upon a crystalline principle, **Vanillin** 87 a (U.S.P.) ($C_6H_3.OH.OCH_3.CHO$, m-methoxy p-oxybenzaldehyde), which does not exist in the green pods, but is

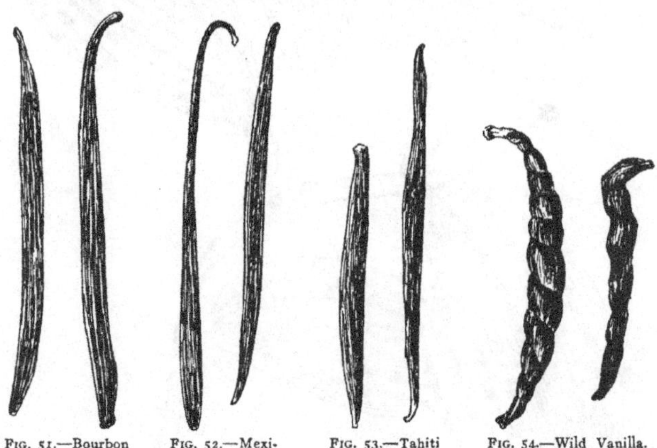

FIG. 51.—Bourbon Vanilla.   FIG. 52.—Mexican Vanilla.   FIG. 53.—Tahiti Vanilla.   FIG. 54.—Wild Vanilla.

developed during the process of curing, and forms the frosty inflorescence upon their surface. It is found in many other plants, being first made artificially from coniferin, a glucoside found in the cambium of the pine; it is now largely made from oil of cloves by reactions upon the eugenol.

Preparation of Vanillin.—Treat alcoholic extract with ether, evaporate, and treat residue with boiling water, when needles of vanillin are deposited. Prepared artificially on large scale from coniferin, $C_{10}H_{22}O_8$ + $2H_2O$, a compound occurring in the sap of the cambium in the Coniferæ. This is first fermented and finally oxidized.

ACTION AND USES.—Carminative, stimulant, aphrodisiac, anti-hysteric.

* An adulteration of benzoic acid crystals can be detected by the latter having rhomboidal form as well as characteristic reactions.

Dose: 5 to 30 gr. (0.3 to 2 Gm.). It is rarely employed medicinally, being principally used as a **flavor.**

87 a. **VANILLINUM** (U.S.P. IX) is described as methylprotocatechnic aldehyde. Should contain not more than 0.05 per cent. of ash.

PIPERACEÆ

Herbaceous or shrubby tropical plants, with jointed stems, and *flowers*, destitute of floral envelopes, arranged in spikes or spicate racemes. The entire order possesses pungent and aromatic properties, due to the presence of volatile oil and resin.

Synopsis of Drugs from the Piperaceæ

A. *Fruits.*
    CUBEBA, 88.
    PIPER, 89.
    Piper Album, 90.
    Piper Longum, 91.

B. *Volatile Oil.*
    OLEUM CUBEBÆ,
    88 a.
C. *Leaves.*
    *Matico, 92.

D. *Roots.*
    Yerba Mansa, 93.
    Jambu Assu, 94.
    Methysticum, 95.

88. CUBEBA.—Cubeb

CUBEBS

The dried unripe but fully grown fruit of Pi'per cube'ba Linné filius.

Botanical Characteristics.—*Stem* climbing, rooting at the joints. *Leaves* 4 to 7 inches long, petiolate, oblong to ovate. *Flowers* diœcious, in spikes opposite the leaves. *Fruit* larger than black pepper, globose, on pedicels about ½ of an inch long.

Source.—Java, Sumatra, Borneo; also in West Indies. It grows extensively in coffee plantations or in grounds reserved for that purpose. The fruit after gathering is sent to Java, thence to Singapore, where it enters the market.

Description of Drug.—The official cubebs are picked while green, becoming **brown or black** and **reticulately wrinkled** on drying; they are about the size of a pea, **still attached to the slender stalk;** this stalk is longer than the fruit, and is formed by the downward lengthening of the pericarp, continuous with the prominent raised ridges on the surface of the berry. The shell or pericarp is hard, almost ligneous, and incloses a central cavity or a black, shrunken seed; **odor and taste** aromatic, spicy, pungent.

Powder.—Characteristic elements: See Part iv, Chap. I, B.

Adulterations.—Frequently adulterated with stems. Black pepper and other piperaceous fruits are often met with, but these are rarely intentional adulterants. *Rhamnus catharticus* (buckthorn berries) is sometimes used as an adulterant and may be readily distinguished by its four-seed fruit.

Constituents.—Volatile oil (5 to 18 per cent.), cubebin, $C_{10}H_{10}O_8$, cubebic acid, $C_{14}H_{16}O_4$, resin, fat, wax, and starch. Cubebin is a colorless principle and forms the greater portion of the sediment which deposits from the official oleoresin on standing. Cubebic acid is the principle upon which depends the diuretic action of cubebs; the volatile oil is stimulating. Ash, not exceeding 8 per cent.

FIG. 55.—*Piper cubeba*—Fruiting branch and fruit (enlarged).

Preparation of Cubebin.—Precipitates from oleoresin, upon standing, in white, crystalline form; inodorous and bitter.

Action and Uses.—Stimulant, carminative, and diuretic. Its especial action is on the mucous membrane of the genito-urinary tract. Dose: 15 gr. to 2 dr. (1 to 8 Gm.).

Official Preparations.

Oleoresina Cubebæ,......................Dose: 5 to 30 ♏ (0.3 to 2 mils).
 Trochisci Cubebæ (⅗ gr. of oleoresin in
  each troche),...................... 1 or 2 troches.

88 a. OLEUM CUBEBÆ, U. S.—OIL OF CUBEB. A greenish volatile oil, becoming yellowish with age (colorless upon rectification), having the odor and taste of cubeb, but less pungent, and a warm, camphoraceous, aromatic taste. It has about the consistence of almond oil and is lighter than water. It is said not to preëxist in the fruit, but to be formed by the prolonged action of the air. The oil consists of dipentene, cadinene, and cubeb camphor. Dose: 5 to 15 ♏ (0.3 to 1 mil).

89. PIPER.—PEPPER

BLACK PEPPER

The dried unripe fruit of **Pi'per ni'grum** Linné.

BOTANICAL CHARACTERISTICS.—Aromatic shrub, with knotted, pointed branches *Leaves* alternate, entire. *Flowers* spicate, perfect, each supported by a scale. *Berry* 1-seeded.

Pepper should not yield less than 6 per cent. of non-volatile ether extract, not less than 25 per cent. of starch.

The yield of total ash should not exceed 7 per cent. The amount of ash insoluble in diluted HCl should not exceed 2 per cent.

Not more than 2 per cent. of stems and foreign matter may be included.

HABITAT.—India and Cochin-China; cultivated in the East Indies.

DESCRIPTION OF DRUG.—**A black, reticulated, berry-like, fruit,** resembling cubebs in size and general appearance, except that it is **destitute of the foot-stalk.** It is hollow inside and contains a single, small, undeveloped seed. Odor aromatic and sternutatory; **taste** sharp, burning, and acrid.

Powder.—Characteristic elements: see Part iv, Chap. I, B.

CONSTITUENTS.—The aromatic and stimulant properties of pepper depend upon its **volatile oil,** $C_{10}H_{16}$, but the pungent taste and medicinal activity are mainly due to a **soft, pungent resin, chavicin;** a neutral principle, **piperine,** is also present which is decomposed by alkalies into piperidine, $C_5H_{11}N$, and piperic acid, $C_{12}H_{10}O_4$. The latter yields piperinal (heliotropine) by oxidation.

Preparation of Piperine.—It is deposited almost pure from freshly made oleoresin; usually has pungent resin associated with it, giving it a biting taste. It is in pale yellow prismatic crystals; odorless, with sharp, bitter taste.

ACTION AND USES.—Stimulant and carminative, its principal use being as a condiment. The principle piperine has been used as an antiperiodic. Dose of pepper: 5 to 20 gr. (0.3 to 1.3 Gm.).

Official Preparation.

Oleoresina Piperis,.....................Dose: ¼ to 2 ♏ (0.016 to 0.13 mil).

Commercial oil of pepper is an oleoresin from which the piperine has crystallized out.

90. **PIPER ALBUM.**—White Pepper. The ripe fruit from which the epidermis has been removed by macerating in water and rubbing off. It is usually somewhat larger than black pepper and has a smooth surface with about ten distinct lines running from base to apex; the seed fills the whole inner cavity. It contains the same principles as black pepper; is seldom used except as a condiment.

FIG. 56.—*Piper nigrum*—Branch and fruit.

91. **PIPER LONGUM.**—Long Pepper. The fruit of Pi′per lon′gum Linné, and of **Pi′per officina′rum** De Candolle. *Habitat:* Southeastern Asia. It consists of cylindrical spikes of the fruits, 25 mm. (1 in.) or more in length; in the market they are of an earthy, grayish-white appearance, but exhibit their deep reddish-brown color when washed. The individual berries are ovoid, about 2.5 mm. (⅒ in.) long, with a nipple-like point at the apex and a bract at the base; they are arranged spirally on the axis. Medical properties same as those of black pepper, but they are inferior and seldom used.

92. MATICO.—Matico, N.F.

MATICO

The leaves of **Pi'per angustifo'lium** Ruiz et Pavon, are readily recognized by the prominent veining of their under surface; upper surface dull green, tessellated or checkered. Odor slight, taste aromatic; contains volatile oil (2 per cent.), resin, tannin, a bitter principle and artanthic acid. Used as an aromatic, stimulant, tonic and styptic. Special action on mucous membrane. Dose: ½ to 2 dr. (2 to 8 Gm.).

93. **YERBA MANSA.**—The root of **Houttuy'nia califor'nica** Bentham and Hooker. Stimulant, tonic, and astringent; used with good results in malarial fevers. Dose of fl'ext.: 15 to 60 ♏ (1 to 4 mils).

94. **JAMBU ASSU.**—The root of Pi'per jaboran'di Vell. Used in its native country, Brazil, as a sudorific like pilocarpus. Dose: 15 to 30 gr. (1 to 2 Gm.).

95. **KAVA, N.F.**—(Kava-kava). The root of Pi'per **methys'ticum,** obtained from a shrub indigenous to the Sandwich Islands. A large, woody, but spongy root, having a thin, grayish-brown bark and a yellowish meditullium which is radiate; usually comes in whitish segments. Odor fragrant, like a perfume rather than a spice; taste pungent, slightly benumbing. Used as a remedy in the treatment of diseases of the mucous membrane, as tonic to the digestive organs, and stimulant to the nerves; also as a diuretic. It perhaps has some reputation as a remedy in gonorrhea.

SALICACEÆ.—Willow family

Diœcious trees or shrubs with both kinds of flowers in catkins; fruit bearing numerous seeds furnished with long, silky down.

96. **SALIX.**—Willow. The bark of Sa'lix al'ba Linné, and of other species of Salix. *Habitat:* Europe; naturalized in North America. The best bark is that collected from the older branches, coming in thin fragments or quills, the thin brownish or yellowish periderm of which overlays a greenish parenchymatous layer. The bark from the trunk is deprived of the outer layer, pale cinnamon-brown, exfoliating; fibrous. Inodorous; taste bitter and astringent. Two varieties—white willow and purple willow, *S. purpurea* (see below).

Constituents.—Tannin about 12 per cent., most abundant in the white willow, and a bitter neutral principle, **salicin,** which is the active glucosidal constituent, occurring and coming into market in silky, shining, white needles or grains; it exists most abundantly in the purple willow, but may be extracted from various other species and from various species of Populus, where it is combined with populin (benzoyl salicin). The degree of bitterness in the barks is probably the best criterion of the value of the several species.

96 a. **SALICINUM** (U.S.P. IX).—It occurs in white, shining, bitter crystals; soluble in 28 parts of water and 68 parts of alcohol. Boiled with sulphuric acid it is converted into saligenin or saligenol, $C_7H_8O_2$, and glucose, according to the following formula: $C_{13}H_{18}O_7 + HO = (C_6H_4)(OH)CH_2OH + C_6H_{12}O_6$. By oxidation with potassium bichromate and sulphuric acid, salicylic aldehyde, $C_6H_4OH.COH$, is formed, having the fragrant odor of the oil of meadowsweet (*Spiræa ulmaris*) and of heliotrope. Tonic, astringent, febrifuge. Dose: 15 to 60 gr. (1 to 4 Gm.). The bark itself is rarely employed, however, salicin being used instead in doses of 10 to 30 gr. (0.6 to 2 Gm.). Ash, not more than 0.05 per cent.

Preparation of Salicin.—Obtained by adding lead subacetate to a decoction of the bark, precipitating the excess of lead with H_2S. Evaporate liquid. Add, near the end of the process, sufficient quantity of animal charcoal to decolorize; filter the liquid while hot. Upon cooling, salicin will deposit in crystalline form.

97. **POPULUS.**—White Poplar. American Aspen. The bark of **Pop'ulus** tremuloi'des Michaux. Tonic and febrifuge. Its active principle, populin, is analogous to the salicin of salix (96). Dose of fl'ext.: 30 to 60 ♏ (2 to 4 mils).

98. **POPULUS BALSAMIFERA.**—Balm of Gilead Buds. The buds of **Pop'ulus** balsamif'era Linné, variety **candicans** Gray. Populi Gemmæ (Balsam Poplar buds, Balm of Gilead buds, N.F.). *Habitat:* Northern North America and Siberia. These buds, as well as those of other species of Populus, are covered with a resinous exudation which is impregnated with a fragrant volatile oil, and is very similar in medicinal action to the turpentine oleoresins. Dose of fl'ext.: 30 to 60 ♏ (2 to 4 mils).

Fig. 57.—*Salix alba*—Branch.

MYRICACEÆ.—Sweet-gale Family

99. **MYRICA, N.F.**—Bayberry Bark. Wax Myrtle. The bark of **Myri'ca** ceri'fera Linné, an indigenous plant growing on seashores, the fruit of which is covered with a layer of white vegetable wax. This bark is occasionally used in medicine as a tonic, and as an astringent gargle in sore throat, etc. Dose of fl'ext.: 15 to 30 ♏ (1 to 2 mils).

100. **COMPTONIA.**—Sweet Fern. The leaves of **Compto'nia** asplenifo'lia Aiton, an indigenous herb. They are linear-lanceolate, with deep, alternate, rounded lobes, and have a spicy odor, especially when rubbed. Stimulant and astringent. Dose: 15 to 30 gr. (1 to 2 Gm.).

JUGLANDACEÆ.—Walnut Family

A small family of trees with monœcious flowers and the fruit a nut.

101. JUGLANS, N.F.—BUTTERNUT. The root-bark of Jug'lans cine'rea Linné, collected in autumn. Off. U.S.P. 1890. Corky layer very thin, smooth, grayish, easily removed, leaving a smooth, deep-brown surface; inner surface pure white when the bark is first removed from the tree, but changes to deep brown on exposure. In the market it is found in flat or curved pieces about ⅕ of an inch (5 mm.) thick, the outer surface dark gray and nearly smooth, or, deprived of the soft cork, deep brown, the inner surface striate. Fracture short, whitish-and-brown checkered; medullary rays somewhat diagonal; odor feeble; taste bitter, somewhat acrid. The leaves and bark of *Juglans nigra* (101 a) (black walnut) have been used as an alterative and deobstruent, and the bark of *Carya alba* (101 b) (shellbark hickory) as a tonic and antiperiodic. The kernels of the nuts of all these trees yield about 25 per cent. of a pale greenish fixed oil (Oleum Juglandis, or nut oil), used as a demulcent. *Constituents:* Bitter oily extractive, in large proportion juglandic acid, $C_{10}H_6O_8$, tannin (?), two other acids, one of them volatile, with potassium, sodium, and other salts. A mild cathartic, especially valuable in habitual constipation. It was much used in the army during the Revolutionary War. Dose: 1 to 2 dr. (4 to 8 Gm.).

CUPULIFERÆ.—Oak Family

An important order on account of its valuable wood. It is characterized by alternate *leaves* and monœcious *flowers*, the *sterile* ones in catkins, the *fertile* in clusters or spikes, and the *fruit* a 1-seeded nut, with or without a woòdy, scaly involucre (cupule).

Synopsis of Drugs from the Cupuliferæ

A. *Bark.*	B. *Excrescence.*	C. *Leaves.*
*Quercus, 102.	GALLA, 105.	*Castanea, 106.
Alnus, 103.	ACIDUM TANNICUM, 105 a.	OLEUM BETULÆ VOLATILE, 107.
Fagus, 104.	ACIDUM GALLICUM, 105 b.	D. *Heart-wood.*
	PYROGALLOL, 105 c.	Ostrya, 108.

102. QUERCUS, N.F.—WHITE OAK

WHITE OAK

The bark of **Quer'cus al'ba** Linné, collected from trunk or branches ten to twenty-five years of age and deprived of the periderm.

DESCRIPTION OF DRUG.—Flat pieces about 6 mm. (¼ in.) thick, deprived of the thick, corky layer; pale brown; coarsely fibrous; inner surface traversed by prominent longitudinal ridges; fracture coarse, fibrous (the tissue contains groups of stone cells and crystals of calcium oxalate); odor faintly tan-like; taste very astringent. It is usually found in the shops as a coarse, fibrous powder.

Powder.—Pale brown. Characteristic elements: Parenchyma of cortex, rather thin-walled, pale brownish rosy hue, some with brown resin or irregular brownish-yellow tannin masses; calcium oxalate, aggregate or prisms (10 to 20 μ in diam.); sclerenchyma with stone cells (25 to 40 μ in diam.), thick-walled; bast fibers 15 to 30 μ thick, long, rather large, thick-walled; crystal fibers with aggregate and prismatic crystals of calcium oxalate (10 to 20 μ in diam.); cork cells, pentagonal or hexagonal (20 to 30 μ in diam.).

CONSTITUENTS.—Quercitannic acid 6 to 11 per cent., a coloring matter, a **bitter** principle (quercin), sugar (quercite), resin, etc. The active principles are soluble in water and alcohol. The amount of tannin varies with the species, the part of the tree, and the season of the year when gathered; the young bark contains a greater proportion than the old.

Quercitannic Acid.—Two forms of this principle exist, according to Lowe —one soluble in water, of the formula $C_{28}H_{28}O_{14}$, and the other scarcely soluble, $C_{28}H_{24}O_{12}$. Both are changed by the loss of water into *oak red*, $C_{28}H_{22}O_{11}$.

FIG. 58.—*Quercus alba*—Branch.

Quercitron.—Under this name large quantities of black oak (*Quercus tinctoria*) bark deprived of its epidermis and reduced to a coarse powder are sent from the United States to Europe as a dye. The coloring principle is called quercitron, $C_{36}H_{38}O_{30}$. This glucoside splits up by hydrolysis into quercetin and isodulcite, or rhammose, $C_6H_{12}O_5(C_5H_9O_5CH_3)$. Quercetron (*Xantho rhamnin*) forms yellowish crystals, odorless and tasteless, but in hot aqueous or alcoholic solution has a bitter taste.

Preparation of Quercin.—Boil bark in acidulated (H_2SO_4) water; add milk of lime to neutralize; filter; add K_2CO_3. Yellow needles slowly form on evaporation of alcoholic solution of above precipitate.

ACTION AND USES.—Astringent and tonic, generally used externally in infusion or decoction as an astringent and tonic bath, injection, etc. Dose: 15 to 60 gr. (1 to 4 Gm.). **Fluidextractum Quercus,** average dose 2 mils.

103. **ALNUS SERRULATA** Willdenow.—Tag Alder. *Habitat:* North America. (Bark.) Tonic, astringent, and alterative. Dose: 30 to 60 gr. (2 to 4 Gm.).

104. **FAGUS FERRUGINEA** Aiton.—American Beech. (Bark and leaves.) Astringent and slightly tonic.

105. GALLA.—Nutgall

GALLS

An excrescence on the young twigs **Quer'cus infectoria** and other species of Quercus produced by the punctures and deposited ova of *Cynips gallæ tinctoriæ* (Fig. 59) Olivier (class, Insecta; order, Hymenoptera). Not more than 5 per cent. of Galls float in water.

BOTANICAL CHARACTERISTICS.—A shrub or small tree 6 to 8 feet high. *Leaves* short-petiolate, obovate-oblong, obtusely toothed, oblique at base. *Acorn* solitary, obtuse, two or three times the length of the cup.

HABITAT.—Levant.

DESCRIPTION OF DRUG.—Hard, heavy, subglobular, from the size of a pea to that of a large cherry, contracted below into a short stipe and covered above with a few or many prominent warts (tuberculated) between which the surface is smooth. Heavy, sinking in water, except the smaller ones which should not be present to a greater extent than 5 per cent. Externally dark bluish or lead color, frequently with a greenish tinge, often with a circular hole near the middle upper part, communicating with the central cavity. They break with a flinty fracture, showing a whitish or brownish interior, with often a central cavity, lined with a thin, hard shell, which contains the insect in all stages of development, or the pulverulent remains of the developed insect mixed with partly eaten fragments of the starchy parenchyma. Odorless; very astringent.

STRUCTURE.—The tissue is chiefly parenchyma, loaded with tannin and chlorophyll; the cavity lining is composed of stone cells containing calcium oxalate crystals; within this cavity, if not eaten out, is a starchy parenchyma.

VARIETIES.—Most of the oaks are occasionally affected as the above species, the resulting excrescence, known as *galls*, developing a tannin which may be employed for various practical purposes. The Aleppo or Syrian, dark colored and heavy (although the designation Aleppo is not wholly applicable to the official galls—"Galla"), are the products of different parts of Asiatic Turkey; still the name is applied to this variety. *Smyrna* galls, grayish-olive color, intermixed with white galls. *Sorian*, size of a pea, blackish. *Japanese* and *Chinese*, from *Rhus simulata*, ½ to 2 inches long, ovate, very irregular, tubercular, grayish downy, inclosing the remnants of numerous insects. The Chinese make use of this product in dyeing and as a medicine.

Powder.—Gray. The microscopic elements consist of: See Part iv, Chap. I, B.

CONSTITUENTS.—Tannin 65 to 77 per cent.   (Acidum Tannicum, gallotan-
nic acid 105 a), chemically known as digallic acid, $C_{14}H_{10}O_9$.   It is a
yellowish-white amorphous substance, insoluble in absolute ether,
chloroform, benzol, benzin, and carbon disulphide, soluble in glycerine,
alcohol, and water; precipitated blue-black by ferric salts, and white by

FIG. 59.—*Quercus lusitanica*—Branch and nutgall.

gelatin.   It appears to exist, in part at least, as a glucoside and digallic
acid.   Digallic acid may be considered as an anhydride of gallic acid,
$C_7H_6O_5$, formed from two molecules of the latter by elimination of one
molecule of water.   Gallic acid also exists in galls.   It is precipitated
blue-black by ferric salts, the color disappearing on boiling, and is
not affected by gelatin when gum is absent.

Preparation of Tannic Acid.—Powdered nutgall is exposed to damp atmosphere for twenty-four hours, then made into paste with washed ether. Allow to stand six hours, then express in canvas cloth between tinned plates. After powdering the pressed cake, again make into paste with washed ether. Repeat the former process and allow the mixed liquid to evaporate spontaneously.

ACTION AND THERAPEUTIC PROPERTIES.—When taken into the digestive tract some of it is changed into gallic acid and absorbed as such; while some may be taken up as a soluble alkaline tannate.

Because of its power and lack of toxicity, tannic acid is one of the most widely used of all the astringents, either in the form of the tannic acid itself or of one of the various vegetables containing it.

Locally applied it may be used to overcome relaxation, as in spongy gums, mercurial sore mouth, hemorrhoids, and chronic sore throat.

To check hemorrhage it may be used whenever the source of flow can be reached directly, as in epistaxis, hæmatemesis, hemorrhage from the bowels, etc.

Tannic is useful as an antidote against a number of poisons including most of the irritant metallic salts, especially those of antimony and iron. Dose: 8 gr. (0.6 Gm.).

OFFICIAL PREPARATIONS.

Preparations commonly employed:
    Unguentum Gallæ (20 per cent.).
    Acidum Tannicum,......................Dose: 10 to 20 gr. (0.6 to 1.2 Gm.).
    Trochisci Acidi Tannici, ...Each 1 gr. (0.06 Gm.).
    Glyceritum Acidi Tannici (20 per cent.), ..Local use.
    Unguentum Acidi Tannici (20 per cent.), ..Local use.
    Collodium Stypticum (2 per cent.),......Local use.

105 b. **ACIDUM GALLICUM,** U.S.—GALLIC ACID. Usually prepared from tannic acid. Also prepared by exposing moistened powdered nutgalls to the action of the air for a month or more; a peculiar fermentation sets in which converts the tannic acid into gallic acid; this is extracted by expression and purified by filtration and crystallization. It is in **light, silky, acicular needles,** colorless when pure, but as usually seen in the shops, of a more or less **pale brownish color;** inodorous; taste sourish and astringent. **It differs from tannic acid** in its sparing solubility in cold water, and in not precipitating gelatin or alkaloids from their solutions. It is less astringent than tannic acid, and inferior to it in all respects except where the astringent effect must be reached through the medium of the general circulation. When applied locally, gallic acid acts as a mild astringent, but does not cause coagulation of the blood, for which reason it is not used locally in the control of hemorrhage. Dose: 5 to 30 gr. (0.3 to 2 Gm.).

105 c. **PYROGALLOL,** U.S.—PYROGALLIC ACID. A triatomic phenol, $C_6H_3(OH)_3$, obtained chiefly by the dry distillation of gallic acid. It is in light, white, shining laminæ, or in fine needles, becoming gray or darker when exposed to the air or light, and should therefore

be kept in amber-colored bottles; inodorous; astringent. Soluble in water and alcohol. **Used exclusively externally** in the form of ointments, in lupus, psoriasis, and other **skin diseases.** Its absorption through abrasions in the skin has caused death by general poisoning.

106. **CASTANEA,** N.F.—Chestnut. The leaves of **Castan'ea denta'ta** Sudworth, collected in September or October while yet green. Off. U.S.P. 1890. Oblong, elliptical, from 150 to 250 mm. (6 to 10 in.) long, and about 50 mm. (2 in.) broad, with a sharply pointed apex and a short petiole; margin somewhat unequally, but strongly, repand-dentate, with prominent parallel veins beneath each tooth (feather-veined); texture firm, flexible; odor slight; taste somewhat astringent. *Constituents:* Tannic acid about 9 per cent., gum, albumen, salts, and traces of resin and fats. Tonic and astringent, used almost exclusively in whooping-cough in the form of infusion or fl'ext. Dose: ½ to 2 dr. (2 to 8 Gm.).

Fluidextractum Castaneæ, U.S.P. 1890, Dose: ½ to 2 fl. dr. (2 to 8 mils).

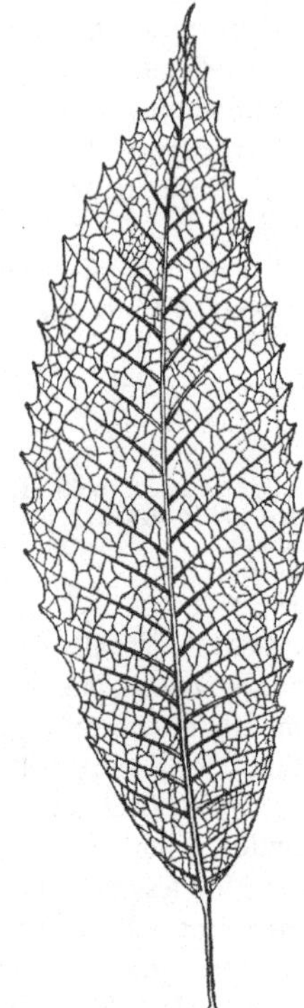

FIG. 60.—Chestnut leaf, natural size.

107. OLEUM BETULÆ.—Volatile Oil of Betula

OIL OF SWEET BIRCH

A volatile oil distilled from the bark of **Betula lenta** Linné (Fam. Betulaceæ, U.S.P. 1900).

BOTANICAL CHARACTERISTICS.—A tree often exceeding 60 feet in height, with a diameter of 2 or 3 feet. The trunk is invested with a dark brown or reddish bark, separating in thin layers. Remarkable for its agreeable fragrance and flavor. *Leaves* cordate, ovate-acuminate, acutely, finely, and doubly serrate, veined beneath. *Flowers* monœcious, *sterile* catkins 2 or 3 inches long, *fertile* much shorter and thicker; *petals* hairy; wood reddish, strong, compact.

DESCRIPTION.—This oil is **identical with methyl salicylate,** $CH_3C_7H_6O_3$, and nearly identical with **oil of wintergreen** (413 a). Its specific gravity is 1.18. In fact, is one of the sources of commercial oil of wintergreen. Dose: 5 to 30 ♏ (0.3 to 2 mils). It should be kept in well-stoppered bottles, protected from the light.

107 a. OLEUM BETULÆ EMPYREUMATICUM RECTIFICATUM, N.F.—Obtained by the dry distillation of the bark and wood of Betula alba and rectified by steam distillation. Is used mainly as an external remedy in cutaneous diseases.

108. **OSTRYA VIRGINICA.**—Iron-wood. Hop-hornbeam. The wood has some reputation as an antiperiodic, tonic, etc. The fl'ext. is used in malaria, in doses of ½ to 1 fl. dr. (2 to 4 mils).

URTICACEÆ.—Nettle Family

A large and very diversified family, consisting of herbs, shrubs, or trees, sometimes with a milky juice yielding caoutchouc; some species have a bark which yields mucilage; the nettleworts are remarkable for the caustic secretion of their glandular stinging-hairs; the juice of the hempworts (suborder Cannabineæ) is bitter and narcotic

Synopsis of Drugs from the Urticaceæ

A. *Bark.*
  ULMUS, 109.
B. *Strobiles.*
  HUMULUS, 110.
C. *Glands.*
  *Lupulinum, 111.

D. *Herb.*
  CANNABIS, 112.

  Urtica, 113.
E. *Seed.*
  Cannabis Semen, 112 a.

F. *Fixed Oil.*
  Oleum Cannabis, 112 b
G. *Fruits.*
  *Ficus, 114.
  Morus, 115.

109. ULMUS.—Elm

SLIPPERY ELM BARK

The dried bark of **Ul'mus ful'va** Michaux (Fam. transferred to Ulmaceæ—U.S.P. 1900), deprived of its periderm.

Botanical Characteristics.—A tree 40 to 60 feet high. *Leaves* ovate-oblong, taper-pointed, doubly serrate, very rough above. *Flowers* nearly sessile, in

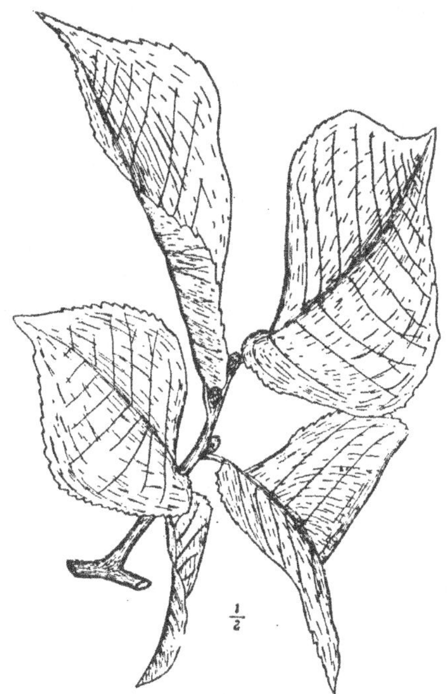

Fig. 61.—*Ulmus fulva*—Branch.

lateral clusters, purplish or brownish. *Fruit* a 1-celled, 1-seeded samara, winged all around.

10

HABITAT.—North America, north of the Carolinas and east of Nebraska.

DESCRIPTION OF DRUG.—Various sized **flat pieces** about 4 mm. (⅙ in.) thick **deprived** of cork, of a uniformly pale brownish-white color, the finely ridged inner surface with a slight reddish tinge; good specimens are **tough and flexible,** capable of being bent double. The texture consists of soft parenchymatous tissue with tangentially arranged bast fibers and numerous medullary rays, giving to a cross-section of the bark a delicately checkered appearance. **Odor** agreeable, resembling fenugreek. Taste highly mucilaginous. It **yields** a fawn-colored powder which is often adulterated with starch. Euro-

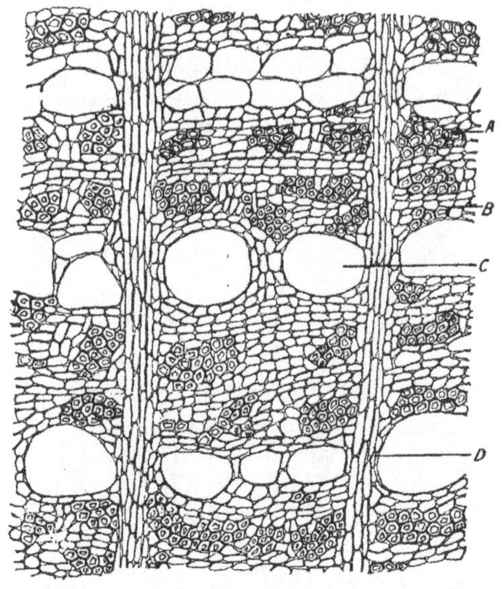

Fig. 62.—Ulmus—Cross-section of bark. *A.* Bast fibers. *B.* Parenchymatous tissue. *C.* Mucilage ducts. *D.* Medullary rays.

pean elm bark, from *U. campestris* and *U. effusa,* cinnamon-colored, nearly inodorous, mucilaginous, but has a bitterish and astringent taste, owing to the presence of a little tannin. A few nearly spherical starch grains (0.005 to 0.01 μ in diameter) are sometimes present.

Powder.—Characteristic elements: See Part iv, Chap. I, B.

CONSTITUENTS.—A large quantity of mucilage (capable of precipitation with alcohol and lead acetate), and some tannin.

ACTION AND USES.—**Demulc**ent—externally as an emollient application, in poultice. Dose: 2 dr. (8 Gm.) or more.

110. HUMULUS.—Hops

HOPS

The strobiles of **Hu'mulus lu'pulus** Linné, carefully dried—bearing the whole of their natural glandular coating (Fam. transferred to Moraceæ, U.S.P. 1900).

BOTANICAL CHARACTERISTICS.—Rough, climbing perennial. *Leaves* palmately 3- to 7-lobed, roughish, ovate. *Flowers* dioecious, the fertile flowers forming a strobile in fruit; *calyx, akene,* etc., thickly studded with yellowish, resinous grains, which give the bitterness and aroma to the hops.

HABITAT.—North Temperate Zone.

DESCRIPTION OF DRUG.—**Strobile** about 30 mm. (1⅕ in.) long, **cone-shaped,** consisting of numerous membranous, greenish-yellow scales attached to a thin, undulating, hairy axis; the scales are oval, leaf-like, translucent, showing delicate veins, and surround a subglobular akene; there are also, covering the surface of the scales at the base and adhering to the zigzag axis, **small yellow grains of lupulin,** upon which the value of hops depends. Odor strong, peculiar, somewhat narcotic; taste bitter, aromatic, slightly astringent.

Powder.—See Part iv, Chap. I, B.

CONSTITUENTS.—**Lupulin** (Lupulinum, U.S.), volatile oil (0.08 per cent.), resin, choline, and tannin. Ash, not exceeding 8 per cent.

ACTION AND USES.—Tonic, anodyne, and slightly narcotic. Dose: ½ to 5 dr. (2 to 20 Gm.), in infusion or tincture. Externally as an anodyne or sedative in fomentation or poultice.

111. LUPULINUM, N.F.—LUPULIN. The granular powder separated from humulus, bright yellow, becoming yellowish-brown with age; mixed with minute scale particles; resinous; odor peculiar, aromatic, like hops, but stronger; taste bitter. Under the microscope each gland is seen to be composed of two reticulated hemispheres, one narrow and one round; the narrow one collapses on drying, giving to the granule a hood-shaped appearance. They are filled with an oleoresin, the volatile oil of which contains a trace of valerianic acid, and valerol, which passes into valerianic acid when kept a long time, causing the valerian-like odor of old hops—lupamaric acid, C_{35}-$H_{35}O_4$.

ACTION AND USES.—Same as hops. Dose: 6 to 15 gr. (0.4 to 0.1 Gm.), in capsules or pills, the latter of which may be made by simply rubbing the powder with warm water until it becomes adhesive.

Fluidextractum Lupulini,..........Dose: 10 to 30 ♏ (0.6 to 2 mils).
Oleoresina Lupulini,.............. 3 to 6 ♏ (0.2 to 0.4 mil).

112. CANNABIS.—INDIAN CANNABIS

INDIAN HEMP. HEMP

The dried flowering tops of the pistillate plant of **Can'nabis sati'va** Linné or of **the** variety indica, Lamarck (Fam. Moraceæ), freed from thicker stems **and** large foliage leaves, and without admixture of more than 10 per cent. of fruits.

Test.—When made into a fl'ext. and assayed biologically, produces inco-
    ördination when administered to dogs in a dose of not more than 0.03 mil
    of fl'ext. per kilogramme of body weight.
Botanical Characteristics.—*Stem* 4 to 8 feet high, annual, tall, and roughish,
    the inner bark consisting of tough fibers. *Leaves* palmately 5- to 7-divided,
    the *leaflets* coarsely serrate. *Flowers* diœcious green, in compound, axillary
    racemes or panicles. *Akene* globose, crustaceous.

Fig. 63.—*Cannabis indica*—Branch.

Source.—The plant is indigenous to Asia, from India northward to
    Western China and Caspian Sea. Its cultivation has extended to
    Central and Southern Europe, Russia, Brazil, and the Western
    United States—in fact, it may be said to grow in all civilized countries
    on the globe.
Description of Drug.—Cannabis indica occurs in commerce as bundles
    of the flowering tops; the branches, digitate leaves, and the numerous
    flower-bracts are more or less compressed, and **agglutinated together**
    with a resinous exudation; color brownish-green; odor peculiar,
    narcotic; taste bitterish, somewhat acrid.   It is sold in Indian bazaars
    for smoking purposes as "gunjah." The leaves, small stalks, and
    capsules, dried separately and mixed with aromatics and fruits, form
    the Arabian confection, "hashish, bhang, or siddhi." "Churrus"

is a brown, earthy-looking resin, brushed off from the plants by leather-clad men running through the field.

Cannabis americana, the plant grown in various parts of the United States, acts similarly to the official plant. See article by author, "Cultivation of Medicinal Plants in U. S." Jour. Amer. Phar. Assoc., 1915.

Powder.—Characteristic elements: See Part iv, Chap. I, B.

CONSTITUENTS.—The resin and a yellow, aromatic volatile oil, $C_{10}H_{16}$, are its most important constituents. The former, **cannabin** (15 to 20 per cent.), is a brown, amorphous powder, soluble in absolute alcohol (but not in cold alcohol of 89 to 90 per cent.), from which solution it is thrown down as a white precipitate by water; it is very potent, $\frac{2}{3}$ of a grain acting as a powerful narcotic; it comes into the market as cannabin tannate; choline, $C_5H_{15}NO_2$, syrupy, soluble in alcohol and water, very sensitive to Mayer's reagent, yielding a yellow, crystalline precipitate, is probably the same as the so-called alkaloid, "tetano-cannabinine." Ash, not exceeding 15 per cent.

Cannabinol.—This principle has been obtained by Wood, Spivey, and Esterfield from the exudate of cannabis indica (*charas*). Several different fractional distillates from the ethereal extract of this exudate were obtained. Among these distillates is *cannabinol*, $C_{18}H_{24}O_2$, boiling at 265°C. It is oleaginous and has a red color. This they have found to largely represent the active principle. A condensed account of the pharmacology of cannabis indica, as contributed by Dr. C. R. Marshall, may be found in "Western Druggist," 1889, pp. 163–166.

Preparation of Cannabin.—Treat drug with water made alkaline with Na$_2$CO$_3$; exhaust dry residue with alcohol; add milk of lime; precipitate with H$_2$SO$_4$; treat filtrate with animal charcoal. From the resulting liquid, concentrated, cannabin is precipitated by water.

ACTION AND USES.—**Powerful narcotic.** The primary effect of the drug is that of exhilaration, intoxication, stimulating the imagination, etc. This is followed by depression, drowsiness, and stupor, the heart becomes weak and slow and the pupil dilated. It has some advantages over opium, it is claimed, in that it is not constipating, and interferes less with digestion; it is more acceptable in certain morbid states of the system and nervous disquietude. Dose: 3 to 5 gr. (0.2 to 0.3 Gm.).

OFFICIAL PREPARATIONS.

112 a. **CANNABIS SEMEN.**—HEMP SEED. These have been used in the form of emulsion as demulcent and anodyne, depending upon the fixed oil which they contain. They are mostly used as a bird-seed, however, and for the extraction of the fixed oil.

112 b. **OLEUM CANNABIS.**—OIL OF HEMP. A greenish fixed oil, becoming lighter and brownish on exposure; odor hemp like; taste mild. Used as a demulcent and protective. Neither it nor the seed are thought to have any narcotic action.

113. **URTICA.**—NETTLE. STINGING NETTLE. The herb of Urti'ca dio'ica Linné. *Habitat:* United States and Europe. Tonic, astringent, and a valuable diuretic. As an astringent it is chiefly used in uterine hemorrhages. Dose: 20 to 40 gr. (1.3 to 2.6 Gm.).

114. FICUS.—Fig, N.F.

FIG

The partially dried fruit of **Fi'cus car**'ica Linné (Fam. Moraceæ, U.S.P. 1900).

BOTANICAL CHARACTERISTICS.—A small tree with palmately lobed, cordate *leaves*. *Flowers* monœcious, inclosed within a pear-shaped receptacle which converges so as to leave only a small orifice at the apex; *style* single; *stigmas* 2.

HABITAT.—Levant; cultivated extensively in the Mediterranean Basin and subtropical regions.

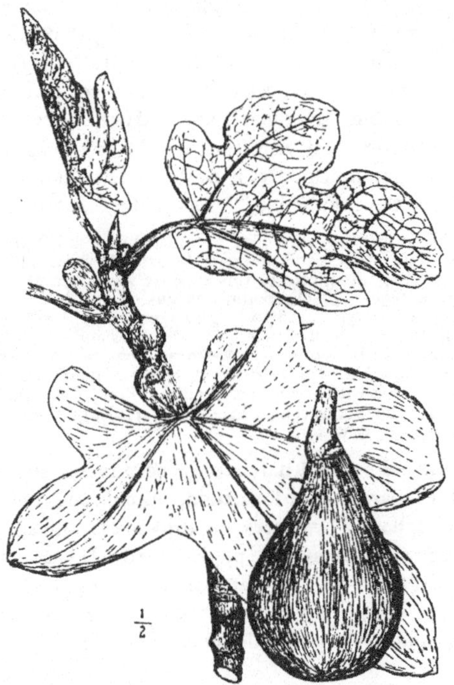

FIG. 64.—*Ficus carica*—Branch and fruit.

COLLECTION.—Figs are either left on the tree to dry or are dried after being gathered by artificial heat or the heat of the sun, and in this condition are called "natural figs," or they are rendered pliant by pulling and kneading. They are then packed in boxes or drums and known as "pulled figs." The largest and best are those of Smyrna and Turkey, the best Smyrna being known as "Eleme figs." The largest amount is imported from Asiatic Turkey, and the remainder from Spain, Portugal, and other countries.

DESCRIPTION OF DRUG.—Figs come into market compressed, and covered with an efflorescence of sugar which melts in warm weather and makes them soft and moist. They are yellowish or brownish, somewhat translucent, and consist mostly of a sweet, viscid pulp, in the center of which are numerous small, yellow ovaries, or akenes, popularly regarded as seeds; odor peculiar; taste sweet, mucilaginous. When soaked in water they may be opened out to their original pear-shaped form, showing the short stalk, or its scar, at the base or pointed end, and scales at the large end surrounding an orifice near which the staminate flowers were situated; the numerous akenes, or ovaries, of the pistillate flowers cover the walls of the hollow interior.

CONSTITUENTS.—Grape sugar (60 to 70 per cent.), gum, fat, and salts.

ACTION AND USES.—Nutrient, laxative, and demulcent. Their principal use medicinally is as a laxative diet in constipation, freely given, which action in dried figs is mainly due to the indigestibility of the seeds and tough skin. Dose: 4 dr. (15 Gm.).

OFFICIAL PREPARATION.

Confectio Sennæ (12 per cent.),....Dose: 1 to 3 dr. (4 to 12 Gm.).

115. **MORUS.**—MULBERRY. The fruit of **Mo'rus ru'bra, M. nigra,** and **M. alba** Linné, indigenous trees. Dense, cylindrical spikes of the small fruit, differing in size, shape, and color in the different species. They are all used,in the fresh state as a refrigerant.

SANTALACEÆ.—Sandalwood Family

116. **SANTALUM ALBUM, N.F.**—SANDALWOOD. The wood of **San'talum al'bum** Linné, and other species of Santalum. It comes in billets from 100 to 150

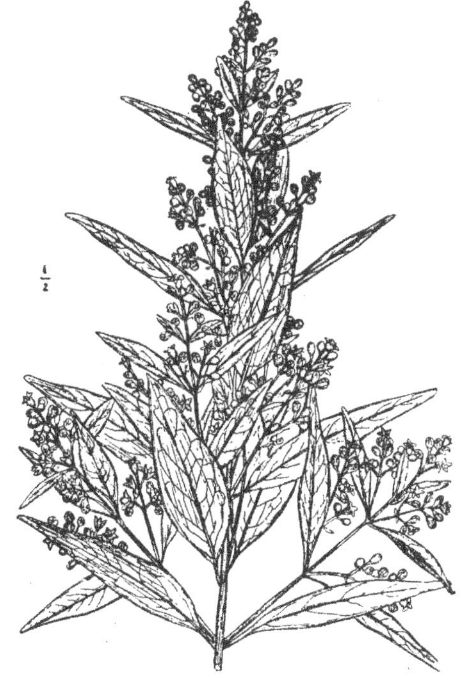

FIG. 65.—*Santalum album*—Branch.

mm. (4 to 6 in.) in diameter, or in split slices; color varying, yellowish, whitish, or brownish; it has only a feeble taste, but an aromatic odor, particularly when rubbed or in powder. Contains from 1 to 4 per cent. of volatile oil.

116 a. **OLEUM SANTALI,** U.S.—Oil of Santal. A yellowish, somewhat thick volatile oil, having a peculiar, strongly aromatic odor, and a pungently aromatic taste. It is a valuable remedy in inflammation of the mucous membrane, used especially in gonorrhea and bronchitis. Its principal use is in the manufacture of perfumery. Dose: 10 to 30 ♏ (0.6 to 2 Gm.) administered usually in capsule.

LORANTHACEÆ.—Mistletoe Family

117. **MISTLETOE.**—The bark of Phoraden'dron flaves'cens Nuttall, a parasitic evergreen growing on various trees, particularly on fruit trees. Laxative, oxytocic, and antispasmodic. As an oxytocic it is claimed to be superior to ergot. Dose: 15 to 60 gr. (1 to 4 Gm.).

ARISTOLOCHIACEÆ.—Birthwort Family

Climbing shrubs, or low herbs, with perfect *flowers*, the lurid *calyx* coherent with the ovary, which forms a 6-celled capsule or berry in fruit. *Leaves* petiolate. Principal constituents are volatile oil and resinous principles.

118. SERPENTARIA.—Serpentaria

VIRGINIA SNAKE-ROOT

The dried rhizome and roots of **Aristolo'chia serpenta'ria** Linné (Virginia), and of Aristolochia reticula'ta Nuttall (Texas).

Botanical Characteristics.—*Stem* 8 to 15 inches high, pubescent. *Leaves* alternate, ovate, or oblong, with a heart-shaped or halberd-shaped base. *Flowers* all next the root, short-peduncled; *calyx-tube* bent like the letter S; *stamens* 6, the sessile anthers adnate to the fleshy style.

Habitat.—United States (**Virginia and Texas**).

Description of Drug.—A rhizome about 25 mm. (1 in.) long, and about the thickness of a quill; contorted, bent up and down; externally light grayish-brown, with short stem-bases on the upper side and numerous long, fibrous, branching rootlets below, interlaced; internally grayish, closely matted. The bark is thin, overlaying quite a large woody zone, and separated into wood-wedges by broad medullary rays; **the pith is not in the center** but is nearer the upper side, making the lower wood-wedges the longest. **Odor family terebinthinate, characteristic;** taste warm, bitter, and camphoraceous. **Virginia and Texas Serpentaria** are both recognized by the U.S.P. The latter is about twice as large as the former, with fewer and thicker rootlets.

ADULTERATIONS.—As found in commerce, serpentaria is frequently adulterated with portions of the stem. *Hydrastis canadensis* has

FIG. 66.—*Aristolochia serpentaria.*

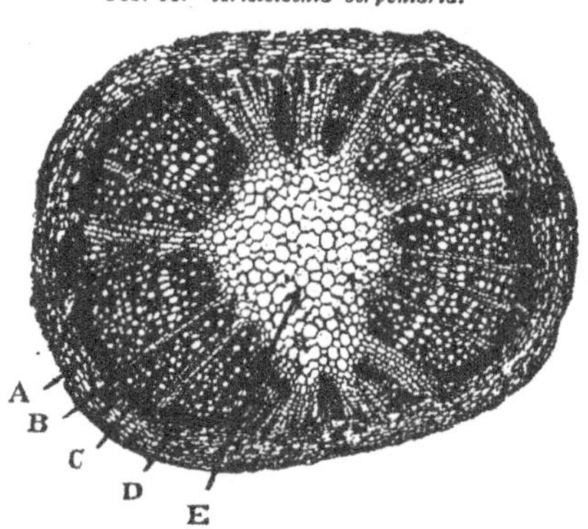

FIG. 67.—Serpentaria—Cross-section of rhizome. (25 diam.) A, Parenchyma of cortex. B, Medullary ray. C, Xylem. D, Phloëm. E, Medulla. (Photomicrograph.)

been used as an intentional adulteration; also spigelia. All of these may easily be distinguished from the genuine by their general characteristics.

Powder.—Characteristic elements: See Part iv, Chap. I, B.

Constituents.—Volatile oil (½ per cent.), containing borneol, aristolochine, $C_{32}H_{22}N_2O_{13}$ (very bitter), tannin, resin, starch, etc.

Preparation of Aristolochine.—Precipitate decoction with lead acetate; exhaust precipitate with hot alcohol; evaporate; dissolve out alkaloid with water. It is bitter, yellow, amorphous, or in needles; soluble in alcohol, water, precipitated by tannin.

Action and Uses.—Aromatic stimulant and tonic. Its only possible therapeutic virtue is as a stimulant to the gastric mucosa—Wood. Dose: 5 to 30 gr. (0.3 to 2 Gm.).

Tinctura Cinchonæ Composita (2 per
    cent. of serpentaria),....   1 to 4 fl. dr. (4 to 15 mils).

119. ASARUM CANADENSE Linné.—Canada Snake-root. Asarum, N.F. Wild Ginger. A long, creeping rhizome, more or less contorted. In commerce broken into pieces from 100 to 150 mm. (4 to 6 in.) long, from the thickness of a straw to that of a goose-quill; somewhat quadrangular or two-edged; externally grayish-brown, longitudinally wrinkled, beset with small fibers, easily broken off; internally nearly white, the small wood-bundles surrounding a large pith; odor peculiar, aromatic; taste aromatic and pungent. It contains a large percentage of volatile oil which is often used in perfumery. This contains asarol, probably identical with linalool, its acetic and valerianic esters, methyl eugenol. Aromatic stimulant and tonic. Dose: 30 gr. (2 Gm.).

POLYGONEÆ.—Buckwheat Family

Herbs or woody plants with alternate, entire *leaves*, and with the stipules in the form of sheaths above the smaller joints of the stem. *Fruit* an akene. The leaves and stem are very rich in crystals of calcium oxalate.

Synopsis of Drugs from the Polygoneæ

A. *Roots.*
    RHEUM, 120.
    Rumex, 121.
    Canaigre, 122.

B. *Rhizome.*
    Bistorta, 124.
C. *Herb.*
    Polygonum, 123.

120. RHEUM.—Rhubarb

RHUBARB

The dried rhizome and roots of **Rhe′um officinale** Baillon, **Rheum palmatum** Linné, and the Var. **Tanguticum** Maximowicz, and probably other species of Rheum, deprived of most of the cortex and carefully dried.

Botanical Characteristics.—Botanical history somewhat obscure. It is known, however, from authentic specimens, that the plant is a herbaceous perennial with acidulous juice, resembling the garden rhubarb, but attaining a larger size than any other species. *Leaves* very large, roundish, cordate at base, and 5- to 7-lobed. The flower-stem, 6 to 8 feet high, bears *flowers* having a greenish perianth; *ovary* (and fruit) triangular, 1-celled.

Source.—Rhubarb is obtained from many species of Rheum, mostly natives of Asia, especially of China, Chinese Tartary, and Thibet. Russian or Turkish rhubarb—so called because all of it imported into these countries from China had to be submitted to official inspection—is now never found in the market. The caravan commerce between Russia and China has been an important one for many generations, and the rhubarb in European commerce was almost entirely carried from China through Persia and Asia Minor; hence the old name of Turkey rhubarb. Later on it was brought through Northern China, Siberia, and European Russia (Kiachta) to St. Petersburg.

The "Russian rhubarb" of early times was evidently what is now known as Shensi variety. That brought into the trade by the port of Canton, known in Europe as Indian rhubarb, is now called Canton. The Chinese rhubarb is the variety recognized in commerce. The root, often attaining a weight of fifty pounds, is cut up into pieces of a suitable size for drying, holes being usually bored through the pieces and a string passed through for hanging them up.

Description of Drug.—In cylindrical, conical, or plano-convex pieces, or pieces with no regular shape, varying in size from 75 to 150 mm. (3 to 6 in.) long, and 50 to 75 mm. (2 to 3 in.) thick; they are usually sorted into "round" and "flat" rhubarb. Externally somewhat shriveled, often with portions of the cortical layer which have not been pared away; usually covered with a bright yellow dust, beneath which it is seen to have a rusty-brown hue; under the lens it is seen to be marked with the medullary rays (innumerable short, broken lines of a deep brown color) crossing a white ground, forming elongated whitish meshes. Well-formed pieces broken transversely display near the cambium zone dark lines arranged as an internal ring of star-like spots, with radiating, reddish medullary rays, marking the internal origin of the leaves. **The tissue is made up** of a white parenchyma, with reddish-brown or brownish-yellow medullary rays, so twisted, however, as to be scarcely recognizable as such, giving a **cross-section a mottled appear**ance of **red, white, and yel**low. The white parenchyma cells are loaded with starch and crystals of calcium oxalate, which cause the grittiness between the teeth; the medullary rays contain the active constituents. Odor characteristics; taste bitter, aromatic, astringent, and gritty. When chewed, it tinges the saliva orange-yellow. It yields a **yellowish powder** with a reddish-brown tinge.

The common pie-plant, Crimean rhubarb, from *Rheum rhaponticum* Linné, is a European variety, having properties similar to that of rhubarb, but the astringent principles predominate. It is fusiform, about 100 mm. (4 in.) long and 20 mm. ($\frac{4}{5}$ in.) thick, with a thick orange-red cork, partially removed; a cross-section shows a

PLATE I.

FIG. 1.—Cross-section of Canaigre root (*Rumex hymenosepalus*). × 5½. (Photomicrograph.)

FIG. 2.—Cross-section of Chinese Rhubarb. × 5½. (Photomicrograph.)

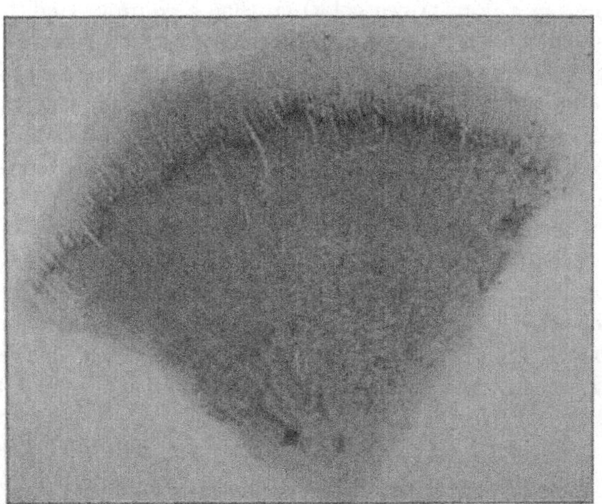

FIG. 3.—Cross-section of *Rheum rhaponticum*. × 5½. (Photomicrograph.)

comparatively regular, radiate structure of red medullary rays traversing a whitish parenchyma and extending into the cortical layer when present; its odor is less aromatic, is less gritty, and its taste more mucilaginous and astringent. *Rumex hymenosepalus*, Canaigre, has been used, in powder, to adulterate powdered rhubarb. For detection, follow general directions for examination of powders, see Part iv, Chap. I.

Choice of Rhubarb.—Select the moderately heavy and compact pieces, which should break with a brittle fracture, presenting a **lively, mottled appearance** of **yellowish and reddish fibers intermingled with white parenchyma**; odor decidedly aromatic; taste bitter, astringent, and gritty, not mucilaginous, tingeing the saliva orange-yellow when chewed. Very light, rotten, or worm-eaten pieces should be rejected. The yield of extractive using dilute alcohol should not be less than 30 per cent. The yield of ash should not exceed 13 per cent. It should be stored in air-tight containers with a few drops of chloroform to prevent the development of insects.

Powder.—Characteristic elements: See Part iv, Chap. I, B.

CONSTITUENTS.—Seemingly a mixture of different coloring principles of a somewhat resinous quality, each having a peculiar solubility of its own: **Chrysophan, $C_{27}H_{30}O_{14}$** (and chrysophanic acid), **emodin, aporetin, phæoretin, erythroretin, rheumic acid, and rheo-tannic acid**; also starch, calcium oxalate, pectin, and arabic acid. Chrysophan is a yellow glucoside yielding, with acidulated water, sugar and chrysophanic acid, $C_{15}H_{10}O_4$, yellow crystals, one of the best solvents for which is hot benzol. According to Hagar, by proper extraction with chloroformic solvent, etc., rhubarb yields not less than 3 per cent. of chrysophanic acid. Chrysophanic acid, or dioxy-methyl-anthraquinone $(C_{14}H_5CH_3(OH)_2O_2)$ is closely related to emodin, which is a trioxy-methyl-anthraquinone $(C_{14}H_4CH_3(OH)_3O_2)$. **Cathartic acid** represents the cathartic principles of rhubarb in a crude but concentrated form. For its preparation, see Senna (240).

EMODIN TEST, in Rhubarb.—Boil 0.100 Gm. of powdered rhubarb with 10 mils of an aqueous solution of potassium hydroxide (1 in 100), allow it to cool, filter, acidulate the filtrate with hydrochloric acid and shake it with 10 mils of ether; on standing, the ethereal layer should be colored yellow. On shaking this ethereal solution with 5 mils of ammonia water, the latter should be colored cherry-red (presence of emodin) and the ethereal layer should remain yellow (presence of chrysophanic acid) U.S.P.

Preparation of Phæoretin.—Wash alcoholic extract with water; dissolve residue in a little alcohol; add ether. This precipitates crude phæoretin.

Preparation of Chrysophanic Acid.—Tincture of rhubarb, after standing for some time, deposits yellow sedimentary crystals. This sediment, dissolved in benzene, deposits the principle on evaporation.

Chrysarobin is a principle easily converted into chrysophanic acid by oxidation. The source of this is Goa powder (from *Andira araroba*). The powder is extracted with hot benzene (benzol), and the liquid allowed to cool. The orange-colored principle separates as the liquid cools.

ACTION AND USES.—**Purgative and astringent.** It has been highly esteemed as an antidysenteric remedy because of the fact that the cathartic principles are accompanied by the antiseptic action of chrysophan, and because catharsis is followed by an astringent and tonic effect upon the mucous lining. Roasting destroys the cathartic quality, when the root becomes simply a bitter astringent. Dose: 15 to 30 gr. (1 to 2 Gm.).

OFFICIAL PREPARATIONS.

Tinctura Rhei (20 per cent., with cardamom),..............................Dose: 1 to 4 fl. dr. (4 to 15 mils).

Tinctura Rhei Aromatica (20 per cent., with cassia cinnamon, cloves, and nutmeg),..................... ½ to 3 fl. dr. (2 to 12 mils).

  Syrupus Rhei Aromaticus (15 per of aromatic tincture),.... 2 to 6 fl. dr. (8 to 24 mils).

Fluidextractum Rhei,.......... 5 to 30 ♏ (0.3 to 2 mils).

  Mistura Rhei et Sodæ (1.5 per cent. with sodium bicarbonate, fl'ext. of ipecac, and spirit peppermint),...

  Syrupus Rhei (Fl'ext. 10 per cent.), 2 to 6 fl. dr. (8 to 24 mils).

Extractum Rhei,..................... 5 to 15 gr. (0.3 to 1 Gm.).

Pulvis Rhei Compositus (25 per cent., with magnesia and ginger),........ 1 to 3 dr. (4 to 12 Gm.).

Pilulæ Rhei Compositæ (each pill containing about 2 gr. of rhubarb, with purified aloes 1½ gr., myrrh, and oil of peppermint),..................... 1 to 3 pills.

121. **RUMEX,** N.F.—YELLOW DOCK. The root of **Ru'mex cris'pus** Linné, and of some other species of Rumex. Off. in U.S.P. 1890. A fusiform root from 100 to 200 mm. (4 to 8 in.) long and 10 to 15 mm. (⅖ to ⅗ in.) thick; **externally reddish-brown,** the upper portion annulate, the lower portion wrinkled; fracture short, exhibiting a rather thick cortical layer and a yellowish or whitish interior, somewhat mottled, the rather porous and horny wood-wedges separated by fine, distinct, reddish medullary rays; inodorous; taste astringent and bitter. Alterative, tonic, and astringent. Dose: 15 to 60 gr. (1 to 4 Gm.). **Extractum Rumicis Fluidum, U.S.P.** 1890. Dose: 15 to 60 ♏ (1 to 4 mils).

122. **CANAIGRE.**—The root of **Ru'mex hymenosep'alus** Torrey, from which a tannin is obtained. This plant resembles common dock, *Rumex crispus,* and flourishes in dry, barren, sandy soil in Southwestern United States and Mexico. It propagates by means of the roots, which grow in clusters of three or four. They are from 50 to 150 mm. (2 to 6 in.) long, and 25 to 50 mm. (1 to 2 in.) thick, reddish-brown to almost black. A cross-section shows a prominent cambium line and a broad radiating center. The tissue is chiefly parenchyma, containing starch, tannin, and a yellowish-red coloring matter. The tannin is yellowish-white, identical with that of rhubarb (rheotannic acid).

123. **POLYGONUM ACRE.**—WATER PEPPER. SMART WEED. (Herb.) Stimulant, diuretic, and emmenagogue. Dose: 1 to 2 dr. (4 to 8 Gm.).

124. **BISTORTA.**—Bistort. The rhizome of **Poly'gonum** bistor'ta Linné. *Habitat:* Europe, Northern Asia, and Northwestern United States, in moist places. An **S**-shaped rhizome (bent upon itself—bistorted), flattened, and transversely striate on upper side, and convex, with depressed root-scars, on lower side; color dark reddish-brown, internally lighter; fracture smoothish, showing a thick bark and a pith of about the same thickness as the bark. Contains tannin, 20 per cent., and starch, with red coloring matter. Tonic and astringent. Dose: 8 to 30 gr. (0.5 to 2 Gm.), in decoction.

CHENOPODIACEÆ.—Goosefoot Family

Weed-like herbs, with minute greenish *flowers;* ovary 2-styled, 1-celled, becoming a 1-seeded thin utricle or caryopsis. Generally bland and innocent.

125. **CHENOPODIUM.**—American Wormseed. The fruit of **Chenopo'dium** ambro'sioi'des Linné, and variety **anthelmin'ticum** Gray. Off. U.S.P. 1890. A small, irregularly globular, seed-like fruit (utricle) not larger than a pin-head and of a grayish-yellow or brownish color. By rubbing the minute grains (fruit) in the hands, the capsular covering to the seeds is broken off, when the shining, lenticular, blackish seeds appear and a peculiar, strong, terebinthinate odor is rendered sensible. Taste pungent and bitter. The variety Anthelminticum gives a similar fruit, but is more aromatic. *Constituents:* Its medical properties depend upon a volatile oil, 3.5 per cent. (125 a), in which it, as well as all the other parts of the plant, abounds. Anthelmintic. Dose: 15 to 30 gr. (1 to 2 Gm.).

125 a. **OLEUM CHENOPODII,** U.S.—Oil of Chenopodium. A thin, yellowish, volatile oil, turning darker or brownish by age, having the peculiar odor and taste of the fruit. It is composed of a hydrocarbon and a heavier oil. Dose: 4 to 8 ℥ (0.25 to 0.50 mil).

PHYTOLACCACEÆ.—Pokeweed Family

Tropical plants represented in the United States by *Phytolacca decandra* and *Rivinia lævis.*

126. PHYTOLACCA, N.F.—Poke Root

The dried root of **Phytolac'ca decan'dra** Linné, collected in autumn.

Botanical Characteristics.—*Stem* red, 3 to 8 feet high, smooth, with an unpleasant odor. *Leaves* large, petiolate, alternate, ovate-lanceolate, entire, cuspidate. *Racemes* lateral, opposite the leaves; *calyx* (perianth) white, lobes ovate, rounded at the apex; *ovary* bright green, *berries* dark purplish-red, pulpy.

Habitat.—North America; naturalized in West Indies and Southern Europe.

Description of Drug.—A large root, often 25 to 75 mm. (1 to 3 in.) in diameter, but cut into various sized **transverse or longitudinal slices** for drying and for the market; externally yellowish-brown, much wrinkled; **internally grayish,** turning yellow on exposure. Structure **loosely fibrous,** almost ligneous, alternating with dark, circular layers; a **transverse slice** shows on its face numerous concentric circles formed by the projecting ends of fibers between which the intervening parenchyma has shrunk; odor slight; taste sweetish, then acrid.

Constituents.—Resin, tannin, starch, gum, sugar, fixed oil, salts, and probably a glucoside. A trace of alkaloid is reported, but the writer has found alkaloidal reaction quite pronounced in concentrated and purified solutions of the drug. Its virtues are imparted to water and alcohol.

Action and Uses.—Alterative, emetic, cathartic. It is not suitable for a cathartic however, because of the narcotic effect often produced. Its most important use is as an alterative in chronic rheumatism, etc., and externally, in the form of ointment, in various skin diseases. Dose: 3 to 30 gr. (0.2 to 2 Gm.). Emetic in the larger dose.

Fluidextractum Phytolaccæ,(U.S.P. 1900), Dose: Emetic, 1.0 mil (15♏).
Alterative 0.2 mil (3♏).

Fig. 68.—Phytolacca—Cross-section of root.   (Photograph.)

127. PHYTOLACCÆ FRUCTUS.—Poke-berries. Globular, purplish or black, berry-like fruits, about 8 mm. (⅓ in.) or less in diameter, adhering together in masses from the exudation and drying of a purplish-red juice. Ten-celled, each containing a single glossy black seed imbedded in a succulent pulp. Inodorous; taste sweetish, slightly acrid, and nauseous. *Constituents:* **Phytolaccin, phytolaccic a**cid, tannin, sugar, gum, and an evanescent coloring matter, turned yellow by alkalies and bleached by sunlight.

CARYOPHYLLEÆ.—Pink Family

Herbs with swollen joints, opposite, entire. and regular flowers; *petals* 4 or 5 mostly removed from the calyx by a short internode. Usually bland herbs; some are highly valued as ornamental plants.

128. SAPONARIA LEVANTICA.—Levant Soapwort. The root of **Gyp′ sophila panicula′ta** Linné. *Habitat:* Italy to Asia Minor. A simple, fusiform root, longitudinally wrinkled, and marked with transverse ridges; used in washing silks and other fabrics. It contains sapotoxin (8.5 per cent.), and the acrid glucoside saponin, yielding by hydrolysis sapogenin, which is used as a detergent.

129. SAPONARIA.—Soapwort. **Sapona′ria officina′lis** Linné. An acrid root, found in Europe and the United States; contains resin, and the glucoside; saponin. The latter is a white powder, soluble in hot water and alcohol, its solution when shaken foams like soap-water. When treated with acids it is split into sugar and a crystallizable principle, sapogenin, soluble in water. Used as an alterative in doses of 15 to 60 gr. (1 to 4 Gm.).

130. **STELLARIA.**—Chickweed. The herb of Stella'ria me'dia Smith. Demulcent and emollient; a poultice is used in ophthalmia, bruises, inflammation, etc.

PORTULACEÆ.—Purslane Family

131. **PORTULACA.**—Garden Purslane. The herb of **Portula'ca olera'cea** Linné. Refrigerant and mild efficient diuretic in ascites; it has a beneficial action in catarrhal affections of the genito-urinary tract. Dose: 1 to 3 dr. (4 to 12 Gm.).

NYMPHÆÆ.—Water Lily Family

Aquatic plants, with peltate or cordate leaves from a prostrate rhizome.

132. **NYMPHÆA.**—Water Lily. The rhizome of Nymphæ'a odora'ta Aiton. *Habitat:* United States, in ponds. About 500 mm. (20 in.) long and 50 mm. (2 in.) thick, usually broken up into grayish, spongy segments, consisting mainly of parenchyma, with a few scattered wood-bundles. Inodorous; taste mucilaginous and astringent. Used as a demulcent and astringent. Dose: 15 to 30 gr. (1 to 2 Gm.).

The rhizome of **Nu'pahr** ad'vena Nuttall, Yellow Pond Lily, has similar properties and uses.

RANUNCULACEÆ.—Crowfoot Family

Herbaceous or somewhat shrubby plants with acrid juice; distinguished by the parts of the flower—sepals, petals, stamens, and pistils—being *free and distinct*—that is, separated and independently situated on the receptacle. The *leaves* are dilated at base, one-half clasping the stem. *Fruit* a pointed or feathery akene, dry pod, or berry. The order has numerous anomalies in the form and structure of the calyx, and corolla in such genera as columbine, aconite, larkspur, ranunculus, anemone, etc., which, nevertheless, agree in the separation of their sepals and petals, the insertion of their numerous stamens, direction of their anthers, structure of seed, etc.

*Synopsis of Drugs from the Ranunculaceæ**

A. *Rhizomes.*	B. *Herbs.*	D. *Tuber.*
CIMICIFUGA, 133.	*Pulsatilla, 140.	**ACONITUM,** 146.
HYDRASTIS, 134.	*Adonis Vernalis, 141.	E. *Leaf.*
Actæa, 135.	Ranunculus, 142.	Hepatica, 147.
*Coptis, 136.	C. *Seeds.*	F. *Root.*
Helleborus Niger, 137.	**STAPHISAGRIA,** 143.	Pæonia, 148.
Helleborus Viridis, 138.	*Delphinium, 144.	
Xanthorrhiza, 139.	Nigella, 145.	

133. CIMICIFUGA.—Cimicifuga

BLACK SNAKEROOT. BLACK COHOSH

The dry rhizome and roots of **Cimicif'uga racemosa** Nuttall.

Botanical Characteristics.—*Stem* 4 to 8 feet high, from a thick rhizome; *leaves* alternate, ternately decompound; *flowers* regular, small, white, in wandlike racemes often 3 feet long; *sepals* 5, petaloid; *petals* from 1 to 8, small, on claws, 2-horned at apex; *stamens* numerous; *pistils* 1 to 3; *fruit* 1 to several dry, dehiscent pods.

11

SOURCE.—This plant is common in rich woodlands of the United States, westward to Iowa and northward to Canada. *Actæ'a racemo'sa* is mentioned by Flückiger as a synonym of this plant.   A similar plant,

FIG. 69.—*Cimicifuga racemosa*—Plant and rhizome.

Actæ'a spicat'a, furnishing a rhizome resembling black snakeroot, is common in Europe; it differs, however, in having juicy berries instead of dry follicles.

DESCRIPTION OF DRUG.—A short horizontal rhizome from 10 to 25 mm. (⅖ to 1 in.) thick, with numerous branches—remains of aërial stems

—each terminated by a deep cup-shaped scar; on the lower side are found **numerous brittle rootlets** from 1 to 2 mm. ($\frac{1}{25}$ to $\frac{1}{12}$ in.) thick; **externally** brownish-black; fracture of rhizome, horny; odor slight (the powder, however, has a heavy odor); taste bitter and acrid.

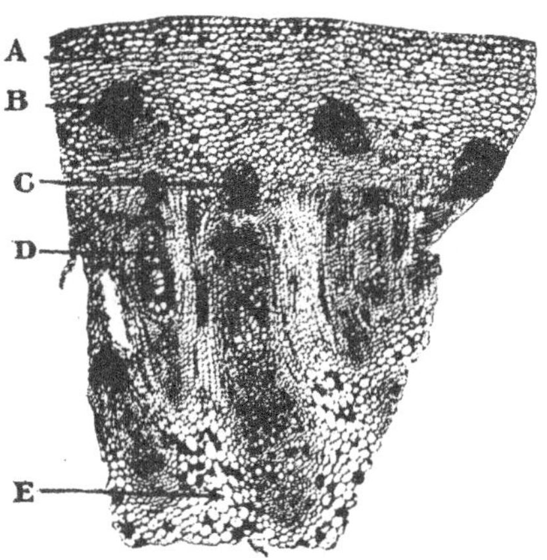

FIG. 70.—Cimicifuga—Cross-section of Rhizome. (18 diam.) A, Parenchyma of cortex. B Vascular bundle. C, Group of bast fibers. D, Xylem. E, Medulla. (Photomicrograph.)

Cross-section of the rhizome exhibits a large, whitish pith, around which, more or less stellately arranged, are wood-wedges separated by

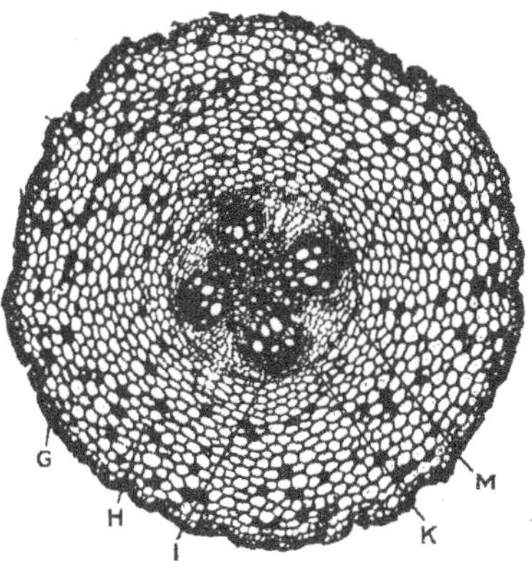

FIG. 71.—Cimicifuga rootlet—Cross-section. (26 diam.) G, Cork. H, Parenchyma of cortex. I, Phloëm. K, Xylem. M, Endodermis. (Photomicrograph.)

medullary rays. Bark hard and thickish. The rootlets display, under the microscope, a thick cortical layer, the space within which contains converging wedges of open, woody tissue, three to five in number, forming a Maltese cross. The stellate arrangement of the woody wedges of the rootlets is one of the best distinguishing characteristics.

Powder.—Characteristic elements: See Part iv, Chap. I, B.

CONSTITUENTS.—Besides the ordinary vegetable principles—fat, sugar, tannin, and starch—there exists a **resin** which has been by some assigned as the active medicinal constituent. This resin, amounting to about 3½ per cent., is contained in the resinoid **cimicifugin** or **marcotin** of the market. An acrid, crystalline principle, soluble in chloroform, ether, and alcohol, and not precipitated by lead acetate, is also said to exist in the root. Ash, not more than 10 per cent.

Preparation of Cimicifugin.—By precipitating the concentrated tincture with water, a crude article is prepared which is known as the resinoid. A purer form is made by precipitating the tincture of the fresh drug with lead subacetate, removing the lead from solution with H_2S, and evaporating. Soluble in alcohol and chloroform.

ACTION AND USES.—Antispasmodic, diaphoretic, and expectorant. It acts like digitalis on the circulation, and as a sedative upon cardiac ganglia; small doses stimulate digestion and secretion; used in rheumatism and disturbances of the menstrual function. It is a powerful uterine stimulant. In large doses cimicifuga causes nausea, headache, vertigo, tremors, muscular relaxation, slowing and weakening of the pulse. Dose: 15 to 30 gr. (1 to 2 Gm.).

OFFICIAL PREPARATIONS.

Fluidextractum Cimicifugæ,............Dose: 5 to 30 ℥ (0.3 to 2 mils).
Extractum Cimicifugæ,.............. 3 to 5 gr. (0.2 to 0.3 Gm.).

134. HYDRASTIS.—HYDRASTIS

GOLDEN SEAL. YELLOW PUCCOON

The dried rhizome and roots of **Hydras'tis canaden'sis** Linné. Yielding not less than 2.5 per cent. of ether soluble alkaloids of Hydrastis.

BOTANICAL CHARACTERISTICS.—Plant about 8 inches high, from a thick, knotty rhizome. The single radical leaf simple, 5-lobed; *stem* 2-leaved at summit; *flowers* terminal, single, greenish; *calyx* of 3-petaloid sepals, regular; *fruit* a head of 1–2-ovuled berries.

SOURCE.—The area of the country over which hydrastis grows in sufficient abundance to be a commercial source of the drug is embraced in Ohio, Indiana, Kentucky, Michigan, and West Virginia. It is also found in other portions of the Eastern United States. Large quantities of the drug are now being cultivated. One of the fields the writer has visited, is located in Douglas, Michigan, "Seal Growers," as they are called, have a coöperative Society to promote their interests in the growing of this plant and ginseng, especially.

Description of Drug.—A knotty, contorted rhizome about 40 mm. (1⅗ in.) long and 5 mm. (⅕ in.) thick; on the upper side are several scars which mark the positions and detachment of former herbaceous stems; these scars (cup-like projections) have given rise to the name "golden seal." **Externally** rough, of a dull yellowish-brown color, annulate, and beset with numerous slender rootlets; **internally** of a lemon-yellow color; breaks with a short, resinous fracture; a

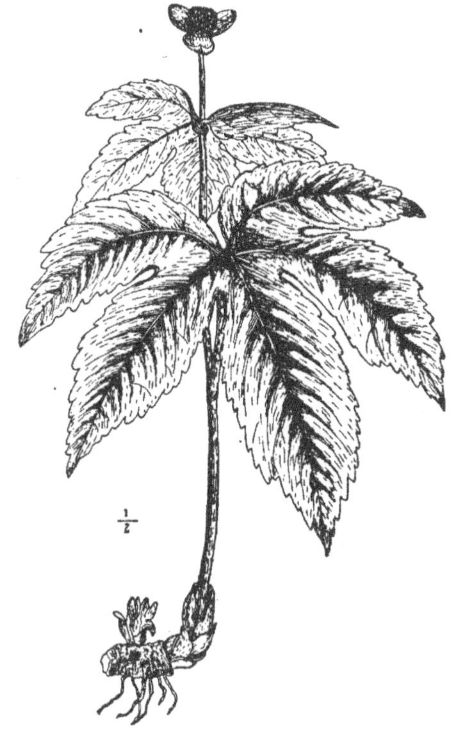

Fig. 72.—*Hydrastis canadensis.*

cross-section shows a thick bark, narrow wood-wedges, and broad medullary rays which radiate from a large pith. The rootlets show a woody center surrounded by a thick parenchymatous cortical tissue which is bordered by an outer row of compressed cells; odor distinct; taste bitter. Two to three hundred thousand pounds of the drug are annually consumed.

Powder.—Characteristic elements: See Part iv, Chap. I, B.

Constituents.—The two alkaloids, hydrastine, $C_{21}H_{21}NO_6$ (colorless and slightly acrid), and **berberine** (yellow and intensely bitter), are the principal constituents. **Berberine, $C_{20}H_{17}NO_4$, is very widely distributed in nature,** being found in drugs from several different families of plants. **Hydrastine, when pure, is in perfectly colorless,** very brilliant, glassy crystals. As a rule, however, they are white and opaque, owing to the presence of numerous fractures. The yellow color of berberine adheres very tenaciously to the hydrastine, so that the absolutely colorless hydrastine is difficult to obtain. **Canadine, $C_{20}H_{21}NO_4$,** tetrahydroberberine, the sulphate of which is soluble in water and alcohol. **The resinoid, hydrastin, should not be confounded with the active alkaloid.** This resinoid is made by precipitating a concentrated alcoholic tincture of hydrastis with acidu-

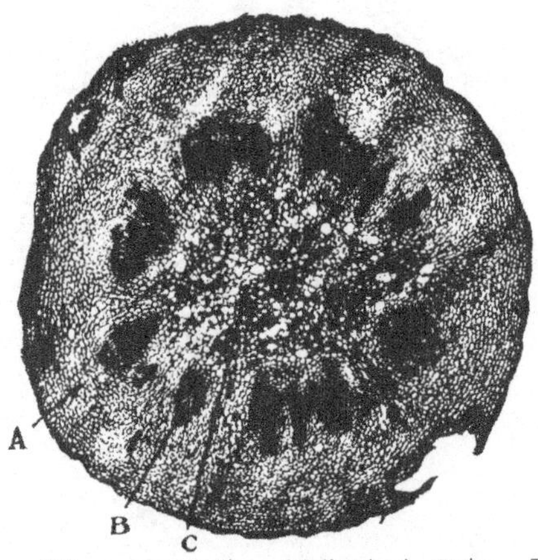

FIG. 73. — Hydrastis — Rhizome, cross-section. (18 diam.) A, Cortex. B, Vascular bundle. C, Medulla. (Photomicrograph.)

lated water, and is probably, in the main, an impure muriate of berberine. Hydrastinine, which Falk regards as a valuable remedy, is made by decomposing the alkaloid, hydrastine, with dilute nitric acid and gentle heat, when opianic acid is also formed.

Preparation of Hydrastine.—Percolate drug with water; precipitate berberine by adding HCl; to filtrate add ammonia in excess. The impure hydrastine which then deposits is dissolved in alcohol, filtered through charcoal, and crystallized.

Preparation of Berberine.—(Obtained also from Berberis vulgaris and allied drugs.) Exhaust powdered root with boiling water, evaporating to soft extract; exhaust this with alcohol; add water. Distil off alcohol; add H_2SO_4 in excess, when berberine sulphate crystallizes in yellow needles.

ACTION AND USES.—Until the introduction of the white alkaloid hydrastine, the drug was used almost exclusively as a local astringent; but of late years, since the many physiological experiments with this alkaloid, it has been used internally in chronic inflammations of the mucous membrane.   Hydrastis is now quite largely employed in the treatment of depraved mucous membranes, as, for example, in chronic rhinitis, the atonic stomach of drunkards, chronic intestinal catarrh, catarrhal jaundice, vaginal leucorrhea, and the later stages of gonorrhea.   It has been recommended in the treatment of uterine hemorrhages resulting from endometritis, and is said to act well in cases in which Ergot has proved useless.

In dyspepsia it has been used as a stomachic stimulant, and has received praise in the vomiting of pregnancy.   Dose: 30 gr. (2 Gm.). Hydrastine is said to have antiperiodic properties and is given in doses of $\frac{1}{32}$ gr. (0.002 Gm.).

OFFICIAL PREPARATIONS.

Extractum Hydrastis,................Dose: 8 gr. (0.5 Gm.).
Fluidextractum Hydrastis, ...   5 to 30 ℥ (0.3 to 2 mils).
Tinctura Hydrastis (20 per cent.),...   10 to 60 ℥ (0.6 to 4 mils).
Glyceritum Hydrastis (each mil contains 1 Gm. of drug).   Used externally.

135. **ACTÆA ALBA.***—WHITE COHOSH.   The rhizome of Actæ′a alb′a Bigelow. *Habitat:* Southern and Eastern United States.   Often found in the European market mixed with black hellebore; its appearance, however, is more like cimicifuga.   Violent purgative, irritant, and emetic.

136. **COPTIS, N.F.**—GOLD THREAD.   The herb of Cop′tis trifol′ia Salisbury. *Habitat:* Northern and Eastern United States.   The drug as found in commerce consists mainly of long, thread-like, yellow rootlets, attached to a slender, terete rhizome, mixed with trifoliate leaves.   Contains *berberine* and a white alkaloid resembling hydrastine.   Tonic.   Dose: 15 to 60 gr. (1 to 4 Gm.) in decoction.

137. **HELLEBORUS NIGER.**—BLACK HELLEBORE.   The rhizome and roots of **Helle′borus ni′ger** Linné.   *Habitat:* Central and Southern Europe.   Irregular and knotty; externally brown-black; internally grayish, with a thick bark; taste sweetish, bitter, and acrid; odor slight, peculiar.   Poisonous; anthelmintic, drastic cathartic, and emmenagogue.   Dose: 5 to 20 gr. (0.3 to 1.3 Gm.).

138. **HELLEBORUS VIRIDIS.**—GREEN HELLEBORE.   The rhizome and roots of **Helle′borus viri′dis** Linné.   This resembles above, but is smaller.   Used as a diuretic, cathartic, and emmenagogue.   Dose: 5 to 20 gr. (0.3 to 1.3 Gm.).   It should not be confounded with veratrum viride (also called green hellebore), a cardiac and nervous sedative.

139. **XANTHORRHIZA.**—YELLOW-ROOT.   The rhizome of **Xanthorrhi′za** apiifol′ia L′Heritier.   *Habitat:* Southern and Central United States.   About 500 to 1,000 mm. (20 to 40 in.) long, and 10 mm. (⅖ in.) thick; externally of a bright yellowish-brown color; internally yellow; inodorous and bitter. Contains *berberine*, the alkaline base of berberis vulgaris; it is a matter of record that in many, perhaps most, berberine-yielding plants, **a colorless alkaloid accompanies berberine**, but, according to Lloyd, a second alkaloid does not exist in this drug.   Used as a tonic.   Dose: ½ to 1 dr. (2 to 4 Gm.).

140. **PULSATILLA, N.F.**—PASQUE FLOWER.   The herb of **Anem′one pulsatil′la** and of **Anem′one praten′sis** Linné, collected soon after flowering.   Off. U.S.P. 1890.   The drug never comes into the market in a condition in which the leaf or other parts are readily recognizable, as they are most always broken or compressed.   The U.S.P. 1890, directed that the herb should be carefully preserved, and not kept longer than one year.   Even the drying of the plant is said to render the drug unreliable.   *Constituents:* A peculiar acrid crystallizable principle exists in the plant known as **anemonin** ($C_{10}H_8O_4$) **an** acrid, unstable principle not well understood.   Some

---

* Drugs treated of in small type and solid paragraphs are unofficial.

authorities state that it undergoes decomposition after its solution, under conditions that are not precisely known, into anemonic acid ($C_{10}H_{10}O_5$) and anemoninic acid ($C_{10}H_{18}O_6$), etc.; others state that it is a volatile, fluid, acrid principle, very susceptible of decomposition.

Preparation of Anemonin.—If aqueous distillate be treated with chloroform, the latter, on evaporation, yields a residue—anemonin. Dose: $1\frac{1}{2}$ to 3 gr. (0.1 to 0.2 Gm.).

Diuretic, diaphoretic, mydriatic, irritant. The action of pulsatilla is said to resemble aconite as a cardiac sedative. One author says it is equivalent to

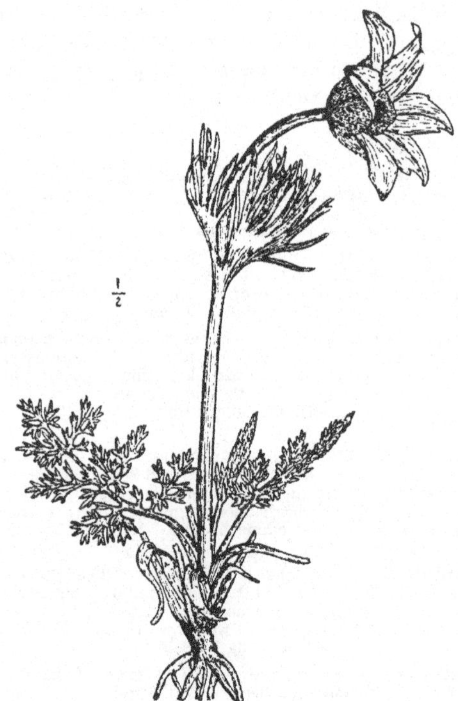

FIG. 74.—*Anemone pratensis.*

senega in convulsive coughs and in bronchitis. The recent tincture, in 5-drop doses (made according to the formula of the tincture of recent herbs, U.S.P. 1890, is highly esteemed by some practitioners. The drug is not infrequently classed among the most useful emmenagogues. Dose: 1 to 5 gr. (0.065 to 0.3 Gm.).

141. **ADONIS VERNALIS, N.F.**—FALSE HELLEBORE. The herb of Adon'is ver-nal'is Linné. This rather obscure drug owes its poisonous quality and medicinal activity to a glucoside, **adonidin**, whose physiological action seems to be almost identical with that of digitalin, except that it is more powerful, and not cumulative. Like digitalis, it is used in heart disease and dropsy, slowing the heart's action, and making it more regular and forcible; it greatly increases urinary secretion. Dose: 2 to 10 gr. (0.12 to 0.6 Gm.), in infusion.

142. **Ranunculus.**—CROWFOOT. BUTTER CUP. The herb of **Ranun'culus bulbo'-sus** Linné. *Habitat:* Europe and North America. Base of stem thick; flowers yellow, the ovaries of which form akenes with a short, curved beak; inodorous, with acrid taste. Used externally as an irritant.

143. STAPHISAGRIA.—STAPHISAGRIA

STAVESACRE

The ripe seed of **Delphin'ium staphisag'ria Linné.**

BOTANICAL CHARACTERISTICS.—*Stem* 3 to 4 feet high, erect, more or less colored purple; *leaves* long petiolate, alternate, palmately 5–9-divided, blotched with purple; *flowers* in loose spoke-like racemes, varying from light-blue to purple; irregular; *sepals* 5, petaloid, upper one prolonged into a spur; *petals* 4, small; *fruit* 3, hairy follicles.

SOURCE.—This herb is a native of Italy, Greece, the Greek Islands, Asia Minor, Mediterranean regions, and Canary Islands. It was introduced into England in 1596.

DESCRIPTION OF DRUG.—About 5 mm. ($\frac{1}{5}$ in.) long, 3 to 4 mm. ($\frac{1}{8}$ to $\frac{1}{6}$ in.) thick; **externally** flattish, tetrahedral, the broadest side convex; testa brownish, with reticulate ridges, rough and deeply pitted; **internally** it contains a whitish, oily albumen, inclosing a small, straight embryo in its sharper end. The outer layer of the testa is made up of thin-walled, narrow cells, which become larger near the edges of the seed and in the superficial wrinkles. They contain a small number of minute starch granules. The interior layer exhibits a single layer of small, densely-packed cells. The albumen is composed of the usual tissue loaded with granules of albuminoid matter and drops of fatty oil. Nearly inodorous; taste bitter and astringent. Dose 1 gr. (0.06 Gm.).

Powder.—Dark greenish. Characteristic elements: The angular cells of the parenchyma of the endosperm with aleurone and oil globules; very large epidermal cells, brown, thick-walled, with irregular thickenings.

CONSTITUENTS.—Besides fixed oil, etc., one of the most prominent constituents **is a poisonous alkaloid, delphinine,** which exists in the form of a malate. This alkaloid, however, is said to be **composed** of **several distinct principles.** Marquis has separated four distinct alkaloids from the seed.

Preparation of Delphinine.—Treat the decoction with magnesia, exhaust the precipitate with alcohol, and evaporate. The crude alkaloid thus obtained consists of three distinct principles—resin, staphisagrine, and delphinine. Pure delphinine is soluble in alcohol and ether.

ACTION AND USES.—Stavesacre is mostly used as a parasiticide to destroy vermin, especially against pediculi vestimentorum—inhabiting the garments next to the skin. A tincture in cologne spirit has been used in some districts as a substitute for tincture of cocculus indicus, applied to the scalp as an antiparasitic.   Internally, the action resembles aconite in its effects upon the heart and respiration.   Dose: 1 to 2 gr. (0.065 to 0.130 Gm.).   Poisonous doses are rapidly diffused, and antidotal measures should be rapidly applied. (Fluidextractum staphisagriæ, used externally as a parasiticide.)

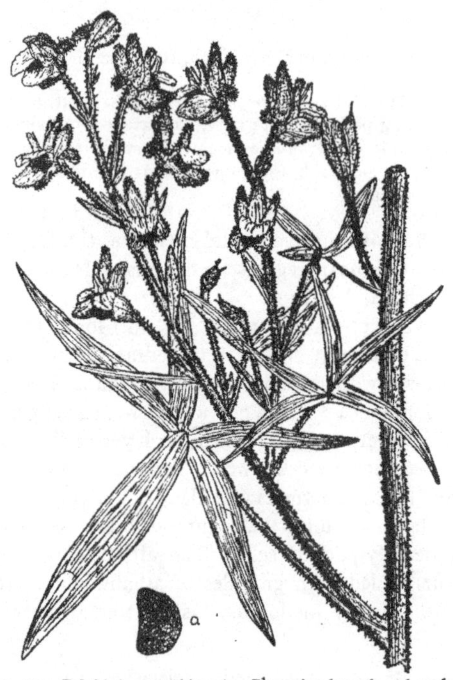

Fig. 75.—*Delphinium staphisagria*—Flowering branch and seed a.

144. DELPHINIUM, N.F.—Larkspur Seed.   The seed of **Delphin'ium consol'ida** Linné. *Habitat:* Central Europe; cultivated and naturalized in some parts of the [United States.   A flattish, tetrahedral seed, 1 to 1.5 mm. ($\frac{1}{25}$–$\frac{1}{16}$ in.) broad; edges sharp, testa black and roughly pitted; internally, it consists of whitish, oily albumen, inclosing a small, straight embryo; inodorous; taste bitter and acrid; contains delphinine.   Used as a diuretic, cathartic, and emetic; poisonous.   Dose: $\frac{1}{2}$ to 3 gr. (0.03 to 0.2 Gm.).

145. NIGELLA.—Nigella.   The seeds of **Nigel'la damasce'na** Linné. *Habitat:* Levant; cultivated.   Triangular-ovate, about 2.5 mm. ($\frac{1}{10}$ in.) long; testa brittle, dull-black; embryo straight and small, with pointed ends.   It has a strawberry-like odor, and bitter taste.   Used as an emmenagogue and diuretic.

146. ACONITUM.—Aconite

MONKSHOOD

The dried tuberous root of Aconi'tum napel'lus Linné.   Yielding, by official assay, not less than 0.5 per cent. of ether soluble alkaloids, also assayed biologically.

The minimum lethal dose of fluidextract should not be greater than 0.00004 mil for each gramme of body weight of guinea-pig.

BOTANICAL CHARACTERISTICS.—*Stem* 3 to 4 feet high, smooth and erect; *leaves* nearly sessile, alternate, palmately 5-divided; *root-leaves* long-petioled; *flowers* deep violet, irregular, very showy, in racemes; *sepals* 5, petaloid, the upper one hooded or helmet-shaped; *petals* 2, concealed.

FIG. 76.—*Aconitum Napellus*—Flowering branch and tuber.

SOURCE AND VARIETIES.—This genus of poisonous herbs, including a number of species, is found throughout cold, mountainous districts of Europe, in the Himalayas, and in Northwestern North America. It is one of the oldest and commonest plants of the English garden, and is often found in dangerous proximity to horseradish (Royle). Hindu writers mention no less than eighteen different kinds of "bish" —the vernacular for aconite. Ten of these are said to be unfit for medicinal use on account of their extremely poisonous nature. The root (tuber) of *A. napellus* is the source of the medicinal preparations of this drug. Nepaul aconite is the source of the extremely active alkaloid, *pseudaconitine* (see below). *A. fischeri* produces Japanese aconite root. It yields japaconitine, stated to be identical with aconitine.

DESCRIPTION OF DRUG.—Almost napiform, abruptly tapering, from 40 to 100 mm. long, about the thickness of a finger at the top, which is tuberculated; **externally** dark-brown, wrinkled longitudinally at lower portion, stem scars visible, rootlets usually detached; **fracture short,** horny or starchy, exhibiting sometimes a spongy or resinous, white, grayish, or brownish tissue; taste at first sweetish, then acrid and tingling, followed by numbness. This peculiar **tingling sensa-**

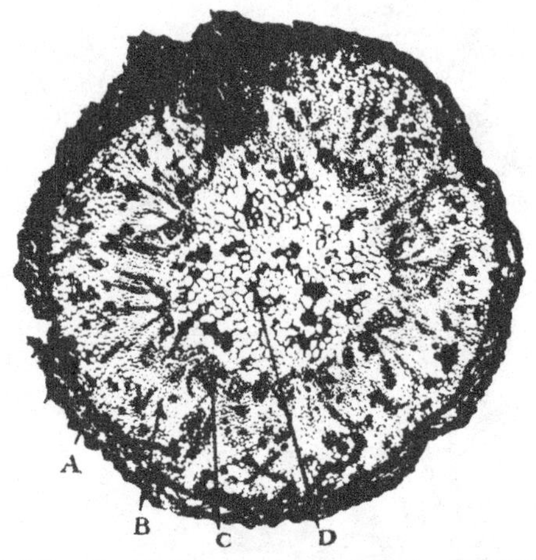

FIG. 77.—Aconite tuber—Cross-section. (14 diam.) A, Cork. B, Parenchyma of cortex. C, Vascular bundle. D, Medulla. (Photomicrograph.)

tion of the tongue is one of the most prominent characteristics upon which the toxicologist depends for the recognition of this drug and its preparations.   At the upper portion of the root there often projects a lateral branch connecting a second tuber, which is an offspring of the other.   A cross-section of the tuber shows a thick bark and **a pith often in the form** of a **star,** the two being separated by a nucleus sheath; the cambium, following the outline of the pith, is also **5- to 7-angled,** and at the terminal and basal extremities of each **ray** are found small groups of vascular bundles; these, however, are inclined to follow the whole cambium line.

Powder.—Microscopical elements of: See Part iv, Chap. I, B.

ADULTERANTS.—With allied aconite roots, defective roots, and horse-radish.   The root of European masterwort resembles aconite root, but it is aromatic and pungent.

CONSTITUENTS.—The principal constituent is **aconitine,** $C_{34}H_{47}NO_{11}$ (0.5 per cent.), forming about one-third the total alkaloid of **the** root. This is white, usually amorphous, but with difficulty may be obtained in rhombic, tabular crystals; almost insoluble in cold water, soluble in alcohol, ether, and diluted acids. Other related principles exist in the drug combined with aconitic acid ($H_3C_6H_3O_6$), but our knowledge of them is not satisfactory. The crystallized alkaloid melts at 189° to 190°C., and yields acetic acid at slightly higher temperature.

Pseudaconitine, $C_{36}H_{49}NO_{12}$, from *Aconitum ferox*, is highly poisonous. Atisine, $C_{22}H_{31}NO_2$ (from *Aconitum heterophyllum*), does not

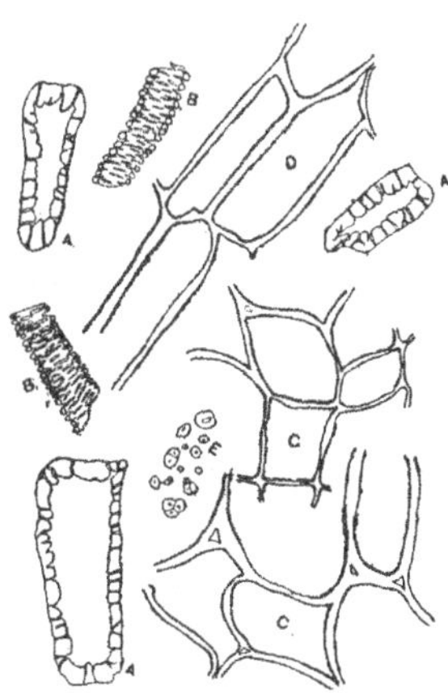

FIG. 78.—Powdered Aconite Tuber. (100 diam.) A, Stone cells. B, Fragments of water tubes. C, Parenchyma, cross-section. D, Parenchyma, longitudinal section. E, Starch.

present any close analogy to the alkaloids of the other and well-known species of aconite (*A. napellus, A. ferox,* and *A. japonicum*). In small doses it is said to be non-toxic, but its action, according to some reports, resembles that of aconite.

Commercial aconitine contains some of the allied principles, which are separated from the alkaloid with difficulty. Ash, not exceeding 6 per cent.

Preparation of Aconitine.—After extracting oil and resin by a suitable solvent, an alcoholic extract is made which is treated with hot water. The aqueous solution is precipitated by adding NH_4OH in excess. This precipitate is exhausted with ether—ethereal solution distilled to dryness. Purify residue by dissolving in acidulated (H_2SO_4) water, again precipitating with NH_4OH, etc. This process yields a commercial product which is not free from pseudoaconitine.

ACTION AND USES.—Antipyretic to a certain extent by reducing circulation; depressant of the sensory nerve-ends, the heart, the respiration, and spinal system. It relaxes the inhibitory apparatus of the heart, and paralyzes the cardiac muscle and its contained ganglia, the respiratory centers, and the spinal cord in all its functions—sensory, reflex,

and motor—but does not affect the cerebrum. Murrell has called
attention to the fact that the English alkaloid is seventeen times
stronger than the German, while the French is variable, but generally
between these; the crystalline variety (Duquesnel's or Merck's aconitine) is therefore to be preferred on account of its uniform strength.
The dose of the commercial aconitine is $\frac{1}{64}$ gr.; the crystallized alkaloid, however, is given in doses of only from $\frac{1}{300}$ to $\frac{1}{250}$ gr.

Dose of drug: 1 gr. (0.06 Gm.).

OFFICIAL PREPARATIONS.

Fluidextractum Aconiti,..............Dose: $\frac{1}{4}$ to 2 ♏ (0.015 to 0.12 mil).
Extractum Aconiti,................... $\frac{1}{6}$ to $\frac{1}{3}$ gr. (0.010 to 0.02 Gm.)
Tinctura Aconiti (10 per cent.),........ $\frac{1}{2}$ to 4 ♏ (0.03 to 0.25 mil)

147. **HEPATICA.**—LIVERWORT. The leaves of **Anem'one hepa'tica** Linné.
Habitat: North America and Europe. Heart-shaped, about 50 mm. (2 in.)
long, slightly leathery; inodorous; astringent and bitter. The more correct
synonym for this plant is liverleaf, as the term liverwort is applied to a family
of cryptogamic, moss-like plants—*Hepaticæ.* Used as a demulcent and
tonic. Dose: $\frac{1}{2}$ to 2 dr. (2 to 8 Gm.) in decoction.

148. **PÆONIA.**—PEONY. The root of **Pæonia officinalis** Linné. Seldom used,
although at one time a popular remedy in epilepsy, diarrhea, and as an
emmenagogue. Occasionally used in chorea, whooping-cough, etc. Dose:
15 to 60 gr. (1 to 4 Gm.), in infusion.

MAGNOLIACEÆ.—Magnolia Family

Trees and shrubs, mostly of subtropical regions. *Leaves* coriaceous; alternate,
simple, usually pellucid-punctate, entire, or rarely dentate; *flowers* axillary or
terminal, usually solitary, perfect, or, in a few genera, unisexual; *sepals, petals,
stamens,* and *pistils* numerous and hypogynous. *Fruit* various, cone-like, or
forming a stellate group of whorl (illicium), or capsular with ventral or dorsal
dehiscence.

Synopsis of Drugs from the Magnoliaceæ

A. *Fruit.*   B. *Barks.*
   Illicium, 149.   Magnolia, 150.
                                                    Wintera, 151.
                                                    Liriodendron, 152.

149. **Illicium.**—STAR ANISE. The dry fruit of **Illi'cium ve'rum** Hooker filius.
Off. U.S.P. 1890. The fruit is pedunculate, and consists of light, stellately-
arranged, one-seeded carpels, which are boat-shaped and united around
a short central column rising from an oblique pedicle. Each carpel is 12
or 15 mm. ($\frac{1}{2}$ to $\frac{3}{5}$ in.) long, woody, wrinkled, with a straight beak; rusty-
brown in color, and split at the ventral suture, exposing the flattish, bright,
glossy-brown, oval seed; odor intermediate between fennel and anise; taste
(residing in the carpel,) aromatic and sweet; seed not aromatic, but oily.
Adulterated with *Illicium religiosum* Siebold (found growing around Buddhist
temples in southwest China, whence its name), a poisonous plant cultivated
in China and Japan, which resembles it in appearance, but is more woody,
has a curved beak, a clove-like odor, and a disagreeable taste. *Constituents:*
A volatile oil resembling the oil of pimpinella anise. The former oil is solidified
at 35°C., and the latter between 50° and 60°C., almost entirely composed of
anethol ($C_{10}H_{12}O$), with small amounts of terpenes, safrol, anisic acid, etc.
It has stimulant, anodyne, diuretic, and carminative properties which
reside exclusively in the volatile oil. Dose: 5 to 30 gr. (0.3 to 2 Gm.).

150. **MAGNOLIA.**—Magnolia. The bark of **Magno'lia glau'ca** Linné. *Habitat:* Middle and Southern United States. A thin-quilled bark of a gray color, or sometimes light brown, fissured, and covered with numerous scattered warts; the inner surface smooth and of a light brown color; fracture short, toward the inner portion somewhat fibrous; nearly inodorous, with a bitter, spicy, and pungent taste. It contains a volatile oil, resin, tannin, coloring matters, gum, and a crystalline glucoside, **magnolin.** Used as a diaphoretic, tonic, and febrifuge. Dose: 10 to 80 gr. (2 to 4 Gm.) in decoction.

151. **WINTERA.**—Winter's Bark. From **Dri'mys winte'ri** Forster, a South American tree. It has an aroma similar to that of canella and cinnamon, for which drugs it has been substituted, and is known in some places as

FIG. 79.—*Illicium verum*—Flowering branch and fruit.

Winter's Cinnamon. The bark of *Drimys granatensis* from New Granada is said to have been offered as Coto bark. It also has an astringent, pungent, as well as aromatic taste. Dose: 15 to 30 gr. (1 to 2 Gm.).

152. **LIRIODENDRON.**—Tulip-tree Bark. From Liroden'dron tulipi'fera Linné. *Habitat:* United States westward to Kansas. In quills and curved pieces obtained from the branches. These quills and pieces are about 2 mm. ($\frac{1}{12}$ in.) thick; outer surface purplish-brown, with thin ridges forming elongated meshes; nearly inodorous; taste pungent and bitter. Tonic, febrifuge and vermifuge. Dose: 1 to 2 dr. (4 to 8 Gm.) in infusion or fluid extract.

. Preparation of Liriodendrin.—Concentrate the alcoholic tincture; add water until a permanent turbidity commences to appear.　Set aside to evaporate spontaneously.　It forms, when purified, white needles or small scales.　Insoluble in water, soluble in ether and alcohol.

FIG. 80.—Flowering branch of *Liriodendron tulipifera.*

CALYCANTHACEÆ.—Calycanthus Family

153. **CALYCANTHUS.**—FLORIDA ALLSPICE.　The bark of **Calycan'thus flor'**-idus.　An aromatic stimulant, used in diarrhea mixtures.　Dose: 10 to 30 gr. (0.6 to 2 Gm.).

MYRISTICACEÆ

A. *Seed.*　　　　　　　　　　　　C. *Fixed Oil.*
　MYRISTICA, 154.　　　　　　Oleum Myristicæ Expressum, 154 b.
B. *Volatile Oil.*　　　　　　　　　D. *Arillode.*
　OLEUM MYRISTICÆ, 154 a.　　　*Macis, 155.

MYRISTICA.—NUTMEG

NUTMEG

The kernel of the ripe seed of **Myris'tica frag'rans** Houttuyn.

BOTANICAL CHARACTERISTICS.—*Tree* about 30 feet high.　*Leaves* oblong-oval, entire, glossy above, whitish beneath, aromatic.　*Flowers* diœcious; *male flowers* in axillary clusters; *female flowers* single, solitary, and axillary, both very small and of a pale yellow color.

HABITAT.—Molucca Islands; cultivated in adjacent East India islands, and especially in the Dutch Banda Islands, whence most of the nutmegs are imported for market.

DESCRIPTION OF DRUG.—A roundish or oval kernel about 25 mm. (1 in.) long; externally light grayish-brown, marked with worm-shaped furrows and covered with lime (done by the Dutch growers to kill the germ, thinking in this way to monopolize its cultivation).　They are hard and not readily pulverizable, but can easily be cut or grated, showing **a waxy luster**; **internally** yellowish, **a cross-section having a mottled appearance,** due to the penetration to the albumen of the inner seed-coat in narrow brown strips; these strips contain oily material; hilum and micropyle on the broad end, chalaza near the upper end, united by a groove corresponding to the raphé; the embryo is small, in a cavity at the base; odor strongly aromatic; **taste** warm and aromatic.

The male, wild, or long nutmeg, as it is variously termed, is occasionally found in market; it is much longer than the official nutmeg,

FIG. 81.—*Myristica fragrans*—Branch and fruit.

elliptical, destitute of the dark brown inner veins, and of a bitter and disagreeable taste. Penang and Singapore nutmegs are unlimed.

California nutmeg, so called, is the seed of *Torrega Californica* (nat. ord. Coniferæ); testa smooth, brownish, internally marbled, resembling nutmeg, but has a terebinthinate odor and taste.

Powder.—Characteristic elements: See Part iv, Chap. I, B.

CONSTITUENTS.—The greater portion of nutmeg (25 to 30 per cent.) consists of a **fixed oil;** this is official in the British Pharmacopœia and

12

is called oil of mace or mace butter; it contains chiefly myristin, with some myristic acid, olein, palmitin, resin, and volatile oil (see 154 b). The aromatic properties of nutmeg depend upon 2 to 8 per cent. of **volatile oil.** Ash, not exceeding 5 per cent.

ACTION AND USES.—Aromatic stimulant and stomachic. Used as a corrective and as a condiment. In large doses it possesses narcotic properties. Dose: 8 to 30 gr. (0.5 to 2 Gm.).

OFFICIAL PREPARATIONS.

154 a. **Oleum Myristicæ,** U.S.—OIL OF NUTMEG. A thin, colorless or pale straw-colored volatile oil, lighter than water, and having the characteristic properties of nutmeg; on standing for a considerable length of time it becomes darker and thicker, and deposits a crystalline fatty glyceride of myristic acid. It contains a hydrocarbon, pinene, myristicin, and an oxygenated compound, myristicol, isomeric with carvol. Action and uses same as nutmeg, but rarely used. Dose: 1 to 3 ♏ (0.065 to 0.2 mil).

154 b. **OLEUM MYRISTICÆ EXPRESSUM.**—EXPRESSED OIL OF NUTMEG. MACE BUTTER (see Myristica Constituents). Unctuous blocks, marbled whitish and brown. Mostly used externally.

155. **Macis,** N.F. (U.S. 1890).—The thick membrane or "arillode" immediately investing the kernel of the nutmeg. It comes in narrow bands, irregularly slit above into somewhat branched and lobed divisions, united at the base in an unbroken band; reddish or orange-yellow in color, with a fatty feeling when scratched or pressed; peculiar aromatic odor and taste. It contains volatile oil (about 8 per cent.), a red fixed oil, gum, resin, sugar, and proteids, but no starch. Aromatic stimulant and tonic; mostly used as a flavoring agent. Dose: 5 to 20 gr. (0.3 to 1.3 Gm.).

MENISPERMACEÆ.—Moonseed Family

Woody climbers, mostly tropical, with peltate or palmate alternate exstipulate *leaves*, and small dioecious, greenish, or whitish *flowers* in axillary panicles. *Sepals* and *petals* alike, in three rows—the petals sometimes wanting. The *stamens* equal or exceed the petals in number. *Pistils* 2 to 6, with nearly straight ovaries, which, however, are incurved in fruiting, so that the seed is either a crescent or a ring.

Synopsis of Drugs from the Menispermaceæ

A. *Roots.*
    CALUMBA, 156.
    *Pareira, 157.

B. *Rhizome.*
    Menispermum, 158.

C. *Fruit.*
    *Cocculus, 159.

156. CALUMBA.—CALUMBA

COLUMBO

The root of **Jateorrhi'za palma'ta** Lamarck, sliced transversely and dried.

BOTANICAL CHARACTERISTICS.—Underground stem a short, irregular rhizome, from which start numerous fleshy fusiform roots 1 to 4 inches in diameter.

Leaves palmate, on long petioles. According to Bentley and Trimen, the blade of the leaf often reaches 14 inches in length. *Flowers* diœcious, *sepals* 6, *petals* 6, *stamens* 6; *anthers* 2-celled; *fruit* about the size of a hazelnut, densely clothed with long, spreading hairs, each tipped with a black, oblong gland.

HABITAT.—East Africa and Madagascar, cultivated in the East Indies.

DESCRIPTION OF DRUG.—**In transverse sections,** circular or oval in outline, 25 to 50 mm. (1 to 2 in.) in diameter; 3 to 12 mm. (⅛ to

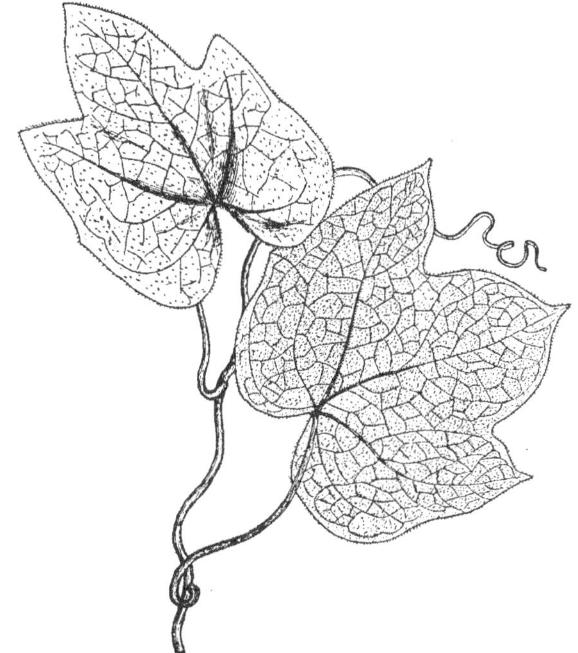

FIG. 82.—*Jateorrhiza palmata*—Portion of Vine.

½ in) thick. The outer edge is covered with a brown wrinkled layer of cork. The bark is about 9 mm. (⅜ in.) thick; a dark, shaded cambium line separates this bark from the **spongy grayish-yellow central portion.** In drying the central portion contracts more than the outer, hence the disks are depressed at this point, where also are found a few interrupted circles of projecting wood-bundles, while the outer portion near the cambium is distinctly radiate. **A microscopic section** shows near the center very distinct **bright yellow wood-bundles,** which are **narrow and radiate near the bark.** The parenchyma is filled with large, oval or circular starch granules.

Odor faint; taste slightly aromatic, very bitter, and mucilaginous. Dose: 30 gr. (2 Gm.).

SUBSTITUTION.—American calumba has frequently been used. It is almost uniformally much smaller, the color is not yellow, it contains no starch and is not mucilaginous. The decoction gives brown precipitate with ferric chloride.

Powder.—Characteristic elements: See Part iv, Chap. I, B.

CONSTITUENTS.—A neutral crystalline principle, **calumbin,** extremely bitter, **berberine, calumbic** acid, and starch, of which it contains

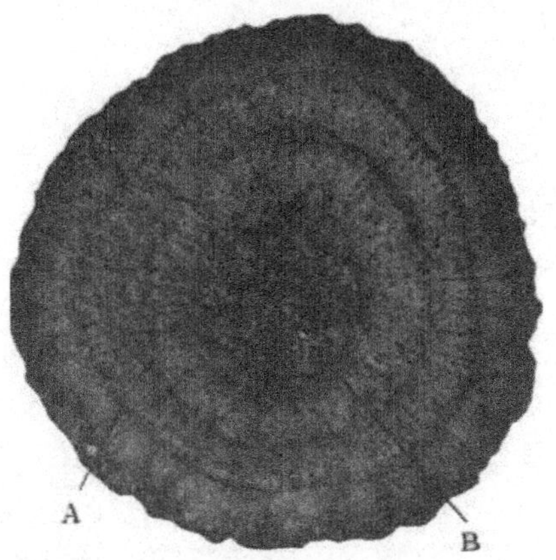

FIG. 83.—Columbo—Cross-section of root. (1.5 diam.) A, Annual ring of growth. B, Medulla. (Photograph.)

33 per cent. **No tannin is present;** it can therefore be compounded with salts of iron. The best solvent for the bitter principle is dilute acetic acid. This liquid, however, is not a good menstruum. Ash, 8 per cent.

Preparation of Calumbin.—Infusion of columbo, made with 3 per cent. of oxalic acid, is neutralized with ammonia. Evaporate to one-third, and when cool, shake out with ether. On evaporation of ethereal solution, white calumbin is obtained.

ACTION AND USES.—A simple tonic, stimulating the appetite through the gustatory nerves, increasing in turn the gastric and salivary secretions. Its special value as a tonic resides in the fact that it has no disagreeable effects, such as nausea, headache, or febrile disorder, like other remedies of its class. Externally, antiseptic, disinfectant, and anthelmintic.

OFFICIAL PREPARATION.

Tinctura Calumbæ (20 per cent.)   Dose: 1 to 4 f℥ (4 to 15 mils).

157. PAREIRA, N.F.—Pareira

PAREIRA BRAVA

The dry root of **Chondoden'dron tomento'sum** Ruiz et Pavon.   With not more
than 5 per cent. of stem bases.

Botanical Characteristics.—A vine with twining stem 4 inches in diameter;
leaves large, cordate, long-petioled, with entire margins; *flowers* diœcious;
fruit purplish, ovoid, 1-seeded, drupaceous, forming thick clusters resembling
bunches of grapes.

Habitat.—Brazil.

Fig. 84.—Pareira—Cross-section of root.   (1.5 diam.)   A, E, F, Xylem.   B, Phloëm.   C, G
Rings of stone cells.   (Photograph.)

Description of Drug.—A long, branching, woody root, found in commerce in
tortuous, subcylindrical pieces, about 100 to 150 mm. (4 to 6 in.) long, and
from 20 to 100 mm. (⅘ to 4 in.) thick.   Externally it varies from brown to
light grayish-brown in color, and is marked with fissures, transverse ridges,
and longitudinal wrinkles.   When cut or sliced it displays a dark brown
interior, leaving under the knife a waxy luster.   A cross-section displays a
thin bark; within this bark circle there are two or more circles (zones) of
radiating wood-wedges.   About 12 of these wood-wedges are found in the
central zone radiating from a common center.   The outer circles (zones)
of wood-wedges are separated from one another by a narrow line of parenchyma, stone cells, and compressed cells, and the short, circular, radiating
wedges of wood are separated from one another by medullary tissue, making
a combination of concentric and radiate arrangement which is quite character-

istic. Sometimes sections of the stem are found in the drug; these have a rather thick bark and a narrow pith. Taste at first mild, then bitter and somewhat acrid; odorless.

Powder.—Brownish-yellow. Characteristic elements: Starch, ellipsoidal, simple or 2 to 4 compound (7 to 15 μ in diam.); sclerenchyma consisting of long bast fibers and numerous isodiametric or elongated stone cells 20 to 50 μ across; wood fibers, simple or bordered pits; cork, dark brown cells (20 to 25 μ in diam.); calcium oxalate, in rosettes, few.

Constituents.—Pelosine (cissampeline), amorphous, insoluble in hot or cold water, soluble in alcohol and chloroform; starch, gum, tannin; taste sweetish-bitter.

Preparation of Pelosine (also known as Cissampeline).—Boil root in acidulated H_2SO_4 water, precipitate with K_2CO_3, purify by redissolving in acidulated water, decolorize with charcoal, again precipitate with K_2CO_3, and purify from solution in ether.

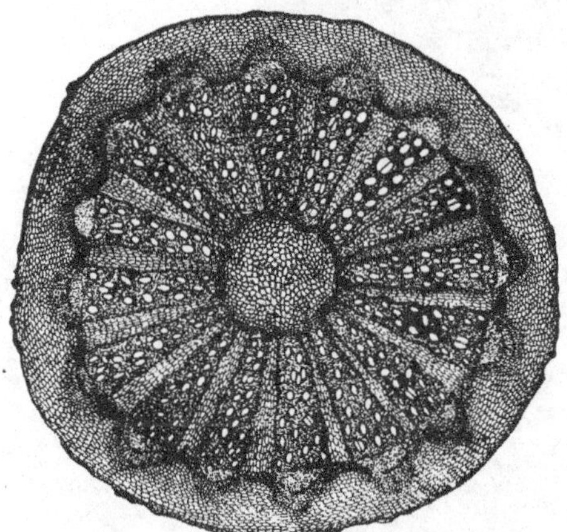

Fig. 85.—Cross-section of Menispermum—Magnified 14 diam.

Action and Uses.—As a remedial agent pareira is generally conceded to be beneficial as a diuretic and tonic in the treatment of cystitis and suppurative kidney diseases, acting in a soothing manner, especially on the bladder. Formerly renowned as a lithontriptic. Dose: 30 to 60 gr. (2 to 4 Gm.). **Fluidextractum Pareiræ**, U.S.P. 1900, Dose: ½ to 2 f ʒ (2 to 8 mils).

158. **MENISPERMUM.**—Yellow Parilla.—The dry rhizome and roots of **Menisper'mum canaden'se** Linné. Rhizome about 1,000 mm. (40 in.) or more long, and 6 mm. (¼ in.) thick; **externally dark yellowish-brown, knotty,** and longitudinally wrinkled; fracture woody and tough; nearly inodorous; taste bitter. Rootlets thin, brittle, yellow. A cross-section of the rhizome displays a thick bark and a yellowish interior. Under the microscope are seen **numerous wood-wedges separated by narrow medullary rays**; at the extremity of each wood-ray there appears a semilunar bundle, which on longitudinal section proves to be composed of bast fibers penetrating the bark. The diameter of the pith varies, not infrequently occupying one-third of the space between the bark. The overground stem, with which the drug is not infrequently mixed, has a very large, porous pith. *Constituents:* **Berberine**

(yellow) in small amount, and **menispine** (white), the principal constituents, with resin, tannin, and starch. Alterative, tonic, diuretic, and laxative; said to resemble sarsaparilla in its action. ·The root was introduced into the market as Texas sarsaparilla. Dose: 5 to 30 gr. (0.3 to 2 Gm.).

159. **COCCULUS.**—FISH BERRIES. Coc'culus In'dicus. N.F. The fruit of **Anamirta cocculus** Wight and Arnott. Obtained from a climbing shrub in Eastern India, native of Malabar coast. The berries are ovoid, **kidney-shaped**, and about the **size of a large pea,** with an obscure ridge around the convex back. Externally wrinkled and blackish-brown in color. **The endocarp** is white, and extends from the concave side deeply into the interior.

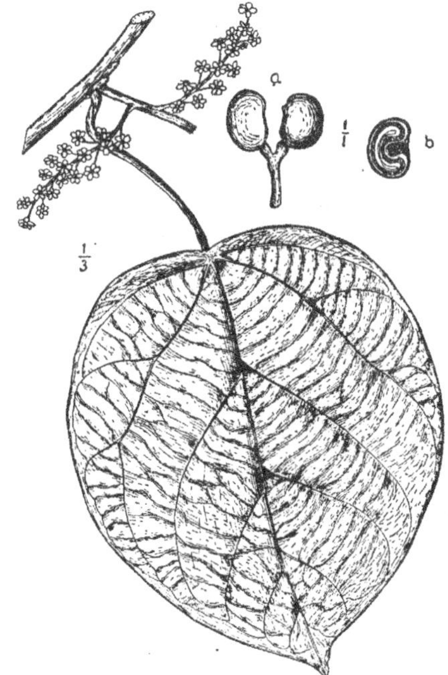

FIG. 86.—*Anamirta cocculus*—Flowering branch. *a*, Fruit. Section of same.

The seed is semilunar, oily, very bitter, but the pericarp is tasteless. The chief constituent is **picrotoxin.**

Preparation of Picrotoxin.—To aqueous extract add MgO; treat this with hot alcohol. ·Evaporate and collect the deposited picrotoxin.

Locally employed in cutaneous affections. The decoction (or tincture added to water, 1 to 4) is used as an insecticide in head lice. **Picrotoxin is an acrid narcotic poison;** in its action on the secretions it is said to resemble pilocarpine. The berries have been used from ancient times for stupefying and capturing fish, but "this unsportsmanlike method of fishing in some parts of the country is now illegal."

Cocculus indicus has been sometimes confounded with the fruit of the *Laurus nobilis*, commonly known as bayberry. The latter is, however, generally larger, distinctly oval in form, and the seeds lie loose within and fill the cavity of the fruit. The seed of the bayberry has an agreeable aromatic taste.

BERBERIDACEÆ.—Barberry Family

Herbs, shrubs, or trees with watery juice. A peculiarity of the **leaves in the** principal genus of the order suggests the name barberry; these are usually **beset with spiny teeth**, occasionally reduced **to simple or branching spines** (barbs). **Inflorescence** various; solitary (Podophyllum), in racemes (Berberis), panicles, cymes, or spikes. **Flowers** greenish (Caulophyllum) or white with outer greenish bractlets (Podophyllum); fruit a berry or capsule (sometimes edible—May apple).

Synopsis of Drugs from the Berberidaceæ

A. *Rhizomes.*
 *Caulophyllum, 160.
 PODOPHYLLUM, 161.
 Jeffersonia, 162.

B. *Roots.*
 Berberis Radix, 163.
 *Berberis, 165.

C. *Bark.*
 Berberis cortex, 164.

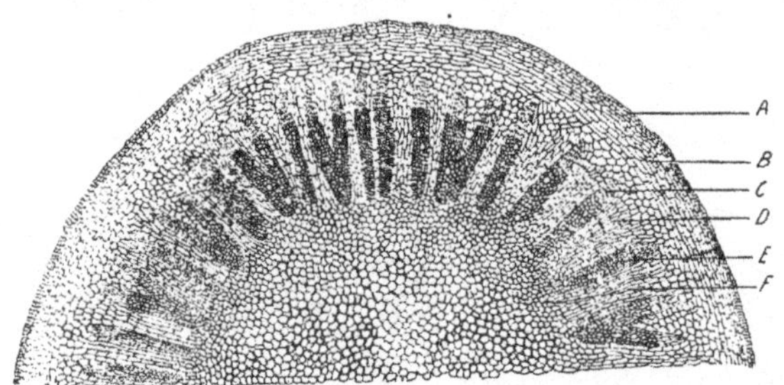

FIG. 87.—*Caulophyllum*—Cross-section of rhizome. *A.* Epidermis. *B.* Parenchymatous tissue. *C.* Phloëm portion of bundle. *D, F.* Medullary rays. *E.* Xylem portion of bundle.

160. **CAULOPHYLLUM.**—Squaw Root. Blue Cohosh. N.F. The rhizome and roots of **Caulophyl'lum** thalictroi'des Linné. Off. in U.S.P. 1890. Rhizome crooked, of horizontal growth, about 100 mm. (4 in.) long, and 6 to 8 mm. ($\frac{1}{4}$ to $\frac{1}{3}$ in.) thick; **on the upper side are** broad **cup-shaped scars** and short bent branches having concave terminations; it is beset with numerous tough and wiry light-brown rootlets matted together. Externally of a dull brown color, internally whitish, with **numerous narrow** wood-wedges, sometimes **in two circles**, inclosing a large pith. The rootlets have a much thicker bark and a thick central woody cord. Nearly inodorous; taste slightly sweetish and somewhat acrid. (Highly magnified starch grains of caulophyllum, see Fig. 87.) *Constituents:* CAULOPHYLLINE. Resins, 12 per cent., tannin, starch, gum, etc. **Caulophylline** is colorless, odorless, and almost tasteless, is not precipitated by alkalies, and crystallizes with difficulty; many of its characteristics make it appear as a proximate principle belonging to a new class of bodies about which little is known.

Preparation of Caulophyllin.—Concentrate alcoholic tincture and add this to a large volume of water. Collect precipitate and dry in current of warm air. *Caulophylline.*—Extract drug with 60 per cent. alcohol. Evaporate tincture to a semi-solid. Add ferric hydrate and sodium bicarbonate to this residue and extract the mixture with chloroform. The principle remains on the evaporation of the solvent. Emmenagogue, diuretic, and antispasmodic; it has some reputation in the treatment of rheumatism and as an expectorant in bronchitis. Dose: 5 to 30 gr. (0.3 to 2 Gm.).

161. PODOPHYLLUM.—Podophyllum

MAY APPLE. MANDRAKE

The dried rhizome and roots of **Podophyl′lum pelta′tum** Linné. Yielding not less than 3 per cent. of resin U.S.P. IX.

Botanical Characteristics.—*Leaf* 7–9-lobed; peltate. *Flowering stem* bearing two one-sided leaves with the stalk thickest near their inner edge. *Flower* large, white, nodding. *Fruit* ovoid, slightly acid, edible.

Fig. 88.—*Podophyllum peltatum*—Plant and rhizome.

Description of Drug.—Rhizome 300 mm. (12 in.) or more long and 5 mm. (⅕ in.) thick, jointed, consisting of nodes and internodes, the length of the internodes being about 50 mm. (2 in.). The rhizome is very much thickened at the nodes, where it is sometimes branched laterally, **each node having a circular scar on the upper side** and about six to ten small brittle rootlets below or scars from broken rootlets; externally smooth, slightly wrinkled longitudinally, of an orange-brown color; fracture short, white and starchy, showing a rather thick bark, and from **sixteen to thirty vascular bundles encircling a broad pith**; the parenchyma contains chiefly starch. Odor faint and characteristic; taste sweetish, slightly acrid, and quite bitter.

Powder.—Characteristic elements: See Part iv, Chap. I, B.

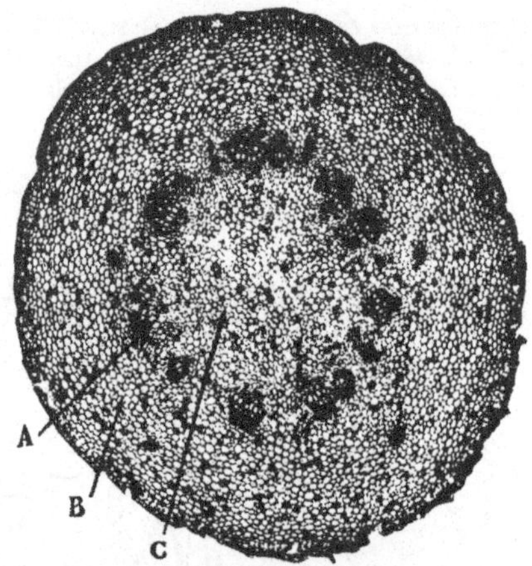

FIG. 89.—Podophyllum. Cross-section of rhizome. (15 diam.) A, Vascular bundle. B, Parenchyma of cortex. C, Medulla. (Photomicrograph.)

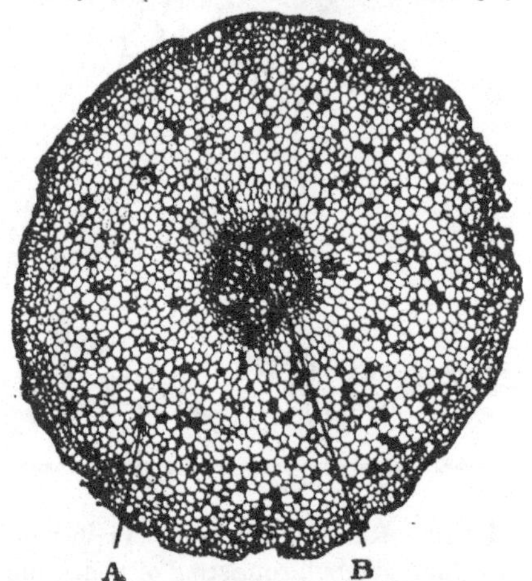

FIG. 90.—Podophyllum. Cross-section of rootlet. (25 diam.) A, Cortex. B, Xylem. (Photomicrograph.)

Preparation of Podophyllin.—Composed of several resinous principles separable by solvents. Ether dissolves out a resin of bright yellow color, leaving a brown, odorless resin of little more prompt activity. A concentrated tincture is precipitated by water containing HCl. The precipitate is collected and dried.

Podophyllin is not found to any extent in the fresh drug, according to Lohman. It is developed to the fullest extent only by storage.

CONSTITUENTS.—Resins associated with other common vegetable principles; **podophyllin** (Resina podophylli, U.S.P.) 4 to 6 per cent., together with amorphous and crystalline principles. Later investigations have given prominence to the following: Podophyllotoxin, $C_{15}H_{14}O_6$ (white crystals), converted by hydration into podophyllic acid, $C_{15}H_{16}O_7$; picropodophyllin, isomeric with podophyllotoxin (inert); quercetin, yellow needles; podophylloresin (purgative). Some authorities state that the purgative principle is closely related to emodin. (See *Rhamnus purshiana*.)

ACTION AND USES.—Classed usually with the drastic cathartics. Dose: 10 to 20 gr. (0.6 to 1.3 Gm.). Podophyllin is an irritant to the mucous membrane; in small doses an active cathartic, having reputed cholagogue properties, hence the name "vegetable calomel." Dose: as a laxative ⅒ gr. (0.006 Gm.), as a purgative ¼ gr. (0.016 Gm.).

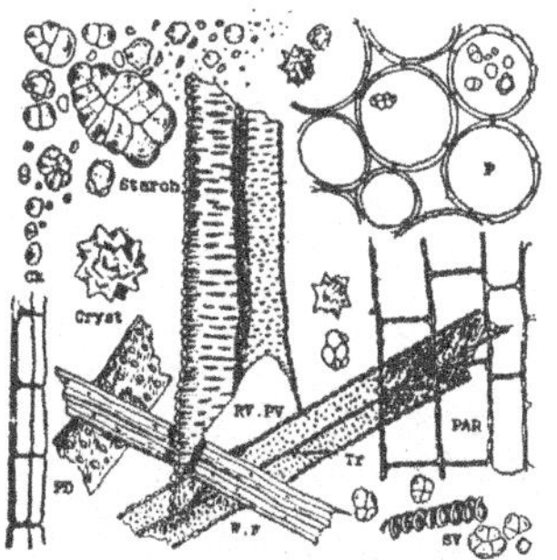

FIG. 91.—Podophyllum in powder. Starch. Cryst, Crystals of calcium oxalate. Ck, Corky tissue, side view. PD, Duct with bordered pores. RV, PV, Reticulated and pitted ducts. Tr, Tracheids. SV, Spiral vessels. W.F, Wood fibers or bast fibers. Par, Parenchyma in longitudinal view. P, Parenchyma in cross-section.—(*After Jelliffe.*)

There is a remarkable difference shown in the medicinal activity of podophyllin, whether precipitated by water alone, whether by acidulated water, or by solution of alum. The one precipitated by water is said to be fifteen to twenty times as active as the one precipitated by acidulated water, and the one precipitated by alum much weaker than either.

OFFICIAL PREPARATIONS.

Fluidextractum Podophylli,....Dose: 5 to 15 ♏ (0.3 to 1 mil).
Resina Podophylli,........... ⅛ to ½ gr. (0.0081 to 0.0324 Gm.).

162. **JEFFERSO'NIA DIPHYL'LA** Persoon.—TWIN-LEAF. (Rhizome.) Has properties somewhat similar to senega; it is also diuretic, alterative, and antispasmodic. Dose: 15 to 60 gr. (1 to 4 Gm.).

163. **BERBERIS RADIX.**—BARBERRY ROOT. The root of **Ber'beris vulga'ris** Linné. *Habitat:* Europe, Western Asia, and North America. Thick, much-branched, from 25 to 50 mm. (1 to 2 in.) in diameter in the thickest part; wood light yellowish, hard, tough, with a very thin bark (see Barberry

Bark below); odor slightly aromatic; taste bitter.   It contains five alkaloids, of which berberine is the most interesting.   Used as a tonic in doses of 30 to 60 gr. (2 to 4 Gm.).

164. **BERBERIS CORTEX**—BARBERRY BARK. The bark of the above root, coming in long, thin pieces, exfoliating, or separating into thin layers; outer surface yellowish-gray; inner surface bright yellow.   It contains the same alkaloids as the root, but in greater proportion.   This species is the host plant for the common wheat rust (*Puccinia graminis*) in its accidio stage.   The leaves when parasiticized by this fungus seem to be covered with yellow spots, the openings of the cups in which the spores are borne.   Dose: 3 to 10 gr. (0.2 to 0.6 Gm.).

165. BERBERIS, N.F.

BERBERIS.   (OREGON GRAPE.)—The rhizome and roots of species of the section *Odostemon* Refinesque of the genus *Berberis* Linné, without the admixture of more than 5 per cent. of the overground parts of the plant or other foreign matter.   Berberis without the bark should be rejected.

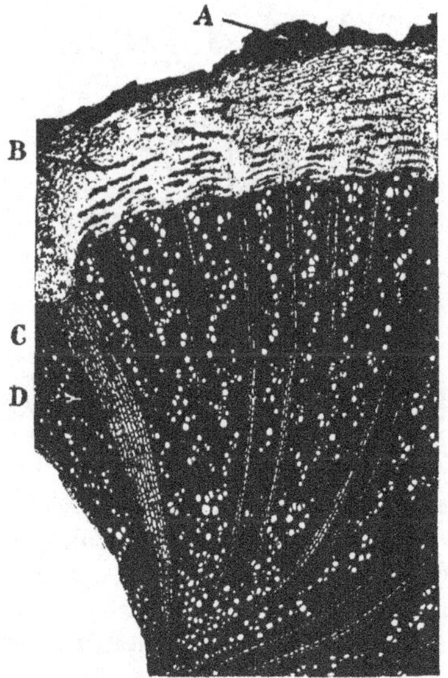

FIG. 92.—*Berberis*.   Cross-section of rhizome.   (22 diam.)   A. Cork.   B. Group of bast fibers. C. Medullary ray.   D. Xylem.   (Photomicrograph.)

   In more or less knotty irregular pieces of varying length and from 3 to 50 mm. in diameter; bark from 0.5 to 2 mm. thick; wood yellowish, distinctly radiate, with narrow medullary rays, hard and tough; rhizome with a small pith; odor distinct; taste bitterish.

Powder.—Yellowish-brown composed chiefly of fragments of wood fibers associated with a few tracheæ and medullary rays. Wood fibers yellow with large simple transverse pores; tracheæ chiefly with bordered pores occasionally reticulate; starch grains single or 2 to 3 compound. The individual grains are irregularly spherical.

CONSTITUENTS.—Contains three alkaloids, berberine, **oxycanthine** and berbamine; the two latter are white. Used as tonic and alterative in doses of 8 to 30 gr. (0.5 to 2 Gm.). (Fluidextractum U.S.P. 1900.)

MONIMIACEÆ

166. **BOLDUS.**—BOLDO, N.F. The leaves of **Peumus boldus** Molina, an evergreen shrub growing in the Chilian Andes. They are broadly oval, about 50 mm. (2 in.) long, with entire margin and rough, reddish-brown surfaces, covered with numerous small glands containing a volatile oil; upper surface glossy, lower surface hairy; midrib prominent; odor fragrant; taste pungent, aromatic, somewhat bitter. They are used as an aromatic stimulant and tonic; in South America in inflammation of the genito-urinary tract. Dose: 15 to 60 gr. (1 to 4 Gm.), in fl'ext., tincture, or infusion.

LAURACEÆ.—Laurel Family

Aromatic trees or shrubs, all parts of which yield volatile oil. *Leaves* simple, alternate, pellucid-punctate.

Synopsis of Drugs from the Lauraceæ

A. *Barks.*
　　CINNAMOMUM ZEYLANICUM, 167.
　　Cinnamomum Cassia, 168.
　　CINNAMOMUM SAIGONICUM, 169.
　　SASSAFRAS, 170.
　　Nectandra, 173.
　　Coto, 174.
　　Lindera, 175.

B. *Leaves.*
　　Laurus, 176.
　　Umbellularia, 177.
C. *Wood.*
　　Sassafras Lignum, 171.
D. *Pith.*
　　*Sassafras Medulla, 172.
E. *Stearopten.*
　　CAMPHORA, 178.

F. *Volatile Oils.*
　　OLEUM CINNAMOMI, 168 a.
　　OLEUM SASSAFRAS, 170 a.
　　OLEUM CAMPHORÆ, 178 a.
G. *Fixed Oil.*
　　Oleum Lauri, 176 a.
H. *Seeds.*
　　Persea, 179.

167. CINNAMOMUM ZEYLANICUM

CEYLON CINNAMON

The dried inner bark of the shoots of **Cinnamo'mum zeylan'icum** Breyne.

BOTANICAL CHARACTERISTICS.—*Tree* about 30 feet high. *Root* with the odor of camphor as well as that of cinnamon. *Leaves* ovate-lanceolate, entire, smooth and shining, tasting of cloves. *Flowers* in panicles, usually unisexual. *Drupe* 1-seeded, the seed large, with oily cotyledons.

HABITAT.—Ceylon.

DESCRIPTION OF DRUG.—Long, cylindrical quills deprived of the corky layer by scraping; compound, consisting of 8 or more thin, papery, light brownish-yellow, quilled layers, inclosed one within the other, their sides curling inward, giving the sticks a flattened appearance on one side; somewhat flexible, with a splintery fracture; the outer surface is marked with shining, wavy bast lines, and occasionally with small scars or perforations indicating the former position of leaves; under the microscope it is seen to be formed by a layer of stone cells.

FIG. 93.—*Cinnamomum zeylanicum*—Branch.

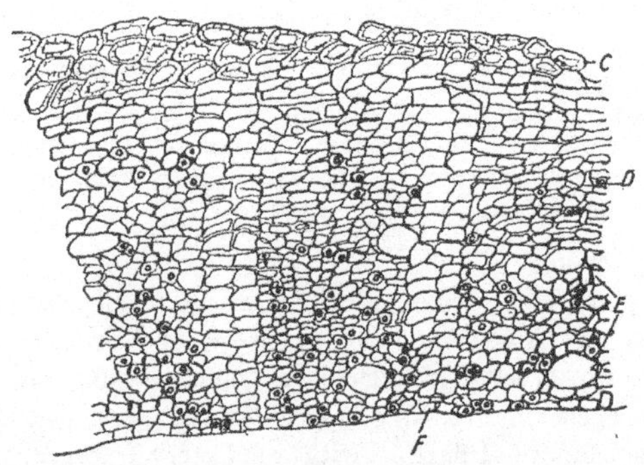

FIG. 94.—Ceylon cinnamon—Cross-section of bark. *C*, Stone cells. *D*, Parenchyma containing numerous bast fibers. *E*, Oil-resin cells. *F*, Medullary rays.

The inner surface is darker and striated. A characteristic, sweet, fragrant odor, and a warm, aromatic, pungent, and sweetish taste run through the different cinnamon barks, but the taste of the Ceylon cinnamon is the more delicate. The broken pieces, caused by re-packing at custom-houses (sorted and sold as "small cinnamon"), are commonly used in pharmacy.

Powder.—Characteristic elements: See Part iv, Chap. I, B.

CONSTITUENTS.—All the cinnamons contain volatile oil, mucilage, resin, tannin, mannite, and bitter substance, in varying relative proportions.

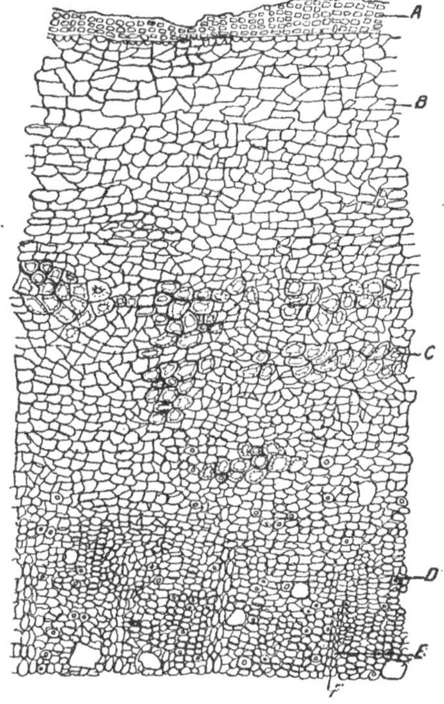

FIG. 95.—Cassia cinnamon—Cross-section of bark. *A.* Cork cells. *B.* Parenchyma cells. *C.* Stone cells. *D.* Bast fibers in parenchyma. *E.* Oil-resin cells (black line from *E* should have been directed to the large cell below and to the left of that letter). *F.* Medullary ray.

In typical samples, the Saigon variety contains the most volatile oil (1 per cent. or more) and mannite, the Cassia variety coming next and the Ceylon last, the oil of the last ranging from 0.50 to nearly 1.00 per cent. Cassia contains the most and Saigon the least, of both tannin and bitter substance. The oil of cinnamon is not identical in the different barks; that of Ceylon cinnamon is recognized as of finer and more perfect flavor, while the Saigon, being sweeter, is more aromatic but the odor is less permanent.   Ash, nor exceeding 6 per cent.; not exceeding 2 per cent. insoluble in HCl.

ACTION AND USES.—Aromatic stimulant and tonic, carminative and astringent.   The different varieties of cinnamon are among the most pleasant and efficient aromatics and form agreeable **adjuvants** to a great many official preparations.   Dose: 8 to 30 gr. (o.5 to 2 Gm.).

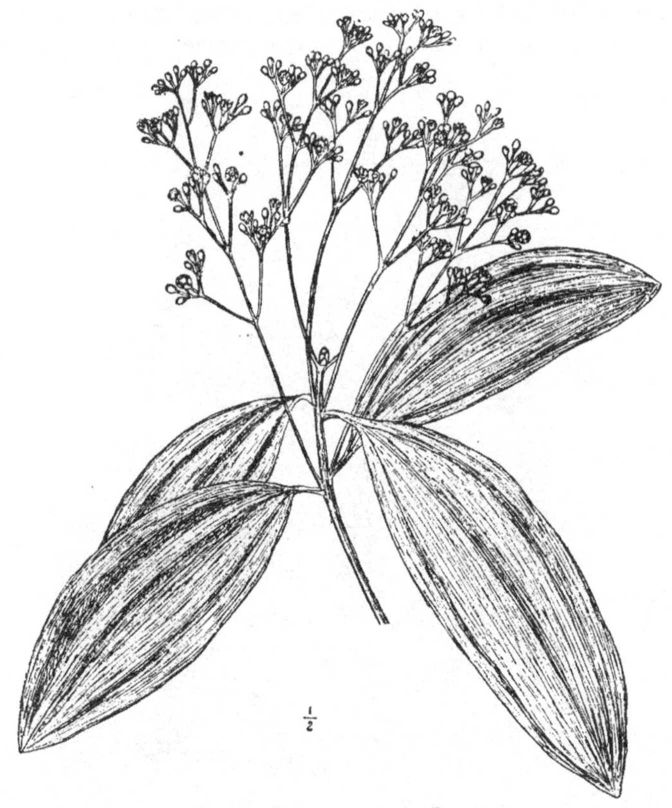

FIG. 96.—*Cinnamomum cassia*—Branch.

168. **CINNAMOMUM CASSIA.**—CASSIA BARK.   The bark of the shoots of one or more undetermined species of **Cinnamo'mum** grown in China (Chinese cinnamon).   Off.· U.S.P. 1890.   Cassia cinnamon is in tubes or curved pieces, of a darker yellowish-brown color than preceding, nearly deprived of the corky layer; these tubes are usually simple, rarely. double, 1 mm. (1/25 in.) or more thick, and break with a rather short fracture; odor and taste similar to, but somewhat less delicate than, that of Ceylon cinnamon. Constituents, the same, the volatile oil being officially recognized as from this source.   This variety has been superseded by Saigon cinnamon in the official preparations containing cinnamon.

168 a. **OLEUM CASSIÆ.**—OIL OF CINNAMON. Contains at least 80 per cent. of **cinnamic aldehyde.** Both the Ceylon oil and that derived from Cassia, and other cinnamon barks are found in commerce, and they are essentially the same. The oil of Ceylon cinnamon has a more delicate odor and flavor. All of the various oils of cinnamon become darker and thicker by age and exposure to the air; they have the characteristic odor of cinnamon, a sweetish, spicy, and burning taste.

CONSTITUENTS.—Oil of cinnamon consists chiefly of **cinnamic aldehyde,** with small quantities of hydrocarbon; when the oil is exposed to the air for a time, the cinnamic aldehyde is oxidized into cinnamic acid, two resins, and water, the oil becoming thicker and darker, and frequently separating out a few crystals of the cinnamic acid.

OFFICIAL PREPARATIONS.

Aqua Cinnamomi (0.2 per cent.),.....Dose: ½ to 1 fl. oz. (15 to 30 Gm.).
Spiritus Cinnamomi (10 per cent.),....   10 to 20 ♏ (0.6 to 1.3 mils).

169. CINNAMOMUM SAIGONICUM.—SAIGON CINNAMON
SAIGON CASSIA

The dried bark of the stem and branches of an undetermined species of
Cinnamo'mum

DESCRIPTION OF DRUG.—It takes its name from Saigon, the capital of French Cochin-China, where it is collected and exported. It is in large quills or broken pieces, 1 or 2 mm. (1/25 to 1/12 in.) thick; the gray or grayish-brown bark, which is not removed, is more or less rough and warty, longitudinally wrinkled and ridged, and covered with whitish patches. Inner bark cinnamon-brown or dark brown, with numerous white striæ near the bark; fracture short, granular; odor aromatic; taste aromatic and pungent. Ash, not exceeding 6 per cent.; not exceeding 2 per cent. insoluble in HCl.

COMPARISON OF THE CINNAMON BARKS.—*Color.*—There is quite a difference in the depth of the color of the three barks. The Ceylon is the lightest, the Saigon is the darkest, and the Cassia intermediate. This difference in shade is shown best in the powder.

Thickness.—The Ceylon is very thin and papery. The Saigon, usually regarded as the thickest, is in the average about the same as Cassia.

Odor.—The odor and taste of the Saigon is the strongest, the Ceylon is the most delicate, the Cassia weakest.

Microscopical.—To distinguish between the barks no difficulty is experienced in cross- and longitudinal sections, which display the oil-cells, stone cells, and other elements. In the powdered condition

13

the Ceylon shows the largest stone cells. In Cassia the stone cells
are less numerous and smaller. In the Saigon the oblong stone
cells are about the same size as those of Cassia, but fewer in number.

Powder.—Elements of: See Part iv, Chap. I, B.

OFFICIAL PREPARATIONS.

Tinctura Cardamomi Composita (2.5 per
cent.),..................... Dose: 1 to 3 fl. dr. (4 to 12 mils).
Tinctura Gambir Composita (2.5 per cent.), ½ to 3 fl. dr. (2 to 12 mils).
Tinctura Lavandulæ Composita (2 per
cent.),..................... ½ to 2 fl. dr. (2 to 8 mils).
Tinctura Rhei Aromatica (4 per cent.),
employed in Syrupus Rhei Aromaticus.
Tinctura Cinnamomi (20 per cent.),... ½ to 2 fl. dr. (2 to 8 mils).
Employed also in Vinum
Opii and Infusum Digitalis.
Pulvis Aromaticus (35 per cent.),......... 15 gr. (1 Gm.).

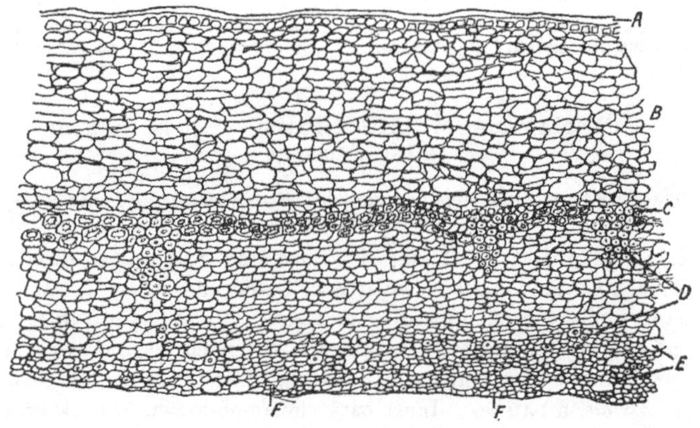

FIG. 97.—Saigon cinnamon—Cross-section of bark. *A.* Corky layer. *B.* Parenchyma cells.
C. Stone cells. *D.* Bast fibers. *E.* Oil-resin cells. *F.* Medullary rays, very inconspicuous.

SASSAFRAS.—SASSAFRAS

Sas'safras variifo'lium O. Kuntze. The various portions used in medicine are the
bark of the root, the volatile oil, and the pith, all official, and the wood,
unofficial.

BOTANICAL CHARACTERISTICS.—Tree with spicy, aromatic bark, 15 to 125 feet
high, with yellowish-green twigs. *Leaves* ovate, entire, or some of them
3-lobed. *Flowers* diœcious, greenish-yellow, in racemes.

HABITAT.—North America, from Kansas eastward.

170. SASSAFRAS.—SASSAFRAS BARK

The dried bark of the root of **Sassafras variifolium** O. Kuntze, collected in early
spring or autumn and deprived of the outer corky layer with not more than 2
per cent. of adhering wood present.

DESCRIPTION OF DRUG.—In small, irregular, rust-brown fragments, deprived of the grayish-brown, fissured, corky layer, leaving a reddish or rust-brown surface; 1 to 5 mm. ($\frac{1}{25}$ to $\frac{1}{5}$ in.) thick. It breaks with a short, corky fracture, exposing a whitish interior dotted with numerous oil-cells; odor highly fragrant, characteristic; taste sweetish, aromatic. Oil is employed in the compound syrup of sarsaparilla.

Powder.—Characteristic elements: See Part iv, Chap. I, B.

CONSTITUENTS.—Volatile oil (about 5 per cent.), camphoraceous matter, tannin (6 per cent.), sassafrid (a derivative of tannin, 9 per cent.), gum, resin, starch, etc. Ash, not exceeding 30 per cent.

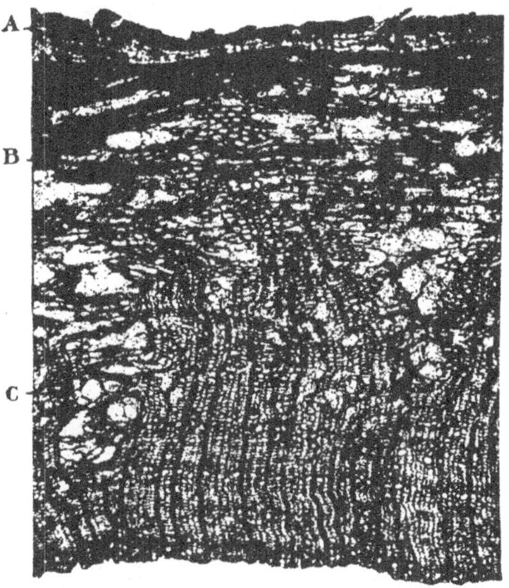

FIG. 98.—Sassafras—Cross-section of bark. (27 diam.) *A*, Cork. *B*, Parenchyma of primary cortex. *C*, Medullary ray. (Photomicrograph.)

ACTION AND USES.—Aromatic stimulant, alterative, and astringent. It is used almost entirely as an adjuvant or corrective. The infusion is used as a popular household remedy for its diuretic and diaphoretic effects in febrile states. Dose: 30 to 120 gr. (2. to 8 Gm.), in infusion.

170 a. **OLEUM SASSAFRAS,** U.S.—A volatile oil usually distilled from the entire root. A colorless or yellow liquid, sp. gr. 1.065–1.075, becoming thicker and of a reddish color by age and exposure, and having the characteristic odor and taste of sassafras. It contains a hydrocarbon (safrene, $C_{10}H_{16}$), and an oxygenated compound, safrol, $C_{10}H_{10}O_2$ (melts at 8.5°C., 47.3°F.), a widely distributed prin-

ciple obtained commercially from oil of camphor, phellandrene, $C_{10}H_{16}$, eugenol, $C_{10}H_{12}O_2$, etc. Generally used as a flavor. Dose: 1 to 5 ♏ (0.065 to 0.3 mil). The oil is sometimes adulterated with the artificial oil and a camphor oil fraction. Virginia is said to be the chief producer of oil of sassafras.

171. **SASSAFRAS LIGNUM** (Unofficial).—Sassafras Wood. The wood of the root, coming in billets, partially or wholly deprived of bark, or in raspings or chips; pale brownish or reddish in color, light and easily cut; medullary rays narrow; odor and taste like the bark, but weaker, there being a smaller proportion of volatile oil. It is used like the bark.

172. SASSAFRAS MEDULLA, N.F.—Sassafras Pith

The dried pith of Sassafras variifolium O. Kuntze.

Description of Drug.—Thin, cylindrical, white pieces, very light and spongy; inodorous; taste insipid and mucilaginous. The tissue is entirely composed of parenchyma. It contains a mucilage (not precipitated by alcohol or lead subacetate) which forms a limpid, ropy, viscid solution with water, but not sufficiently tenacious to hold insoluble substances in suspension. **Demulcent,** often used as an application to inflamed eyes.

Preparation.
Mucilago Sassafras Medullæ (2 per cent.).

173. **NECTANDRA.**—Bebeeru Bark. Greenheart Bark. From **Nectan'dra ro'diæi** Schomburgk. *Habitat:* South America. Large, flat, heavy pieces, from 250 to 300 mm. (10 to 12 in.) long, 50 to 150 mm. (2 to 6 in.) broad; usually deprived of the cork, leaving longitudinal depressions in the grayish-brown outer surface similar to the digital furrows of flat calisaya bark; internally pale brown, roughly striate. Its structure is chiefly short liber cells filled with secondary deposit, causing it to break with a short fracture. Inodorous; intensely bitter, somewhat astringent. It contains tannin, beberine (identical with buxine and pelosine), and sipirine.
　　Action and Uses.—Tonic, astringent, and febrifuge, introduced as a substitute for cinchona as an antiperiodic, but much inferior. Dose: 15 to 60 gr. (1 to 4 Gm.), commonly used in the form of beberine sulphate.

174. **COTO.**—Coto Bark. Origin undetermined. *Habitat:* Bolivia. Very large, flat pieces, about 5 to 15 mm. (⅕ to ⅗ in.) thick, usually deprived of cork; the outer surface cinnamon-brown, rough, having the appearance of having been shaved or split off; inner surface darker brown, rough from numerous close ridges of longitudinally projecting bark fiber; a fresh cross-section shows numerous small, yellowish spots (groups of stone cells). Odor aromatic, cinnamon-like, stronger when bruised; taste hot, bitter.
　　PARACOTO BARK, N.F.—Which occasionally enters our market from Bolivia, very much resembles the above, but is marked with whitish fissures, and has a fainter, somewhat nutmeg-like odor.
　　Constituents.—Cotoin, in true coto bark, paracotoin in the other; both barks contain volatile oil, resin, and piperonylic acid. They have established quite a reputation in diarrhœa. Dose: 5 to 10 gr. (0.3 to 0.6 Gm.).

175. **LINDERA BENZOIN** Meissner.—Spice Bush. (Bark, berries, and leaves.) Aromatic stimulant, tonic, and diaphoretic. The berries have been used as a substitute for allspice. Dose: 15 to 60 gr. (1 to 4 Gm.).

176. **LAURUS.**—Laurel. Sweet Bay. The leaves of **Lau'rus nobi'lis** Linné. Oval-oblong, about 50 to 100 mm. (2 to 4 in.) long, brownish, pellucid-punctate; margin entire, wavy; taste aromatic, bitter, somewhat astringent; odor fragrant, due to a volatile oil. The chief constituent, however, is a fixed oil (see below) present to the extent of about 30 per cent. Stimulant and astringent, quite popular as an astringent injection.

176 a. **OLEUM LAURI.**—Laurel Oil. A green, granular, semi-solid of the consistence of butter. It consists mainly of laurostearin, but contains a small quantity of volatile oil which makes it a very aromatic base for liniments and ointments.

177. **UMBELLULARIA CALIFORNICA** Nuttall.—California Laurel. (Leaves.) They contain a volatile oil which seems to be a strong local anæsthetic, used in neuralgic headache, cerebro-spinal meningitis, intestinal colic, and atonic dyspepsia. Dose: 15 to 30 gr. (1 to 2 Gm.).

178. CAMPHORA.—Camphor

GUM CAMPHOR

A ketone obtained from **Cinnamo'mum cam'phora** Nees et Ebermaier, and purified by sublimation. It is dextrogyrate.

Botanical Characteristics.—A large and handsome tree. *Leaves* evergreen, shining, alternate, ovate-lanceolate. *Flowers* small, perfect, in corymbose panicles; *anthers* 4-celled, opening by terminal pores.

Fig. 99.—*Cinnamomum camphora*—Branch and flower.

Source.—The camphor tree grows in Japan and China, especially in the island of Formosa. This island alone furnishes half of the total product of the globe, or 5,200,000 pounds. Japan grows 1,560,000 pounds. The rest comes from China, Java, Sumatra, and Florida. It should be mentioned that the camphor of Malaysia is not extracted from *Cinnamomum camphora*, but from *Dryobalanops aromatica*. The United States alone uses 2,000,000 pounds of camphor yearly. The trunk, root, and branches are cut into chips and exposed to vapors of boiling water. The camphor volatilizes and condenses in small granules on the straw with which the head of the still is lined. It is freed from the volatile oil by draining or expressing, and is purified by resubliming with lime from a vessel into which the steam is allowed to escape through a small aperture. The camphor condenses in a compact cake, with a circular hole in the center corresponding to the aperture. Camphor has had to compete with rivals which are cheaper. In the manufacture of celluloid, the substitution of naphthalin for camphor has produced a considerable effect in controlling the high price resulting from the Japanese monopoly of the industry.

DESCRIPTION OF DRUG.—Refined camphor comes in **white, translucent masses,** tough and somewhat flexible, breaking with a shining, crystalline fracture; reduced to a powder only by the addition of a few drops of alcohol, ether, chloroform, glycerin, volatile or fixed oils, or other volatile liquids for which it has an affinity, by triturating with an equal weight of sugar, by precipitating the alcoholic solution with water, or by sublimation. It is very volatile, even at ordinary temperatures, giving out a **characteristic penetrating** odor. Taste pungent, aromatic, leaving a cooling sensation in the mouth. Lighter than water, small pieces taking up a circulatory motion therein, which ceases upon the addition of a drop of oil. Very inflammable, burning with a dense smoke, and leaving no residue. When triturated with about molecular proportions of thymol, phenol, or chloral hydrate, it liquefies. It melts at 175°C. (347°F.) and boils at 204°C. (399.2°F.).

Borneo or Sumatra camphor is an allied camphor. By oxidation it yields ordinary camphor. Borneol Valerates have been introduced as useful in various neuroses. See "New and Non-official Remedies."

CONSTITUENTS.—Camphor has the composition $C_{10}H_{16}O$, and is considered as a ketone yielded indirectly by the oxidation of borneol, a secondary alcohol having the composition $C_{10}H_{18}O$. By treatment with various reagents camphor yields a number of interesting compounds, as cymol, camphoric acid, etc. With iodine and bromine it forms compounds, one, **the monobromated camphor** ($C_9H_{15}BrCO$), being used as a nerve sedative in doses of 3 gr. (0.2 Gm.); it is made by heating equal portions of bromine and camphor at 172°F.; one-half the bromine goes off as hydrobromic acid. One H of the camphor molecule, is replaced by Br in the reaction. Camphoric acid,

$C_{10}H_{16}O_4$, and camphronic acid, $C_9H_{12}O_5$, are produced by oxidation with nitric acid. Ash, not more than 0.05 per cent.

ACTION AND USES.—Stimulant and antispasmodic. Externally anodyne and rubefacient. Dose: 3 to 10 gr. (0.2 to 0.6 Gm.), in pill or emulsion.

OFFICIAL PREPARATIONS.

Aqua Camphoræ (0.8 per cent.),	Dose: ½ to 2 fl. oz. (15 to 30 mils)
Spiritus Camphoræ (10 per cent.),........	5 to 40 ♏ (0.3 to 2.6 mils)
Tinctura Opii Camphorata (0.4 per cent.),	1 to 4 fl. dr. (4 to 15 mils)
Linimentum Camphoræ (20 per cent.).	
Linimentum Saponis (4.5 per cent.).	
Linimentum Chloroformi (70 per cent.).	
Linimentum Belladonnæ (5 per cent.).	

178 a. **OLEUM CAMPHORÆ.**—OIL OF CAMPHOR. Obtained in the sublimation of camphor from the wood. It is a reddish liquid with a slight yellowish tint, and is probably a mixture of a hydrocarbon and camphor. It resembles the latter in medical properties, but is more of a stimulant, and is especially applicable to those cases of bowel complaint or spasmodic cholera in which an anodyne and stimulant effect is wanted. This volatile oil must not be confounded with Linimentum Camphoræ, the common name for which, with many, is oil of camphor. Dose: 2 or 3 ♏ (0.13 to 0.2 mil).

179. **PERSEA GRATISSIMA** Gaertner.—ALLIGATOR PEAR. (Seeds.) Used by the Mexicans as an anthelmintic, and, in the form of liniment, in intercostal neuralgia. Dose of fl'ext.: 30 to 60 ♏ (2 to 4 mils).

PAPAVERACEÆ.—Poppy Family

Herbs, with milky, narcotic juice. *Leaves* alternate. *Flowers* large, with caducous calyx. *Ovary* one-celled, with parietal placentæ. The genus Papaver, a description of which is given under Opium, is typical of the order. See also illustrations below.

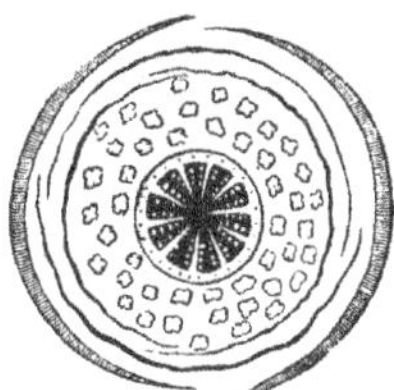

FIG. 100.—Transverse section of flower of Poppy.

FIG. 101.—Gynæcium of Poppy, with one stamen remaining.

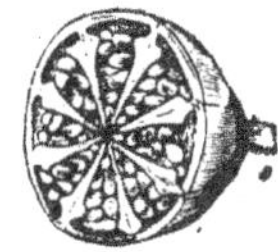

FIG. 102.—Transverse section of ovary of Poppy.

Synopsis of Drugs from the Papaveraceæ

A. *Concrete Juice.*	C. *Seed, and Fixed Oil.*	E. *Rhizome.*
OPIUM, 180.	Papaver, 182.	**SANGUINARIA,** 185.
B. *Capsule.*	D. *Herbs.*	F. *Flower-petals.*
*Papaver, 181.	Chelidonium, 183.	Rhœas, 186.
	Eschscholtzia, 184.	

180. OPIUM.—Opium

OPIUM

The concrete milky exudation obtained by incising the unripe capsules of **Papa′ver somnif′erum** Linné, and its variety, **album**, DeCandolle. Containing not less than 9.5 per cent. of anhydrous Morphine.

Botanical Characteristics.—*Leaves* large, sessile, wavy, cut, or toothed; *flowers* large and terminal, drooping before expansion; *petals* 4, large, roundish, white or purplish with a darker colored spot near the claws; *stigmas* 4 to 20,

Fig. 103.—*Papaver Somniferum*—Flowering branch and fruit.

radiating, sessile upon the disk, which covers the ovary. *Capsule* obovate, 1-celled; *placentæ* extended so as to almost divide the cavity into several cells; dehiscence by small chinks or pores beneath the crown formed-by the radiating stigmas; *seeds* numerous, reniform.

Source.—Western Asia; cultivated in the elevated plains of India,·in Egypt, Persia, Asia Minòr, and in some parts of Europe.   Varieties: (1) Smyrna, Levant, Turkey, or Constantinople; opium generally in flattish masses—the most abundant in the market, to which descriptions in text-books usually apply; (2) Egyptian, in flattened, roundish cakes; (3) Persian, in cylindrical sticks or cakes of a black color; (4) Indian, in flat squares, covered with layers of mica, and further protected by a coating of wax or an oiled-paper wrapper; (5) Chinese, in flat, globular cakes; (6) European.

DESCRIPTION OF DRUG.—**In irregular or subglobular lumps weighing from four ounces to two pounds,** enveloped in remnants of poppy leaves and with chaffy fruits of a species of *Rumex* adhering; **when fresh it is plastic,** breaking or tearing apart, showing an irregular, chestnut-brown surface, shining when rubbed; odor peculiar, narcotic; taste bitter. When examined with a pocket lens, it is seen to be composed of yellowish, agglutinated tears. **The value of the gum,** however, **is determined only by assay.** Opium should yield not less than 9 per cent.; powdered opium not less than 12, nor more than 12.5 per cent., of crystallized morphine when assayed by the official process.

Granulated opium, or coarsely powdered opium, is an article of commerce, and is especially recommended as a form of the drug best adapted to the preparation of the tinctures.

Powder.—Characteristic elements: See Part iv, Chap. I, B.

ADULTERATIONS.—To increase the weight various articles are used, such as sand, clay, scrapings of poppy capsules, and various mucilaginous, albuminous, and saccharine matters. The writer has taken from the interior of about a two-pound lump of opium over a quarter of a pound of lead bullets.

A mixture sold for opium was analyzed and found to be mostly aloes which after suitable mixing, had been buried in the ground until the odor of aloes was gone.

Factitious opium has occasionally been met with, of soft consistence, blackish-brown color, less odorous than the genuine. It is probably an aqueous extract of the poppy plant.

Alkaloidal assay, and microscope, easily betray adulteration.

CONSTITUENTS.—Opium contains a mixture of sixteen or more different alkaloids, with meconic acid, coloring matters, and various inert substances. The principal constituents are the following alkaloids: Morphine, $C_{17}H_{19}NO_3 + H_2O$; codeine, $C_{18}H_{21}NO_3 + H_2O$ (both official); narcotine, narceine, paramorphine, papaverine, meconidine, pseudomorphine, codamine, laudanine, and oxynarcotine; these are in combination with meconic and thebolactic acids. Mineral constituents average about 6 per cent.

Preparation of Morphine.—To the concentrated infusion of opium add three volumes of a mixture composed of one part of alcohol, two volumes of ether, and one-third volume of ammonia; shake, and set aside for crystals to form.

Preparation of Codeine.—The mother liquor, from which morphine has separated, yields crude codeine on evaporation. Obtained artificially by heating morphine with methyl iodide and soda or potassa.

Preparation of Narceine.—The concentrated infusion of opium is shaken with ether. This removes narcotine. If alkali be added in excess, codeine is deposited. From the filtrate morphine can be crystallized, and from the mother liquor narceine may be obtained upon evaporation.

Preparation of Meconic Acid.—Add $CaCl_2$ to an infusion of opium, which precipitates calcium meconate; decompose the latter by dilute HCl at 180°F. This deposits the calcium bimeconate, which is dissolved in warm concentrated HCl, from which the pure meconic acid deposits in cooling.

ACTION AND USES.—Stimulant, narcotic, anodyne, antispasmodic, and intoxicant. It restrains the movements and checks the secretions of the stomach and intestinal canal. The dominant action of opium, however, is upon the brain, first producing **mental and emotional exhilaration, then hypnotic depression.** It is a powerful respiratory depressant, death usually resulting from paralysis of the respiratory center in the medulla. Toxic doses, also, finally paralyze both the heart and vagi, and produce a rapid and feeble pulse. While the effects are due to the morphine present, the drug is not fully represented by this alkaloid. Codeine is also hypnotic, but affects the cerebrum less. Narcotine is antiperiodic. Thebaine is sudorific and excitant. Dose of opium: 1 to 2 gr. (0.065 to 0.13 Gm.).

POISONING shows three stages or degrees as follows:

1. Rather slow respiration, slow heart but good blood pressure, much contracted pupils. The patient is sluggish or inattentive. There may be nausea perhaps retching or vomiting.

2. A stupor which supervenes in from fifteen to thirty minutes. The face is cyanotic flushed, the skin warm, the respirations regular, only 4 to 10 per minute, slow heart but blood pressure remains good, pupils pin point, the patient in a state of unconsciousness from which he can be aroused with difficulty.

3. This stage is manifested by coma and collapse. The skin is cyanotic, cold and clammy, the pulse is weak, patient cannot be aroused, respirations are very infrequent and shallow about 3 or 4 per minute.

ANTIDOTES.—Emetics, apomorphine subcutaneously injected, strong coffee and stimulants, evacuation by mechanical means (stomach-pump, etc.), or rousing and walking the patient. Atropine is the physiological antagonist.

OFFICIAL PREPARATIONS.

Opii Pulvis (10 to 10.5 per cent. of anhydrous Morphine),........Dose:	¼ to 2 gr. (0.016 to 0.13 Gm.).
Extractum Opii (20 per cent. Morphine),......................	¼ to 1 gr. (0.016 to 0.065 Gm.).
Opium Deodoratum (12 to 12.5 per cent. Morphine),..............	¼ to 2 gr. (0.016 to 0.13 Gm.).
Mistura Glycyrrhizæ Composita,...	(2¼ fl. dr.)
Opium Granulatum,..............	(1 Gm.).

Pulvis Ipecacuanhæ et Opii (1 gr.
Opium and 1 gr. Ipecac in every
10 gr.),................   5 to 10 gr. (0.3 to 0.6 Gm.).
Tinctura Opii (10 per cent.)   5 to 15♏ (0.3 to 1 mil).
Tinctura Opii Camphorata (Opium,
Camphor, Benzoic Acid, and Oil
of Anise, each 0.4 per cent.), ...   1 to 4 fl. dr (4 to 15 mils).
Mist. Glycyrrhizæ. Co (From
Camphorated Tr.),
Tinctura Opii Deodorati (10 per
cent.),....................   5 to 15♏ (0.3 to 1 mil).
Alkaloids:
  Morphina, ...   ⅛ to ½ gr. (0.008 to 0.032 Gm.).
  Morphinæ Hydrochloidum,........   ⅛ to ½ gr. (0.008 to 0.032 Gm.)
  Morphinæ Sulphas,..............   ⅛ to ½ gr. (0.008 to 0.032 Gm.)
  Codeina, ...   ¼ to 2 gr. (0.016 to 0.13 Gm.)

181. **PAPAVER.**—Poppy Capsules. (Papaveris Fructus, N.F.) The nearly ripe capsules, free from seeds, of **Papa'ver somnif'erum** Linné. There are two varieties, distinguished by the color of their seeds. The white poppy is usually considered the true opium plant; its capsule is smooth, of various shapes, but usually **subglobular and somewhat flattened at the extremities;** it is of **a gray or a light yellowish-brown color,** 50 to 100 mm. (2 to 4 in.) in diameter, **crowned with the sessile stigmas arranged in a** circle; *placentæ* parietal, projecting toward the center; odor slight; taste bitter.

    Constituents.—Morphine, codeine, narcotine, narceine, papaverosine, and rhœadine, united with organic acids, of which meconic is the most important.

    Action and Uses.—Hypnotic and sedative in syrup or extract; local anodyne in decoction. Dose: 15 to 30 gr. (1 to 2 Gm.).

182. **PAPAVERIS SEMEN.**—Poppy Seed. Maw Seed. The seed of **Papa'- ver somnif'erum,** remarkable for containing so large a per cent. of fixed oil, which is very useful in the arts, and is also demulcent and anodyne. The seeds are less than a millimeter in length, kidney-shaped, with the surface regularly pitted, giving them a beautiful appearance under a lens. There is a black-seeded and a white-seeded variety under cultivation.

    Fifty per cent. of oil is obtained from the seeds by warm and 30 per cent. by cold pressure. It is pale yellow, with a bland and slightly sweetish taste, totally destitute of narcotic properties. Poppy-seed oil is used for salads, paints, soaps, illumination, and to adulterate olive and almond oils.

183. **CHELIDONIUM.**—Celandine. The entire plant of **Chelido'nium ma'jus** Linné. Off. in U.S.P. 1890. Stem hairy, arising from a reddish-brown, branching root, and bearing light green, lyrate-pinnatified leaves about 200 mm. (8 in.) long; odor slight; taste acrid. Cathartic, diuretic, diaphoretic, and expectorant. In certain sections it is used in the treatment of jaundice. Dose: 15 to 60 gr. (1 to 4 Gm.).

    Alkaloids and Principles of Chelidonium and Allied Plants.—Important researches of J. O. Schlotterbeck have shown that chelerythrine, yielding lemon-colored salts, exists also as a prominent alkaloid in sanguinaria and other plants of the same family. Protopine, $C_{20}H_{19}NO_5$, a frequently occurring alkaloid in the poppy family, occurs also in the plants of the fumariaceæ. In physical properties protopine agrees, in every particular with fumarine. Protopine has been found in *Papaver somniferum, Eschscholtzia californica, Sanguinaria canadensis, Stylophorum diphyllum,* and *Adlumium cirrhosa;* it constitutes two-thirds of the entire alkaloidal content of *Bocconia cordata.* ("Proc. Amer. Phar:," 1900, p. 131.) Wintgen found the constituent, chelidonine, to be $C_{20}H_{19}NO_{50}$ + H_2O. Schlotterbeck finds its more exact formula as, $C_{20}H_{18}(OH)NO_4 + H_2O$. ("Proc. Am. Ph.," 1903, p. 321.) It occurs in colorless monoclinic prisms, melting at 135° to 136°C. The coloring-matter, known as chelidoxanthin, found in chelidonium and stylophorum diphyllum, has been found to be identical with the alka-loid berberine. ("Pharm. Rev.," Jan., 1902, pp. 4, 5.)

184. **ESCHSCHOLTZIA CALIFORNICA** Chamisso.—(Herb.) A valuable calmative, soporific, and analgesic, "free from the disadvantages of opium." Dose of alcoholic extract: 10 gr. (0.6 Gm.), gradually increased to 3 dr. (12 Gm.) in a day.

185. SANGUINARIA.—Sanguinaria

BLOOD ROOT

The dried rhizome of **Sanguina'ria canaden'sis** Linné.

BOTANICAL CHARACTERISTICS.—A low perennial, common in rich woods, having a thick, prostrate root-stock, surcharged with an orange-red, acrid juice, and sending up in earliest spring a rounded, palmately lobed leaf and a one-flowered naked scape. *Flower* white, handsome; *sepals* 2; *petals* 8 to 12; *stamens* about 24; *style* short; *stigma* two-grooved; *pod* oblong, turgid, one-celled.

HABITAT.—Rich woods of North America.

DESCRIPTION OF DRUG.—A horizontal cylindrical rhizome about 50 mm. (2 in.) long and 10 mm. (⅖ in.) thick, slightly tapering and branched;

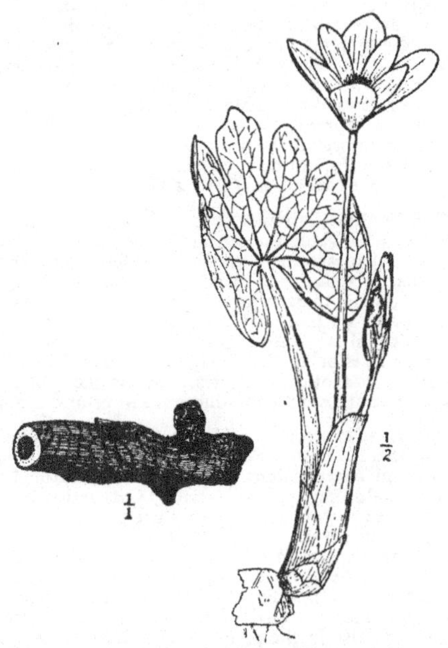

FIG. 104.—*Sanguinaria canadensis*—Plant and rhizome.

externally reddish-brown, rough, wrinkled, and annulate; **internally** spongy, **dotted with small resin cells of a ruby color.** The color of a cut surface varies from a light to a very dark red, and presents a glossy, dotted appearance; bark thin, with resin cells scattered in the parenchyma; frequently the transverse surface shows either a uniform dark blood-red color, or a whitish, starchy surface scattered with numerous red dots; **odor slight; taste** bitter **and** acrid; the powder is sternutatory. The infusion of the drug becomes blood-red with sulphuric or hydrochloric acid.

Powder.—Characteristic elements: See Part iv, Chap. I, B.

ADULTERATION.—E. M. Holmes calls attention to an adulteration of *Helionas rhizome* (q.v.), false unicorn, a rather expensive admixture amounting, in one case, to 40 per cent. This root has a different transverse surface, being of a dirty white hue and horny texture, and exhibits a well-defined central column, occupying about one-third of the diameter, and containing irregularly placed vascular bundles.

CONSTITUENTS.—**Sanguinarine**, $C_{20}H_{15}NO_2$, a colorless **alkaloid yielding red** salts, chelerythrine yielding lemon-yellow salts, homochelidonine and protopine. See Alkaloids, under Chelidonum (183). "A careful analysis of sanguinaria shows that the yield of sanguinarine scarcely reaches 1 per cent." Schlotterbeck believes that "the name Sanguinarine should be applied to the predominating alkaloid, to chelery-

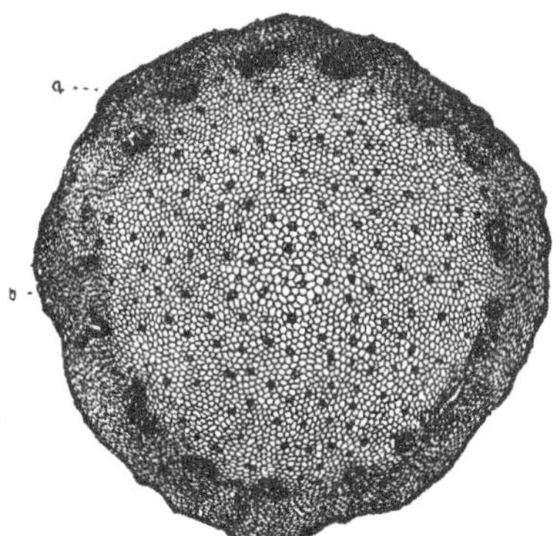

FIG. 105.—Sanguinaria. *a*. Wood-bundle. *b*. Pith.

thrine which forms yellow salts. Sanguinarine nitrate is becoming recognized more and more by the medical profession as a remedy in respiratory disorders and throat troubles." Ash, not exceeding 3 per cent.

Preparation of Sanguinarine.—Treat infusion of the powdered rhizome with dilute HCl or acetic acid, add NH_4OH, collect precipitate, redissolve in alcohol, decolorize, and evaporate. It is white, soluble in alcohol, ether, benzene; yields bright red salts of an acrid taste.

ACTION AND USES.—An acrid emetic, stimulant, narcotic. Moderate doses produce nausea and circulatory depression, and in large doses it inflames the stomach, causing intense burning, thirst, vomiting, dimness of vision, vertigo, great prostration, and collapse.

Powdered sanguinaria snuffed up the nostrils is sternutitory, and applied locally it acts as a stimulant to indolent ulcers and as an escharotic to fungous granulations. The physiological action of sanguinaria bears no relation to its principal therapeutic application, namely, as a stimulating expectorant in subacute and chronic bronchitis. Dose: Expectorant, o.2 Gm. (3 gr.); emetic, 1 Gm. (15 gr.).

OFFICIAL PREPARATION.

Tinctura Sanguinariæ (10 per cent.),. . .Dose: 15 to 30♏ (1 to 2 mils).

186. **RHŒAS.**—RED POPPY. The petals of **Papa'ver rhœ'as** Linné, the red or corn poppy of our gardens, growing abundantly as a wild plant in Europe. Nearly round, 50 mm. (2 in.) broad, contracted below into a short blackish claw; when fresh, they are of a scarlet-red color, but become brownish-purple on drying, and have an opium-like odor and a somewhat bitter taste. All parts of the plant contain the alkaloid rhœadine, which produces interesting reactions with acid and alkalies. It does not appear to be poisonous. Acid solutions produce a purple color, which disappears when neutralized. One part of the alkaloid produces a deep purple with 10,000 parts of water, rose with 20,000, and a perceptible redness with 800,000 parts. According to Hesse, the milky juice also contains meconic acid. Red poppy is a weak and uncertain opiate; **used in pharmacy almost wholly in the fresh state for coloring preparations.**

FUMARIACEÆ.—Fumitory Family

Erect or climbing herbs with alternate leaves. Slightly bitter, innocent plants. Bocconia cordata (= Macleya cordata), Tree Celandine, belongs to this order (see Chelidonium). Yields protopine.

187. **CORYDALIS, N.F.**—TURKEY CORN. Tubers of **Dicen'tra canaden'sis** De Candolle. *Habitat:* Canada and the mountains of the United States south to Kentucky. Small, heavy, **pebble-like tubers,** often united, three around a common center; of a dull yellowish to a dull black color, semitranslucent; inodorous; bitter. They contain four alkaloids, the chief of which is corydaline ($C_{18}H_{19}NO_4$), four-sided prisms, inodorous, tasteless, insoluble in water, soluble in ether, alcohol, and chloroform. This interesting alkaloid has been found in other species of corydalis, as *C. cava.*
Preparation of Corydaline.—Treat the residue from evaporated tincture with dilute HCl. Precipitate with ammonia and dissolve precipitate in boiling alcohol; on evaporation of this solution four-sided prisms of the alkaloid are deposited.

CRUCIFERÆ.—Mustard Family

Herbs with pungent, watery juice; *sepals* and *petals* 4 each, cruciform; *stamens* 6, tetradynamous; *capsule* usually spuriously 2-celled; *fruit* a silique or silicle.

Synopsis of Drugs from the Cruciferæ

A. *Seeds.*	B. *Herb.*	C. *Root.*
SINAPIS ALBA, 188.	Bursa Pastoris, 190.	Armoracia, 191.
SINAPIS NIGRA, 189.		

188. SINAPIS ALBA

WHITE MUSTARD

The seed of **Sina'pis al'ba** Linné.

BOTANICAL CHARACTERISTICS.—Stem 1 to 2 feet high, round, smooth. *Leaves* lyrate-pinnatifid. *Flowers* yellow. *Silique* hispid. *Seeds* whitish, with the *embryo* folded upon the surface of one of the cotyledons, which is also folded so as to inclose it.

HABITAT.—Asia and Southern Europe; cultivated.

DESCRIPTION OF DRUG.—The principal difference between this and black mustard is that of color and size, being 1 to 2 mm. in diam., of a yellowish color, and less pungent. **The oily embryo consists of a curved caudicle and two cotyledons, one folded over the other.** Both the black and white seeds are practically free from starch. Commercial ground mustard is an unctuous yellowish powder which cakes on pressure; it is usually a mixture of the ground white mustard (dull yellow) and the black mustard (yellowish-green). The mixture is, however, often rendered brighter by the addition of turmeric; when this is the case, it will respond to the test for starch, and will acquire a red-brown color with a solution of borax or boric acid. The "limit of starch" test of the U.S.P., page 394 (8th ed.), will betray dilution with starchy substances. Moistened with water the powder quickly develops a pungent odor.

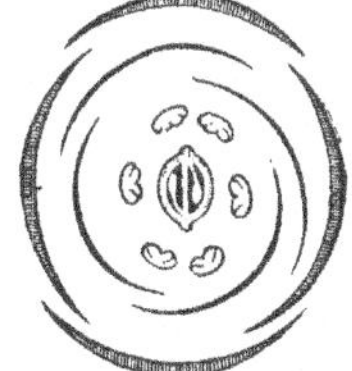

FIG. 106.—Diagram of a flower of the *Cruciferæ*.

FIG. 107.—The embryo of *Brassica pumias orientalis*.

Powder.—Characteristic elements: See Part iv, Chap. I, B.

CONSTITUENTS OF BLACK AND WHITE MUSTARD.—Both contain a fixed oil, 22 to 23 per cent.; mucilage about 19 per cent. Both seeds contain the ferment myrosin, the white having usually the larger quantity. The quantity of myrosin in these seeds is quite variable, sometimes being as low as 2 per cent., then as high as 18 per cent. A glucoside exists in the white mustard, *sinalbin* ($C_{30}H_{44}N_2S_2O_{16}$), which, decomposed by myrosin, yields glucose, sinapine sulphate, and a fixed oil, which is the *sulphocyanate of acrinyl*, and is found to be identical with *para oxyphenylacetic acid*. H. Salkowski manufactured this principle synthetically. The black mustard contains *sinigrin* ($C_{10}H_{18}KNS_2O_{10}$), which yields, when decomposed by the ferment myrosin, glucose, potassium sulphate, and a volatile oil, *allyl isothiocyanate* ($CS : N.C_3H_5$), the common mustard oil. Ash, not exceeding 9 per cent.

ACTION AND USES.—Same as Sinapis Nigra. Average dose: 2 dr. (8 Gm.).

Preparation of Sinalbin.—Extract powdered white mustard with benzene (CH) to remove oil. Treat the dried dregs with four times its weight of boiling alcohol. Filter the alcoholic liquid while hot. On standing in a cool place the liquid deposits crystals of sinalbin.

Preparation of Sinigrin.—Oil is removed, as in the case of sinalbin.   The oil cake is then boiled in alcohol and evaporated to dryness.   Repowder and extract with cold water.   Treat the resulting liquid with barium carbonate and evaporate on a waterbath to dryness.   Extract the residue with strong boiling alcohol and filter while hot.   On cooling and standing the solution deposits silky needles of sinigrin, or potassium myronate.

189. SINAPIS NIGRA

BLACK MUSTARD

The ripe seed of **Bras'sica ni'gra** Linné.

BOTANICAL CHARACTERISTICS.—Similar to *S. alba* (see above), but has larger flowers, a longer hispid silique, and a smaller blackish seed.

HABITAT.—Asia and Southern Europe; cultivated.

DESCRIPTION OF DRUG.—A globular seed about 1 mm. ($\frac{1}{25}$ in.) in diameter, with a circular hilum and a short beak not filled with albumen; **testa hard,** black, or reddish-brown, **finely pitted.**   The yellow embryo and cotyledons are folded and bent along the midrib. Inodorous when dry, but **pungent and penetrating when moist;** taste hot, acrid.   The powder should give only a faint reaction for starch by the iodine test.   Ash, not exceeding 9 per cent.

Powder.—Characteristic elements: See Part iv, Chap. I, B.

ACTION AND USES.—Externally a **powerful** rubefacient and **counter-irritant,** internally **emetic,** especially valuable in cases of poisoning by narcotics from its reflex stimulation of the heart and respiration. Dose: 1 to 4 dr. (4 to 15 Gm.).

OFFICIAL PRODUCTS.

189 a. **Oleum Sinapis Volatile.**   U.S.P. IX.   A product yielding not less than 92 per cent. of "allyl isothiocyanate."   It is produced synthetically or obtained from the seed of Brassica Nigra by maceration with water and subsequent distillation, and must conform in name to the source from which it is derived. Great caution should be exercised in smelling this oil.   It should not be tasted except when highly diluted.

DESCRIPTION AND SOURCE.—Volatile oil of mustard is not contained as such in seeds but is formed by the decomposition of "sinigrin" or "potassium myronate" in the presence of emulsin.   The ground mustard seed is deprived of its fatty oil with the aid of hydraulic presses.   The press cakes are mixed with tepid water, allowed to undergo fermentation, and then distilled with water vapor.   The yield varies between 0.5 to 0.75 per cent. of the original seed. At a temperature exceeding 70°C. (158°F.) no fermentation takes place because the myrosin is coagulated and rendered inactive.

PROPERTIES.—Oil of mustard is a colorless or yellowish, limpid and refractive liquid with an exceedingly pungent and acrid odor.   Inasmuch as it draws blisters when in contact with the skin, it should not be tasted.

COMPOSITION.—In addition to "mustard oil," C_3H_5SCN, or allyl isosulphocyanate, the oil from black mustard contains variable amounts of "allyl cyanide," C_3H_5CN, and carbon disulphide, CS_2.

ACTION AND USES.—Volatile oil of mustard is rarely given internally.   Locally it may be employed as a counter-irritant.   Diluted with olive oil, it may be used as a substitute for mustard papers and as a stimulating liniment.   Dose: $\frac{1}{125}$ mil ($\frac{1}{8}$ ♏).

189 b. **OLEUM SINAPIS EXPRESSUM** (Unofficial).—Crushed seeds of the black and white mustard yield, by cold expression, about 22 per cent. of a bright

yellow (white) or brownish-yellow (black) oil, of a bland taste. This oil is a commercial oil and not infrequently used for the adulteration of other oils. Rapeseed, or colza, oil is obtained from the seeds of different varieties of the genus Brassica, rape (*Brassica napus*) in particular. In Europe the term rapeseed oil is sometimes applied to the product of rape alone, colza being restricted to the oil obtained from the ruta-baga, or Swedish turnip (*B. campestris*), while "Rubsen" oil is furnished by the common turnip (*B. rapa*). There is great confusion among authors in the use both of the common names of the oils and the scientific names of the varieties of Brassica which produce them. The seeds of rape contain from 33 to 43 per cent. of oil, which, when crude, is a dark yellow-brown and used for lubricating. Refined and freed from albumen and mucilage the oil becomes bright yellow. Rape oil is extensively used for lamps, lubricating machinery, and for adulterating both almond and olive oils.

FIG. 108.—*Sinapis nigra*—Branch.

190. **BURSA PASTO'RIS.**—SHEPHERD'S PURSE. The herb of Capsel'la bursa-pastoris Moench, a small plant very common along our roadsides. It derives its name from its inversely heart-shaped fruit in elongated racemes. The small white flowers are in corymbose racemes. Nearly inodorous; taste acrid, pungent, and bitter. Contains a little volatile oil of mustard. An active diuretic, also tonic and stimulant. Dose: 15 to 60 gr. (1 to 4 Gm.).

191. **ARMORACIA.**—HORSERADISH. The root of Cochlea'ria armora'cia Linné. Indigenous to Europe, but cultivated in our gardens as a condiment. A cylindrical root 300 mm. (12 in.) long, 12 to 25 mm. (½ to 1 in.) thick; externally pale yellowish-brown, warty; internally white; fracture short; odor when crushed pungent; taste sharp and acrid. Contains a volatile oil similar to oil of mustard. Used only in fresh state as a stimulant to digestion, as a diuretic, and externally as a rubefacient. Dose: 1 to 2 dr. (4 to 8 Gm.).

14

SARRACENIACEÆ.—Pitcher-plant Family

192. **SARRACE'NIA FLA'VA** and **S. PURPU'REA** Linné.—The curious pitcher-plant, fly-trap, or side-saddle plant of our Southern States, where their rhizomes are much used in dyspepsia. They are tonic and diuretic. Dose: 15 to 30 gr. (1 to 2 Gm.).

DROSERACEÆ.—Sundew Family.

193. **DROSERA, N.F.**—Sundew. The herb of **Drose'ra rotundifo'lia** Linné. (See Conspectus.) *Habitat:* North America and Europe. Used principally as a pectoral in bronchitis, coughs, etc. Dose: 5 to 15 gr. (0.3 to 1 Gm.).

CRASSULACEÆ.—Orpine Family

194. **SEDUM ACRE.**—Biting Stone-crop. English Mass. The whole plant, Se'dum a'cre Linné. *Habitat:* Europe; cultivated in New England gardens. It is said to be very successful in the treatment of diphtheritic sore throat, by dissolving and expelling the false membrane. Dose: 15 to 30 gr. (1 to 2 Gm.).

195. **PENTHORUM.**—Virginia Stone-crop. The herb of **Pentho'rum sedoi'des** Linné. Astringent, demulcent, and laxative, in diseases of the mucous membranes. Dose: 15 to 30 gr. (1 to 4 Gm.).

SAXIFRAGEÆ.—Saxifrage Family

196. **HEUCHERA.**—Alum Root. The root of **Heu'chera america'na** Linné. (See Conspectus.) *Habitat:* United States. It contains about 14 per cent. of tannin, and is a powerful astringent in doses of 15 to 30 gr. (1 to 2 Gm.).

197. **HYDRANGEA, N.F.**—The root of **Hydran'gea arbores'cens** Linné. (See Conspectus.) *Habitat:* United States. It consists of several bent, branched roots, arising from a thick, knotty head, or, as usually seen, of pieces of these roots cut up into various lengths. The rather thick, light gray, or pale brown bark is longitudinally ridged and covered with rust-colored patches, and separates easily from the tough, white, tasteless wood; wood-wedges long, narrow; odorless; taste of bark sweetish, afterward pungent. Used as a diuretic and as an antilithic in those cases where there is an alkalinity of the urine and a tendency toward the deposition of phosphatic calculi. Dose: 30 to 60 gr. (2 to 4 Gm.).

198. **MITELLA NUDA** Linné.—Coolwort. (Leaves.) Diuretic; used in inflammatory and catarrhal affections of the bladder and kidneys.

HAMAMELIDACEÆ.—Witchhazel Family

Shrubs or trees with alternate, simple *leaves* and deciduous stipules. *Flowers* in heads or spikes, often polygamous or monœcious. *Fruit* a woody capsule, 2-beaked, 2-celled, 2-seeded. A family which contains but few species, but is dispersed over both hemispheres. The wood of a tree, Parrolin, is extremely hard, and in Persia is called iron-wood.

Synopsis of Drugs from the Hamamelidaceæ

A. *Leaves.*
  ***HAMAMELIDIS Folia, 199.**
B. *Bark.*
  HAMAMELIDIS CORTEX, 200.

C. *Balsam.*
  STYRAX, 201.
  Liquidambar, 202.

199. HAMAMELIDIS FOLIA, N.F.—Hamamelis Leaves

WITCHHAZEL

The dried leaves of **Hamame'lis** virgin'iana Linné, collected in autumn before the flowering of the plants. Not more than 10 per cent. of stems and foreign matter permitted.

Habitat.—North America.

Description of Drug.—Leaves broadly elliptical to obovate, more or less unequal, 3.5 to 12 cm. long, 2.5 to 7 cm. broad; apex rounded, acute or acuminate; base obliquely cordate; margin sinuate or sinuate-dentate. Upper surface dark green, midrib and veins prominent, veins of the first order running nearly parallel to the margin; under surface light green, texture coarse, brittle; odor slight; taste astringent.

Twigs with nodes 2-ranked giving the younger portions frequently a zig-zag outline; externally yellowish-brown, with a purplish tinge, nearly smooth, faintly longitudinally wrinkled and with small circular lenticels; fracture tough, fibrous bark easily separable from the whitish or green white, finely radiate wood, in which the annular rings are not very distinct; odor slight and characteristic.

Powder.—Dull green. Characteristic elements: The trichomes, one-celled, in groups of 8 to 15, radiating from a center; crystal fibers, calcium oxalate prisms, and stomata. Seldom employed as powder.

Constituents.—Gallic acid; hamamelo-tannic acid, $C_{14}H_{14}O_9 + 5H_2O$, resin, and extractive. **Distilled Extract of Witchhazel, Hamamelis Water, Aqua Hamamelidis,** is prepared from hamamelis bark by macerating the bark in water for twenty-four hours, then distilling the product until the distillate reaches 85 per cent. of the bark used; then add 15 per cent. of alcohol. It has a peculiar odor, a somewhat saccharine taste, is quite stable, and presents no pharmaceutical, chemical, or therapeutical incompatibility. Its mode of preparation has been to some extent a trade secret, but the above formula furnishes a good preparation. This preparation has built up quite an industry along the Connecticut Valley, where the distillation of the liquid is performed almost exclusively.

Action and Uses.—It has come into extensive use as an astringent in hemorrhoids and internal hemorrhages, and as a general vulnerary. The distillate, known as "Extract of Witchhazel," is alleged to have properties which are not professionally recognized. Average dose: 30 gr. (2 Gm.).

Fluidextractum Hamamelidis Foliorum, Dose: 10 to 60 ♏ (0.6 to 4 mils).

200. **HAMAMELIDIS CORTEX.**—Witchhazel Bark. Thin pieces covered with an easily separable grayish or grayish-brown cork, more or less covered with blackish dots and scars. When deprived of this layer, the bark is pale cinnamon-brown, fibrous. Odorless; taste astringent, bitter, and somewhat pungent. Its medical properties are the same as those of the leaves. The bark and twigs are official under the above title.

201. STYRAX.—Storax

LIQUID STORAX

A balsam obtained from the wood and inner bark of Liquid'ambar orienta'lis Miller.

Source and Description.—This balsam is not a natural part of the plant but is produced as a result of the stimulus from wounds in the bark.

The outer bark is bruised, then the inner bark becomes saturated with this pathological exudation.   The outer bark is removed and the inner is boiled in sea water, the storax is skimmed off the surface as it rises, then afterward the boiled bark is pressed.   The bark which yet contains some balsam is dried and used chiefly as incense.   Good storax should not contain over 30 per cent. of water and 60 per cent. should be soluble in alcohol.

The *Liquidambar orientalis*, growing in the southwest districts of Asia Minor, produces the balsam also, it is said.   It is a gigantic tree

FIG. 109.—*Liquidambar orientalis*—Branch.

"like the great oak, having clusters (of berries) like those of the oak, but its berries are larger."   The inner bark of the tree is boiled in water and the balsam pressed out.   A superior kind is said to be obtained by simply pressing the bark before it is boiled.   Another kind of liquid storax is mentioned—that which exudes naturally.

HABITAT.—Asia Minor.

DESCRIPTION OF DRUG.—It is a viscid, gray semi-liquid, with an agreeable odor, and a balsamic, somewhat acrid taste; a heavier dark brown layer separates on standing.

CONSTITUENTS.—Containing a volatile oil and a resin, and cinnamic and benzoic acids, storax is rightly classed as a balsam. Its most abundant constituent is **storesin,** $C_{36}H_{58}O_3$, existing both free and as a cinnamic ether. **Cinnamic acid** exists to the extent of 6 to 12 per cent., various ethers of it occurring, **styracin** being the cinnamate of cinnamyl. Storax also contains a liquid hydrocarbon, **styrol,** C_8H_8, or **cinnamene,** having the storax odor and taste, and another fragrant constituent, **vanillin,** not more than 1 per cent. of ash.

ACTION AND USES.—Stimulant expectorant. Dose: 5 to 20 gr. (0.3 to 1.3 Gm.).

OFFICIAL PREPARATION.

Tinctura Benzoini Composita (8 per cent.),..Dose: ½ to 2 fl. dr. (2 to 8 mils).

202. **LIQUIDAMBAR.**—SWEET GUM. A balsam exuding spontaneously or from incisions made in the trunk of Liquidam'bar styraci'flua. *Habitat:* Southern United States and Central America. It is a pale yellowish, opaque liquid of honey-like consistence, or thick, golden brown, solidifying on exposure to a transparent, amber-colored mass, which softens at the heat of the hand; odor storax-like; taste aromatic and pungent. Stimulant expectorant, mostly used in the manufacture of chewing-gum.

ROSACEÆ.—Rose Family

Herbs, shrubs, or trees, with pinnate, palmate, or simple, alternate *leaves.* *Flowers,* regular, *sepals* usually 5, united *petals* 5, perigynous; *stamens* numerous distinct, perigynous; *pistils* 1 to many. The different tribes are characterized by the fruit—a drupe in Pruneæ, follicles in Spiræeæ, druples in Rubeæ, dry akenes in Potentilleæ and Poterieæ, bony akenes in Roseæ, and pomes in Pomeæ. Except in the seeds of the drupe-fruits, which develop the poison hydrocyanic acid, this order is destitute of noxious qualities.

Synopsis of Drugs from the Rosaceæ

A. *Barks.*
  PRUNUS VIRGIN-
    IANA, 203.
  Choke Cherry, 204.
  Malus, 207.
  *RUBUS, 217.
  *Quillaja, 212.

B. *Seeds.*
  AMYGDALA AM-
    ARA, 209.
  AMYGDALA DUL-
    CIS, 210.
  Cydonium, 208.

C. *Fruits.*
  *Prunum, 205.
  *Rubus Idæus, 218.
  Crategus, 219.
  Pyrus Malus, 207

D. *Leaves.*
  Persica, 206.
  Laurocerasus, 220.
  Fragaria, 221.

E. *Flowers and Petals.*
  ROSA GALLICA, 213.
  Rosa Centifolia, 214.
  *Cusso, 222.

F. *Rhizomes.*
  Tormentilla, 223.
  Geum Urbanum, 224.
  Geum Rivale, 225.
  Gillenia, 226.

G. *Herbs.*
  Agrimonia, 227.
  Potentilla, 228.
  Spiræa, 229.
  Rosa Canina, 215.

H. *Volatile Oils.*
  OLEUM ROSÆ, 216.
  OLEUM AMYGDA-
    LÆ AMARÆ, 209 a.

I. *Fixed Oils.*
  OLEUM AMYGDA-
    LÆ EXPRESSUM, 211.

203. PRUNUS VIRGINIANA.—Wild Cherry

WILD CHERRY BARK

The bark of **Pru'nus sero'tina** Ehrhart, collected in autumn and carefully dried and preserved.

BOTANICAL CHARACTERISTICS.—A large forest tree. *Leaves* oval-oblong or lance-oblong, brilliant green, smooth on both sides, unequally serrate; *flowers* white, in racemes; *drupes* purplish-black and shining; bitter.

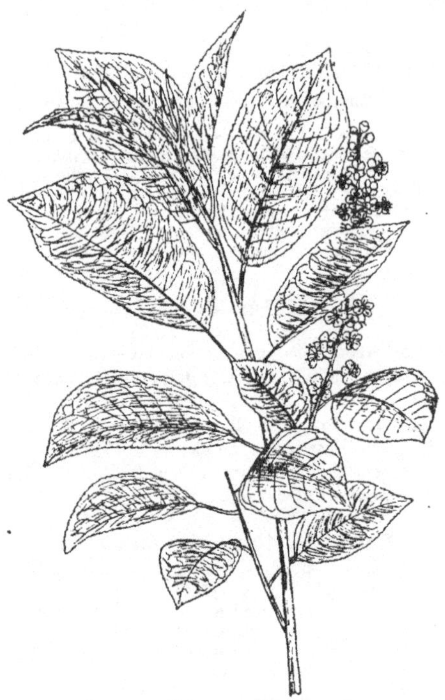

FIG. 110.—*Prunus serotina*—Branch.

SOURCE.—United States and Canada. Although the name *Prunus virginiana* has been held as the official and medicinal name, the botanical name is *P. serotina*. This leads to confusion among botanists, who strongly urge the discontinuance of the above official title. *Prunus virginiana* is the botanical name of the common choke cherry, not of the black wild cherry. *Prunus Pennsylvanica*, the wild red cherry, growing in rock woods and along the lake shores, is frequently mistaken for the *P. serotina*.

DESCRIPTION OF DRUG.—About 2 mm. ($\frac{1}{12}$ in.) or more in thickness, curved or flat. The newer bark is covered with a smooth, **greenish** periderm, but bark collected from the older parts usually has the corky layer removed, leaving a rough, rust-brown surface, inner surface lighter colored, finely striate; fracture granular. Almost inodorous, but emits the characteristic odor of bitter almonds when moistened; taste astringent, aromatic, and bitter, at the last bitter almond-like.

STRUCTURE.—Beneath the corky layer are found numerous clusters of stone cells, forming an interrupted zone. Just beneath this layer the medullary rays, which in the whole bark are wavy, terminate very obliquely. Between the medullary rays are found masses of stone cells and more elongated bast fibers.

The bark of the root is thought to be the most active, but that of the whole tree is collected indiscriminately.

RELATIVE VALUE OF THE OLD AND NEW BARK.—Experiments by Dohme and by Stevens have been made to decide whether the green bark is richer in hydrocyanic acid than the older, thick, brown bark. The results of the experiments of these gentlemen are somewhat contradictory. Dohme obtains 0.216 and 0.183 per cent. of HCN respectively, while the older bark assays 0.167 and 0.159 per cent. Stevens found in the older bark 0.335 per cent., while the younger assayed only 0.25 per cent. It is probably safe to say that the older thick bark is not so unworthy of recognition as some believe.

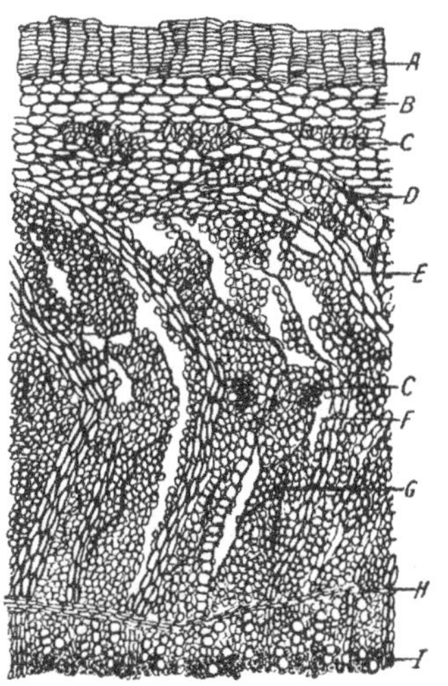

FIG. 111.—Cross-section of bark of stem of *Prunus Virginiana* (*P. serotina*). *A.* Cork. *B.* Middle or green layer of bark. *C.* Clusters of stone cells. *D.* Compressed sieve tissue. *E.* Medullary ray. *F.* Fissure between medullary ray and bast. *G.* Bast tissue. *H.* Cambium zone. *I.* Ducts in mature wood.

Powder.—Characteristic elements: See Part iv, Chap. I, B.

CONSTITUENTS.—Tannin, a bitter glucoside, resin, starch, etc. The volatile oil and hydrocyanic acid, to which the sedative action is due, do not preëxist in the bark, but, as in the bitter almond, are formed by the action on amygdalin, in the presence of water, of a ferment analogous to, if not identical with, emulsin; the action of this ferment is destroyed at a boiling temperature, and therefore heat should never be used in making preparations of this bark.

ACTION AND USES.—Tonic and sedative. Owing to the bitter principle it is a stomachic and bitter tonic. Useful in gastric atony and general debility. The syrup forms the basis of many of the cough syrups. Dose: 30 to 60 gr. (2 to 4 Gm.).

OFFICIAL PREPARATION.

Syrupus Pruni Virginianæ,. . . .   Dose: 1 to 4 fl. dr. (4 to 15 mils).

204. CHOKE CHERRY.—The bark of **Pru′nus virginia′na** Linné, a small tree growing in the Northern and Western States. Tonic and antiperiodic.

205. PRUNUM, N.F.—PRUNE

PRUNE

The partly dried ripe fruit of **Pru′nus domes′tica** Linné.

BOTANICAL CHARACTERISTICS.—The French variety, or *Juliana*, the principal commercial prune, bears ovate-oblong, deep-purple *drupes*, not depressed

FIG. 112.—*Prunus domestica*—Fruiting branch and flowering branch.

at the insertion of the stalk, and with a scarcely visible suture and no furrow; pulp greenish and rather austere. The tree is small, with smooth branches and elliptical *leaves; flower-buds* formed of one or two flowers; *petals* white, oblong-ovate.

HABITAT.—Western Asia; cultivated in temperate regions. Most of the prunes come from France, the best from Bordeaux.

DESCRIPTION OF DRUG.—Dried shriveled, oblong, almost globular, about 30 mm. (1⅕ in.) long; externally brownish-black. The sarcocarp (the medicinal portion) consists of a brownish-yellow pulp having a sweet, acidulous taste, and surrounds a single stone (putamen), which is very hard, smooth or ridged, and incloses a white, bitter weed.

CONSTITUENTS.—Sugar 12 to 25 per cent., pectin, malic acid, and salts. The seeds contain fixed oil, amygdalin, and emulsin.

Preparation of Amygdalin.—Obtained by solvent action of boiling alcohol upon the "oil cake," evaporating off alcohol, fermenting residue by yeast, and precipitating amygdalin and gum. Boiling alcohol takes up the principle which is deposited on cooling.

ACTION AND USES.—Laxative and nutrient, as an article of food or in laxative confections. **Confectio Sennæ** (U.S.P. VIII). Dose: 1 to 3 dr. (4 to 12 Gm.).

206. **PERSICA.**—PEACH LEAVES. From **Pru'nus per'sica** Linné. Mild sedative, generally administered in infusion. Dose: 15 to 30 gr. (1 to 2 Gm.).

207. **MALUS.**—APPLE TREE. The bark of **Pyr'us ma'lus** Linné. Tonic and febrifuge. Dose of fluidextract: 15 to 60 ♏ (1 to 4 mils).
SUCCUS POMARUM, N.F.—The freshly expressed juice of sound, ripe, sour apples, of cultivated varieties.

208. **CYDONIUM.**—QUINCE SEED. **Pyr'us cydo'nia** Linné. *Habitat:* Western Asia; cultivated. About 6 mm. (¼ in.) long, ovate, somewhat triangularly compressed, with the hilum near the pointed end; testa dark brown, covered with a thin, mucilaginous membrane or epithelium, causing the seeds to adhere in masses. The two cotyledons are thick and oily, veined, with a short conical radicle. Taste and odor of the embryo like bitter almonds, of the unbroken seed mucilaginous and insipid. The testa contains a large amount of mucilage; the embryo, fixed oil. A decoction is often used as a demulcent, and as an addition to eye-lotions.

209. AMYGDALA AMARA.—BITTER ALMOND (U.S.P. VIII)

BITTER ALMOND

The ripe seed of **Pru'nus Amyg'dalus**, var. **Amara**, De Candolle.

This is an oblong-ovate flattened seed with marked numerous longitudinal lines. Inodorous, bitter. Constituents: Fixed oil, 45 per cent. and amygdalin a crystalline glycoside, which by the action of emulsin, a ferment existing in the seed in the presence of water, splits up into glucose, HCN and benzaldehyde. Used as a sedative. From the seed is extracted the fixed oil by expression, and, from the residue, the volatile oil by distillation.

209 a. **OLEUM AMYGDALÆ AMARÆ**, U.S.—OIL OF BITTER ALMOND. A pale yellowish volatile oil obtained by macerating in water the residue left from bitter almonds after the fixed oil has been expressed, and distilling. It has a bitter, acrid taste, and a **strong odor of hydrocyanic acid.** It consists chiefly of benzoic aldehyde, to the oxidation of which is due the sediment, benzoic acid, thrown down on long exposure to air. The source from which it is derived in every case to be stated on the label. It should yield when assayed by the U.S.P. process not less than 85 per cent. of benzaldehyde and not less than 2 per cent. nor more than 4 per cent. of HCN. This oil is intended for medicinal use and not for flavoring foods. Sedative. Dose: ¼ to 1 ♏ (0.0164 to 0.0650 mil), in emulsion.

Official Preparations.

    Aqua Amygdalæ Amaræ (0.1 per cent.), . . . Dose: ½ to 2 fl. dr. (2 to 8 mils).
    Spiritus Amygdalæ Amaræ (1 per cent.), . .   5 ℥ (0.3 mil).

Fig. 113.—*Prunus amygdalus*—Branch, flower, and fruit.

210. AMYGDALA DULCIS.—Sweet Almond

SWEET ALMOND

The ripe seed of **Pru'nus Amyg'dalus**, var. Dulcis, De Candolle.

Botanical Characteristics.—Like Amygdala Amara, except that the *style* is much longer than the *stamens*, and the *seed* is sweet.

Source.—Western Asia and Barbary; extensively cultivated in Southern Europe, Spain and Southern France chiefly supplying the market.

Description of Drug.—Closely resembles the bitter almond, but is somewhat larger, with more convex sides, and has a **bland, sweetish taste,** free from rancidity. When triturated with water, if forms a milk-white emulsion, free from the odor of hydrocyanic acid.

Constituents.—Fixed oil from 50 to 55 per cent., nitrogenous compounds 25 per cent. (**myrosin, vitellin, conglutin**) precipitated by acetic acid, **emulsin,** mucilage, and sugar amounting to about 6 per cent. Ash, not exceeding 4 per cent.

ACTION AND USES.—Nutrient and demulcent; being free from starch, sweet almonds are often used as a diet in diabetes.

OFFICIAL PREPARATION.

Emulsum Amygdalæ (6 per cent.),....Dose: 2 to 8 fl. oz. (60 to 240 mils).

211. OLEUM AMYGDALÆ EXPRESSUM.—EXPRESSED OIL OF ALMOND

ALMOND OIL

A fixed oil expressed from Bitter or Sweet Almond.

DESCRIPTION.—A thin, clear, colorless or straw-colored liquid, with a mild, sweet taste and slight odor.

CONSTITUENTS.—Chiefly olein, with a slight quantity of palmitin.

ACTION AND USES.—Lenitive in pulmonary affections, in the form of emulsion.   Dose: 1 to 4 fl. dr. (4 to 15 mils).

OFFICIAL PREPARATION.

Unguentum Aquæ Rosæ (56 per cent., with spermaceti, white wax, stronger rose-water and borax).

212. QUILLAJA, N.F.—QUILLAJA

SOAPBARK

The dried bark derived of the periderm of Quilla'ja sapona'ria Molina.

DESCRIPTION OF DRUG.—In rather thick, flattish pieces of various sizes, deprived of the corky layer; outer surface brownish-white, sometimes with patches

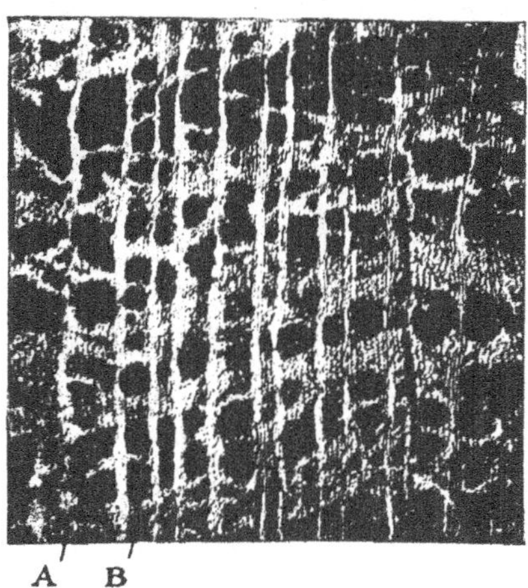

FIG. 114.—Quillaja—Cross-section of bark.   (25 diam.)   A, Medullary ray.   B, Bast fibers and stone cells.   (Photomicrograph.)

of the reddish-brown cork adhering; when held up to the light it shows numerous glistening crystals of calcium oxalate, which are scattered throughout the tissue.   Fracture tough and is fibrous, a transverse section showing a checkered arrangement of pale brown bast fibers imbedded in the white wood.   Odorless; taste persistently acrid.   The powder is sternutatory. The powder of quillaja has been suspected as an adulterant of senega.   It is not at all difficult to detect its presence in such admixtures, as in quillaja powder there are found elements not at all represented in senega.   In quillaja there is a considerable amount of sclerotic tissue, numerous bast fibers, and prismatic crystals of calcium oxalate.   Any and all of these clearly mark the

FIG. 115.—*Quillaja saponaria*—Branch.

powder of quillaja, and would at once betray its presence in the powder of senega.

Powder.—Grayish.   Inner parenchyma of cortex colorless (15 to 25 μ by 50 to 150 μ in diam.), mostly with large, long prisms of calcium oxalate; parenchyma of cortex with starch (3 to 10 μ in diam.); sclerenchyma with bast fibers (20 to 30 μ in diam.), thick-walled, porous, occasionally branched; stone cells (50 to 150 μ in diam.).

CONSTITUENTS.—Its irritant property is due to the presence of **saponin**, $C_{19}H_{30}O_{10}$, a mixture of the two glucosides, **quillaiac acid** and **sapotoxin**.

Preparation of Saponin.—Exhaust quillaja with hot alcohol, from which it separates upon cooling.   Saponin is regarded as a mixture of two glucosides, quillaiac acid and sapotoxin.

ACTION AND USES.—Containing about the same principles as senega, it has been recommended as a substitute for that drug as an expectorant in pulmonary affections. Dose: 15 to 30 gr. (1 to 2 Gm.).

213. ROSA GALLICA.—RED ROSE

RED ROSE

The dried petals of **Ro'sa gal'lica** Linné, collected before expanding.

BOTANICAL CHARACTERISTICS.—A dwarfish bush, with odd-pinnate *leaves* and adnate *stipules; leaflets* elliptical, rugose. *Flowers* large, red; *stamens* many. *Carpels* several, becoming bony akenes in *fruit*. Receptacle urn-shaped, with *styles* rising from inner surface.

HABITAT.—Asia and Europe; cultivated.

DESCRIPTION OF DRUGS.—The buds are collected before expanding, the petals being loosely imbricated in the form of cones, or separate and crumpled. They are roundish-obovate, with a dark red, velvety appearance, which they retain after drying, during which process the fresh petals lose 90 per cent. of their weight; claws yellow; odor fragrant; taste bitter and astringent.

Powder.—Elements in: See Part iv, Chap. I, B.

CONSTITUENTS.—The astringency is due principally to **quercitrin,** with which their color is also doubtless connected. They contain some tannin, fat, and volatile oil. Boiling water extracts their virtues.

Not more than 3.5 per cent. of ash.

ACTION AND USES.—Mild tonic and astringent; chiefly employed as a vehicle for tonic and astringent preparations. Dose: 15 to 60 gr. (1 to 4 Gm.).

OFFICIAL PREPARATIONS.

Fluidextractum Rosæ,...............Dose: 15 to 60 ♍ (1 to 4 mils).
Mel Rosæ (12 per cent.).

214. **ROSA CENTIFOLIA.**—PALE ROSE. HUNDRED-LEAVED OR CABBAGE ROSE. The petals of **Ro'sa centifo'lia** Linné. Off. U.S.P. 1890. The full-blown flower is picked off just below the calyx, and the petals separated. They are a beautiful pink when fresh, dull brown when dry; thin and delicate, roundish-obovate, sometimes obcordate, with a fragrant odor, and a bitter, faintly astringent taste. They may be preserved fresh for a considerable time by packing them in half their weight of common salt. These petals were formerly used in making the compound syrup of sarsaparilla, but wisely have been dropped as one of the ingredients. *Constituents:* Malic and tartaric acids, tannin, etc. Their odor depends upon a volatile oil existing in small quantity, about 0.04 per cent. Seldom, if ever, used medicinally. In pharmacy used principally for preparing rose-water.

215. **ROSA CANINA.**—HIPS. DOG ROSE. The fruit of **Ro'sa cani'na** Linné, common in Europe. Ovoid, or pitcher-shaped, about 18 mm. (¾ in.) long, with a smooth, shining, red surface. It consists of the ripened fleshy calyx, surmounted by the five calyx teeth; its cavity is hairy inside, and contains numerous hard, hairy akenes, but these akenes and hairs are removed before the hips are used. Taste acidulous, slightly astringent, due to the malic and citric acids and slight quantity of tannin contained; odorless. Refrigerant, mild astringent, and diuretic. Confection of hips is a familiar preparation abroad.

216. OLEUM ROSÆ.—OIL OF ROSE

ATTAR OF ROSES

A volatile oil distilled from the fresh flowers of **Ro'sa damasce'na** Miller.

SOURCE.—District of Kisanlik, in southern slope of the Balkans.

DESCRIPTION.—A pale yellow liquid having a specific gravity of 0.87, an agreeable rose odor, and sweetish taste. It solidifies between 16° and 21°C. into a transparent solid, containing numerous slender, iridescent crystals of the stearopten, which float on the surface when the solid is melted, as by the heat of the hand.

CONSTITUENTS.—It consists of two parts, one of which is fragrant and the other comparatively inodorous. The fragrant principles are mainly geraniol and citronellol; the other a white crystalline stearopten, $C_{16}H_{34}$, melting at 36.5° to 38°C. Used as a perfume in ointments, pomades, etc.

FIG. 116.—*Rubus villosus*—Branch and fruit.

217. RUBUS, N.F.—RUBUS

BLACKBERRY ROOT

The dried bark of the rhizome of **Ru'bus villo'sus** Aiton, **Rubus Nigrobaccus** Bailey, and **Rubus cuneifolius** Pursh.

DESCRIPTION OF DRUG.—In thin, tough, pliable bands 1 to 2 mm. ($\frac{1}{25}$ to $\frac{1}{2}$ in.) thick, having a blackish-gray outer surface, longitudinally wrinkled, and a pale brown inner surface; bast layers tangential, the fibers easily removed. Odorless; taste astringent and somewhat bitter. The root of *Rubus canadensis* Linné (dewberry) very closely resembles that of blackberry in medical properties.

Powder.—Light brown. Characteristic elements: Parenchyma of cortex, thin-walled, with starch, spherical (3 to 7 μ in diam.), thick, porous, elongated; bast fibers, walls of medium thickness, with some starch; wood fibers, ducts and tracheids, numerous with simple pores; cork considerable (20 to 30 μ in diam.); calcium oxalate crystals, aggregate (25 to 30 μ in diam.).

CONSTITUENTS.—The virtues of the bark depend chiefly upon the tannin present, about 10 to 15 per cent.

ACTION AND USES.—Tonic and astringent. From a popular domestic remedy it has come into extensive use in the treatment of diarrhea, dysentery, and relaxed conditions of the bowels generally. Dose: 15 to 30 gr. (1 to 2 Gm.).

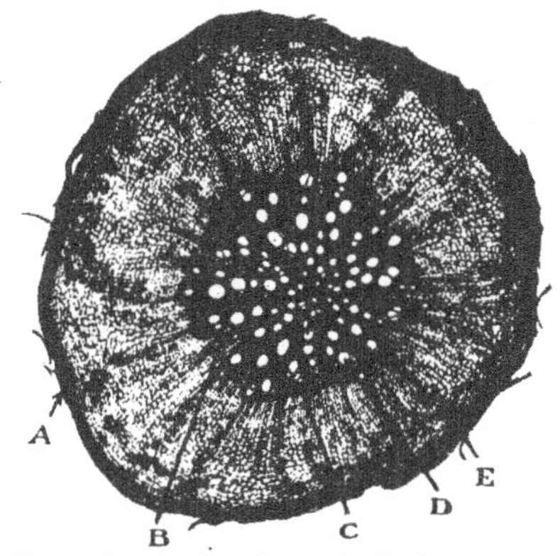

FIG. 117.—*Rubus villosus*—Cross-section of root, showing bark attached to wood. (17 diam.) A, Cork. B, Medullary ray. C, Xylem. D, Water tube. E, Groups of bast fibers. (Photomicrograph.)

PREPARATIONS: Fluidextractum Rubi, Syrupus Rubi, N.F.

RUBI FRUCTUS, N.F.—Includes two varieties of ripe fruits: *Nigrobaccus* and *Villosus.* A Syrup is recognized in the N.F.

218. **RUBUS IDÆUS.**—RASPBERRY. The fruit of **Ru'bus idæ'us** Linné. Off. U.S.P. 1890. A collective fruit, hemispherical, about 12 mm. (1½ in.) broad; it consists of numerous small, red, hairy drupes united at the base around the receptacle, from which the coalesced fruits are easily removed, leaving a conical cavity. Contains a bright red, acidulous juice; odor agreeable. Used only in the fresh state. The purplish-black fruit of *Rubus occidentalis* Linné may be substituted for it.

SYRUPUS RUBI IDÆI.—Rubi Idæi Fructus, N.F., includes two varieties: *Idæus and Strigosus.*

219. **CRATÆGUS.**—The fruit of **Cratæ'gus oxyacan'tha,** English Hawthorn. Heart tonic. Its value as a cardiac stimulant and tonic has recently come to the medical profession through Dr. M. C. Jennings, of Chicago. Dose of fluidextract: 10 to 15 ♏ (0.6 to 0.9 mil).

220. **LAUROCERASUS.**—CHERRY LAUREL. The leaves of **Pru'nus laurocera'sus** Linné, an ornamental shrub native to Western Asia. They contain an amygdalin-like principle, **laurocerasin,** and a ferment. Odor bitter, almond-like;

taste aromatic and bitter.   Used in making cherry-laurel water, a preparation much employed in Europe as a sedative narcotic, much as the dilute hydrocyanic acid is used here.

221. **FRAGARIA VESCA** Linné.—Strawberry.   (Leaves.)   Mild astringent **and** diuretic.   Dose: 1 dr. (4 Gm.), in infusion.

Fig. 118.—*Rubus idæus*—Branch.

222. BRAYERA, KOOSO, N.F.

The dried panicles of the pistillate flowers of **Hage'nia abyssin'ica** Gmelin, with-. out the presence of more than 10 per cent. of the staminate flowers, other parts of the tree, or other foreign matter.   Reject any portions of the stem over 3 mm. in diameter and any binding material before the drug is powdered or used.

Habitat.—Abyssinia.

Description.—Small, reddish, pistillate flowers, consisting of two reddish bracts and a calyx of five reddish, hairy sepals inclosing one or two nutlets.   They come into market in cylindrical bundles of the compressed panicles, or detached, on short, hairy peduncles; odor tea-like; taste bitter and nauseous. In trade the "brown" and "red" kusso are known.   The former are mixed with male flowers.   In the "red," the best variety, the sepals are reddish; in the "brown" they are greenish or brownish and smaller.

Powder.—Light brown.   Characteristic elements: These are to be found in the glandular trichomes consisting of stalks, 2 to 3 celled, head 1, 2, 4 celled; non-glandular trichomes, one-celled, curved: few ellipsoidal pollen grains with 3 pores. Powder seldom dispensed.

CONSTITUENTS.—The chief constituents are **kosotoxin** (amorphous), a muscle poison, and **protokosin** (crystalline), inactive. Kosotoxin with baryta water yields a neutral body said to be identical with commercial **kosin**, an active principle soluble in alkalies; a neutral principle, Koussein (dose: i5 to 30

FIG. 119.—*Hagenia abyssinica*—Flowering branch, and male and female flowers.

gr.) is marketed; **tannin** 24 per cent., and a tasteless and an acrid **resin**. Ash, not more than 9 per cent.

Preparation of Kosin.—Heat cusso repeatedly with alcohol to which calcium hydrate has been added, boil residue with water, mix liquids, filter, and distil. Kosin is then precipitated by treating the solution with acetic acid. Is in flocculent form, soon becoming dense and resin-like. Purified by crystallization.

ACTION AND USES.—Tænifuge. Dose: 15 Gm. (240 gr.).

Fluidextractum Cusso (U.S.P. 1890),..Dose: 1 to 4 fl. dr. (4 to 15 mils).

223. **TORMENTILLA.**—TORMENTIL. The rhizome of **Potentil'la tormentil'la** Sibthorp. *Habitat:* Europe. Large, somewhat fusiform, longitudinally

15

wrinkled, and rough from numerous stem and rootlet scars; externally dull reddish-brown; fracture smooth, showing a pale reddish interior, consisting of one or two distinct circles of wood-fiber around a large central pith; inodorous; taste astringent. Used as a tonic and astringent. Dose: 10 to 30 gr. (0.6 to 2 Gm.), in powder or decoction.

224. **GEUM URBANUM.**—AVENS. EUROPEAN AVENS. The rhizome of Ge'um urba'num Linné. *Habitat:* Europe. Short, oblong, hard, with a dark-brown, warty, and scaly surface; a cross-section shows a thin bark, and a large, reddish pith surrounded by a circle of whitish wood. The rootlets are long and fibrous, light brown in color, and have a comparatively thicker bark. Odor aromatic, slightly clove-like when fresh, but nearly absent when dry; taste aromatic, bitter, and astringent. Used as an astringent and tonic. Dose: 15 to 45 gr. (1 to 3 Gm.), in powder or decoction.

225. **GEUM RIVALE** Linné.—WATER AVENS. (Rhizome.—See Conspectus.) Astringent and tonic. Dose: 15 to 45 gr. (1 to 3 Gm.).

226. **GILLENIA.**—AMERICAN IPECAC. The rhizome of Gille'nia stipula'cea Nuttall. *Habitat:* Western United States. A knotty rhizome, with numerous tortuous, annulate rootlets, the thick bark of which is in two reddish layers and incloses a tough, whitish, finely-rayed wood. *Gillenia trifoliata* Moench, growing east of the Allegheny Mountains, is a smaller and less knotty rhizome, and the rootlets are nearly straight and smooth. Both rhizomes are similar in medical properties, being mildly emetic and cathartic, somewhat resembling ipecac in action. Dose: 15 to 30 gr. (1 to 2 Gm.).

227. **AGRIMONIA.**—AGRIMONY. The herb of Agrimo'nia eupato'ria Linné. Common in the United States west to the Rocky Mountains, and in Europe. Tonic and astringent. Dose: 30 to 60 gr. (2 to 4 Gm.).

228. **POTENTILLA CANADENSIS** Linné.—CINQUEFOIL. *Habitat:* North America. (Herb.) Astringent. Dose: 30 to 60 gr. (2 to 4 Gm.) in infusion.

229. **SPIRÆA TOMENTOSA** Linné.—HARDHACK. An indigenous herb used as an astringent and tonic in doses of 30 to 60 gr. (2 to 4 Gm.). As found in market it consists of the slender, reddish-brown stems, broken leaves covered below with a rust-brown wool, and a few of the dull reddish flower-petals. Odor slight, aromatic; taste astringent and bitter.

LEGUMINOSÆ.—Pulse Family

Herbs, shrubs, or trees with alternate and usually compound leaves. *Flowers* papilionaceous, or rarely regular. *Stamens* usually ten and mostly monadelphous or diadelphous. *Pistil* becoming in fruit a legume, from which the order takes its name. Most of the plants are innoxious; the marked exception to the rule, however, is the calabar bean.

Synopsis of Drugs from the Leguminosæ

I. *Cellular.*

GLYCYRRHIZA, ...	Root, 230.	Stylosanthes, ..	Herb,	244.
Abri Radix, .	" 231.	Galega,*	"	245.
Baptisia, ..	" 233.	Trifolium Pratense,*	"	243.
Erythrophlœum.	Bark, 234.	Trifolium Repens, ...	"	243 a.
Cercis, ...	" 235.	Scoparius,*	"	246.
Saraca, ...	" 236.	Cassia Fistula,*	Fruit,	247.
Piscidia, ...	" 237.	Ceratonia, ...	"	248.
HÆMATOXYLON ..	Wood, 238.	Tamarindus,*	"	249.
SANTALUM **RUBRUM,** }	" 239.	Dipteryx, ...	"	250.
		Abri Semen, ...	Seed,	232.
SENNA, ...	Leaves, 240.	Fœnum Græcum, ...,	"	251.
Cassia Marilandica, .	" 241.	**PHYSOSTIGMA,**	"	252.
Melilotus, ...	Herb, 242.	Mucuna, ...	Hairs,	253.

II. Non-cellular.

<table>
<tr><td>Araroba, ... Powder, 254.</td><td>OLEUM } . . . .Volatile Oil, 259 a.
COPAIBÆ, }</td></tr>
<tr><td>ACACIA, ... Gum, 255.</td><td>Pongamia, ... Fixed Oil, 260.</td></tr>
<tr><td>TRAGACANTHA, . . . " 256.</td><td>Copal, ... Resin, 261.</td></tr>
<tr><td>CATECHU } . . .Extractive, 257.
(Gambir), }</td><td>BALSAMUM } . . . . Balsam, 262.
PERUVIANUM, }</td></tr>
<tr><td>KINO, ... " 258.</td><td>BALSAMUM }
TOLUTANUM, } . . . " 263.</td></tr>
<tr><td>EXTRACTUM } " 230 a.
GLYCYRRHIZÆ, }</td><td></td></tr>
<tr><td>COPAIBA. . . . . . . Oleo-resin, 259.</td><td></td></tr>
</table>

230. GLYCYRRHIZA.—Glycyrrhiza

LICORICE ROOT

The dried rhizome and root of Glycyrrhi'za gla'bra typica Regel et Herder, and Glycyrrhiza glabra glandulif'era Regel et Herder. Spanish and Russian respectively.

Fig. 120.—*Glycyrrhiza glabra*—Branch.

BOTANICAL CHARACTERISTICS.—Plants 4 to 5 feet high. *Leaves* impairipinnate; *leaflets* about 13, oval. *Racemes* axillary, *flowers* distinct, pale blue. *Legume* ovate, compressed.

SOURCE.—Russia exports the largest amount, Syria the smallest. Partiality for the Spanish root is now unwarranted; the close digging, and the limited and practically exhausted fields of Spain are the causes of its deterioration. Russia, with its new and almost unlimited fields, furnishes roots rich in glycyrrhizin and extractive, much better suited for commercial purposes because better and cheaper than the Spanish root. Anatolian root ranks between the Spanish and Russian in the quality of sweetness. In commerce no attention is paid to the botanical varieties of licorice root. From the root alone it is quite impossible to determine its true botanical origin, the usual designation being from the countries of growth, as Spanish, Russian, Anatolian, etc., although all varieties except the Spanish are often classified as "Greek root." Peeled root may now be prepared in Russia, but Syria formerly prepared it for shipment to Europe, some of which found its way into the market as "peeled Russian."

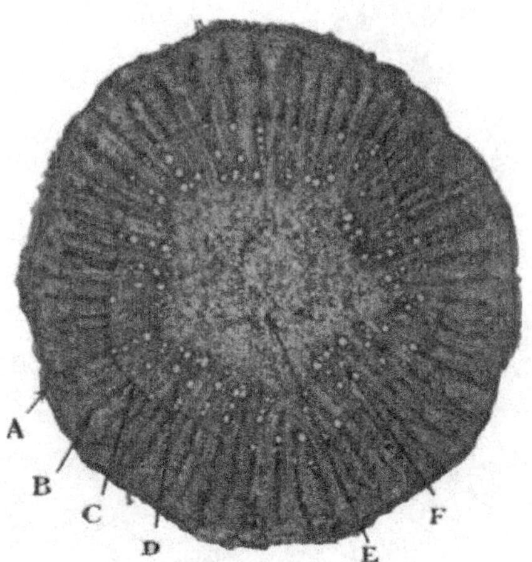

FIG. 121.—Glycyrrhiza—Cross-section of root. (13 diam.) A, Cork. B, Parenchyma of cortex C, Medullary ray. D, Xylem. E, Medulla. F, Water tube. (Photomicrograph.)

DESCRIPTION OF DRUG.—**Long, cylindrical pieces** from 5 to 25 mm. (⅕ to 1 in.) in diameter; **externally dark-brown, longitudinally wrinkled; internally of a light-yellow** color; pliable, fibrous, tough, readily **tearing into long,** fibrous **strips.** Odor peculiar, earthy, taste sweetish, afterward acrid. A **cross-section** shows a rather thick bark, the inner layer of which is composed principally of bast fibers. The **meditullium is made up of three kinds of cells,** ligneous, with oblique ends, parenchymatous, almost cubical, and large pitted ducts giving to the wood a porous appearance. Wood-wedges narrow, separated by distinct medullary rays.

Glycyrrhizal glabra glandulifera, so-called Russian, is thicker, less sweet, and more acrid than *G. glabra typica* (Spanish).

Powder.—Characteristic elements: See Part iv, Chap. I, B.

CONSTITUENTS.—Glycyrrhizin, asparagin, glycyramarin, an acrid resin, starch, etc. Glycyrrhizin is a glucosid, sparingly soluble in alcohol and ether, splitting up by hydrolysis into sugar and a brownish-yellow bitter substance, glycyrrhetin; it probably exists in combination with ammonia. Ash, not to exceed 7 per cent.

Preparation of Glycyrrhizin.—Obtained from the cold infusion (from which albumen has been removed by heat) by precipitating with H_2SO_4. Purify precipitate by dissolving in very weak ammonia water 1 to 10, filtering, and evaporating.

ACTION AND USES.—Expectorant and demulcent in bronchial affections. Frequently used to disguise the disagreeable taste of other medicines, and as a sweetening ingredient for medicinal preparations. Dose: 15 to 60 gr. (1 to 4 Gm.).

OFFICIAL PRERAPATIONS.

Fluidextractum Glycyrrhizæ,.	Dose: 15 to 60 ♏ (1 to 4 mils).
Extractum Glycyrrhizæ Purum,......	5 to 60 gr. (0.3 to 4 Gm.).
Mistura Glycyrrhizæ Composita (3 per cent. of extract, with wine of antimony, paregoric, sweet spirits of niter, syrup, and mucilage of acacia),...................	2 to 6 fl. dr. (8 to 24 mils).
Glycyrrhizinum Ammoniatum,.......	5 to 15 gr. (0.3 to 1 Gm.).
Pulvis Glycyrrhizæ Compositus (23.6 per cent., with senna, washed sulphur, oil of fennel, and sugar), ...	½ to 2 dr. (2 to 8 Gm.).
Elixir Glycyrrhizæ.	

230 a. **EXTRACTUM GLYCYRRHIZÆ**—Extract of Licorice. Made by evaporating the aqueous extract of the root. It is found in market in black, brittle, cylindrical rolls about 150 mm. (6 in.) long; flexible when warm, but when dry breaks with a brittle, conchoidal fracture, showing a glossy surface; odor characteristic; taste sweet. It yields a brown powder. It contains glycyrrhizin, both free and combined with ammonia, to which combination its sweetness is due, glycyrrhizin itself being almost tasteless. It is an excellent demulcent, the presence of a small piece in the mouth often allaying cough by coating and thus protecting the irritated membrane. Not less than 60 per cent. of the extract of glycyrrhiza should be soluble in cold water. Dose: 15 to 60 gr. (1 to 4 Gm.). Ash, not more than 6 per cent.

OFFICIAL PREPARATIONS.

Trochisci Ammonii Chloridi (each troche containing about 3 grains each of glycyrrhiza and 1½ of ammonium chloride, with sugar, tragacanth, and syrup of tolu), ...Dose: 1 or 2 troches.
Trochisci Cubebæ.

231. **ABRI RADIX.**—Indian Licorice.   The root of **A′brus precato′rius** Linné, indigenous to India, naturalized in most tropical countries.   Reddish-brown, twisted pieces, having a thin bark, and a meditullium composed of alternating zones of porous wood-bundles and parenchyma, traversed by medullary rays.   Inodorous; taste bitter, afterward sweetish.   It is thought to contain glycyrrhizin, and is used as a demulcent like glycyrrhiza.

232. **ABRI SEMEN.**—Prayer Beads.   Jequirity.   The seeds of **A′brus precato′rius** Linné.   Subglobular, about 5 to 8 mm. ($\frac{1}{5}$ to $\frac{1}{3}$ in.) long, scarlet-red, glossy, with a black spot at the hilum; inodorous; taste bean-like.   They contain two proteids, paraglobulin, and albumose, which are irritating to the eyes.   A weak infusion of the seed is used in granular ophthalmia.

233. **BAPTISIA**, N. F.—Wild Indigo.   The root of Bapti′sia tincto′ria R. Brown. *Habitat:* United States.   It contains baptisine (acrid, poisonous), baptisin (a bitter glucoside), and baptin (a purgative glucoside).   Chiefly used for its antiseptic properties, in lotion and ointment, although it acts also as an emetic and cathartic.   Dose: 5 to 15 gr. (0.3 to 1 Gm.).

234. **ERYTHROPHLŒUM.**—Sassy Bark.   A poisonous bark from Erythrophlœ′um guineens′e Don, used as an ordeal in Africa, where the tree grows, and therefore sometimes called doom-bark.   It is in thick, warty, curved pieces, reddish-brown, fissured.   Inodorous; taste astringent and bitter.   It contains an alkaloid, **erythrophleine, which gives it an action on the** heart similar to digitalis; also astringent, emetic, diaphoretic, and analgesic.   Dose: 5 to 15 gr. (0.3 to 1 Gm.).

Preparation of Erythrophleine.—Treat concentrated aqueous solution of the alcoholic extract of the bark with ammonia and exhaust the mixture with acetic ether.   The alkaloid is yielded on evaporation.

235. **CERCIS CANADENSIS** Linné.—Redbud.   The bark of this indigenous tree has been recommended as a mild, non-irritating, but active astringent in diarrhea and dysentery.   Also used as a local application to mucous membranes.   Dose of fluidextract: 15 to 60 ♏ (1 to 4 mils).

236. **SARACA INDICA** Linné.—Asoca.   (Bark.)   Much employed by the Hindoo physicians as a sedative in the treatment of uterine affections; it is also astringent.   Dose of fluidextract: 15 to 60 ♏ (1 to 4 mils).

237. **PISCIDIA.**—Jamaica Dogwood.   The bark of Piscid′ia erythri′na Jacquin. *Habitat:* West Indies.   Quills or curved pieces about 4 mm. ($\frac{1}{6}$ in.) thick; externally of a dark, yellowish-gray color, ridged longitudinally.   Odor **opium-like when broken.**   Taste bitter, acrid, producing a burning sensation in the mouth.   Used as a mild **soporific for children and aged persons,** and for those not able to bear a strong narcotic like opium.   Dose: 15 to 45 gr. (1 to 3 Gm.).

238. HÆMATOXYLON, N.F.—Hæmatoxylon

LOGWOOD

The heart-wood of **Hæmatox′ylon campechia′num** Linné.   Usually found in commerce in the form of **deep, brownish-red** chips.—When the surface has a greenish metallic luster, the wood has undergone fermentation and should be rejected.   Odor slight; taste sweetish, astringent.

Constituents.—**Hæmatoxylin,** $C_{16}H_{14}O_6$, sweet, colorless crystals, giving to the wood its characteristic colors by the combined action of the oxygen of the air and the alkaline bases existing in the wood; it is readily soluble in hot water and alcohol, sparingly in cold water; by the action of ammonia and oxygen in the air dark purple scales of hæmatein, $C_{16}H_{12}O_6$, are formed, often observable as the fine greenish hue upon logwood chips.   This principle gives a blue color with alkalies.   Hæmatoxylon also contains tannin, fat, resin, and

FIG. 122.—*Hæmatoxylon campechianum*—Branch.

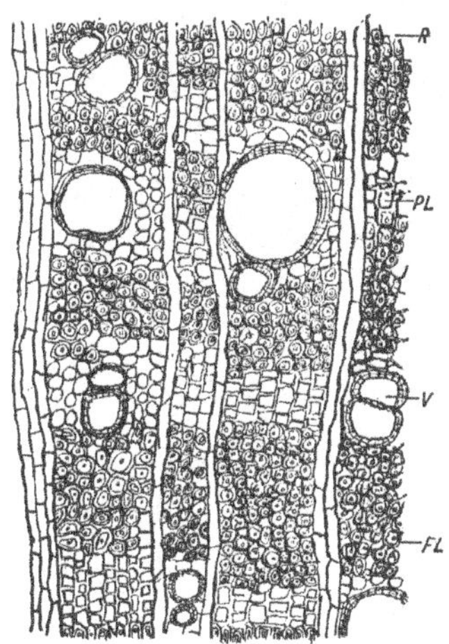

FIG. 123.—*Hæmatoxylon*—Cross-section of wood.   *P*, Medullary ray, consisting of two vertical rows of cells, to which the black line from *R* should be extended.   *V*, Pitted vessels.   *FL*, Ligneous fibers.   *PL*, Wood parenchyma.

a trace of volatile oil. With an alkali hæmatoxylon gives a purple color, brazil-wood a red color, and red saunders is not affected.

Preparation of Hæmatoxylin.—To ethereal extract add water and allow to crystallize; add a little H_2SO_3 or sulphite to prevent oxidation. Yellowish prisms of sweetish taste, violet-blue, with alkalies. Soluble in alcohol and water. Sunlight causes a red color.

ACTION AND USES.—A mild astringent. Dose: 30 to 60 gr. (2 to 4 Gm.), in decoction or extract. A solution of hæmatoxylon as a staining fluid in microscopy is one of the most useful, as it stains both lignified and cellulose tissue, but not suberin or cutin. It is also one of the very best nuclear stains.

PREPARATION: Ext. Hæmatoxyli, N.F. Dose, 1 Gm. (15 gr.).

239. SANTALUM RUBRUM.—RED SAUNDERS

RED SANDALWOOD

The heart-wood of **Pterocar'pus santali'nus** Linné.

BOTANICAL CHARACTERISTICS.—A large tree with dark red, heavy, and compact wood; a reddish juice exudes from its bark. *Racemes* axillary; *flowers* yellow, streaked with red. *Legumes* orbicular.

FIG. 124.—*Pterocarpus santalinus*—Branch.

HABITAT.—Madras.

DESCRIPTION OF DRUG.—In commerce usually in deep reddish-brown raspings or small chips, or a coarse powder; tasteless and nearly odorless. The wood consists mostly of the lower parts of the stem, and thick roots, imported in irregular logs of various sizes, usually deprived of the bark, and externally of a dark-brown color; internally of a rich red color, showing in transverse sections circles of a lighter tint. Used in Compound Tincture of Lavender.

Powder.—Microscopical elements of: See Part iv, Chap. I, B.

CONSTITUENTS.—The most important constituents are the **red coloring-matter, santalin,** in needles, soluble in alcohol, ether, acetic acid, and alkaline solutions, but insoluble in water, and only slightly soluble in boiling water and santalic acid, $C_{15}H_{14}O_5$. The yellow ethereal solution is turned to violet by alkalies. Santol, pterocarpin, and homopterocarpin are also constituents. Ash, not to exceed 3 per cent.

Preparation of Santalin.—Precipitate alcoholic tincture with lead acetate; decompose this precipitate with H_2S in presence of alcohol and evaporate. Red needles are obtained, which are inodorous, tasteless, resinous; soluble in the alkalies with violet, and in ether with yellow color.

ACTION AND USES.—Of no value medicinally. Used in pharmacy for coloring preparations.

OFFICIAL PREPARATION.
Tinctura Lavandulæ Composita.

240. SENNA.—SENNA

SENNA

The dried leaflets of **Ca'ssia acutifo'lia** Delile and **C. angustifolia** Vahl.

BOTANICAL CHARACTERISTICS.—The acute-leaved senna, *C. acutifo'lia,* is a leafy shrub 2 to 5 feet high, bearing axillary racemes of yellow *flowers. Legume* flat, broadly oblong, very slightly curved inward, rounded at the extremities, terminating in an indurated and nearly obsolete style.

SOURCE.—**Alexandria senna,** exported by the way of Alexandria, is derived from *Ca'ssia acutifo'lia,* a species growing wild abundantly in upper Egypt, Nubia, etc. **India senna** (*C. angustifo'lia*) is obtained **chiefly in Arabia,** reaching western ports by way of Bombay and other Indian ports; sometimes called Mocha senna, as originally from that port. **The same plant in cultivation yields Tinnevelly senna.** The plant yields two annual crops, the best at the close of the rainy season (September), and the other during the dry season. Prepared for market by the natives, who carry it there on camels, where it is cleaned (garbled) and sold.

DESCRIPTION OF DRUG.—Both the Alexandria and the India senna consist of leaflets, **a prominent distinction between the two being their size;** the former, the acutifolia, is described as follows: Lanceolate or ovate-lanceolate, 1.5 to 3 cm. long, 5 to 8 mm. broad; apex acute, mucronate; base unequal, acute; margin entire; upper surface light green, nearly glabrous, midrib sometimes depressed, veins of first order more or less prominent; under surface light grayish-green, midrib prominent, minutely pubescent, especially near the veins; petiole about 1 mm. long; texture coriaceous, fibrous; odor slight; taste somewhat bitter. *Powder:* Light green; non-secreting hairs 0.1 to 0.2 mm. long, one-celled, thick-walled, the wall of the upper part strongly cuticularized; calcium oxalate crystals rosette-shaped or in monoclinic prisms. The powder of Indian senna (*C. angustifolia*) is dark green and has relatively few non-secreting hairs. (For fuller particulars of the microscopical distinction of the two powders, see article by the author, "Amer. Jour. Pharm.," June, 1897, p. 298.) The India senna is by far a cleaner senna; senna should be free from stalks and other inert materials, and from Argel leaves (*Solenostem'ma ar'-gel*, N. O. Asclepiadeæ), which are thick, even at the base, and one-veined.

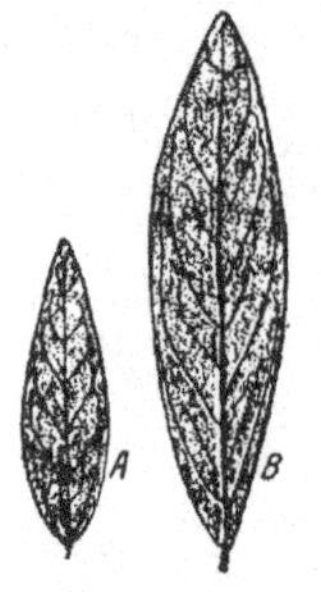

FIG. 125.—*A,* Alexandria senna. *B,* India senna.

Powder.—Characteristic elements: See Part iv, Chap. I, B.

CONSTITUENTS.—The purgative action of senna depends upon a sulphuretted glucoside, **cathartic acid,** insoluble in alcohol, soluble in water, but rendered partially or wholly inert by prolonged evaporation or boiling of its solution. Senna **also contains chrysophan,** phæoretin, sennacrol, and glucosennin, $C_{22}H_{18}O_8$; this latter is probably an emodin glucoside. The emodin is said to be identical with that found in Barbadoes and Cape Aloes. The principles giving the odor and taste to senna, also its griping action, are extracted by alcohol, somewhat affecting the cathartic action, however. Ash, not more than 12 per cent. not less than 3 per cent.; insoluble in HCl.

EMODIN TEST.—This test is applied to the emodin-bearing drugs such as Rhubarb, Aloes, Senna, etc. The tests as applied are practically the same. For Senna it is as follows: Mix 0.5 Gm. of powdered Senna with 10 mils of an alcoholic solution of potassium hydroxide (1 in 10), boil the mixture for about two minutes, dilute it with 10 mils of water and filter. Now acidify the filtrate with hydrochloric acid, shake it with ether; remove the ethereal layer and shake it with 5 mils of ammonia water; the latter is colored yellowish-red.

Preparation of Cathartic Acid.—Rhubarb or senna may be treated separately as follows: Moisten the drug with alcohol. Macerate 48 hours and percolate with strong alcohol till exhausted, to remove chrysophanic acid, resin, etc. Exhaust the marc with 60 per cent. alcohol. Evaporate the percolate at 50°C. to syrup, with constant stirring. Precipitate extract with 85 per cent. alcohol and filter to remove gum. The filtrate, after evaporating to a syrupy consistence, is added to a large excess of absolute alcohol. The brown precipitate thus produced is spread on glass to dry. It is then in light, shining scales.

ACTION AND USES.—A prompt and efficient cathartic. Its griping action may be prevented by combining it with an aromatic and one of the alkaline salts, or, as before stated, by first extracting the griping principle with alcohol. Dose: 2 to 8 dr. (8 to 30 Gm.).

OFFICIAL PREPARATIONS.

Infusum Sennæ Compositum (6 per cent., with manna and Epsom salts, each, 12 per cent., and fennel 2 per cent.), ...	Dose: 1 to 2 ½ fl. oz. (30 to 75 mils).
Syrupus Sennæ (25 per cent. of Fl'ext.),	1 to 4 fl. dr. (4 to 15 mils).
Fluidextractum Sennæ, ...	½ to 4 fl. dr. (2 to 15 mils).
Pulvis Glycyrrhizæ Compositus (18 per cent.), ...	1 to 2 dr. (4 to 8 Gm.).
Syrupus Sarsaparillæ Compositus,	4 fl. dr. (15 mils).

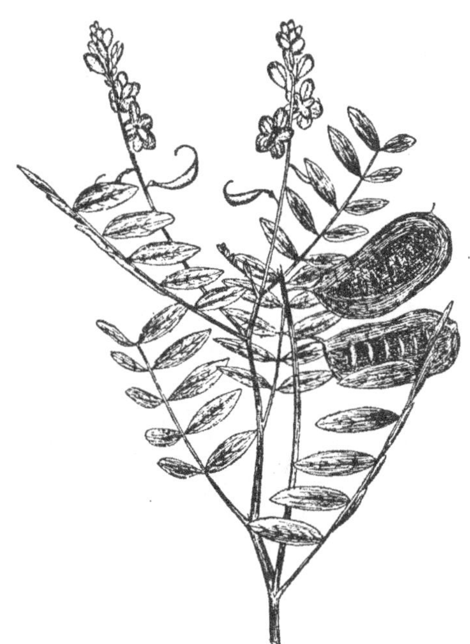

FIG. 126.—*Cassia acutifolia*—Branch showing flower and fruit.

241. CASSIA MARILANDICA Linné.—AMERICAN SENNA. (Leaflets.) Oblong-lanceolate, about 25 mm. (1 in.) in length, mucronate at the apex and uneven and short-stalked at base; lower surface lighter green than upper surface. They have a weaker odor and taste than senna, but have similar medicinal properties, their action depending upon the same principle, cathartic acid.

242. MELILOTUS, N.F.—SWEET CLOVER. The flowering tops of **Melilo'tus officina'lis** Willdenow. The small yellowish or white flowers are in a close, rounded raceme on an angular stem; leaves serrate, trifoliate; odor fragrant, honey-like; taste aromatic and bitter. They contain melilotol (a fragrant volatile oil), coumarin (the aromatic principle of tonka), cumaric acid, and melilotic (hydrocumaric) acid, having a honey-like odor. An infusion is used as a stimulant and antispasmodic in whooping-cough, but it is generally used as a local anodyne in poultices.

243. **TRIFOLIUM PRATENSE** Linné (Trifolium, N.F.).—The flowering tops of this, our common red clover, are now being used quite extensively as an alterative; they are also deobstruent and sedative in whooping-cough.

243 a. **TRIFOLIUM REPENS.**—WHITE CLOVER. The tops are used in whooping-cough and other spasmodic affections, in the form of infusion.

244. **STYLOSANTHES ELATIOR** Swartz.—PENCIL FLOWER. This herb is much used in domestic practice as a uterine sedative and tonic. The fluidextract is not miscible with water. Dose of fluidextract: 10 to 20 ♏ (0.6 to 1.3 mils).

245. **GALEGA, N.F.**—GOAT'S RUE. The herb of Galega officinalis Linné. Europe. Recently introduced. An erect glabrous perennial, about three feet high. Leaves alternate, oddly pinnate, and stipulate; stipules lanceolate; leaflets smooth, lanceolate, and mucronate. Flowers in loose, axillary racemes longer than the leaves; blue, appearing in June or July. *Preparation:* Fluidextract. *Properties:* Vermifuge, nervous stimulant, galactagogue. In typhoid conditions diuretic and tonic. Dose: 15 to 20 minims.

FIG. 127.—*Cytisus scoparius*—Flowering branch and pod.

246. SCOPARIUS, N.F.—SCOPARIUS

BROOM

The dried tops of **Cyti′sus scopa′rius** (Linné) Link. *Habitat:* Europe and Asia.

DESCRIPTION OF DRUG.—Thin, flexible, branched twigs, pentangular and winged, nearly smooth, and of a dark greenish-brown color; as found in the market they are usually free from the small trifoliate leaves. Odor slight, stronger when bruised; taste very bitter.

Powder.—Greenish-brown. Characteristic elements: Sclerenchyma with bast fibers, long, thick-walled, associated with crystal fibers containing calcium oxalate prisms; ducts, spiral, annular, and reticulate; trichomes, non-glandular (0.5 to 0.7 μ in diam.), thick-walled, yellowish, one-celled; pollen, brownish; grains, oval.

CONSTITUENTS.—A neutral crystalline principle, **scoparin**, $C_{20}H_{20}O_{10} + 5H_2O$, to which the diuretic action is due, and the colorless, volatile, liquid alkaloid, **sparteine**, $C_{15}H_{26}N_2$, acting as a powerful cardiac tonic; this is oily, very bitter, soluble in alcohol, chloroform, and ether; it has been made official as the salt, sparteinæ sulphas. Prisms freely soluble in water. Oxidation products, such as oxysparteine, $C_{15}H_{24}N_2O$, produce an increase of heart activity, while dioxysparteine, $C_{15}H_{26}N_2O$, produces an inverse effect upon the heart. Sparteine has an aniline-like odor.

Preparation of Scoparin.—Allow a concentrated decoction of broom-tops to gelatinize; express and purify the jelly-like mass by repeated solution in hot water, and finally in hot alcohol.

Preparation of Sparteine.—Extract plant with acidulated water and distil concentrated liquid with NaOH. A colorless oily liquid, forming crystalline salts. Sulphate official.

ACTION AND USES.—Scoparius is a reliable diuretic and laxative in small doses of 10 to 30 gr. (0.6 to 2 Gm.), and is an efficient remedy in dropsy. Dose of sparteinæ sulphas: ⅛ to 1 gr. (0.0081 to 0.065 Gm.). Used to regulate heart action.

247. CASSIA FISTULA, N.F.—CASSIA FISTULA

PURGING CASSIA

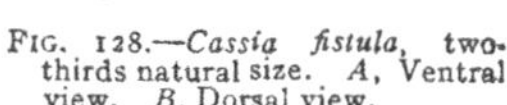

FIG. 128.—*Cassia fistula,* two-thirds natural size. *A,* Ventral view. *B,* Dorsal view.

The dried fruit of Cas'sia fist'ula Linné.

BOTANICAL CHARACTERISTICS.—Tree from 20 to 50 feet high, with showy *racemes* 1 to 2 feet long, of bright yellow, fragrant *flowers,* followed by cylindrical pods of the same length. *Legume* woody, indehiscent. Tropical, extensively cultivated.

DESCRIPTION OF DRUG.—Cylindrical pods or legumes 450 to 600 mm. (18 to 24 in.) long and about 25 mm. (1 in.) in diameter, with a blackish-brown, woody pericarp; indehiscent, but with two smooth sutures or bands on opposite sides running the whole length of the pod, and showing the union of the two valves. The dorsal band is marked with a fine ridge, while the ventral band is seem_ ingly divided into two by a shallow, longitudinal groove. **The interior of the pod consists of numerous (25 to 100) transverse cells, each containing a** single, flattish, glossy, red-brown seed, imbedded in a sweet, blackish-brown pulp; odor prune-like.

CONSTITUENTS.—The pulp, which is the part used, consists mainly of sugar (about 60 per cent.), with mucilage, pectin, albuminoids, and organic salts.

ACTION AND USES.—A mild laxative, generally combined with other mixtures. Dose: 1 to 8 dr. (4 to 30 Gm.).

248. **CERATONIA.**—St. John's Bread. The fruit of **Cerato'nia sil'iqua Linné.** *Habitat:* Southern Europe. Broad, flat pods, brown and glossy, divided into six to twelve transverse cells, in each of which is a sweet, black pulp having a single seed imbedded in it. This pulp is used as a laxative and demulcent, but chiefly as an ingredient in expectorant mixtures.

FIG. 129.—*Trigonella fœnum græcum.*—Branch.

249. TAMARINDUS, N.F.—Tamarind

TAMARIND

The preserved pulp of the fruit of **Tamarin'dus in'dica Linné** (the Indian date).

A tough, reddish-brown mass, made adhesive by the syrup in which the fruit is preserved. This preserved pulp **consists of a fibrous or stringy** mucilaginous mass, the thin membranous epicarp (the pericarp being re-moved), and numerous large, somewhat quadrangular, brown seeds, each inclosed in a tough membrane; inodorous; taste sweetish and acidulous.

CONSTITUENTS.—Tartaric acid and acid potassium tartrate, with traces of citric and malic acids. These organic salts amount to about 10 per cent.

ACTION AND USES.—Laxative and refrigerant, in confection of senna. Dose: 1 to 8 dr. (4 to 30 Gm.).

250. **DIPTERYX.**—TONKA BEAN. The fruit of a large tree, **Dip'teryx odora'ta** Willdenow, growing in Guiana. Oblong, flattened, rounded at each end, 37 to 50 mm. (1½ to 2 in.) long; pericarp thin, wrinkled, of a dark-brown color, somewhat glossy, and often covered with small, white crystals of cou-marin; internally oily, pale brown; odor fragrant, similar to vanilla; taste aromatic and bitter. Its odor is due to the aromatic, crystalline principle **coumarin.** Used as a flavor, as an adulterant of vanilla, and to flavor cigars.

250 a. **COUMARINUM.**—COUMARIN. The anhydride $(C_6H_4(CH)_2OCO = 146.05)$ of ortho-oxycinnamic acid, occurring naturally in Touka, Melilot and other plants, or prepared synthetically, N.F.

251. **FŒNUM GRÆCUM.**—FENUGREEK. The seeds of Trigonel'la **fœnum græ'cum** Linné. *Habitat:* India and the Mediterranean Basin. Brownish or yellowish, rhomboid seeds, about 3 mm. (⅛ in.) in diameter, often wrinkled or distorted. They are divided into two equal lobes by a deep furrow run-ning from the hilum on the sharper edge, diagonally across the sides. Odor peculiar, characteristic; taste mucilaginous and bitter. Used mostly as a demulcent in condition-powders.

252. PHYSOSTIGMA.—PHYSOSTIGMA

CALABAR BEAN

The ripe seed of **Physostig'ma veneno'sum** Balfour, yielding, by official assay, not less than 0.15 per cent. of alkaloids of Physostigma.

BOTANICAL CHARACTERISTICS.—A lofty, half-shrubby, twining plant, obtaining its name from its peculiar footed stigma. *Leaves* trifoliate, *leaflets* ovate. *Flowers* purplish-pink, in axillary racemes. *Legume* about 7 inches long.

HABITAT.—Africa.

DESCRIPTION OF DRUG.—**About the size of a pecan nut,** oblong, some-what flattened, and **kidney-shaped,** invested with a light to **deep chocolate-brown testa.** Along its entire convex edge there extends a prominent black furrow, bordered on each side by a reddish ridge, and traversed the entire length by the raphe as a little ridge in the center. This raphe is **terminated at one end** by a **small funnel-shaped depression, the micropyle.** Exalbuminous, embryo large, the cotyledons are concavo-convex, the concave surfaces inclosing a rather large cavity, thus enabling the bean to float upon water. Nearly odorless; taste bean-like, afterward acrid. Spurious cala-bar beans have been called "calibeans" in European commerce, hose occurring the most frequently belonging to the following species: *Entada scandens, E. gingalobium* D. C., *Mucuna urens* D. C., and

seeds of oil palms, *Elæis Guineensis*. E. H. Holmes called attention to certain specimens of calabar beans of commerce bearing a close
resemblance to the genuine beans. They were longer, of circular cross-
section, and the hilum did not extend the full length of the beans.
They also differ chemically, as upon touching the cotyledons with a
solution of potassa a permanent yellow tint was produced, and upon
treating the spurious article similarly a deep, almost orange, color
is formed, turning to a greenish hue. It has been found that the
ordinary test-reagents for alkaloids are so sensitive for physostigmine
(eserine) that one one-millionth part of a gram may be recognized.
The poisonous qualities reside in the seeds, especially in the cotyledons.
It has been ascertained that the leaves and stems are not poisonous.

Powder.—Characteristic elements: See Part iv, Chap. I, B.

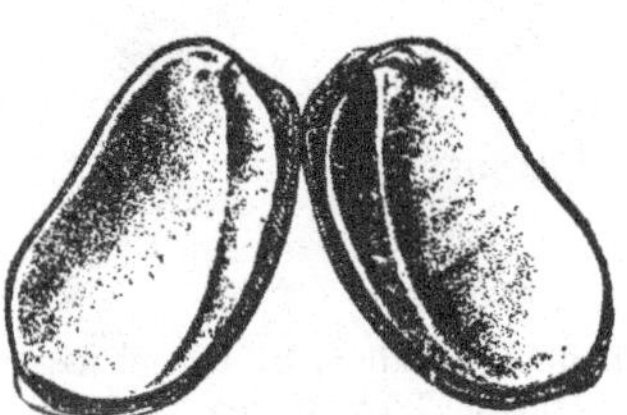

FIG. 130.—Calabar bean.

CONSTITUENTS.—**Physostigmine**, $C_{15}H_{21}N_3O_2$ (also known as **eserine**), contracting the pupil of the eye; calabarine, a tetanizing principle, a derivative of physostigmine; eseridine, $C_{15}H_{23}N_3O_3$ (producing purgation); and physosterin, a neutral principle closely related to cholesterin. These principles are soluble in alcohol. Physostigmine is amorphous, tasteless, reddened by potassa, soda, and lime when exposed to the air, due to absorption of oxygen. The drug sometimes contains over 0.15 per cent. of the alkaloid Physostigmine. Ash, not exceeding 3 per cent.

Preparation of Physostigmine (Eserine).—Treat powdered drug (mixed with
1 per cent. tartaric acid) with water. Shake out coloring matter with ether, make
aqueous solution alkaline with an alkaline bicarbonate, and shake out alkaloid
with ether. Evaporate ethereal solution.

Preparation of Eseridine (Calabarine).—Precipitate the alkaloid from the liquid
from which physostigmine has been separated by lead subacetate and ammonia;
evaporate the filtrate, treat the residue with alcohol, precipitate with phospho-
tungstic acid, and decompose with baryta. It is converted into physostigmine by
hydrolysis.

Preparation of Physosterin.—Exhaust beans with petroleum ether and evaporate
solvent.

ACTION AND USES.—Physostigmine is used in medicine chiefly for three
purposes: as a depressant for the spinal cord; as a stimulant to the
intestinal muscles; and to contract the pupils. As a motor depressant
physostigmine is useful in the treatment of tetanus and strychnine

poisoning. Its greatest value in internal medicine is as a stimulant to intestinal muscles in paralytic forms of colic, but especially in chronic constipation in conjunction with cathartic drugs.

Physostigmine stimulates the secretory nerve-endings of glands and the nerve-endings of striated and smooth muscle. It therefore antagonizes the effects of atropine and curare.

Fig. 131.—*Physostigma venenosum*—Portion of plant and fruit.

If a drop of 1 : 200 aqueous solution of eserine is placed in the eye, contraction of the pupil begins in one or two minutes and reaches its maximum in one-half to one hour.

When the alkaloid calabarine is present in excess in the drug, and is taken in overdose, convulsions develop. Dose of drug: 1 to 4 gr. (0.065 to 0.25 Gm.).

OFFICIAL PREPARATIONS.

Physostigminæ Salicylas,......Dose: $\frac{1}{120}$ to $\frac{1}{30}$ gr (0.0005 to 0.00216 Gm.).
Extractum Physostigmatis,.... $\frac{1}{10}$ to $\frac{1}{2}$ gr (0.0064 to 0.0324 Gm.).
Tinctura Physostigmatis (10 per cent.),................   10 to 40 ♏ (0.6 to 2.6 mils).

16

253. **MUCUNA.**—Cowage, or Kiwach, the Hindustan name, vulgarly corrupted
into cow-itch. The hairs from the pods of **Mucu'na pru'riens** De Candolle,
a high-climbing plant growing in tropical Africa, America, and India. These
hairs are about 3 mm. (⅛ in.) long, stiff, brown-red, and readily penetrate
the skin, causing violent itching. Detached from the pod (which forms an
article of diet in India) by dipping it in honey and then scraping. An electuary is used in doses of a teaspoonful to a tablespoonful. Cowage acts as
an **anthelmintic mechanically,** penetrating the bodies of the worms and thus
irritating and dislodging them.

254. **ARAROBA.**—Goa Powder. A mixture of neutral principles obtained from
radial fissures in the wood of a Brazilian tree, Andi'ra araro'ba Aguiar. This
powder is of a light yellow color, with a somewhat earthy appearance, turning
dark brown or purplish on exposure; somewhat crystalline, rough, and mixed
with pieces of wood-fiber; inodorous and very bitter. It consists chiefly of
chrysarobin (Chrysarobinum). Used externally, in ointments, in skin diseases caused by fungi.

255. ACACIA.—Acacia

GUM ARABIC

A gummy exudation from **Aca'cia sen'egal** Willdenow and of other species of
Acacia.

Botanical Characteristics.—A small tree about 20 feet high, with a gray bark.
Leaves bi-pinnate. *Flowers* pale yellow, in dense spikes. *Legumes* broad,
three to four inches long.

Habitat.—The acacia tree forms dense scrubby forests in the sandy regions watered by the Senegal, and in Abyssinia and Kordofan.

Description of Drug.—In roundish, brittle **tears or broken fragments
about the size of a pea,** or larger, with an opaque appearance, due to
the numerous fissures. Inodorous; **taste mucilaginous and insipid.**
Soluble in water, forming a thick mucilaginous liquid; **insoluble in
alcohol.** The aqueous solution has an acid reaction and yields gelatinous precipitates with subacetate of lead, ferric chloride, and concentrated solution of borax. Oxalates precipitate the calcium base.
There are two kinds of "powdered acacia" on the market, the "granulated" and the "finely dusted." The former is more soluble and less
liable to form lumps, and is, therefore, preferable for pharmaceutical
purposes.

Varieties and Grades.—The Kordofan and Senegal gums are the product of *A. Senegal*. The former has been described above. Gum
Senegal, deriving its name from the river Senegal, comes in larger
tears than the former, varying in color between yellow and yellowish-
brown, being less fissured and more transparent. As to the grades
of gum, it may be said that the quality entering the market varies
exceedingly in its solubility, viscosity of its mucilage, and its color.
In the market the grades are designated by numbers, No. 1 being
the best carefully selected tears, No. 2 the next best, and so on until

several selections have been made, the remaining colored pieces containing impurities being termed "sorts;" but this term is sometimes applied to unsorted gum arabic, often consisting of a mixture of the lower grades. The terms "strong" and "weak" have been applied, designating the quantity of moisture, the strong being the drier and probably the most soluble; the weak being that which possibly swells in water, does not completely dissolve, and hence yields a relatively small percentage of mucilage.

Mesquite gum is obtained from *Prosopis juliflora*, found in Southwestern America and South America. Quite abundant in some portions of Texas and New Mexico. It occurs in colorless or amber-brown tears; resembles gum arabic somewhat in fissures, specific gravity, solubility, its behavior to nitric acid, and the amount of ash yielded upon incineration (2.1 to 3 per cent.). Its aqueous solution is not precipitated by subacetate of lead, ferric salts, or borax. Acetate of lead, with ammonia added subsequently, yields a gelatinous precipitate. These reactions, however, differ to some extent in different samples.

CONSTITUENTS.—Arabic acid, $C_{12}H_{22}O_{11}$, **combined with calcium, magnesium, and potassium,** to the presence of which its solubility is due; boiled with dilute acid it yields arabinose or arabin sugar. A solution of the gum is **unaffected** by **neutral lead acetate.** The gum contains about 14 per cent. of moisture and some sugar. Ash, not exceeding 4 per cent.

Preparation of Arabic Acid.—Obtained by adding alcohol to acidified (HCl) mucilage, and drying the precipitate. It yields arabiose in prismatic crystals when boiled with acids and possibly also galactose.

Powder.—Not more than 1 per cent. should be insoluble in water (limit of dirt, etc.), nor should the powder contain more than 15 per cent. moisture.

ACTION AND USES.—Demulcent. Used in pharmacy for suspending insoluble matters in water, as in emulsions, and as an excipient.

Powder.—Elements of: See Part iv, Chap. I, B.

OFFICIAL PREPARATIONS.

Mucilago Acaciæ (34 per cent.).
Syrupus Acaciæ (10 per cent. of acacia),..Dose: 1 to 8 fl. dr. (4 to 30 mils).
Pulvis Cretæ Compositus (20 per cent.), used as an excipient.

256. TRAGACANTHA.—TRAGACANTH

GUM TRAGACANTH

The spontaneously dried gummy exudation from **Astra'galus gum'mifer Labillardiere,** or from other Asiatic species of Astragalus.

BOTANICAL CHARACTERISTICS.—A small, tangled, spiny bush of compact growth, the *petioles* being converted into long spines. *Flowers* yellow, in axillary clusters. *Legume* partially two-celled.

HABITAT.—Western Asia.

DESCRIPTION OF DRUG.—The flake tragacanth comes in transversely lined, curved, and contorted bands, somewhat resembling fragments of oyster shell, but tough and horny; color whitish or yellowish, translucent. Taste insipid, sometimes faintly bitterish; inodorous. It is difficult of pulverization, made less so, however, by the use of a warm mortar. It does not dissolve in water, but swells up and forms a thick, gelatinous mass.

VARIETIES.—Very narrow bands or strings variously coiled. Tragacanth in sorts—stratified or nodular, conical and subglobular pieces, more or less brown, often adulterated with the gum of the almond and plum trees.

Powder.—Elements of: See Part iv, Chap. I, B.

CONSTITUENTS.—**Traganthin or bassorin,** $C_6H_{10}O_5$, constituting about 43 per cent., swelling up in water, but not dissolving; and **arabin,** the calcium salt of gummic acid, soluble in water, but not identical with the arabin or arabic acid of acacia. Ash, not more than 3.5 per cent.

ACTION AND USES.—Used as a demulcent, but rarely, however, on account of its insolubility. Chiefly used in pharmacy to give consistence to lozenges, etc.

OFFICIAL PREPARATION.

Mucilago Tragacanthæ (6 per cent.).

257. CATECHU.—CATECHU

An extract prepared from the heart-wood of **Aca′cia cat′echu** Linné.

BOTANICAL CHARACTERISTICS.—Small tree with straggling, thorny branches, and compact, dark red wood. *Leaves* bipinnate; *petiole* angular, with prickles on its under side. *Flowers* pale yellow. *Legume* about three-seeded.

SOURCE.—The tree is common in most parts of India and Burmah, where the export of cutch forms, next to the sale of timber, the most important item of forest revenue. It abounds in the forests of tropical Eastern Africa, but in many places where the tree abounds it is only valued for its wood. In comparatively few regions is any extract manufactured. From *Acacia suma*, a nearly related species growing in Southern India, catechu is also made. The extract from these two species of acacia furnishes a variety of catechu, but a catechu formerly prescribed as *Catechu pallidum* (pale catechu), gambir, is official in the present Pharmacopœia and is described as follows:

GAMBIR

GAMBIR (CATECHU)

An extract prepared from the leaves and twigs of **Ourouparia Gambir** (Hunter) Baillon (Fam. Rubiaceæ).

Irregular masses of cubes about 25 mm. in diameter; externally reddish-brown, pale brownish-gray or light brown; fracture dull-earthy, friable, crystalline; inodorous, bitterish, very astringent with a sweetish after-taste.

Not less than 70 per cent. should be soluble in alcohol; the ash should not be more than 5 per cent., and starch should not be present.

CONSTITUENTS.—Mainly **catechu-tannic acid,** 45 to 55 per cent., which does not produce gallic acid on exposure to air as does the tannin of galls; it is turned blackish-green by ferric salts. Catechin is an interesting principle which, by dry distillation, yields pyrocatechin, or catechol, $C_6H_6O_2$, which, with ferric chloride, gives a dark green color by ammonia changing to violet. Ash, not more than 9 per cent.

Preparation of Catechin.—On allowing the decoction of catechu to stand several days, crude catechin is deposited. This deposit is purified to white silky needles by dissolving in dilute alcohol, washing with ether, and evaporating from hot aqueous solution. It has a sweetish taste, is precipitated by albumen, but not by gelatin.

ACTION AND USES.—**A powerful astringent** like kino. Dose: 8 to 30 gr. (0.5 to 2 Gm.).

OFFICIAL PREPARATION.

Tinctura Gambir Composita (5 per cent.,
     with saigon cinamnon 2.5 per cent.),..Dose: 15 to 60 ♏ (1 to 4 mils).

258. KINO.—KINO

KINO

The spontaneously inspissated juice of **Pterocar'pus marsu'pium** (Roxburgh).

BOTANICAL CHARACTERISTICS.—A leafy tree 40 to 80 feet high, with reddish-brown bark. *Leaflets* 5 to 7, coriaceous, dark green, shining, 3 to 5 inches long. *Flowers* yellowish-white. *Legume* woody, indehiscent.

SOURCE.—East Indies. We have several varieties other than the Malabar (East India), the official kind as described above—namely, African or Gambia kino (*P. erinaceus*), Palas or Bengal kino (*Butea frondosa*), Botany Bay or Eucalyptus kino (*E. amygdalina*), from Australia, and West Indian or Jamaica kino (*Coccoloba uvifera*). These all furnish extractives known as kino.

A new kind of kino from the juice of the bark of several kinds of Asiatic Myristica has been noticed, differing from the Malabar by containing, in the crude state, calcium tartrate. By this characteristic it may easily be distinguished from the official and other kinos of the market.

DESCRIPTION OF DRUG.—Small, **dark** reddish-brown, **shining, angular fragments,** much lighter and nearly transparent in thin layers. Adheres to the teeth when chewed, and colors the saliva a deep red;

odorless; taste sweetish and **astringent.**   The powder is of a brownish-red color.

Powder.—Elements of: See Part iv, Chap. I, B.

CONSTITUENTS.—Kino-tannic acid (colored black-green by ferric salts, in neutral solution; violet by ferrous salts), kinoin, neutral crystal-

FIG. 132.—*Pterocarpus marsupium*—Branch.

line prisms, pyrocatechin, kino-red, pectin, and ash.   Ash, not exceeding 3 per cent.

Preparation of Kinoin.—Boil kino with dilute HCl and agitate clear solution with ether. Evaporate off the ether. Heating this to 266°F., an insoluble amorphous kino-red is obtained.
   Pyrocatechin results from the dry distillation of kino, or is obtained by treating kino with ether.

ACTION AND USES.—A powerful astringent. Dose: 8 to 30 gr. (0.5 to 2 Gm.).

OFFICIAL PREPARATION.

   Tinctura Kino (5 per cent.),.........Dose: 1 to 2 fl. dr (4 to 8 mils).

259. COPAIBA.—Copaiba

BALSAM COPAIBA

The oleoresin of **Copai'ba langs'dorffii**[1] O. Kuntze, and of other species of Copaiba.

BOTANICAL CHARACTERISTICS.—Lofty forest trees, natives of Central America. bearing alternate, pinnate leaves. The wood of the trees is *replete with oleoresin*, sometimes even to bursting.

SOURCE AND COLLECTION.—This oleoresin is derived from several species of copaiba, as *C. officinalis* (Carthagena), *C. langsdorffii* (São Paulo),

FIG. 133.—*Copaiba langsdorffii*—Branch.

C. multifuga (Para). These furnish the several commercial varieties. Obtained by making large augur holes, square or wedge-shaped boxes, into the center of the trunk, where the oleoresin collects. Sometimes these openings are closed or sealed with wax, and often the pressure from the high liquid column is said to burst the trunk with a very loud report. A tree may yield from 10 to 12 gallons.

[1] Sometimes written, incorrectly, lansdorffii (Lloyd).

If 4 fluidrams of the above varieties of copaiba be mixed with
1½ fluidrams of aqua ammonia and shaken in a test-tube, the mixture will be clear, but milky if more alkali or fixed oil be present.
Maracaibo (Colombia copaiba) is thicker, darker, not always clear.
It solidifies, however, with magnesia and contains from 20 to 40 per
cent. of the volatile oil.

DESCRIPTION OF DRUG.—A more or less viscid, yellow or light brown,
transparent liquid, of about the consistence of olive oil; specific gravity, 0.950 to 0.955 at 25°C. (77°F.); it becomes thicker and darker
with age, the volatilization and the oxidation of the volatile oil leaving
a greater proportion of the soft resin.   Odor peculiar, aromatic; taste
bitter, acrid, and nauseous.

Para copaiba is a pale, limpid liquid containing from 60 to 90 per cent.
of volatile oil.   Maranham and Rio Janeiro copaiba are of the consistence of olive oil, and contain a somewhat smaller proportion of
volatile oil—40 to 60 per cent.   Maracaibo copaiba is dark yellow
or brownish, thick, somewhat turbid.   It contains from 20 to 40 per
cent. of oil of copaiba.

CONSTITUENTS.—**Volatile oil,** upon which its value mostly depends; a
bitter principle, **and two resins,** copaibic acid, $C_{20}H_{30}O_2$ (soluble in
ammonia and absolute alcohol), and a viscid, non-crystalline resin.
Para copaiba contains oxycopaivic acid, $C_2H_{28}O_3$; Maracaibo copaiba,
metacopaivic acid, $C_{22}H_{34}O_4$.   Copaiba contains no benzoic nor cinnamic acids, hence the term balsam is a misnomer.

Preparation of Copaibic Acid.—Mix nine parts of copaiba and two parts
of ammonia (sp. gr. 0.95); lower the temperature to 10°C.; crystals of copaibic
acid are then obtained, which agree with abietic acid in composition, but not in
properties.

ACTION AND USES.—Stimulant, diuretic, laxative.   Its principal action,
however, is on mucous membranes.   Dose: 15 ♏ (1 mil), in emulsion.

259 a. **OLEUM COPAIBÆ.**—OIL OF COPAIBA.   A volatile oil distilled
from copaiba.   A pale yellowish liquid of an aromatic, bitterish taste,
and having the general properties of the oleoresin.   It is a pure
hydrocarbon having the formula $C_{20}H_{32}$.   Dose: 5 to 15 ♏ (0.3 to
1 mil), in emulsion.

260. **PONGAMIA OIL.**—KURUNG OIL.   A deep yellow, or reddish-brown, fixed
oil expressed from the seeds of an East Indian tree, **Ponga'mia gla'bra** Ventenat.   It is used by the natives as a local application in skin diseases and rheumatism; especially recommended in pityriasis versicolor, and other cutaneous
diseases due to fungous growth.

261. **COPAL.**—GUM COPAL.   A resin found as a fossil in Zanzibar, or exuding
from various species and genera of trees of the natural order **Leguminosæ,**
growing in South America, West Indies, and Africa.   Yellowish or brownish,
irregular masses, often with a wrinkled surface; breaks with a glossy conchoidal
fracture; odorless and tasteless.   Used in making varnishes.

262. BALSAMUM PERUVIANUM.—Balsam of Peru

BALSAM OF PERU

A balsam exuded from the bruised trunk of Tolui′fera perei′ræ Baillon.

Botanical Characteristics.—A leafy tree, with wood containing a liquid balsam. *Leaves* imparipinnate; *leaflets* 5 to 11, alternate. *Racemes* 6 to 7 inches long. *Fruit* a one-celled, one-seeded pod about 3¼ inches long; *mesocarp* fibrous, the inner part with receptacles of oleoresin.

Fig. 134.—*Toluifera pereiræ*—Flowering branch and fruit.

Source and Collection.—This valuable tree grows in the wild forests of San Salvador, singly or in groups. The trees, owned by individuals, are carefully guarded. The balsam is collected by loosening the bark with a blunt mallet for some distance in four alternate sections so as not to kill the tree. The loosened bark soon splits; it is set on fire and charred, leaving the wood bare. Pockets thus made are covered with rags to $abso_{rb}$ the exuding balsam. These, when saturated, are thrown into boiling water, as a means of separating the balsam, which collects at the bottom of the vessel. The annual yield per tree is about twenty pounds. The fruit yields by expression a white balsam (balsam blanco, white Peru balsam), having a tonka-like odor, which contains a crystallizable resin. The name Myroxylon, as sometimes applied to the balsam, suggests the fact that for a long time it was supposed to be derived from a species of Myroxylon (*M. peruiferum*).

DESCRIPTION OF DRUG.—A brownish-black, oleoresinous, non-viscous liquid, transparent in thin layers, and, by transmitted light, a bright red-brown; heavier than water; odor balsamic and vanilla-like; taste warm, bitterish, afterward acrid.

CONSTITUENTS.—Benzoic and cinnamic acid, cinnamein (the cinnamate of benzyl alcohol) constituting the greater part, about 60 per cent.; resin 32 per cent., and small quantities of benzyl alcohol, $C_6H_5CH_2CH_2OH$; benzylic benzoate, $C_7H_5(C_7H_7)O_2$; stilbene, $C_{14}H_{12}$; styrol, C_8H_8; styracin; toluol, C_7H_8.

ACTION AND USES.—Stimulant, expectorant, and stomachic. Externally in ointment. Dose: 8 to 30 gr. (0.5 to 2 Gm.).

263. BALSAMUM TOLUTANUM.—BALSAM OF TOLU

BALSAM OF TOLU

A balsam exuding from incisions in the trunk of Tolui'fera **Balsamum** Linné.

BOTANICAL CHARACTERISTICS.—A lofty evergreen tree with warty branches; the wood contains a liquid balsam, which exudes when incisions are made. *Leaflets* 7 to 8, ovate-oblong. *Legume* indehiscent, with winged expansions and a winged stalk; very broad at apex.

HABITAT.—Venezuela and New Granada.

COLLECTION.—The balsam is obtained by making **V**-shaped incisions through the bark and collecting the exudate in small cups or calabashes. It is imported from Venezuela in tins holding from ten to twenty-five pounds. This tapping of the tree continues for eight months, causing the tree to become partially exhausted, showing itself in the lessened foliage. A spurious article has been found on the market. It has a soft consistence, is very sticky, especially when chewed, and under the microscope shows only an occasional crystal. On distilling a portion of this balsam with water, it was observed to contain more of a fragrant volatile oil and less cinnamic acid than the genuine drug.

DESCRIPTION OF DRUG.—A very viscid, yellowish-brown semi-solid, with a sweet, fragrant odor, and feebly aromatic taste. Long kept, it gradually hardens into a more or less solid mass, which is brittle in the cold. Soluble in volatile oils, alcohol, chloroform, glacial acetic acid, and solution of potassa. Readily fusible, and burns with an aromatic odor.

Constituents.—A volatile oil (chiefly toluene, $C_{10}H_{16}$), a resin, free acids (cinnamic and benzoic), and benzylic ethers of these, principally of the former. If a thin layer of the balsam be viewed under the microscope, numerous crystals of the free cinnamic acid are seen.

Action and Uses.—Stimulant expectorant, similar in action but weaker than balsam of Peru. The syrup is used as an agreeable basis for cough mixtures. Dose of the balsam: 8 to 30 ℥ (0.5 to 2 mils).

Fig. 135.—*Toluifera balsamum*—Branch and fruit.

Official Preparations.

Syrupus Tolutanus (5 per cent. of tr.),..Dose: 2 to 6 fl. dr. (8 to 24 mils).
Tinctura Tolutani (20 per cent.),....... ½ to 2 fl. dr. (2 to 8 mils).
Tinctura Benzoini Composita (4 per cent.),............................. 15 to 60 ℥ (1 to 4 mils).

LINACEÆ.—Flax Family

Stems herbaceous; annual or perennial, rarely woody plants closely allied to the mallows, remarkable, however, in having the inner fiber of the bark very tenacious, and for the mucilaginous covering of the seed, in which there is an abundance of drying fixed oil. A few are bitter.

  LINACEÆ

264. LINUM.—Linseed

FLAXSEED

The ripe seed of Li'**num** **usitatis'simum** Linné, including not more than 3 per cent.
of other harmless fruits, seeds or foreign matter.

Botanical Characteristics.—The common flax is an annual; stem corym-
bosely branched at top. *Leaves* sessile, linear-lanceolate, smooth. *Flowers*
in a corymbose *panicle*, with sky-blue petals. *Pod* about the size of a pea, of
5 united carpels (into which it splits in dehiscence), and 5-celled, with two
seeds hanging from the summit of each cell, which is partly or completely

Fig. 136.—*Linum usitatissimum.*

divided into two by a false partition projecting from the back of the carpel,
the pod thus becoming 10-celled.

Habitat.—All temperate countries.

Description of Drug.—**Oblong-ovate, flat,** obliquely pointed at one
end and blunt at the other. The brown, glossy, polished surface is
seen, under the lens, to be marked with fine pits, and to be covered
with a transparent **mucilaginous epithelium** that swells in water.
The hilum occupies the slight hollow just below the apex. The embryo is oily, whitish, and inodorous. Taste mucilaginous, oily, and

slightly bitter. Flaxseed meal is of a brownish-gray color, and has a slight odor.

Powder.—Characteristic elements: See Part iv, Chap. I, B.

CONSTITUENTS.—A viscid yellow fixed oil, 30 to 35 per cent., proteids 25 per cent., resin, wax, a small quantity of amygdalin. The powder upon extraction with petroleum should yield not less than 30 per cent. of fixed oil, 98 per cent. of which should be saponifiable. An althæa-like mucilaginous substance resides in the epithelial layer, which swells considerably in water. This gummy matter from the investing coat is rapidly imparted to hot water, forming a thick, viscid mucilage, precipitated by alcohol and lead subacetate. The gummy principle is considered as transformed starch, which latter

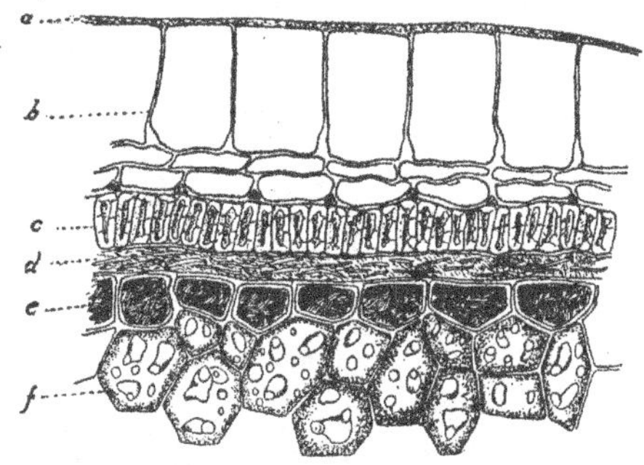

FIG. 137.—Cross-section of Flaxseed. *a.* Epithelium. *b.* Epidermal cells in swollen condition. *c.* Stone cells. *d.* Layer of collapsed cells for building stone cells. *e.* Pigment coat. *f.* Containing oil.

exists in the immature seed, but is absent in the ripe seed. Ash, not exceeding 6 per cent.

264 a. **OLEUM LINI.**—A yellowish fixed **oil expressed** (for medicinal use) **from the seed** without heat, having a slight, pleasant odor, and a bland taste; on exposure to the air it gradually thickens and acquires a strong odor and taste. The oil used in the arts is obtained on a large scale by roasting the seeds before being pressed, in order to destroy the gummy constituents of the coating. It does not congeal above $-20°C.$ ($-4°F.$). The most characteristic principle in the oil is linolein, $C_{12}H_{28}O_2$, a glyceride of linoleic acid, and considered to be a mixture of two acids—linolic, $C_{18}H_{32}O_2$, and linolenic acid, $C_{18}H_{30}O_2$. The drying property of the oil resides in this constituent.

ACTION AND USES.—The whole seed is used in decoction as a demulcent; ground flaxseed is a favorite farina for poultices; the expressed oil is laxative, and, in combination with lime-water (Linimentum Calcis), is much employed as a protective in burns, etc.

OFFICIAL PREPARATION.

From Oleum Lini.
Linimentum Calcis (equal parts of linseed-oil and lime-water).

265. COCA.—COCA (U.S.P. VIII)

ERYTHROXYLON

The dried leaves of **Erythrox'ylon Co'ca** Lamarack (Fam. Erythroxyllaceæ,) known commonly as **Huanuco (Bolivian) Coca,** or of **E. Truxillense** Rusby, known commercially as Truxillo (Peruvian) **Coca,** yielding, when assayed by U.S.P. process, not less than 0.5 per cent. of ether-soluble alkaloids of coca.

BOTANICAL CHARACTERISTICS.—Shrub about 6 feet high, with bright green leaves, size and shape similar to those of tea, and white blossoms, which are succeeded by small scarlet berries. When the leaves mature, the branches are stripped and the leafless plant is soon again covered with verdant foliage. The plant is propagated in nurseries from the seed.

SOURCE.—The shrub bearing coca leaves is extensively cultivated on the slopes of the Andes about 2,000 to 5,000 feet above the sea level, in **Peru and Bolivia.** The province of La Paz in Bolivia produces about the largest crops. That of Bolivia is considered superior to the Peruvian, although the latter country produces double the quantity. In this latter country, especially owing to the European demand, the cultivation has considerably increased. The annual production reaches the enormous figures of about one hundred million pounds. Two varieties, "Truxillo" and "Huanuco," having different characteristics, come to this market, the former named after the port Trujillo in the northern part of Peru, and the latter from the city of Huanuco, in the central part of Peru. The culture of coca leaves has been tried in other countries, but with questionable results, except, perhaps, on the Island of Java. The plant yields its first crop when eighteen months old, and continues to bear about forty years. There are two pickings yearly—April and September; the latter is considered the best and most abundant. The leaves are laid out in a paved drying yard and afterward pressed in drums (tambors) of plantain leaves, the tambor weighing forty pounds net.

DESCRIPTION OF DRUG.—*Huanuco Coca.*—Greenish-brown to clear brown, smooth and slightly glossy, thickish and slightly coriáceous, stoutly and very short petioled; blade 2.5 to 7.5 cm. long and nearly elliptical, with a very short and abruptly narrowed basal portion and a short point, the margin entire; midrib traversed above by a slight ridge, very prominent underneath, the remaining venation obscure, especially above; underneath, two conspicuous lines of collenchyma tissue run longitudinally on either side of the midrib and about one-third of the distance between it and the margin, the enclosed areola being of a slightly different color from the adjacent surface; odor characteristic; taste bitterish, faintly aromatic, followed by a numbness of the tongue, lips, and fauces.

Truxillo Coca.—Pale green, thin, brittle and usually much broken, smooth but not shining, shortly and stoutly petioled; blade 1.6 to 5 cm. long and one-third to one-half as broad, obovate to oblanceolate, narrowed from near the middle into the petiole, usually with a slight projecting point at the summit, the margin entire; underneath two irregular lines of collenchyma tissue, usually incomplete or obscure, and frequently wanting, run beside the midrib; odor more tea-like than that of Huanuco Coca; taste and numbing effect similar.

Powder.—Greenish. Characteristic elements: Calcium oxalate of parenchyma in prisms, 3 to 10 μ in diam.; sclerenchyma, bast, and crystal fibers; small papillæ on under epidermal cells.

CONSTITUENTS.—A volatile liquid alkaloid, hygrine, and **cocaine** $(C_{17}H_{21}NO_4)$, which has been found to be a compound body represented in a methyl benzoyl compound of another organic base, **ecgonine** $(C_9H_{16}NO_2)$. There **are** also present in the leaves **benzoyl ecgonine, a methyl compound of which constitutes the alkaloid cocaine.** This complex body cocaine is readily decomposed

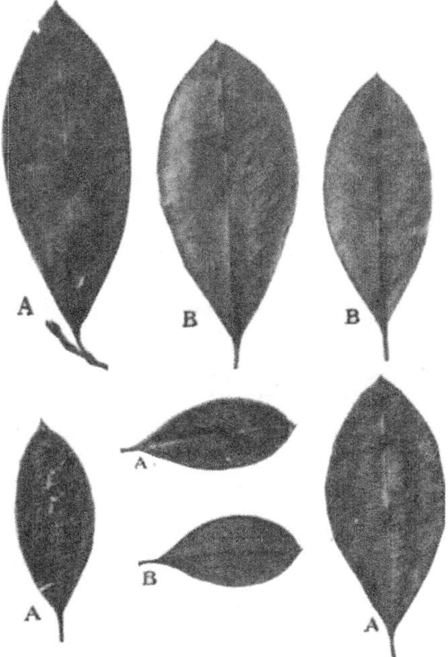

FIG. 138.—Coca Leaves—A, Upper side. B. Under side. (Photograph. Natural Size.)

into its component parts, methyl alcohol, benzoic acid, and ecgonine, by heating with HCl. Hydrochloric acid is, therefore, unsuitable for the extraction of cocaine in the process of its manufacture. The percentage of cocaine varies greatly, hence it is important to assay the leaves and its preparations. Assay shows an average of 0.5 per cent. of ether-soluble alkaloids of the leaf.

Preparation of Cocaine.—Exhaust the powdered drug by repercolation with water acidulated with 5 per cent. H_2SO_4. Agitate the concentrated liquid with pure coal oil and an excess of Na_2CO_3. The oily liquid is then shaken with acidulated water and again precipitated by Na_2CO_3 in the presence of ether. From the ethereal solution the alkaloid can be obtained on evaporation.

COCAINA (U.S.P. IX).—Cocaine. Average dose: 0.015 Gm. (¼ gr.).

ACTION AND USES.—Stimulant to digestion, the brain, and respiration. Checks the process of wasting, enabling the laborer to endure a greater amount of physical exertion with a small amount of food. For this purpose the leaves are habitually chewed by the natives. Dose: 15 to 60 gr. (1 to 4 Gm.). Cocaine is a valuable local anæsthetic. Applied to mucous surfaces and injected subcutaneously. Dose: ½ to 1 gr. (0.0324 to 0.064 Gm.).

Solutions of the alkaloid in olive and castor oil are stable. Cocaine hydrochloride ointment should not be made with lard or vaseline, as it is insoluble in these fats. If the hydrochloride be dissolved in a little water before admixture, a stable ointment is effected.

COCA PRÆPARATA, N.F.,...................1 to 4 fl. dr. (4 to 15 mils).

GERANIACEÆ.—Geranium Family

Herbs with opposite or alternate leaves, usually stipulate, simple or compounds. *Flowers* regular or irregular; *carpels* prolonged above into beaks terminated by the styles, which give rise to the name Cranesbill, applied to the principal genus.

266. GERANIUM.—GERANIUM, N.F.

CRANESBILL

The dried rhizome of Gera′nium macula′tum Linné.

DESCRIPTION OF DRUG.—**Rough, knotty, cylindrical, horizontal, rhizome, 50 to 75 mm. (2 to 3 in.) long, and 10 mm. (⅖ in.) thick;** longitudinally wrinkled, tuberculated, very hard, and sometimes beset with shriveled, brittle rootlets; externally dark brown; fracture short, reddish-gray, showing a thin bark, several small, yellowish wood-wedges forming a circle near the cambium line, and a large pith; medullary rays broad. The rootlets have a thick bark and a thin central column of fibrovascular tissue. Inodorous; taste astringent.

Powder.—Grayish-brown. Characteristic elements: Large aggregate crystals of calcium oxalate; ducts porous and reticulate; parenchyma with crystals and starch. (Highly magnified starch grains, see Fig. 139).

CONSTITUENTS.—Tannic (12 to 37 per cent.) and gallic acids, with resin, starch, gum, pectin, and a red coloring matter. Both alcohol and water extract its virtues.

ACTION AND USES.—A valuable and pleasant astringent. It has been claimed that the rhizome contains mucilaginous material which, acting as a demulcent, makes a decoction a much more desirable preparation than a simple solution of tannin. The fluidextract is said to be useful in buccal ulcer, etc. Dose: 15 to 30 gr. (1 to 2 Gm.).

267. IMPATIENS PALLIDA.—Jewel Weed. Indigenous herb occasionally used as an alterative and diuretic in infusion.   Dose: 1 dr. (4 Gm.).   *Impa'-tiens balsam'ina*, the touch-me-not of the gardens, has the same properties.

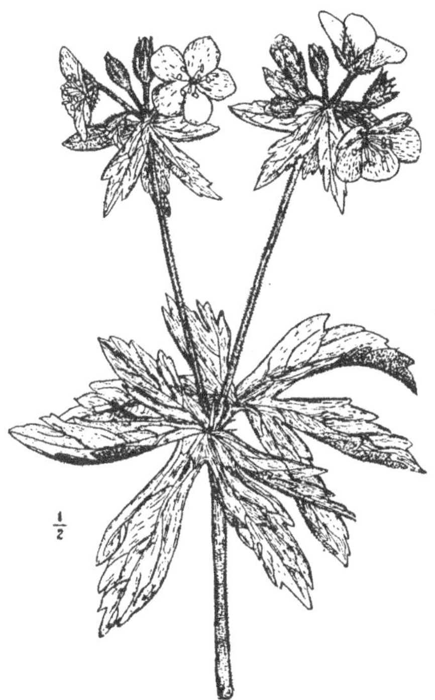

Fig. 139.—*Geranium maculatum*—Flowering branch.

ZYGOPHYLLACEÆ

The wood of many species of this order is remarkable for its excessive hardness.   The two official drugs from the order are the wood, 75, and resin, 76, of guaiacum.

268. GUAIACI LIGNUM, N.F.—Lignum Vitæ.   The heart-wood of **Gua'iacum officina'le and G. sanctum** Linné.   Greenish-brown, resinous raspings or chips, mixed with yellowish particles of the sap-wood; odor slight, agreeable, increased by heating or rubbing; taste slightly aromatic, but irritating and persistent after chewing some time.   The heart-wood of guaiac is imported in billets or logs and used for turning out various instruments and utensils, the shavings from these being used in pharmacy.   The **sap-wood** is yellowish, the heart-wood dark greenish-brown, hard and heavy, remarkable in that its specific gravity is such as to sink in water.   *Constituents:* **The resin** (soluble in alcohol and alkaline fluids) **is the most important constituent,** of which it contains about 26 per cent.; it also contains 0.8 per cent. of bitter, pungent extractive.   The wood or chips are turned a bluish-green by the action of nitric acid fumes.

17

Stimulant, diaphoretic; also a reputed antirheumatic and antisyphilitic. Generally given in the form of compound decoction of sarsaparilla. Dose: 15 to 60 gr. (1 to 4 Gm.).

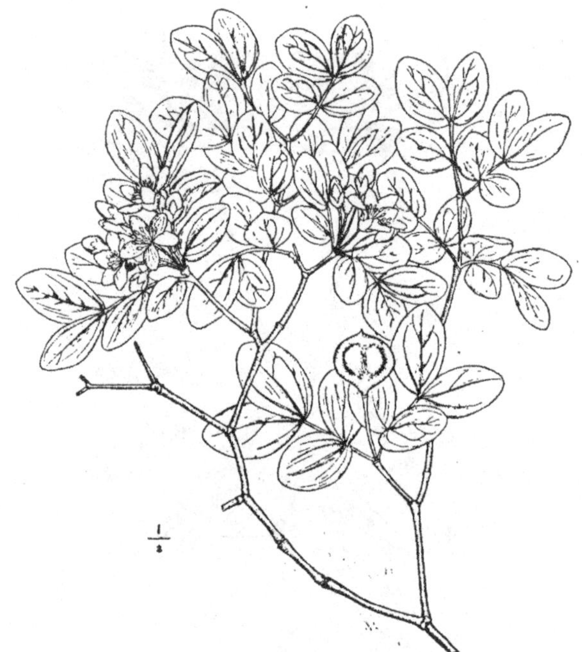

FIG. 140.—*Guaiacum sanctum*—Flowering branch.

269. GUAIACUM.—Guaiac

GUM GUAIAC

The resin from the wood of **Gua'iacum officina'le** Linné and of G. **sanctum.**

SOURCE.—Obtained from natural exudation or from incisions into the trunk, occasionally by boring longitudinally through a billet, placing one end in the fire, and catching the melted resin as it exudes from the hole in the other end; more commonly, however, by extracting the chips or raspings with a boiling solution of common salt.

DESCRIPTION OF DRUG.—Greenish-brown, irregular masses, containing fragments of wood and bark; brittle, breaking with a glossy fracture; in thin pieces, transparent. The powder is gray when fresh, but **becomes green on exposure,** and blue when in contact with oxidizing agents. Odor slight, balsamic, when heated resembling benzoin; taste slightly irritating.

CONSTITUENTS.—**Guaiacic** acid, β-resin (11.75 per cent.), and guaiac yellow, $C_{20}H_{20}O_7$, soluble in milk of lime; guaiaretic acid, $C_{20}H_{24}O_4$, 11.15 per cent.; **guaiaconic acid,** 50 per cent., and gum and ash in small quantity. Guaiacene, guaiacol, cresol, and pyroguaiacin are obtained by dry distillation. The coloring matter crystallizes in pale yellow or quadratic octahedra having a bitter taste. Ash, not exceeding 4 per cent.

The so-called "guaiacum oil" is obtained by boiling guaiacum resin with solution of sodium carbonate, allowing to cool, filtering,

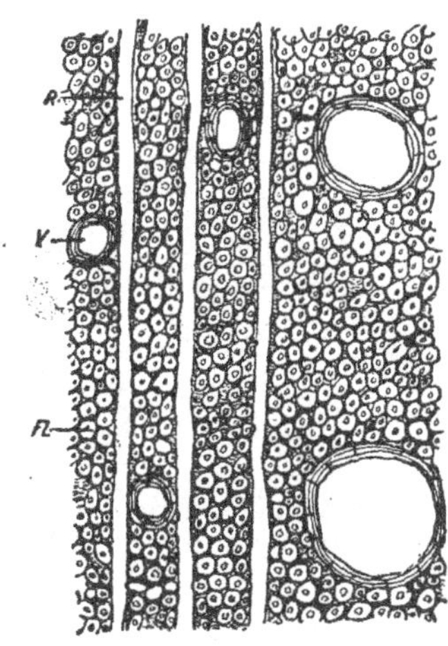

FIG. 141.—*Guaiacum*—Cross-section of wood. *R*. Medullary rays, composed of one, two, and three ranges of cells. *V*. Closed vessels. *F*. Ligneous fibers, very much developed and forming concentric zones.

saturating the filtrate with carbon dioxide, again filtering, extracting the oil with ether, and allowing the solvent to evaporate. The product is soluble in water, alcohol, and ether. From the alkaline liquid acids precipitate the yellow coloring matter ("guaiacum yellow"), which imparts a blue color to strong sulphuric acid.

The blue color which guaiacum resin produces with certain oxidizing agents is due to an oxidation product of guaiaconic acid.

ACTION AND USES.—Stimulant, diaphoretic, and alterative; also a mild purgative. Dose: 5 to 30 gr. (0.3 to 2 Gm.).

OFFICIAL PREPARATIONS.

Tinctura **Guaiaci** (20 per cent.),........Dose: 30 to 60 ♏ (2 to 4 mils).
Tinctura **Guaiaci Ammoniata** (20 per
cent.),........................... 30 to 60 ♏ (2 to 4 mils).

RUTACEÆ.—Rue Family

To facilitate study, this order has been divided, one of the subdivisions being the sub-order Aurantieæ (see below). The rueworts are remarkable for yielding acrid and resinous principles and volatile oil. *Ruta montana*, growing in Spain, is so extremely acrid that it raises pustules on the skin of those who gather it. The peduncles and flower of the European Dittany are so laden with volatile oil that the plant ignites at the approach of a lighted candle.

Synopsis of Drugs from the Rutaceæ

A. *Barks.*
XANTHOXYLUM, 270.
Angustura, 272.
Ptelia Trifoliata, 273.

B. *Leaves.*
BUCHU, 274.
PILOCARPUS, 275.
Ruta, 276.

C. *Fruits.*
Bela, 277.
*Xanthoxyli Fructus, 271.

(Products of the sub-order Aurantieæ, p. 267.)

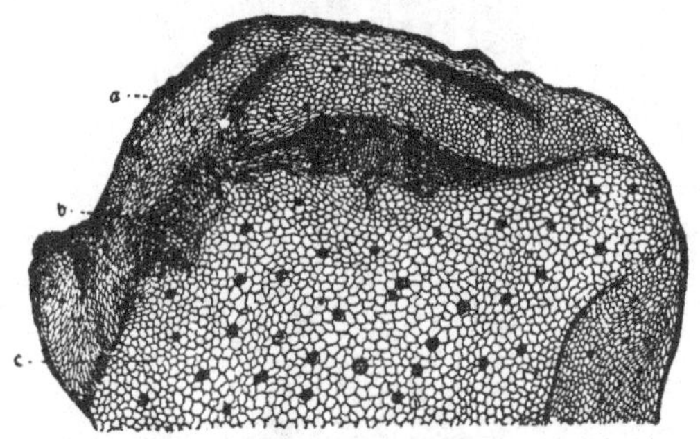

FIG. 142.—Cross-section of Cranesbill. *a.* Bark. *b.* Wood-wedge. *c.* Pith. (12 diam.)

270. XANTHOXYLUM.—Xanthoxylum

PRICKLY-ASH BARK

The bark of **Xanthox'ylum america'num** Miller, and of **Fagara clava-her'culis** Linné, known in commerce respectively as Northern Prickly-ash and Southern Prickly-ash.

BOTANICAL CHARACTERISTICS.—The northern prickly-ash, *X. americanum*, bears its *leaves* and *flowers* in sessile, axillary, umbellate clusters; *leaflets* 2 to 4 pairs, and an odd one, obovate-oblong, downy when young. The southern prickly-ash, *F. clava-herculis*, bears its *flowers* in an ample terminal cyme, appearing after the leaves; *leaflets* 3 to 8 pairs, and an odd one, ovate or ovate-lanceolate, oblique, shining above.

HABITAT.—United States.

DESCRIPTION OF DRUG.—Northern prickly-ash (*X. americanum*), as found in commerce, is in curved or quilled pieces about 1 mm. (1/25 in.) thick; the outer surface is of a brownish-gray color, longitudinally furrowed and **showing a few yellowish-gray patches of foliaceous** lichens, also numerous black dots and a few straight spines. Inner surface is light brown or yellowish; fracture uneven, short; inodorous; taste bitter, pungent, and acrid. Southern prickly-ash (*F. clava-herculis*) is somewhat thicker and has conical corky projections, with a few spines rising from corky bases. Inner surface free from acicular crystals.

Powder.—Characteristic elements: See Part iv, Chap. I, B.

CONSTITUENTS.—An acrid green oil, a colorless crystalline resin, sugar, ash 11 to 12 per cent., tannin (small quantity), and a bitter principle which is turned brown by H_2SO_4.

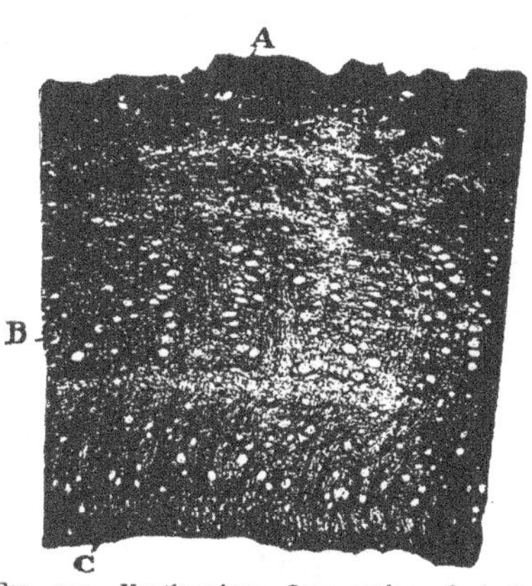

FIG. 143.—Xanthoxylum—Cross-section of bark. (15 diam.) A, Cork. B, Resin cells. C, Medullary ray. (Photomicrograph.)

ACTION AND USES.—Alterative, sialagogue, stimulant, and tonic, its action being similar to that of guaiac and mezereum. The bark chewed is a popular remedy for toothache, giving rise to the synonym, toothache-tree. The fluidextract is frequently combined with such alteratives as stillingia, lappa, etc. The berries are used in compound syrup of stillingia (see National Formulary). Dose: 15 to 45 gr. (1 to 3 Gm.).

OFFICIAL PREPARATION.

Fluidextractum Xanthoxyli,.   Dose: 30 to 60 ♏ (2 to 4 mils).

271. **XANTHOXYLI FRUCTUS,** N.F.—PRICKLY-ASH FRUIT. Consists of brownish-red capsules about 4 to 5 mm. (⅕ to ¼ in.) in diameter, sessile on the thin receptacle (*X. clava-herculis*), or borne on short stalks (*X. americanum*); the two valves open when ripe and expose the one or two shining, more or less wrinkled, black seeds; odor aromatic; taste very pungent and somewhat bitter. Stimulant, tonic, and alterative; used in fluidextract of stillingia, N.F. Dose: 15 to 30 gr. (1 to 2 Gm.).

272. **ANGUSTURA.**—CUSPARIA BARK. The bark of **Galipe'a cuspa'ri** St. Hillaire. *Habitat:* Northern South America. Found in the market in flattish, quilled, or channeled pieces about 3 mm. (⅛ in.) thick, and not longer than 150 mm. (6'in.), but usually shorter; externally it is covered with a yellowish-gray, corky layer, which is marked by shallow longitudinal fissures, and in most cases easily removed by the nail; inner surface light cinnamon-brown, often with adhering strips of wood; internally reddish-brown, showing white points due to deposits of calcium oxalate. The tissue of the bark is loaded with oil cells. Odor musty, due to volatile oil; taste bitter and nauseous. Besides volatile oil and resin, the bark contains a bitter principle, **angusturin, and four alkaloids,** the most important of which is **cusparine.** Used as an aromatic bitter. Dose: 8 to 30 gr. (0.5 to 2 Gm.).

273. **PTELIA TRIFOLIATA** Linné.—WATER ASH.   Shrub growing in the United States east of the Mississippi.   (Root-bark.)   It contains berberine.   Used as a tonic and antiperiodic, "its mild, non-irritating properties rendering it especially valuable in low fevers attended with gastro-intestinal irritation; this soothing influence causes it to be retained when other tonics would be rejected."   Dose of fluidextract: 15 to 30 ♏ (1 to 2 mils).

FIG. 144.—*Barosma betulina*—Branch and flower.

FIG. 145.—*Barosma crenulata*—Flowering branch.

274. BUCHU.—BUCHU

SHORT BUCHU

The dried leaves of **Barosma Betulina** (Thunberg) Bartling and Wendland, known commercially as short buchu, or of **Barosma Serratifolia** (Curtis), Willdenow, known commercially as long buchu, with which may be mixed not more than 10 per cent. of the stems of the plants or other foreign matter.

BOTANICAL CHARACTERISTICS.—Shrubby plant.   The characteristics common to the buchus are opposite *leaves*, small, simple, coriaceous, dotted with pellucid glands.   *Flower* pink (*betulina*), white (*crenulata*), solitary on axillary or terminal peduncles.   *Fruit* composed of five follicles, adherent at the axis and dehiscing at the summit.

HABITAT.—Southern Africa, Cape of Good Hope.

DESCRIPTION OF DRUG.—About 15 mm. long, varying between oval and obovate, yellowish-green, apex obtuse, margin crenate or serrate with a gland at the base of each tooth, base more or less wedge-shaped; coriaceous, both surfaces beset with numerous slight projections; odor strong and characteristic; taste somewhat mint-like, pungent and bitterish. *B. serratifolia* (very narrow, linear-lanceolate) con‑ stitute the "long buchu" of commerce. The long buchu (off. in U.S.P. 1890) contains less of the volatile oil. Transverse sections show a subcuticular layer of thickened cells, rich in mucilage, and containing sphæro-crystals. Both kinds usually require careful garb‑ ling, as they are often mixed with branchlets, fragments of capsules, and with leaves of allied species. The long buchu is sometimes

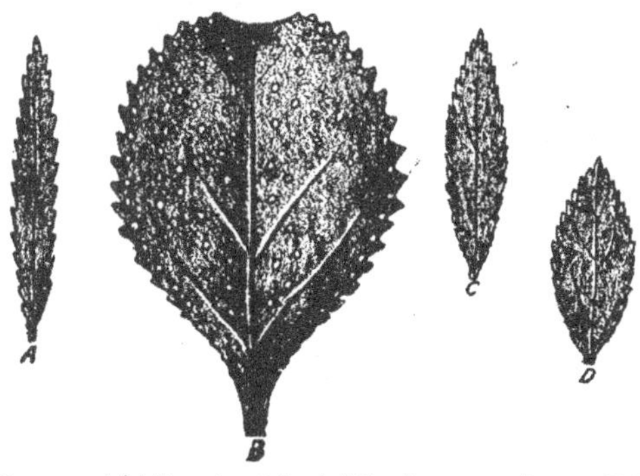

FIG. 146.—Buchu Leaves and Adulterant. *A*, Leaf of *Empleurum serrulatum*. *B*, Leaf of *Barosma betulina* (enlarged). *C*, Leaf of *Barosma seratifolia*. *D*, Leaf of *Barosma crenulata*.

mixed with the leaves of *Empleurum serrulatum*, but these are still narrower, often longer, and terminate in an acute point, without an oil duct.

Powder.—Characteristic elements: See Part iv, Chap. I, B.

CONSTITUENTS.—**Volatile oil** is contained in large circular cells just be‑ neath the epidermis of the under surface of the leaf; the short buchu yields the greater per cent. (1 to 1.56 per cent.). On exposure to cold it separates out barosma camphor, which existed in the oil dis‑ solved in a hydrocarbon. The upper surface of the leaves swells up in water, due to a layer of mucilage cells just beneath the surface. The bitter principle is **rutin**; **resin** is also present.

ACTION AND USES.—A mild diuretic in disorders of the urinogenital organs, its action depending upon the volatile oil. In Cape Colony the leaves are employed as a stimulant and stomachic. Dose: 15 to 45 gr. (1 to 3 Gm.).

OFFICIAL PREPARATION.

Fluidextractum Buchu,................Dose: 15 to 60 ♏ (1 to 4 mils).

275. PILOCARPUS.—Pilocarpus

JABORANDI

The leaflets of **Pilocar'pus jaboran'di** Holmes or of **Pilocarpus microphyllus**
Stapf.   Yielding when assayed by U.S.P. process not less than 0.6 per cent.
of alkaloids.

BOTANICAL CHARACTERISTICS.—A shrub 4 to 5 feet high.   *Leaflets* 1 to 4 pairs,
petiolate.   *Flowers* in long racemes.   *Ovary* with 5 carpels.   *Seeds* black,
angular.

FIG. 147.—*Pilocarpus selloanus*—Branch from flowering top.—(*After Kohler.*)

SOURCE, VARIETIES, AND ADULTERATIONS.—The name Jaborandi is a
generic one, applied in South America to several plants possessing
diaphoretic properties.   The shrub, *Pilocarpus jaborandi*, grows in
Brazil in the neighborhood of Pernambuco, known commercially as
Pernambuco Jaborandi.   *P. microphyllus* (which yields a large percentage of alkaloid), differs from this in absence of oil from their
tissues, by their reticulated venation, etc., is known commercially as

Maranham Jaborandi.   It has been adulterated with species of Piper, which are not pellucid-punctate, with *Laurus nobilis*, etc.

DESCRIPTION OF DRUG.—Leaves nearly sessile, pinnate, with a terminal leaflet; the leaflets, which come into market separate, are **ovate-oblong, entire,** about 100 mm. (4 in.) long, and 50 mm. (2 in.) broad; short-petiolate; uneven at the base; slightly **revolute at margin,** near which **the anastomosing veins form one or two distinct wavy lines;**

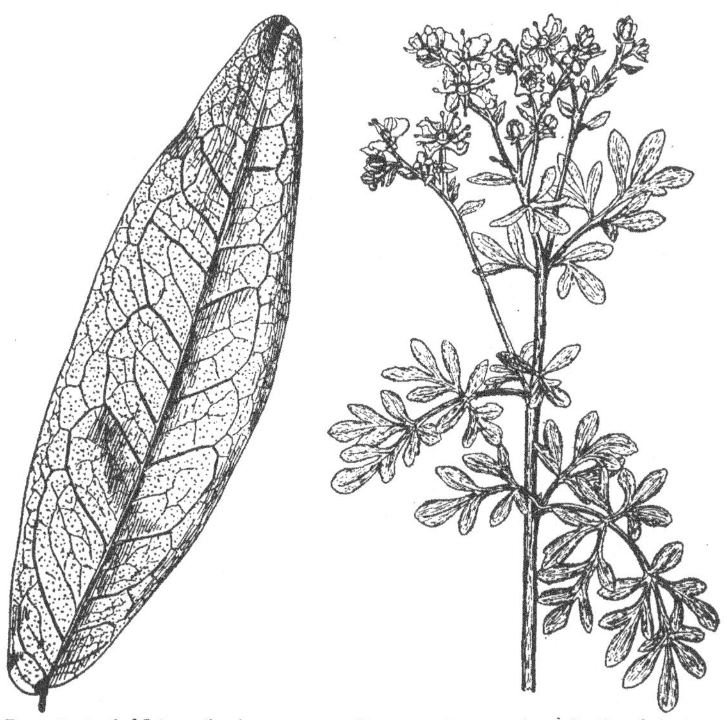

FIG. 148.—Leaf of Jaborandi as it appears in the market.

FIG. 149.—*Ruta graveolens.*—Portion of plant.

coriaceous; dull green, finely marked with small, transparent dots or oil-cells, plainly visible when held up to the light; texture coriaceous, brittle; when bruised a peculiar, rather unpleasant odor is emitted; this odor is predominant in the fluidextract. Taste disagreeable, slightly pungent, and bitter. The leaflets of *P. microphyllus* (Maranham jaborandi) are smaller (2.5 to 4 cm. in length), usually ovate in outline, deeply emarginate at apex. **Aklaloidal content (chiefly Pilocarpine) of best leaf ranges from 0.5 to 1 per cent.**

Powder.—Yellowish-brown.   Characteristic elements: See Part iv, Chap. I, B.

Constituents.—A volatile oil, and two alkaloids, pilocarpine ($C_{11}H_{17}N_2O_2$), deliquescent, crystalline, inodorous, and slightly bitter, and jaborine, chemically isomeric with, but directly antagonistic to, the first named in physiological action. Pilocarpine is the most active, and yields jaborine and pilocarpidine ($C_{10}H_{14}N_2O_2$) when heated with HCl; its salts are readily soluble in water; their action is similar to that of nicotine. Jaborine ($C_{22}H_{32}N_4O_4$) is yellow, amorphous, and resembles atropine in action; its presence in the commercial pilocarpine explains the different effects following the use of the latter when improperly made. It is therefore very necessary, in using pilocarpine or any of its preparations, to obtain them free from jaborine.

Preparation of Pilocarpine.—To an aqueous solution of acidulated alcoholic extract add alkali and shake with chloroform. From the chloroformic solution the alkaloid is separated by shaking with acidulated (HCl) water, filter, and allow it to crystallize.

Action and Uses.—Powerfully diaphoretic and sialagogue by stimulating the nerves supplying the glands and involuntary muscular fiber; cardiac depressant. The most important effects of pilocarpine are due to the stimulation of certain nerve terminations. It stimulates the peripheral endings of all the autonomous nerves. The most important effect of the ingestion of a therapeutic dose of pilocarpine is an increase in the secretory activity of nearly all the glands of the body, especially of the salivary and sweat-glands. Dose: of drug 5 to 60 gr. (0.3 to 4 Gm.). Pilocarpine is used as a myotic in ophthalmic practice. It has acquired some reputation in the treatment of diphtheria and croup; frequently administered hypodermically; poisonous. Dose of pilocarpinæ hydrochloridum, $\frac{1}{8}$ to $\frac{1}{12}$ gr. (0.008 to 0.005 Gm.). Ash, not exceeding 7 per cent.

Official Preparation.

Fluidextractum Pilocarpi,................Dose: 5 to 60 ♏ (0.3 to 4 mils).

276. RUTA.—Rue. The leaves of **Ru'ta graveo'lens** Linné. *Habitat:* Mediterranean region; cultivated. The whole plant is active, but the leaves are the portion generally employed. They are ternate, the leaflets being obovate-oblong, yellowish-green, thickly dotted with minute, transparent oil-vesicles; odor strong, disagreeable, increased by rubbing; taste bitter, hot, and acrid.

Their medicinal value depends chiefly upon the volatile oil, but there is also present a peculiar coloring matter, rutinic acid, found also in other plants, and an acrid principle, the activity of which is diminished in the dried leaves; the fresh leaves will inflame or even blister the hands if much handled.

Action and Uses.—Emmenagogue, vermifuge, and diaphoretic. Dose: 5 to 20 gr. (0.3 to 1.3 Gm.) in infusion. The Romans used rue as a condiment, as the Germans still do.

Oleum Rutæ.—A yellowish-green volatile oil, powerfully irritant; used as a uterine stimulant, emmenagogue, etc. Dose: 2 to 5 ♏ (0.13 to 0.3 mil).

277. BELA.—Bael Fruit. Bengal Quince. From Æ'gle marme'los Correa. *Habitat:* Himalaya Mountains; cultivated in India, where it is employed

and considered as a valuable remedy in dysentery and diarrhœa, relieving without causing constipation. Dose: 1 to 2 dr. (4 to 8 Gm.). It is collected when half ripe and dried; usually enters commerce in segments having a smooth, grayish rind, and a hard, reddish, gummy pulp; whitish internally and divided into cells, each of which contains four or five woolly seeds; taste mucilaginous, slightly bitter; nearly inodorous.

AURANTIEÆ.—SUB-ORDER OF RUTACEÆ.—The Orange Family

The trees and shrubs which compose this sub-order of Rutaceæ are distinguished from others of the order merely by the character of the fruit. In the Aurantieæ the fruit is an indehiscent, juicy, berry-like fruit, botanically known as hesperidium (lemon, orange, and lime), having a leathery rind, containing numerous oil-glands. The capsular fruit of the rueworts proper is usually dehiscent. The leaves and fruit of both sub-orders abound in minute receptacles of volatile oil. These attain their maximum development in the rind of the orange, lemon, etc. (see Figs. 150, 151, 152).

The Official and Unofficial Products of the Aurantieæ

I. *The Products of the Orange.*
  A. Official.
    The Peel, 278.
    The Oil, 279.
    Oleum Aurantii Florum, 281 a.
  B. Unofficial.
    The Leaf, 280.
    The Flower, 281.

II. *The Products of the Lemon.*
  A. Official.
    *The Juice, 282.
    The Rind, 283.
    The Oil, 284.
    Oil of Bergamot, 285.
  B. Unofficial.
    White Zapote, 286.

THE ORANGE PRODUCTS

Source.—Universally cultivated in India and widely in tropical regions. The *sweet orange* was introduced from China by the Portuguese. It has been much improved by cultivation. There are now some fifty varieties in different parts of the globe, these taking the name of the places where cultivated, the sweetest coming from Havana, Florida, and California. *Bitter oranges* were introduced into Europe from India by the Arabians and were used medicinally from very early times, the bitter fruit being usually termed the Seville or Bigarade orange.

278. **AURANTII AMARI CORTEX.**—The Rind. Bitter Orange Peel, the dried rind of the unripe fruit of *Citrus vulgaris* Risso. Ash, not to exceed 7 per cent.

AURANTII DULCIS CORTEX.—Sweet Orange Peel, the undried outer rind of the ripe fruit of *Citrus Aurantium* Linné. The orange tree is cultivated in the south of Europe, in the Azores, and in the United States—Southern States and California. It is said to be one of great longevity; thus, a tree in Versailles, known as the "Grand Bourbon," planted in 1421, is still in existence (Mueller).

DESCRIPTION OF DRUG.—*Bitter:* In narrow, thin bands or in quarters, epidermis brownish-yellow color, outer layer with numerous oil reservoirs, inner layer spongy, light yellowish-brown; odor fragrant; taste aromatic, bitter.   The Curacao orange peel is obtained from a variety of the orange cultivated in the island of Curacao.   *Sweet:* Outer surface orange-yellow with numerous oil reservoirs, odor highly fragrant, taste pungently aromatic.

Powder.—Microscopical elements of: See Part iv, Chap. I, B.

FIG. 150.—*Citrus Aurantium*—Branch.

CONSTITUENTS.—Volatile oil (contained in vesicles of the epidermis), hesperidin, ash, and a white principle which turns black with ferric salts.

ACTION AND USES.—Tonic, carminative, and stomachic; a valuable addition to preparations of the bitter tonics like gentian.   Dose: 15 to 30 gr. (1 to 2 Gm.).

OFFICIAL PREPARATIONS.

Bitter Orange Peel.
    Fluidextractum Aurantii Amari,   Dose: 15 to 60 ♏ (1 to 4 mils).
    Tinctura Aurantii Amari (20 per cent.), ...   1 to 2 fl. dr. (4 to 8 mils).
    Tinctura Cinchonæ Composita (8 per cent.),   8.0 mils to 2 fl. dr.
    Tinctura Gentianæ Composita (4 per cent.),   4.0 mils. to 1 fl. dr.
Sweet Orange Peel.
    Syrupus Aurantii (5 per cent. of Tinct.),.   ¼ to 1 fl. oz. (8 to 30 mils).
    Tinctura Aurantii Dulcis (50 per cent.),.   Flavoring.

279. **OLEUM AURANTII.**—The Oil. Obtained from the fresh peel of either the bitter or sweet orange. A pale yellow liquid, having a characteristic aromatic odor. Optical rotation should not be more than 95° to the right in a 100 mm. tube, and at a temperature of about 25°C. (77°F.). It contains some hesperidin, and an aldehyde geranial.

Oil of Petit-grain is obtained from the small, fragrant, immature oranges (berries about the size of a cherry). Recently, however, the leaves and shoots have been used for this purpose.

Manufacture.—The oils of the fruit of the Aurantieæ are manufactured by subjecting the outer rind to expression, distillation, or, preferably, to the *écuelle* process. This instrument (the écuelle) is described in most works on pharmacy.

Official Preparations.

Spiritus **Aurantii Compositus** (contains 25 per cent. oil and the oils of lemon, coriander, and anise).
Elixir Aromaticum (1.2 per cent.).

Fig. 151.—*Citrus vulgaris*—Flowering branch.

280. **AURANTII FOLIA.**—The Leaf. From Cit'rus vulgar'is Risso. Oval, from 50 to 100 mm. (2 to 4 in.) long, on a broadly-winged petiole, pellucid-punctate; odor aromatic; taste bitter. It is the principal source of *essence de petit-grain*, used to adulterate Oleum Neroli. Stimulant and tonic

281. **AURANTII FLORES.**—The Flower. Orange Flowers. The flowers Cit'rus vulgar'is and **C. Aurantium,** collected before they are expanded, solely for the volatile oil, which is then most fragrant. Generally used while fresh, in which state they may be preserved for some time by mixing

with half their weight of common salt. They are about 12 mm. (½ in.) long, with small, cup-shaped calyx and white, rather fleshy petals. Occasionally used as a stimulant and antispasmodic, but principally for preparing orange-flower water and the volatile oil.

281 a. OLEUM AURANTII FLORUM, U.S. VI—OLEUM NEROLI. A thin, yellowish, or brownish-yellow volatile oil, very fragrant. Used as a flavor and as a perfume. Neroli is the predominant odor in Farina Cologne.

THE LEMON PRODUCTS

282. LIMONIS SUCCUS—THE JUICE.—LEMON JUICE (Succus Citri, N. F.). The freshly expressed juice of the ripe fruit of Cit'rus **medica** Linné (C.

FIG. 152.—*Citrus limonum*—Branch.

limonum Risso, U.S.P. 1900). A slightly turbid, yellowish liquid having the odor of lemon, due to the presence of some of the volatile oil from the rind; taste acid, often slightly bitter. It contains about 7 per cent. of free citric acid, also phosphoric and malic acids. Refrigerant and antiscorbutic; used in the form of lemonade, or in effervescing draughts. Dose: 1 fl. oz. (30 mils).

Lemon juice should contain from 7 to 9 per cent. of citric acid. It should be free from added preservatives; preserved by sterilization. For tests see U.S.P. VIII. Lemon juice contains from 0.5 to 1 per cent. of gum and sugar.

283. LIMONIS CORTEX—THE RIND.—Lemon Peel. The undried outer rind of the ripe fruit of Citrus medica Linné (C. limonum Risso, U.S.P. IX), removed by grating. The fruit comes from the Mediterranean and tropical regions (see Orange). The outer surface is of a light yellow color and ruggedly glandular from the oil-cells; odor fragrant; taste aromatic and bitterish.

Microscopically, the rind of the lemon resembles that of the orange.

Powder.—Microscopical elements of: See Part iv, Chap. I, B.

CONSTITUENTS.—A pale yellow volatile oil (sp. gr. 0.87) consisting mainly of hydrocarbons, citrene ($C_{10}H_{16}$), cymene ($C_{10}H_{14}$), also citral ($C_{10}H_{16}O$), and a compound ether. Hesperidin ($C_{22}H_{26}O_{12}$), a bitter principle, produces with ferric salts a black color.

Used as a flavoring agent.

OFFICIAL PREPARATION.

Tinctura Limonis Corticis (50 per cent.).

284. OLEUM LIMONIS.—Oil of Lemon Peel, or Rind. A volatile oil obtained by expression from the fresh lemon peel. It is a pale yellow, limpid liquid, having a lemon taste and a fragrant odor. . It should be protected from light in well-stoppered bottles. Oil of citral, used in perfumery, is obtained from *Cit'rus med'ica* Risso, a large oblong fruit with rough surface—known in England as the citron.

Oil of lemon consists of two isomeric oils, chiefly citrene or limonine, $C_{10}H_{16}$, with citral (an aldehyde) and a crystalline product which fuses at 143° to

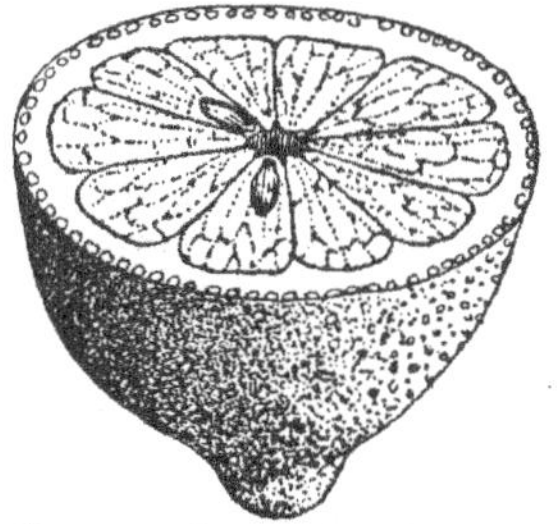

FIG. 153.—Cross-section of Lemon fruit.

144°C. (289° to 291°F.), colored yellow by H_2SO_4, and green by HNO_3. Used principally as a flavor assayed by the official process not less than 4 per cent. of the aldehydes from oil of lemon calculated as citral.

ADULTERATION OF THE OIL OF LEMON.—It is adulterated with the volatile oil of other fruits of the genus Citrus. These are difficult to detect; odor and taste must be chiefly relied upon.

OFFICIAL PREPARATIONS:

Spiritus Aurantii Compositus (5 per cent.).
Spiritus Amonniæ Aromaticus, ...Dose: 30 ♏ (2 mils).

285. OLEUM BERGAMOTTÆ.—Oil of Bergamot. A volatile oil obtained by expression from the rind of the fresh fruit of Cit'rus berga'mia Risso et Poiteau, the fruit being collected in November or December, still greenish, unripe, but full grown. By some, the bergamot orange is supposed to be an established hybrid—a product of cultivation. A greenish or greenish-yellow, thin liquid, having a peculiar, very fragrant odor, and an aromatic, bitter taste. The color is due to chlorophyll. It is distinguished from the oils of orange and lemon by forming a clear solution with solutions of potassium. This oil, so valuable in perfumery, was official in the U.S.P. of 1890, but was dropped from the list in 1900.

CONSTITUENTS.—By fractional distillation there comes over as the first fraction at 60° to 65° about 40 per cent. of the oil. This has a lemon odor and consists of almost pure **limonine.** The second fraction (10 per cent.), distilling at 77° to 82°, consists principally of dipentene, $C_{10}H_{16}$. The third fraction of about 25 per cent., distilling between 87° and 91°, consists of linalool, $C_{10}H_{18}O$. The fourth fraction, 90° to 105° (approximately 20 per cent.), having the pronounced bergamot odor, consists of **linalool (linalyl) acetate,** $C_{10}H_{17}OC_2H_3O$. It is to this that the peculiar odor of bergamot is probably due.

286. **WHITE ZAPOTE.**—The seeds of **Casimuroa edulis,** growing in Mexico. Used as a hypnotic in the hospitals of the City of Mexico. Recently introduced in United States. Dose of fl'ext.: ½ to 9 ℥ (0.1 to 0.6 mils).

SIMARUBACEÆ

Shrubs and trees with scentless foliage; almost confined to the tropics. *Leaves* generally compound and alternate. The bitter bark and wood are employed in medicine.

287. QUASSIA

QUASSIA

The wood of **Picrasma excelsa** (Swartz) Planchon, known commercially as **Jamaica Queen,** or of **Quassia amara** Linné (Surinam Quassia).

BOTANICAL CHARACTERISTICS.—A tree resembling the common ash, attaining a height of 50 or 60, even 100, feet. *Leaves* pinnate, with an odd leaflet; leaflets opposite, 4 to 8 pairs. *Flowers* small, pale yellowish-green, in loose panicles, polygamous. *Fruit* drupaceous, globose, glossy, black.

HABITAT.—Jamaica and other West India islands.

DESCRIPTION OF DRUG.—Imported in dense, **tough billets,** often 300 mm. (12 in.) thick, freed from the thick, tough bark. The yellowish-white or white **raspings or chips** are usually employed in pharmacy. The tissue consists mostly of prosenchyma, associated with long wood-fibers with tapering ends, and ducts which, on transverse sections of the wood, appear as pores; inodorous; taste intensely bitter. Quassia **tonic drinking cups** are made from the wood on a turning lathe; water poured into them acquires a bitterness, in a few minutes, of which the wood seems inexhaustible.

Quas'sia amar'a Linné, Surinam Quassia, comes in much thinner billets, and has a thin, brittle bark, it seldom reaches our market. It may be distinguished from the *Picrasma excelsa* (Jamaica quassia) by the fact that the medullary rays in the former consist of single rows of cells, while those of the latter consist of three rows each. The cells composing the rays in the *Q. amara* are of equal size, and their radial walls appear wavy in tangential section; whereas the corresponding cells in *P. excelsa* are of variable size and exhibit regular walls in tangential section. The true source of Quassia is said, by some authorities, to be a *simaruba*.

Powder.—Characteristic elements: See Part iv, Chap. I, B.

Fig. 154.—*Picrasma excelsa*—Branch.

CONSTITUENTS.—*Picras'ma excel'sa* contains a bitter neutral principle, picrasmin, *Quas'sia amar'a*, an analogous principle, quassin, both soluble in water, alcohol, and chloroform. The principles can easily be obtained from the precipitated tannate by mixing it with lead carbonate, drying, and extracting with alcohol. They crystallize from alcoholic solution in needles; purified by recrystallization. Quassia **contains no tannin,** and therefore can be prescribed with salts of iron.

18

Preparation of Quassin.—Neutralize infusion with NaOH; add tannin to precipitate the neutral principle; heat with lead oxide or lime to decompose precipitate, and dissolve out with alcohol. White, opaque, very bitter. Soluble in hot alcohol, chloroform; slowly in water.

Preparation of Picrasmin.—Precipitate tannate with lead acetate, the former obtained by precipitating the neutral infusion with tannin. In needles; very soluble in hot alcohol, chloroform, acetic acid, but sparingly in water.

ACTION AND USES.—A valuable simple bitter tonic. Dose: 15 to 60 gr. (1 to 4 Gm.). It is poisonous to insects, a strong infusion being often used as a parasiticide on animals.

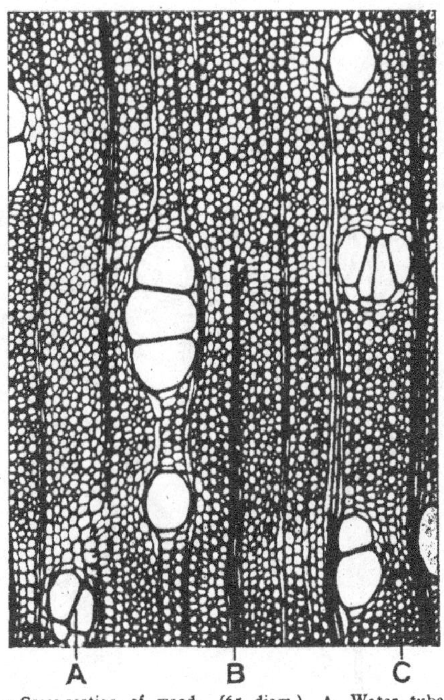

FIG. 155.—Quassia—Cross-section of wood. (65 diam.) A, Water tube. B, Medullary ray. C, Wood fibers. (Photomicrograph.)

OFFICIAL PREPARATION.

Tinctura Quassiæ (20 per cent.),........Dose: 30 to 60 ♏ (2 to 4 mils).

288. QUASSIÆ CORTEX.—QUASSIA BARK. The bark of **Picræ'na excel'sa** Lindley. In flat or curved pieces 5 mm. (⅕ in.) or more thick. The outer surface is of a dark gray color and longitudinally furrowed; inner surface yellowish-white and smooth; inodorous; very bitter. The bark of *Surinam Quassia* is much thinner. These barks have the same constituents and are used for the same purposes as the wood—as tonics.

289. SIMARUBA.—The root-bark of Simaru'ba officina'lis De Candolle. *Habitat:* Northern South America and West Indies. In curved or quilled pieces about 50 to 100 mm. (2 to 4 in.) long, and 3 mm. (⅛ in.) thick; it is of a yellowish-white color, generally deprived of the yellowish or brownish peri-

derm; inner surface light brown, finely striate; bast coarsely fibrous, tough, flexible, the fibers easily separable; inodorous; very bitter. It contains probably quassin or picrasmin, some resin, and a trace of volatile oil. Tonic, used in dysentery and chronic diarrhea. Dose: 8 to 30 gr. (0.5 to 2 Gm.), in infusion or decoction.

290. **CEDRON.**—CEDRON SEED. From Sima'ba ce'dron Planch, a South American tree. These seeds are used by the natives as a remedy for the bite of poisonous serpents and insects. Cerebral sedative, antispasmodic, and antiperiodic; poisonous. Dose of fluidextract: 1 to 8 ♏ (0.065 to 0.5 mil).

291. **AILANTHUS.**—TREE OF HEAVEN. CHINESE SUMAC. The bark of **Ailanth'- us** glandulo'sa Des Fontaines, a common shade tree. The powder is of a greenish-yellow color, and has a strong, narcotic, nauseating odor. A powerful nerve-depressant and antispasmodic, used in asthma, hiccough, twitching of the muscles, epilepsy, etc. When chewed, it produces a general sense of uneasiness, weakness, dazzling, cold sweats, shivering, nausea, etc., similar to that produced by tobacco. These effects depend upon a volatile oil, which is so powerful that persons preparing the extract are often thus affected by the vapor. Dose: 15 to 30 gr. (1 to 2 Gm.).

292. **CASCARA AMARGA.**—HONDURAS BARK. From undetermined species of **Picram'næa.**—A valuable alterative, claimed to be almost a specific in syphilitic affections; it contains an alkaloid, picramnine. The use of tobacco and alcohol is said to counteract its action. Dose: 30 to 60 gr. (2 to 4 Gm.).

293. **CHAPARRO AMARGOSO.**—Castela Nicholsoni Hook. AMARGOSA. Bark of root. West Indies and along the Rio Grande in Mexico and Texas. *Preparation:* Fluidextract. *Properties:* Tonic, antiperiodic, astringent. *Uses:* Considered a specific for diarrhea. Dose: 30 to 60 minims. In pieces about ten inches long by one in width and a quarter in thickness. Outer portion composed of a corky layer $\frac{1}{24}$ inch thick and ash-gray in color. Inner side of a dirty yellowish color; slightly striated. Fracture brittle but fibrous. Slightly aromatic. Very bitter.

BURSERACEÆ

Tropical trees and shrubs abounding in resinous and oily secretions. Drugs of the order are: *Myrrha* (294); *Olibanum* (295); *Bdellium* (296), and *Elemi* (297).

294. MYRRHA.—MYRRH

MYRRH

A gum-resin obtained from one or more species of **Commiph'ora myr'rha** Engler and other species.

BOTANICAL CHARACTERISTICS.—A shrub forming the chief underwood of the Arabian and African forests along the shores of the Red Sea. Squamose, spinescent branches, with pale, ash-gray, odorous bark; *leaves* ternate; *flowers* solitary, greenish; *fruit* drupaceous, with the persistent calyx attached.

SOURCE.—Myrrh is now imported from the East Indies, where it is brought from Arabia and the northeastern coast of Africa. It is usually imported in chests containing from one hundred to two hundred pounds. The terms Turkish and Indian myrrh are now obsolete. Up to recent times most of the myrrh came from India but now it chiefly comes direct from Aden.

DESCRIPTION OF DRUG.—Irregular masses of agglutinated tears, varying from small grains up to pieces about the size of an egg, or sometimes much larger; of a reddish-yellow to a reddish-brown color, dusty, opaque, waxy, and unctuous. Freshly broken, the shining surface often shows characteristic white marks or streaks. Odor pleasant, balsamic; taste bitter, aromatic. This description applies to the best *Turkey*-official myrrh. The *India* variety comes in darker pieces, more opaque, less odorous, and abounding in impurities. Bdellium and other gummy or resinous substances are often mixed with it. *False myrrh* is the name sometimes given to these other gummy and resinous substances. As it is difficult to detect adulteration when it is in the powdered form, it is best purchased in mass. The best variety yields a brownish-yellow tincture, which acquires a purple tint upon the addition of nitric acid. A tincture which does not show this color reaction betrays an impure article, which should be rejected.

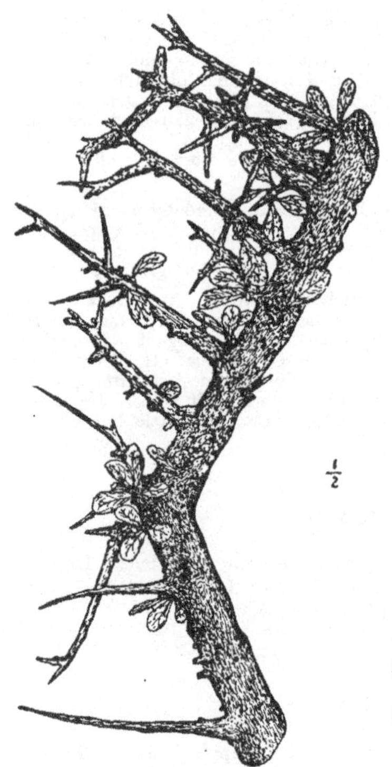

FIG. 156.—*Commiphora myrrha*—Branch.

Powder.—Microscopical elements of: See Part iv, Chap. I, B.

CONSTITUENTS.—A volatile oil, myrrhol (3 to 4 per cent.); a bitter principle; a **resin,** 35 per cent., and **gum,** 60 per cent., forming with water a yellowish or brownish emulsion, which deposits a sediment upon standing. Recent investigations of Tschirch and others, have cleared up many obscure points regarding the chemistry of the resins in such drugs as myrrh. An excellent classification of the resins is found in a volume entitled "Pharmacopedia," by White and Humphrey, London (pp. 400, 403) and in Allen's "Commercial Organic Analysis," (pp. 1–103, vol. iv, 4th edition). Myrrh of good quality should contain not more than 70 per cent. of matter insoluble in alcohol. Ash, not more than 8.5 per cent.

ACTION AND USES.—A stomachic, carminative, and emmenagogue. Used mostly in mouth-washes. Dose: 2.5 to 15 gr. (0.15 to 1 Gm.), in pills and emulsion.

OFFICIAL PREPARATIONS.

Tinctura Myrrhæ (20 per cent.),....... Dose: 10 to 60 ♏ (0.6 to 4 mils).
Pilulæ Rhei Compositæ.

295. **OLIBANUM.**—FRANKINCENSE. A gum-resin exuding from incisions into the bark of **Boswel'lia carte'rii** Birdwood. *Habitat:* Eastern Africa and Southern Arabia. In tears of various shapes, generally rounded; yellowish or pale brown, thickly covered with a white dust; fracture dull, waxy, pale yellowish or reddish; softens when chewed; odor agreeably aromatic, stronger on heating; taste terebinthinate, somewhat bitter, but not unpleasant. Contains a volatile oil, a gum resembling gum arabic, and a resin, forming with water a pure white emulsion. Rarely used medicinally; mostly used for fragrant fumigations and pastilles, and as an altar incense.

296. **BDELLIUM.**—A gum-resin obtained from **Commi'phora mu'kul** Hooker and from **C. africana** Engler. *Habitat:* (1) East India; (2) Western Africa. (1) Dusty pieces breaking with a dark brown, conchoidal fracture; translucent in thin sections; (2) irregular, dusty tears, breaking with a yellowish to brown-red, waxy, angular fracture. Contains resin, volatile oil, and gum. Odor and taste resemble myrrh. Used for the same purposes.

297. **ELEMI.**—MANILA ELEMI. An oleoresin exuding from incisions in **Cana'-rium commu'ne** (?) Linné. *Habitat:* Philippine Islands. A soft, unctuous substance, colorless when pure, becoming firmer and yellow with age; often contaminated with carbonaceous matter, which renders it grayish or blackish. It has a strong, pleasant odor, like lemon and fennel; taste bitter, disagreeable, and pungent. Contains volatile oil, resin, clemic acid, and breidin, a crystalline principle, soluble in water. Used in plasters and ointments as a stimulant and irritant.

MELIACEÆ

Tropical trees, rarely undershrubs, with mostly pinnately compound leaves. The order contains many plants which have acrid, bitter, and astringent properties. None official.

298. **MAREGAMIA ALATA.**—GOANESE IPECAC. (Root.) *Habitat:* Western India. Expectorant and emetic. Dose: 1 to 3 gr. (0.065 to 0.2 Gm.); as an emetic, 5 to 10 gr. (0.3 to 0.6 Gm.).

299. **COCILLANA, N.F.**—The bark of an undetermined species of **Guarea**, a large Bolivian tree. Expectorant and emetic properties similar to ipecac. Dose of fluidextract: 10 to 30 ♏ (0.6 to 2 mils). A popular compound expectorant, syrupy, preparation furnishes a much used remedial agent.

300. **AZEDARACH.**—MARGOSA BARK. The root-bark of **Me'lia azed'arach** Linné. *Habitat:* China and India; cultivated in Southern United States. Fibrous pieces about 5 mm. (⅕ in.) thick, and 50 to 75 mm. (2 to 3 in.) wide. The outer surface is reddish-brown, with irregular, blackish, longitudinal ridges. The inner surface is yellowish-white to brown, and striated longitudinally; fracture fibrous; inodorous; taste sweetish, acrid, and bitter. If collected from old roots, the bark must be freed from the corky layer. The active principle is a yellowish-white resin. Azedarach was once extensively used in the Southern States as an anthelmintic. Dose: 15 to 60 gr. (1 to 4 Gm.), in decoction.

POLYGALEÆ.—Milkwort Family

Plants often with milky juice in roots, low herbs in temperature regions, with *leaves* mostly simple, entire, dotted, exstipulate. *Flowers* irregular; *sepals* 5, the two inner large, petaloid, *petals* 3, the anterior one larger.   Properties: generally bitter (polygala), acrid (senega), or astringent (krameria).

Synopsis of Drugs from the Polygaleæ

A. *Roots.*   B. *Herb.*
    KRAMERIA, 301.   Polygala, 303.
    SENEGA, 302.

FIG. 157.—*Krameria triandra*—Flowering branch.

301. KRAMERIA, N.F.—Krameria

RHATANY

The dried root of **Krame'ria trian'dra** Ruiz et Pavon, and of **Krame'ria ixi'na Linné** and other undetermined species of Krameria.   (Fam. transferred to Krameriaceæ U.S.P. 1900.)

BOTANICAL CHARACTERISTICS.—A low, woody shrub, with grayish leaves and red flowers.   The *flowers* are solitary in the axils of the upper leaves, short-stalked.   The *fruit* is globular, leathery, indehiscent, about the size of a pea, and covered with reddish-brown, hooked prickles.

SOURCE.—*Krameria triandra* (Red rhatany) is a native of Peru, the commercial supply being obtained from the southern provinces; abundant about the cities of Huanuco and Lima; shipped from Paytu. *Krameria ixina* (Savanilla or New Granada rhatany) is yielded by several varieties, as *K. tomentcsa*. St. Hil., an extremely wooly form growing in Colombia, British Guiana, and Northern Brazil; shipped from Carthagena, Santa Marta, etc. Para rhatany, described by Berg, is said to be from *K. argentea;* grayish-brown color.

DESCRIPTION OF DRUG.—From 10 to 30 mm. (⅖ to 1⅕ in.) thick, knotty, and with several thick heads above, and branches below, from which emanate cylindrical roots about 6 to 12 mm. (¼ to ½ in.) thick and from 100 to 400 mm. (4 to 16 in.) long. In commerce the more woody pieces, with short stumpy branches, constitute the largest proportion; the bark is tough and fibrous, dark reddish-brown, scaly, rugged, and about 1 to 2 mm. (⅟₂₅ to ⅟₁₂ in.) thick; the wood is hard and compact, light reddish-brown in color, and when cut with a knife, presents a shining surface, marked with concentric circles and fine medullary rays. Inodorous; taste very astringent, the bark more so than the wood. *Krameria ixina* (Savanilla rhatany) is more slender and less knotty, dull purplish-brown, with smooth, closely adhering bark. The roots are less

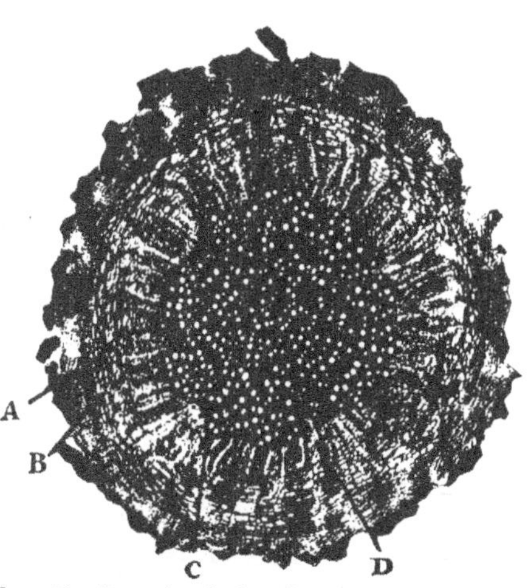

FIG. 158.—*Krameria triandra*—Cross-section of root. (18 diam.) A, Cork. B, Thin-walled parenchyma of cortex. C, Xylem. D, Medullary ray. (Photomicrograph.)

flexuous and less tapering than the Peruvian rhatany and are usually separate, not usually exceeding 12 cm. in thickness, externally purplish-brown or chocolate-colored and marked with numerous fissures, the fracture less tough than that of Peruvian rhatany, the bark and wood darker. The bark is more astringent than that of Peruvian rhatany. The yield of aqueous extract should not be less than 9 per cent. The yield of ash should not exceed 5 per cent.

Powder.—Deep red. Characteristic elements: Parenchyma cells of cortex with reddish-brown coloring-matter; starch grains, 20 to 30 μ in diam., 1 to 4 compound; calcium oxalate in prisms and pyramids; sclerenchyma with few short, thick-walled bast fibers. In Savanilla variety the sclerenchymatous fibers, the parenchyma, bast, and ducts, are larger.

CONSTITUENTS.—**Kramero-tannic** acid (20 per cent.), rhatanin, and rhatanic-red (a coloring matter). The tannic acid in a state of purity is perfectly colorless, but accompanying it is phlobaphene, an' extractive which gives its solutions a reddish-brown color. Gives a dark green precipitate with ferric salts, a flesh-colored precipitate, with gelatin, and none with tartar emetic. **Extracts of krameria should be made with cold water, the solution being** evaporated at a low temperature. Boiling water extracts apothem, the presence of which is a detriment to the astringent principle.

ACTION and USES.—A powerful astringent, with some tonic properties. Dose: 5 to 30 gr. (0.3 to 2 Gm.).

302. SENEGA.—SENEGA

SENEKA. SENEGA SNAKEROOT

The dried root of **Polyg'ala sen'ega** Linné.

BOTANICAL CHARACTERISTICS.—*Stems* several, from a thick and hard, knotty root-stock; *leaves* lanceolate, with rough margins; *calyx* with 3 *sepals*, small, greenish, and 2 larger (called wings), colored; *flowers* white, in a solitary, close spike.

SOURCE.—Almost all parts of the United States east of the Rocky Mountains. It is collected for market in Kentucky and in the states west and southwest of it, and in Wisconsin, and in immense quantities in northern Minnesota. This latter variety is known as northern senega. It is, as a rule, a larger root than the southern; the anatomical and structural differences between the two roots are probably very slight. *Polygala alba*, Nutt., inhabits Western Texas and Western Kansas, but this variety of senega is not systematically collected for the market as are the roots of Minnesota and Kentucky.

DESCRIPTION OF DRUG.—A contorted root, about 100 mm. (4 in.) long, with a **knotty crown bearing numerous remnants of scaly leaves.** The main root is from 5 to 10 mm. (⅕ to ⅖ in.) thick, fleshy, but void of starch. It varies in color from a light yellow to a dark brown externally; much-branched, the branches spreading, tortuous, longitudinally wrinkled, annulate near upper end; **bark thickish,** inclosing a porous, yellowish wood, but easily separable from it; it consists of three layers, the inner one excessively **developed on one side, forming a prominent cord or keel on drying,** fracture short when dry. Odor faint, sometimes wintergreen-like; taste sweetish, afterward acrid and nauseating. The liquid preparations of it have a characteristic nauseous odor.

Powder.—Characteristic elements: See Part iv, Chap. I, B.

CONSTITUENTS.—The acrid principles to which its medicinal action is entirely due, are **polygalic acid,** $C_{19}H_{30}O_{10}$, and **senegin,** $C_{17}H_{26}O_{10}$—two homologues. The distinction between polygalic acid and senegin is mainly one of solubility in alcohol (the former more soluble). **Lead** acetate precipitates polygalic acid, but does not precipitate senegin. The root also contains a fixed oil, and a small quantity of volatile oil, which is a mixture of valerianic ether and methyl salicylate, resin, malic acid, and sugar.   Liquid preparations of senega are apt to become gelatinous, which is ascribed to the presence of pectin compounds; but is very likely, at least in part, due to sapogenin, generated under the influence of acids or other compounds; the jelly is rendered

FIG. 159.—*Polygala senega*—Plant and rhizome.

soluble again on the addition of an alkali.   The above proximate principles are similar to the saponins.   Ash, not exceeding 5 per cent.

ACTION AND USES.—A valuable stimulating expectorant, for which it is generally used; also diuretic, and in large doses emetic and cathartic. It affects the heart like digitalis.   Dose: 10 to 30 gr. (0.6 to 2 Gm.).

OFFICIAL PREPARATIONS.

Fluidextractum Senegæ,..............Dose: 10 to 30 ♏ (0.6 to 2 mils).
  Syrupus Senegæ(20 per cent. of fl'ext.),   30 to 60 ♏ (2 to 4 mils).

Syrupus Scillæ Compositus (Fl'ext.
  senega 8 per cent., Fl'ext. squill 8 per
  cent., Tartar emetic, 0.2 per cent.),   10 to 60 ♏ (0.6 to 4 mils).

303. **POLYGALA RUBELLA** Willdenow.—Bitter Polygala. A North American herb, used for its tonic properties. The bitter principle is easily extracted by water and alcohol.

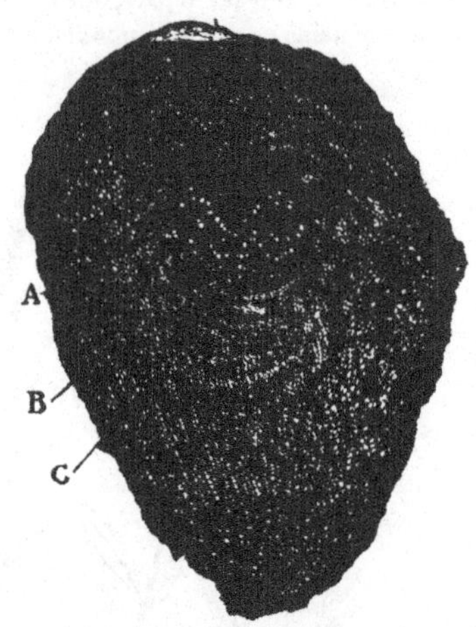

Fig. 160.—Senega—Cross-section of Root. (21 diam.) A, Xylem. B, Parenchyma of cortex.
              C, Cork. (Photomicrograph.)

EUPHORBIACEÆ.—Spurge Family

Herbs, shrubs, or trees, usually with an acrid, milky juice, which in some
cases yields rubber. A volatile oil is found in the bark of a few species, and a
fatty oil is found abundantly in the seeds of other plants, as *tiglium* and *ricinus*.

Synopsis of Drugs from the Euphorbiaceæ

A. *Roots.*
    STILLINGIA, 304.
    Euphorbia, 305.
    Euphorbia Corollata,
      305 a.
    Euphorbia Ipecacu-
      anha, 305 b.
B. *Herbs.*
    Euphorbia Pilulifera,
      305 c.
    Mercurialis, 308.

C. *Gum-resins.*
    Euphorbium, 306.
D. *Concrete Juices.*
    Alveloz Milk, 307.
    ELASTICA, 309.
E. *Resin.*
    Lacca, 310.
F. *Bark.*
    *Cascarilla, 311.

G. *Seeds.*
    Ricinus, 312.
    Tiglium, 313.
    Curcas, 314.
H. *Fixed Oils.*
    OLEUM RICINI,
      312 a.
    OLEUM TIGLII,
      313 a.
I. *Glands.*
    Kamala, 315.

304. STILLINGIA.—Stillingia

QUEEN'S ROOT.   QUEEN'S DELIGHT

The dried root of Stillin'gia sylvat'ica Linné.

Botanical Characteristics.—*Stem* herbaceous, 1 to 3 feet high. *Leaves* alternate, nearly sessile, oblong-lanceolate, finely serrate. *Flowers* monœcious,

in a terminal spike (the fertile flowers at the base), with saucer-shaped glands at the base of each; *stamens* 2 or 3; *style* 1; *stigmas* 3. *Capsule* 3-celled, 3-lobed, 3-seeded.

HABITAT.—United States, from Virginia to Florida, in sandy soil.

DESCRIPTION OF DRUG.—A subcylindrical root, 300 mm. (1 ft.) long, 25 to 50 mm. (1 to 2 in.) or more thick, slightly tapering and sparingly branched; compact; fracture fibrous; odor distinct, peculiar,

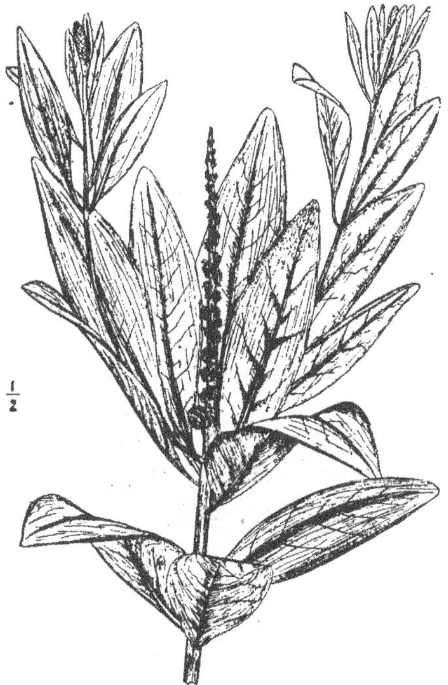

FIG. 161.—*Stillingia sylvatica*—Branch.

stronger and disagreeable when fresh; taste bitterish and pungent, persistently acrid.

The color of the exterior surface varies considerably, due, probably, to the varied character of the soils in which the plants grow. Roughly speaking, the roots would thus be classified into *light* and *dark* stillingias. By the accidental removal of their outer bark the pinkish inner bark is exposed. Transversely the woody cortex is seen to occupy about one-half of the diameter of the root. Around this is disposed the thick bark containing numerous bast fibers separately imbedded in the parenchyma. The cambium line is composed of distinctly marked flat cells. Woody center radiate, through which numerous tracheids, arranged in four or five radiating rows that are quite regular in their disposition.

Powder.—Characteristic elements: See Part iv, Chap. I, B.

CONSTITUENTS.—The active principle has not yet been determined; it is probably a volatile principle, as old roots are nearly inert. An acrid resin (sylvacrol, soluble in alcohol and chloroform, insoluble in benzene), volatile oil, fixed oil, resin, starch, tannin, and gum have been separated. The so-called oil of stillingia, as found in the market, is intended to be the ethereal extract, but sometimes possesses very little of the persistent acrimony of the root. Ash, not to exceed 5 per cent.

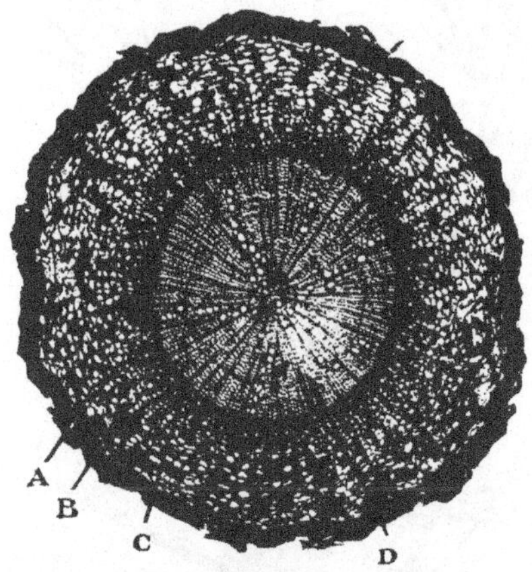

FIG. 162.—Stillingia—Cross-section of root. (17 diam.) *A,* Cork. *B,* Parenchyma of cortex. *C,* Medullary ray. *D,* Xylem. (Photomicrograph.)

ACTION AND USES.—An efficient **alterative** and antisyphilitic, usually given in combination, often with sarsaparilla, but more generally in the compound syrup of stillingia. Dose: 15 to 30 gr. (1 to 2 Gm.).

OFFICIAL PREPARATION.

Fluidextractum Stillingiæ,..............Dose: 15 to 20 ♏ (1 to 2 mils).

305. **EUPHORBIA.**—There are a number of species of this genus yielding medicinal products:

305 a. **EUPHORBIA COROLLATA** Linné.—LARGE FLOWERING SPURGE. (Root.) Long, branched; externally purplish-black, wrinkled; internally whitish or yellowish. The medical virtues reside in the very thick, internally whitish bark, which constitutes about two-thirds of the whole root. Inodorous; taste sweetish, somewhat bitter and acrid. Emetic in doses of 10 to 20 gr. (0.6 to 1.3 Gm.); diaphoretic, expectorant, and cathartic in smaller doses.

305 b. **EUPHORBIA IPECACUANHA.**—IPECACUANHA SPURGE. (Root.) Has medical properties similar to the above. It is of a light brown color externally, with a thick bark inclosing a yellowish or whitish wood. The action of these two drugs is due to a resinous matter. Both are indigenous.

305 c. **EUPHORBIA PILULIFERA, N.F. Linné.**—A common herb along the road-sides in Australia, where it enjoys a great reputation for the prompt and complete relief it gives in asthma and pectoral complaints generally. Dose: 15 to 60 gr. (1 to 4 Gm.).

306. **EUPHORBIUM.**—EUPHORBIUM. A gum-resin exuding from one or more undetermined species of **Euphorbia,** ascribed to some leafless, cactus-like plants of Egypt, Arabia, and the East Indies. It occurs in dull brownish-yellow or reddish, rounded pieces of the size of a pea or larger, often pierced with, or inclosing, the spines around which it has hardened on the stem of the plant; almost inodorous, the powder sternutatory; taste mild at first, but afterward intensely acrid and burning. Only used externally, mostly in veterinary practice as a vesicant.

307. **ALVELOZ MILK.**—The milky juice of a Brazilian lan , **Euphor'bia hete-rodox'a** Müller. It has an action resembling that of papain, and is used in eating out cancerous and other ulcers.

308. **MERCURIALIS ANNUA** Linné.—MERCURY WEED. A European herb, employed from the most ancient times as a purgative and emmenagogue.

309. ELASTICA.—INDIA-RUBBER (U.S.P. VIII)

CAOUTCHOUC

The prepared milk-juice of **He'vea Braziliensis** Mueller and other species, known in commerce as Para rubber. Large trees containing a milky juice which, on hardening, forms india-rubber. **Ficus elastica,** producing the greatest quantity, has its seeds germinate in the forks of the tree, giving off aërial roots which descend to the ground and form a great many trunks.

HABITAT.—South America and India, the finest quality coming from Brazil.

DESCRIPTION.—Large, flat pieces, or molded into various shapes—balls, hollow, bottle-shaped pieces, etc. When the juice first hardens it is yellowish-brown externally and yellowish-white within, but in the processes of molding and drying it acquires a smoky, blackish appearance; very elastic; odor peculiar. Insoluble in water and alcohol, but soluble in chloroform, carbon bisulphide, and benzin. The common adulterants are the carbonates of zinc and lead; when pure or nearly pure, india-rubber should float in water.

CONSTITUENTS.—The elastic principle has been termed caoutchoucin; it, or a similar principle, is contained in a great number of milky-juiced plants.

USES.—On account of its insolubility it has no therapeutic application, but is extensively used in the arts. Employed in some of the pharmaceutical plasters, *e.g.*, Emp. Elasticum. U.S. IX.

310. **LACCA.**—LAC. GUM-LAC. A resinous exudation from punctures, made by insects, in the bark of several East Indian trees, and also in plants growing in Arizona and other Western States. The twigs, with their deep reddish-brown incrustations, are called **stick-lac. Seed-lac** consists of the small, irregular fragments broken off from the twigs. **Lump-lac** is made by melting the stick-lac, and, after it has hardened, breaking the brown, translucent mass into lumps. **Shell-lac** or gum-shellac, the most common form, is prepared by spreading the melted lac out in thin layers, which, on drying, form thin, brittle sheets, glossy, more or less transparent, varying from amber to dark brown in color; in packing, these sheets are broken into fragments, in which form shellac is commonly met with in market; odorless and tasteless. Lac contains several resins, laccin (a peculiar principle insoluble in alcohol), and a coloring matter varying in quantity in the different forms; this coloring matter, "lac dye," is equal to cochineal dyes: it is soluble in water, being obtained from the washings in making the different forms of lac. Lac is not used medicinally, but is extensively employed in the arts for making varnishes and sealing-wax.

311. CASCARILLA, N.F.—CASCARILLA BARK. The bark of **Cro′ton elute′ria** Bennet. Small broken quills having a grayish fissured cork, more or less covered with white lichen patches, but often pa all or wholly removed, showing the dull brown inner bark; inner surface smooth; bast fibers few; fracture short, resinous; odor·feeble, stronger when rubbed; when ignited, it emits a strongly aromatic odor, somewhat resembling musk, but weaker and more agreeable; taste warm, aromatic, very bitter. Copalchi bark (see also Aspidosperma, 353) has a cascarilla-like odor, and melambo bark, from *Croton Melambo*, Venezuela, and other species of Croton, are similar to cascarilla. *Constituents:* Volatile oil (1.5 to 3 per cent.); **cascarillin** (a

FIG. 163.—*Croton eluteria*—Branch.

bitter crystalline principle), tannin, fat, resin, etc. Aromatic, stimulant, and tonic. Once used as a febrifuge as a substitute for cinchona. Dose: 15 to 30 gr. (1 to 2 Gm.).

312. RICINUS.—CASTOR-OIL SEED. The seeds of **Rici′nus commu′nis** Linné (Palma Christi), a herbaceous plant about 4 to 6 feet in height, native to India, but cultivated in tropical and warm temperature countries; stems hollow, purplish-red; leaves large, palmately 9-divided, on long petioles, with glands at the apex of the petiole; flowers monœcious, in terminal panicles, the lower ones male, the upper female; male flowers—stamens numerous; female flowers—style 1, stigma 3, colored red; capsule covered with prickles, 3-celled, each cell containing one seed.

The seeds are about the **size of a bean,** oval-oblong, flattened on one side; at one end is a yellowish caruncle from which runs an obscure, longitudinal ridge (raphé) to the opposite end; externally smooth, of a **glossy grayish** color, **mottled with reddish-brown** from the removal, in places, of the thin, white pellicle investing the black, brittle testa. Embryo and albumen very oily; cotyledons broad and veined. Inodorous; taste sweetish, then acrid. They contain a **fixed oil,** 45 to 50 per cent. (Oleum Ricini),

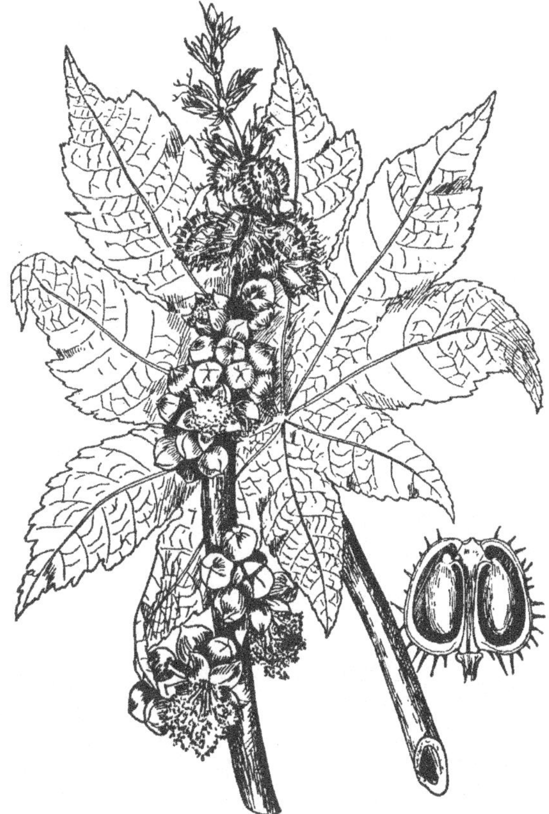

FIG. 164.—*Ricinus communis*—Flowering stem, leaf, and section of fruit.

and a poisonous principle, ricin, which is left behind in the extraction of the oil, some cases of poisoning have occurred from the ingestion of the whole seeds, symptoms are violent gastroenteritis and collapse. They are more active, weight for weight, than the oil.

312 a. **OLEUM RICINI.**—CASTOR OIL. The commercial fixed oil is extracted in several ways, the finest product being yielded by the process known as cold expression. It is a thick, viscid, transparent liquid with a feeble odor, and a mild, somewhat acrid and nauseous taste, soluble in its own weight of strong alcohol. On standing, it becomes

thicker, and deposits a white, crystalline substance. Ricinolein (the glyceride of ricinoleic acid) constitutes the bulk of castor oil, with small quantities of palmitin, stearin, myristin and an acrid principle. A mild and efficient **cathartic.** Dose: ¼ to 2 fl. oz. (8 to 60 mils). Formerly employed in making flexible collodion, 3 per cent.

ADMINISTRATION.—Various methods of administration to hide the nauseating taste have been devised. The three-layer method in which the

FIG. 165.—*Croton tiglium*—Flowering branch; flower (enlarged); seed, entire and in section (enlarged).

oil is suspended between two layers of flavored watery or alcoholic liquid, is the favorite. For this purpose compound tincture of cardamon, spirit of peppermint, whisky, orange juice, lemon juice, lemonade or beer may be used. "The layers should *not* be stirred together." The favorite drug store method is to place some syrup of sarsaparilla in a glass and cause it to foam by adding carbonated water from the soda fountain or by a little tartaric acid and sodium carbonate. Then the oil is poured in without allowing any to get on the edge of the tumbler. "The mixture must not be stirred." The oil floats between some of the syrup below and foam above, and the whole is drunk without stopping. The oil is not tasted at all. The principle of these methods is to have the mouth and tongue moistened with a pleasant flavored liquid (the top layer), upon which the oil will readily slip down. For infants and children, an emulsion made with acacia and flavored syrup may be employed.—Bastedo.

313. **TIGLIUM.**—CROTON SEED. The seed of **Cro'ton tig'lium** Linné, a small tree indigenous to China, but extensively cultivated in India. The fruit is a smooth capsule about the size of a large hazelnut, 3-celled, each containing a single seed. **The seeds** are from 12 to 15 mm. ($\frac{1}{2}$ to $\frac{3}{5}$ in.) long, oval-oblong, resembling castor-oil seeds in shape but somewhat larger, and **quadrangular,** and with the caruncle usually absent; the testa is soft, dull yellowish-brown, generally partially, but occasionally wholly, rubbed off from the black tegmen by friction, giving the seeds a **mottled** or **nearly** black appearance; albumen and embryo yellowish-brown; odor feeble; taste acrid. It yields about 50 to 60 per cent. of an **acrid fixed oil.**

313 a. **OLEUM TIGLII.**—CROTON OIL. A rather viscid, pale yellowish to brown fixed oil, with a peculiar, faint odor, and an exceedingly hot, acrid taste, continuing in the mouth for several hours. It consists of the glycerides of lauric, myristic, palmitic, stearic, formic, acetic, crotinic, $C_4H_6O_2$, isobutyric, isovalerianic, and tiglinic, $C_5H_8O_2$, acids. Saponification value 200 to 215, iodine value 104 to 110. The vesicating properties are due to a croton resin. Purgative principle is insoluble in alcohol. **Drastic purgative,** capable of causing death in excessive doses. Dose: $\frac{1}{2}$ to 2 ♏ (0.0324 to 0.13 mil), in emulsion. Applied externally in liniment, it is a powerful rubefacient.

314. **CURCAS.**—PURGING NUTS. The seeds of **Cur'cas pur'gans** Adanson. *Habitat:* Brazil, West Indies, and Africa. They resemble croton seeds, but have a dull black, fissured surface and are somewhat milder in action. The purgative principle is ricinoleic acid; they also contain about 40 per cent. of an acrid, colorless fixed oil.

315. **KAMALA.**—ROTTLERA. The glands and hairs from the capsule of **Mallo'- tus philippinen'sis** Mueller Arg. Official U.S.P. 1890. A brick-red, mobile, finely granular powder, almost odorless and tasteless, with a gritty feeling between the teeth; excessive grittiness, however, indicates a probable adulteration with earthy matter, which may be detected by floating it in water. It is inflammable, flashing up like gunpowder, with a red flame. Under the microscope the powder is seen to consist of depressed globular, transparent sacs, containing numerous red, hood-shaped vesicles, and mixed with colorless hairs. Almost insoluble in water; soluble in alcohol, imparting a deep red color to the solution, from which water throws down a resinous precipitate. *Flemingia rhodocarpa* Baker or Warrus, a leguminous plant indigenous to Eastern Africa, has been employed as substitute. The powder is coarser than kamala, is deep purple, in a water-bath becomes black, and has a slight odor. The glands are cylindrical or subconical. *Constituents:* Resins (supposed to be the active principle) and resinous coloring matters, one of which has been isolated and termed **rottlerin,** $C_{22}H_{20}O_6$. Vermifuge. Dose: 1 to 2 dr. (4 to 8 Gm.).

Preparation of Rottlerin.—Obtained by exhausting with ether or carbon disulphide, evaporating and crystallizing; occurs in yellowish needles; soluble in hot alcohol, ether, benzene, or carbon disulphide; changes on exposure.

19

ANACARDIACEÆ.—Cashew Family

Trees or shrubs with gummy, milky or resinous juice, often poisonous. *Leaves* usually compound. *Fruit* drupaceous, not infrequently having a strong turpentine odor and taste. The *seeds* of many species yield an abundance of bland oil. Drugs from the order: *Rhus Toxicodendron*, 316;* *Rhus Glabra*, 317; *Rhus aromatica*, 318; *Mastiche*, 319; *Terebinthina Chia*, 320; *Anacardium*, 321; *Semecarpus*, 322.

316. **RHUS TOXICODENDRON.**—Poison Ivy. Poison Oak. The fresh leaves of **Rhus rad'icans** Linné. Off. U.S.P. 1890. The leaves are trifoliate, the terminal leaflet ovate, stalked, the lateral ones sessile, obliquely ovate. These leaflets are about 100 mm. (4 in.) in length, with margins entire, or coarsely toothed or indented; odorless; taste bitter, acrid, and astringent. The dried leaves are brittle and papery, of a pale green color. *Constituents:* The fresh leaves abound in an acrid, milky juice, which blackens on exposure to the air, and in contact with the skin causes inflammation and swelling. The acridity is due to what was formerly termed toxicodendric acid, the vapor of which was said to be the cause of vesicular eruptions, but this principle has been found to be, by Pfaff and Balch, an oil, which was given the name, "toxicodendrol." It is said by some authorities (Bessey) that it is volatile. A. B. Stevens shows the principle to be a resin, soluble in a mixture of ether and alcohol, which solvent removes completely the poison from the parts affected. Bessey has shown by test upon himself that, to sensitive persons, the poison may be communicated without handling the plant, and concludes that the principle is volatile. They also contain tannin producing greenish precipitates with iron salts, wax, fixed oil, resin, etc.

Preparation of Toxicodendric Acid.—To bruised leaves add $Ca(OH)_2$; macerate with water; express; add H_2SO_4; distil. The condensed vapor is a very acrid liquid (see above), which causes the characteristic vesicular eruption of ivy-poison.

Local irritant and rubefacient. Used in treatment of eczema, but is no longer in vogue. Dose: 1 to 5 gr. (0.065 to 0.3 Gm.).

317. RHUS GLABRA, N.F.—Rhus Glabra

SUMAC

The dried fruit of **Rhus gla'bra** Linné.

Description of Drug.—Berries (drupes) about 3 mm. (⅛ in.) in diameter, densely covered with a dark-red down. The sarcocarp (the outer portion of a stone fruit) is composed of two layers, the outer being crimson, and the inner whitish; putamen (stone) flattish, ovoid, smooth. Inodorous; taste acidulous and astringent.

Powder.—Dark reddish-brown. Characteristic elements: Thick-walled cells of testa, porous; many celled trichomes deep red in color; seldom dispensed as powder. •

Constituents.—The acidity of the fruit is due to the acid calcium and potassium malates present; there are also tannic and gallic acid, a red coloring-matter, etc.

Action and Uses.—Astringent and refrigerant. Used as a gargle in the form of decoction or fluidextract. Dose: 30 gr. (2 Gm.).

Official Preparation.

Fluidextractum Rhois Glabræ, (U.S.P. VIII). Dose: 1 to 2 fl. dr. (4 to 8 mils).

318. **RHUS AROMATICA** Aiton (Var. Trilobata Gray).—Sweet Sumach. An indigenous bush, with leaves smaller than those of *R. glabra*, and un-

pleasantly scented. (Root-bark.) It acts as an excitant to the unstriped muscular fiber, particularly of the bladder, and is therefore an efficient remedy in incontinence of urine. Dose: 5 to 30 gr. (0.3 to 2 Gm).

319. MASTICHE, N.F.

MASTIC

A concrete resinous exudation from Pista'cia lentis'cus Linné. A shrub about 12 feet high. *Fruit* a small, roundish drupe, brownish-red, produced chiefly in the island of Scio.

FIG. 166.—*Pistacia lentiscus*—Branch.

DESCRIPTION OF DRUG.—A handsome-appearing resin, globular, somewhat elongated, yellowish, translucent tears about the size of a pea, brittle, and dusty from powder derived from attrition; plastic when chewed; odor balsamic; taste slight turpentine-like and faintly bitter. Soluble in ether and nearly so in alcohol.

CONSTITUENTS.—Volatile oil 1 to 2 per cent., and two resins, mastichic acid (alpha-resin), soluble in alcohol, and masticin (beta-resin), insoluble in alcohol, but soluble in ether.

ACTION AND USES.—Mild stimulant, but rarely used internally. Dose: 30 gr. (2 Gm.). Used as a filling for carious teeth, and for making paints, varnishes, etc., and formerly official in **Pilulæ Aloes et Mastiches.**

320. **TEREBINTHINA CHIA.**—CHIAN TURPENTINE. An oleoresin from **Pista'cia terebin'thus** Linné, a tree growing on the island of Scio. Incisions are made and the exuding juice is allowed to fall upon smooth stones. It is a greenish-yellow, pellucid, syrupy liquid, hardening to a transparent mass when exposed by the evaporation of its volatile oil; odor fennel-like; taste bitterish.. It is used for destroying cancerous growths in which it is claimed to be very efficient. Dose: 5 to 20 gr. (0.3 to 1.3 Gm.), in emulsion.

321. **ANACARDIUM.**—CASHEW NUT. The fruit of **Anacar'dium occidenta'le** Linné. *Habitat:* North America. Kidney-shaped, about 25 mm. (1 in.) long, invested with a grayish-brown, finely punctate pericarp containing cardol (a reddish-yellow fixed oil, very active and poisonous). The seed is white and consists principally of a bland fixed oil. Vermifuge and escharotic.

322. **SEMECARPUS.**—ORIENTAL CASHEW NUT. The fruit of **Semecar'pus anacar'dium** Linné, growing in Eastern India, a heart-shaped, somewhat flattened nut, about 20 mm. (⅘ in.) long, invested with a blackish-brown pericarp containing a brown, acrid, vesicating oil. Used as a local irritant.

ILICINEÆ.—Holly Family

Trees and shrubs indigenous to tropical and temperate climates. Leaves coriaceous, evergreen.

323. **ILEX OPACA** Aiton.—HOLLY. (Leaves.) Petiolate, about 50 mm. (2 in.) long, leathery, smooth; inodorous; taste mucilaginous, bitter, and astringent. They contain a bitter principle, ilicin, and tannin. Demulcent, tonic, and emetic. Dose: 15 to 30 gr. (1 to 2 Gm.).

324. **ILEX PARAGUAYENSIS** Lambert.—PARAGUAY TEA. (Leaves.) *Habitat:* Brazil and Argentine Republic. Lance-oblong, about 50 mm. (2 in.) long, on a short petiole; surface smooth; margin few-toothed. The maté of the market is a coarse, dark powder, slightly roasted, with a tea-like odor and a bitter, astringent taste. Contains caffeine, giving it properties differing only slightly from tea, for which it is used as a substitute by the natives.

325. **PRINOS.**—BLACK ALDER. WINTERBERRY. The bark of **I'lex verticilla'ta** Gray. *Habitat:* North America, in swampy thickets. Thin, yellowish-green fragments, usually deprived of the grayish or brownish periderm, which, when present, is marked with whitish patches and black lines and dots; inodorous; taste bitter and slightly astringent. It contains tannin, wax, sugar, resin, starch, chlorophyll, and a yellow, amorphous, bitter principle. Used as a tonic, antiperiodic, and astringent. Dose: 15 to 60 gr. (1 to 4 Gm.).

CELASTRINACEÆ.—Staff-tree Family

Small trees and shrubs, sometimes climbing. *Leaves* alternate, rarely opposite, often coriaceous. A peculiarity of the *flowers* is that the perigynous stamens are inserted on the disk which fills the bottom of the calyx and sometimes covers the ovary. *Fruit* a capsule, an indehiscent drupe, or a samara. *Seeds* furnished with a pulpy, colored, cupular aril.

326. EUONYMUS, N.F.—EUONYMUS

WAHOO

The dried bark of the root of **Euon'ymus at.opurpu'reus** Jacquin.

BOTANICAL CHARACTERISTICS.—Tall, ornamental shrub, 6 to 14 feet high; *leaves* petiolate, oval-oblong; *flowers* dark purple, in fours; *pods* smooth, deeply lobed; *seeds* inclosed in a red aril.   Ornamental in autumn from its copious crimson fruit, drooping in long peduncles.

DESCRIPTION OF DRUG.—In quilled or curved pieces about 2 mm. ($\frac{1}{12}$ in.) thick. The **periderm is of an ash-gray color,** covered with blackish patches **or ridges,** and removable in scales from the whitish **or yellowish-brown inner bark;** fracture, smooth and short.   It contains a hygroscopic tissue, which readily absorbs moisture, thus becoming less brittle; odor distinct; taste sweetish, bitter and somewhat acrid.   It is sometimes mixed with branches and pieces of the wood.

Powder.—Light brown.   Characteristic elements: Sclerenchyma consisting of long, thin-walled bast fibers; ducts and wood fibers sometimes present; spherical starch grains and rosette-shaped calcium oxalate crystals also present.

CONSTITUENTS.—Its chief constituent of therapeutic value, **euonymin,** is bitter, amorphous, and precipitated from its solution by phosphomolybdic acid and lead subacetate.   This product is not to be confounded with a resinoid of the same name (see below).   The bark also contains atropurpurin, asparagin, cuonic acid, fixed oil, and albumen.

Preparation of Euonymin.—Add chloroform to a dilute alcoholic tincture and shake; separate chloroformic solution and evaporate; treat residue with ether, then alcohol, and lead acetate; add H_2S to precipitate lead; finally evaporate.   Soluble in ether, alcohol, and water.   The eclectic resinoid, by this name, is a dried precipitate, resulting when concentrated alcoholic tincture is added to water.

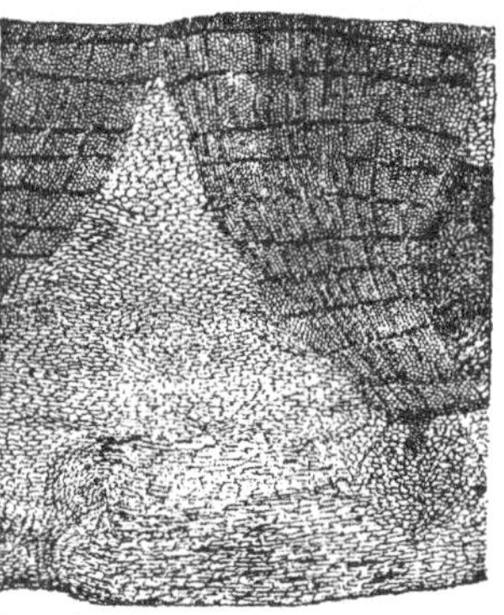

FIG. 167.—Cross-section of Wahoo bark. Magnified 15 diam.

ACTION AND USES—A cholagogue cathartic in doses of 0.8 to 30 gr. (0.5 to 2 Gm.); also tonic and laxative.

OFFICIAL PREPARATIONS.

Extractum Euonymi (From Fl'ext.), ... Dose: 1 to 5 gr. (0.065 to 0.3 Gm.)
Fluidextractum Euonymi,............. $\frac{1}{2}$ to 2 fl.dr. (2 to 8 mils).

327. **CELASTRUS SCANDENS** Linné.—CLIMBING STAFF-TREE.   FALSE BITTER-SWEET.   *Habitat:* North America.   (Root-bark.)   Alterative, diaphoretic, diuretic, and emetic; has been used in chronic affections of the liver. Dose of fluidextract: 1 to 2 fl. dr. (4 to 8 mils).

328. **ACER RUBRUM** Linné (Aceraceæ).—RED OR SWAMP MAPLE.   The bark of this indigenous maple was the favorite remedy of the Indians for sore eyes; it is a mild astringent.

SAPINDACEÆ.—Soapberry Family

Trees or shrubs, rarely herbs.   Stem with watery juice, erect or climbing. The members of the order are called soapworts because of the *fruit of many species containing a saponaceous principle.*   The flowers are unsymmetrical, racemed, or panicled, the pedicels often changed into tendrils.   The order furnishes a variety of dissimilar products, as will be seen in *Guarana,* 329; *Æsculus glabra,* 330; *Æsculus hippocastanum,* 331; *Acer rubrum,* 328; and *Macassar oil,* 332.

329. GUARANA

GUARANA

A dried paste consisting chiefly of the crushed or pounded seeds of **Paullin'ia cupan'a**
Kunth, yielding, by the official process, 4 per cent. of caffeine.

BOTANICAL CHARACTERISTICS.—A climbing shrub with alternate, imparipinnate
leaves on long stalks, with five oblong-oval, irregularly sinuate-dentate
leaflets 5 to 6 in. long and 2 to 3 in. broad, contracted into a shortly attenuated blunt point. *Flowers* in axillary spicate panicles. *Fruit* ovoid or
pyriform, about the size of a grape, with a short, strong beak, and six longitudinal ribs. *Pericarp* thin, leathery, hairy inside, inclosing lenticular,
thorny seeds resembling small horse-chestnuts, and each invested with an
easily removed, flesh-colored aril.

HABITAT.—Brazil.

DESCRIPTION OF DRUG.—In **cylinders, cakes,** or **balls** of a dark reddish-
brown color, not infrequently met with in the form of a light **reddish-**
brown powder. In preparing the cylinders, etc., above referred to,
the seeds deprived of arilode (papery shell) of the plant are first
roasted, then ground, kneaded with water in a heated mortar into
a pasty and pliable dough, made into forms, and dried. The forms
thus made break with an uneven fracture, black-mottled from fragments of seeds. The drug has a peculiar characteristic chocolate-like
odor and a bitter, astringent taste afterward sweetish. Guarana
constitutes the habitual beverage of thousands of people in the
Amazon valley.

Powder.—Characteristics: See Part iv, Chap. I, B.

CONSTITUENTS.—Tannic acid, not precipitated by tartar emetic or copper,
gum, albumin, starch, a trace of volatile oil, saponin, a greenish
fixed oil, and **guaranine,** an alkaloid identical with caffeine or theine.
Of this it contains a much larger percentage as compared with other
caffeine-yielding drugs. For example, good black tea gives an average yield of 2.13 per cent.; coffee, 1 per cent.; Paraguay tea (324), 1.2
per cent., and guarana, 4.5 per cent.

Preparation of Guaranine.—Treat the powder with boiling water. Evaporate the decoction on a water-bath to dryness, and exhaust the residue with chloroform. Distil off chloroform, treat residue with boiling water, filter, and evaporate
the liquid to obtain caffeine (guaranine). Tea and kola can be treated in the same
way for their active constituents.

ACTION AND USES.—Stimulant, especially beneficial in nervous headache,
and used like tea, coffee, and other drugs containing caffeine-like
principles. Dose: 15 to 60 gr. (1 to 4 Gm.).

OFFICIAL PREPARATION.

    Fluidextractum Guaranæ,..   .Dose: 15 to 60 ♏ (1 to 4 mils).

330. ÆSCULUS GLABRA Willdenow.—OHIO BUCKEYE. (Bark.) It has an
especial action on the portal circulation and the liver, and promotes the
biliary secretions.   Dose of fluidextract: 3 to 5 ♏ (0.2 to 0.3 mil).

331. ÆSCULUS HIPPOCASTANUM Linné.—HORSE-CHESTNUT. (Bark and
Fruit.)   *Habitat:* Asia; cultivated as an ornamental tree in Europe and
North America.   The bark contains a bitter glucosid, esculin, isomeric
with quinovin in cinchona bark, for which it is used as a substitute in Europe.
It is tonic, astringent, antiperiodic, narcotic, and antiseptic.   The nuts have
a similar action, but in addition are antispasmodic, used chiefly in neuralgic
affections.   The administration of the fluidextract has been recently recommended as a palliative in hæmorrhoids.   Dose of bark: ½ to 2 dr. (2 to 8 Gm.);
of the nuts: 5 to 15 gr. (0.3 to 1 Gm.), generally in fluidextract.

Preparation of Esculin.—Precipitate a decoction of the bark with lead acetate,
treat the filtrate with H_2S, evaporate and recrystallize.

332. MACASSAR OIL.—A fixed oil expressed from the seeds of **Schlerche'ra
triju'ga** Willdenow, a small East Indian tree which is also a source of lac.
This oil has a great reputation in its native country as a stimulating application to promote the growth of the hair, and also as a remedy in skin diseases, especially eczema.

RHAMNACEÆ.—Buckthorn Family

Shrubs or small trees with simple leaves; branches somewhat spinescent.
Flowers somewhat diœcious.   *Fruit* an indehiscent, fleshy, winged drupe, with a
hard, woody endocarp, or a pod not arilled.

333. FRANGULA.—FRANGULA

BUCKTHORN

The dried bark of **Rham'nus fran'gula** Linné, collected at least one year before
using.

BOTANICAL CHARACTERISTICS.—An elegant arborescent shrub, known as the
berry-bearing alder.   *Leaves* entire, with about 7 pairs of nearly opposite
parallel veins.   *Flowers* perfect, style simple; the fleshy berry is round, red,
and on ripening becomes black and juicy.

HABITAT.—Europe and Northern Asia.

DESCRIPTION OF DRUG.—Quilled, about 1 mm. (¹⁄₂₅ in.) thick; outer surface grayish-brown, or blackish-brown, with numerous small, whitish,
transversely-elongated lenticels and occasional patches of foliaceous
lichens; inner surface smooth, pale brownish-yellow; fracture in the
outer layer short, of a purplish tint; in the inner layer fibrous and pale
yellow; when masticated, coloring the saliva yellow; odor distinct;
taste sweetish and bitterish.

Medullary rays not converging at the outer ends (distinction from
Rhamnus Purshiana); stone cells absent (distinction from *Rhamnus
Purshiana* and *Rhamnus Californica*).

Powder.—Characteristic elements: See Part iv, Chap. I, B.

CONSTITUENTS.—Frangulin, or rhamno-xanthin, $C_{20}H_{20}O_{10}$, is a crystalline, lemon-yellow, odorless, tasteless glucoside; and emodin, a red-

dish principle, exists in the old bark; these develop by age. Two products are obtained from frangulin by hydrolysis—emodin, $C_{15}H_{10}O_5$, and rhamnose, $C_6H_{12}O_5$. Frangula-emodin differs from the rhubarb-emodin in melting-point, and in some color reactions. Senna and aloes also contain an isomeric emodin. (See *Rhamnus Purshiana*.) Ash, not exceeding 6 per cent.

FIG. 168.—*Rhamnus frangula*—Branch.

Preparation of Frangulin.—Macerate the bark for four days in carbon disulphide. Evaporate; exhaust residue with alcohol; evaporate alcoholic solution to dryness; crystallize from ethereal solution. Forms sublimable yellow crystals; becomes purple when treated with alkalies. Dyes cotton, silk, wool, etc., yellow.

ACTION AND USES.—A mild laxative or cathartic, acting like senna and often used in its stead. Dose: ½ to 2 dr. (2 to 8 Gm.).

OFFICIAL PREPARATION.

Fluidextractum Frangulæ, . . . Dose: ⅓ to 2 fl. dr. (1.3 to 8 mils).

334. CASCARA SAGRADA.—Cascara Sagrada

CHITTEM BARK

The dried bark of the trunk and branches of **Rham'nus** purshian'a De Candolle.

Botanical Characteristics.—Plants of this species of Rhamnus attain a height of 10 to 20 feet. The *leaves* are ovoid, 3 to 5 in. in length, and about ½ in. in their greatest width, serrate except at base. *Flowers* are small and white, appearing after the leaves have matured. The *fruit* is a plain, round, black berry about ¼ in. in diameter, and contains three seeds. This species differs from other species of Rhamnus in that it is a larger tree and bears a larger fruit.

Source.—Several allied species growing in the cascara district in California seem to contribute the cascara sagrada bark of the market. The official species grows abundantly in Northern California, Oregon,

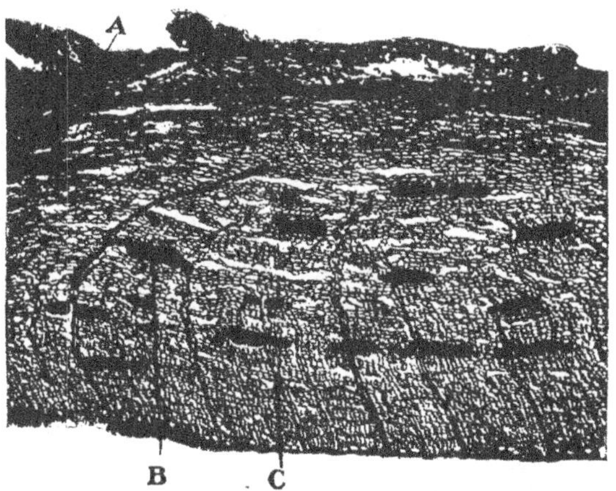

Fig. 169.—*Rhamnus frangula*—Cross-section of bark. (37 diam.) A, Cork. B, Group of bast fibers. C, Medullary ray. (Photomicrograph.)

and Washington. "If the bark comes and is actually collected from Northern California, it is presumptive evidence that it is genuine. The probabilities of adulteration increase with its southward sources, and if collected in, or south of, Central California, it is to be looked upon with greatest suspicion" (Rushy).

Description of Drug.—Curved pieces or quills 1 to 4 mm. (⅟₂₅ to ⅙ in.) thick, and about 100 mm. (4 in.) long. The outer surface is reddish brown, frequently more or less covered with grayish or whitish lichens, the young bark having numerous rather broad, pale-colored warts; sometimes mottled or figured; inner surface smooth and finely striate, yellowish, turning brown or nearly black on exposure; fracture short, yellowish, of the inner layer somewhat fibrous and thick. A cross-section shows numerous thin, almost straight, broadening medullary rays, which run on an average about three-fourths of the distance across the bark.　Medullary rays in groups converging at their outer ends (distinction from *Rhamnus Californica*); stone cells present (distinction from *Rhamnus frangula*).　If to a small quantity of the powdered barks an alkaline solution be added, the color developed in the *Rhamnus Californica* is a deep red, while that of *R. Purshiana* is orange.　Odor distinct; taste bitter and slightly acrid.

Powder.—Characteristic elements of: See Part iv, Chap. I, B.

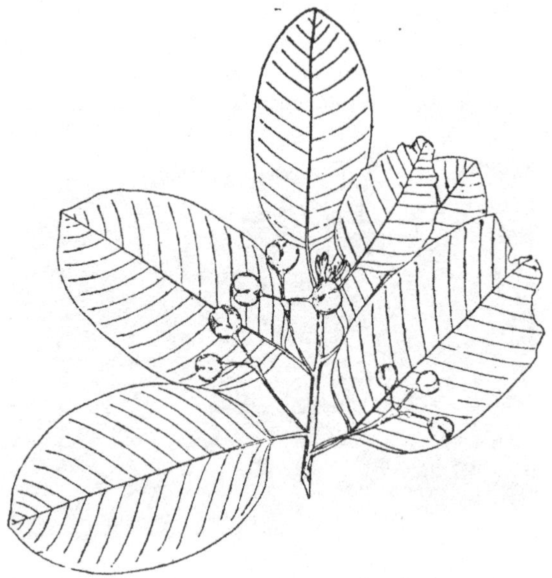

FIG. 170.—*Rhamnus purshiana*—Branch.

CONSTITUENTS.—Emodin and frangulic acid; frangulin and purshianin—the two latter being glucosides, yielding, on hydrolysis, emodin and sugar.　The principle, emodin, is found in many purgative drugs. Its composition, and its relation to several carbon compounds are shown in the following:

$C_{14}H_{10}$	$C_{14}H_8O_2$	$C_{14}H_6(OH)_2O_2$	$C_{14}H_4CH_3(OH)_3O_2$
Anthracene	Anthraquinone	Chrysophanic Acid	Emodin

Emodin is, therefore, said to be a trioxy-methyl-anthraquinone.　It is contained in rhubarb, senna, aloes, etc.　See emodin test under Rhubarb (120).　The resins are turned a vivid purple-red by caustic potash. The fresh bark is active as a purgative, causing much griping.　By keeping and properly curing, however, this griping principle is destroyed, and the bark becomes more accurate in action and less likely to cause this discomfort. Ash, usually about 8 per cent.

Purshianin is a glucoside reported by Dohme and Englehardt. Obtained by first removing oil, etc., from the drug by means of chloroform, then extracting the residue with alcohol, etc. It crystallizes from acetone and ethylic acetate in dark brown-red needles, melting at 237°. On heating with alcoholic hydrochloric acid it yields sugar and emodin.

ACTION AND USES.—A valuable laxative in chronic constipation. Dose: 30 to 60 gr. (2 to 4 Gm.).

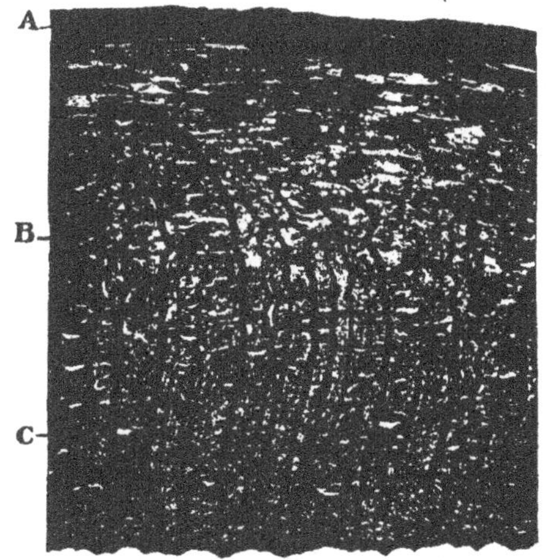

FIG. 171.—*Rhamnus purshiana*—Cross-section of bark. (20 diam.) A, Cork. B, Group of bast fibers and stone cells. C, Medullary ray. (Photomicrograph.)

OFFICIAL PREPARATIONS.

Extractum Cascara Sagradæ,................Dose: 4 gr. (0.25 Gm.).
Fluidextractum Cascara Sagradæ,..... 15 ♏ (1 mil).
Fluidextractum Cascara Sagradæ Aromaticum,⁎ 30 ♏ (2 mils).

335. **RHAMNUS CATHARTICA,** N.F.—BUCKTHORN. The fruit of **Rham′nus** cathar′tica Linné. *Habitat:* Europe, Northern Asia, and naturalized in North America. Small, berry-like fruits about the size of a pea, borne on a receptacle at the end of a slender stalk; apex tipped with the style remnants. Smooth, purplish or black when fresh, in which state they are generally used; wrinkled on drying; four-celled, each containing a single triangular seed, surrounded by a brownish-green pulp; odor unpleasant; taste sweetish, afterward bitter and nauseous. They contain rhamnocathartin, rhamnin, sugar, gum, and tannin. A syrup is made from the juice, having strong purgative properties. Dose of syrup: 2 to 5 fl. dr. (8 to 20 mils). The green fruit treated with lime yields a pigment, sap-green.

336. **CEANOTHUS.**—NEW JERSEY TEA. RED ROOT. The root of **Ceano′-** thus america′na Linné. *Habitat:* North America. About 300 mm. (12 in.) long, and 12 to 25 mm. (½ to 1 in.) thick, contorted and knotty; bark reddish-

brown, thin, inclosing a tough, light brown wood, finely rayed; odor none; taste astringent and bitter. It contains ceanothine, tannin, mucilage, etc. Astringent and expectorant. Dose: 10 to 30 gr. (0.6 to 2 Gm.).

337. **GOUANIA.**—Chewstick. The stems of Goua'nia domingen'sis Linné. *Habitat:* West Indies. Brownish-gray, wrinkled pieces of the stems, with a thin bark, and a yellowish-gray, fibrous, porous wood. It contains a bitter principle and is used as a tonic.

Fig. 172.—*Theobroma cacao*—Branch and fruit.

AMPELIDEÆ

Mostly climbing shrubs. *Stems and branches* nodose; tendrils and flower clusters opposite the leaves. *Fruit* a two-celled berry. Plants abounding in the Tropics.

338. **UVA PASSA.**—Raisin. The dried fruit of Vi'tis vinif'era Linné. *Habitat:* Western Asia, Europe, and California; the Valencia raisins are the kind generally used in pharmacy. Shriveled and pressed; brown, slightly translucent; internally pulpy, two-celled, with two seeds in each cell; taste sweet. Chiefly used as an agreeable saccharine addition to preparations.

339. **AMPELOPSIS QUINQUEFOLIA** Michaux.—AMERICAN IVY. WOODBINE. (Root-bark.) Alterative, tonic, astringent, and expectorant. Dose of fluid-extract: 30 to 60 ♏ (2 to 4 mils).

TILIACEÆ.—Linden Family

Mostly tropical trees, some of the species of the genus Tilia, yielding tenacious fibers for cordage. Flowers balsamic, furnishing infusions which are antispasmodic and diaphoretic.

340. **TILIA AMERICANA** Linné.—LINDEN FLOWERS. BASSWOOD LIME TREE. *Habitat:* North America. Flowers yellowish; petals notched at base; odor pleasant; taste sweet and mucilaginous. Stimulant, diaphoretic, and lenitive. Dose: 15 to 30 gr. (1 to 2 Gm.). The bark is used as a demulcent, emollient and vulnerary.

MALVACEÆ.—Mallow Family

Mucilaginous, innocent plants, with tough bark and palmately-veined *leaves; stamens* monadelphous, in a column, and united with the short claws of the petals; *pistils* several, the ovaries united in a ring, or forming a several-celled pod.

Synopsis of Drugs 'from the Malvaceæ

A. *Root.*
    ALTHÆA, 341.

B. *Flowers.*
    Althæa Rosea, 342.
    Malva, 343.

    344. DERIVATIVES OF THE COTTON PLANT.
*Bark, 344 a. Filamentous Hairs, 344 b. Oil, 344 c.

FIG. 173.—Vertical section of Mallow flower.

341. ALTHÆA.—ALTHÆA

MARSHMALLOW

The dried root of **Althæ'a officina'lis** Linné, deprived of the brown corky layer and small roots.

BOTANICAL CHARACTERISTICS.—Stem 2 to 4 feet high. *Leaves* ovate, or slightly heart-shaped, toothed, downy. *Flowers* pale rose color.

HABITAT.—Europe, Asia, United States, and Australia.

DESCRIPTION OF DRUG.—Whitish, cylindrical, or conical pieces deprived of the outer corky layer, from 75 to 150 mm. (3 to 6 in.) long, and about 10 mm. (⅖ in.) or more in diameter; longitudinally wrinkled, and marked with numerous brownish scars; somewhat **hairy externally from loosened bast fibers**; it breaks with a short mealy fracture, with projecting fiber-ends near the outer edge; odor faint, but characteristic, stronger in infusion; taste sweetish and **mucilaginous.** A cross-section shows small wood-bundles of scalariform and pitted vessels scattered throughout the prevailing parenchymatous tissue, but

with an indistinctly radiate arrangement near the edge. The cells
of the **parenchyma** contain starch and mucilage, with a few stellate
rhaphides. Most of this drug now appears cut into fine pieces or
granules. This often looks beautifully white, but on scrutiny it is
found coated with lime. (Rushy.)

Powder.—Characteristic elements: Microscopical elements of: See Part iv,
Chap. I, B.

CONSTITUENTS.—**Asparagin,** $C_4H_8N_2O_3H_2O$, 1 per cent. (a colorless,
nearly tasteless, crystalline principle), **bassorin,** $C_{12}H_{20}O_{10}$, 25 per
cent. (althæa mucilage, a turbid, slimy, non-adhesive mucilage, which when dried forms a very coherent mass), sugar 8 per cent., **pectin** 10 per cent., ash 5 per cent., starch 35 per cent., a fixed oil, and a trace of tannin. Ash, not to exceed 8 per cent.

FIG. 174.—*Althæa officinalis*—Flowering branch.

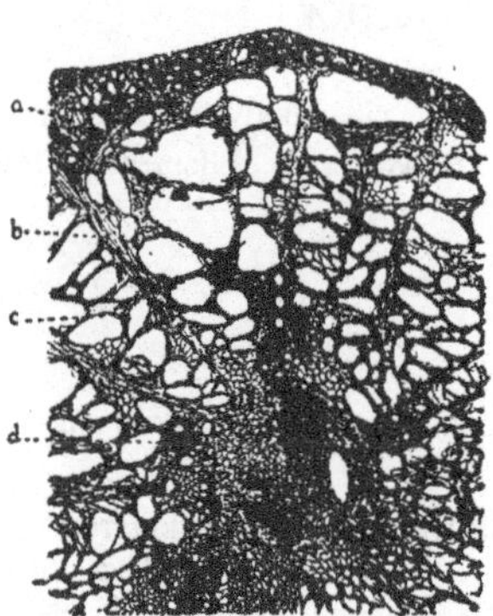

FIG. 174a.—Cross-section of a portion
of the trimmed branch of Althæa. *a.*
Bark. *b.* Medullary ray. *c.* Parenchyma. *d.* Wood-bundle. Magnified
18 diam.

ACTION AND USES.—Used as a demulcent application to inflamed mucous
tissues, as in bronchitis. Powdered marshmallow root being exceedingly absorbent, is used advantageously to impart consistency to
soft pill-masses. (In Mass. Hydrarg., 15 per cent. In Blaud's
Pills and Pil. Phosphorus.)

ALTHEA FOLIA—recognized by the N.F.

342. ALTHÆ'A RO'SEA Cevanilles.—HOLLYHOCK. (Petals.) Indigenous to Western Asia, but cultivated in gardens for its large, purple, ornamental flowers. Petals broadly obovate, notched above and with a claw at base; odor slight; taste sweetish, mucilaginous, and astringent. They contain tannin, mucilage, and a coloring matter. An infusion is occasionally used as a demulcent.

343. MALVA.—MALLOW. The flowers of **Mal'va sylves'tris** Linné, an herbaceous plant growing abundantly in Europe. When fresh, of a rose-red or purple color, becoming blue when dried; odor slight; taste sweetish and mucilaginous. Emollient and demulcent.

MALVÆ FOLIA—recognized by the N.F.

344. DERIVATIVES OF THE COTTON PLANT

Bark, Hairs of Seed, and the Oil of **Gossyp'ium herba'ceum** Linné, and other species of Gossypium.

BOTANICAL CHARACTERISTICS OF GOSSYPIUM HERBACEUM.—Large herbs with alternate *leaves*, which are more or less palmately-lobed. *Flowers* are large, showy, more or less yellow or red; *pistils* 5, united at their base. *Stamens* numerous, united below and adhering to the petals. *Capsule* roundish, 3- to 5-celled, opening at the top by as many valves. The numerous *seeds* are glossy, covered with long, woolly hairs, which constitute the cotton.

HABITAT.—Asia and Africa; cultivated in the United States.

344 a. ROOT BARK.—Gossypii Cortex, N.F. COTTON-ROOT BARK.—Long bands or curved pieces, sometimes in quills. The outer surface is of a yellowish-brown color, dotted with a few small black spots, and, from the abrasion of the thin cork, numbers of orange-brown patches; the inner surface is whitish and has a silky luster; **the bast fibers are long and tough, and may easily be separated into papery layers**; inodorous; taste very slightly acrid and astringent.

Powder.—Light brown. The microscopical elements are: The simple and compound starch grains, the aggregate calcium oxalate crystals, colored resin, and tannin masses; the numerous long, slender, and thick-walled bast fibers (8 to 15 μ thick), large cork cells, etc.

CONSTITUENTS.—A yellow resin, fixed oil, tannin (small quantity), sugar, starch, and, in the fresh bark, a yellow chromogen, which becomes red and resinous on exposure to the air. To this change is due the red color of old specimens, and old preparations, of the bark.

ACTION AND USES.—Emmenagogue and oxytocic, stimulating uterine contractions probably by direct action on the uterine center in the spinal cord; said to be as efficient and more safe than ergot. Dose: 15 to 60 gr. (1 to 4 Gm.).

344 b. HAIRS OF SEED.—Gossypium Purificatum. PURIFIED COTTON. Fine, white, soft filaments, which, under the microscope, appear as hollow, flattened, and twisted bands; unacted upon by ordinary solvents. Ordinary raw cotton contains among other impurities fatty substances, which, when removed by chemical means, such as alkaline or ethereal solvents, changes its character so that the fiber, which formerly was almost impenetrable by aqueous liquids, now becomes so absorbent that it no longer floats on water, but when placed on the surface of that liquid will readily absorb it and sink.

CONSTITUENTS.—Almost pure cellulose; by the action of nitric acid this is converted into soluble gun-cotton.

ACTION AND USES.—Employed as a dressing for burns, scalds, and excoriated surfaces, and for making antiseptic cottons, such as salicylated cotton, benzoinated cotton, iodoform cotton, etc.

Pyroxylinum (Soluble Gun-cotton), the basis of the various official collodions.

344 c. **OIL.**—OLEUM GOSSYPII SEMINIS.   A fixed oil expressed from the seeds. Pale yellowish, odorless, with a bland, nut-like taste; specific gravity 0.920 to 0.930 at 15°C. (59°F.), solidifying at about 0° to   5°C. (32° to 23°F.);

FIG. 175.—*Gossypium herbaceum*—Branch.

very sparingly soluble in alcohol.   Brought into contact with concentrated sulphuric acid, the oil at once assumes a dark reddish-brown color.   Color reactions with nitric acid and silver nitrate (see U.S.P. tests) distinguish this oil from other similar oils.   The oil is used as a basis for Linimentum Ammoniæ, Linimentum Camphoræ, etc.   Processes have been invented for purifying the crude oil to abstract its acrid resin, and so leave it bland and as palatable as the olive oil, for which it is oftentimes substituted as a table or salad oil.

CONSTITUENTS.—Palmitin, olein, and a pale-yellow coloring-matter that is non-saponifiable.

STERCULIACEÆ

Trees or shrubs with soft wood; sometimes climbing. *Fruit* dry, rarely fleshy (Theobroma, 346); *seeds* globose or ovoid, with coriaceous or crustaceous testa. The two plants of interest of the order are the one mentioned and Cola, 70.

345. **COLA** N.F.—COLA (KOLA). The dried kernel of the seed of **Cola acumi-nata** R. Brown (Fam. *Sterculiaceæ*), yielding by assay 1 per cent. of total alkaloids. Occurring in irregular somewhat plano-convex pieces; cotyledons from 15 to 30 mm. long and 5 to 10 mm. thick; dark brown or reddish-brown'; fracture short, tough; odor faintly aromatic, taste astringent and somewhat aromatic. The drug contains alkaloids consisting mostly of **caffeine and theo-bromine,** about 40 per cent. of starch, a little volatile oil, fat, and tannin. The *kolanin* of Knebel is simply a kolatannate of caffeine. Kolatannic acid

FIG. 176.—Cola (Kola Nut). Showing longitudinal section of fruit × ⅓; cross-section of red seed × ⅓; longitudinal section of red seed showing embryo × ⅓; cross-section of red seed × ⅓; longitudinal section of white seed. (*After Kohler.*)

differs from caffeotannic acid in being free from sugar. Tonic, stimulant, and nervine; used as a beverage by the natives of Africa as is coca by the natives of South America. Dose: 10 to 30 gr. (0.6 to 2 Gm.).

"Bissey nuts" are the seed of the Cola naturalized and cultivated in the West Indies. It should be said with regard to the many preparations of Cola that they seem to lack a certain degree of permanence: the fluidextract of the Cola, for example, is an unsatisfactory preparation, because of the immense precipitation which goes on for a long time after the preparation is made.

346. **THEOBROMA.**—CACAO. CHOCOLATE NUT. The seed of **Theobro'ma caca'o** Linné. *Habitat:* Mexico; cultivated in the West Indies. About the size of an almond, flattened, invested with a thin, longitudinally wrinkled testa, varying from reddish to grayish-brown in color; somewhat ovate in shape, the hilum being situated on the broader end. The cotyledons are brown, oily, somewhat ridged. Odor agreeable when bruised; taste bitterish, oily. Contains 45 to 53 per cent. of fixed oil (Cacao Butter), and 1.5 per cent. of theobromine, an alkaloid similar to caffeine. Chocolate is made by roasting the seed, removing the testa, then powdering the kernels, forming the powder into cakes with water, and flavoring with vanilla or other substances.

20

Theobromine and Its Compounds.—Theobromina, $C_7H_7N_4O_2$.—3,7—dimethyl-xanthine, occurs also in Kola (Cola, 345), etc., also made synthetically, action and uses same as caffeine.

Theobromine Sodium Salicylate ("Diuretin").—A white powder, odorless, soluble in water. Dose: 15 gr. (1 Gm.).

Theobromine Sodium Acetate (Agurin), has great solubility and is well tolerated by the stomach. Dose: 15 gr. (1 Gm.).

Preparations of Theobromine.—Obtained from an infusion of cacao, precipitating it with lead acetate, removing excess of lead by H_2S, evaporating, and exhausting the residue with boiling alcohol. The alkaloid separates on cooling. Sparingly soluble in cold water, alcohol, and ether.

346 a. **OLEUM THEOBROMATIS,** U. S.—Cacao Butter. A fixed oil expressed from the seed. A yellowish-white, brittle, fatty solid, of tallow-like consistence, melting at 30° to 33°C. (86° to 91.4°F.), about the temperature of the body; has a faint, chocolate-like taste and agreeable odor. Should respond to the various important official tests (see U.S.P.). Contains palmitin, stearin, laurin, olein (small quantity), theobromine, and glycerides of formic, acetic, and butyric acids. Employed largely in making suppositories.

TERNSTRŒMIACEÆ.—Tea or Camellia Family

Trees or shrubs with simple, usually alternate, leaves, often fascicled at the tops of the branches.

347. **THEA.**—Tea. The leaves of **Camel'lia the'a** Link. *Habitat:* Southern Asia; cultivated. From 25 to 75 mm. (1 to 3 in.) long, petiolate, acute at both ends, irregularly serrate except at base, and with anastomosing veins near the margin; bluish-green or blackish. The green color of tea is not infrequently intensified by a mixture of Prussian-blue and gypsum. Odor peculiar, taste bitter and astringent. Contains volatile oil and an alkaloid, theine, which is analogous to, if not identical with, caffeine. Much of the caffeine of commerce is made from tea siftings. Astringent, tonic, stimulant, and nervine; one of the most valuable stimulating and restorative agents.

GUTTIFERÆ

Trees or shrubs with opposite or whorled coriaceous leaves; *stamens* indefinite; *stigmas* sessile, radiant. Many species, like the gamboge, yield a yellow juice; the *seeds* of others are oily. Among the edible *fruits* of the order is the mangosteen, regarded as the most delicious fruit in the world.

348. CAMBOGIA.—Gamboge

GAMBOGE

A gum-resin from **Garcin'ia hanbu'rii** Hooker filius.

Botanical Characteristics.—The gamboge tree has diœcious flowers and a foliage resembling that of laurel. *Flowers* yellow; *male flowers* in axillary clusters, on short, one-flowered pedicels. *Female flowers* sessile. *Fruit* a berry, about the size of a large cherry, reddish-brown, containing a sweet pulp.

HABITAT.—Anam, Camboja, Siam, and Cochin-China.

DESCRIPTION OF DRUG.—**Lumps,** or **cylindrical sticks** (pipes), 25 to 50 mm. (1 to 2 in.) in diameter, and 100 to 200 mm. (4 to 8 in.) in length, striated lengthwise by impressions from the bamboo in which it is

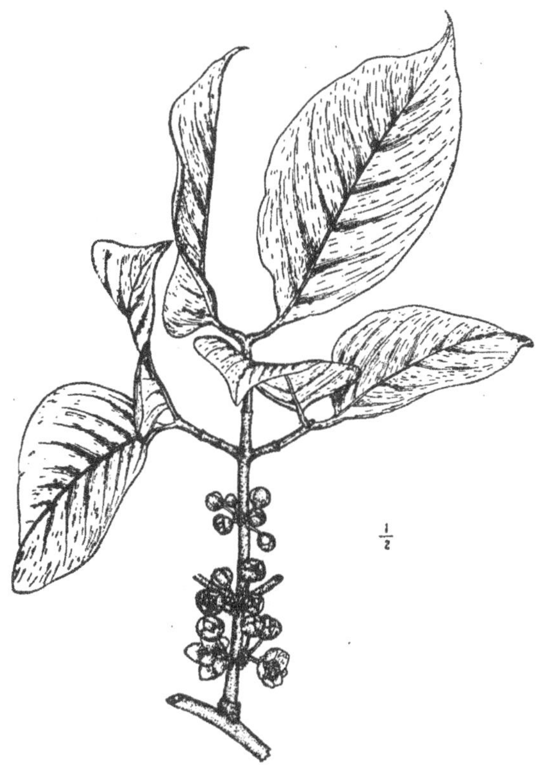

FIG. 177.—*Garcinia hanburii*—Branch.

collected. Externally, grayish-brown. It has a smooth, conchoidal fracture of **a waxy luster and orange-red in color.** The powder is bright yellow and sometimes adheres to the drug, giving it a yellow appearance. Taste at first mild, afterward very acrid; odor irritating, sternutatory. The cake or lump gamboge is sold in masses weighing two or three pounds. It is less uniform, less brittle, and is sometimes called "coarse gamboge." Adulterated specimens are easily recognized by their general inferior appearance, by the grayish or bronze appearance of a broken surface, and by giving a blue or green color with iodine when starch is one of the impurities. Pure gamboge is completely soluble by successive treatment with ether or alcohol and then water.

CONSTITUENTS.—A bright yellow resin (gambogic acid) 73 per cent., soluble in alcohol and ether, turned to a red color by alkalies, and black-brown by ferric chloride; **gum** 16 to 26 per cent., which, with the resin and hot water, forms a yellow emulsion; wax 4 per cent. and ash not more than 2 per cent.

ACTION AND USES.—A drastic **hydragogue** cathartic, but so liable to produce vomiting and griping that its action is usually modified by combining it with other milder purgatives. Dose: ½ to 5 gr. (0.0324 to 0.3 Gm.), generally in pill form.

OFFICIAL PREPARATION.

Pilulæ Catharticæ Compositæ,....................Dose: 2 to 5 pills.

349. **MANGOSTANA.**—MANGO FRUIT. MANGOSTEEN. The pericarp of the fruit of **Garcin'ia mangosta'na** Linné, of India. Astringent; used in various diseases of the mucous membrane, in injections, etc. Mangostin has been isolated from the pericarp. It is golden-yellow in color, crystallizes in scales, soluble in alcohol and ether. The fruit yields a fatty oil, concrete oil of mangosteen, called kokum butter, used in soap-making. It is well adapted for pharmaceutical preparations and candle-making. Dose: 15 to 60 gr. (1 to 4 Gm.).

HYPERICINEÆ.—St. John's-wort Family

350. **HYPERICUM.**—ST. JOHN'S-WORT. The herb of **Hype'ricum perfora'tum** Linné. *Habitat:* Europe, Asia, and North America. The drug as it appears in market is composed of a mixture of oblong-ovate, pellucid-punctate leaves, thread-like branches, and less slender, brittle stems, with occasionally black-dotted flower petals, the whole having a greenish-brown appearance. Constituents: Resin, tannin, and a red coloring matter. Used as a stimulant, diuretic, and astringent. Dose: 30 to 60 gr. (2 to 4 Gm.).

DIPTEROCARPEÆ

Trees often gigantic, exuding a resinous juice; rarely shrubs.

351. **GURJUN.**—GURJUN BALSAM. WOOD-OIL. An oleoresin exuding from **Dipterocar'pus turbina'tus** Gaertner, and other species of Dipterocarpus. *Habitat:* India and the East Indies. A thick, viscid balsam with uses and properties similar to copaiba. Opaque, and grayish, greenish or brownish in reflected light; transparent and reddish-brown or brown in transmitted light; odor copaiba-like; taste bitter. It contains a volatile oil, 40 to 70 per cent., which is similar to oil of copaiba in composition, and produces a red or violet color with a drop of H_2SO_4 and HNO_3 mixed; also gurjunic acid (crystalline), resin, and a bitter principle. Owing to its close resemblance to copaiba it has been used in considerable quantities for the purpose of adulterating the latter.

352. **BORNEO CAMPHOR.**—SUMATRA CAMPHOR AND BORNEOL. A stearopten, or camphor, $C_{10}H_{18}O$, obtained in solid crystalline form from fissures and cavities in a gigantic forest tree, **Dryoba'lanops aroma'tica** Colebrook, growing in the Malay Archipelago. It occurs in masses some pounds in weight. Differs from the ordinary camphor in having a higher specific gravity (heavier than water) and in being less volatile. With nitric acid it yields the Japan (laurel) camphor, $C_{10}H_{16}O$.

FRANKENIACEÆ

353. **FRANKENIA.**—Yerb'a Reum'a. (Herb.) A California plant, **Franken'ia grandiflo'ra** Chamisso et Schlechtendal. A valuable topical application in catarrhal affections, and in diseases of the mucous membranes generally. Dose of fluidextract: 10 to 30 ℥ (0.6 to 2 mils), diluted.

CISTINEÆ.—Rock-rose Family

354. **HELIANTHEMUM,** N.F.—Frostwort. The herb of **Helian'themum canaden'se** Michaux. *Habitat:* North America. As found in commerce it consists of broken branches or stems not longer than 1 to 1½ inches, mixed with a few broken roots, crushed, woolly leaves, and, occasionally, yellow petals; the stems are red-brown, thread-like, slightly pubescent, internally whitish, with a large pith; taste astringent and bitter. It contains a bitter glucoside, soluble in water, alcohol, and benzol, and 11 per cent. of tannin, with sugar and gum. Tonic, astringent, and alterative, in the treatment of scrofulous diseases. Dose: 5 to 20 gr. (0.3 to 1.3 Gm.).

BIXINEÆ

Trees and shrubs with alternate simple leaves and regular, symmetrical flowers. The fruits of some species are edible, and gums are obtained from a few others.

355. **GYNOCARDIA.**—Chaulmoo'gra. The seed of **Gynocar'dia odora'ta** R. Brown. *Habitat:* Malayan Peninsula. Contains an acrid, whitish fat, known in market as **chaulmoogra oil,** separated from the kernels by expression or by boiling water, then taken up by ether or chloroform, which, when evaporated, leaves the oil almost pure. Gynocardic acid, a constituent, is sometimes employed in medicine. "The oil is a very successful remedy in eczema of the third stage." The oil is esteemed in India for the treatment of all manner of skin diseases. Its unctuous smoothness has been compared to that of goose-grease. Dose (of oil): 10 to 20 ℥ (0.6 to 1.3 mils), in gelatin capsules or in emulsion.

356. **ANNATO.**—A coloring substance obtained from a tropical American tree, **Bix'a orella'na.** The seeds steeped in water and allowed to ferment, and this liquid evaporated to a paste, becomes the **anna'to of commerce, used as a cheese and** butter color. By the natives the fragrant reddish pulp of the seeds is used as an astringent in diarrhea. It is also used as a dyestuff for silks and other fabrics.

CANELLACEÆ

An order furnishing mostly aromatic trees.

357. **CANELLA.**—Canella, N.F. The bark of **Canel'la al'ba** Murray. A native of Florida, West Indies, etc. In quills or broken pieces deprived of the corky layer; outer surface orange-red, marked with small scars and depressions; inner surface whitish; odor slight, **aromatic; taste bitter and very pungent and biting.** It contains a reddish volatile oil (about 2 per cent.), a portion of which is closely related to eugenol of oil of cloves, with resin, ash, mannite, a bitter principle, cellulose, albumen, and starch. Aromatic and stimulant, used as an adjuvant. The powder is used in making "hiera picra," Pulv. aloes et canellæ, at one time recognized as an official preparation.

358. **CINNAMODENDRON.**—The bark of **Cinnamoden'dron cortico'sum** Miers. An aromatic bark from Jamaica, coming in curved or quilled pieces. Odor cinnamon-like; taste bitter, biting, giving a suggestion of canella, but this bark contains tannin, which canella does not. Used as an aromatic stimulant. Enters commerce solely from the Bahamas, where it is known as cinnamon bark, or as white wood bark.

VIOLARIEÆ.—Violet Family

Herbs with alternate or radical leaves; *corolla* of 5 unequal petals, one being spurred; *stamens* 5, connivent, alternate with the petals; *fruit* a 3-valved capsule.

359. **VIOLA TRICOLOR.**—Pansy. Heart's-Ease. The herb of Viola tricolor Linné. *Habitat:* Europe, North America, and Northern Asia; cultivated. The drug consists of the herbaceous upper portion of the plant, including green leaves, straw-colored, broken stems, and the variegated flowers. Odor slight, pleasant; taste somewhat bitter. It contains salicylic acid 1 per cent., sugar, mucilage, a bitter principle, resin, and violin (in small quantity). Mucilaginous, emollient; much used in Europe as an alterative in skin diseases, especially eczema. Dose: ½ to 2 dr. (2 to 8 Gm.).

TURNERACEÆ

360. **TURNERA.**—Damiana, N.F. The leaves of a Mexican plant, **Turnera aphrodisiaca (T. diffu'sa** Willdenow). About 8 to 16 mm. (⅓ to ⅔ in.) long, obovate or lanceolate, with a few-toothed margin; surface smooth or with a few hairs on the under side along the ribs. They generally have mixed with them pieces of the slender, woody stem, which is reddish-brown and hairy, the branches being terminated by hairs; odor somewhat aromatic, due to the presence of about 0.5 per cent. of volatile oil. Damiana leaves form the basis of a number of the quack aphrodisiacs. It is not known as a *drug* in Mexico, but as a general tea-like beverage. Dose: about 1 dr. (4 Gm.), in infusion.

PASSIFLOREÆ.—Passion-flower Family

361. **CARICA PAPAYA.**—Melon-tree. True Papaw (wholly different from the common papaw, **Asim'ina trilo'ba,** of our Southern States). *Habitat:* Tropics; cultivated. Although the inspissated juice (papain) of the unripe fruit has been for a long time known as a medicinal agent, having a reputation in its native country as a remedy for hæmoptysis, bleeding piles, and ulcers of urinary passages, and for ringworm, etc., it has only comparatively recently attracted attention as a digestive agent. Dymock, in his treatise on the drugs of British India, says: "Its digestive action on meat was probably known in the West Indies at a very early date. * * * It has long been the practice to render meat tender by rubbing it with the juice of the unripe fruit or by rubbing it with the leaves. Its therapeutic value, in the form of papain, is specially commended in aggravated symptoms of dyspepsia." Its constituents are mainly globulin, albumin, and albumoses. Dose: 1 to 3 gr. (0.065 to 0.2 Gm.).

362. **PASSIFLORA,** N.F.—Passion Flower. The herb of **Passiflo'ra incarna'ta** Linné; indigenous. Said by eclectic and homœopathic practitioners to be a somnifacient, useful in neuralgia, sleeplessness, dysmenorrhœa, etc. Dose of a saturated tincture: 15 to 30 ♍ (1 to 2 mils).

CACTEÆ.—Cactus Family ·

363. **CACTUS GRANDIFLORUS,** N.F. Linné.—Night-Blooming Cereus. *Habitat:* Tropical America; cultivated as an ornamental herb. The fleshy, hexagonal flowering branches are used in the fresh state. Sedative and diuretic; useful in diseases of the heart when there is an irregularity of action. The tincture and fluidextract have of recent years been growing in popularity, but the supply of the drug seems difficult to obtain, and for this reason, partly, the drug is not official. Dose: 5 gr. (0.3 Gm.).

364. **ANHALONIUM LEWINI,** Henning.—A Mexican cactus, acting powerfully as a cardiac and respiratory stimulant; it has been used to a slight extent in medicine in angina pectoris and asthmatic dyspnea. A source of **mescal buttons.** A powerful habit-forming narcotic and intoxicant.

THYMELEACEÆ.—Mezereum Family

Shrubby plants, with the bark containing strong bast fibers, and very bitter.

365. MEZEREUM.—Mezereum

MEZEREON BARK

The dried bark of **Daph'ne** meze'reum Linné, or **Daphne** **guidium** Linné or of Daphne Laureola.

BOTANICAL CHARACTERISTICS.—A small shrub with smooth, evergreen, lanceolate *leaves*. *Flowers* spicate, appearing before the leaves, rose-colored, 4-lobed. *Berry* bright red, fleshy, 1-seeded.

FIG. 178.—*Daphne mezereum*—Fruiting branch and flowers.

HABITAT.—Mountainous regions of Europe, Siberia, Canada, and New England.

DESCRIPTION OF DRUG.—This bark comes to us in **tough, pliable strips,** from 2 to 4 feet long, 25 mm. (1 in.) or less broad, always rolled into bundles or balls; the very thin periderm is of **a greenish-orange** or purple color, marked with transverse scars and minute black dots; beneath it is a soft, greenish parenchymatous layer, from which it separates easily. The inner surface is whitish, covered with irregular layers of white silky bast fibers, tangentially arranged. Fracture tough. Odorless; taste exceedingly acrid.

Powder.—Characteristic elements: See Part iv, Chap. I, B.

CONSTITUENTS.—It contains a crystalline glucoside, daphnin, $C_{15}H_{16}O_9$, which is not the active principle, however, the medical virtues depending upon an acrid resin termed **mezerein**.

ACTION AND USES.—Sialagogue, stimulant, and alterative. Externally vesicant, in ointment or applied in the form of a small square, moistened. Dose: 1 to 8 gr. (0.065 to 0.6 Gm.).

OFFICIAL PREPARATION.

Fluidextractum Sarsaparillæ Compositum
(3 per cent.),......................Dose: ½ to 1½ fl. dr. (2 to 6 mils).

PUNICACEÆ.—Pomegranate Family

366. GRANATUM.—POMEGRANATE

POMEGRANATE

The stem-bark and root-bark of Pu'nica grana'tum Linné, without more than 2 per cent. of adhering wood and other foreign matter.

FIG. 179.—*Punica granatum*—Branch with flowers.

BOTANICAL CHARACTERISTICS.—Tree shrubby, 20 feet in height; branches numerous, sometimes bearing thorns. *Leaves* opposite, entire, oblong, pointed at each end. *Flowers* large, rich scarlet, terminal. *Fruit* a berry about the size of an orange; rind thick, having a reddish-yellow exterior; pulp many-seeded, acidulous.

HABITAT.—Mediterranean Basin and various portions of Asia; cultivated in all warm climates for its ornamental flowers.

DESCRIPTION OF DRUG.—The stem bark comes occasionally in quills, more frequently in curved pieces 20 to 80 mm. long, 5 to 20 mm. in diameter; bark 0.5 to 2 mm. thick, outer surface yellowish-brown, with grayish patches; longitudinally wrinkled; small lenticels.   Inner

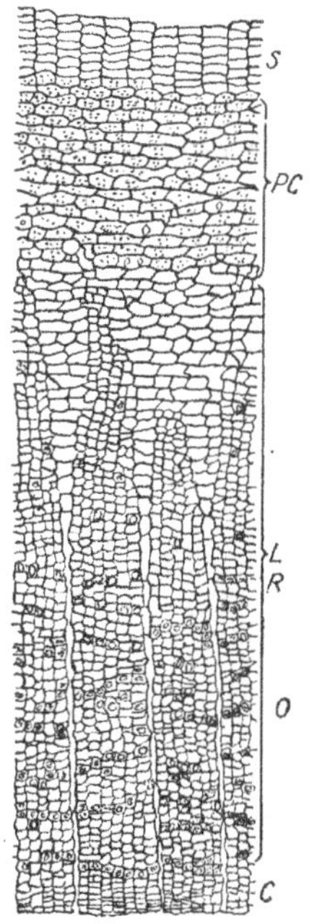

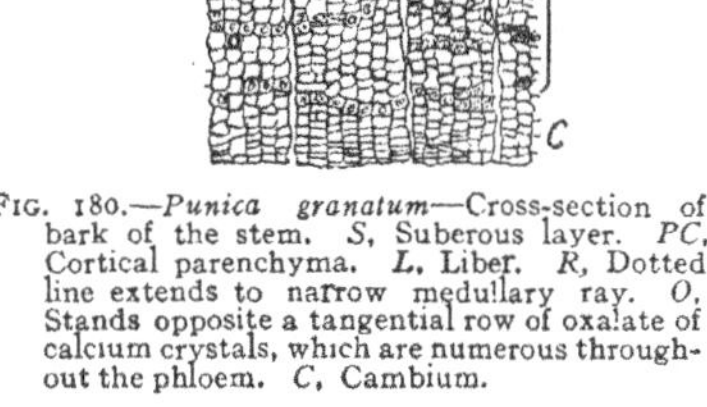

FIG. 180.—*Punica granatum*—Cross-section of bark of the stem. *S*, Suberous layer. *PC*, Cortical parenchyma. *L*, Liber. *R*, Dotted line extends to narrow medullary ray. *O*, Stands opposite a tangential row of oxalate of calcium crystals, which are numerous throughout the phloem. *C*, Cambium.

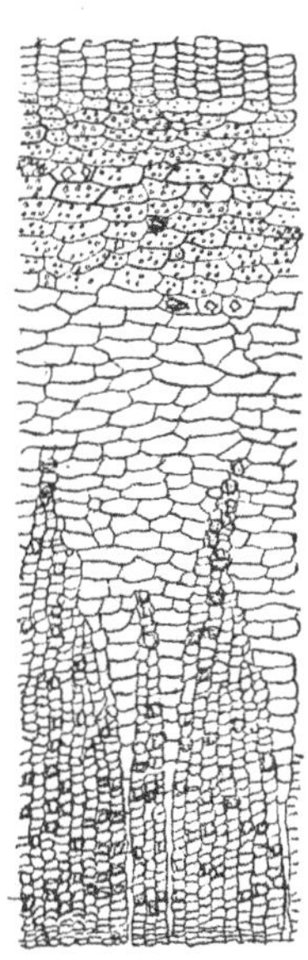

FIG. 181.—*Punica granatum*—Cross-section of the bark of the root.

surface light yellow or brownish-yellow, finely striate, smooth.   Fracture short, smooth, inner layer of bark (phelloderm) dark green, inner bark light brown, odor slight; taste astringent, somewhat bitter.

The root bark has a rough, yellowish-gray to brown outer surface, marked with more or less longitudinal patches of cork, green inner layer of bark absent.   Medullary rays extending nearly to the outer layer; inner surface smooth and yellowish with irregular brownish blotches.

Assay of the drug consists in the extraction and separation of the alkaloid from the drug by acidulated water, washing out the aqueous solution of the salt (after neutralization) with chloroform, again washing the latter solution with $\frac{N}{10}$ hydrochloric acid and titrating final solution in the usual way. No authoritative standard has been fixed.

STRUCTURE.—The tissue consists chiefly of large-celled parenchyma, traversed by one-rowed medullary rays of quadratic cells, each ray accompanied by a single row of crystal cells. The inner bark steeped in water and then rubbed **on paper produces a yellow stain,** which **is rendered blue by ferrous sulphate, and rose-red by nitric acid,** soon vanishing. These properties distinguish it from the bark of the box-root and the barberry, with which it is sometimes adulterated.

Powder.—Characteristic elements: See Part iv, Chap. I, B.

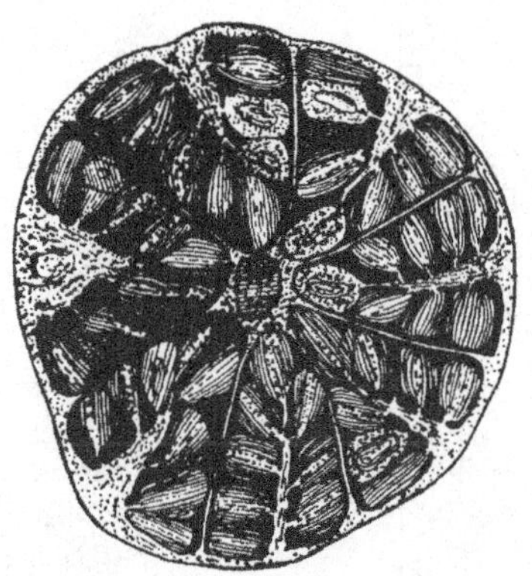

FIG. 182.—*Pomegranate*—Cross-section of fruit.

CONSTITUENTS.—Mannite, punico-tannic acid, 22 per cent. (resolved by hydrolysis into sugar and ellagic acid), and the active constituent, pelletierine, $C_8H_{13}NO$, with its three allied alkaloids, **methyl-pelletierine,** $C_9H_{17}NO$, pseudo-pelletierine, $C_9H_{15}NO$, and **iso-pelletierine.** Pelletierine is a liquid alkaloid, readily soluble in water, alcohol, and ether. Several salts of it are made, but the **tannate** is the **official** one. This is yellowish, hygroscopic, and pulverescent, with a pungent astringent taste, soluble in 700 parts of water and 80 parts of alcohol. Ash, not exceeding 16 per cent.

Preparation of Pelletierine.—Displace powder with water mixed with lime' exhaust percolate with chloroform, etc. It is claimed by Tanret to be the anthelmintic constituent. Is probably a mixture of several alkaloids.

ACTION AND USES.—Astringent, **tæniafuge.** Dose: ½ to 1½ dr. (2 to 6 Gm.). The alkaloid pelletierine is a tæniafuge in extensive use; it is given in the form of tannate in doses of about 5 gr. (0.3 Gm.).

Preparation. Official

Fluidextractum Granati,.........Dose: 1 to 2 fl. dr. (4 to 8 mils).

367. **GRANATI FRUCTUS CORTEX.**—Pomegranate Rind. Irregular fragments, of a yellowish or reddish-brown color; outer surface rough from tubercles; inner surface marked with small depressions; hard; brittle. It contains a greater proportion of tannin than the bark, but has the same medical properties.

MYRTACEÆ

Trees and shrubs, without stipules. *Leaves* opposite, entire, pellucid-punctate, usually with a vein running close to the margin; they are usually fragrant and pungent, due to volatile oil residing chiefly in the pellucid dots or glands.

Synopsis of Drugs from the Myrtaceæ

A. *Leaves.*
   EUCALYPTUS, 368.
   Myrcia, 369.
   Chekan, 370.
B. *Flower.*
   CARYOPHYLLUS, 371.
C. *Fruits.*
   Caryophylli Fructus, 371 a.
   *PIMENTA, 372.

D. *Seed.*
   Jambul, 373.
E. *Volatile Oils.*
   OLEUM EUCALYPTI, 368 a.
   OLEUM MYRCIÆ, 369 a.
   OLEUM CARYOPHYLLI, 371 b.
   OLEUM PIMENTÆ, 372 a.
   OLEUM CAJUPUTI, 374.

368. EUCALYPTUS.—Eucalyptus

EUCALYPTUS

The dried leaves of Eucalyp'tus glob'ulus Labillardierre, collected from the older parts of the tree with not more than 3 per cent. of the stems and fruit of the tree or other foreign matter.

Botanical Characteristics.—Rapid-growing trees, attaining the height of 200 to 300 feet. *Flowers* solitary, or in clusters of 2 or 3, axillary; peduncles broad, somewhat hemispherical in shape, prolonged into a cone, and united with the petals and 4- or 5-celled ovary, making a peculiar hard, brittle, floral envelope, which is quite aromatic. Wood exceedingly hard, remarkable for toughness and durability.

Source.—This is an Australian tree, but is cultivated extensively, especially in malarial districts in various subtropical portions of the world. In California the tree is abundant. At the State Forestry Station at San Monica forty-four species are cultivated. Among these, the Globulus is the most valuable. The Amygdalina possesses the best emollient properties. *E. rostrata* Schlecht (red gum) furnishes an inspissated juice, which is used for the same purpose as kino.

It has been stated that the anti-malarial property attributed to these trees is probably due to their power of absorbing moisture rather than from emanations from them. They probably act in a dual capacity.

Description of Drug.—Petiolate, **scythe-shaped,** from 150 to 300 mm. (6 to 12 in.) long, 20 to 40 mm. ($\frac{4}{5}$ to $1\frac{3}{5}$ in.) broad, tapering from near the base to the apex; pale grayish-green, smooth, and of a leathery texture; margin entire, with a **parallel** vein a short distance from it, running from base to apex of the leaf; odor camphoraceous; taste cooling, bitter, astringent, and aromatic.

Powder.—Characteristic elements: See Part iv, Chap. I, B.

FIG. 183.—*Eucalyptus globulus*—Branch.

CONSTITUENTS.—The virtues of the leaves depend upon a volatile oil (which contains the valuable antiseptic, **Eucalyptol**) existing to the extent of 2 to 6 per cent.; the freshly-dried leaves yield the greatest proportion.

ACTION AND USES.—Used as a febrifuge, stimulant, and astringent. Its principal action, however, is that of the volatile oil, or rather its chief constituent, eucalyptol, $C_{10}H_{18}O$, antiseptic. Dose: ½ to 2 dr. (2 to 8 Gm.). Dose of eucalyptol cineol: 5 ♏ (0.3 mil).

OFFICIAL PREPARATION.

Fluidextractum Eucalypti,..   Dose: 5 to 60 ♏ (0.3 to 4 mils).

368 a. **OLEUM EUCALYPTI.**—A colorless or yellowish volatile oil, distilled from the fresh leaves. It has a spicy, cooling taste, and somewhat camphoraceous odor. Consists of two hydrocarbons (cymene, $C_{10}H_{14}$, and eucalyptene, $C_{10}H_{16}$), a terpene, and **Eucalyptol,** $C_{10}H_{18}O$, upon which its value depends; it is obtained as one of the fractions

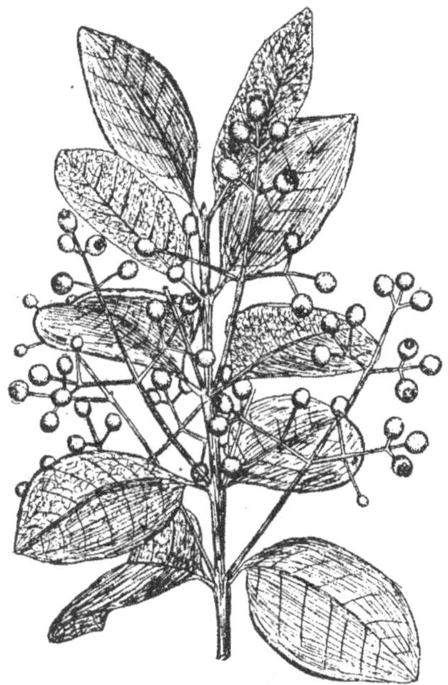

FIG. 184.—*Myrcia acris*—Branch with fruit.

in the distillation of the oil, coming over between 170° to 178°C. It should contain not less than 70 per cent. of Eucalyptol when assayed by the official process. Should be kept protected from light in cool place. It is a nearly colorless liquid, with a strong, aromatic, camphoraceous odor; slightly soluble in water, but very soluble in alcohol, carbon disulphide, and glacial acetic acid. Dose: 5 to 10 ♏ (0.3 to 0.6 mil). Antiseptic. Used frequently as an inhalant in respiratory diseases either with atomizer or with steam. Dose of the oil: 5 to 10 ♏ (0.3 to 1 mil).

368 b. **EUCALYPTUS ROSTRATA** Schlecht.—Red Gum.   The resin or inspissated juice.   *Synonyms:* Creek Gum, Murray Red Gum, Red Gum Kino, Eucalypti Gummi.   *Habitat:* Australia.   Small, angular, ruby-red, shining pieces; in thin layers transparent.   Resembles kino, but has a brighter appearance and is less astringent.   The taste is bitter.   Almost entirely dissolved by alcohol.   *Properties:* A good astringent, similar to kino.   *Preparations:* Fluid and lozenges.   *Uses:* Checks the purging of mercurial pills administered for syphilis.   Has been recommended for seasickness.   Dose: 5 to 20 minims of the fluid.

369. **MYRCIA.**—Bay Leaves. Wax Myrtle. Wild Clove Leaves.   The leaves of **Myr′cia ac′ris** De Candolle, a West Indian tree.   These leaves are aromatic and spicy, containing a volatile oil, which, when distilled, forms the Oleum Myrciæ, and when distilled over with rum, forms bay rum.

369 a. **OLEUM MYRCIÆ** (1890).—Oil of Bay.   A volatile oil distilled from bay leaves.   It is a brownish-yellow, slightly acid liquid, having an agreeable, somewhat clove-like odor, and a warm, spicy taste; sp. gr. 0.96 to 0.98.   It consists of a light and a heavy oil— the light a hydrocarbon identical with that of cloves and allspice, the heavy composed chiefly of **eugenol.**

Preparation.

Spiritus Myrciæ (U.S.P. 1890) (8 per cent., with the oils of orange-peel and pimenta).   Artificial Bay Rum.

370. **CHEKAN.**—Cheken.   The leaves of a Chilian evergreen shrub, **Eu ge′nia che′kan** Molina.   Tonic, expectorant, with some diuretic action.   Dose of the fluidextract: 30 to 60 ♏ (2 to 4 mils).

371. CARYOPHYLLUS.—Cloves

CLOVES

The dried flower buds of **Euge′nia aroma′tica** O. Kuntze (Iambosa caryophyllus (Sprengel) Niedenzu) with not more than 5 per cent. of peduncles, stems and other foreign matter.

Botanical Characteristics.—A shrubby evergreen, with hard wood, covered with a smooth, gray bark.   *Leaves* opposite, ovate-lanceolate, coriaceous.   *Petals* 4, globular in bud, afterward spreading, whitish, aromatic.   *Ovary* 2-celled; *fruit* a large, elliptical berry.

Source.—The original habitat of the clove tree was the Molucca Islands, but they have been introduced into other East Indian Islands, into Zanzibar (which now forms the principal source), and into Cayenne. They are picked singly while green and are dried in the sun.   Commercial: There are several varieties, as Molucco, Sumatra, and South American, the latter being rather inferior.

Description of Drug.—Cloves are about 15 mm. (⅗ in.) long, of a dark brown or reddish-brown appearance; the calyx tube is long, nearly cylindrical, crowned with the four stiff teeth (clasping the unexpanded corolla); **corolla** of four lighter colored, **unexpanded petals, forming a hollow ball on the** top of **the calyx-tube,** inclosing the numerous curved stamens and the single style; the ovary is inferior, situated near the top of the calyx-tube, and consists of two cells, each containing many ovules.   A **cross-section** of the lower part of the calyx-tube under the microscope shows a thin outer layer surrounding a darker zone; this outer layer contains a double ring of oil cells; the inner darker zone contains an outside circle of about thirty fibrovascular bundles, with a larger bundle running through the center. Odor highly aromatic, especially when scratched; **taste** pungent and aromatic, followed by slight numbness.

Powder.—Characteristic elements: See Part iv, Chap. I, B.

FIG. 185.—*Eugenia aromatica.*

CONSTITUENTS.—About **18 per cent.** of volatile oil, 17 per cent. of tannin, a little fixed oil, gum, resin, etc. Two crystalline principles have been separated, caryophyllin, $C_{10}H_{16}O$, white, odorless, and tasteless, resinous, and eugenin, $C_{10}H_{12}O_2$, isomeric with eugenol of the volatile oil, soluble in boiling alcohol and ether, as is also caryophyllin, but differing from the latter in turning red with nitric acid. Water extracts the volatile oil with scarcely any of the pungency of taste. Ash, not exceeding 8 per cent.

Preparation of Caryophyllin.—Treat ethereal extract of cloves with water, collect precipitate, and purify with ammonia.

ACTION AND USES.—Stimulant and carminative, used mostly as a **syner gist.** Dose: 5 to 10 gr. (0.3 to 0.6 Gm.).

OFFICIAL PREPARATION.

Tinctura **Lavandulæ** Composita (0.5 per
  cent.),..........   ....Dose: ½ to 2 fl. dr. (2 to 8 mils).
Tinctura **Rhei** Aromatica.

371 a. **CAROPHILLI FRUCTUS.**—The ripe fruit, or Mother Cloves, resembles cloves in appearance, but is thicker and somewhat lighter in color and less aromatic; the corolla is absent, but the calyx-teeth still adhere.

371 b. **OLEUM CARYOPHYLLI.**—OIL OF CLOVES. A pale yellowish-brown, thin liquid, becoming reddish-brown on exposure. It has a specific gravity of 1.060, and boils at about 250°C.; slightly acid; taste aromatic and hot; odor characteristic, aromatic. Oil of cloves consists of two oils—one lighter than water, the other heavier; the light oil, caryophyllene, $C_{15}H_{24}$, sp. gr. 0.91, is a pure hydrocarbon, and is thought to be inactive; the heavy oil is a phenol-like liquid termed **eugenol,** or eugenic acid, $C_{10}H_{12}O_2$, sp. gr. 1.064 to 1.070.

ACTION AND USES.—Used for the same purposes as cloves, more commonly, however, for introduction into an **aching, carious tooth.** Dose: 1 to 5 ℥ (0.065 to 0.3 mil).

372. PIMENTA, N.F.—PIMENTA

ALLSPICE

The nearly ripe dried fruit of Pimen'ta officina'lis Lindley, including not more than 5 per cent. of stems and foreign matter.

BOTANICAL CHARACTERISTICS.—An elegant tree about 30 feet high, evergreen. *Leaves* pellucid-punctate, petiolate. *Flowers* in racemes, white. *Calyx* and *petals* 4-fold, the latter greenish-white. *Fruit* a berry, covered by the roundish, persistent base of the calyx. After ripening, they lose their aromatic warmth and acquire a somewhat juniper-like taste; hence they are gathered in the unripe state.

SOURCE.—West Indies, Mexico, and South America, the principal source being Jamaica—from which it has received the name of Jamaica pepper.

DESCRIPTION OF DRUG.—Globular, about the size of a large pea; picked while yet green, becoming wrinkled and brownish on drying, with the four calyx-teeth and the short style still adherent to the apex, or a raised ring marking the position of the calyx-teeth; it is divided into two cells, each of which contains a single, brownish, plano-convex seed. The pericarp is finely tuberculated with numerous oil tubercles. Odor spicy and agreeably pungent; taste clove-like.

Powder.—Reddish-brown. Characteristic elements: Parenchyma of endosperm, with starch and resin; parenchyma of pericarp, with starch, resin, and calcium oxalate in aggregate crystals about 10μ in diam.; sclerenchyma with stone cells, having simple, branching pores; trichomes, short, one-celled; large oil and resin ducts; starch grains, spherical, 10μ simple or compound. See Fig. 301.

CONSTITUENTS.—The properties depend upon a volatile oil and a green, acrid fixed oil, existing to the extent of 10 per cent. and 8 per cent. respectively in the pericarp, and in considerably less quantities in the embryo. The yield of total ash should not exceed 6 per cent. of which the amount soluble in dilute HCl should not exceed 0.5 per cent.

ACTION AND USES.—Stimulant and carminative, as an adjuvant to tonic and purgative mixtures. Dose: 5 to 30 gr. (0.3 to 2 Gm.).

FIG. 186.—*Pimenta officinalis*—Branch and flower.

372 a. **OLEUM PIMENTÆ** (U.S.P. IX).—A colorless, or pale yellow, volatile oil, becoming thick and reddish-brown by age. Specific gravity 1.02 to 1.05. It closely resembles oil of cloves (*q v.*), but has a more pleasant and less pungent odor; taste aromatic. Consists, like oil of cloves, of a light and a heavy oil, the heavy oil being identical with eugenol.

ACTION AND USES.—Same as the other stimulant aromatic oils. Dose: 1 to 5 ℥ (0.065 to 0.3 mil).

OFFICIAL PREPARATION.

Spiritus Myrciæ (U.S.P. 1890) (0.05 per cent.).

373. JAMBUL.—JAVA PLUM. A large tree, Eugen'ia jambola'na, growing in the East Indies, where its fruit is eaten as a food. All parts are astringent, but the bark, and especially the seeds, possess, in addition, the peculiar property of arresting the formation of sugar in diabetes, and hence are "likely to prove a valuable remedy in this disease." Dose: 5 to 10 gr. (0.3 to 0.6 Gm.).

374. OLEUM CAJUPUTI.—OIL OF CAJUPUT

OIL OF CAJUPUT

A volatile oil distilled from the leaves of Melaleu'ca leucaden'dron Linné.

BOTANICAL CHARACTERISTICS.—A tree with crooked stem and scattered branches, the branchlets drooping like those of the weeping willow; bark whitish.

FIG. 187.—*Melaleuca leucadendron*—Branch.

Leaves lanceolate, deep green, entire, from 3 to 4 inches long. *Flowers* small, white, inodorous, in axillary spikes.

HABITAT.—East Indies.

DESCRIPTION OF DRUG.—A light bluish-green (probably due to copper), limpid liquid having a penetrating, agreeable odor, and a warm, camphoraceous, bitter, afterward saline or cooling, taste. Specific gravity 0.912 to 0.925. It has a slightly acid reaction.

CONSTITUENTS.—The principal constituent is the hydrate of the hydrocarbon, cajuputene, $C_{10}H_{16}$ (**Cajuputol, $C_{10}H_{16}H_2O$**), said to be identical with eucalyptol, or cineol, from eucalyptus. The commercial oil often contains a trace of copper, not in large enough quantities to be dangerous, however.

ACTION AND USES.—Highly stimulant, carminative, and a counter-irritant in rheumatism. Dose: 1 to 10 ♏ (0.065 to 0.65 mil).

COMBRETACEÆ

375. **MYROBOLANUS.**—MYROBOLANS. The fruit of Termina'lia chebu'la Retzia, and of other species of Terminalia growing in the East Indies. Oblong, pyriform, or roundish-oval, from 30 to 50 mm. (1⅕ to 2 in.) in length, dark brown or orange color. Several varieties of the fruit are used occasionally as a mild laxative and astringent, but now principally in the arts for tanning, etc.

ONAGRARIEÆ.—Evening Primrose Family

376. **EPILOBIUM.**—WILLOW-HERB. The herb of Epilo'bium angustifo'lium Linné. *Habitat:* Northern Hemisphere. It has a smooth, reddish stem, branching above, arising from a long, yellowish-white root, and bearing the purplish-pink flowers in a raceme resembling those of the willow; hence the name willow-herb. Demulcent and astringent. Dose: 30 to 60 gr. (2 to 4 Gm.).

377. **ŒNOTHERA** BIENNIS Linné.—EVENING PRIMROSE. *Habitat:* North America. Astringent, alterative.

ARALIACEÆ.—Ginseng Family

Synopsis of Drugs from the Araliaceæ

A. *Root.*
    Panax, 378.

B. *Rhizome.*
    Aralia Nudicaulis, 379.
    *Aralia Racemosa, 379 a.
    Aralia Hispida, 380.

378. **PANAX.**—GINSENG. (Official, 1840–1880). The root of **Pa'nax quinquefo'lium** Willdenow. Cultivated in Ohio, West Virginia, Minnesota, and quite extensively and profitably in Michigan, and exported to China, where, from its fancied resemblance to the human figure, it is supposed to possess miraculous powers in preventing and curing diseases, and where at one time it was valued at its weight in gold. It has, however, little medicinal properties except as a demulcent and aromatic stimulant; not used extensively in medicine. It is a soft, yellowish-white, fusiform root, about the thickness of the finger, with two or three equal branches below. A **cross-section** shows a hard central portion, surrounded by a thick, soft, white inner cortical layer; with thin bark, containing numerous reddish resin-cells; wood-wedges narrow; medullary rays broad; odor feeble; taste sweet, slightly aromatic. The sweet principle is **panaquilon,** $C_{12}H_{25}O_9$.

Preparation of Panaquilon.—Concentrate the cold infusion to a syrup, precipitate by concentrated solution of sodium sulphate, wash the precipitate thoroughly with the saline solution, then treat with alcohol, which dissolves the principle; evaporate to dryness.

379. **ARALIA NUDICAULIS** Linné.—FALSE SARSAPARILLA. WILD LICORICE. *Habitat:* North America. (Rhizome.) Horizontal, often 300 mm. (12 in.) in length, and about the thickness of the little finger; it has a yellowish-brown, wrinkled, and annulate bark, inclosing a yellow wood and spongy pith; somewhat aromatic; taste warm, aromatic, and sweetish. The rhizome of Ara'lia racemo'sa, N.F. Linné (American Spikenard) is short and from 25 to 50 mm. (1 to 2 in.) thick, marked above by prominent stem-scars and beset below with long, branching rootlets; externally pale brown, internally whitish; more aromatic and spicy than *A. nudicau'lis.* Both rhizomes are used extensively in domestic practice as stimulant, diaphoretic, and alterative. Dose: 30 to 60 gr. (2 to 4 Gm.), in infusion.

380. **ARALIA HISPIDA** Ventenat.—DWARF ELDER. *Habitat:* United States. (Rhizome.) Diuretic; used in dropsy, etc. Dose of fluidextract: 1 to 2 fl. dr. (4 to 8 mils).

UMBELLIFERÆ.—Parsley Family

Herbs with hollow stems. The umbellate inflorescence—the general character of the order—gives rise to its name. The *fruit,* called a cremocarp (from *cremao,* to support, and *karpos,* fruit), is perhaps the most marked characteristic of the order; it originates from one ovary surmounted by 2 styles and often crowned by the limb of the calyx, and has 2 cells and 2 seeds. The entire fruit is usually ellipsoidal, but in the case of the coriander it is spherical; it divides itself into two *mericarps* (half-fruits) suspended by their summits from a slender axis (carpophore), usually 2-forked; each mericarp has 5 to 10 more or less prominent ridges (juga), in the furrows or grooves between which are several oil-tubes (vittæ), usually visible in cross-section; in anise there are usually 15, in coriander 2. The *roots* contain an abundance of aromatic resin.

Synopsis of Drugs from the Umbelliferæ

A. *Fruits.*
  ANISUM, 381.
  FŒNICULUM, 382.
  *Conium, 383.
  CARUM, 385.
  CORIANDRUM, 386.
  Anethum, 387.
  *Apium, 388.
  Ajowan, 389.
  *Petroselinum, 391.
  Phellandrium, 392.
  Cuminum, 393.
  Carota, 394.
E. *Searopten.*
  THYMOL, 390.
F. *Roots.*
  SUMBUL, 400.
  Imperatoria, 401
  Laserpitium, 402
  *Angelica Atropurpurea, 395.
  Angelica, 396.

B. *Leaves.*
  Conii Folia, 384.
C. *Volatile Oils.*
  OLEUM ANISI, 381 a.
  OLEUM FŒNICULI, 382 a.
  OLEUM CARI, 385 a.
  OLEUM CORIANDRI, 386 a.
  Oleum Anethi, 387 a.
D. *Gum Resins.*
  ASAFŒTIDA, 397.
  Galbanum, 398.
  Ammoniacum, 399.

F. *Roots.*—(Continued.)
  Levisticum, 403.
  *Petroselinum, 391.
  *Pimpinella, 404.
  Thapsia, 405.
  Cicuta, 406.
  Eryngium, 407.
  Osmorrhiza, 408.

381. ANISUM.—ANISE

ANISE

The ripe fruit of **Pimpinel'la an'isum** Linné, with not more than 3 per cent. of foreign seeds and other vegetable matter.

BOTANICAL CHARACTERISTICS.—Stem about 1 foot high. Umbels on long stalks without involucre; *flowers* small, white; *calyx* obsolete; *carpels* 5, with filiform ridges.

HABITAT.—Levant and Egypt; extensively cultivated in Europe.

DESCRIPTION OF DRUG.—Two or three varieties have been produced by cultivation, the Spanish being the smallest, and usually preferred. In general appearance **anise resembles conium very much**, but it is distinguished from the latter in being usually longer and more ovate, the mericarps, which usually adhere together, having their five ribs more or less hairy and not jagged, and having about 15 oil tubes, of which conium has none; odor fragrant; **taste** aromatic, sweetish. The fruit is often accompanied with its adhering short peduncle.

Powder.—Characteristic elements: See Part iv, Chap. I, B.

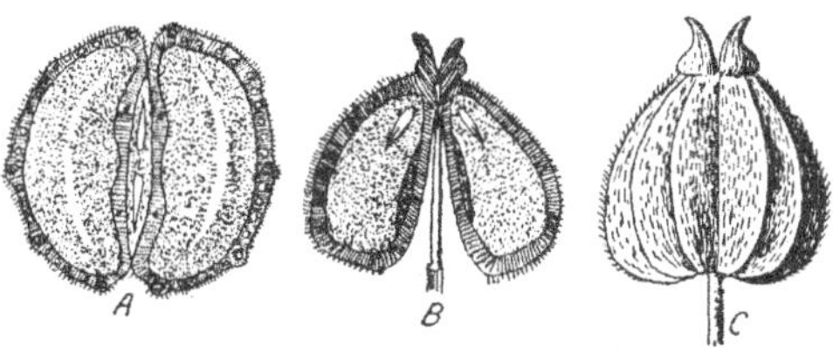

FIG. 188.—*Anisum.* A. Transverse section. B. Longitudinal section. C. Fruit (enlarged).

CONSTITUENTS.—Volatile oil (1½ to 3 per cent.). Ash, not exceeding 9 per cent.

ACTION AND USES.—Stimulant and aromatic carminative. Dose: 8 to 30 gr. (0.5 to 3 Gm.).

381 a. **OLEUM ANISI,** U.S.—A colorless or pale yellow volatile oil, having the aromatic odor and taste of the fruit; neutral in reaction; sp. gr. 0.98 to 0.99, depending upon age. Dose: 5 ♏ (0.3 mil).

CONSTITUENTS.—It contains a slight quantity of a light hydrocarbon oil, but principally **anethol,** $C_{10}H_{12}O$, which is present in both liquid (liquid anethol) and solid form (anise camphor); by oxidation this anethol is converted into **anisic acid;** anethol is the principal constituent also of fennel and star anise, the most of the commercial anise oil being derived from the last-named fruit. Anethol is recognized in the National Formulary.

Preparation of Anethol.—Obtained by fractional distillation; by oxidation is converted into anisic acid.

OFFICIAL PREPARATIONS.

Aqua Anisi (0.2 per cent.),.................Dose: 4 fl. dr. (16 mils).
Spiritus Anisi (10 per cent.),...............   90 ♏ (6 mils).
Spiritus Aurantii Compositus (0.5 per cent.),
Tinctura Opii Camphorata (0.4 per cent.),.....   2 fl. dr. (8 mils).

382. FŒNICULUM.—Fennel

FENNEL

The dried nearly ripe fruit of **Fœnic′ulum** vulga′re Miller with not more than 4
per cent. of foreign matter.

Botanical Characteristics.—Stem somewhat furrowed, 3 feet high. *Leaves*
much compounded, cut into fringe-like segments.   Umbels with 6 to 8 rays,
without involucre or involucel.

Habitat.—Chiefly imported from Germany, although the cultivated
plants in the gardens of this country partially supply the market.

Description of Drug.—Varying in size, the longest often being 12 mm.
(½ in.) in length; oblong, terete, a cross-section showing a nearly
circular surface; the mericarps are usually separated, however, and
slightly curved, their surface dark brown and smooth, with the exception of the five prominent, filiform, lighter colored ribs, the two
lateral ones rather broader; in each depression is one oil tube, and
on the flat side or commissure there are two.   There are two prominent varieties: Saxon, or German, about 4 mm. (⅙ in.) long,
dark brown, usually in half-fruits without foot-stalks. The other
(Roman) is about 12 mm. (½ in.) in length, lighter brown, with more
prominent ribs, and often in the whole state and furnished with foot-
stalk. Both, however, are about the same in aromatic properties,
and have a warm, sweet, aromatic taste. Bitter fennel, from a wild
plant of Southern France, is a small fruit, bitter and spicy. Indian
fennel (6.7 mm. in length), anise-like odor; used in the preparation of
compound infusion of senna (2 per cent.).

Powder.—Characteristic elements: See Part iv, Chap. I, B.

Constituents.—From 2.5 to 4 per cent. of volatile oil, almost chemically
identical with that of anise.   It contains phellandrene, $C_{10}H_{16}$.   Ash,
not exceeding 9 per cent.

Action and Uses.—Stimulant, carminative, stomachic, corrective. Dose:
8 to 30 gr. (0.5 to 2 Gm.), in infusion or powder.

Official Preparation.

Infusum **Sennæ Compositum**,...............Dose: 4 fl. dr. (120 mils).

382 a. **OLEUM FŒNICULI.**—A colorless or pale yellow volatile oil,
having a specific gravity. of 0.96.   It usually solidifies at from 5°
to 10°C. (41° to 50°F.).   It has essentially the same constituents
as the oil of anise. Stimulant and carminative, and a corrective
of harsh, purgative preparations.   Dose: 1 to 5 ℥ (0.06 to 0.3 mil).

OFFICIAL PREPARATIONS.

Aqua Fœniculi (0.2 per cent.),...	Dose: ¼ to 1 fl. oz. (8 to 30 mils).
Pulvis Glycyrrhizæ Compositus (0.4 per cent.),......................	½ to 2 dr. (2 to 8 Gm.).
Spiritus Juniperi Compositus (0.05 per cent.),..................	1 to 4 fl. dr. (4 to 15 mils).

FIG. 189.—*Fœniculum capillaceum*—Branch and fruit entire and in cross-section.

383. CONIUM.—CONIUM

POISON HEMLOCK. *Ger.* SCHIERLINGSFRCÜHTE

The full-grown, but unripe fruit of **Con'ium macula'tum** Linné, carefully dried and preserved, should yield by assay, not less than 0.5 per cent. of coniine. It should not be kept longer than two years.

DESCRIPTION OF DRUG.—Gathered when full grown but yet green, the yield of alkaloid being greatest at this time. Small, roundish-ovate, laterally compressed, grayish-green. The mericarps, which are often separated, have five jagged ribs but no oil-tubes; the flat side or commissure is deeply furrowed, giving to a transverse cut surface a reniform outline. Almost odorless; taste disagreeable and somewhat acrid; when triturated with a solution of KOH, conium emits the peculiar, mouse-like odor characteristic of the volatile alkaloid, coniine, which is liberated thereby. The total alkaloids in the fruit may reach as high as 3.5 per cent., rapidly diminishing as it ripens.

Powder.—Pale yellowish-brown. Characteristic elements: Parenchyma of endosperm, rather thick-walled with oil globules and aleurone (4 to 7 μ in diam.); aggregate calcium oxalate (1 to 2 μ in diam.); other parenchyma with starch and chloroplastids; sclerenchyma, from fruit and stalk with bast fibers, long and thin-walled, with numerous pores; collenchymatous cells from mericarp, yellowish, nearly isodiametrical, irregularly thickened.

Constituents.—The liquid alkaloid, coniine, $C_8H_{17}N$ (the active constituent), methyl coniine, $C_8H_{16}(CH_3)N$ (also liquid), conhydrine, and its isomer, pseudo-coniine. Coniine is a yellowish, oily, volatile liquid (sp. gr. 0.88), very acrid, and of a strong, mouse-like odor; it is strongly basic, and is combined

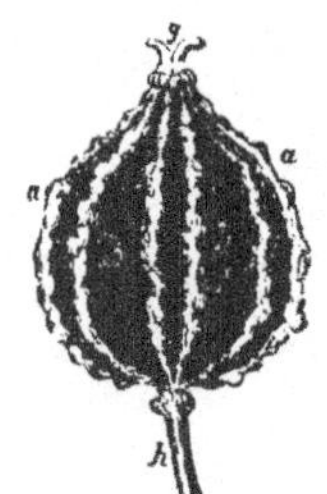

FIG. 190.—Fruit ot Hemlock (*Conium maculatum*).

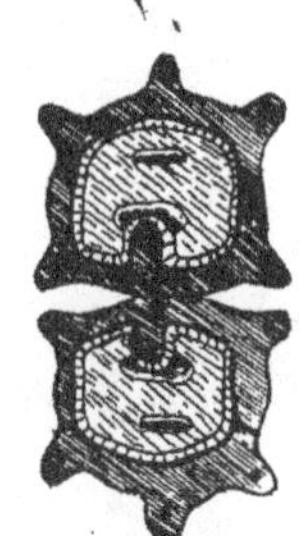

FIG. 191.—Cross-section of the fruit of *Conium maculatum*, 20 diam.

in the fruit with conhydric acid, from which it may instantly be freed and its odor developed in the fruit by rubbing with potassa, as noted above; its action is that of a paralyzant to the motor nervous system. Methyl coniine resembles it in action. Conhydrine is in iridescent scales, melting at 120.6°C.

Preparation of Coniine.—Liberated from drug by distilling it with alkali. Methyl coniine and conhydrine is likely to come over with it.
Separation of Conhydrine from Coniine.—Reduce the temperature of the oily liquid containing the two by a freezing mixture. Recrystallize from ether. Occurs in iridescent scales, less poisonous than coniine.

Action and Uses.—Conium is narcotic and sedative; its principal action is as a paralyzant to the motor nerves. Dose: 3 to 5 gr. (0.2 to 0.3 Gm.). The alkaloid coniine is an active poison, the dose being from ¼ to ½ ♏ (0.0164 to 0.0324 mil); dose of the hydrochlorate is probably about ⅙ gr. (0.01 Gm.).

384. CONII FOLIA.—Hemlock Leaves. Grayish-green, thin, smooth, from 100 to 300 mm. (4 to 12 in.) long, twice or thrice decompound, with oblong-lanceolate, acute, sharply serrate divisions; petiolate, the petiole hollow; odor mouse-like; taste disagreeable. They contain coniine in very small quantity, and are less active than the fruit, but used for the same purposes —as an anodyne and antispasmodic for controlling maniacal excitement and spasmodic affections, such as whooping-cough, etc. Dose: about 5 gr. (0.3 Gm.).

385. CARUM.—Caraway

CARAWAY

The dried fruit of **Car'um car'vi** Linné prevented from attacks of insects by chloroform or carbon tetrachloride.

BOTANICAL CHARACTERISTICS.—A biennial 2 feet in height, with bipinnate *leaves*. The *umbel* rarely involucrate, *flowers* consisting of 5 obcordate, small, white *petals; carpels* with 5 filiform ridges; stylopodium (the disk-like expansion of the receptacle) depressed. *Fruit* brownish, oblong, slightly curved.

HABITAT.—Asia; introduced into America.

DESCRIPTION OF DRUG.—The mericarps, which are usually separated, are about 4 to 5 mm. ($\frac{1}{6}$ to $\frac{1}{5}$ in.) in length, tapering somewhat at the ends. Surface dark brown, smooth, with the exception of the **five** lighter colored, filiform **ribs,** between which are the six large, easily visible oil-tubes. A cross-section shows the pentangular seed and oil-tubes. **Odor and taste** aromatic, agreeable. "Drawn fruits:" This name has been applied to a form of adulterated caraway—a partially exhausted fruit, whereby they have been deprived of a portion of the volatile oil. It is said that "Dutch seed" of fair quality should give over 5 per cent. of volatile oil. Exhausted fruits have been found to contain but 1.5 to 1.9 per cent. of oil. They are of much darker color than the genuine. The American seed is slightly smaller than the German. The seed cultivated in Northern Germany is too deficient in essential oil for profitable distillation, but it has a fine appearance.

Powder.—Characteristic elements: See Part iv, Chap. I, B.

CONSTITUENTS.—Volatile oil 4 to 5 per cent., consisting of carvone and carvene, see 385 a; readily soluble in alcohol, slightly soluble in water. Ash, not more than 8 per cent.

ACTION AND USES.—Stimulant, stomachic, and carminative, and an adjuvant. Dose: 15 to 30 gr. (1 to 2 Gm.).

OFFICIAL PREPARATION.

Tinctura Cardamomi Composita (1.2
per cent.),......................Dose: 1 to 4 fl. dr. (4 to 15 mils).

385 a. **OLEUM CARI,** U.S.—A limpid, colorless or pale yellow volatile oil, specific gravity 0.92, with an aromatic odor and taste, becoming acrid and of a higher specific gravity when exposed. It consists of two portions, a light hydrocarbon, **carvene,** identical with limonene, and a heavy oil, **carvone,** isomeric with thymol.

ACTION AND USES.—Stimulant, stomachic, carminative, and adjuvant. Dose: 1 to 10 ♏ (0.065 to 0.6 mil).

OFFICIAL PREPARATION.

Spiritus Juniperi Compositus (0.05
per cent.),......................Dose: 2 to 4 fl. dr. (8 to 15 mils).

386. CORIANDRUM.—Coriander

CORIANDER

The dried ripe fruit of **Corian'drum sati'vum** Linné without admixture of more than 5 per cent. of other fruit, seeds or other foreign matter.

BOTANICAL CHARACTERISTICS.—An annual herb about two feet high, with an offensive, bedbug-like odor, with smooth stem and bipinnate *leaves*. *Calyx* 5-toothed; *petals* obcordate (the exterior ones bifid), white, often with a pink tinge. *Capsules* with primary ridges obsolete, the four secondary ones prominently keeled. *Fruit* globose; *seed* covered with a loose membrane.

HABITAT.—Italy; cultivated in all parts of Europe and United States.

DESCRIPTION OF DRUG.—**Almost globular,** about 3 mm. (⅛ in.) in diameter, slightly pointed at the apex (style) and with the persistent calyx-teeth around the pedicel-scar at the base. The two concave, hemispherical **meri-carps are closely united at the edge** by the woody pericarp; their outer surface is pale yellowish-brown, sometimes purplish-tinted, with five primary ribs merely indicated by wavy, slightly raised lines,

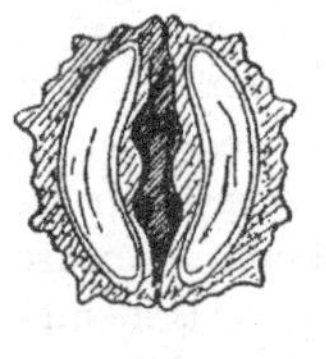

FIG. 192.—*Coriandrum sativum*—Whole fruit and cross-section.

and four more prominent secondary ribs. The interior of the fruit is a lenticular cavity. Odor fragrant (the odor of the fresh plant and fruit is fœtid, resembling bedbugs); **taste** aromatic.

Powder.—Characteristic elements: See Part iv, Chap. I, B.

CONSTITUENTS.—**Volatile oil,** ½ to 1 per cent., containing coriandrol, $C_{10}H_{18}O$, also dextropinene, fat, mucilage. Ash, not exceeding 7 per cent. Soluble ether extract, 0.5 per cent.

ACTION AND USES.—Feeble aromatic and stimulant; mostly used as an aromatic addition to, or a **corrective** of, purgative preparations. Dose: 8 to 30 gr. (0.5 to 2 Gm.).

386 a. **OLEUM CORIANDRI.**—An almost colorless or yellowish volatile oil with the characteristic aromatic odor and taste of the fruit; specific gravity 0.863 to 0.875; neutral in reaction. It is one of the most stable of the volatile oils in its power of resisting oxidation when exposed. It consists mainly of d-linalool or coriandrol, $C_{10}H_{18}O$. Stimulant and carminative, like the other aromatic oils. Dose: 1 to 5 ♏ (0.065 to 0.3 mil).

OFFICIAL PREPARATIONS.

Spiritus **Aurantii Compositus** (2.0 per
    cent.),. . . . .   Dose: 1 to 4 fl. dr. (4 to 15 mils) linalool.
Syrupus **Sennæ** (0.5 per cent.),. . . . . .   1 fl. dr. (4 mils).

387. **ANETHUM.**—DILL FRUIT OR DILL SEED. The fruit of Ane'thum **graveo'-**
lens Linné, an herb of Levant and Southern Europe. Oval-oblong, usually
separated into the two thin mericarps; these have a smooth brown surface,
with five ribs, the two lateral ones expanded into a lighter colored, **membranous wing surrounding the** fruit; oil-tubes six, two on the concave inner face
and one in each interval between the ribs; odor and taste caraway-like, depending upon the volatile oil, the heavy portion of which is doubtless carvol.
Stimulant, carminative, and stomachic. Dose: 8 to 30 gr. (0.5 to 2 Gm.).

387 a. **OLEUM ANETHI.**—Pale yellow, with the characteristic odor of the fruit,
and a pungent, sweetish, acrid taste. It is official in the British Pharmacopœia, where it is sometimes used to prepare dill-water.

388. **APIUM.**—CELERY FRUIT. From A'pium graveo'lens Linné, N.F., the common celery of our gardens, native to Levant and Southern Europe. Roundish-ovate, very small, brown cremocarps, generally separated into the two
mericarps, which have five ribs and about six oil-tubes. They contain a
volatile oil and a yellowish liquid principle, **apiol,** an oleoresinous substance,
but somewhat analogous to the fixed oils; this apiol is chiefly extracted for
medicinal use from parsley, however; it is used as an emmenagogue in doses
of 10 to 15㎜ (0.6 to 1 mil).

Preparation of Apiol.—The simplest process for its separation is to exhaust
the fruit with petroleum-benzene, evaporate the solvent, and treat the residue with
strong alcohol. On evaporation, the apiol remains. A process resulting in a pure,
almost colorless apiol is published in "Pharm. Archiv," Feb., 1899. Dose: 7½ to
23 gr. (0.5 to 1.5 Gm.).

Celery is stimulant, antispasmodic, and carminative. Dose of fl'ext.:
5 to 15 ㎜ (0.3 to 1 mil).

389. **AJOWAN.**—The fruit of Ca'rum ajow'an Bentham and Hooker. *Habitat:*
Southern Asia and Egypt. Ovate, somewhat compressed laterally, about
2 mm. (1/12 in.) long, with a rough, grayish-brown surface; mericarps usually
separated, containing six oil-tubes. The large fruits much resemble those
of common parsley, but are readily distinguished from them and other small
umbelliferæ by their odor and very rough surface. Odor thyme-like; taste
pungent and aromatic, due to a volatile oil, 5 to 6 per cent., which consists of
a terpene, cymene, and the stearopten, thymol. Ajowan is one of the commercial sources of this stearopten. Oil of ajowan, when freshly distilled, is colorless, but soon acquires a slightly yellow tinge. It has an acrid, burning taste.
Carminative, stomachic, having the same properties as thymol (see below).
Dose: 10 to 30 gr. (0.6 to 0.2 Gm.).

390. THYMOL

THYMOL

A phenol, $C_{10}H_{13}OH$, obtained by fractional distillation of oils from Thymus **vul**
garis, **Carum ajowan,** and **Monarda punctata.** That portion coming over at
392°F. 260°C.) is separately collected and subjected to freezing, when thymol
crystallizes out; or by distilling off a greater part of the light oils or hydrocarbons and obtaining the thymol from the remaining heavier liquid by the
use of caustic soda and HCl.

DESCRIPTION.—Small, colorless scales or large, translucent crystals of the
hexagonal system having a **thyme-like** odor and pungent taste,
somewhat caustic to the lips. It melts at about 50°C. (122°F.), but

does not crystallize again until a much lower temperature is reached. Sparingly soluble in water (1:1200), but dissolves in less than its own weight of alcohol, ether, or chloroform. The crystals have a specific gravity of 1.069, but the melted liquid is lighter than water. Chemically, thymol is considered as isopropyl-meta-cresol ($C_6H_3.-CH_3.OH.C_3H_7$), and is closely related to carvacrol, which is regarded as isopropyl-ortho-cresol; the two differing in the relative position of the hydroxyl group. When 2 Gm. are volatilized on waterbath not more than 0.05 per cent. of residue should remain. It should melt from 48° to 51°C.

As a solid it is heavier than water but when liquefied by fusion is lighter than water.

ACTION AND USES.—Stimulant and powerful antiseptic, generally applied externally in ointment or lotion, or in a spray, considered almost as a specific in Hookworm disease. Aristol.—A name applied to thymol iodide (*q.v.*). Internal dose: 1 to 2 gr. (0.065 to 0.13 Gm.).

391. **PETROSELINI RADIX**, N.F.—PARSLEY. The root of **Petroseli'num sati'-vum** Hoffman, native to Southern Europe, but cultivated extensively as a common garden plant. A tapering root from 100 to 200 mm. (4 to 8 in.) long, and about 12 mm. (½ in.) thick; externally yellowish or light brown, marked with close annular rings above and longitudinal wrinkles at the lower end; fracture short, showing a thick bark dotted with resin cells, and a porous, pale yellow wood, with very irregular, white medullary rays. When fresh, it has a strong, aromatic odor, but is only faintly so when dry; taste sweetish, slightly aromatic. It is the chief source of apiol (also found in celery), a yellowish liquid somewhat analogous to the fixed oils, given as an emmenagogue in doses of 10 to 15 ♏ (0.6 to 1 mil). The root is given in infusion as a carminative, and as a laxative and diuretic in nephritic and dropsical affections. Dose: 30 to 60 gr. (2 to 4 Gm.).

391 a. **PETROSELINUM**, U.S.P. IX, applies this term to the fruit which is ovate, about 2 mm. (½₂ in.) long, with a greenish or brownish-gray surface, the mericarps usually separated. It contains the same principal ingredients, and is used for about the same purposes as the root. Dose: 8 to 30 gr. (0.5 to 2 Gm.). See Apiol 391 b.

Powder.—Microscopical elements of: See Part iv, Chap. I, B.
Official Preparation.—Oleoresina Petroselini.

391 b. **APIOL** (L. *apinum*, parsley, + ol), an oleoresinous liquid, heavier than water, of a persistent odor, distinct from the plant, and an acrid, pungent taste; from certain umbelliferous fruits, chiefly parsley "seed" (fruit). A crystalline compound, $C_{12}H_{14}O_4$, a purified *apiol* (parsley camphor) is obtainable. Dill oil yields a liquid *apiol* which has the same composition as the crystallizable *apiol* from the parsley. (See also 388.)

392. **PHELLANDRIUM.**—WATER DROPWORT. FIVE-LEAVED WATER HEMLOCK. The fruit of a European aquatic plant, **Œnan'the phellan'drium** Lamarck. From 2 to 3 mm. (½₂ to ⅛ in.) in length, terete, oblong, narrowed at one end, and crowned with the stylopodium; yellowish-brown or blackish-brown in color; taste aromatic, slightly acrid; odor strong, somewhat caraway-like, but disagreeable. Its aromatic properties depend upon a volatile oil, but there are indications of a narcotic alkaloid, possibly coniine, as the characteristic mouse-like odor is developed when the powdered seeds are rubbed with a solution of potassa. Slightly narcotic, stimulant, but more particularly used in chronic affections of the air-passages, as bronchitis, etc. Dose of powder about 5 gr. (0.3 Gm.), cautiously increased.

393. **CUMINUM.**—Cumin Seed. The fruit of **Cumi′num cym′inum** Linné. *Habitat:* Egypt; cultivated in Southern Europe. Resembles caraway, but may be distinguished by its entirely different, peculiar, heavy odor, and in being whole fruits and not half-fruits, as in the latter; surface brown, rough, and hairy; ribs 18, oil-tubes 6; taste aromatic, bitterish, disagreeable. It contains a volatile oil, often used as a carminative, which consists of three different oils (two hydrocarbons and cuminol). Cumin is much stronger as a stimulant than the other umbelliferous fruits. Dose: 8 to 30 gr. (0.5 to 2 Gm.).

394. **CAROTA.**—Carrot Fruit. From wild plants of **Dau′cus caro′ta** Linné. *Habitat:* United States and Europe. Light, oval-oblong fruits, dorsally compressed; mericarps usually united, brownish, each with five hairy primary ribs and four more prominent secondary ones beset with long, white bristles; odor aromatic; taste warm, bitterish. Aromatic stimulant, diuretic. Dose: 8 to 30 gr. (0.5 to 2 Gm.).

395. **ANGELICA ATROPURPUREA.**—American Angelica. (Root.) This highly aromatic root was official in the U.S.P., 1860-′70. It is similar to—

396. **ANGELICA, A. OFFICINALIS.**—European or Garden Angelica. (Root.) The aroma is due to a fragrant volatile oil. Also contains **angelic** acid (also found in sumbul), which has an action on the nerves. *Description:* Root-stock 5 to 10 cm. (2 to 4 in.) long, 2.5 to 5 cm. (½ in.) thick, crowned with remnants of leaf-bases, rather thick bark, curved yellowish, porous wood-wedges, a whitish pith, spongy, especially in root-branches, radiating lines of large resin-ducts in the bark, bast rays destitute of bast fibers. Aromatic stimulant, stomachic, and carminative. Dose: 30 to 60 gr. (2 to 4 Gm.).

Angelica Fructus, the ripe fruits of Angelica Archangelica, Linné, and Angelica Radix, the rhizome and roots of Angelica Atropurpurea, Linné, are recognized in the National Formulary.

397. ASAFŒTIDA.—Asafetida

ASAFETIDA

A gum-resin obtained by incising the rhizomes and roots of Ferula asafœtida, Linné, of Feru′la fœ′tida Regel, and some other species of Ferula.

Botanical Characteristics.—A gigantic herbaceous plant, 10 feet high, with radical *leaves* 18 inches long, bipinnate; *calyx* nearly obsolete, consisting of 5 minute points. *Fruit* broadly elliptical, thin, foliaceous, with dilated border; vittæ inconspicuous.

Source.—This plant, and other species from which commercial asafetida is procured, grows in Western Thibet, Kashmir, Persia, Turkestan, and Afghanistan. The plant is cut off at the root, and the milky juice exuding is allowed to harden, the sun being excluded by branches and leaves thrown over the cut surface; when it has solidified it is scraped off, and another slice of the root is cut off to expose a fresh surface, this operation being continued until the root is exhausted.

Description of Drug.—Masses composed of white tears of various shapes and sizes, imbedded in a brown, sticky mass, along with vegetable trash and earthy impurities. These masses are at first soft, but harden on exposure, the tears breaking with a conchoidal fracture, at first milk-white, but gradually turning pink, and at last brown. It resembles galbanum very much in appearance, but is easily distinguished by its strong, disagreeable, alliaceous odor, due to a sulphuretted volatile oil present to the extent of 3 to 9 per cent. On adding ammonia to a decoction of the sublimated resin, a blue fluorescence is exhibited. Taste acrid, bitter, and alliaceous.

When assayed by the official process asafœtida should contain not less than 60 per cent. of alcohol soluble constituents.

VARIETIES.—Besides the above-described variety, the amygdaloid, which is the most common, there are other forms in which it enters the market:

Liquid asafœtida is a permanent, syrupy liquid, white, turning brown on exposure.

Asafœtida in tears is the purest variety.

Stony asafœtida, never used medicinally, consists of pieces of gypsum or other earthy material coated with a thin layer of the milk-juice.

CONSTITUENTS.—The greater part of asafœtida consists of a gum (20 to 30 per cent.) and resin (50 to 70 per cent.). These, with the volatile oil (3 to 9 per cent.), form with water a milky emulsion. The resin is regarded by Tschirch as the ferulic ester of asaresino-tannol, $C_{24}H_{35}O_5$, which, by sublimation, yields umbelliferone. There is also contained in the drug vanillin 0.06 per cent., ferulic acid, $C_{10}H_{10}O_4$, 1.28 per cent. The resin, when fused with KOH, yields resorcin and protocatechuic acid. The mineral impurities often amount to 40 %, especially in that imported from Herat, where it is adulterated with red clay. Ash (of Resin), not to exceed 15 per cent.; (Powder), not to exceed 30 per cent.

For an exhaustive treatise on Gum Resins, etc., the student is referred to "Analysis of Resins, Balsams and Gum Resins, Their Chemistry and Pharmacognosis," by Carl Dietrich (Scott, Greenwood & Co., London).

ACTION AND USES.—Asafœtida combines the properties of a stimulating antispasmodic with those of an efficient expectorant, making it a valuable remedy in spasmodic affections of the respiratory tract, as whooping-cough, asthma, etc. It is also a laxative, especially useful in cases of flatulence. Dose: 5 to 8 gr. (0.3 to 0.5 Gm.).

OFFICIAL PREPARATIONS.

Emulsum Asafœtidæ (4 per cent.),......Dose: 2 to 4 fl. dr. (8 to 15 mils).
Tinctura Asafœtidæ (20 per cent.),...... 10 to 40 ℥ (0.6 to 2.6 mils).
Pilulæ Asafœtidæ (each pill containing
  about 3 gr. of asafœtida, with soap as
  an excipient),................... 2 to 5 pills.

398. GALBANUM.—GALBANUM. A gum-resin imported from Persia, but the botanical source of which is not definitely decided; it is generally considered, however, as a spontaneous exudation from Feru'la galbani'flua Boissier et Buhse, and other species of Ferula, large plants growing in that region. It is usually met with in pale yellow or brownish tears, ranging in size from a pea to a hazelnut, occasionally separate and with a shining, varnished surface, but

more generally agglutinated into a more or less hard mass by means of a darker, yellowish-brown, sometimes greenish, substance. In winter this mass has the consistence of firm wax, but in the heat of summer it becomes soft and sticky; odor balsamic; taste acrid and bitter.

CONSTITUENTS.—Besides gum and resin, it contains the interesting principle, **umb**elliferone (common to many umbelliferous plants), acicular crystals, producing a brilliant blue fluorescence on the addition of an alkali.

ACTION AND USES.—Stimulant, expectorant, and antispasmodic. Dose: 5 to 8 gr. (0.3 to 0.5 Gm.).

399. **AMMONIACUM.**—GUM AMMONIAC. A gum-resin exuding from **Dore'ma** ammoni'acum Don. Off. U.S.P., 1890. Roundish tears varying in size from 1.5 to 12 mm. ($\frac{1}{16}$ to $\frac{1}{2}$ in.) in diameter, externally yellow or pale yellowish-brown. When warm it is of the consistence of wax, but it becomes brittle when cold, breaking with a milk-white, waxy fracture, translucent at the edges; odor balsamic, stronger on heating; taste acrid, bitter, and nauseous. Lump ammoniac is an inferior quality in which the tears are agglutinated. Cake ammoniac is a very impure, dark-colored, resinous mass exuding from the roots; imbedded in it are a few tears and much vegetable and earthy trash; it is not used internally. *Constituents:* Volatile oil, gum resembling acacia, resin (about 70 per cent. composed of two, one acrid resin and one indifferent resin); it yields no umbelliferone. By fusing with KOH, yields protocatechuic acid and resorcin, $C_6H_6O_2$. Among the derivatives of the acid resin are salicylic acid, ammoresinotannol, etc. Similar to asafœtida— stimulating expectorant, antispasmodic and laxative—but less powerful. Dose: 10 to 30 gr. (0.6 to 2 Gm.).

Emulsum Ammoniaci (4 per cent.),
   U.S.P. 1890,....................Dose: $\frac{1}{2}$ to 1 fl. oz. (15 to 30 mils).
Emplastrum Ammoniaci cum **Hydrargyro** (72 per cent., with mercury, oleate of mercury, dilute acetic acid, and lead plaster), U.S.P. 1890.

400. SUMBUL.—SUMBUL

MUSK ROOT

The rhizomes and roots of **Feru'la** sum'bul (Kauffmann) Hooker filius.

BOTANICAL CHARACTERISTICS.—*Root* fusiform; perennial stem 8 to 10 feet high. *Fruit* oblong-ovate, monocarpous. When punctured, the branches yield an angelica-flavored milk-juice.

HABITAT.—Regions north and east of British India.

DESCRIPTION OF DRUG.—**Transverse segments** about 10 to 50 mm. ($\frac{2}{5}$ to 2 in.) long, and 25 mm. (1 in.) thick. They have a **dusky-brown,** wrinkled bark, just beneath which is a whitish, spongy, parenchymatous layer, under the microscope dotted with brown, translucent, resinous exudations from large resin-ducts. **The brownish-yellow interior** is a **spongy mass** consisting of coarse fibers, easily separable, and indiscriminately mixed and twisted with the medullary rays; fracture short and fibrous. **Odor musk-like;** taste sweetish at first, becoming bitter and balsamic, and leaving a sensation of warmth in the mouth and throat. E. M. Holmes recommends that the true root be cultivated, which he thinks possible in temperate and mountainous districts in the colonies or in ordinary gardens and fields of England. The true root has a strong, persistent, musky odor.

Powder.—Characteristic elements: See Part iv, Chap. I, B.

Constituents.—**Sumbulic or angelic acid,** $C_5H_8O_2$, a small quantity of **valerianic acid,** $C_5H_{10}O_2$, and a small percentage of bluish volatile oil, to which, however, its odor is not due, but to two balsamic resins, or probably to some principle connected with them not yet isolated. The oil contains umbelliferone, $C_9H_6O_3$.

Action and Uses.—**Antispasmodic** (due to the angelic and valerianic acids contained), stimulant, and tonic. Dose: 15 to 30 gr. (1 to 2 Gm.).

Official Preparations.

Fluidextractum Sumbul, ...Dose: 30 ♏ (2 mils).
Extractum Sumbul,...............   5 to 15 gr. (0.3 to 1 Gm.).

401. **IMPERATORIA.**—Masterwort. The root of Imperato'ria ostru'thium Linné. *Habitat:* Southern Europe. A conical root with a dark brownish-gray, annulated and tuberculated bark, inclosing a whitish wood-circle and a resin-dotted central pith; odor angelica-like; taste pungent and bitter. It is a stimulant aromatic, but is rarely used in this country.

402. **LASERPITIUM.**—White Gentian. The root of Laserpi'tium latifo'lium Linné. *Habitat:* Central Europe. Somewhat conical, wrinkled and annulated above, branched below; wood whitish, porous, deprived of the brown, corky layer; aromatic and bitter. Used as a tonic and stimulant. Dose: 15 to 60 gr. (1 to 4 Gm.).

403. **LEVISTICUM.**—Lovage. The root of an aromatic European herb, **Ligus'-ticum levis'ticum** Linné. This is thick, sparingly beset with fibers, and has an annulate, reddish-brown bark, inclosing a porous yellow wood; it has an aromatic odor resembling that of angelica, and a sweetish, aromatic, and pungent taste, somewhat bitter. Its medicinal properties are similar to those of angelica, being used as an aromatic stimulant and carminative, and as an adjuvant to tonic mixtures. Dose: 8 to 30 gr. (0.5 to 2 Gm.), in infusion.

The root of Ligus'ticum [filici'num, Osha or Colorado Cough Root, has enjoyed some notoriety as an expectorant.

404. **PIMPINELLA.**—N.F. Pimpernel. The root of **Pimpinel'la saxifra'ga** Linné. *Habitat:* Europe. Diaphoretic, diuretic, and stomachic. It has also been employed in chronic catarrh, asthma, dropsy, amenorrhœa, etc., and as a masticatory in toothache. Dose: 15 to 30 gr. (1 to 2 Gm.), in infusion or powder.

405. **THAPSIA GARGANICA** Linné.—(Root.) Used chiefly as a counter-irritant in rheumatism, gout, bruises, etc.

406. **CICUTA MACULATA.**—American Water-hemlock. Wild Parsnip. The root and leaves of **Cicu'ta macula'ta** Linné. Poisonous, sedative, narcotic; **resembles conium in action** and has been used in its stead, but the two drugs should not be confounded when conium is prescribed, as it sometimes is, by its old name, cicuta. Dose: 3 to 5 gr. (0.2 to 0.3 Gm.). Children have been poisoned by eating the fresh root, which resembles parsnip in taste and smell.

407. **ERYNGIUM AQUATICUM** Linné.—Water Eryngo. Rattlesnake's Master. *Habitat:* United States. (Root.) Diaphoretic and expectorant, and has been used as a substitute for senega. Dose of fluidextract: 20 to 40 ♏ (1.3 to 2.6 mils).

408. **OSMORRHIZA LONGISTYLIS** De Candolle.—Sweet Cicely. *Habitat:* United States and Canada. (Root.) Aromatic, stomachic, carminative, and expectorant. It contains a volatile oil identical with oil of anise. Dose: 1 to 2 dr. (4 to 8 Gm.).

CORNACEÆ.—Dogwood Family

409. CORNUS FLORIDA.—Dogwood. Cornus, N.F. The root-bark of **Cor'nus flori'da** Linné. *Habitat:* North America. Appears in pieces of various sizes, generally broken up and more or less curved; about 2 mm. ($\frac{1}{12}$ in.) in thickness when deprived of its brownish-gray cork, as it generally is, with a fawn-colored outer surface; inner surface red, due to the tannin contained, plainly radially striate; fracture short, whitish, showing numerous striæ of brownish-yellow stone cells. Inodorous; taste astringent and bitter, the bitter principle being termed cornin. It yields a grayish powder. tinged with red. Tonic and astringent, and almost equal to cinchona as an **antiperiodic** in intermittent fevers. Dose: 10 to 30 gr. (0.6 to 2 Gm.). The barks of two other dogwoods, *Cor'nus circina'ta* (green osier bark or round-leaved dogwood bark) and *Cor'nus serice'a*, are often used.

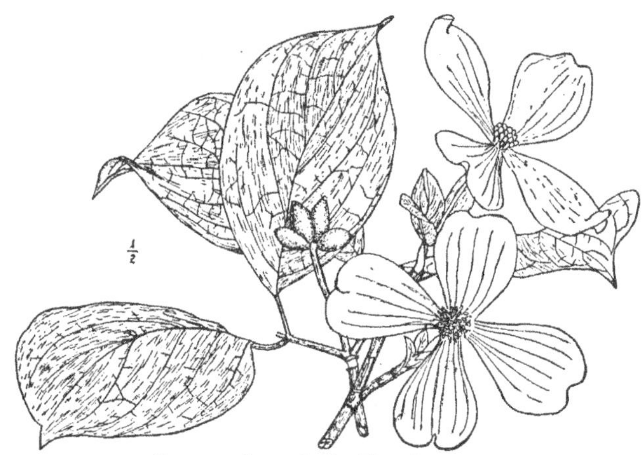

Fig. 193.—*Cornus florida*—Flowering branch.

410. GARRYA FREMONTII Torrey.—California Fever Bush. (Leaves.) Used as a tonic and antiperiodic in chills and fevers. They contain a bitter principle similar to quinine in therapeutic action. Dose: 15 to 30 gr. (1 to 2 Gm.).

ERICACEÆ.—Heath Family

Trees or shrubs, rarely herbs; *leaves* generally foliaceous; *flowers* regular, gamopetalous, usually bell-shaped or urn-shaped; *anthers* two-celled, with porous dehiscence. A large order, with *leaves* astringent and bitter, because of the presence of glucosides. Some species contain a poisonous principle, andromedotoxin.

Synopsis of Drugs from the Ericaceæ

A. *Leaves.*
  UVA URSI, 411.
  Arctostaphylos, 412.
  Gaultheria, 413.
  *Chimaphila, 414.
  Epigæa, 415.
  Vaccinium, 416.

  Kalmia, 417.
  Ledum, 418.
  Oxydendrum, 419.
  Rhododendron, 420.
B. *Volatile Oil.*
  OLEUM GAULTHERIÆ, 413 a.

22

411. UVA URSI.—Uva Ursi

BEARBERRY

The dried leaves of Arctostaph'ylos u'va ur'si (Linné) Sprengel, with not more than 5 per cent. of stems or other foreign matter.

BOTANICAL CHARACTERISTICS.—Shrubs with trailing stems. *Leaves* alternate, coriaceous, evergreen, obovate or spatulate, entire. *Flowers* in terminal racemes, nearly white; *corolla* urn-shaped. *Fruit* a red drupe.

SOURCE.—In dry, sandy, or rocky soil from Hudson's Bay to New Jersey, in some parts of which it grows in abundance.

RELATED SPECIES.—*Arctostaphylos glauca*, indigenous to California (412).

DESCRIPTION OF DRUG.—**Short-stalked, rather thick, coriaceous, obovate leaves,** about 20 mm. ($\frac{4}{5}$ in.) in length, rounded at the apex and narrowed at the base; margin entire; surface smooth, glossy, grayish-green above, lighter colored and reticulated below; taste astringent, bitter; odor slight. (The powder has a hay-like odor.)

They are sometimes adulterated with the leaves of *Vaccinium vitis ideæ* (European uva ursi), distinguished from the genuine by their rounder shape, their revolute margin, which is sometimes toothed, and the dotted appearance of their under surface. Chimaphila leaves, which are occasionally mixed with uva ursi, may be readily distinguished by their greater length, their cuneiform-lanceolate shape, and their serrate edges. *Leiophyllum buxifolium* (sand myrtle) and *Epigæa repens* (trailing arbutus, 415) are also used as adulterants.

Powder.—Characteristic elements: See Part iv, Chap. I, B.

CONSTITUENTS.—**Tannic and gallic acids,** and the three principles, **arbutin,** $C_{12}H_{16}O_7$, **ericolin,** $C_{34}H_{56}O_{21}$, and **ursone,** $C_{10}H_{16}O$, which are common to the plants of the natural order Ericaceæ. Arbutin is a bitter glucoside, occurring in colorless crystals; it is resolved by hydrolysis into glucose and hydroquinone or arctuvin, $C_6H_6O_2$. Ericolin is a yellow, crystalline, bitter glucoside. Ursone is in tasteless needles.

Preparation of Arbutin.—Precipitate decoction with lead acetate; filter; **add** H₂S; evaporate; evaporate slowly, when needles crystallize out. Dilute Fe₂Cl₆ gives blue color. Dose: 5 to 15 gr. (0.3 to 1 Gm.).

Preparation of Ursone.—Obtained by exhausting drug with ether. The alcoholic solution of the ethereal residue yields the crystals on slow evaporation. Occurs in tasteless needles; sparingly soluble in alcohol and ether. Insoluble in water.

ACTION AND USES.—Astringent, tonic, and **diuretic;** valuable in ulcerations of the kidneys, bladder, or urinary passages. It has been recommended in cystitis, its action being due to the decomposition of arbutin in the system and the excretion of the hydroquinone, which is a powerful disinfectant and antiferment. Dose: 15 to 60 gr. (1 to 4 Gm.).

Official Preparation.

Fluidextractum Uvæ Ursi,.   Dose: 15 to 60 ♏ (1 to 4 mils).

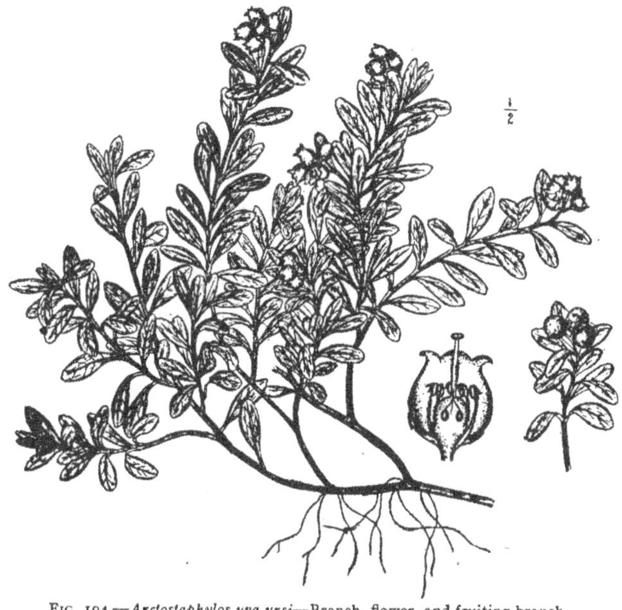

Fig. 194.—*Arctostaphylos uva ursi*—Branch, flower, and fruiting branch.

412. ARCTOSTAPHYLOS GLAUCA Lindley.—Manzanito. This is a small California evergreen tree or shrub whose leaves are there highly esteemed as an astringent, like uva ursi.

413. GAULTHERIA.—Wintergreen. Checkerberry. Partridge Berry. The leaves of Gaulthe'ria **porcum'bens** Linné. *Habitat:* Northern Hemisphere. This is a small evergreen shrub, consisting of slender, erect, reddish stems, bare below, leafy at top, rising at intervals from a creeping root to the height of eight or ten inches. Fruit a scarlet-red, berry-like, fleshy capsule. **Leaves** roundish, oval or obovate, about 37 mm. (1½ in.) long, on a short pedicel; coriaceous; margin serrate, with a few appressed teeth; somewhat revolute at the edges; odor fragrant, especially when chewed; taste aromatic, astringent. The fragrance is due to a volatile oil (413 a). Stimulant, astringent, and diuretic. Dose: 15 to 60 gr. (1 to 4 Gm.).

413 a. OLEUM GAULTHERIÆ.—Oil of Wintergreen. A nearly colorless volatile oil, distilled from the fresh leaves, "consisting almost entirely of methyl salicylate, $CH_3C_7H_5O_3$, and nearly identical with volatile oil of betula." The latter, according to Power, is composed entirely of methyl salicylate and is optically inactive, while the former is lævogyrate. As it comes into market it is of a brownish-yellow or reddish color and has a very agreeable, characteristic odor and taste. **Specific gravity of 1.172–1.182.** An aqueous solution gives, with ferric salts, a purplish color. It yields, with 6 parts of 70 per cent. alcohol at 20°C., a perfectly clear solution—a property serving to detect adulterations. (Official as Methyl Salicylas, 413 b).

413 b. METHYL SALICYLAS.—A product yielding not less than 98 per cent. of methyl salicylate ($CH_3C_7H_5O_3$). *It is produced synthetically* or obtained by distillation from Betula lenta Linné, or from Gaultheria procumbens Linné, and the source from which it is derived must be stated on the label.

Specific gravity at 25°C.: Synthetic 1.18 to 1.185; when from Sweet Birch or Gaultheria 1.172 to 1.182.

Most of the so-called "true" oil of wintergreen is made by distilling a mixture. of wintergreen leaves and the bark of the sweetbirch.

Dose of Methyl Salicylas: 12 minims (0.75 mils), U.S.P. IX.

414. CHIMAPHILA.—Chimaphila N.F.

PIPSISSEWA. PRINCE'S PINE

The dried leaves of Chimaph'ila umbella'ta Nuttall.

BOTANICAL CHARACTERISTICS.—A low, nearly herbaceous plant, with long, running, underground shoots. *Leaves* evergreen, thick, and shining, whorled, wedge-lanceolate, sharply serrate, not spotted. *Flowers* umbelled, on a terminal peduncle; *petals* rose-color; *anthers* violet. *Capsule* 2- to 5-celled.

HABITAT.—Northern Hemisphere, in dry woods.

DESCRIPTION OF DRUG.—Oblanceolate, about 25 to 50 mm. (1 to 2 n.) in length· sharply serrate, with pointed apex, cuneiform and entire at base; coriaceous; surfaces smooth, upper dark green, glossy, lower lighter in color; odor slight; taste astringent, slightly bitter.

Chimaphila maculata (spotted wintergreen or pipsissewa) has the same medicinal qualities, but differs physically in being oval-lanceolate, with a paler upper surface, and in being dotted with small white holes along the midrib.

RELATED PLANTS.—*Pyrola rotundifolia* (known as wintergreen or shin leaf), *P. elliptica* and *P. chlorantha* are used similarly to the above.

Powder.—Brownish-green. Characteristic elements: Parenchyma, mesophyll with irregular reddish-brown tannin masses, other cells with few starch grains, simple or compound, calcium oxalate crystals, aggregate (40 to 60 μ in diam.); stomata and few tracheids present.

CONSTITUENTS.—Same as uva ursi (411) with the addition of chimaphilin, $C_{24}H_{21}O_4$ (yellow, odorless, tasteless, volatile crystals), and several white crystalline principles.

Preparation of Chimaphilin.—When the leaves are distilled with water, yellow crystals are deposited in the neck of the retort. These, dissolved out with chloroform, will deposit from this solution on evaporation. Shaking out the tincture with chloroform will also dissolve out the principle.

ACTION AND USES.—Like uva ursi (411). Also used in scrofula and other cutaneons eruptions. Dose: 15 to 60 gr. (1 to 4 Gm.).

[FIG. 195.—*Chimaphila umbellata.*

415. **EPIGÆA.**—TRAILING ARBUTUS. GRAVEL PLANT. The leaves of **Epigæ'a re'pens** Linné. *Habitat:* North America, on woody hillsides. Ovate, about 50 mm. (2 in.) long, with heart-shaped base and mucronate apex; coriaceous; margin entire. They contain the same three principles that uva ursi does and have the same general medicinal properties, but are particularly valuable in those cases of local irritation of the urinary organs in which they have often given relief when uva ursi and buchu had failed. They are also claimed to be highly beneficial in lithic acid gravel. Dose: 15 to 60 gr. (1 to 4 Gm.), in decoction or fluidextract.

416. **VACCINIUM CRASSIFOLIUM** Andrzejowski.—The leaves of this indigenous shrub have properties very much resembling uva ursi and may be used in its stead.

417. **KALMIA.**—MOUNTAIN LAUREL. SPOONWOOD. The leaves of **Kal'mia latifo'lia** Linné, an evergreen shrub common on the hills and mountains

of North America. They are lance-oval, acute at both ends, about 50 to 75 mm. (2 to 3 in.) in length; petiolate; coriaceous; both surfaces smooth, green. In medicinal doses kalmia is astringent, sedative to the heart, and antisyphilitic; also used externally in skin diseases. From its affirmed effect upon sheep and other small animals it is supposed to have toxic, narcotic properties, but no such principle has yet been found. Dose: 10 to 30 gr. (0.6 to 2 Gm.).

418. **LEDUM.**—LABRADOR TEA. The leaves of Le'dum latifo'lium Aiton. *Habitat:* Canada and Northern States. Elliptic-oblong, covered beneath with a rust-colored wool. Besides the tannin and other principles common to the Ericaceæ, they contain a poisonous principle, andromedotoxin, rendering them poisonous in large doses. Astringent, tonic, and alterative. Dose: 15 to 30 gr. (1 to 2 Gm.), in infusion.

419. **OXYDENDRUM ARBOREUM** De Candolle.—SOURWOOD. The leaves of this North American tree are tonic, diuretic, and refrigerant, used in dropsy. Dose of fluidextract: ½ to 2 fl. dr. (2 to 8 mils).

420. **RHODODENDRON MAXIMUM** Linné.—GREAT LAUREL. (Leaves.) Tonic, diuretic, astringent, expectorant. Dose of fluidextract: 5 to 15 gr. (0.3 to 1 Gm.).

MYRSENEÆ

421. **EMBELIA RIBES** Burman.—The pepper-corn-like, aromatic fruit of this East Indian plant is said to be an efficient tæniafuge.

PLUMBAGINEÆ.—Leadwort Family

422. **STATICE.**—MARSH ROSEMARY. The root of Stat'ice limo'nium Linné, growing in flat marshes along the Atlantic coast of the United States. Spindle-shaped, from 300 to 600 mm. (12 to 24 in.) long, and about 25 mm. (1 in.) thick; externally rough, purplish-brown; bark thick; wood yellowish, in narrow wood-wedges; inodorous; bitter and strongly astringent. It contains about 12 per cent. of tannin and is used like catechu and kino in diarrhea, but more particularly as an astringent gargle in ulcerations of the mouth and throat, and as an injection. Dose: 10 to 30 gr. (0.6 to 2 Gm.).

423. **BAYCURU.**—The root of Statice brasilien'sis Boissier. *Habitat:* Brazil. One of the most powerful of astringents, chiefly used locally in gargle, injection, and lotion.

PRIMULACEÆ.—Primrose Family

424. **ANAGALLIS ARVENSIS** Linné.—SCARLET PIMPERNEL. This plant, growing in the United States and Europe, is applied locally to ulcers and employed internally in consumption, dropsy, etc. It contains a pepsin-like ferment

SAPOTACEÆ.—Sapodilla Family

425. **GUTTA-PERCHA, N.F.**—The concrete juice of large trees, **Isonandra** (or *Palaquium oblongifolium*), **Dichop'sis gut'ta,** and other species, growing in the Malay Peninsula and the East Indies. In grayish or yellowish masses, often streaked with red; hard and tenacious at ordinary temperatures, with a somewhat unctuous feeling, but at a higher temperature, or when immersed in hot water, it becomes plastic, retaining, when hard and dry, the form into which it has been molded. Upon this property its uses in the arts chiefly depend. In medicine it is used as a surgical dressing in the formation of splints, supports, etc. A Liquor Guttæ Perchæ is often applied as a protective, the evaporation of its solvent, carbon disulphide, leaving a thin, flexible coating over the wounded surface.

426. **MONESIA.**—An extract obtained from a South American tree, Lucu'ma glycyph'læa Martius et Eiohler. Dark brown, almost black, cakes, about 25 mm. (1 in.) in thickness; very brittle, often coming into the market in

broken fragments; inodorous; taste sweetish, astringent, and then acrid, its acrimony being very persistent, especially in the fauces. This acridity is due to **monesin,** a principle identical with saponin. Monesia also contains tannin, glycyrrhizin, and lucumin (silky needles). Stomachic stimulant, alterative, and astringent. Used in diarrhea, hemorrhages, in astringent gargles, and in powder or ointment applied to scrofulous ulcers. Dose: 5 to 20 gr. (0.3 to 1.3 Gm.).

EBENARCEÆ.—Ebony Family

427. **DIOSPYROS.**—Persimmon. The unripe fruit of Dio'spyros virginia'na Linné. (Official, 1820–'80.) Very astringent. Used in uterine hemorrhage, leucorrhœa, and sore throat. Dose: 15 to 60 gr. (1 to 4 Gm.) in infusion, syrup, or vinous tincture.

STYRACEÆ.—Storax Family

428. BENZOINUM.—Benzoin

BENZOIN

A balsamic resin obtained from Sty'rax ben'zoin Dryander, and other species of styrax.

Botanical Characteristics.—A large tree with tomentose branches. *Leaves* alternate, oblong, the under surface tomentose. *Inflorescence* compoundly racemose; *calyx* 5-toothed; *corolla* 5-parted, gray; *stamens* 10, their filaments coherent at the base into a short tube.

Source and Varieties.—Sumatra and Java. Sumatra-Penang, grayish-brown with many white tears, odor storax-like; Siam, reddish-brown, odor vanilla-like; Palembang resembles Sumatra, but yields more benzoic acid; false benzoin, catappa benzoin (*Terminalia angustifolia*), whitish brown.

A deciduous shrub of the Lauraceæ, a native of Virginia, and called spice-wood or Benjamin tree, was at one time thought to be a source of benzoin. The berries of this tree are aromatic, and have been used as a substitute for allspice.

Collection.—In Sumatra the benzoin is collected by making incisions in the tree during its seventh year, only the unhealthy trees yielding resin. The milky juice which flows first is the purest and most fragrant, but soon hardens upon exposure to the air. That which flows subsequently is brownish, and some is scraped out when the tree is cut down and split open, as it is soon killed by the process of tapping. These varieties are in common called head, belly, and feet benzoin, and have the relative value to each other of 105, 45, and 18, being esteemed according to their whiteness, semi-transparency, and freedom from admixture (Royle). A product of the younger tree furnishes a variety known as amygdalina benzoin, which contains whitish, almond-like tears diffused through its substance.

DESCRIPTION OF DRUG.—It exudes from incisions in the bark, hardening on exposure into agglutinated shining tears of a yellowish-brown or reddish-brown color; internally milk-white; usually, however, it is in various-sized pieces, having a resinous fracture, showing a **mottled surface** of smooth, shining white spots, tears, imbedded in the somewhat rough and porous, reddish-brown mass. It has a very agreeable odor and a slightly aromatic taste, leaving an irritating sensation in the mouth and fauces.

FIG. 196.—*Styrax benzoin*—Branch.

CONSTITUENTS.—Benzoin has the constitution of a balsam and is by some authors considered as a solid balsam; it contains resin, benzoic acid, $C_7H_6O_2$, 20 to 24 per cent., which comes off in dense white vapor when benzoin is heated and melted, and cinnamic acid, $C_9H_8O_2$ (in some varieties), detected by boiling in milk of lime, decomposing with HCl, and adding permanganate of potassium, when the odor of bitter almonds is given off. Siam benzoin contains vanillin, $C_8H_8O_3$, and has a vanilla-like odor. *Sumatra Benzoin.* Ash, not more than 2.5 per cent. *Siam Benzoin.* Ash, not more than .2 per cent.

Preparation of Benzoic Acid.—Obtained by simple sublimation of benzoin. Is also prepared artificially from toluol, but sometimes from phthalic acid or hippuric acid. Contamination with cinnamic acid is detected by mild oxidation when it yields the odor of oil of bitter almond.

ACTION AND USES.—Stimulant and diaphoretic, but seldom used as such except in the compound tincture of benzoin. It is used locally as a stimulant and irritant, and in tooth powders and fumigations. Dose: 8 to 30 gr. (0.5 to 2 Gm.).

OFFICIAL PREPARATIONS.

Tinctura Benzoini Composita (10 per
 cent., with aloes, storax, and tolu),..Dose: 15 to 60 ♏ (1 to 4 mils).
Tinctura Benzoini (20 per cent.),...... 10 to 40 ♏ (0.6 to 2.6 mils).
Adeps Benzoinatus (2 Gm. digested in 100 Gm. of lard).

OLEACEÆ.—Olive Family

Trees or shrubs with simple leaves—illustrated by the olive, the ash, the lilac, and the privet. The olive fruit contains mannite, which is converted into olive oil on ripening.

429. MANNA.—MANNA

MANNA

The concrete saccharine exudation from Frax'inus or'nus Linné.

BOTANICAL CHARACTERISTICS.—A tree about 25 feet high. *Leaves* pinnate, *leaflets* 7 to 9, serrate. *Panicles* dense; *calyx* 4-cleft; *corolla* white, divided to the base into linear segments.

SOURCE.—The tree yielding the manna is a native of Sicily, Calabria, and Apulia. The juice exudes spontaneously, or its flow is hastened by incision. Although this is the only manna officially recognized, saccharine substances known as mannas are yielded by many other trees and plants, and are obtained from the cocoons of some insects. The manna of Scripture was doubtless a lichen which grows extensively in the Sahara and Western Asia, and which occasionally falls like rain over the adjacent country.

DESCRIPTION OF DRUG.—In stalactiform pieces from 1 to 6 inches long, or irregular fragments, yellowish or brownish-white, internally white and porous; very friable. Manna in tears is a pure kind, but manna in flakes is chiefly valued and mostly met with. Manna in sorts, minute tears, internally crystalline, and fat manna, brownish viscid, non-crystalline masses, are also met with. Odor honey-like; taste sweetish, afterward nauseous. Soluble in water and alcohol. When long kept, manna darkens and deliquesces into a liquid.

OTHER MANNAS.—Manna occurs in irregular masses, consisting of brittle and soft resin-like fragments from yellowish-white to yellowish-gray color. The quantity of the yellowish-white fragments should not be less than 40 per cent. of whole. The varieties of manna generally distinguished in our commerce are large flake, small flake and sorts.

INFERIORITY.—Inferior manna may have a greenish color due to froxin, a fluorescent glucoside resembling æsculin. A new variety of manna from Rhodesia, probably derived from Gymnosporia deflexa is on the market.

ADMINISTRATION.—May be given to very young children as a gentle laxative. Given by dissolving in milk. When administered to adults it is combined with senna, rhubarb, and more energetic laxatives.

CONSTITUENTS.—Chiefly mannite (75 per cent.), a sweet principle which separates out from the boiling alcoholic solution in crystals, also sugar, dextrin, mucilage, and a nauseous principle, to which its laxative action is doubtless due.

ACTION AND USES.—Gentle laxative, usually given in combination with other purgatives. Dose: ½ to 2 oz. (15 to 60 Gm.).

PREPARATION.—Infusum Sennæ Compositum,...Dose: 1 to 4 fl. oz. (30 to 120 mils).

430. OLEUM OLIVÆ.—OLIVE OIL

SWEET OIL

A fixed oil expressed from the ripe fruit of O'lea europæ'a Linné:

BOTANICAL CHARACTERISTICS.—A small evergreen with hard wood. *Leaves* short-petiolate, opposite, ovate-lanceolate, mucronate. *Flowers* white, in axillary clusters. *Fruit* a drupe, ½ to 1 in. long, ovoid, purple, sarcocarp firm, fleshy, filled with oil.

HABITAT.—Levant and the Mediterrànean Basin and California.

DESCRIPTION OF OIL.—A pale yellow or greenish-yellow, unctuous liquid when pure, having a bland, sweetish taste, but scarcely any odor. Specific gravity 0.915 to 0.918 at 15°C. (59°F.). On exposure it absorbs oxygen and becomes thick and rancid and loses its color, but does not dry as does linseed oil.

The oil is obtained by crushing the ripe fruit and subjecting the pulp to strong pressure. The expressed oil is run into water and the floating oil is skimmed after a few days' subsidence (virgin oil); the expressed cake is now broken up, mixed with hot water, and again subjected to pressure, resulting in a second-grade oil. The remaining marc yields by solvents, such as carbon disulphide, or by a third expression after fermentation, a very inferior oil.

The oil is adulterated sometimes with cotton-seed oil chiefly, with oil of benne, and with peanut oil.

Preparations:
> Emplastrum plumbi,
> Unguentum diachylon.
> *Sapo*, soap, is employed in:
> Linimentum saponis.

From the olive is obtained the wood so famous for its capability of receiving a fine polish; used in cabinet work of various kinds. The unripe fruit is served at the table. It is prepared by repeatedly steeping it in water containing lime and ashes, then bottling in a slightly aromatic salt solution; the small French or Provence, the finest, and the large Spanish are both used for this purpose.

Fig. 197.—*Olea europæa*—Branch.

CONSTITUENTS.—At about 5°C. (41°F.) white crystalline granules separate out, which consist of palmitin with possibly some stearin and arachin. The liquid portion remaining consists almost entirely of olein, $C_3H_5(OC_{18}H_{33}O)_3$, which forms about 72 per cent. of the oil. The green color is due to chlorophyll.

ACTIONS AND USES.—Nutritive and laxative, a common ingredient in laxative enemata; externally **protective and emollient.** Its chief use in pharmacy is in liniments, cerates, and plasters. Dose: 1 fl. oz. (30 mils).

431. **FRAXINUS AMERICANA** Linné.—(Fraxinus, N.F.) WHITE ASH. (Bark.)
Quills·or curved pieces, having an ash-gray periderm and a white inner bark,
and breaking with a splintery, coarsely fibrous fracture. Emmenagogue.
Dose: about 15 gr. (1 Gm.).

432. **FRAXINUS SAMBUCIFOLIA** Lambert.—BLACK ASH. (Bark.) *Habitat:*
United States. Tonic and astringent. Dose: 1 to 4 dr. (4 to 15 Gm.).

433. **CHIONANTHUS**, N.F.—FRINGE-TREE. The root-bark of **Chionan'thus**
virgin'ica Linné. *Habitat:* United States. Tonic, aperient, and diuretic.
Dose: 15 to 60 gr. (1 to 4 Gm.).

434. **LIGUSTRUM**.—PRIVET. The leaves of Ligus'trum vulga're Linné, a
shrub growing wild in the United States and Europe. Astringent; the decoction is used in sore throat, ulcerations of the mouth, stomach, and intestines.

LOGANIACEÆ.—Logania Family

Herbs, shrubs, or trees, with opposite, entire *leaves* connected by stipules or
a stipular line, and with regular 4–5-merous, 4–5-androus *flowers*, the ovary free
from the calyx. Many of the plants belonging to this order are extremely
poisonous.

Synopsis of Drugs 'from the Loganiaceæ

A. *Seeds.*
  NUX VOMICA, 435.
  *Ignatia, 436.
B. *Bark.*
  Hoang-nan, 437.

C. *Rhizomes.*
  GELSEMIUM, 438.
  SPIGELIA, 439.
D. *Extractive.*
  Curara, 440.

435. NUX VOMICA.—NUX VOMICA

DOG BUTTON. QUAKER BUTTON

The dried ripe seed of Strych'nos **nux** vom'ica Linné yielding, by assay, not less
than 2.5 per cent. of alkaloids of Nux Vomica.

BOTANICAL CHARACTERISTICS.—A small tree with a crooked *stem* resembling
a dogwood. *Leaves* short-petiolate, smooth, oval, mucronately, palmately,
3- to 5-nerved. *Flowers* small, greenish-white, in terminal corymbs; *corolla*
funnel-form. Fr*uit* round, orange-like.

SOURCE.—Indigenous to the coasts of most parts of India, Burmah, Siam,
and northern parts of Australia. Large quantities of the drug are
brought into the London market from British India. The export
from Bombay is considerable. Madras and Calcutta are also
shipping points.

DESCRIPTION OF DRUG.—**Orbicular disks** from 18 to 25 mm. ($\frac{3}{4}$ to 1
in.) in diameter, and about 4 mm. ($\frac{1}{6}$ in.) thick; flat or slightly
convex on one side and concave on the other, with a slightly raised
margin on the concave side. On one side is a ridge (raphe) extending from a raised point in the center (hilum) to a point on the edge
where the radicle is situated (chalaza). Both surfaces have a **grayish
or a grayish-green, shiny, silky appearance, due to a large number
of silky hairs, closely pressed to the seed** and forming a tuft around
its edge. Testa thin, fragile, somewhat soft, inclosing two disks of

horny, translucent or opaque, yellowish or white albumen around **a** large central cavity. The embryo is contained in this cavity, and consists of a short radicle and two flat, heart-shaped, veined cotyledons extending about one-fourth the distance across it. Inodorous; taste extremely bitter. **Powdered nux vomica** is yellowish-gray and has a faint, sweetish odor. Should contain 1.25 per cent. of strychnine.

Fig. 198.—*Strychnos nux vomica*—Flowering branch and seeds.

Considerable difficulty has been experienced in keeping nux vomica and it is recommended that the container be kept in a second container containing a layer of unslaked lime.

Powder.—Characteristic elements: See Part iv, Chap. I, B.

Constituents.—The total alkaloids amount from 2.5 to 3.5 per cent. They consist principally of **strychnine,** $C_{21}H_{22}N_2O_2$, 1.25 per cent., and brucine, $C_{23}H_{26}N_2O_4$, the former being in excess. These are combined in the seed with igasuric acid. A third alkaloid, igasurine, has been claimed, but it is probably simply a mixture of the other two. A glucoside, loganin, $C_{25}H_{34}O_{14}$, has been found in the seeds, but it exists in greater quantity in the pulp surrounding the seed of the fruit. Other constituents are a concrete fixed oil, gum, wax, phosphates, and a yellow coloring matter.

Strychnine.—As usually found in commerce, strychnine is a white or grayish-white powder. When rapidly crystallized from an

FIG. 199.—Cross-section of the fruit of *Strychnos nux vomica.*

alcoholic solution, it has the form of a white granular powder; when slowly crystallized, that of an elongated octahedra, or rhombic prisms with pyramidal capping. It is officially described as "in colorless, transparent, octahedral, or prismatic crystals," etc.

The test usually employed for its recognition is sulphuric acid with potassium bichromate; gives a deep violet or blue color. A physiological test is usually employed by toxicologists as confirmatory to the chemical tests.

Brucine.—Brucine occurs in rectangular octahedra containing $4H_2O$, readily soluble in alcohol; nitric acid colors blood-red, changing to orange and yellow, the yellow liquid becoming violet upon the addition of stannous chloride or ammonium or sodium sulphide. Ash, not to exceed 3.5 per cent.

Preparation of Strychnine.—Boil powdered seeds with acidulated (HCl or H_2SO_4) water. Decompose solution of alkaloidal salts by adding milk of lime, which precipitates strychnine and brucine. Wash precipitate; treat with dilute alcohol to dissolve brucine, or with alcohol or benzene to take out strychnine, thus leaving brucine in the residue. Purify with animal charcoal and reprecipitate with ammonia. Occurs in four-sided rhombic prisms; very bitter; soluble in boiling alcohol 5 parts chloroform, 110 alcohol.

ACTION AND USES.—Nux vomica is a tonic, spinal nervine, and a poison. In small doses it stimulates the appetite and digestion and the respiration. Dose: ½ to 5 gr. (0.0324 to 0.3 Gm.). Strychnine represents its action fully. Brucine has the same physiological action as strychnine, but is only about one-twelfth as strong.

OFFICIAL PREPARATIONS.

Fluidextractum Nucis Vomicæ,..Dose: ½ to 5 ♍ (0.0324 to 0.3 mil).
Extractum Nucis Vomicæ,....... ½ to 1 gr. (0.0324 to 0.0650 Gm.).
Tinctura Nucis Vomicæ,........ 5 to 15 ♍ (0.3 to 1 mil).

Druggists should never make the tincture from the fluidextract of nux vomica regardless of whether the latter is assayed or not.

STRYCHNINE AND ITS OFFICIAL PREPARATIONS.

Strychnina,....................Dose: $\frac{1}{60}$ to $\frac{1}{20}$ gr. (0.001 to 0.003 Gm.).
Strychninæ Sulphas, $\frac{1}{64}$ gr. (0.001 Gm.).
Strychninæ Nitras,.............. $\frac{1}{40}$ gr. (0.0015 Gm.).

436. IGNATIA, N.F.—ST. IGNATIUS' BEAN. The seeds of **Strych'nos igna'tia** Lindley, a tree growing in the Philippine Islands, where they are much esteemed as a medicine, and whence they were introduced to the medical world by the Jesuits, who conferred upon them the name of the founder of their order. The fruit is pear-shaped, and contains 10 to 15 of these hard, heavy seeds lying one upon the other and imbedded in a dry medullary mass, but the seeds come into market separate. Their shapes are various, owing to the manner in which they were situated in the fruit; but their general form is ovate, somewhat flattened, and more or less **angular.** They are about 25 mm. (1 in.) long, but considerably narrower, and have at one end a small depression indicating their point of attachment (hilum). Their testa is of a less silky nature than that of nux vomica, and of a gray-brown color. In commerce they are perfectly smooth, the testa and hairs being removed by the rubbing of the seeds against one another, and therefore the outer surface consists of dull brown or blackish horny albumen, translucent when fresh. The embryo is oblong, situated in the broad end of the seed, the cotyledons extending only about half the distance across the irregular cavity. Inodorous; taste excessively bitter.

CONSTITUENTS.—Same as nux vomica (435) but in different proportions, the strychnine existing to the extent of about 1.2 per cent. against $\frac{1}{3}$ to $\frac{1}{2}$ per cent. in nux vomica. Ignatia was once used for the preparation of this alkaloid, strychnine, but rarely at the present day, as nux vomica is imported in such large quantities and is a much cheaper source. Dose: $\frac{1}{2}$ to 5 gr. (0.0324 to 0.3 Gm.).

437. **HOANG-NAN** or **HWANG-NAO.**—TROPICAL BINDWEED. The bark of **Strych'nos malaccen'sis** Bentham, a creeping vine growing in the mountains of Tonquin. This bark is in general use among the natives of Tonquin, Cochin-China, Venezuela, etc., as a remedy in leprosy and hydrophobia, and as an antisyphilitic and alterative. First brought to the notice of the medical profession by the missionary fathers. It contains strychnine and brucine in about equal proportions, and probably has about the same range in medicine as nux vomica. Dose: $\frac{1}{2}$ to 5 gr. (0.0324 to 0.3 Gm.).

438. GELSEMIUM.—GELSEMIUM

YELLOW JASMINE

The dried rhizome and roots of Gelse'mium sempervir'ens Aiton.

BOTANICAL CHARACTERISTICS.—*Stem* smooth, climbing. *Leaves* short-petiolate, shining, ovate. *Flowers* in short axillary clusters, very fragrant; *corolla* bright yellow, funnel-form, 5-lobed.

HABITAT.—Southern United States, notably Florida.

DESCRIPTION OF DRUG.—Generally in very light and fibrous cylindrical sections, 90 to 200 mm. long, 4 to 15 mm. in diameter; externally of a brownish-yellow color, slightly wrinkled; tough, breaking with a fibrous, splintery fracture; bark thin, with silky bast fibers, adhering to the light-yellowish, porous, broadly rayed wood; the wood-cells are more or less indurated and free of starch-grains; medullary rays contain few starch-granules; pith small; odor characteristic; taste persistently bitter.

ADULTERATION.—Mixed with the true gelsemium there are sometimes found the roots of the jessamine or jasmine; as an adulterant this has become known as false gelsemium. The true yellow jasmine (*Jasminum fructicans* Linné) is called *Gelsemium officinale* in Europe. In cross-section the false root, according to Dohme, has no indurated cells in the medulla. Its medullary rays are full of starch-grains, and the sieve-ducts at the outer end of the woody cylinder are, in

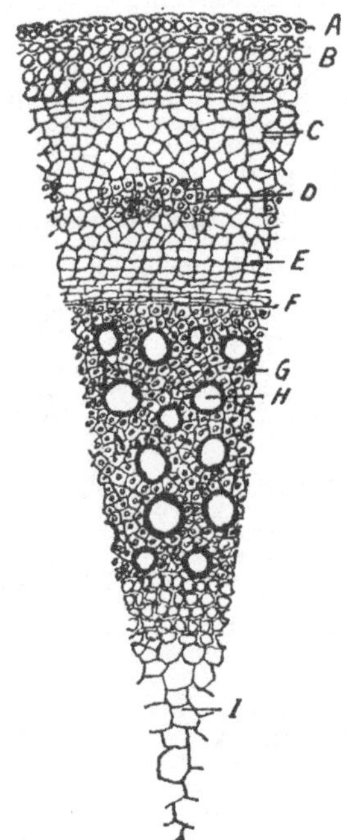

FIG. 200.—Cross-section of the stem of Gelsemium. *A*, Epidermis. *B*, Collenchyma. *C*, Parenchyma of cortex. *D*, Bundles of sclerenchymatous cells. *E*, Sieve tissue. *F*, Cambium. *G*, Wood fibers. *H*, Tracheal tubes. *I*, Pith cells.

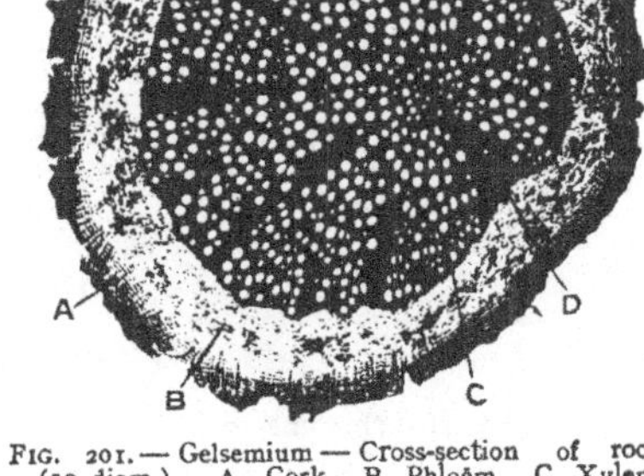

FIG. 201.—Gelsemium—Cross-section of root. (19 diam.). A, Cork. B, Phloëm. C, Xylem. D, Medullary ray. (Photomicrograph.)

the case of every woody wedge, surrounded and protected by several rows of bast fibers. The true gelsemium has no such bast fibers. Accidental admixture of stems may be detected by the latter having bundles of bast fibers near the cortex. In the rhizome the fibers are not in bundles, but in a more or less interrupted ring.

Powder.—Microscopical elements of: See Part iv, Chap. I, B.

CONSTITUENTS.—**Gelsemine**, $C_{54}H_{69}N_4O_{12}$, gelseminine, gelseminic acid, volatile oil, resins, gallic acid, etc. Gelsemine is a brittle, white, transparent solid, soluble in alcohol, from which it crystallizes with difficulty. Gelseminine (uncrystalline) and Sempervirine, yellow-ish-red crystals. The latter has been separated in minute quantities (Sayre). The root yields from o.3 to o.5 per cent. of alkaloids. Gelseminine is rapidly growing in favor as presenting most of the benefits occurring from the use of morphine without any of the disadvantages of the latter (Ellingwood). The activity seems to reside principally in the gelseminine.

Preparation of Gelsemine.—Add acetic acid to concentrated tincture; precipitate resin with water; concentrate the aqueous filtrate; remove gelsemic acid with chloroform or ether. The acid liquid yields impure alkaloid when precipitated by Na_2CO_3. This is purified by solution in chloroform and slow evaporation. It is a white, amorphous, very bitter alkaline alkaloid; with HCl or HNO_3 forms crystalline salts.

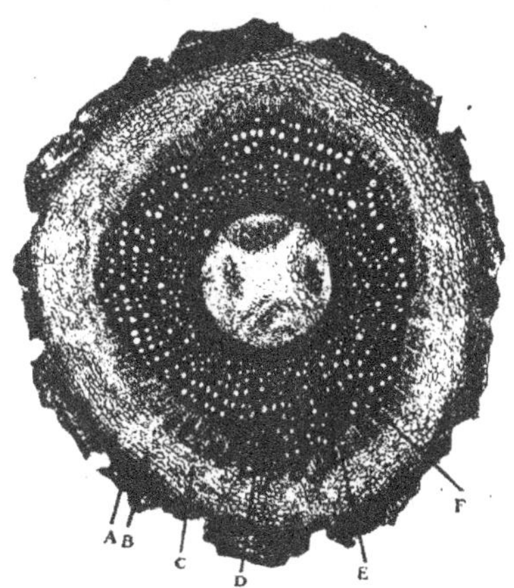

FIG. 202.—Gelsemium—Cross-section of rhizome. (25 diam.) A. Cork, B. Parenchyma of cortex. C, Bast fibers. D, Xylem. E, Medullary ray. F, Medulla. (Photomicrograph.)

ACTION AND USES.—**Antispasmodic, sedative,** and diaphoretic. Dose: 2 to 10 gr. (o.13 to o.6 Gm.).

OFFICIAL PREPARATIONS.

Fluidextractum Gelsemii,........Dose: 2 to 10 ℥ (0.13 to 0.6 mil).
Tinctura Gelsemii (10 per cent.),.. 2 to 15 ℥ (0.14 to 1 mil).
Extractum Gelsemii,............ 0.01 Gm. (⅙ gr.).

439. SPIGELIA.—SPIGELIA

PINK ROOT. CAROLINA PINK

The dried rhizome and roots of Spige'lia marilan'dica Linné.

BOTANICAL CHARACTERISTICS.—*Root* perennial; *stem* simple and erect. *Leaves* sessile, ovate-lanceolate, acute. *Flowers* in a short spike; *corolla* red externally, yellow within, four times the length of the calyx; *stamens* and *pistil* exserted.

23

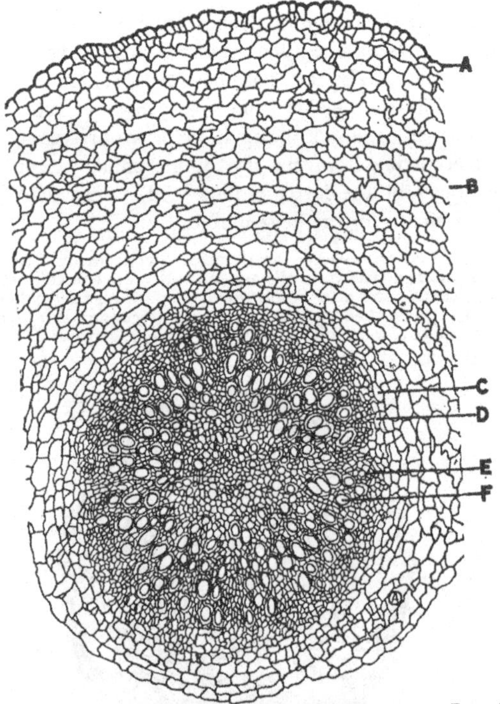

FIG. 203.—Cross-section of root of Spigelia (×64). A, Epidermis. B, Parenchyma of primary cortex. C, Endodermis. D, Pericycle. E, Phloëm. F, Xylem.

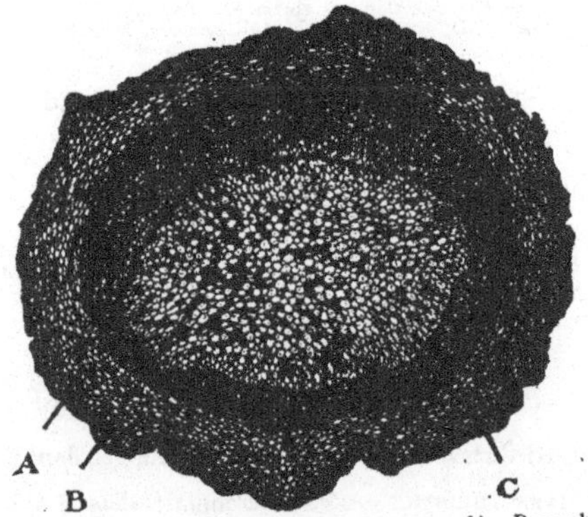

FIG. 204.—Spigelia—Cross-section of rhizome. (21 diam.) A, Parenchyma of cortex. B, Medulla. C, Xylem. (Photomicrograph.)

HABITAT.—United States, Maryland southward and westward in rich woods.

DESCRIPTION OF DRUG.—Rhizome **thin, bent, purplish-brown,** on the upper side marked with stem scars, on the lower side beset with numerous lighter colored, slender, branching rootlets. Fracture short. Odor slight, aromatic; taste sweetish, bitter, and pungent.

RELATED SPECIES.—Another species which has attracted attention as an anthelmintic is *Spigelia anthelmia* of South America and the West Indies, which in that country is said to have greater medicinal properties than the official.

A wholesale adulteration of this drug was discovered a short time ago. To the surprise of pharmacognosists this adulterant (a species of Ruellia) had completely replaced the official article in commerce. A cross-section of the authentic drug (rootlets and rhizome) is given in Figs. 203 and 204.

Powder.—Characteristic elements: See Part iv, Chap. I, B.

CONSTITUENTS.—A volatile alkaloid, **spigeline,** is the active principle. Ash, not more than 10 per cent.

Preparation of Spigeline.—Distil the powdered drug over a paraffin bath with milk of lime; collect the distillate in HCl and evaporate to dryness; crystallize from alcoholic solution.

ACTION AND USES.—A **powerful anthelmintic.** Dose: 15 to 60 gr. (1 to 4 Gm.).

OFFICIAL PREPARATION.

    Fluidextractum Spigeliæ,.....   Dose: 15 to 60 ♏ (1 to 4 mils).

440. **CURARA.**—CURARE. WOORARI. From Strych'nos castelnæa'na and other species of Strychnos growing in South America, where an extract is prepared by the natives as an arrow-poison. This extract is a blackish, friable solid or of extract-like consistence, having a somewhat resinous appearance, and very hygroscopic. It contains a very bitter and poisonous alkaloid, curarine. As a remedial agent curara has probably little value, although it has been used in tetanus, hydrophobia, epilepsy, and chorea. It is a strong depressant of the motor nerves, causing a gradual loss of muscular power, deepened respiration, and death by asphyxia. Dose: $\frac{1}{10}$ to $\frac{1}{2}$ gr. (0.006 to 0.02 Gm.).

Curarine ($C_{18}H_{35}N$).—From the drug Roulin obtained this principle by a very intricate process. The alkaloid is extremely deliquescent and crystallizes in prisms, soluble in water, and changes litmus feebly.

GENTIANEÆ.—Gentian Family

Smooth herbs with a colorless, bitter juice, and containing little or no tannin.

Synopsis of Drugs from the Gentianeæ

A. *Roots.*
    GENTIANA, 441.
    Frasera, 442.

B. *Herbs.*
    *Chirata, 443.
    Sabbatia, 444.
    *Menyanthes, 445.

441. GENTIANA.—Gentian

GENTIAN

The dry rhizome and roots of Gentia′na lu′tea Linné.

Botanical Characteristics.—Root perennial, large; stem 2 to 3 feet high, *Leaves* opposite, sessile, 5- to 7-nerved, ovate-acute, more or less clasping. *Flowers* in whorls, bright yellow; *corolla* with 5 or 6 green glands at its base; *stigmas* 2.

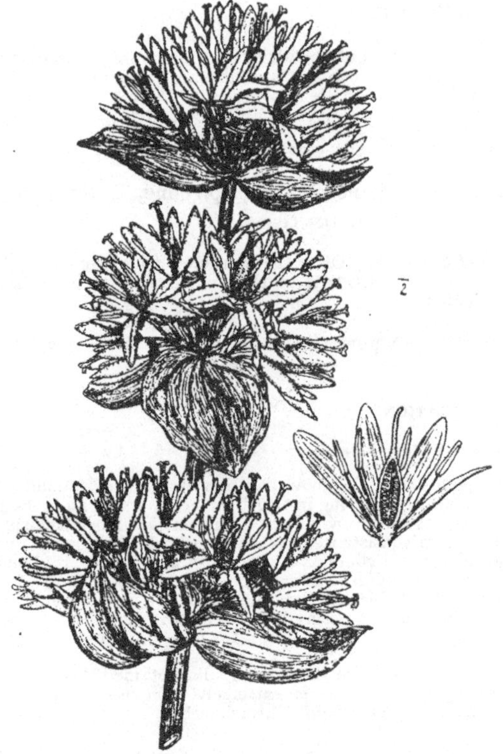

Fig. 205.—*Gentiana lutea*—Flowering head and dissected flower.

Habitat.—Mountainous portions of Central Europe.

Description of Drug.—Cylindrical, fleshy, and very long, often 3 feet or more; it is generally cut longitudinally about 100 to 200 mm. (4 to 8 in.) long, and 5 to 40 mm. ($\frac{1}{5}$ to $1\frac{3}{5}$ in.) thick; in drying, these slices are depressed in the center and the bark overlaps; **yellowish-brown,** much wrinkled longitudinally and marked transversely, especially in the upper portion, with numerous rings. Transversely the bark is rather thick, wrinkled, and contorted, separated by a black cambium line from the yellowish-brown, porous, and spongy meditullium marked with indistinct medullary rays. Fracture

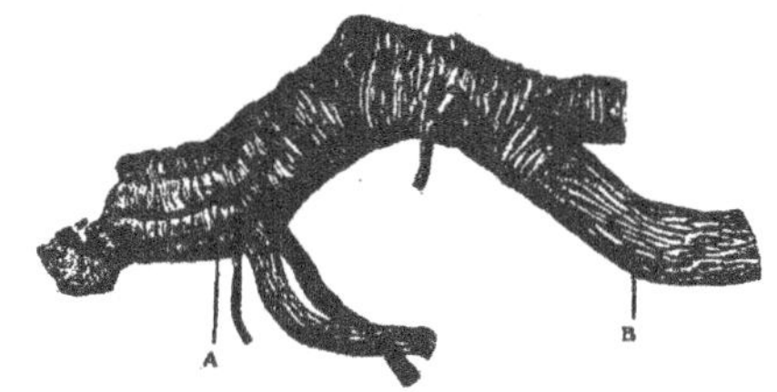

Fig. 206.—Gentian—A, Rhizome portion.   B, Root portion.   (½ natural size.)

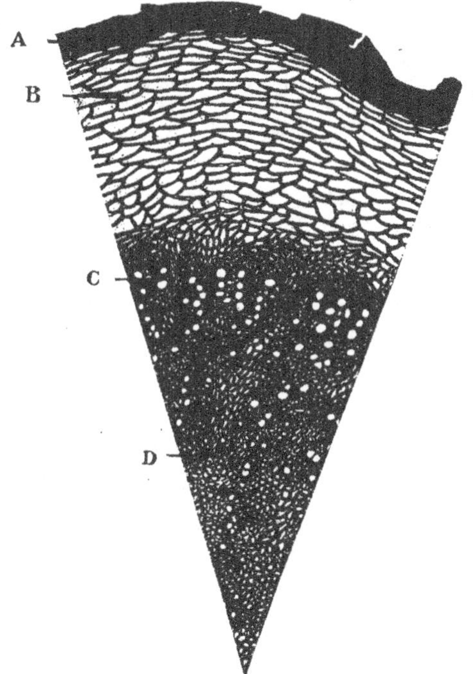

Fig. 207.—*Gentiana lutea*—Cross-section. A, Cork. B, Parenchyma of cortex. C, Water tube. D, Parenchyma of xylem.

irregular, brittle when dry, flexible and tough when damp; odor pronounced and characteristic; **taste** intensely bitter, sweetish, and not disagreeable.

Gentiana catesbæi, the blue gentian of the Southern States, growing in mossy swamps, is said to be little inferior to the official species.

It is sometimes used to adulterate senega.   Other indigenous species, as *G. purpurea* and *G. punctata*, have about the same properties as the official gentian and are used similarly.   The herb *G. quinqueflora* is used in liver affections, chronic ague, jaundice, etc.

Powder.—Characteristic elements: See Part iv, Chap. I, B.

CONSTITUENTS.—The bitter principle is a neutral principle, **gentiopicrin,** $C_{20}H_{30}O_{12}$, the yellow color is due to **gentisin,** $C_{14}H_{10}O_5$, or **gentisic acid** (tasteless yellow prisms).   The root also contains from 12 to 15 per cent. of glucose (gentianose), $C_{16}H_{66}O_{31}$, but is remarkable in that it contains **no starch, calcium oxalate,** or **tannin.**   Ash, not to exceed 6 per cent.

Preparation of Gentisic Acid.—The alcoholic extract is washed with water, then with ether.   The residue dissolved in alcohol yields the acid on evaporation. It is in yellow, tasteless crystals, partially soluble in alcohol and ether; with ferric salts gives dark brown color.

Preparation of Gentiopicrin.—Obtained by making aqueous solution of alcoholic extract.   This solution is subjected to the absorptive action of charcoal. Charcoal is then boiled with alcohol, tincture evaporated, and treated with lead oxide to remove color.   Lead removed by H_2S; solution agitated with ether. Set solution aside to crystallize.   Yellowish-brown, soluble in water and dilute alcohol.

ACTION AND USES.—**Simple bitter tonic,** long known and very valuable.   Dose: 5 to 30 gr. (0.3 to 2 Gm.).

OFFICIAL PREPARATIONS.

Fluidextractum Gentianæ, ...Dose: 5 to 30 ♏ (0.3 to 2 mils).
Extractum Gentianæ, ...   5 to 10 gr. (0.3 to 0.6 Gm.).
Tinctura Gentianæ Composita (10 per
    cent., with orange-peel and cardamom),   1 to 2 fl. dr. (4 to 8 mils).

442. FRASERA.—AMERICAN COLUMBO.   The root of **Fra'sera wal'teri** Michaux, a plant growing extensively in Southern and Western United States, especially in Arkansas and Missouri.   Its root is long and spindle-shaped, but comes into market in transverse slices, irregularly circular, about 25 mm. (1 in.) in diameter; these disks consist of a central medullary matter, yellowish-brown, shrunken in the middle, and a reddish-brown exterior; inodorous; taste at first sweet, then bitter.   It may be distinguished from columbo by its greater uniformity of internal structure, the absence of concentric and radiating lines, and its purer yellow color without the green tinge.   It occasionally comes into the market in longitudinal slices under the name of American gentian.   It contains gentiopicrin and gentisic acid, but no starch or tannin.   Simple bitter tonic like columbo and gentian.   Dose: 15 to 30 gr. (1 to 2 Gm.).

443. CHIRATA.—CHIRATA, N.F.

CHIRETTA

The dried plant **Swer'tia chira'yita** Hamilton.

HABITAT.—Nepal and other parts of Northern India.

DESCRIPTION OF DRUG.—Chirata of the market consists principally of **short sections** of the stem and branches, **orange-brown** or dark **purple** in color, **generally pressed and split,** showing the yellow pith, and mixed with a few leaves and flower panicles.   These stems when entire are about 4 mm. (⅙ in.)

in thickness, round at base and quadrangular toward the top, jointed, the internodes being from 37 to 100 mm. ($1\frac{1}{2}$ to 4 in.) in length; branches opposite. Inodorous when dry, but when moistened it has a perceptible odor; taste very bitter, persistent.

In the Indian bazaars there are a number of species of Ophelia, known by the name of *Chiretta*, which possess, to a greater or less degree, the bitter properties of that drug. Flückiger states: "We have frequently examined the chiretta found in the English market, but have never met with any other than the legitimate sort." Bentley noticed, in 1874, the substitution of *O. angustifolia*, which he found by far to be less bitter than the true chiretta. J. S. Ward ("Pharm. Jour.," 4th Series, 1, 1897) calls attention to a false chirata entering the eastern market. He recognized it as the product of *Andrographia paniculata*, nat. ord. Acanthraceæ, a plant distributed throughout India from Lucknow and Assam to Ceylon, and cultivated in the West Indies—a domestic remedy for fevers, debility, etc. Sold by herbalists in the fresh state.

Powder.—Grayish-brown. Characteristic elements: Parenchyma of medulla, slightly lignified with simple pores; sclerenchyma with fibers, long, narrow, and thick-walled; tracheids, numerous; ducts with spiral or scalariform markings; yellowish-brown pollen and stomata present.

CONSTITUENTS.—**Chiratin,** $C_{26}H_{48}O_{15}$ (yellow, hygroscopic powder, very bitter), **ophelic acid,** $C_{13}H_{20}O_{10}$ (a syrupy liquid, very bitter), resin, coloring matter, bitter extractive, gum, and salts. Water and alcohol extract its virtues.

ACTION AND USES.—Bitter tonic like the other plants of the order Gentianæ.

Dose: 15 gr. (1 Gm.).

OFFICIAL PREPARATIONS.

Tinctura Chiratæ (10 per cent.) (1890), Dose: $\frac{1}{2}$ to 2 fl. dr. (2 to 8 mils).
Fluidextractum Chiratæ,.   15 ♏ (1 mil).

444. **SABBATIA.**—CENTAURY. (Centaurium, the dried flowering plant of Centaurium Contanrium, N.F.) Three species of this indigenous herb are more or less used in this country as tonic and antiperiodic. These are *Sabba'tia angular'is* Pursh (American centaury), *S. paniculata* Pursh, and *S. Ellio'tti* Steudel (quinine flower); the whole plant of the two first-named species is used, the root of the last-named; they probably all contain the same principle, erythrocentaurin, $C_{27}H_{24}O_8$. Dose: about 1 dr. (4 Gm.).

445. **MENYANTHES.**—WATER SHAMROCK. BUCKBEAN (N.F.) The herb of **Menyan'thes** trifolia'ta Linné, an aquatic plant growing in bogs in the temperate zone of the Northern Hemisphere. Leaves ternate, rising out of the water on long petioles from a rhizome; leaflets obovate, about 50 mm. (2 in.) long, with entire margin, and smooth, green upper surface, paler beneath. It has no odor, but a very bitter taste, due to a bitter principle, menyanthin, $C_{33}H_{64}O_{16}$(?). Bitter tonic, in large doses cathartic. Employed in the preparation: Vinum Aurantii Compositum, N.F. (*Elix. Aurantiorum Compositum*, Germ. Pharm.). Dose: 15 to 45 gr. (1 to 3 Gm.).

APOCYNACEÆ.—Dogbane Family

Herbs, shrubs, or trees, mostly tropical, with a milky juice which is often drastic or poisonous. *Leaves* mostly opposite, exstipulate. *Flowers* regular, 5-merous and 5-androus, with the pollen cohering into granular, waxy masses. *Fruit* a pair of follicles; seeds often comose.

Synopsis of Drugs from the Apocynaceæ

A. *Roots.*
    * **Apocynum,** 446.
    Apocynum Androsæmifolium, 446 a.
B. *Barks.*
    Aspidosperma, 447.
    Alstonia Constricta, 448.
    Alstonia Scholaris, 449.
    Conessi, 450.

C. *Seeds.*
    STROPHANTHUS, 451.
D. *Leaves.*
    Oleander, 452.
E. *Herb.*
    Urechites, 453.

446. APOCYNUM.—Apocynum, N.F.

CANADIAN HEMP

The dried rhizome of **Apocy'num canna'binum** Linné without the presence of more than 5 per cent. of stems and foreign matter.

Botanical Characteristics.—*Stems* much branched, 2 to 3 feet high. *Leaves* from oval to oblong or lanceolate, short petiolate or sessile. *Inflorescence* cymose; *corolla* greenish-white, with nearly erect lobes, the tube not longer than the calyx tube.

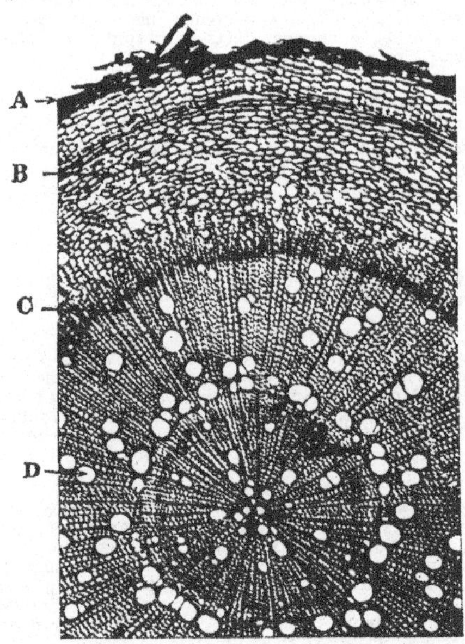

Fig. 208.—*Apocynum cannabinum*—Cross-section. (28 diam.) A, Cork. B, Parenchyma of cortex. C, Medullary ray. D, Water tube. (Photomicrograph.)

Habitat.—United States.

Description of Drug.—A long, cylindrical root, somewhat contorted, about 8 mm. (⅓ in.) thick, with a rather thick light brown bark, longitudinally wrinkled and transversely fissured, and a yellowish, porous wood divided by fine medullary rays into very narrow wood-wedges; fracture short. The thick inner cortical layer has numerous lactiferous vessels scattered through it, which in the fresh state secrete a milky juice which hardens into a caoutchouc-like substance. Odor slight, or none: taste bitter, nauseous.

Apocynum androsæmifolium Linné, dogbane, resembles the above, but has a relatively thicker bark inclosing a white, porous wood, and contains, in the outer portion, stone-cell groups. By applying the phloroglucin test to a section, the groups of stone-cells are revealed, stained red. Two species sold indiscriminately.

CONSTITUENTS.—**Apocynein,** a yellowish glucoside (acting like digitalin); **apocynin,** a bitter, resin-like extractive; tannin, resin, starch, etc.

ACTION AND USES.—**A** valuable **diuretic** in moderate doses, in large doses emetic and cathartic, producing considerable diaphoresis and expectoration; most used and most beneficial **in dropsy.** Recently the drug has attracted some attention as a most valuable deobstruent in relieving renal congestion in the second stage of tubular nephritis. It is also a decided heart tonic. Dose as a diuretic, 4 to 5 gr. (0.3 to 0.324 Gm.); as an emetic and cathartic, 15 to 30 gr. (1 to 2 Gm.).

OFFICIAL PREPARATION.

Fluidextractum Apocyni, ...Dose: 15 ♏ (1 mil).

447. **ASPIDOSPERMA.**—QUEBRACHO. (U.S.P. IX.) The bark of **Aspidosper'ma quebra'cho blancho** Schlechtendal. Thick, flat pieces (from ⅖ to 1 in. in thickness), with a very thick, yellowish-gray cork, which constitutes more than one-half of its entire substance, and is separated from the lower layer by a more or less sharply defined outline, deeply fissured, and traversed by parallel yellowish lines; between these lines are whitish dots visible in a cross-section scattered through both the outer and inner layers. Internally reddish-brown to yellow; odor slight; taste aromatic and bitter. *Constituents:* Aspidosperma is very rich in alkaloids, six having been discovered thus far; the most important are aspidospermine, $C_{22}H_{30}N_2O_2$, and quebrachine, $C_{21}H_{26}N_2O_2$. A peculiar sugar, quebrachite, is also present, and tannin, 3 to 4 per cent. Cardiac tonic. Its special action, however, is upon the respiration, lessening the rate and increasing the amplitude of the respiratory movements; it is chiefly used in asthmatic dyspnœa. Dose: 5 to 30 gr. (0.3 to 2 Gm.).

Powder.—Microscopical elements of: See Part iv, Chap. I, B.

Preparation of Aspidospermine.—Treat alcoholic extract with alkaline chloroform; dissolve chloroformic extract in acidulated (H_2SO_4) water and precipitate with NaOH; dissolve precipitate (mixed alkaloids) in boiling alcohol and cool, when alkaloids will crystallize.

To separate aspidospermine, crystallize from dilute HCl, when this alkaloid will remain in the mother liquor, from which it may be removed by neutralization and recrystallization. As found in commerce, this alkaloid is a mixture of this and the other associated principles, among which quebrachine is the most important. Crude aspidospermine sulphate is a commercial article, is deliquescent and unstable; it is much more soluble in water than the alkaloid.

Fluidextractum Aspidospermatis,......Dose: 5 to 30 ♏ (0.3 to 2 mils).

448. **ALSTONIA CONSTRICTA** F. Mueller.—AUSTRALIAN FEVER BARK. Tonic, antiperiodic. Dose of fl'ext.: 2 to 8 ♏ (0.13 to 0.5 mil).

449. **ALSTONIA SCHOLARIS** R. Brown.—DITA. A tree growing in the Philippine Islands, the bark of which is used in India as a substitute for cinchona. Dose of fl'ext.: 2 to 8 ♏ (0.13 to 0.5 mil).

450. **CONESSI.**—The bark of **Holar'rhena antidysenter'ica** Wallr. Has been used in Europe and is still extensively employed in India in dysentery. Its alkaloid, conessine, enters commerce.

451. STROPHANTHUS.—STROPHANTHUS

STROPHANTHUS

The ripe seed of **Strophan'thus Kombé** Oliver or of Strophanthus hispidus De Candolle, deprived of its long awn.

BOTANICAL CHARACTERISTICS.—A woody climber, ascending to the tops of high trees, from which it hangs in festoons. *Flowers* in terminal cymes, gamopetalous, the lobes prolonged into long, tail-like points, often 8 or 9 inches long. *Fruit* two long follicles.

SOURCE.—The genus Strophanthus contains about 20 species, native of Africa and Asia, where it is probable that more than one of them are used for the preparation of arrow-poison.

FIG. 209.—*Strophanthus hispidus*—Branch and seed with comose awn.

DESCRIPTION OF DRUG.—Lance-ovoid, flattened and obtusely edged; from 7 to 20 mm. in length, about 4 mm. in breadth and about 2 mm. in thickness; externally, of a light fawn color, with a distinct, greenish tinge, silky lustrious from a dense coating of closely appressed hairs, (S. Kombe); or light to dark brown, nearly smooth and sparingly hairy (S. hispidus), bearing on one side a ridge running from about the center to the summit; fracture short and somewhat soft, the fractured surface whitish and oily; odor heavy when the seeds are crushed and moistened; taste very bitter. U.S.P. IX.

Powder.—Characteristic elements: See Part iv, Chap. I, B.

TEST U.S.P.—If made into the official tincture and assayed biologically the minimum lethal dose should not be greater than 0.00006 mil of tincture, or the equivalent in tincture of 0.0000005 Gm. of ouabain, for each gramme of body weight of frog. Preserve Strophanthus in tightly closed containers, adding a few drops of chloroform or carbon tetrachloride, from time to time, to prevent attack of insects.

CONSTITUENTS.—Its medical properties depend upon an intensely bitter glucoside, **strophanthin,** $C_{32}H_{48}O_{16}$ (anhydrous), 2 to 2.5 per cent., choline, trigonelline, kombïc acid, resin, mucilage, and a fixed oil are also present. Ash, not to exceed 5 per cent.

OUABAIN, CRYSTALLIZED.—Crystallized Strophanthin.—G. Strophanthin Thoms. $C_{30}H_{46}O_{12} + 9H_2O$. A glucoside, obtained from *Acocanthera ouabaio* by Arnaud, or, as now commonly prepared, from *Strophanthus gratus*, in which case it is also called crystallized strophanthin, or g-strophanthin Thoms. (The official strophanthin is methyl ouabain $C_{31}H_{48}O_{12}$.) Recent investigation shows that this alkaloid varies in proportion to water of crystallization.

Preparation of Strophanthin.—Treat powdered seeds with acidulated (HCl) alcohol; evaporate to soft extract; treat with water. The aqueous solution containing tannate is treated with lead oxide, and from the purified aqueous solution white crystals are obtained.

ACTION AND USES.—Used in all forms of cardiac disease to supplant digitalis, but is not generally regarded as its equal. It has a diuretic action similar to digitalis through its action on the circulation, and also by direct promotion of urinary secretion, and is especially indicated in cardiac dropsy as being superior to digitalis; given in the form of tincture. Dose: 1 gr. (0.065 Gm.).

OFFICIAL PREPARATION.

Tinctura Strophanthi (10 per cent.),   Dose: 4 to 8 ℥ (0.25 to 0.50 mil).

452. OLEANDER.—The leaves of Ne'rium odor'um, a heart stimulant belonging to the digitalis group. Oleandrin is a cardiac poison.

453. URECHITES.—YELLOW-FLOWERED NIGHTSHADE. A poisonous plant growing in the West India Islands. A cardiac poison not very unlike digitalis in effect. Dose of fl'ext.: 2 to 10 ℥ (0.13 to 0.6 mil).

ASCLEPIADEÆ.—Milkweed Family

Herbs, usually milky-juiced, with opposite or whorled entire *leaves.* *Anthers* connected to the stigma and the pollen, cohering into waxy masses which hang in pairs from the glands of the stigma. The *juice* contains caoutchouc.

454. ASCLEPIAS TUBEROSA (N.F.).—The root of Ascle'pias tubero'sa Linné. Off. in U.S.P. 1890. Enters the market in transverse or longitudinal sections about 20 mm. (⅘ in.) in thickness, and of various lengths; externally pale orange-brown or grayish, wrinkled longitudinally; internally it consists of a grayish or yellowish porous wood with broad, white medullary rays; fracture tough, uneven, showing the two distinct layers of the thin bark,

the inner one white; odorless; taste bitter, somewhat acrid. Diaphoretic expectorant. Dose: 15 to 60 gr. (1 to 4 Gm.). Fl'ext., off. U.S.P. 1890, dose: 15 to 60 ♏ (1 to 4 mils).

455. **ASCLEPIAS CORNUTI** Decaisne.—Common Silk-weed or Milk-weed. (Rhizome.) Cylindrical sections, from 6 to 25 mm. (¼ to 1 in.) thick, beset with a few simple rootlets; externally grayish-brown, finely wrinkled, and rough from stem-scars and undeveloped branches. It breaks with a short or splintery fracture, showing a thick bark containing lactiferous vessels, and a yellowish, porous wood in narrow wood-wedges. Odorless; taste bitter and nauseous. Diuretic, alterative, and expectorant; recommended in pectoral affections and in dropsy. Dose: 15 to 60 gr. (1 to 4 Gm.), in decoction.

456. **ASCLEPIAS INCARNATA** Linné.—Swamp Milk-weed. *Habitat:* North America. An oval or globular, yellowish-brown rhizome, with a tough, white wood, and a central pith; rootlets smooth, light yellowish-brown, brittle; odorless; taste sweetish, bitter, and acrid. It contains an emetic principle, **asclepiadin**; it is also alterative and cathartic. Dose: 15 to 45 gr. (1 to 3 Gm.).

457. **ASCLEPIAS CURASSAVICA** Linné.—Blood Flower. A West Indian herb used as an emetic, in smaller doses cathartic and vermifuge. Dose of fl'ext.: 1 to 2 fl. dr. (4 to 8 mils).

458. **HEMIDESMUS.**—Indian Sarsaparilla. The root of a climbing East Indian plant, **Hemides'mus in'dicus** R. Brown. Long, cylindrical, slender, and tortuous; externally wrinkled and fissured, dark brown; wood yellowish, separated from the thin bark by a dark, wavy cambium line. Odor sweetish, tonka-like; taste sweetish and acrid. It is used in India as an alterative, and also in Great Britain, where it is official. Dose: 30 to 60 gr. (2 to 4 Gm.), in infusion or decoction.

459. **CONDURANGO** (N.F.).—The bark of Gonolo'bus conduran'go Triana, a South American vine, largely used there as an alterative. It was first introduced as a medicine here as a specific in cancer, but experience has shown it to be of no value in that trouble. It is from 2 to 6 mm. (½₂ to ¼ in.) thick, the outer surface or periderm ash-gray, with greenish or blackish lichen patches scattered over it; odor slight; taste bitter and acrid. It is given in doses of about 30 gr. (2 Gm.).

CONVOLVULACEÆ.—Convolvulus Family

Chiefly twining or trailing herbs, sometimes with milky juice, with alternate *leaves*, and regular, 5-androus *flowers*.

Tuber.	*Resin.*
JALAPA, 460.	SCAMMONIUM, 462.
Root.	
Ipomœa, 461 a.	
SCAMMONII RADIX, 462 a.	

460. JALAPA.—Jalapa

JALAPA

The dried tuberous root of **Exogo'nium pur'ga** (Wenderoth) Bentham, yielding, by assay, not less than 7 per cent. of resin.

Botanical Characteristics.—*Stem* brownish, smooth. *Leaves* long-petiolate, cordate-ovate, acuminate, entire, smooth. *Peduncles* axillary, 2-flowered; *corolla* crimson or light red, four times the length of the calyx.

Habitat.—Mexico; now successfully cultivated in India.

Description of Drug.—A compact, heavy, hard, pear-shaped tuber, varying in size, but never larger than the fist; the larger ones are longitudinally incised to facilitate the drying, which is done over the hearths of the Indian huts, hence **externally** brown, smoky, more or less wrinkled, covered with thick, round warts of a somewhat lighter color; **internally** gray to dark brown; fracture horny and resinous; odor peculiar, smoky, partly due to the manner of drying; taste starchy, afterward slightly acrid. .**Powdered jalap** is yellowish-gray, and when inhaled causes sneezing and coughing.

Structure.—Cortical layer thin, with a dense circle of resin cells near the cambium line; interior composed chiefly of parenchymatous

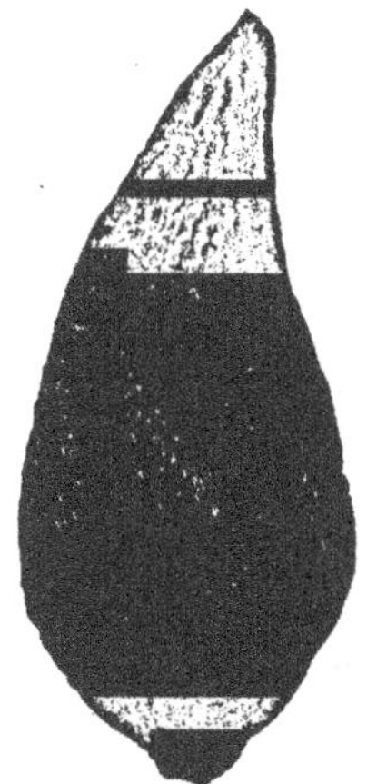

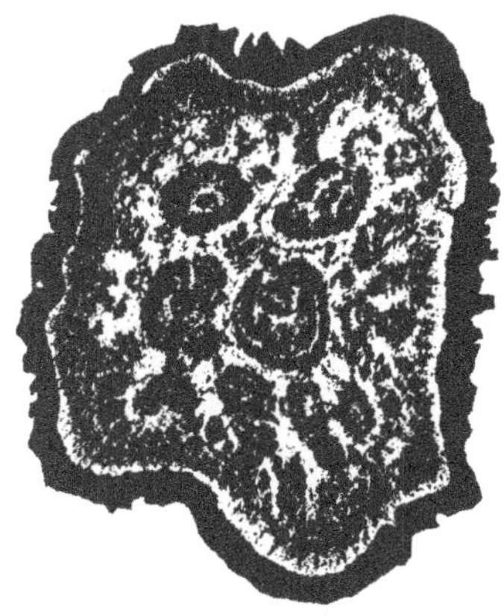

Fig. 210.—Jalap tuber. (Photograph.)   Fig. 211.—Jalap—Cross-section of tuber. (2 diam.) (Photograph.)

tissue containing starch and calcium oxalate, arranged in which are concentric zones of resin cells, the broader, darker, alternate zones being formed by a closer packing of the cells; medullary rays small, but plainly visible. The false jalaps which frequently adulterate the drug in market may usually be detected by the difference in internal structure.

Adulterations.—Immature roots, roots partially deprived of resin by treatment with alcohol. These are sticky, internally darker than the genuine and other species of Ipomœa.

Powder.—Characteristic elements: See Part iv, Chap. I, B.

Constituents.—Besides starch, calcium oxalate, etc., jalap contains a resinous substance which consists of two portions, a soft resin, **jalapin,** soluble in ether, and a hard resin, constituting nine-tenths of the mixture, termed **jalapurgin or convolvulin** (a glucoside, $C_{31}H_{50}O_{16}$); this latter is supposed to be the active principle. The U.S. Pharmacopœia (IX) has fixed the lowest limit of resin at 7 per cent. (which includes both resins).

The varying strength in jalap may be accounted for by the fact that the roots are dug at all seasons of the year. In the fall, when the aërial stem has decayed, it is better than in the spring, at the sprouting season. Ash, not to exceed 6.5 per cent.

ACTION AND USES.—**Hydragogue cathartic**, generally used in dropsy in the compound powder of jalap. Dose: 15 to 30 gr. (1 to 2 Gm.).

OFFICIAL PREPARATIONS.

Pulvis Jalapæ Compositus (35 per cent.
  with potassium bitartrate),   Dose: 15 to 60 gr. (1 to 4 Gm.).
Resina Jalapæ,...............   2 to 5 gr. (0.13 to 0.3 Gm.).
Pilulæ Catharticæ Compositæ,.   2 to 5 pills.

461 a. **IPOMŒA PANDURATA.**—WILD JALAP. MAN-ROOT. MAN OF THE EARTH. The root of **Ipomœ'a pandura'ta** Meyer. Occasionally met with in commerce, in the form of longitudinal slices with an irregularly wrinkled, brownish-gray bark overlapping the white wood. The woody center is divided into narrow wood-wedges by medullary rays dotted with resin cells. Nearly inodorous; taste sweetish and bitter. Contains panaquilon (the sweet principle found in panax), mucilage, starch, resin, etc. Diuretic and cathartic. Dose: 15 to 60 gr. (1 to 4 Gm.).

461 b. **FALSE JALAPS.**—*Ipomœa simulans* (Tampico jalap), a somewhat globular root yielding a resin (tampicin), very similar to jalapin, nearly soluble in ether. *I. orizabensis* (fusiform or male jalap), a spindle-shaped, large, woody root, often in sections, the resin orizabini (unfortunately named jalapin) entirely soluble in ether.

462. SCAMMONIUM.—SCAMMONY

SCAMMONY

A gum resin obtained by incising the living root of **Convol'vulus scammo'nia** Linné.

BOTANICAL CHARACTERISTICS.—*Root* perennial, tapering, 3 to 4 feet long, from 9 to 12 in. in circumference at the crown, and abounding in a milky, acrid juice. *Stem* annual, smooth. *Leaves* petiolate, sagittate, entire. *Peduncles* cymose, 3-flowered, twice the length of the leaves; *calyx-lobes* with a reflexed point; *corolla* pale yellow. *Capsule* 2-celled, 4-seeded.

HABITAT.—Western Asia. Obtained in the same manner as asafœtida.

DESCRIPTION OF DRUG.—The pure, or, as it is called, the "genuine" scammony is scarce in the market, the ordinary article being impure from flour, chalk, ashes, sand, etc., mixed with the exuded milk-juice before it has entirely hardened. It usually comes in hemispherical cakes, convex on one side, about 100 to 150 mm. (4 to 6 in.) in diameter; externally dark gray or nearly black; fracture brittle, shining, somewhat rough, exhibiting a usually porous interior, lighter colored and tinged with yellow or green. It yields a light-gray powder having a peculiar odor resembling cheese or putty; taste slight, but leaves an acrid sensation in the throat.

CONSTITUENTS.—Gum, resin, starch, **scammonin,** $C_{34}H_{66}O_{16}$, etc. Not less than 75 per cent. of the drug should be soluble in ether; ash not more than 3 per cent.

ACTION AND USES.—**Hydragogue cathartic,** on account of its harshness, generally given in combination. Uncertain on account of frequent impurities. Dose: 1 to 8 gr. (0.065 to 0.5 Gm.), in emulsion.

FIG. 212.—*Convolvulus scammonia*—Branch.

462 a. SCAMMONII RADIX.—SCAMMONY ROOT

The dried root of Convolvulus scammonia Linné yielding, when assayed by the official process, not less than 8 per cent. of the total resins of scammony root.

SOURCE AND DESCRIPTION.—This is the root of a morning glory-like plant, a native of Levant. The root is cylindrical or somewhat tapering from 10 to 25 cm. in length and 1 to 4.5 cm. in thickness. Externally, it is grayish to reddish-brown usually distinctly twisted, deeply longitudinally furrowed and marked by distinct root scars. Fracture tough, irregular and with projecting wood-fibers. Internally somewhat mottled showing yellowish, porous wood-wedges, separated by whitish parenchyma containing starch and resin. Bark, thin;

odor, slight, resembling that of jalap; taste, slightly sweet, becoming slightly acrid.

ACTION AND USES.—For its action it depends on the gum resin. Hydrogogue, cathartic, on account of its harshness it is generally given in combination. Its action is often uncertain due to adulteration.

Powder.—Characteristic elements: See Part iv, Chap. I, B.

OFFICIAL PREPARATIONS.

Resina Scammonii,.............................Dose: 3 gr. (0.2 Gm.).
Extractum colocynthidis Compositum,.   7½ gr. (0.5 Gm.).

POLEMONIACEÆ.—Polemonium Family

463. POLEMONIUM REPTANS Linné.—ABSCESS ROOT. The root of this American plant has been used as an alterative, astringent, diaphoretic, and expectorant. Dose: 30 to 60 gr. (2 to 4 Gm.).

HYDROPHYLLACEÆ.—Waterleaf Family

464. ERIODICTYON.—ERIODICTYON

YERBA SANTA.   MOUNTAIN BALM.   CONSUMPTIVE'S WEED

The dried leaves of Eriodic'tyon Californicum Greene.

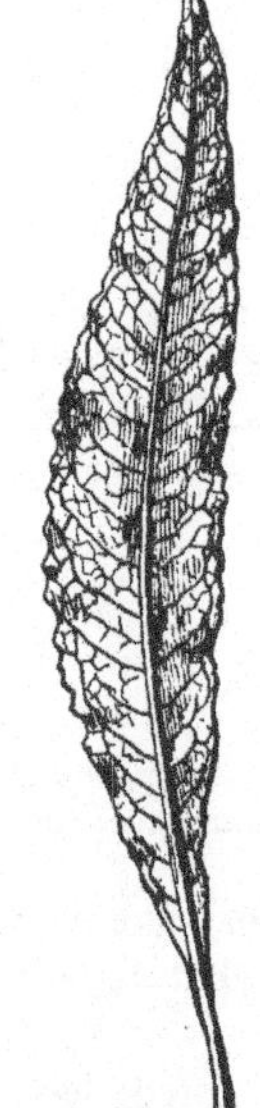

FIG. 213.—Leaf of Yerba Santa. Under side.

BOTANICAL CHARACTERISTICS.—Low shrubs with alternate *leaves*. *Calyx* of narrow sepals; *corolla* violet or purple, occasionally white, with the filaments adherent to it; *ovary* 2-celled. *Fruit* a small capsule.

HABITAT.—California, and in mountains of Northern Mexico.

DESCRIPTION OF DRUG.—Oblong-lanceolate, from 50 to 100 mm. (2 to 4 in.) long, 10 to 30 mm. (⅖ to 1⅕ in.) broad, with a sharp apex, and narrowed at the base into a short foot-stalk; margin sinuate or almost entire; upper surface brownish-green and **varnished with a resinous coating; under surface greenish-white, hairy,** with a prominent midrib and **distinct reticulations;** brittle, odor aromatic; taste balsamic, sweetish, free from bitterness.

RELATED SPECIES.—*Eriodictyon tomentosum*, growing with the other, is large and has a dense coat of short, villous hairs, becoming whitish or musty-colored with age.

Powder.—Characteristic elements: See Part iv, Chap. I, B.

CONSTITUENTS.—Volatile oil, an acrid resin, tannin, ericolin, $C_{34}H_{56}O_{21}$.

ACTION AND USES.—Long used in California as a stimulant balsamic expectorant. Its preparations are principally used, however, as

vehicles to disguise the taste of disagreeable medicines like quinine. Dose: 15 to 30 gr. (1 to 2 Gm.).

OFFICIAL PREPARATION.

Fluidextractum Eriodictyi, ...Dose: 15 to 30 ♏ (1 to 2 mils).

BORRAGINACEÆ.—Borage Family

465. **ALKANNA.**—ALKANET. The root of **Alkan'na tincto'ria** Tausch. *Habitat:* Grecian Archipelago and Southern Europe. Fusiform, about 100 mm. (4 in.) long, from the thickness of a quill to that of the little finger, often crowned with soft, white, hairy root-stocks; the bark is of a dark-purple color, friable, and separates easily in thin, papery layers from the yellowish, twisted ligneous column; the wood is composed of distinct, slender wood-fibers cohering together and cleft by purple, friable, medullary rays; in the commercial samples, however, it is generally more or less decayed, loose, and spongy. Odorless and tasteless. Alkanna is employed exclusively for coloring oils, ointments, and plasters, which is accomplished by suspending it, tied up in a rag, into the melted fat. Its coloring principle has been termed alkannin; it is a red, resin-like substance, soluble in alcohol, ether, and fats, but insoluble in water.

Preparation of Alkannin.—Obtained by evaporation of ethereal tincture, or precipitating a weak alkaline aqueous solution of alkanet by an acid.

466. **SYMPHYTUM.**—COMFREY. The root of **Sym'phytum officina'le** Linné. *Habitat:* United States and Europe; cultivated. About 150 mm. (6 in.) or more long, and from the thickness of a quill up to an inch in diameter, often split; externally black, wrinkled; internally whitish, and horny when dry; inodorous; taste sweetish, astringent, and very mucilaginous, containing as much mucilage as, or more than, althæa, for which it may often be substituted. It is chiefly used as a demulcent in domestic cough remedies, and has been highly esteemed as a vulnerary. Dose: 2 to 4 dr. (8 to 15 Gm.).

467. **BORAGO OFFICINALIS** Linné.—BORAGE. *Habitat:* Europe. (Leaves.) They contain a large quantity of mucilage, with potassium nitrate and other salts, upon which their virtues depend. Diuretic, refrigerant, demulcent, etc. Dose of fluidextract: 1 fl. dr. (4 mils).

468. **PULMONARIA OFFICINALIS** Linné.—LUNGWORT. *Habitat:* Europe. (Leaves.) Pectoral and demulcent. Dose: 30 to 60 gr. (2 to 4 Gm.).

VERBENACEÆ.—Vervain Family

469. **LIPPIA MEXICANA.**—The leaves of **Lip'pia dul'cis** Treviranus. Demulcent and expectorant. Dose: 8 to 15 gr. (0.5 to 1 Gm.).

470. **VERBENA HASTATA** Linné.—AMERICAN BLUE VERVAIN. (Root and Herb.) (Verbena, N.F., is the dried overground portion of the plant, collected when flowering.) The hot infusion is used as a sudorific in colds, etc. Also tonic and expectorant. Dose of fl'ext.: 30 to 60 ♏ (2 to 4 mils).

471. **VERBENA URTICÆFOLIA** Linné.—WHITE VERVAIN. *Habitat:* Tropical America. (Root.) Febrifuge. Credited with the cure of the opium-habit. Dose of fl'ext.: 30 to 40 ♏ (2 to 2.6 mils).

472. **TONGA.**—A drug introduced under this name has been found to be a mixture of bark, leaves, and woody fibers, tied into bundles by means of the inner bark of the cocoanut tree. The bark comes from **Premna taitensis** (nat. ord. Verbenaceæ), a shrubby tree having a sweet and slightly astringent inner bark, containing little volatile oil, etc. The fibrous material comes from **Rhaphidophora vitiensis** (nat. ord. Araceæ), a creeper having a stem about the size of a quill, containing potassium chloride, a volatile alkaloid, tongine, etc. From this mixture a fl'ext. is prepared which has proved efficient in neuralgia. Dose of fl'ext.: 1 fl. dr. (4 mils).

24

LABIATÆ.—Mint Family

One of the most natural groups of plants in the vegetable kingdom. Its members being so uniform, it would seem as if all of its species could be comprehended in a single genus; hence the characteristics of its different genera are very difficult to make out.

DESCRIPTION.—*Herbs* with opposite or whorled leaves. *Flowers* in axils of leaves or bracts, solitary or clustered cymes, scattered or crowded into spikes. *Calyx* sometimes 2-lipped, upper lip bifid, lower trifid, sometimes subregular. *Corolla* monopetalous, bilabiate, the upper lip entire or emarginate, the lower 3-lobed, sometimes bell- or funnel-shaped, with 4 subequal lobes (Mentha). *Stamens* 4, inserted on the corolla tube, didynamous (2 long and 2 short), or 2 by the abortion of the 2 upper (Lycopus, Salvia, Rosmarinus). *Ovary* 4-lobed. *Ovules* 4. *Style* simple, rising from the base of the ovarian lobes. *Fruit* separating into 4 akenes. *Stem* quadrangular, with volatile oil secreted in vascular glands.

GENERAL DESCRIPTION OF DRUGS OF THE ORDER

In most instances the drug consists of dry herbs composed of leaves, or leaves and tops, with portions of stem, branches, and flowers. These are usually broken and intermixed. Odor aromatic, due to the secreted volatile oil; some species hold in solution a solid hydrocarbon (stearopten) analogous to camphor. Taste aromatic, pungent, cooling, and bitterish (marrubium). The odor and taste are frequently sufficient to distinguish the different drugs, but a knowledge of the size, shape, and marginal character of the leaves and their texture, and the character of the stem and branches is sometimes quite useful in the identification of the various drugs derived from the order.

Synopsis of Drugs from the Labiatæ

A. *Herbs.*
MENTHA PIPER-
  ITA, 473.
MENTHA VIRIDIS,
  474.
Hedeoma, 475.
Marrubium, 476.
Melissa, 477.
*Scutellaria, 478.
Origanum, 479.
Cunila, 480.
Glechoma, 481.
Lycopus, 482.
Majorana, 483.
Serpyllum, 484.
Leonurus, 485.
Monarda, 486.
Hyssopus, 488.
*Cataria, 489.

Teucrium, 490.
Lamium, 491.
B. *Leaves.*
  Salvia, 492.
  Rosmarinus, 493.
  *Thymus, 494.
  Orthosiphon, 495.
  Pycnanthemum, 496.
  Satureia, 497.
  Yerba Buena, 498.
  Ocimum, 499.
  Monarda Fistulosa, 487.
  Betonica, 500.
C. *Flowers.*
  Lavandula, 501.
D. *Volatile Oils.*
  OLEUM MENTHÆ
    PIPERITÆ, 473 a.

OLEUM MENTHÆ
  VIRIDIS, 474 a.
Oleum Hedeomæ
  475 a.
Oleum Origani, 479 a.
Oleum Monardæ, 486 a.
OLEUM ROSMARINI,
  493 a.
OLEUM THYMI, 494 a.
OLEUM LAVANDULÆ
  FLORUM, 501 a.
E. *Stearopten.*
  MENTHOL, 473 b.
F. *Rhizome.*
  Collinsonia, 502.

473. MENTHA PIPERITA.—PEPPERMINT

¶ The dried leaves and tops of **Men′tha piperi′ta** Linné.

DESCRIPTION.—Leaves petiolate, ovate, lanceolate, about 2 inches (50 mm.) long, acute, sharply serrate, glandular, nearly smooth; light or dark green flowers in terminal spikes, purplish; odor strong and characteristic; taste pungent and cooling. Statistics show that 300,000 pounds of peppermint are annually consumed by the world, and that more than 90 per cent. of this is grown within 25 miles of Kalamazoo, Mich. A few miles from Fenville, Mich., there are two famous mint farms, one section covers about 1400 acres, the other about 2100 acres. The former tract is known as the "Campania Farm" the other "Mentha Farm." A distilling plant is on the

FIG. 214.—*Mentha piperita*—Flowering branch.

ground. An average yield of oil is about 20 pounds per acre. The "mint" industry is a specialty with peculiar features, combining farm and factory—agriculture in growing the plant, and the manufacture in separating the oil by distillation. There are about 80 "stills" in southwestern Michigan, and since there are 4000 acres of the plant under cultivation, one "still" is required for every 50 acres of peppermint.

Powder.—Microscopical elements of: See Part iv, Chap. I, B.

CONSTITUENTS.—Volatile oil, consisting of a terpene of complex composition (liquid) and menthol, $C_{10}H_{20}O$.

ACTION AND USES.—Carminative and diffusive stimulant. Dose: 15 to
60 gr. (1 to 4 Gm.).

OFFICIAL PREPARATION.

Spiritus Menthæ Piperitæ (1 per cent.),...Dose: 15 to 30 ♏ (1 to 2 mils).

473 a. OLEUM MENTHÆ PIPERITÆ, U.S.

A volatile oil distilled from peppermint. A colorless, or yellowish, or
greenish-yellow liquid, turning darker and thicker by age and exposure to the air, having a strongly aromatic, pungent taste, followed
by a sensation of cold when air is drawn into the mouth. Its composition is very complex, consisting of a number of terpenes, aldehydes,
and acids: pinene, phellandrene, cineol, dipentene, limonene, men-
thone, and menthol, etc. In a freezing mixture the oil becomes
cloudy and thick, and will separate crystals of menthol (473 b). The
oil yields not less than 5 per cent. of esters calculated as methyl
acetate and not less than 50 per cent. of total menthol.

OFFICIAL PREPARATIONS.

Aqua Menthæ Piperitæ (0.2 per cent.),...Dose: 4 fl. dr. (15 mils).
Spiritus Menthæ Piperitæ (10 per cent.),   15 to 30 ♏ (1 to 2 mils).

473 b. MENTHOL

A secondary alcohol from the official oil of peppermint (from *Mentha
piperita* Smith), or from Japanese or Chinese oil of peppermint (from
Mentha arvensis Linné, variety *piperascens* Holmes, and *Mentha canadensis* Linné, variety *glabrata* Holmes). Colorless, acicular or prismatic crystals, having a strong and pure odor of peppermint, and a
warm, aromatic taste, followed by a sensation of cold when air is
drawn into the mouth. It is slightly soluble in water, freely soluble
in olive-oil, and very soluble in alcohol, ether, chloroform, and in
petroleum benzin. When menthol is triturated with about an
equal part by weight of camphor, thymol or hydrated chloral, the
mixture becomes liquid.

Lubulinski recommends the use of a solution of menthol in liquid
paraffine for acute coryza. Dose: 0.06 Gm. (1 gr.).

474. MENTHA VIRIDIS.—SPEARMINT

The dried leaves and flowering tops of **Men'tha spica'ta** Linné.

DESCRIPTION.—The leaves of the spearmint resemble those of the peppermint, but the former are rather subsessile. The branches of the
spearmint are mostly light green, while those of the peppermint are
often purplish. The stamens of the spearmint are exserted, while

those of the peppermint are short; odor and taste mint-like, but less cooling, .quite characteristic.

Powder.—Microscopical elements of: See Part iv, Chap. I, B.

CONSTITUENTS.—Volatile oil containing carvone, $C_{10}H_{14}O$, limonine, etc.

ACTION AND USES.—Carminative; an antispasmodic of milder property than peppermint, often preferred in infantile cases.   Dose: 15 to 60 gr. (1 to 4 Gm.), in infusion, employed in Spiritus Menthæ Viridis.

OFFICIAL PREPARATION.—**Spiritus Menthæ Piperitæ.**

FIG. 215.—*Mentha viridis*—Flowering branch.

474 a. OLEUM MENTHÆ VIRIDIS

A volatile oil distilled from the flowering plant of Mentha Spicata Linné (Mentha Viridis Linné) and yielding when assayed by the U.S.P. process not less than 40 per cent. by volume of carvone. It is a colorless, yellow or greenish-yellow liquid having characteristic odor and taste of spearmint.

Michigan is the chief producer of this oil in U.S.

OFFICIAL PREPARATIONS.

Aqua Menthæ Viridis (0.2 per cent.).
Spiritus Menthæ Viridis (10 per cent.),.....Dose: 30 ♏ (2.0 mils).

475. HEDEOMA.—American Pennyroyal U.S.P. VIII

The dried leaves and tops of **Hedeo'ma pulegioi'des** Persoon.

DESCRIPTION.—*Stem* hairy; *leaves* ½ inch (12 mm.) long, short-petioled, oblong-ovate, slightly serrate; *flowers* in small axillary cymules, with small, pale blue, spotted, pubescent stamens; odor mint-like. Taste aromatic and pungent.

CONSTITUENTS.—Volatile oil containing hedeomol, $C_{10}H_{18}O$, and pulegone, $C_{10}H_{16}O$. The oil obtained from *Mentha pulegium* Linné has about the same specific gravity and optical rotation, and contains pulegone.

FIG. 216.—*Hedeoma pulegioides*—Flowering branch.

475 a. OLEUM HEDEOMÆ, U.S. VIII—Oil of Pennyroyal

A volatile oil distilled from the flowering plant of Hedeoma pulegioides Persoon.

SOURCE AND DESCRIPTION.—Most of the oil of pennyroyal is reported as being distilled in North Carolina and in the southern and eastern parts of Ohio.

It is a pale yellow liquid, having the characteristic odor and taste of Hedeoma. Its specific gravity is 0.920 to 0.935 at 25°C. It is soluble in 2 volumes of 70 per cent. alcohol forming a solution showing not more than a slightly acid reaction with litmus.

The principal and only constituent known definitely to exist in the oil is "pulegone," a ketone which can be identified by its hydrated oxime.

ACTION AND USES.—Oil of pennyroyal possesses stimulant, carminative, and emmenagogue properties.

The dose is from 2 to 10 minims (0.10 to 0.60 mil).

476. MARRUBIUM.—HOREHOUND U.S.P. VIII

The dried leaves and flowering tops of **Marru'bium vulga're** Linné.

DESCRIPTION.—*Stem* white, tomentose; *leaves* roundish-ovate, about 1 inch (25 mm.) long, obtuse, crenate, downy; *flowers* whitish, aromatic, and bitter; *odor* distinct and agreeable; *taste* aromatic and bitter.

CONSTITUENTS.—Volatile oil in minute quantity, marrubiin, crystalline prisms soluble in hot water and ethereal solvents, insoluble in benzene.

Preparation of Marrubiin.—Treat the infusion with charcoal; exhaust latter with Alcohol, which dissolves marrubiin and tannin; precipitate tannin with lead oxide; exhaust precipitate with alcohol. This leaves behind insoluble tannate of lead and dissolves the bitter principle.

ACTION AND USES.—A bitter tonic, laxative when given in large doses; employed in catarrh and chronic affections of the lungs attended by copious expectoration. Dose: 15 to 30 gr. (1 to 2 Gm.), in infusion or decoction.

477. **MELISSA.**—BALM. The leaves and tops of **Melis'sa officina'lis** Linné. Official U.S.P. 1890. Margin of leaves crenate, serrate, base rounded or rather heart-shaped, somewhat hairy; flowers in about four-flowered cymules, whitish or purplish; aromatic, astringent, and bitterish. *Constituents:* Volatile oil about 0.1 per cent., containing citral and citronellal. Carminative, stimulant, diaphoretic. Dose: 15 to 60 gr. (1 to 4 Gm.), in infusion or decoction.

478. SCUTELLARIA, N.F.—SKULLCAP

The dried plant of **Scutella'ria lateriflo'ra** Linné.

DESCRIPTION.—*Leaves* 2 inches (50 mm.) long, somewhat ovate, serrate; *stem* smooth and branched; corolla pale blue; odor slight, taste bitterish. The other species of Scutellaria are sometimes collected as *S. integriafolia, S. pilosa,* and *S. galericulata.*

CONSTITUENTS.—A bitter crystalline glucoside, trace of volatile oil, tannin.

ACTION AND USES.—Tonic and antispasmodic. Dose: 30 to 60 gr. (2 to 4 Gm.), in infusion or fl'ext.

Fluidextractum Scutellariæ,..........Dose: 30 to 60 ℔ (2 to 4 mils).

479. **ORIGANUM.**—WILD MARJORAM. The herb of Ori'ganum vulga're Linné, formerly used in making the Vinum Aromaticum, U.S.P. 1880.

479 a. **OLEUM ORIGANI.**—OIL OF ORIGANUM is a favorite among some practitioners as an ingredient in various liniments.

480. **CUNILA.**—DITTANY. The herb of **Cunil'a marian'a** Linné. Carminative and sudorific. Dose: 15 to 60 gr. (1 to 4 Gm.).

481. **GLECHOMA.**—GROUND IVY. The herb of Glecho'ma hedera'cea Linné. Pectoral, tonic, and diuretic. Dose: 30 to 60 gr. (2 to 4 Gm.).

482. **LYCOPUS.**—BUGLE. The herb of Ly'copus virgin'icus Linné, and of L. sinuatus Elliott. Astringent, sedative. Dose: 8 to 30 gr. (0.5 to 2 Gm.).

483. **MAJORANA.**—SWEET MARJORAM. The herb of Ori'ganum majora'na. Carminative, stimulant, and emmenagogue. Dose: 15 to 60 gr. (1 to 4 Gm.).

484. **SERPYLLUM.**—WILD THYME. The herb of Thy'mus serpyl'lum. Carminative, stimulant, tonic, and emmenagogue. Dose: 15 to 60 gr. (1 to 4 Gm.).

485. **LEONURUS.**—Motherwort. The herb of Leonu'rus cardia'ca. Tonic and expectorant. Dose: 30 to 60 gr. (2 to 4 Gm.).

486. **MONARDA.**—Horsemint. The herb of Monar'da puncta'ta Linné. Carminative, emmenagogue, and nervine. Dose: 15 to 60 gr. (1 to 4 Gm.).

486 a. **OLEUM MONARDÆ.**—Oil of Horsemint. Used as an embrocation and as an addition to stimulating liniments.

487. **MONARDA FISTULOSA** Linné.—Wild Bergamot. Indigenous. (Leaves.) Introduced as a substitute forq uinine; in large doses diaphoretic. Dose: 15 to 60 gr. (1 to 4 Gm.).

488. **HYSSOPUS.**—Hyssop. The herb of Hysso'pus officina'lis Linné. Carminative, sudorific, and stimulant. Dose: 15 to 60 gr. (1 to 4 Gm.).

489. **CATARIA,** N.F.—Catnip. The herb of Nep'eta cata'ria Linné. Carminative, stimulant, tonic, and diaphoretic. Dose: 15 to 60 gr. (1 to 4 Gm.).

490. **TEUCRIUM.**—Germander. The leaves and tops of Teu'crium chamæ'drys. Aromatic stimulant; noted as an ingredient in the famous gout remedy known as Portland Powder.

491. **LAMIUM ALBUM** Linné.—Dead Nettle. (Herb.) An active hemostatic.

492. SALVIA.—Sage, U.S.P. VIII

The dried leaves of **Sal'via officin'alis** Linné.

Description.—About 2 inches (50 mm.) long, ovate, obtuse, base narrow to the long petiole, thickish, wrinkled, grayish-green, soft, hairy, and reticulated and glandular beneath; odor aromatic, taste bitterish and astringent. Salvia is said to be adulterated with other species, closely resembling the official in late summer.

Constituents.—Volatile oil (0.5 to 0.75 per cent.), resin, tannin, etc. The volatile oil contains pinene, cineol, and salviol, $C_{10}H_{18}O$.

Action and Uses.—Stimulant, tonic, astringent, vulnerary, in infusion or decoction. Dose: 15 to 60 gr. (1 to 4 Gm.).

493. **ROSMARINUS.**—Rosemary. The leaves of Rosmari'nus officina'lis Linné. Rigid, linear, obtuse at summit, margin entire; odor strong, balsamic, and camphoraceous.

Action and Uses.—Carminative, stimulant, diaphoretic, emmenagogue. Dose: 3 to 15 gr. (0.2 to 1 Gm.)

493 a. **OLEUM ROSMARINI,** U.S.—Oil of Rosemary. A volatile oil distilled from the fresh flowering tops of Rosmarinus officinalis Linné, yielding, when assay by official process, not less than 2.5 per cent. of ester, calculated as bornyl acetate ($C_{10}H_{17}C_2H_3O_2$) and not less than 10 per cent. of total borneol ($C_{10}H_{17}OH$).

Description.—It is a colorless or pale yellow liquid, having the characteristic odor of rosemary and a camphoraceous taste.

Action and Uses.—In moderate amounts acts as stimulant, aromatic and carminative. In local application, it is said to do good in the treatment of chronic rheumatism, sprains, etc.

Official Preparations.

Tinctura Lavandulæ Composita (0.2 per cent.),............................Dose: ½ to 2 fl. dr. (2 to 8 mils).
Linimentum Saponis (1 per cent.).

494. **THYMUS,** N.F.—Garden Thyme. The leaves of Thy'mus vulga'ris Linné. Carminative, tonic, antispasmodic. Dose: 30 to 60 gr. (2 to 4 Gm.).

494 a. **OLEUM THYMI,** U.S.—Oil of Thyme. Used as an antiseptic, etc. A volatile oil distilled from the flowering plant of Thymus vulgaris Linné, containing about 20 per cent. by volume of phenols. It is a colorless red liquid

having a characteristic odor and taste. Specific gravity is from 0.894 to 0.929. It is soluble in 2 volumes of 80 per cent. alcohol.

PROPERTIES.—Commercially "red" and "white" oil are distinguished. The latter, however, is not obtained by simple rectification of the ordinary kind. "White" thyme oil, offered at a lower price than the "red," is apt to contain much turpentine oil.

THYMOL.—(See Ajowan, 389.)

495. ORTHOSIPHON STAMINEUS Bentham.—JAVA TEA. (Leaves.) Used as a diuretic and in gravel. Dose of fl'ext.: 20 to 30 ♏(1.3 to 2 mils).

496. PYCNANTHEMUM MONTANUM Michaux.—MOUNTAIN MINT. (Leaves.) Stimulant, tonic, and carminative. Dose: 15 to 60 gr. (1 to 4 Gm.).

497. SATUREIA HORTENSIS Linné.—SUMMER SAVORY. *Habitat:* Southern Europe; cultivated in our gardens. (Leaves.) Stimulant, carminative, and emmenagogue. Dose: 1 to 4 dr. (4 to 15 Gm.).

498. YERBA BUENA.—The leaves of a California plant, **Microme'ria dougla'sii** Bentham. A grateful aromatic stimulant, tonic, and emmenagogue. Dose of fl'ext.: ½ to 2 fl. dr. (2 to 8 mils).

499. OCIMUM BASILICUM Linné.—SWEET BASIL. (Leaves.) Aromatic, stimulant, and tonic.

500. BETONICA.—The leaves of **Sta'chys beto'nica** Bentham. Used in atonic dyspepsia, rheumatism, hepatic diseases, etc. Dose: 15 to 60 gr. (1 to 4 Gm.).

501. LAVANDULA.—GARDEN LAVENDER. The flowers of **Lavan'dula ve'ra** De Candolle. Calyx tubular, blue-gray, hairy, 5-toothed; corolla violet-blue, hairy, and glandular on the outside, tubular and 2-lipped; odor characteristic, somewhat camphoraceous. Stimulant and carminative. Dose: 15 to 30 gr. (1 to 2 Gm.).

501 a. OLEUM LAVANDULÆ FLORUM, U.S.—OIL OF LAVENDER FLOWERS. A volatile oil distilled from the fresh flowers of **Lavan'dula officina'lis** Chaix. French oil contains linalool, geraniol, partly free and partly as ester, principally as acetate, but in small part. Also as propionate, butyrate and valerianate. English oil contains linaloyl acetate and free linalool, also limonene and sesquiterpene, and cineol.
U. S. P. IX gives quantitative test for *esters*.

ACTION AND USES.—Used as perfumery and as flavoring agent in certain pharmaceuticals.

OFFICIAL PREPARATIONS.

Spiritus Lavandulæ (5 per cent. of the oil).
Tinctura Lavandulæ Composita (0.8 per cent. of the oil, with oil of rosemary. Saigon cinnamon, cloves, nutmeg and red saunders). Dose, 30 ♏(2.0 mils).

OIL OF SPIKE, used as an embrocation in rheumatic affections, is obtained by distillation of the leaves, tops, etc., of *Lavandula spica*.

502. COLLINSONIA.—STONE ROOT. The rhizome of **Collinso'nia canaden'sis** Linné. Long, with short, knotty branches and numerous stem-scars; hard; internal whitish; nearly inodorous; taste bitter and nauseous. Contains resinous matter. Diaphoretic, diuretic, and irritant.

SOLANACEÆ.—Nightshade Family

Herbs or, rarely, shrubs, with rank-scented, often poisonous, foliage, and colorless juice. *Leaves* alternate. *Stamens* five, equal, inserted on the corolla. *Fruit* a capsule or berry. This order owes its poisonous qualities to the presence of alkaloids such as atropine.

Synopsis of Drugs from the Solanaceæ

A. *Roots.*	HYOSCYAMUS, 509.	*Dulcamara, 514.
BELLADONNÆ	Tabacum, 511.	F. *Fruits.*

RADIX, 503.	Duboisia, 512.	**CAPSICUM,** 516.
Manaca, 505.	C. *Stems and Branches.*	Lycopersicum, 517.
B. *Leaves.*	Pichi, 513.	G. *Herb.*
BELLADONNÆ	D. *Seeds.*	*Solanum Carolinense,
FOLIA, 504.	Stramonii Semen, 508.	515.
STRAMONII FOLIA,	Hyoscyami Semen, 510.	H. *Rhizome.*
507.	E. *Branches.*	Scopola, 506.

BELLADONNA.—Deadly Nightshade

The dried root and the dried leaves official.

Botanical Characteristics.—At′ropa Belladon′na Linné. *Root* perennial, fleshy, white within; *stem* 3 to 5 feet high, with a tinge of red. *Leaves* short-petiolate, ovate, acute, entire, more or less hirsute. *Flowers* solitary, drooping; *calyx* campanulate; *corolla* campanulate, twice the length of the calyx, greenish at the base, varying to dark purple at the border. *Berry* 9-lobed, violet-black; seeds uniform.

503. BELLADONNÆ RADIX

The dried root of **Atropa Belladonna** Linné, yielding, when assayed by U.S.P. process, not less than 0.45 per cent. of its alkaloids.

Description of Drug.—Rough, irregular, longitudinally wrinkled, somewhat tapering pieces, from 12 to 25 mm. (½ to 1 in.) thick, of a dirty-gray color externally, internally whitish; fracture short, mealy when dry, tough when damp; odor narcotic; taste slightly sweetish, afterward bitter and acrid. Tough, woody roots, breaking with a splintery fracture, should be rejected, also the hollow stem-bases sometimes present.

Structure.—The bark is rather thick, free from bast fibers, composed almost entirely of parenchymatous tissue containing starch-grains and calcium oxalate raphides; directly underneath the periderm is a darker line composed of six to eight tabular cells. In the center of the root is a small pith, surrounded in the younger root by distant wood-fibers scattered throughout the parenchymatous tissue; in older roots the wood-bundles are more numerous, and traversed by broad medullary rays.

Belladonna is sometimes mistaken for, or adulterated with, althæa, from which it may be distinguished by the smoothness of its outer layer (althæa has projecting fibers), by its fracture, which does not show protruding fiber-ends, and by the wood-bundles, which are readily discernible in the former, but not in the latter.

Adulterations.—Certain species of Mandragora yield very nearly allied roots both in external appearance and structure, but they are not likely to be confounded with belladonna roots.

The rhizomes of *Scopola carniolica* are very similar to the root of belladonna; the bark, however, of the former, is less thick, starch-grains smaller, and shape less distinct. *Scopola Japonica* (Japanese belladonna) is found to be similar to *S. carniolica*.

Powder.—Microscopical elements of: See Part iv, Chap. I, B.

CONSTITUENTS.—The active principles are alkaloids, the chief of which are atropine, hyoscyamine and hyoscine. Atropine is a compound of equal amounts of the isomers, dextro- and levo-hyoscyamine into which it separates and is readily changed to dextro-hyoscyamine. In the growing belladonna the hyoscyamine is said to form in the young leaves, to be later changed to atropine.

According to the predominance of one or other of these alkaloids, and to the amounts present, the drugs of this group fall into a regular pharmacologic series, as follows:

FIG. 217.—*Atropa belladonna*—Branch.

1. Belladonna (root and leaves), the leaves containing 0.35 per cent., and the root, 0.5 per cent., of alkaloid, which is nearly all atropine. It has, therefore, a typical atropine action.

2. Scopola (root) contains 0.5 per cent. of alkaloid, about equally hyoscyamine and atropine. It acts like belladonna, but with somewhat less strength.

3. Stramonium (leaves) contains 0.35 per cent. of alkaloid, mostly hyoscyamine but with small amounts of atropine and hyoscyine. It is less stimulating to the cerebrum and may be narcotic.

4. Hyoscyamus (leaves) contain 0.08 per cent. of alkaloid, mostly hyoscyamine, with a fair amount of hyoscine, and only traces of atropine. It is rather narcotic but is weaker than the other drugs of the group (Bastido).

Ash, root, not more than 7 per cent.; leaves, not more than 20 per cent.

ACTION AND USES.—Applied externally belladonna is anodyne and anesthetic. Internally the activity of the peripheral terminations of all the secretory nerves in the body is depressed. Dropped into the eye, solutions of belladonna or atropine quickly dilate the pupil and accommodation is paralyzed. Upon the heart it has a stimulating action; toxic doses abolish the function of the cardiac muscles and the heart stops in diastole. When a 1 per cent. solution of

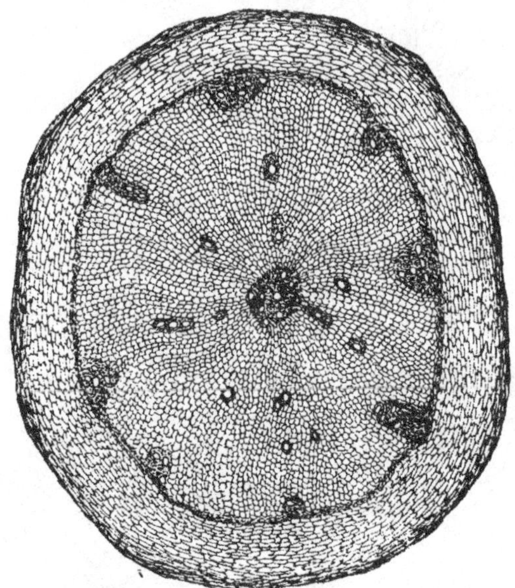

FIG. 218.—Cross-section of Belladonna root.

atropine sulphate is dropped into the eye, the pupil dilates in about fifteen or twenty minutes, but takes two hours to reach the maximum dilation. The pupil gradually regains its power but is not fully restored to normal for one or two weeks.

An antagonist of atropine is physostigmine, which stimulates the ends of the third nerve. It is not powerful enough to remove the effects of atropine at once, but greatly lessens the time which the eye takes to return to normal.

Dilated pupils, dry throat, and wild cerebral symptoms are the regular warnings of overdosage. In full poisoning there is a stage of central stimulation followed by collapse. Dose: 1 to 3 gr. (0.065

to 0.2 Gm.); of atropine, $\frac{1}{64}$ to $\frac{1}{100}$ gr. According to Cushney, hyoscyamine is twice as active as atropine in checking secretions and in pupil dilatation.

OFFICIAL PREPARATIONS.

Fluidextractum Belladonnæ Radicis, Dose: 1 to 3 ♏ (0.065 to 0.2 mil).
Linimentum Belladonnæ (95 per cent., with camphor 5 per cent.).

504. BELLADONNÆ FOLIA

The dried leaves of Atropa Belladonna, yielding not less than 0.3 per cent. of total alkaloids.

DESCRIPTION OF DRUG.—As they come into market, these leaves are crumpled and broken, of a dull brownish-green tint, the under surface paler than the upper, and with a prominent woody midrib prolonged below into a petiole, margin entire; one of the characteristics is the small, circular holes puncturing the leaves by the dropping off of corky excrescences. This, however, applies, but in a less degree, to the other narcotic leaves. It should be observed that the margins of the three narcotic leaves, belladonna, stramonium, and hyoscyamus, are quite different.

Powder.—Microscopical elements of: See Part iv, Chap. I, B.

CONSTITUENTS.—The alkaloids hyoscyamine and atropine (0.3 to 0.7 per cent.) are present. Belladonnine (oxyatropine) and other alkaloids of less importance exist, with chrysatropic acid. Ash, not exceeding 20 per cent.

ACTION AND USES.—Same as the root. Dose: 1 gr. (0.065 Gm.). The extract is employed in: Pil. Laxative Co. and Pil. Podophyl., Bellad. et Capsici, and in the following:

OFFICIAL PREPARATIONS.

Tinctura Belladonnæ Foliorum (10
    per cent.),....................Dose: 5 to 15 ♏ (0.3 to 1 mil).
Extractum Belladonnæ Foliorum
    (1.4 per cent. alkaloid),......... $\frac{1}{8}$ to $\frac{3}{4}$ gr. (0.008 to 0.048 Gm.).
Unguentum Belladonnæ (10 per cent.).

505. MANACA.—Portions of the root and stem of Brunfel'sia hopia'na Bentham, a Brazilian plant. Strongly recommended in chronic subacute rheumatism as a powerful alterative. Dose: 15 to 60 gr. (1 to 4 Gm.).

506. SCOPOLA.—SCOPOLA, U.S.P. VIII

The dried rhizome of Scopola Carniolica Jacquin, yielding by former U.S.P. process not less than 0.5 per cent. of its alkaloids.

DESCRIPTION OF DRUG.—From 25 to 100 mm. (1 to 4 in.) long and from 10 to 20 mm. ($\frac{3}{8}$ to $\frac{4}{5}$ in.) thick, frequently sliced. The upper surface is beset with cup-shaped stem scars; externally, yellowish-brown to dark brown; wrinkled longitudinally, obscurely annulate, rough and nodular; fracture short, showing a yellowish-white bark, its corky layer dark-brown or pale brown, indistinctly radiate wood; pith rather hard, but becoming soft and spongy when macerated in water. As compared to belladonna root, Coblentz concludes that scopola rhizome is more constant in alkaloidal content; that it is to be preferred to belladonna root in securing preparations of uniform standard.

CONSTITUENTS.—(See Belladonna.)

ACTION AND USE.—The action of scopola is about the same as that of belladonna, but preparations of the rhizome have not been professionally recognized until recently. The extract has been used as a substitute for the extract

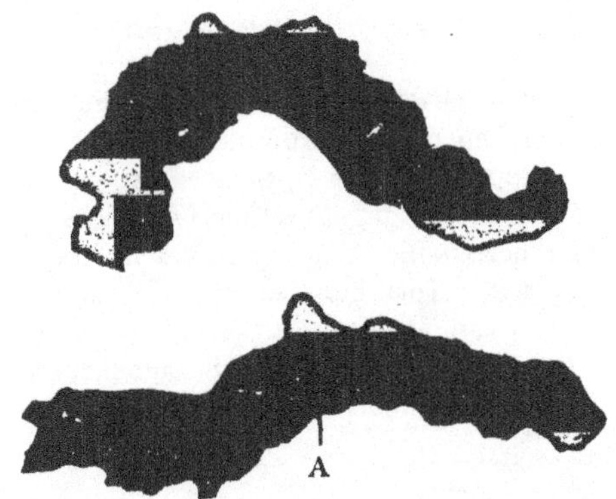

FIG. 219.—Scopola rhizome. A, Leaf scar. (½ natural size.) (Photograph.)

of belladonna in making of plasters. It has been stated that scopola costs about forty dollars per ton, while belladonna costs three hundred dollars per ton.

STRAMONIUM.—THORNAPPLE, JIMSON WEED

The leaves and the seed are medicinal.

BOTANICAL CHARACTERISTICS.—**Datu'ra stramo'nium** Linné. Rank, narcotic, poisonous annuals. *Leaves* ovate, sinuate-toothed. *Corolla* white, funnelform, the border 5-toothed. *Fruit* a 4-valved, 2-celled capsule, the outer side covered with prickles, longer toward the apex.

HABITAT.—Europe, Asia, and North America; almost universally distributed.

507. STRAMONIUM.—LEAVES

The dried leaves of **Datu'ra Stramo'nium** or of D. tatula Linné.

DESCRIPTION OF DRUG.—These leaves, in the dried and broken state resemble somewhat those of belladonna, but are lighter in color; odor distinct, heavy, and narcotic; taste nauseous. Admixture of more than 10 per cent. of stems or other foreign matter not permitted. **The drug should yield not less than 0.25 per cent. of total alkaloids.**

Powder.—Characteristic elements: See Part iv, Chap. I, B.

CONSTITUENTS.—Daturine o.2 per cent., which, according to Ladenburg, is a mixture of atropine and hyoscyamine, with the latter usually predominating; it is said to be stronger than atropine. Ash, not to exceed 20 per cent.

ACTION AND USES.—Stramonium acts similarly to belladonna in every particular, but more strongly, and chiefly on the sympathetic system,

FIG. 220.—*Datura stramonium*—Branch showing flower and fruit.

without affecting the motor or sensory nerves. **Its chief use is in asthma,** the powdered leaves being sprinkled with a solution of potassium nitrate, dried, and smoked in a pipe, or ignited and the smoke inhaled. Dose: 1 to 5 gr. (0.065 to 0.3 Gm.)

OFFICIAL PREPARATIONS.

Tinctura Stramonii,...............	Dose: ♏ 8 (0.5 mil).	
Extractum Stramonii { Pilular Extract,	0.01 Gm. (⅙ gr).	
{ Powder, Extracted........	0.01 Gm. (⅙ gr).	
Unguentum Stramonii.		

508. **STRAMONII SEMEN.**—Off. in U.S.P. 1890. Small, somewhat reniform, flattened seeds, with a blackish testa covered with small indentations; the embryo, curved parallel with the convex edge of the seed, is imbedded in a whitish, oily albumen. Inodorous in the whole state, but with a peculiar disagreeable odor when crushed; taste oily, slightly acrid, bitter, and nauseous. *Constituents:* Daturine o.3 per cent., combined with malic (daturic)

acid, scopolamine, fixed oil, etc.   Dose: 1 to 3 gr. (0.065 to 0.2 Gm.).   A
tincture, extract, and fluidextract were official in the U.S.P. 1890.

509. HYOSCYAMUS.—Henbane

The dried leaves and flowering tops of **Hyoscy'amus ni'ger** Linné, collected from
plants of second year's growth, yielding by official assay not less than 0.65 per cent.
of the alkaloids of Hyoscyamus.

Botanical Characteristics.—Clammy, pubescent, fœtid, narcotic annuals or
biennials. *Leaves* clasping, sinuate-toothed, and angled. *Flowers* sessile,
in one-sided, sessile spikes in the axils of the leaves; *calyx* urn-shaped;
corolla dull yellow, reticulated with purple veins. *Fruit* a 2-celled capsule.

Fig. 221.—*Hyoscyamus niger*—Flowering branch.

Source.—Europe and Asia; from biennial plants growing wild or cultivated in Britain, when about two-thirds of the flowers are expanded.
The plant is found in the northeastern section of the United States
in wet grounds, growing in great abundance about Detroit and in
other parts of Michigan.

Description of Drug.—The fresh leaf is from 2 to 10 inches long, 1
to 4 inches broad, ovate to ovate-oblong in shape. On each side
3 to 5 coarse, sinuate teeth or lobes, which are rather acute and oblong or triangular. On drying, the leaves shrivel and crumple up

around the very large, light-colored midribs, and generally have the large petiole still attached; they are grayish-green, and of a coriaceous texture; leaves, in the market, are very much broken; odor heavy, narcotic; taste bitter and nauseous.

Powder.—Microscopical elements of: See Part iv, Chap. I, B.

CONSTITUENTS.—By distillation the leaves yield a very poisonous volatile oil, but the active principles are **hyoscyamine,** $C_{17}H_{23}NO_3$ (crystalline), and **hyoscine,** $C_{17}H_{21}NO_4$ (amorphous). They also contain about 2 per cent. of potassium nitrate, which causes them to crackle when thrown in the fire. Ash, not exceeding 30 per cent.

Preparation of Hyoscyamine from Seed.—First extract fatty matter; acidulate with HCl; evaporate; wash acid solution with benzene. Neutralize solution with NH₄OH, shake out with chloroform, and evaporate latter solvent.

ACTION AND USES.—Anodyne, hypnotic, narcotic. The action of hyoscyamus is that of its alkaloid, hyoscyamine, which acts like atropine but is less irritant and more calmative and hypnotic. Hyoscine is a decided anodyne and hypnotic. The extract in the form of a suppository is frequently employed to relieve the pain of hemorrhoids, and a poultice made from the leaves may be employed to allay the pain of cancerous and other ulcers. Dose of leaves: 5 to 15 gr. (0.3 to 1 Gm.); Hyoscyamine salts, $\frac{1}{100}$ gr. (0.0006 Gm.); Hyoscine hydrobromate (Scopolamine hydrobromate), $\frac{1}{100}$ gr. (0.0006 Gm.).

RELATED SPECIES.—*Hyoscyamus pallidus* (flowers pale yellow), *H. agrestis* (flowers few, leaves smaller), and *H. albus* (flowers white). The latter is used indiscriminately in France with the *niger*, with which it appears to be identical in medicinal properties.

VARIETIES.—There are two varieties of henbane, the annual and biennial. The former when properly grown are not devoid of active properties. The official plant is susceptible of considerable diversity of character. causing varieties which have been considered by some as distinct species, and by cultivation differing somewhat in medical properties.

OFFICIAL PREPARATIONS.

Tinctura **Hyoscyami** (10 per cent.), .Dose: 10 to 60 ♏ (0.6 to 4 mils).
Fluidextractum **Hyoscyami,**........ 1 to 3 gr. (0.065 to 0.2 Gm.).
Extractum **Hyoscyami,**........... · 5 to 15 ♏ (0.3 to 1 mil):

510. **HYOSCYAMI SEMEN** (unofficial).—Used for the same purposes as the leaves and contain the same alkaloids, but in larger proportion, together with a large quantity of fixed oil and a bitter glucoside, hyoscyopicrin. They are small, reniform, and have a peculiar gray-brown surface, much wrinkled and finely pitted; near the raised portion of the testa they are rather sharp (distinction from stramonium seed). The embryo is curved so as to form a figure 9, the lower part of which is the radicle, and is surrounded by a whitish, oily albumen. Odorless in entire state, but when rubbed, of a distinctly narcotic odor; taste oily and bitter.

511. **TABACUM.**—TOBACCO. The leaves of **Nicotia'na taba'cum** Linné. Off. U.S.P. 1890. Large, oval, or oval-lanceolate leaves, often 500 mm. (20 in.) long when entire, but they are more generally somewhat broken; brown;
25

thin; friable; the glandular hairs, so thick on the leaves when fresh, are scarcely discernible; short-petiolate; odor peculiar, heavy, narcotic; taste strong, bitter, and acrid. *Constituents:* **Nicotine,** $C_{10}H_{14}N_2$, nicotianine (a camphor), bitter extractive, salts, resin, etc. Nicotine is a volatile liquid alkaloid and a virulent poison; there is hardly any of it contained in Turkish tobacco; by heat it is decomposed, yielding various pyridine compounds, hydrocyanic and acetic acids, etc.; these pass off in the smoke; the chief of these compounds are pyridine (in smoke from pipes), collidine (from cigars), lobeline, coniine, piperidine, sparteine, trimethylamine, etc.

Preparation of Nicotine.—Concentrated infusion made with acidulated water is treated with KOH and shaken with ether. The ethereal solution is precipitated with oxalic acid; the oxalate of the alkaloid thus precipitated is dissolved in boiling alcohol; evaporate to a syrup, agitate with ether, and make alkaline with KOH. On fractional distillation the colorless, oily alkaloid remains. It is very unstable.

Narcotic, sedative, diuretic, and emetic. It is rarely used in medicine. Dose: ½ to 2 gr. (0.0324 to 0.13 Gm.). Oil of tobacco is a pharmaceutical product, official in the U.S.P. in 1870, obtained by destructive distillation of coarsely powdered tobacco; it is a tarry liquid of offensive odor. Considerable oil is obtained by distilling the leaves with water. It contains nicotine (a dark, oily liquid).

512. **DUBOISIA.**—Duboisia. The leaves of **Duboi'sia myoporoi'des** R. Brown, a tall Australian shrub or small tree. The medicinal qualities of the leaves make the plant related to hyoscyamus and other narcotic plants of this order Lanceolate, 75 to 100 mm. (3 to 4 in.) long and 12 to 25 mm. (½ to 1 in.) broad, tapering below into a short petiole; midrib prominent; margin entire; they are generally seen, however, in broken fragments of a brownish-green color; inodorous; taste bitter. They contain **duboisine** (a mixture of hyoscyamine and atropine), and their action is, therefore, nearly identical with that of belladonna, except that they are less of a cerebral excitant and more calmative and hypnotic.

513. **PICHI.**—The stems and leafy branches of a Chilian shrub, **Fabia'na imbrica'ta** Ruiz et Pavon. A terebinthinate diuretic, used in gravel, cystitis, and diseases of the genito-urinary tract when the kidneys are not inflamed. Dose of fluidextract: 30 ♏ (2 mils).

514. **DULCAMARA,** N.F.—Bittersweet. Woody Nightshade. The young branches of **Sola'num dulcama'ra** Linné. Off. U.S.P. 1890. Very small cylindrical pieces (branches cut in sections) about the thickness of a quill; externally longitudinally striate and marked with alternate leaf-scars; periderm light greenish-brown or greenish-gray, thin, overlaying a uniformly green, rather thick, inner bark. Wood whitish or yellow, with greenish spots, surrounding a central pith, or, as is generally the case, a hollow; it is in one or two circles, with large ducts and numerous one-rowed medullary rays. The bark consists principally of parenchymatous tissue. Inodorous; taste at first bitter, afterward sweet. *Constituents:* **Solanine,** the active alkaloid, and a glucoside termed **dulcamarin,** $C_{22}H_{34}O_{10}$, to which the taste of the drug is due; also resin, wax, gum, starch, and calcium lactate. Commercial Solanin is a mixture of Solanin and Solanidin. Solanidin is soluble in alcohol. Solanin is practically insoluble, excepting in boiling alcohol.

Preparation of Dulcamarin.—Digest aqueous infusion of the drug with animal charcoal; treat charcoal with alcohol. Precipitate aqueous solution of alcoholic extract with lead subacetate, wash, digest with alcohol, and decompose with H_2S. Evaporate resulting solution. Purify product by resolution, filtration and evaporation.

Dulcamara is feebly narcotic and anodyne, but is chiefly employed as an alterative and resolvent in skin diseases, particularly those of a scaly character. Dose: 1 to 2 dr. (4 to 8 Gm.).

Extractum Dulcamaræ Fluidum, U.S.
P. 1890,......   ....Dose: 1 to 2 fl. dr. (4 to 8 mils).

515. SOLA'NUM Carolinensis, N.F. Linné.—HORSE NETTLE. A 20 per cent. tincture of this herb has been recommended in epilepsy in doses of 30 to 60 ℥ (2 to 4 mils).

516. CAPSICUM.—CAPSICUM

CAYENNE PEPPER. RED PEPPER

The dried ripe fruit of **Cap'sicum** frutescens Blume, deprived of its calyx.

BOTANICAL CHARACTERISTICS.—A small, rough, branched shrub, 1 to 2 feet high. *Leaves* ovate or lanceolate, entire, hairy. *Flowers* small, white, solitary, axillary, drooping. *Capsule* deep red, very pungent.

SOURCE.—Tropical America and Asia; cultivated.

DESCRIPTION OF DRUG.—The fruits vary in shape and size, but those most generally used are oblong, wrinkled, pendulous, pod-like berries,

FIG. 222.—*Capsicum fastigiatum*—Branch.

the largest (American), about the thickness of a finger, with a long, recurved apex; pericarp bright red, sometimes yellow, thin, translucent; it incloses two or three cells and contains numerous flat, reniform, whitish seeds, which are surrounded by a dry, loose parenchyma, and fastened to a slender placenta; odor peculiar, very irritating, especially in powder or in the fresh state; taste fiery.

Powdered capsicum of the market consists of several species of capsicum ground up together. It is of a reddish color. This is especially true of the American capsicum, which is grown to a limited extent in Texas and Mexico, where it is ground and called "paprika." The African (Zanzibar) pod yields a powder of a greenish- or brownish-yellow color. The commercial variety known as Bombay yields a powder of a more yellowish color than the African, but is not at all like the reddish-orange powder resulting from the American pod. This color fades and disappears on long exposure to the light. It is often adulterated with sawdust and red lead; the former may be detected with the microscope, the latter by digesting the powder in dilute nitric acid, filtering, and adding a solution of sodium sulphate, which will throw down a white precipitate if any lead oxide is present.

STRUCTURE.—A microscopical examination for the distinction of the above varieties has been suggested. This test is based upon the size and character of the cells of the outer layer of the epidermis, the American having, in dimension, the largest and the African the smallest cell in the outer layer of the pericarp. The value of capsicum can be estimated only by assay.

Powder.—Characteristic elements: See Part iv, Chap. I, B.

CONSTITUENTS.—**Capsaicin,** $C_9H_{14}O_2$, an exceedingly active pungent principle existing principally in the pericarp; **a volatile alkaloid** having an odor like coniine, supposed to be the result of a decomposition process during ripening of the fruit, as it does not exist in the unripe fruit; fixed oil, fat acids (oleic, palmitic, and stearic), and a red coloring matter (a cholesterin ester of the fat acids). Ash, not exceeding 7 per cent.; insoluble in HCl 1 per cent.

Capsicum should yield not less than 15 per cent. of non-volatile ether extract, soluble in ether, U.S.P. IX.

Preparation of Capsaicin.—Treat petroleum ether extract with alkali; pass CO_2 through the solution; collect crystals after standing. Soluble in ether, alcohol, benzene, and fixed oils.

ACTION AND USES.—Externally **rubefacient.** Internally a powerful stimulant. Its chief value medicinally is in the treatment of malignant sore throat and scarlet fever, used internally and as a gargle. Dose: 1 to 5 gr. (0.06 to 0.3 Gm.).

OFFICIAL PREPARATIONS.

Tinctura Capsici (10 per cent.),........Dose: 15 to 30 ℞ (1 to 2 mils).
Oleoresina Capsici,.......   ¼ to 1 ℞ (0.0162 to 0.065 mil).

517. **LYCOPERSICUM ESCULENTUM** Miller.—TOMATO. The ripe fruit is said to exert a curative action on ulcerated mucous membranes, given internally and applied locally. Dose of fluidextract: 30 to 60 ℞ (2 to 4 mils).

SCROPHULARIACEÆ.—Figwort Family

Herbs or rarely trees with didynamous *stamens,* and an irregular, usually 2 lipped, *corolla; fruit* a capsule. A large order of plants, containing a bitter glucoside.

Synopsis of Drugs 'from the Scrophulariaceæ

A. *Leaves.*
    DIGITALIS, 518.
    Euphrasia, 519.
    *Verbascum, 520.
B. *Rhizome.*
    * Leptandra, 521.

C. *Herbs.*
    Veronica Officinalis, 522.
    Scrophularia, 523.
    Chelone, 524.

518. DIGITALIS.—DIGITALIS

FOXGLOVE

The carefully dried leaves of **Digita'lis purpurea** Linné, without admixture of more than 2 per cent. of stems, flowers, or other foreign matter.

BOTANICAL CHARACTERISTICS.—Biennial, hoary-pubescent. *Leaves* alternate, ovate-lanceolate, crenate, rugose. *Racemes* terminal, loose; *flowers* purple, sometimes white, hairy, and spotted within.

SOURCE.—The plant is indigenous to Southern and Central Europe, particularly in the western section, and grows wild as far north as Norway, also in Madeira and the Azores, and is cultivated in the United States. It is found on the edges of woody land and prefers sandy soil.

It is claimed by some investigators that Digitalis leaves of the first and second year's growth have proved identical in their activity, and the cultivated leaves are at least as active as those wild grown.

DESCRIPTION OF DRUG.—The margin of this leaf is rather irregularly double crenate. In the market it comes in wrinkled, velvety fragments, the lower surface paler green than the upper, softly pubescent, especially along the midrib and veins; the midrib is prominent, but not so much so as in hyoscyamus; the venation forms prominent meshes on the under surface of the leaf, the principal veins joining the midrib at a very acute angle; odor slight and characteristic; taste strongly bitter.

ADULTERATIONS.—Other dried leaves are sometimes mixed with digitalis; the commonest of these are: *Inula conyza* (*Conyza squarrosa*), spikenard, and *Inula helenium*, both having entire, instead of crenate or serrate, margins, and the latter having its veins branching off at about right angles to the midrib; accidental impurities, such as comfrey leaves, *Symphytum officinale*, have been found. These are lanceolate and bear isolated stiff hairs.

Powder.—Characteristic elements: See Part iv, Chap. I, B.

CONSTITUENTS.—The exact chemical composition of digitalis is a vexed question, but the latest analysis shows it to be composed of at least five principlès: **digitalin,** $C_5H_8O_2$ (soluble in alcohol, insoluble in water), **digitalein** (soluble in water and alcohol), **digitonin,** $C_{27}H_{44}O_{13}$ (readily soluble in water, insoluble in alcohol, the diuretic principle), **digitin** (inert), and digitoxin, $C_{31}H_{50}O_{10}$, the most active ingredient, crystalline (insoluble in water, and sparingly soluble in alcohol, deposited as a sediment from the alcoholic preparations of the leaf). Digitoxin, by recent experimentation, is found to yield with hydrochloric acid digitoxigenin, $C_{22}H_{32}O_4$, and a glucose, digitoxose, $C_9H_{18}O_6$, the former in colorless crystals.

DIGITALIS PRINCIPLES.—The search for pure principles representing the complete action of the drug seems to be hopeless, but many pro-prietary preparations have been countenanced, in a measure, by the Council of the A.M.A. These are: Digitalein, Crude; Digitalin, True; Digitalin, "French;" Digitalin, "German;" Digitoxin; Digitoxin-Merck. These principles are all described in "New and Non-official Remedies." Ash not to exceed 15 per cent.

TEST.—If made into a fluidextract and assayed biologically the minimum lethal dose should not be greater than 0.0006 mil of fluidextract, or the equivalent in fluidextract of 0.0000005 Gm. of ouabain, for each gramme of body weight of the frog.

Preparation of Digitalin.—A concentrated fluidextract is first treated with water acidulated with acetic acid and charcoal. The filtrate is neutralized with ammonia, then precipitated with tannin. The washed precipitate is then rubbed with lead oxide, boiled with alcohol, decolorized, and filtered. Evaporate to solid and wash with ether. In this way a digitalin of indefinite composition is obtained, consisting of such glucosides as digitin, digitonin, etc.

FIG. 223.—*Digitalis purpurea*—Flowering branch.

ACTION AND USES.—**Cardiac tonic and stimulant and diuretic.** It slows the heart's action and increases its force, and by stimulating the vascular nervous system causes contraction of the arterioles and therefore greatly increases arterial tension. Its efficient diuretic action in cardiac diseases is due to its peculiar effects upon the general and renal circulations. Dose: 1 to 2 gr. (0.065 to 0.03 Gm.). Dose of digitalin: 1/10 gr. (0.006 Gm.), much depends on the quality. Digitalin, "French." Homolle's Digitalin, for example: Dose:

$\frac{1}{250}$ to $\frac{1}{35}$ gr. (0.00025 to 0.002 Gm.), Digitoxin $\frac{1}{120}$ gr. (0.0005 Gm.), Digitalein, crude $\frac{1}{60}$ gr. (0.001 Gm.).

OFFICIAL PREPARATIONS.

Infusum Digitalis (1.5 per cent.),........Dose: 1 to 4 fl. dr. (4 to 15 mils).
Tinctura Digitalis (10 per cent.),........ 5 to 30 ♏ (0.3 to 2 mils).
Fluidextractum Digitalis,........ 1 to 2 ♏ (0.065 to 0.13 mil).

519. **EUPHRASIA OFFICINALIS** Linné.—EYEBRIGHT. The leaves of this common plant have been stated to be almost a specific in acute nasal catarrh, given in the form of infusion.

520. **VERBASCUM THAPSUS** Linné.—MULLEIN. Both the flowers and leaves of this field weed are used. (V. Flores and V. Folia, N.F.). Mullein contains a large proportion of mucilage, which makes it a good demulcent and emollient. Anodyne properties are also ascribed to it. Popularly used in pectoral complaints, especially consumption, in which it is said to relieve the cough and also to improve the nutrition. Dose: 2 to 3 dr. (8 to 12 Gm.), in infusion. The dried leaves are sometimes smoked for nasal catarrh.

521. LEPTANDRA, N.F.—LEPTANDRA

CULVER'S ROOT. CULVER'S PHYSIC

The dried rhizome and roots of **Veron'ica virgin'ica** Linné.

BOTANICAL CHARACTERISTICS.—*Stem* erect, 2 to 6 feet high. *Leaves* whorled, in 4's or 7's, very smooth, or sometimes slightly downy, lanceolate, serrulate. *Spikes* panicled; *corolla* small, pinkish, or nearly white; *stamens* much exserted.

HABITAT.—United States, east of the Mississippi.

DESCRIPTION OF DRUG.—Horizontal rhizome, 4 to 6 inches long, somewhat flattened, about the thickness of a quill, branched, generally broken into pieces an inch or more long; very hard and firm; from a light to a dark brown color; upper side marked with broad stem-scars, under side beset with the remnants of the thin, fragile, wrinkled rootlets. **Fracture** woody—bark thin, blackish, wood-circles one or two, yellowish, pith 6-rayed; tissue surrounding pith irregular and angular; inodorous; taste bitter and acrid.

Powder.—Brown. Characteristic elements: Parenchyma of cortex, isodiametrical or elongated with spherical starch grains (2 to 4 μ in diam.) and brown resin; sclerenchyma with bast fibers and narrow, thick-walled stone cells; ducts with scalariform, spiral, simple pores; tracheids, numerous; outer wall of epidermal rootlets very thick; considerable cork from rhizome.

CONSTITUENTS.—Besides tannin, gum, and a small quantity of volatile oil; it contains a crystalline glucoside, the active principle, which should be termed leptandrin instead of the resin or resinoid called by that name; this resinoid is obtained by precipitating a concentrated alcoholic tincture with water; its action is probably due to a small amount of the crystalline glucoside mixed with it.

Preparation of Leptandrin.—Remove coloring matter from infusion by basic acetate of lead, excess of lead removed by Na_2CO_3. Treat resulting liquid with animal charcoal. Extract washed charcoal with boiling alcohol; evaporate; dissolve in ether to purify. Upon evaporation needle-shaped crystals are obtained which are bitter; soluble in water, alcohol, and ether. The eclectic leptandrin is made by precipitating concentrated alcoholic tincture with water, and is a mixture of inert matter with pure leptandrin.

ACTION AND USES.—Cholagogue cathartic.　Dose: 15 to 60 gr. (1 to 4 Gm.).
The fluidextract, extract and vegetable cathartic pills formerly represented the drug (U.S.P. VIII).

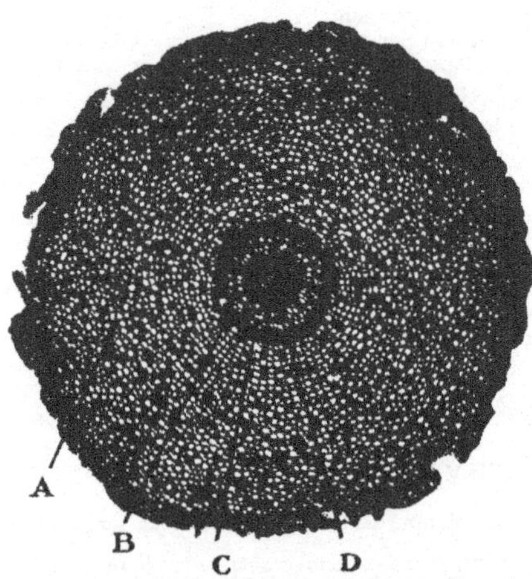

FIG. 224.—Leptandra—Cross-section of rootlet. (25 diam.) A, Parenchyma of cortex. B,
Phloëm. C, Xylem. D, Medulla. (Photomicrograph.)

522. VERON'ICA OFFICINA'LIS Linné.—SPEEDWELL. Indigenous. (Herb.)
Alterative, diuretic, and expectorant, in infusion.

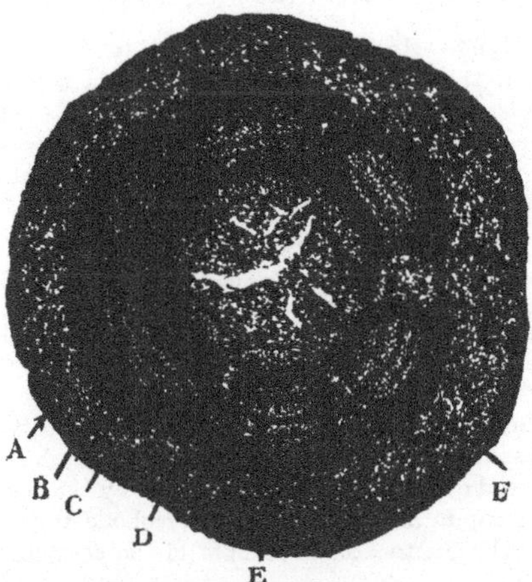

FIG. 225.—Leptandra—Cross-section of rhizome. (13 diam.) A, Cork. B, Parenchyma of
cortex. C, Phloëm. D, Medullary ray. E, Xylem. F, Medulla. (Photomicrograph.)

523. SCROPHULA'RIA NODO'SA Linné.—FIGWORT. This indigenous herb is
peculiar from the rank, fœtid odor of the leaves, especially when fresh. It
has alterative, diuretic, and anodyne properties, and is used in hepatic
diseases, scrofula, cutaneous diseases, dropsy, and as a depurative. Dose
of fluidextract: 30 to 60 ♏ (2 to 4 mils).

524. **CHELONE.**—BALMONY. SNAKE-HEAD. The herb of **Chelo'ne gla'bra** Linné. *Habitat:* United States. Tonic, anthelmintic, and laxative, with a supposed peculiar action on the liver. It has been largely employed in domestic practice as an external application in diseases of the skin. Dose: 30 to 60 gr. (2 to 4 Gm.).

OROBANCHACEÆ.—Broom-rape Family

525. **EPIPHEGUS.**—BEECH-DROP. CANCER-ROOT. The herb of **Epiphe'gus virginia'na** Barton, growing in all parts of North America as a parasite on the roots of the beech tree. It is a fleshy plant with a scaly, tuberous root, and smooth, yellowish or purplish stem, about 400 mm. (16 in.) tall, covered with small scales instead of leaves; taste bitter, astringent, and nauseous. It receives its name, cancer-root, from the popular belief that the powder was beneficial in the treatment of cancerous ulcers. It is often given as an astringent. Dose: 30 to 60 gr. (2 to 4 Gm.).

BIGNONIACEÆ.—Bignonia Family

526. **NEWBOULDIA.**—The root-bark of **Newboul'dia læ'vis** Seeman, introduced from tropical Africa as an astringent in diarrhea and dysentery. Dose of fl'ext.: 15 to 60 ♏ (1 to 4 mils).

527. **CAROBA.**—The leaves of **Jacaran'da proce'ra** Sprengel. *Habitat:* South America. A valuable alterative and antisyphilitic. Dose of fl'ext.: 15 to 60 ♏ (1 to 4 mils).

PEDALINEÆ

528. **SESAMUM.**—BENNÉ. From **Se'samum in'dicum** Linné, a plant growing to the height of 4 or 5 feet, native to the East Indies, but long cultivated in Asia and Africa; from the latter country it was introduced by the negroes into Southern United States. Both the leaves and the seeds are used, and a fixed oil expressed from the latter.

528 a. **The leaves** are oblong-lanceolate, from 75 to 125 mm. (3 to 5 in.) long, heart-shaped at base; pubescent, prominently veined beneath. They abound in a gummy matter to such an extent that two leaves stirred in a cup of water will make a sufficiently thick mucilage for use as a demulcent.

528 b. **The seeds** are used chiefly as a source of the fixed oil, of which they contain from 50 to 60 per cent. They are used by the southern negroes as food. Ovate, flattened, about 3 to 4 mm. (⅛ to ⅙ in.) long; externally yellowish-white to pale brown (in one species, *S. orientale*, purplish-brown), with four longitudinal ridges, and, on the pointed end, a somewhat prominent hilum; internally whitish, oily; taste bland.

528 c. **Oleum Sesami.**—TEEL OIL. BENNÉ OIL. A yellowish, limpid, transparent fixed oil, thinner at ordinary temperatures than most of the fixed oils; odor slight; taste bland, nut-like. It has a specific gravity of 0.919 to 0.923, and congeals to a yellowish-white mass at $-5°$C. ($-23°$F.). It is often used to adulterate olive and almond oils, in which it may be detected by shaking a portion of the suspected oil with an equal weight of concentrated hydrochloric acid; a bright emerald-green color will usually be produced, changing to blue, then violet, and finally to deep crimson on the addition of about one-tenth its weight of cane-sugar and agitating.

CONSTITUENTS.—Contains olein (76 per cent.), myristin, palmitin, stearin-resinoid compound, higher alcohol, $C_{25}H_{44}O$, sesamin, $C_{11}H_{12}O_3$, crystalline.

PLANTAGINEÆ

529. PLANTAGO.—Plantain. The herb of **Planta'go ma'jor** and other species. Used principally in domestic practice, the leaves being externally applied as a stimulant application to sores, frequently in the form of a poultice, not infrequently applied whole.

RUBIACEÆ.—Madder Family

Herbs, shrubs, or trees, with opposite, simple, and entire *leaves*, connected with interposed stipules, or in whorls without stipules. A very large family in tropical regions, represented by the coffee plant (Arabia and Africa) and by the cinchonas (South America).

Synopsis of Drugs from the Rubiaceæ

A. *Root.*
    IPECACUANHA, 530.
B. *Rhizome.*
    Rubia, 531.
C. *Bark.*
    CINCHONA, 532.
    CINCHONA RUBRA, 532 a.
    Remijia, 533.
    Cephalanthus, 534.

D. *Herb.*
    Mitchella, 535.
    Galium, 536.
E. *Seed.*
    *Coffea, 537.
F. *Extractive.*
    Catechu Pallidum (Gambir), 538.

530. IPECACUANHA.—Ipecac

IPECAC

The dried root, of **Cephæ'lis Ipecacuan'ha** (Brotero) A. Richard (Fam. *Rubiaceæ*), known commercially as *Rio Ipecac*, **C. acuminata** Karsten, known commercially as *Cartagena Ipecac*. The value is dependent upon the percentage of alkaloidal constituents, should yield not less than 1.75 per cent. of ether soluble alkaloids of Ipecac, U.S.P.

Two important alkaloids (**emetine and cephaëline**) are present in ipecac; the proportion in which these exist seems to vary, and this variation seems to depend upon the accidents of growth and the surroundings of the individual plant. —See Constituents.

Botanical Characteristics.—The root perennial, knotty, with transverse rings; *stems* suffruticose, ascending, somewhat pubescent toward the apex. *Leaves* opposite, oblong, roughish above, finely pubescent beneath. *Inflorescence* capitate, inclosed by a large one-leafed involucre; *flowers* bracteate; *corolla* white, funnel-form, the limb with reflexed segments; *stamens* 5, slightly exserted. *Fruit* a dark violet berry, crowned by the limb of the calyx, 2-celled, 2-seeded.

Source and Varieties.—Grows in the damp woods of the Brazilian valleys, notably in the provinces of Para, Rio Janeiro, Pernambuco, etc. This variety is known in commerce as Rio ipecac, while that from Colombia is called Carthagena ipecac. The former is usually preferred, but the latter is now more common. The plant *Psychotrin medica* is sometimes termed and sold as Carthagena ipecac, but it is devoid of alkaloid. The Brazilian plant is quite hardy, appearing as a creeping vine or bush. The roots usually grow thicker as they penetrate the ground and then taper off again to a point or thin root-let. Collectors usually leave a part of every other plant in the ground, so that in about three years another crop may be harvested. "Wiry root," consisting of about 75 per cent. of woody portion and 25 per cent. cortex, is, according to Dohme, richest in alkaloids. It has a rather rough, uneven appearance, and is popularly less esteemed than

FIG. 226.—*Cephælis ipecacuanha*—Plant and dried root.

the so-called "fancy" root consisting of 75 per cent. cortex. This prejudice, according to Dohme, is difficult to overcome.

DESCRIPTION OF DRUG.—**Rio Ipecac.**—In pieces of irregular length, rarely exceeding 25 cm.; stem portion 2 to 3 mm. thick, light gray-brown, cylindrical and smoothish; root portion usually red-brown, occasionally blackish-brown, rarely gray-brown, 3 to 6 mm. thick, curved and sharply tortuous, nearly free from rootlets, occasionally branched, closely annulated with thickened, incomplete rings, and usually exhibiting transverse fissures with vertical sides, through the bark; fracture short, the very thick, easily separable bark whitish, usually resinous, the thin, tough wood yellowish-white, without vessels; odor very slight, peculiar, the dust sternutatory; taste bitter and nauseous, somewhat acrid. It is stated by Rusby that the Rio variety has almost ceased to arrive in the market, the Carthagena variety being supplied. This is now mostly what is known as Panama Ipecac.

Carthagena ipecac is of a dull gray color, thicker, less frequently and sharply crooked, and lacks the constrictions characteristic of Rio ipecac, although it bears the annular thickenings, or merging

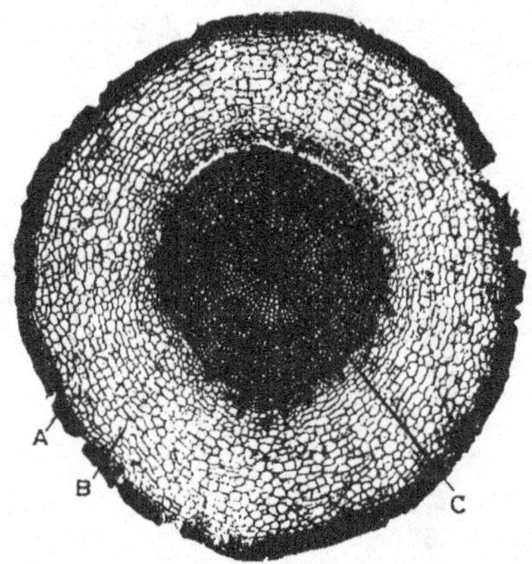

FIG. 227.—Ipecac—Cross-section of root. (25 diam.) A, Cork. B, Parenchyma of cortex. C, Xylem. (Photomicrograph.)

annulæ. The thick bark, on cross-section, has rather a grayish color, the medullary rays are more prominent and more numerous.

STRUCTURE.—The thin outer layer of cork cells contains a brownish-red deposit, thought by some to be emetine in combination with ipecacuanhic acid. The thick inner cortical layer consists of starchy parenchyma, free from medullary rays, but containing a circle of stone cells filled with calcium oxalate crystals. Transverse sections show rather a small layer of cork cells, a thick cortical portion consisting of parenchyma, loaded with starch and rich in alkaloid. The woody portion, radiate, contains little or no alkaloid.

Powder.—Characteristic elements: See Part iv, Chap. I, B.

CONSTITUENTS.—**Emetine** (1 to 2 per cent.), cephaëline, psychotrine, and a peculiar tannic acid called ipecacuanhic or cephaëlic acid, starch, resin, etc. The active principles exist only in the bark of the root, and probably in the thin, outer layer of cork cells. Recently considerable light has been thrown on emetine, $C_{15}H_{22}N_2O_5$, and cephaëline, $C_{14}H_{20}NO_2$, which were formerly supposed to be one body. According to Paul and Cownley ("Pharm. Jour.," 1896) cephaëline is the emetic principle and emetine the expectorant principle of the drug. This naming is unfortunate, and should be reversed. Emetine is amorphous; cephaëline crystalline. Ash, not less than 1.8 per cent. nor more than 4.5 per cent.

KRYPTONINE.—This is the name of a new alkaloid of ipecac, discovered by J. U. Lloyd. The principle itself, as well as its acid compounds, are colloidal in character. It belongs apparently to a new group of principles awaiting further investigation. Filter paper shows a marked adsorptive property to this alkaloid. It is black in mass but of varying color in different solvents. See Proc. Amer. Ph. Asso., 1916. Condensed description Amer. Druggist, Oct., 1916, p. 28.

Preparation of Emetine.—A very simple process is to exhaust the drug with boiling chloroform made slightly alkaline with solution of ammonia. Upon distilling off the chloroform the emetine is left in a very pure condition, and, when dried at 100°C., gives a residue which, when weighed, gives one a rough estimate of the value of the drug. Cephaëline is extracted usually with emetine in most of the processes for assay. It is less soluble in ether than emetine.

Preparation of Ipecacuanhic Acid (Cephaëlic Acid).—Precipitate decoction with lead acetate, dissolve precipitate with acetic acid, and precipitate solution with lead subacetate; wash and dry. Resembles caffeotannic acid.

ACTION AND USES.—When locally applied, acts as counter-irritant. Small doses are diaphoretic and expectorant. In large doses a **systemic emetic,** in minute doses stomachic, aiding digestion. Ipecac has been used, since its introduction into medicine, as a remedy in dysentery, when there is said to be a peculiar tolerance of the drug; but the fact is the stomach almost invariably rejects large doses. Recent experiments prove that ipecac, when deprived of its emetine, possesses its full antidysenteric properties, without the drawbacks of depression, nausea, etc. Accordingly there appears in the market to meet this peculiar demand a preparation made from **de-emetinized** bark. **Emetine** has recently been highly praised in the treatment of pyorrhea, Riggs' disease. Hypodermic tablets of the hydrochloride, containing from 0.016 to 0.032 Gm. are prepared. Used in the form of injections in diseases due to pathogenic amebas. Also administered internally, "Alcresta Ipecac" when thus administered is decomposed in the alkaline fluid of the intestines with liberation of alkaloids and produce amebacidal action. Tablets of same, representing 10 gr. of ipecac are dispensed. Dose 2 or 3 tablets three times a day at first period of few days, then discontinued for a day or two, if laxative effect is produced. Dose of ipecac as expectorant, 1 gr. (0.06 Gm.); emetic, 10 to 15 gr. (0.6 to 1 Gm.).

Official Preparations.

Fluidextractum Ipecacuanhæ,...Dose: 3 to 8 ♏ (0.2 to 0.5 mil); 15 to 60 ♏ (1 to 4 mil).

Syrupus Ipecacuanhæ (7 per cent.),   Adult exp. 30 ♏ (2 mils), Emetic 6 fl. dr. (24 mils).

Pulvis Ipecacuanhæ et Opii (10 per cent. of each),..   5 to 15 gr. (0.3 to 1 Gm.).

531. RUBIA.—Madder. The rhizome of **Ru'bia tinc'torum** Linné. *Habitat:* Levant and Southern Europe, chiefly supplied from Holland, where it is cultivated. Usually comes into market in a coarse, red powder. Its most important constituent is **alizarin,** a red coloring-matter soluble in water and alcohol. Chiefly used as a dye.

532. CINCHONA.—Cinchona

PERUVIAN BARK

The dried bark of **Cincho'na Ledgeriana** Moens, **Cincho'na calisa'ya** Weddell, **Cinchona officinalis** Linné, and of hybrids of these with other species of Cinchona, yielding, when assayed, not less than 6 per cent. of cinchona alkaloids.

Source, Varieties, History, Etc.—The genus Cinchona is composed of over three dozen species, but few furnish the commercial barks. It is well known that the original source of the drug is South America (10° N. lat. to 19° S. lat., from about 3000 to 12,000 feet above sea-level), the area of the growth of the various species being confined exclusively to the Andes, chiefly on the eastern face of the Cordilleras—occasionally on the western face, which is covered by forests. The best known varieties from South America were the dark brown Loxa bark and the pale yellow-gray Huanuco. The cinchonas seldom form an entire forest, but rather groups interspersed among tree-ferns, gigantic climbers, bamboos, etc., sometimes growing separately in exposed situations, but under peculiar climatic conditions, such as a great humidity of atmosphere and a mean temperature of about 62°. Shade seems to favor the development of alkaloids. Dymock calls attention to the fact that "the north or shaded side of a tree has a richer bark than that on the south side," a fact which explains the success of the "mossing system."*

Cultivated trees in recent years have been the chief source of the commercial barks. To some extent the cultivation has been carried on in South America, but great success has attended the persevering

---

* There are four methods of collecting or harvesting the bark: (1) By taking it in longitudinal strips from the standing tree and leaving the bark to renew over the exposed wood; (2) by scraping and shaving off the bark; (3) by coppicing; and (4) by uprooting. The first is most in use . . . The trees are barked preferably in the rainy season, when the bark "lifts," or is more easily removed from the wood. The coolie inserts the point of a knife in the tree as far as he can reach and draws it down, making an incision in the bark straight to the ground; he then makes another cut parallel to the first, about an inch and a half distant, and, loosening the bark with the back of the knife, the strip or ribbon is taken off. If the operation is performed carefully and the cambium cells are not broken, a new layer of bark will be formed in place of that which is taken away. The tree is then covered with moss, grass, or leaves, bound on by strings of fiber. All this is done to foster the growth of the new bark (renewed bark) from the cambium and to thicken the untouched layers of natural bark, which is now termed **mossed bark.**—*Pharmacographia Indica.*

Outline map of South America, showing the native distribution of cinchona bar and certain other prominent drugs of that country. See also Geographical Class fication of Drugs.

efforts of the Dutch Government and the Government of British India. Extensive plantations of cinchona are now flourishing, to the extent of several million trees of the more important species, on the Neilgherry Hills and in the valleys of the Himalaya in British Sikkin. The tree is also cultivated in Ceylon, Java, Jamaica (Blue Mountains), and other countries.

VARIETIES.—There are about twenty varieties of cinchona barks, and it is a very difficult matter to distinguish them, since they have been and

FIG. 228.—*Cinchona officinalis*—Branch.

are now changed so much by grafting and crossing. The varieties generally used and best known are: *C. succirubra* Pavon, *C. calisaya* Weddell, *C. ledgeriana* Moens, *C. lancifolia* Mutis, and *C. officinalis* Hooker.

The success of the Dutch planters of Java has been so pronounced that the greater portion of cinchona bark comes from this place, leading varieties being ledgeriana and succirubra bark. In Java great care is exercised in the cultivation. The trees are allowed to reach the age of twelve years before the bark is collected. The cultivation is largely confined to the variety Ledgeriana. Over 500,000 pounds are collected annually from Java plantations.

DESCRIPTION.—In quills or curved pieces of variable size, usually 2 or 3, sometimes 5 mm. thick; externally gray, rarely brownish-gray, with numerous intersecting transverse and longitudinal fissures, having nearly vertical sides; the outer bark may be wanting, the color externally being then cinnamon brown; the inner surface light cinnamon brown, finely striate; fracture of the outer bark short and granular,

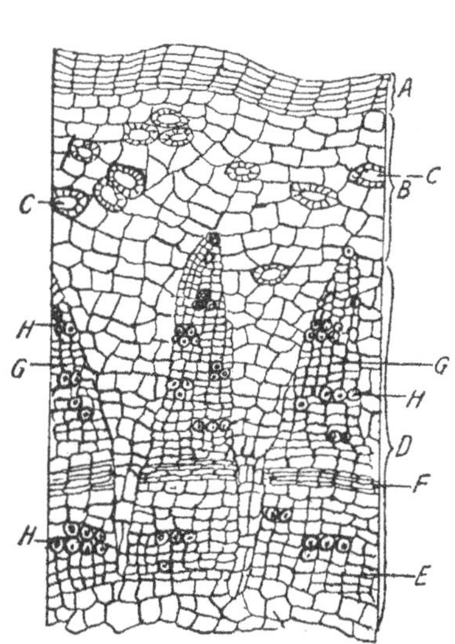

FIG. 229.—*Cinchona calisaya*—Cross-section of bark. *A*, Cork cells. *B*, Cortical parenchyma. *C*, Stone cells. *D*, Phloëm portion. *E*, Soft bast. *F*, Phellogen forming bark. *G*, Medullary rays. (The black line from *G* should be extended to the parenchyma cells between the phloëm portions.) *H*, Bast fibers.

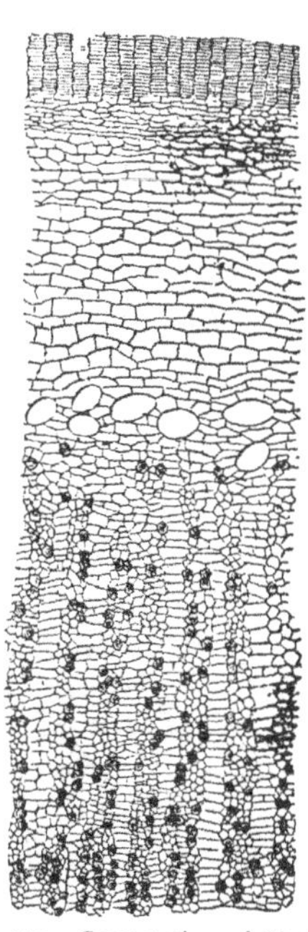

FIG. 230.—Cross-section of *Cinchona calisaya* bark (var. Micrantha).

of the inner finely splintery; powder light brown or yellowish-brown; odor slight, aromatic; taste bitter and somewhat astringent.

MICROSCOPICAL.—The calisaya (variety Micrantha) transversely shows milk-vessels in the cortical parenchyma and absence of stone cells, which are present in Lancifolia. The rays of the woody portion are more elongated and the medullary rays larger in size. Bast fibers comparatively small and less numerous, but are spindle-shaped, as they are in all true cinchona barks showing longitudinal section. In *C. rubra* the stone cells and milk-ducts are both wanting, while the

26

bast fibers are more numerous and stouter. The bark is richer in coloring matter. In cuprea bark the cork cells are thicker and the cortical parenchyma cells smaller, stone cells present, milk-ducts absent, few bast fibers, but the woody portion contains indurated cells, which simulate them. The ligneous cells are very numerous and extend even down into the medulla. They are smaller than the bast fibers of true cinchona barks, but much more numerous.

These barks are thoroughly saturated with pigments, principally cinchona red, the phlobaphen of all cinchona barks. Before microscopical examination these pigments must be removed by a weak alcoholic solution of ammonia. This requires considerable practice (Dohme). Compared with other barks, the fibers of the liber of cinchona are shorter and more loosely arranged, being for the most part separated into simple fibers imbedded in the bast parenchyma, or united into very short bundles.

Grahe's test for the distinction of cinchona bark is as follows: On heating about 0.1 Gm. (1½ gr.) of the powdered bark in a dry test-tube a tarry distillate of a red color is obtained.

Powder.—Microscopical elements of: See Part iv, Chap. I, B.

OFFICIAL PREPARATIONS.

    Fl. Ext. Cinchonæ,....................Dose: 15 ♏. (1 mil).
    Tr. Cinchonæ,........................   1 fl. dr. (4 mils).

532 a. **CINCHONA RUBRA.**—The dried bark of *Cinchona Succirubra* Pavon or its hybrids, yielding not less than 6 per cent. of the total alkaloids of Cinchona. "In quills or incurved pieces, varying in length, and from 2 to 4 or 5 mm. (1/12 to 1/6 or 1/5 in.) thick; the outer surface covered with a grayish-brown cork, more or less rough from warts and longitudinal, warty ridges, and from few, mostly short and not frequently intersected transverse fissures, having their sides sloping; inner surface more or less **deep reddish-brown** and distinctly striate; fracture short, fibrous in the inner layer; outer layer, granular. For years practically all of the red cinchona bark, so called, was only a hybrid, but recently, and especially for a year past, fine quill bark of pure succirubra has frequently been received.

Powder.—Microscopical elements of: See Part iv, Chap. I, B.

CONSTITUENTS.—Upon **quinine,** $C_{20}H_{24}N_2O_23H_2O$, the bark almost exclusively depends for its value. This alkaloid is colorless, amorphous, or in acicular crystals; inodorous, very bitter; soluble in 1670 parts water, 6 parts alcohol, 26 parts ether. Aqueous solutions of the salts have a blue fluorescence, and when treated with chlorine water and ammonia a beautiful green color is produced—"Thalleoquin test." The solutions deviate the plane of polarization to the left. The tartrate is soluble in water. A cold aqueous solution of the sulphate remains unaffected by potassium iodide T. S. (difference from quinidine). The other prominent principles are:

CINCHONIDINE, $C_{19}H_{22}N_2O$—isomeric with cinchonine, non-fluorescent; forms colorless, anhydrous crystals, soluble in 20 parts alcohol (80 per cent.), 1680 of water, and 188 of ether. The sulphate is more soluble in water than quinine, and the tartrate very insoluble. The

FIG. 231.—*Cinchona succirubra*—Branch.

Thalleoquin test (see above) gives a white precipitate. Represented in Cinchonidinæ Sulphas.

CINCHONINE, $C_{20}H_{24}N_2O$—white lustrous prisms, soluble in 3760 parts water, 116 parts alcohol, and 526 parts ether; has exactly the opposite action to cinchonidine and quinine upon polarized light.

QUINIDINE, $C_{20}H_{24}N_2O_2$—isomeric with quinine; crystallizes in prisms soluble in 2000 parts water, 0.8 part alcohol, about 30 parts ether; turns the plane of polarization to the right. A cold aqueous solution of the sulphate yields a white precipitate with potassium iodide T. S. (difference from sulphate of quinine). Represented in Quinidinæ Sulphas.

Among the **unofficial alkaloids and principles** found in the bark are the following: Isomeric with quinine and quinidine is **quinicine**; with cinchonine and cinchonidine, are **cinchonicine, homocinchonine, homocinchonidine, homocinchonicine**, and **apoquinamine**; a brown amorphous alkaloid is obtainable from the mother-liquor known as chinoidine (quinoidine), a mixture of various not well-defined alkaloidal substances; kinic acid, $C_7H_{12}O_6$, and kinovic acid; kinovin; bitter cinchonic acid (derived from preceding); volatile oil, a minute quantity.

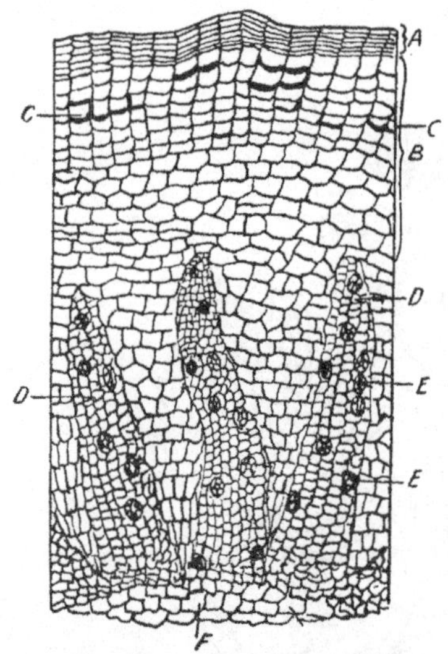

FIG. 232.—*Cinchona rubra*—Cross-section of bark. *A*. Cork cells. *B*. Cortical parenchyma. *C*. Coloring matter, deposits between and within the cells. *D*. Medullary ray. (Black line from *D* should not extend beyond the first two rows of cells of the figure.) *E*. Bast fibers.

Separation of Total Alkaloids.—Moisten powdered cinchona with ammonia water and allow it to stand for an hour, then hot water is added. To the mixture, after cooling, milk of lime is added and the whole evaporated to dryness. This is placed in an extraction apparatus and exhausted with ether. Water acidulated with HCl is added to neutralize the alkaloids and the ether distilled off. The cooled liquid is filtered and decinormal solution of soda is added. Finally, sodium hydrate is added to complete the precipitation of the alkaloids. There are numerous other processes, but this seems a simple and satisfactory one to use for assay purposes.

YIELD OF ALKALOID.—The richest government bark brought to the market until recently has not exceeded 9½ per cent. of sulphate of quinine; 7 to 8 per cent. is a good average in government plantations. Barks taken from the trees in the government gardens at Pioeng Goenoeg, Java, have recently been analyzed and found to equal respectively 12.66 and 16.04 per cent. of quinine sulphate.

ACTION AND USES.—The action of cinchona bark is due almost entirely to the alkaloids therein contained. Quinine is a powerful antiseptic, destructive, in weak solution, to infusorial and vegetable life. Internally it stimulates the muscular fibers of the stomach, acting as a bitter tonic, invigorating the vital functions and aiding digestion. In large doses the brain is affected, giving rise to symptoms such as fullness, frontal headache, deafness, ringing in the ears, and mental dullness. This effect is called "cinchonism" attributed to partial anæmia of the brain, contraction of blood-vessels, etc. Heart action is depressed. Reflex excitability of the spinal cord is lowered. In the blood, quinine arrests the migration of the white corpuscle and checks its amœboid movement; the oxygen-carrying function of the red corpuscle is impaired; infectious micro-organisms in the blood and tissues are probably rendered inactive or destroyed. The toxic symptoms produced by quinine and allied salts are spoken of collectively as cinchonism, which ordinarily is not allowed to go further than tinnitus aurium.

Dose of cinchona: 15 to 60 gr. (1 to 4 Gm.), in powder. fluidextract, or its equivalent in the salts of the alkaloids

OFFICIAL PREPARATION.

Tinctura Cinchonæ Composita (10 per cent., with bitter orange-peel 8 per cent., and serpentaria 2 per cent.),.... 1 to 4 fl. dr. (4 to 15 mils).

533. **REMIJIA.**—CUPREA BARK. The bark of **Remij'ia peduncula'ta** Triana and of **Remijia purdiea'na** Weddell, resembling cinchona in physical properties and constitution. A copper-red bark from the United States of Colombia, grown at an altitude of from 3000 to 6000 feet, usually in flat or curved pieces; odor slight; taste bitter. Quinine is contained in this bark to the amount of 0.5 to 2.5 per cent., but no cinchonidine is found; homo-quinine—a compound of quinine and cupreine—is also a constituent. Remijia bark is largely imported by manufacturers; it was said that the importations of this bark at one time exceeded in amount the entire importations of all the cinchona barks, by reason of its cheapness for the manufacture of quinine. Cinchonamine, $C_{19}H_{24}N_2O$, is one of the principal products of *R. purdieana*, the bark from which does not respond to Grahe's test.

534. **CEPHALANTHUS OCCIDEN-TALIS** Linné.—BUTTON BUSH. POND DOGWOOD. *Habitat:* United States. (Bark.) Tonic, febrifuge, laxative, and diuretic. It has an indirect action on the lungs, and is much used in consumption, coughs, and colds generally. Dose: 30 to 60 gr. (2 to 4 Gm.).

535. **MITCHELLA.**—SQUAW VINE. PARTRIDGE BERRY. The herb of **Mitchell'a re'pens** Linné, a creeping evergreen growing in the woods of this country east of the Mississippi. Stem branching, bearing about 12 mm. (½ in.) long, sometimes purplish, the ovary ripening into a small, roundish-ovate, entire, evergreen leaves, marked with white lines; flowers pale scarlet-red berry. Tonic, astringent

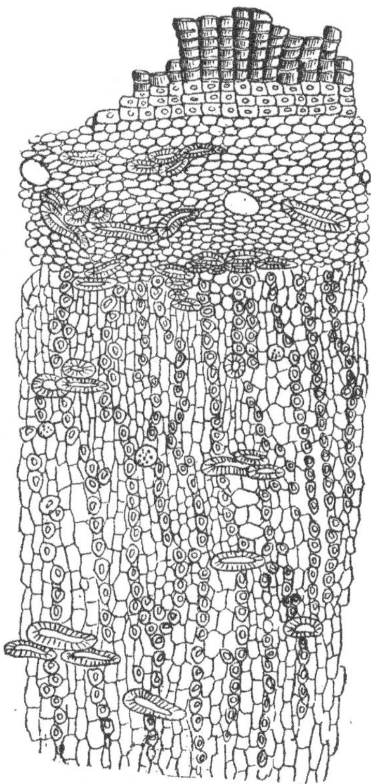

FIG. 233.—Cross-section of Cuprea bark.

and diuretic, resembling pipsissewa in action and often substituted for it. It is frequently combined with black haw. Dose: 30 to 60 gr. (2 to 4 Gm.).

536. **GALIUM.**—CLEAVERS. LADY'S BEDSTRAWS. The herb of **Ga'lium apari'ne** Linné. *Habitat:* Northern Hemisphere. Stem weak, quadrangular, prominently winged, and covered with retrorse prickles; leaves linear-lanceolate, borne in whorls. Flowers small, white, axillary, the single ovary ripening into a two-seeded, bristly fruit. Aperient, diuretic, and alterative; also used in psoriasis and other skin diseases. Dose: 30 to 60 gr. (2 to 4 Gm.), in infusion.

G. ve'rum (Yellow Lady's Bedstraw) has a smooth stem, bearing yellow flowers. *G. triflo'rum* contains coumarin, and has a fragrant odor when dry.

537. **COFFEA.**—COFFEE. The seeds of **Cof'fea arab'ica** Linné. *Habitat:* Southern Arabia and Abyssinia; cultivated in South America, Java, and various tropical countries. The fruit is a roundish berry, about the size of a large cherry, becoming dark purple, and containing two seeds, which are inclosed within a membranous covering, and a purplish pulp. These seeds, when freed from the pericarp, form the coffee of the market. They are brownish-green or bluish-gray, planoconvex, the flat surface being elliptical, with a longitudinal groove curving deeply into the horny albumen; odor peculiar, faint, growing stronger by age; taste sweetish, somewhat astringent. Good berries are hard and sink readily in water. Soft, light, dark-colored berries should be rejected.

CONSTITUENTS.—Its properties depend upon the alkaloid caffeine (2 to 8 per cent.), the constituent common to most of the stimulating beverages. It also contains sugar, tannic acid, caproic acid, fat, etc. When roasted, the sugar is converted into caramel, the caffeic acid partially into methylamine, and several volatile and empyreumatic substances (caffeone) are formed. Pyridine has been separated from these mixed products due to roasting, giving to coffee its peculiar aroma. It loses from 15 to 18 per cent. of moisture in drying.

Preparation of Caffeine (Theine).—Precipitate infusion of tea or coffee with lead acetate; remove lead from filtrate with H_2S; concentrate second filtrate, neutralize with NH_4OH, and allow it to cool, when caffeine will crystallize out. An aqueous solution of caffeine does not form a precipitate with Mayer's reagent.

ACTION AND USES.—Cerebrospinal stimulant, tonic; aids digestion and allays hunger and fatigue by lessening tissue waste.

537a. **COFFEA TOSTA, N.F.**—Yielding not less than 1 per cent. of caffeine.

538. **CATECHU PALLIDUM.**—TERRA JAPONICA. GAMBIR. An extract obtained from a climbing plant of the East Indies, **Ourouparia Gambir** (Hunter) Baillon, by boiling the leaves, twigs, etc., in water. It is in about one-inch cubes, or in irregular pieces, reddish-brown or yellowish, breaking with a dull, earthy, pale yellowish fracture, showing under the microscope numerous crystals; inodorous; taste astringent and bitter, leaving finally a sweet taste in the mouth. It is mostly used in this country in tanning, dyeing, etc.; in its native country it is chewed with betel-nuts.

CAPRIFOLIACEÆ.—Honeysuckle Family

Shrubs, as viburnum, or twining plants, as the honeysuckle, with opposite, exstipulate *leaves*, a gamopetalous *corolla*, and the *fruit* a berry, pod, or drupe. The *calyx-tube* is adherent to the 2- to 5-celled ovary.

Synopsis of Drugs from the Caprifoliaceæ

A. *Flowers.*
    * Sambucus, 539.
B. *Bark.*
    * Viburnum Opulus, 540.
      VIBURNUM PRUNIFOLIUM, 541.

C. *Root.*
    Triosteum, 542.

539. **SAMBUCUS, N.F.—**Elder. The dry flowers of **Sambu'cus canaden'sis** Linné. Collected when in full bloom and rapidly dried, the commercial drug being composed of the small, yellowish, somewhat wheel-shaped and shriveled flowers, mixed with a few expanded ones; usually detached from their peduncles, which are mixed with them. They have a sweetish, somewhat bitter taste, and a slight, peculiar, agreeable odor, due to a very small quantity of volatile oil. The European elder (*S. nigra*) resembles *S. canadensis.* *Con-*

Fig. 234.—*Viburnum opulus*—Flowering branch and fruit.

stituents: Besides volatile oil, they contain sugar, mucilage, fat, wax, resin, pectin, albuminoids, and probably a little tannin. Stimulant, carminative, and diaphoretic. Dose: 30 to 60 gr. (2 to 4 Gm.).

540. VIBURNUM OPULUS, N.F.—Cramp Bark

HIGH BUSH CRANBERRY

The dried bark of Vibur'num op'ulus Linné.

Habitat.—North America.

Description of Drug.—Very thin pieces or occasionally quills, outer surface, light gray, with purplish-brown stripes and very small brown lenticels; thicker pieces purplish-red, or occasionally blackish; odor strong and characteristic; taste bitter; the inner surface is yellowish or brownish; fracture short. The bark of the mountain maple (Acer Spicatum) was an adulterant formerly described by *misled* authorities, as Viburnum opulus.

Powder.—Light brown. Characteristic elements: Parenchyma of inner cortex, with rosettes of calcium oxalate; middle bark bearing reddish-brown coloring-

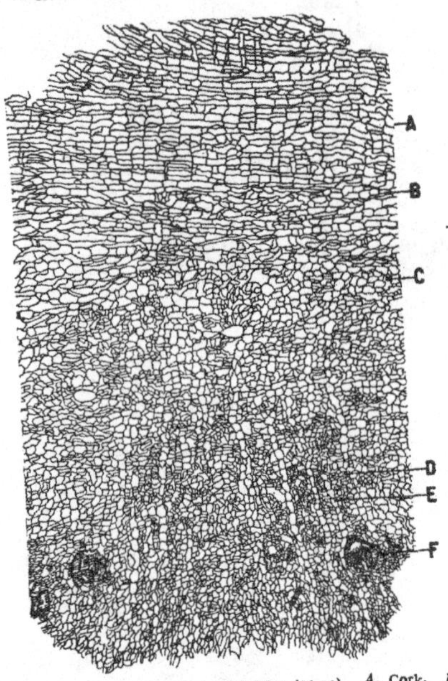

FIG. 235.—Cross-section of bark of *Viburnum Opulus* (× 43). *A,* Cork. *B,* Cork cambium. *C,* Parenchyma of cortex. *D,* Phloëm. *E,* Medullary ray. *F,* Group of stone cells.

FIG. 235 a.—*Viburnum Opulus* (×137). Group of stone cells taken from *F* in Fig. 235.

matter, starch (5 to 12 μ in diam.); tracheal fragments with lignified wood fibers; few stone cells; crystals of calcium oxalate, few aggregate (15 to 30 μ in diam.); polygonal cork cells, thin-walled.

ACTION AND USES.—Claimed to be **antispasmodic**, hence the name cramp bark. Dose: 30 gr. to 2 dr. (2 to 8 Gm.).

541. VIBURNUM PRUNIFOLIUM

BLACK HAW

The dried bark of the root of Vibur'num prunifo'lium Linné or of V. lentago Linné, without admixture of more than 5 per cent. of wood or other foreign matter.

BOTANICAL CHARACTERISTICS.—A tall shrub or small tree. *Leaves* oval, obtuse, or slightly pointed, finely serrate. *Cymes* compound, sessile. *Fruit* an oval, black, sweet drupe.

HABITAT.—Middle and Southern United States, east of the Mississippi.

DESCRIPTION OF DRUG.—In irregular, transversely curved or quilled pieces from 1.5 to 6 cm. in length, and from 0.5 to 1.5 mm. in thickness;

FIG. 236.—*Viburnum prunifolium*—Branch with flowers.

outer surface, grayish-brown, or, where the outer cork has scaled off, brownish-red, longitudinally wrinkled; inner surface reddish-brown, longitudinally striated; fracture short but uneven, showing in bark which is young or of medium thickness, a dark brown cork, a brownish-red outer cortex, and a whitish inner cortex in which are numerous light yellow groups of sclerenchymatous tissues; odor slight; taste distinctly bitter and somewhat astringent. U.S.P. IX.

Powder.—Characteristic elements: See Part iv, Chap. I, B.

CONSTITUENTS.—A brown resin, a bitter principle (viburnin), valerianic acid, tannic acid, oxalic, malic, and citric acids, sulphates, and chlorides.

ACTION AND USES.—Diuretic, and a **tonic and sedative to the uterine and ovarian nerve centers;** used in threatened abortion. Dose: 30 to 60 gr. (2 to 4 Gm.).

OFFICIAL PREPARATIONS.

Extractum Viburni Prunifolii, ...Dose: 0.5 Gm. (8 gr.).
Fluidextractum Viburni Prunifolii,. . .   30 to 60 ♍ (2 to 4 mils).

542. **TRIOSTEUM.**—FEVER ROOT. BASTARD IPECAC. The root of Trios'teum perfolia'tum Linné, common in most parts of the United States. (See Conspectus.) Cathartic and emetic in large doses. Dose: 15 to 30 gr. (1 to 2 Gm.).

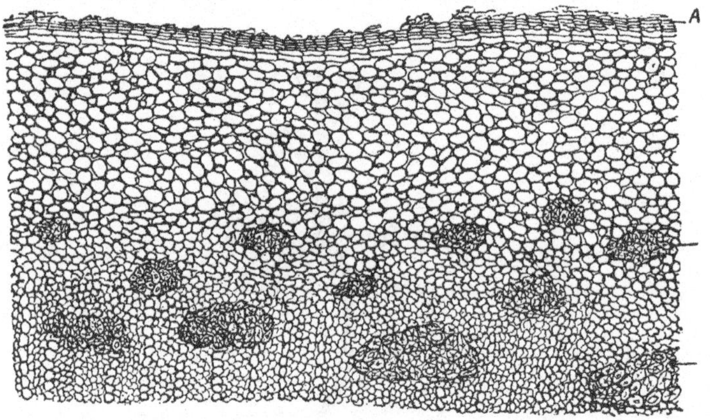

FIG. 237.—*Viburnum prunifolium*—Bark of root, cross-section. *A.* Cork. *B.* Group of stone cells in cortical parenchyma.

VALERIANEÆ

Herbs with opposite, exstipulate *leaves. Flowers* in panicled or head-like cymes. Many of the species possess antispasmodic properties, due to the presence of a volatile oil, from which is developed valerianic acid.

543. VALERIANA.—VALERIAN

VALERIAN

The rhizome and roots of Valeria'na officina'lis Linné.

BOTANICAL CHARACTERISTICS.—*Root* perennial, tuberous. *Leaves* pinnate or pinnately cut. *Corolla* roseate, funnel-form, 5-lobed; *stamens* 3. *Fruit* a feathery akene.

SOURCE.—Europe, especially in Holland, Belgium, England, and Germany as well as Japan. The Japanese root is said to be richer in volatile oil than the Belgian. The fresh rhizomes and roots are preferred for distilling the oil, as there is a loss of nearly 50 per cent. of the oil in drying the rhizome and root for medicinal use.

DESCRIPTION OF DRUG.—Obconical, from 6 to 75 mm. (¼ to 3 in.) in length, with stem-remnants above, and beset with numerous rootlets; those rhizomes grown in dry localities are smaller, nearly globular,

FIG. 238.—*Valeriana officinalis*—Plant and rhizome.

with lighter colored, thinner, and less shriveled rootlets, and contain a greater proportion of volatile oil than those grown in moist ground; the latter are generally sliced longitudinally. Externally brown, internally pale brownish; odor strong, disagreeable, increasing with age; taste camphoraceous and bitter. A cross-section shows a rather thin bark, and a wood-circle, narrow, white, inclosing a large pith. Nucleus sheath mostly indistinct; branches have a similar structure but a thicker bark. The rootlets have a thick bark and a slender, woody column, distinctly radiate, and contain a small pith inclosed in a nucleus sheath.

Powder.—Characteristic elements: See Part iv, Chap. I, B.

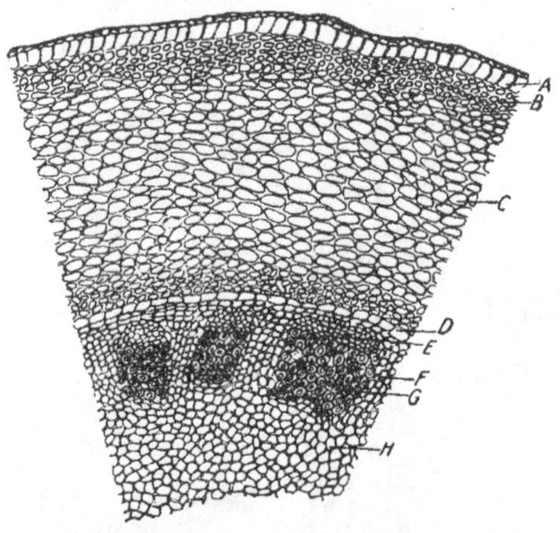

Fig. 239.—Valerian—Cross-section of rhizome. *A*, Cork cells. *B*, Collenchyma. *C*, Cortical parenchyma. *D*, Endodermis. *E*, Small irregular liver-cells. *F*, Medullary rays. *G*, Punctuated vessels of wood-rays. *H*, Pith-cells.

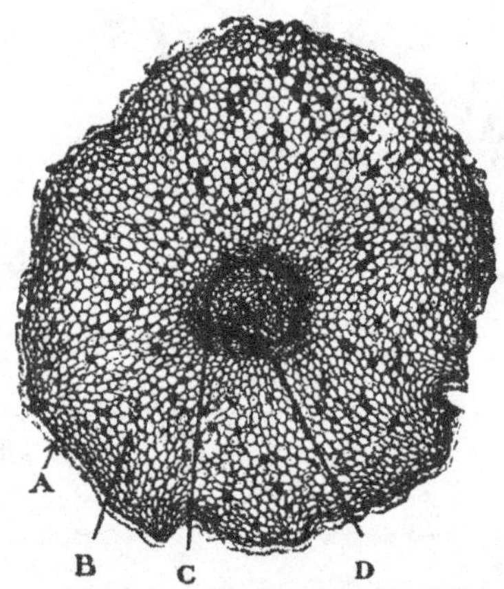

Fig. 240.—Valerian—Cross-section of rootlet. (17 diam.) A, Epidermis, B, Parenchyma of cortex. C, Phloëm. D, Xylem. (Photomicrograph.)

CONSTITUENTS.—Besides the common vegetable principles, it contains a terpene, isovaleric acid, $C_6H_{10}O_2$ (distilling at 300°C.), and a volatile oil of complex constitution, consisting mainly of an alcohol, borneol; its ether, and its formic, acetic, and valerianic acid esters, which are gradually decomposed on exposure, liberating the acids. This oil (Oleum Valerianæ, U.S.P. VI) is of a pale greenish color, becoming yellow and viscid on exposure, and has the peculiar odor of the root. Ash, not exceeding 20 per cent.

ACTION AND USES.—Gentle nerve stimulant and antispasmodic, employed in hysterical disorders. Dose: 15 to 60 gr. (1 to 4 Gm.).

OFFICIAL PREPARATIONS.

Tinctura Valerianæ (20 per cent.),......Dose: 1 to 2 fl. dr. (4 to 8 mils).
Tinctura Valerianæ Ammoniata
   (20 per cent.),...............   30 to 60 ♏ (2 to 4 mils).

CUCURBITACEÆ.—Gourd Family

Succulent herbs, creeping or climbing by tendrils. *Leaves* alternate. *Flowers* monœcious and polygamous; *stamens* with long and wavy or twisted anthers. *Fruit* a pepo.

Synopsis of Drugs from the Cucurbitaceæ

A. *Root.*
  * Bryonia, 545.
B. *Fruits.*
   COLOCYNTHIS, 544.
   Luffa, 546.
   Momordica, 547.

C. *Seeds.*
   PEPO, 548.
   Citrullus, 549.
   Cucumis, 550.
D. *Resin.*
   Elaterium, 551.

544. COLOCYNTHIDIS PULPA.—COLOCYNTH

BITTER APPLE. *Ger.* KOLOQUINTEN

The dried pulp of the fruit, **Citrul′lus colocyn′this** Schrader, containing not more than 5 per cent. of seeds nor more than 2 per cent. of epicarp.   U.S.P. IX.

BOTANICAL CHARACTERISTICS.—Stem procumbent, angular, hispid; *leaves* cordate-ovate, lobate; tendrils short. *Flowers* axillary, *female flowers* solitary, *petals* yellow with greenish veins. *Fruit* globose, smooth, 6-celled, with very bitter pulp; seeds whitish, sometimes brownish.

HABITAT.—Asia, Europe, and Africa.

DESCRIPTION OF DRUG.—The fresh fruit has a marbled green surface, not very unlike the watermelon. It has a thick rind inclosing a white, spongy pulp, imbedded in which are numerous light-colored seeds. The fruit on drying loses about 90 per cent. of water, leaving a very light, spongy, white or yellowish-white pulp, which, deprived of the seed, constitutes the official drug. Colocynth "apples," as they appear in the market, contain the seeds, but are deprived of the rind; 50 to 100 mm. (2 to 4 in.) in diameter. A cross-section of the spherical pulp ("apples") makes apparent three distinct wedges, each of which has two branches; this structure is due to the parietal placentæ, which project to the center of the fruit, then divide and turn back, making convoluting branches directed one toward the other. Inodorous; so **intensely bitter** that the bitterness is imparted to **any** object brought in contact with it.

Powder.—Microscopical elements of: See Part iv, Chap. I, B.

CONSTITUENTS.—Resin, gum, and amyloid principles. Colocynthin, $C_{56}H_{84}O_{23}$, a yellowish, somewhat translucent, bitter, and friable

FIG. 241.—Colocynth—Portion of vine and whole fruit.

glucoside, is, perhaps, the most important constituent; it is contained in the pulp to the extent of about 2 per cent. Colocynthin is a taste-

 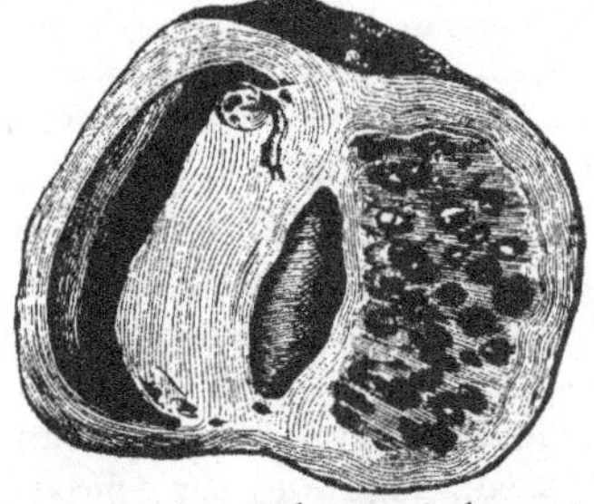

FIG. 242.—Transverse section of colocynth fruit.　FIG. 243.—Longitudinal section of colocynth fruit.

less crystalline principle left after treating the alcoholic extract with cold water in preparing colocynthin. Ash, not to exceed 15 per cent. The powder should not yield more than 2 per cent. of fixed oil when treated with petroleum benzin—a check test on the 5 per cent. limit of seeds. U.S.P. IX.

Preparation of Colocynthin.—Exhaust alcoholic extract with water, precipitate with lead acetate and subacetate, remove lead from liquid by treating with H_2S, filter, then precipitate with tannin; suspend the tannate in alcohol, decompose with lead hydroxide, remove excess of lead by H_2S, filter and evaporate, and wash the residue with ether.

ACTION AND USES.—**A powerful hydragogue cathartic,** given in combination with weaker purgatives. Dose: 3 to 10 gr. (0.2 to 0.6 Gm.).

OFFICIAL PREPARATIONS.

Extractum Colocynthidis,........ ....Dose: ½ to 2 gr. (0.0324 to 0.13 Gm.).
Extractum Colocynthidis Compositum (Extract Colocynth 16 per cent., with aloes, scammony, cardamon and soap),.... 5 to 25 gr. (0.3 to 1.6 Gm.).
Pilulæ Catharticæ Compositæ (8 per cent. of compound extract), ... 2 to 5 pills.

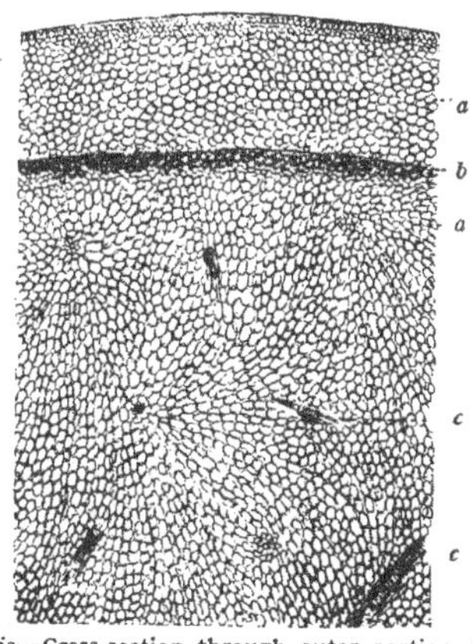

FIG. 244.—*Citrullus colocynthis*—Cross-section through outer portion of the fruit. *a, a,* Parenchyma. *b,* Stone cells. *c, c,* Fibro-vascular bundles.

545. **BRYONIA, N.F.**—BRYONY. The root of **Bryo′nia al′ba** and of **Bryonia′ dioi′ca** Linné. Off. in U.S.P. 1890. A dull reddish-brown, longitudinally wrinkled root, usually appearing in the market in transverse disks about 50 to 100 mm. (2 to 4 in.) in diameter, of a white or yellowish-white color; bark thin, with a thin, friable cork; the bark is separated by a brown cambium line from the meditullium, in which the wood-bundles are arranged radically and concentrically; the wood-wedges and zones are separated by rather broad rays and concentric circles of parenchymatous tissue; fracture short. Inodorous; taste disagreeably bitter. The active principle is bryonin, $C_{48}H_{80}O_{19}$, an intensely bitter glucoside, soluble in water, but best extracted with strong alcohol. Obtained by precipitating the hydro-alcoholic percolate with tannin. The moist tannin compound is mixed with lead oxide and then digested with alcohol. The alcoholic solution yields bryonin on evaporation. Drastic hydragogue cathartic, formerly much used in the treatment of dropsy, but now superseded by jalap. Dose: 10 to 30 gr. (0.6 to 2 Gm.).

Tinctura Bryoniæ (10 per cent.)
(U.S.P. 1890),........ ..Dose: 1 to 4 fl. dr. (4 to 15 mils).

546. **LUFFA.**—VEGETABLE SPONGE. WASH-RAG SPONGE. GOURD TOWEL. The gourd-like fruit of **Luf′fa ægyp′tiaca,** a vine growing in Arabia and Egypt. The layer of tissue next the epidermis is composed of interwoven woody

fibers, and, when deprived of the epidermis, makes a good substitute for sponge. The fruit of *Luffa echinata*, growing in India, contains a principle related to, if not identical with, colocynthitin.

547. **MOMORDICA BALSAMINA** Linné.—BALSAM APPLE. This is a climbing East Indian plant, cultivated in our gardens for the sake of its cucumber-like fruit, which is often used in domestic practice as a vulnerary.

548. PEPO.—PUMPKIN SEED

PUMPKIN SEED

The ripe seed of **Cucur'bita pe'po** Linné.

BOTANICAL CHARACTERISTICS.—Stem hispid, procumbent; tendrils branched. *Leaves* very large, cordate, palmately 5-lobed. *Fruit* yellow, very large (sometimes two feet in diameter), roundish or oblong, smooth, and furrowed.

HABITAT.—Tropical Asia and America.

DESCRIPTION OF DRUG.—Flat, broadly ovate seeds, about 20 mm. ($\frac{4}{5}$ in.) long, and 2 mm. ($\frac{1}{12}$ in.) thick, with a **flat ridge and shallow groove** around **the edge;** testa **dull white,** inclosing two flat, white, oily cotyledons and a short radicle; inodorous; taste bland and oily.

Powder.—Microscopical elements of: See Part iv, Chap. I, B.

CONSTITUENTS.—From 30 to 40 per cent. of a thick, red fixed oil, **an acrid resin,** considered to be the tæniafuge principle, starch, sugar, fatty acids, and the proteids, myosin and vitellin, the myosin precipitating from an infusion saturated with NaCl, and the addition of CO_2 separating out the vitellin, apparently identical with that of egg yolk.

ACTION AND USES.—Tæniafuge. Dose: 1 to 2 oz. (30 to 60 Gm.), in emulsion.

549. **CITRULLUS.**—WATERMELON SEED. The seed of **Cucu'mis citrul'lus** Seringe. Indigenous to Southern Asia, but cultivated extensively in the United States. Differs from the pumpkin seed in being blackish-marbled or brownish in color, somewhat smaller, and with a blunt, ungrooved edge. They are used like pumpkin seeds as a tæniafuge, and also have diuretic and demulcent properties. Dose: 2 dr. to 2 oz. (8 to 60 Gm.).

550. **CUCUMIS SATIVUS** Linné.—CUCUMBER SEED. Flat and thin, lance-oblong, from 8 to 12 mm. ($\frac{1}{3}$ to $\frac{1}{2}$ in.) long, acutely edged, ungrooved, dull white in color. Resembles above in properties.

551. **ELATERIUM.**—A peculiar resinous substance obtained from the fruit of Ecbal'lium elate'rium A. Richards (squirting cucumber), a vine growing in the Mediterranean regions of Europe, Africa, and Asia. The fruit, when ripe, separates suddenly from its stalk, and the internal pressure forces the juice out of the aperture thus made in a stream; in collecting, therefore, the fruits are gathered green, sliced, and the juice expressed by slight pressure; on standing it deposits a sediment, which, when dried, forms the commercial Elaterium.

Elaterium is in flat pieces of varying sizes, light **and friable,** pale green when fresh, but taking on a **gray or** light **buff** color as it becomes older; the surface is covered with small crystals of elaterin; odor somewhat tea-like; taste acrid and intensely bitter, due to the active ingredient, **elaterin,** which constitutes from 25 to 30 per cent. of the drug. This principle is insoluble in water, readily soluble in chloroform and hot alcohol; it is a violent irritant poison; its alcoholic solution is colored red by warm sulphuric acid; its carbolic acid solution, crimson, rapidly changing to scarlet. There is also present ecballin (soft, yellow, acrid), hydroelaterin, and elaterid.

ELATERINUM (U.S.P. IX).—*Elaterin.*—Exhaust elaterium with chloroform; add ether; white crystals deposit immediately. Wash with ether and recrystallize from chloroform. This principle is odorless and crystalline, is bitter and acrid in taste. No weighable ash remains on incinerating 0.1 Gm. of Elaterin.

ACTION AND USES.—Elaterin is a powerful hydragogue cathartic, used in the treatment of dropsy. Dose: ⅟₂₀ to ⅟₁₂ gr. (0.003 to 0.005 Gm.). *Preparation:* **Trituratio Elaterini** (10 per cent.). Dose: ½ gr. (0.030 Gm.).

FIG. 245.—*Ecballium elaterium*—Branch with flowers.

CAMPANULACEÆ.—Campanula Family

Herbs or shrubbery plants, with acrid, milky juice, alternate *leaves*, and scattered flowers, *corolla* 5-lobed. *Fruit* a one- to several-celled capsule. Many species of the tribe Lobeliæ are acrid-narcotic poisons.

552. LOBELIA.—LOBELIA

INDIAN TOBACCO

The dried leaves and tops of Lobe'lia infla'ta Linné (fam. Lobeliaceæ U.S.P. IX), without the presence or admixture of more than 10 per cent. of stems or other foreign matter.

BOTANICAL CHARACTERISTICS.—*Stems* much branched from an annual root, pubescent; *leaves* ovate or oblong, gradually diminishing into leaf-like bracts. *Capsule* inferior.

RELATED SPECIES.—*Lobelia syphilitica* (great lobelia), *Lobelia cardinalis* (cardinal plant).

HABITAT.—United States.

27

D ESCRIPTION OF DRUG.—In the market the herb is broken up, but the fragments of green leaves, small pieces of the longitudinally ridged stem, the rather elongated, dried flowers, and the inflated, membranous capsules serve to identify it; odor irritating when inhaled; taste very pungent, persistently acrid, and tobacco-like.

Powder.—Characteristic elements: See Part iv, Chap. I, B.

FIG. 246.—*Lobelia inflata*—Portion of plant and flower.

CONSTITUENTS.—**Lobeline** (a poisonous, acrid, yellowish, aromatic liquid alkaloid), **lobelic acid, lobelacrin** (an active principle, probably lobelate of lobeline), inflatin (a tasteless, colorless, and odorless, probably inert, neutral principle), resin, fixed oil, gum, probably volatile oil, salts, etc.

Preparation of Lobeline.—Evaporate the acetic alcoholic tincture to syrup; triturate this with MgO in excess; agitate filtrate with ether. Evaporate ether and concentrate over sulphuric acid. It is quite volatile.

Preparation of Lobelacrin.—Obtain by concentrating tincture of lobelia in persence of animal charcoal; exhaust charcoal with boiling alcohol. It is the acrid principle, lobelate of lobeline. Ash, not more than 8 per cent.

ACTION AND USES.—Poisonous; diaphoretic and expectorant; used in **asthma,** whooping-cough, and other spasmodic pulmonary affections. In large doses it is a cathartic and emetic, but, being a violent gastro-irritant, **it should not be used for these purposes on account of its danger.** Dose: 1 to 15 gr. (0.065 to 1 Gm.). The latter dose as an emetic. The two species, *syphilitica* and *cardinalis*, are used medicinally, the former antisyphilitic and diaphoretic and the latter anthelmintic. Both were used by the Indians.

OFFICIAL PREPARATIONS.

Fluidextractum Lobeliæ,..... .Dose: 1 to 5 ♏ (0.065 to 0.3 mil).
Tinctura **Lobeliæ** (10 per cent.),....... Expectorant 15 ♏ (1 mil).
Emetic 1 fl. dr. (4 mils).

COMPOSITÆ.—Composite Family

Herbaceous or woody plants, rarely shrubs, with the *flowers* in close heads on a common receptacle, and surrounded by a common imbricated involucre. *Stamens* 5, their anthers united into a tube surrounding the pistil. *Flowers* of two sorts, strap-shaped or ligulate, and tubular, and hence the family is divided into three tribes: Tubulifloræ (flowers tubular in all the perfect flowers, and ligulate in the marginal or ray-flowers), Ligulifloræ (all the flowers of the head being strap-shaped, ligulate), and Labiatifloræ (with tubular flowers more or less labiate). *Fruit* an akene.

Synopsis of Drugs from the Compositæ

A. *Roots.*
  TARAXACUM, 553.
  Cichorium, 554.
  PYRETHRUM, 555.
  Pyrethrum Germani-
    cum, 555 a.
  Inula, 557.
  *Lappa, 558
  Polymnia, 560.
  Laciniaria, 561.
  Helianthella, 562
  * Echinacea, 563.
B. *Rhizomes.*
  Arnicæ Radix, 564.
  Cnicus Arvensis, 566.
C. *Leaves.*
  Erechthites, 567.
  Trilisa, 568.
  Pertocaulon, 569.
  Guaco, 570.
  Ambrosia, 571.
  Strumarium, 572.
  Spinosum, 573.
  Eupatorium Purpure-
    um, 575.

D. *Herbs.*
  *Eupatorium, 574.
  GRINDELIA, 576.
  Tanacetum, 577.
  *Absinthium, 578.
  Artemisia, 579.
    A. Frigida (a).
    A. Vulgaris (b).
    A. Abrotanum (c).
  Erigeron, 580.
  Erigeron Canadense,
    581.
  Gnaphalium, 582.
  Helenium, 583.
  Achillea, 584.
  Tussilago, 585.
  Carduus Benedictus,
    586.
  Silphium, 587.
  Mutisia, 588.
  Elephantopus, 589.
  Rudbeckia, 590.
  Bidens, 591.
  *Senecio, 592.
  Solidago, 593.

Lactuca Sativa, 595.
Lactuca Canadensis,
  596.
Parthenium, 597
Cotula, 598.
E. *Flowers.*
  MATRICARIA, 599.
  Anthemis, 600.
  Santonica, 601.
  ARNICA, 565.
  *Calendula, 602.
  Carthamus, 603.
  Pyrethri Flores, 556.
F. *Concrete Juice.*
  LACTUCARIUM, 594.
G. *Volatile Oil.*
  Oleum Erigerontis,
    581 a.
  Oleum Anthemidis,
    600 a.
H. *Seeds.*
  Helianthus, 604.
I. *Fruit.*
  Lappæ Fructus, 559.

553. TARAXACUM.—TARAXACUM

DANDELION

The dried rhizome and roots of **Tarax'acum officina'le** Weber. Preserve the thoroughly dried drug in tightly closed containers, adding a few drops of chloroform or carbon tetrachloride from time to time, to prevent attack by insects.

BOTANICAL CHARACTERISTICS.—*Root* perennial; *leaves* radical, runcinate, pinnatifid or lyrate; scape hollow. *Flower-head* solitary, many flowered, yellow. After blossoming, and while the fruit is forming, a pappus raises which soon exposes to the wind the naked fruit, which is blown about.

SOURCE.—A plant of very extensive geographical distribution, native to Europe, but very abundant in the United States, where, in some parts, it is a troublesome weed.

Dandelion root may be dug from July to September, during which time the juice it contains becomes thicker and more bitter. The roots should be washed and carefully dried, and care should be taken to avoid maggots, which attack the well-dried roots.

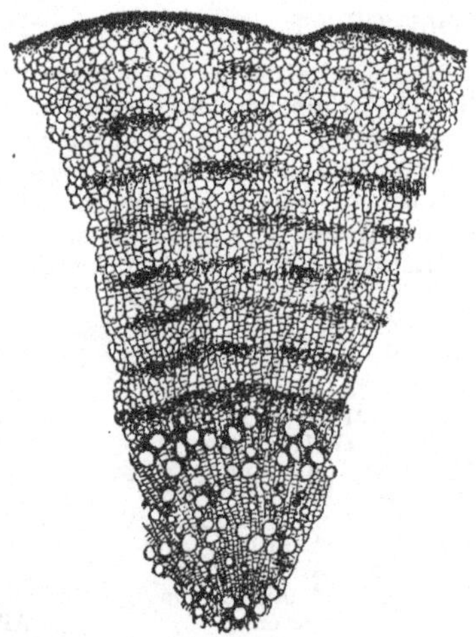

FIG. 247.—Cross-section of Taraxacum root.

DESCRIPTION OF DRUG.—The dry root is fleshy, long, and tapering, seldom branching; 5 to 25 mm. (⅕ to 1 in.) thick at the top, surmounted by several heads. Externally brownish, soon darkening by exposure. In the fall, about November, the root acquires a deep orange color throughout. **Internally** white, abounding in a bitter, inodorous, milky juice. A **cross-section** displays a **thick, white bark** with numerous **concentric** circles of laticiferous vessels surrounding a **yellow woody center.** The central column is easily separated from the thick bark, when the former is found to have along its exterior at intervals minute knotty projections; a cross-section of the root at this point shows woody fibers branching from the ligneous cord, penetrating, and passing through, the bark. Inulin spherules are plainly discernible under the microscope if, before sectioning, the fresh root be macerated in alcohol. The root loses in drying from 78 to 88 per cent. of moisture. The dried root is longitudinally and spirally wrinkled; when quite dry, has a brittle fracture, showing a dark brown exterior and a thick, white bark.

Powder.—Characteristic elements: See Part iv, Chap. I, B.

CONSTITUENTS.—Taraxacin (a bitter principle), **taraxacerin,** $C_9H_{15}O$, resin, inulin, sugar, and mucilaginous substances. The percentage of sugar varies with different seasons and with condition of soil; it is said to diminish in the summer. Recent investigations have shown the existence of an alkaloid. But this has been found to be exceedingly minute—a mere trace. Ash, not more than 10 per cent.

Preparation of Taraxacin.—Treat decoction with animal charcoal, wash the latter with water, and dissolve out bitter principle with boiling alcohol; evaporate. It has not been proven that this is crystalline. Composition uncertain.

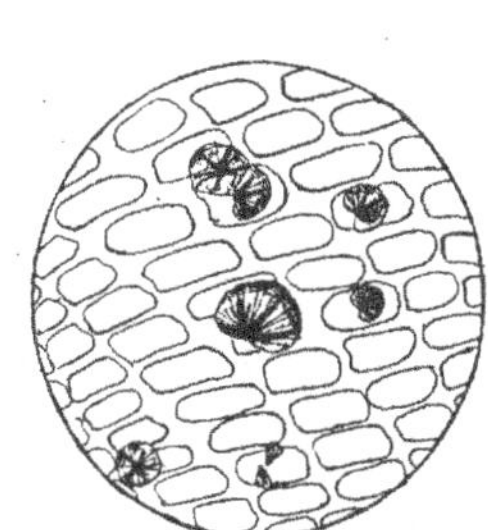

FIG. 248.—Inulin spherules in Taraxacum.

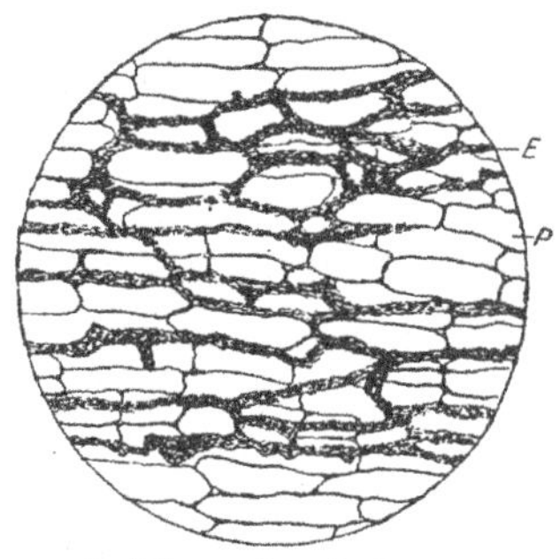

FIG. 249.—Laticiferous tissue in Taraxacum root. E, Laticiferous tissue. P, Parenchyma.

ACTION AND USES.—Deobstruent, tonic. As a remedial agent dandelion root has not been properly appreciated, possibly because it is such a common weed. It is worthy of more study on the part of pharmaceutical chemists and clinicians. The fluidextract and extract are used in **hepatic** disorders. Dose: 1 to 4 dr. (4 to 15 Gm.).

OFFICIAL PREPARATIONS.

Fluidextractum Taraxaci,............... Dose: 1 to 4 fl. dr. (4 to 15 mils).
Extractum Taraxaci,.................. 5 to 60 gr. (0.3 to 4 Gm.).

554. **CICHORIUM.**—CHICORY. The root of Cichor'ium in'tybus Linné. *Habitat:* Europe; naturalized in the United States. Nearly cylindrical, resembling dandelion, but lighter in color, more woody, with a thinner bark, and with the laticiferous vessels of the woody column and the bark arranged *radially;* very bitter. It contains inulin and a bitter principle. Bitter tonic in doses of 15 to 60 gr. (1 to 4 Gm.), in decoction. Its greatest demand is as an adulterant of coffee. It should be stated, however, that roasted chicory has become a favorite in many parts as a coffee substitute. The cultivation of the plant for this purpose and as a forage plant has grown to be a permanent agricultural industry in nearly every country of Europe and in many parts of the United States.

555. PYRETHRUM.—Pyrethrum

PELLITORY. ROMAN PELLITORY

The root of **Anacy′clus pyre′thrum** (Linné) De Candolle. Preserve in tightly closed containers, adding a few drops of chloroform or carbon tetrachloride, to prevent attack by insects.

BOTANICAL CHARACTERISTICS.—*Root* long, fusiform. *Stems* numerous, branched, pubescent. *Radical leaves* pinnatifid, *stem-leaves* sessile. *Florets* of the ray pistillate, white above and purplish beneath; of the disk, yellow, tubular, 5-toothed. *Akene* flat, winged; pappus short.

SOURCE.—Mediterranean Basin, coming solely from Algeria, thence to Mediterranean points.

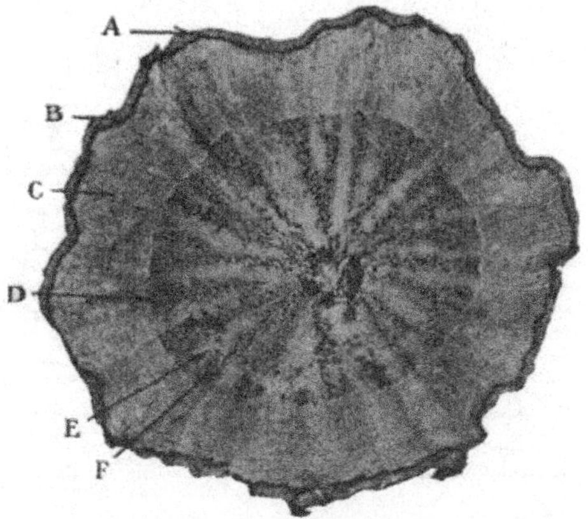

FIG. 250.—Pyrethrum—Cross-section of root. (11 diam.) A, Cork. B, Ring of stone-cells. C, Parenchyma of primary cortex. D, Cambium. E, Medullary ray. F, Xylem. (Photomicrograph.)

DESCRIPTION OF DRUG.—**A hard, compact,** somewhat fusiform root, about the size of the little finger, with sometimes leaf-remnants at the top, and beset with few or no hair-like rootlets; externally brownish, **deeply fissured** longitudinally. It breaks with a short fracture, showing a rather thick bark adhering closely to the pale brown wood, from which it is separated by a narrow cambium line. This woody column is traversed by broad, distinct medullary rays, and contains, as does also the bark, large scattered resin ducts. Odor very slight; taste slight at first, but afterward **persistently acrid,** leaving a singular tingling sensation in the mouth and throat, and exciting a remarkable flow of saliva.

555 a. *Pyrethrum Germanicum,* from *Anacyclus officinarum* Hayne, is of a grayish color, about half as thick as above, tapering to filiform at the lower end; has long been cultivated near Magdeburg and in Saxony. It resembles the above in foliage and flowers.

CONSTITUENTS.—A very acrid resinous substance, two acrid oils—pyrethrine, extracted by ether (crystalline, bitter, burning taste), which under action of alcoholic KOH decomposes into piperidine. Most of the parenchymatous cells are loaded with inulin, which forms about 35 per cent. of the root. Ash, not more than 5 per cent.

ACTION AND USES.—Used almost exclusively as a **sialagogue** in headache, neuralgic and rheumatic affections of the face, toothache, etc., or as a local stimulant in palsy of the tongue or throat, or relaxation of the uvula. Dose when chewed: 30 to 60 gr. (2 to 4 Gm.).

OFFICIAL PREPARATION.

Tinctura Pyrethri (20 per cent.),....................Used externally.

556. PYRETHRI FLORES.—INSECT FLOWERS. The flowers of (1) **Pyre'thrum carne'um** and **Pyrethrum rose'um** Weber, yielding, when powdered, Persian or Caucasian Insect Powder, and (2) **Pyrethrum cinerariæfo'lium** Visiani, yielding Dalmatian Insect Powder, which is more powerful than the Persian powder; this latter is now produced of very superior quality in California by cultivation. The plants resemble matricaria and bear flower-heads about 38 mm. ($1\frac{1}{2}$ in.) in diameter, surrounded by an imbricate involucre, (1) having brownish scales with a white scarious (membranous) edge, whitish ray-florets, and yellow disk-florets, and (2) having greenish involucral scales with scarious edge, rose-colored ray-florets, and yellow disk-florets. The flowers seldom come in market, but are in the form of a yellowish-brown or yellowish-green powder, which is used either as a powder or in tincture as an insecticide. It is not actively poisonous to human beings. Its strength or purity, and the variety from which obtained, may be ascertained by microscopical examination. A deficiency of pollen and presence of sclerenchymatous tissue would show a scarcity of flowers and the presence of stems in the powder, and consequent inferiority in strength.

557. INULA, N.F.—ELECAMPANE. The root of **In'ula Hele'nium.** Off. in U.S.P. 1890. Found in the market in slices cut in various directions. Externally grayish-brown, wrinkled, with overlapping bark. Internally gray. When dry, breaks with a horny fracture. Odor aromatic, suggestive of orris and camphor; taste slightly bitter, warm, aromatic. Gentle stimulant and tonic, supposed also to have diaphoretic, diuretic, expectorant, and emmenagogue properties. Chiefly used in this country for dyspepsia and pulmonary troubles. Dose: $\frac{1}{2}$ to 2 dr. (2 to 8 Gm.), in powder or decoction.

558. LAPPA, N.F.—LAPPA

BURDOCK ROOT

The dried root of **Arc'tium lappa** Linné, and possibly of other species of Arctium, collected from plants of the first year's growth.

BOTANICAL CHARACTERISTICS.—*Root* biennial, fusiform; *stem* 1 to 3 feet high. *Leaves* strong-smelling, ovate, with cordate and crenate base, or lanceolate, with cuneate base. *Involucre* composed of imbricated coriaceous scales, the stiff, needle-like points of which are hooked. Heads solitary or clustered; *flowers* white or light purple, all tubular. *Akenes* oblong, flattened.

DESCRIPTION OF DRUG.—A fusiform, fleshy root several inches in length and about 25 mm. (1 in.) thick, sometimes sliced longitudinally; grayish-brown, longitudinally wrinkled from drying, and having withered scales near the top; internally lighter colored, spongy, **a cross-section** showing a thick bark (in young roots, thin in old), the inner layer of which, and the meditullium, is traversed by broad medullary rays. Fracture horny. It has a slight unpleasant odor, and a sweetish, somewhat bitter taste.

Powder.—Brownish-gray. Characteristic elements: Parenchyma of cortex, thin-walled, elongated with glassy masses and sphæro-crystals of inulin; ducts large and small, with reticulate, simple pores; wood fibers and resin ducts, few.

CONSTITUENTS.—Mucilage, sugar, fat, a little tannin, a bitter glucoside, and inulin.

ACTION AND USES.—Diuretic, diaphoretic, and alterative. Dose: ½ to 2 dr. (2 to 8 Gm.). **Fluidextractum Lappæ,** Dose: ½ to 2 fl. dr. (2 to 8 mils).

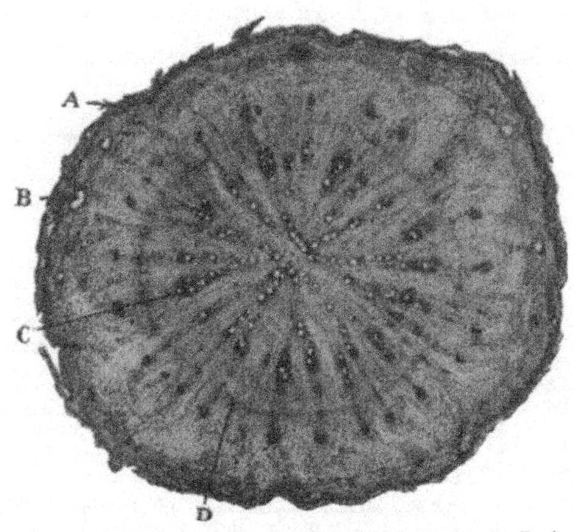

FIG. 251.—Lappa—Cross-section of root. (10 diam.) A, Cork. B, Resin cell. C, Xylem. D, Cambium. (Photomicrograph.)

559. **LAPPÆ FRUCTUS.**—BURDOCK FRUIT. A somewhat angular fruit, about 6 mm. (¼ in.) long, rough and wrinkled, and covered with short, stiff hairs, which are easily rubbed off. Very bitter. A tincture is used in psoriasis and other skin diseases.

560. **POLYMNIA UVEDALIA** Linné.—BEARSFOOT. An indigenous plant, the root of which, in ointment form, has had virtues ascribed to it as a discutient and anodyne, particularly in the treatment of malarial splenic enlargements.

561. **LACINIARIA SPICATA** Willdenow.—BUTTON SNAKEROOT. *Habitat:* United States. (Root.) Diuretic; also used as a gargle and injection. Dose: ½ to 2 fl. dr. (2 to 8 mils).

562. **HELIANTHELLA TENUIFOLIA** Torrey and Gray.—The root of this plant has the properties of an aromatic expectorant and antispasmodic, used as an addition to cough mixtures.

563. **ECHINACEA, N.F.**—The root of Echina'cea angustifo'lia De Candolle. *Habitat:* Western United States. This plant has grown into considerable importance, especially among the eclectic practitioners, in the treatment of phagedenic ulcerations, boils, various forms of septicæmia, etc. The common name of the plant is "nigger-head." The flower-head has from twelve to fifteen rays, 2 inches long, rose-colored or red, drooping; receptacle conical, with finely tipped chaff, longer than the disk-florets; disks purplish. The root has a brownish-black color, the epidermis shrunken causing longitudinally twisted wrinkles. Over 200,000 pounds were consumed in 1903. Since that time the demand has been kept up quite regularly at the same figure.

In cross-section are seen wood-wedges and medullary rays, colored dark gray or blackish; fracture short and rough; taste peculiar and somewhat acrid and biting, reminding one of pyrethrum; odor heavy, mousey, accompanied by a peculiar pungency. The root contains a very small percentage of alkaloid and a crystalline principle soluble. in carbon· disulphide. Active principle contained, apparently, in an oleoresin which represents the medicinal properties. Allied species: *Echinacea purpurea.*

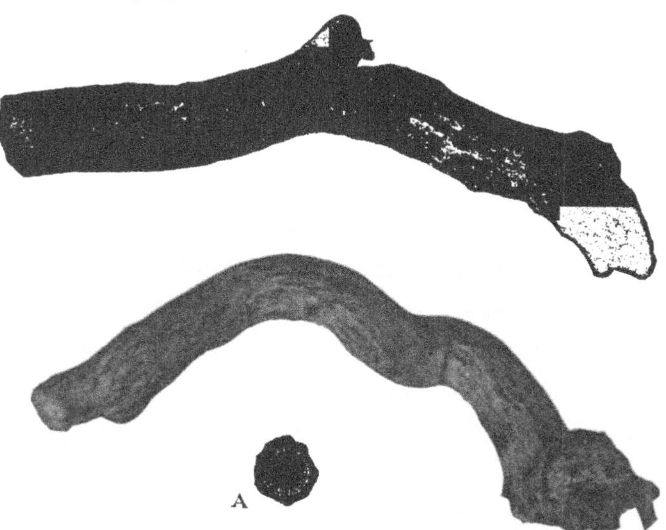

FIG. 252.—*Echinacea angustifolia*—Root. A, Cross-section of root. (Photograph.)

564. ARNICÆ RADIX.—ARNICA ROOT. A horizontal, contorted rhizome about 50 to 75 mm. (2 to 3 in.) long, and 3 to 4 mm. (⅛ to ⅙ in.) thick; externally dark brown, rough from scars, longitudinally wrinkled, and beset with numerous thin, fragile rootlets. Fracture short, showing a rather thick bark containing a circle of resin cells near the cambium line, a circle of short, yellowish wood-bundles, and a very large, whitish pith. Odor slightly aromatic; taste pungent and bitter. Adulterated with other roots of the Compositæ, also with *Geum urbanum* roots and *Frageria vesca* Off. in. U.S.P. 1890. Stimulant and tonic. Dose: 5 to 30 gr. (0.3 to 2 Gm.).

565. ARNICA.—ARNICA FLOWERS

The dried flower heads of **Ar'nica montana** Linné.

DESCRIPTION OF DRUG.—About 25 mm. (1 in.) in length and 15 to 20 mm. (⅗ to ⅘ in.) in diameter, surrounded by lanceolate, involucral scales; the receptacle is flat, and bears about 15 to 20 bright yellow, ligulate ray-florets, 3-toothed, striate, about 25 mm. (1 in.) long, and numerous shorter, tubular disk-florets; **pappus long and hairy,** giving the heads a characteristic appearance; odor peculiar and agreeable; taste persistently acrid and bitter. The powder is sternutatory. Adulterated with many flowers of the Compositæ, such as calendula, anthemis, inula, senecio, etc.

Powder.—Characteristic elements: See Part iv, Chap. I, B.

CONSTITUENTS.—Four per cent. of arnicin, and 0.04 to 0.07 per cent. of
butyraceous volatile oil. A bitter alkaloid arnicine with crystallizable salts was reported, but has not since been confirmed. Ash, not
more than 9 per cent.

ACTION AND USES.—Same as the root. Dose: 15 to 30 gr. (1 to 2 Gm.).
The tincture is used externally as a **vulnerary.**

OFFICIAL PREPARATION.

Tinctura **Arnicæ** (20 per cent.),..   .Dose: 10 to 30 ♏ (0.6 to 2 mils).

566. **CNICUS ARVENSIS** Hoffmann.—CANADA THISTLE. An indigenous plant,
the rhizome of which is popularly used for its astringent properties.

567. **ERECHTHITES HIERACIFOLIA** Rafinesque.—FIREWEED. *Habitat:*
United States. (Leaves.) The name (fireweed) comes from the fact that
the plant springs up spontaneously in burned districts. Tonic and astringent in dysentery, etc. Dose: 30 to 60 gr. (2 to 4 Gm.). The volatile oil
of this plant has been used to adulterate the oil of erigeron.

568. **TRILISA ODORATISSIMA** Cassini.—DEER TONGUE. VANILLA LEAF.
This plant contains coumarin, and the leaves are used in the Southern States
to flavor tobacco. Aromatic, stimulant, and tonic; used as a corrective.
Dose 30 to 60 gr. (2 to 4 Gm.).

569. **PTEROCAULON PYCNOSTACHYON** Elliott.—BLACK ROOT. Leaves
used by the Indians as an alterative. Dose: 15 to 30 gr. (1 to 2 Gm.).

570. **GUACO.**—By this name are known the leaves and roots of various herbs
belonging to the genus **Mikania,** growing in Central and South America,
where they are used as a febrifuge, anthelmintic, alterative, and alexiphar-
mic. They at one time gained considerable attention in Europe in the
treatment of epidemic cholera and chronic diarrhea. Dose: 15 to 30 gr. (1
to 2 Gm.).

571. **AMBROSIA ARTEMISIÆFOLIA** Linné.—RAGWEED. The leaves of this
common weed have been used in domestic practice as an astringent, styptic,
and hemostatic.

571 a. **AMBROSIA.**—RAGWEED. The staminate flowers of Ambrosia artemisiæfolia Linné, North America. Staminate flowers very small, yellowish;
surrounded by the cup-like, green involucre. *Preparation:* Fluidextract.
Properties: Tonic and astringent. *Uses:* In treatment of inflammation
from wounds and injuries; in hemorrhoidal tumors and ulcers; internally for
hay-fever. **See Pollen Extract, under Serotherapy.** Also in the treatment of
dysmenorrhea.

572. **STRUMARIUM.**—CLOTBUR. COCKLEBUR. The leaves of **Xan′thium stru-
ma′rium** Linné. Hemostatic and styptic.

573. **SPINOSUM.**—SPINY CLOTBUR. The herb of **Xan′thium spino′sum** Linné.
Diaphoretic, sialogogue, and diuretic. It is asserted that it has been used
with success in warding off hydrophobia. Dose of fluidextract: 15 to 30 ♏
(1 to 2 mils).

574. EUPATORIUM, N. F.—EUPATORIUM

BONESET. THOROUGHWORT

The dry leaves and flowering tops of **Eupato′rium perfolia′tum** Linné.

DESCRIPTION OF DRUG.—As it appears in the market, the drug consists of broken,
wrinkled fragments of the dark green leaves and corymbs of the numerous
white florets. The leaves have a rough upper surface, and downy, resin-
dotted lower surface. Odor faintly aromatic; taste strongly bitter and slightly
astringent.

Powder.—Yellowish-green. Characteristic elements: Sclerenchyma with bast fibers, thin-walled, very slightly or not at all lignified; ducts, spiral, annular, with bordered pits; trichomes, glandular and non-glandular present, 2- to 12-celled, of different shapes; stomata present; pollen, ellipsoidal (10 to 20 μ diam.); pappus, multicellular axis, unicellular branches.

CONSTITUENTS.—A peculiar, bitter, crystallizable glucoside (**eupatorin**), soluble in boiling water, alcohol, ether, and chloroform; resin, gum, tannin, and an undetermined wax-like, crystalline matter.

FIG. 253.—*Eupatorium perfoliatum*—Portion of plant and flower (enlarged).

ACTION AND USES.—Stimulant and tonic, in large doses emetic and cathartic, and as a diaphoretic often used in warding off a cold and in fevers. Dose: 30 to 60 gr. (2 to 4 Gm.), in infusion, powder, or fluidextract, which was formerly official.

575. EUPATORIUM PURPUREUM Linné.—QUEEN OF THE MEADOW. GRAVEL ROOT. The leaves and root of this indigenous plant are an excellent diuretic. Also tonic, stimulant, and somewhat astringent. Dose: 30 to 60 gr. (2 to 4 Gm.).

576. GRINDELIA.—Grindelia

GRINDELIA

The dried leaves and flowering tops of Grindelia camporum Greene, or Grindelia
cuneifolia Nuttall, or Grindelia squarrosa (Pursh) Nutall, without the presence of admixture of more than 10 per cent. of stems and other
foreign matter.

BOTANICAL CHARACTERISTICS.—Woody herbs; *leaves* clasping, resinous, somewhat cuneate. *Involucre* hemispherical or globular, coated with resin; rays
fertile, yellow; disk-florets yellow, tubular, and perfect. *Akenes* compressed, the outermost somewhat triangular; pappus awned. *Grindelia
robusta* is found in rather elevated regions, while *G. squarrosa* is found in
the plains. The former is more woody than the latter.

SOURCE.—This genus inhabits the western part of both North and South
America. A resinous exudation is common to the various species of
the genus, being most abundant in the flower-heads, and it is possible
that medicinal properties are common to the genus. Besides the
official species, there are found the *hirsutula* and the *glutimosa*, similar
species growing in the western part of the United States, often cultivated and mixed with the official.

DESCRIPTION OF DRUG.—Rough, grayish-green fragments of the leaves,
mixed with brownish-yellow stem fragments, and with flower-heads
about 15 mm. (⅗ in.) in diameter, usually destitute of florets, leaving
the bare **receptacle surrounded by the stiff, varnished, resinous**
bracts of the involucre; odor balsamic; taste aromatic and bitter.

 Distinction of the Two Species.—It may be said that the two species,
squarrosa and *robusta*, resemble each other very much. *Robusta* is
said to have a more leafy involucre and the leaves to be more coarsely
serrate. The *squarrosa* in general is said to be less leafy and bushy,
but on close examination of numerous specimens it is a question
whether the distinction will hold.

 Powder.—Characteristic elements: See Part iv, Chap. I, B.

CONSTITUENTS.—The medicinal properties of grindelia seem to reside in
the resinous exudation. An alkaloid principle has been claimed by
some investigators and termed grindeline.

ACTION AND USES.—**Antispasmodic and sedative, in asthma.** Dose: 15
to 60 gr. (1 to 4 Gm.). The fluidextract is said to be an efficient
application in rhus poisoning.

OFFICIAL PREPARATION.

 Fluidextractum Grindeliæ,   .Dose: 15 to 60 ♏ (1 to 4 mils).

577. **TANACETUM.**—TANSY. The leaves and tops of **Tanace'tum vulga're**
Linné. Off. in U.S.P. 1890. Leaves pinnate, the lobes sharply serrate, in
wrinkled, broken pieces mixed with the reddish stems; midrib heavy and prominent on under side; odor strong, fragrant, diminished by drying; taste bitter,

somewhat mint-like. *Constituents:* **Tanacetin,** $C_{11}H_{16}O_4$ (a bitter principle), malic acid, volatile oil (0.25 per cent.), tannin, resin, etc. Stimulant, tonic, emmenagogue, and anthelmintic. The dose of the volatile oil is from 1 to 5 ♏; used also as a domestic abortifacient and as a remedy for amenorrhea. Its use should be prohibited except upon physician's order, as it is a dangerous drug. Dose: 15 to 60 gr. (1 to 4 Gm.), in infusion.

578. **ABSINTHIUM.**—Wormwood, N.F. The leaves and tops of **Artemis'ia absin'thium** Linné. Off. U.S.P. 1890. Consists of the grayish, softly, hairy, longitudinally ribbed or furrowed stems with the petiolate, pinnatifid, pubescent leaves mostly broken beyond recognition; flower-heads in racemes, hemispherical, about 3 mm. (⅛ in.) broad; receptacle small, hairy, convex, with all yellow, tubular florets; akenes obovoid, without pappus; odor strongly aromatic; taste intensely bitter and nauseous. *Constituents:* Tannin, resin, malates, absinthin, $C_{15}H_{20}O_4$ (a bitter glucoside), **absinthic acid** (probably succinic acid), and a dark green volatile oil, about 1 per cent. (mainly absinthol), which has the odor of the drug, and when mixed with alcohol and oil of anise constitutes the absinthe of the French. Stomachic, tonic, anthelmintic and febrifuge. Dose: 15 to 60 gr. (1 to 4 Gm.).

Isolation of Absinthin.—Obtained by precipitating infusion, previously deprived of color, with tannin. The alcoholic extract of this precipitate is mixed with lead oxide and again extracted with alcohol. Absinthia deposits on evaporation of this tincture.

579. **ARTEMISIA.**—Nearly all the varieties of **Artemis'ia** seem to have similar properties—anthelmintic. Besides absinthium and santonica, some common indigenous plants of this genus are more or less used in medicine:

579 a. **ARTEMISIA ABROTANUM.**—Southernwood. Old Man.

579 b. **ARTEMISIA VULGARIS.**—Mugwort. Also alterative and emmenagogue, and externally as a vulnerary.

579 c. **ARTEMISIA FRIGIDA.**—Mountain Sage. Antiperiodic; first introduced as a substitute for quinine.

579 d. **ARTEMISIA TRIDENTATA.**—Sage Brush—of the Rocky Mountains. **A.** trifolia, the dwarf variety of the above, and **A. dracunculus** Terragon, are well known, but only used locally in making domestic remedies of aromatic, bitter, and tonic character.

580. **ERIGERON.**—Fleabane. Daisy Fleabane. The herb of **Erig'eron an'nuus** Persoon, **E. philadelphicus** Linné, and **E. strigosus** Muhlenberg. *Habitat:* North America and Europe. All resemble one another and are indiscriminately employed in medicine. They have erect stems, much branched at the top, bearing terminal corymbs of wheel-shaped flowers having delicate, thread-like, white or purple ray-florets and yellow disk-florets; all parts of the plant are pubescent. Taste bitterish; odor feebly aromatic, due to a small quantity of volatile oil. Diuretic and stomachic, sometimes used in the treatment of gravel and dropsy. Dose: 30 to 60 gr. (2 to 4 Gm.), in infusion.

581. **ERIGERON CANADENSE** Linné.—Canada Fleabane. *Habitat:* North America. (Herb.) This differs from the other species principally in having a bristly stem and flowers with very inconspicuous ray-florets and straw-colored disk-florets. Odor aromatic; taste bitterish, somewhat acrid. It contains a bitter principle, and a volatile oil which is official in the U.S.P. VIII. Properties and dose about the same as preceding.

581 a. **OLEUM ERIGERONTIS,** U.S.P. VIII.—(Canada Fleabane.) A limpid, straw-colored liquid becoming thick and dark on exposure; odor aromatic, persistent; taste characteristic. Adulterated with the oil of fireweed, *Erechthites hieracifolia* (567). Stimulant and diuretic, resembling oil of turpentine in action, especially as a hemostatic, but is less irritating and stimulating. Dose: 10 to 30 ♏ (0.6 to 2 mils).

582. **GNAPHALIUM.**—Life Everlasting. The herb of **Gnapha'lium polyceph'alum** Michaux. *Habitat:* North America. Leaves lanceolate, entire, woolly, sessile on the erect stem, which is branched, and bears dense terminal clusters of small obovate flower-heads surrounded by dry, whitish in-

volucres; florets yellow, tubular; odor pleasant, taste aromatic, bitterish. It probably possesses little medicinal value, but is a popular domestic remedy, used as a tea in diarrhea, hemorrhages, etc., and externally in a fomentation as a vulnerary.   Dose: 30 to 60 gr. (2 to 4 Gm.).

583. **HELENIUM.**—Sneezewort. The herb of **Helen'ium autumna'le** Linné. *Habitat:* North America. A square-stemmed herb, the leaves and flowers of which, when powdered and snuffed up the nose, produce violent sneezing, hence the name sneezewort. It has been used as an errhine.

584. **ACHILLEA.**—Yarrow. Milfoil. The herb of **Achille'a millefo'lium** Linné, common in Europe and North America. Stem hairy, branched at top bearing the large corymbs of white flower-heads, each composed of five pistillate ray-florets, and greenish-white, perfect disk-florets; leaves lanceolate, thrice pinnatifid, the divisions linear. In market, however, the leaves are broken or crumpled, and the flower-heads destitute of florets; odor chamomile-like; taste aromatic, bitterish, and astringent. Used as a vulnerary and occasionally as an internal remedy for hemorrhages and mucous discharges, as in consumption. Dose: 30 to 60 gr. (2 to 4 Gm.), in infusion.

585. **TUSSILAGO.**—Coltsfoot, N.F. The herb of **Tussila'go farfar'a** Linné. *Habitat:* Europe, and Middle and Northern United States, along the banks of streams. Demulcent, popularly used in the treatment of coughs (hence the name, from *tussis*, cough). Its expectorant properties are not pronounced, however. Dose: 30 to 60 gr. (2 to 4 Gm.), in decoction.

586. **CARDUUS BENEDICTUS.**—Blessed Thistle. The herb of **Cni'cus benedic'tus** Gaertner. *Habitat:* Levant and Europe. The drug consists of the woolly stems, with the soft, spiny leaves and a few of the large, ovate, yellow flower-heads; it has a slight, unpleasant odor and a very bitter taste. In cold infusion it is a bitter tonic, in hot infusion in large quantities diaphoretic and emetic. *Cnicus marianus* Gaertner has been used for the same purposes, and in Europe as a depurative.

587. **SILPHIUM LACINIATUM** Linné.—Rosin Weed. *Habitat:* United States. (Herb or root.) It has given good results in intermittent fevers, and in dry, obstinate coughs, its action being somewhat like grindelia.

588. **MUTISIA VICLÆFOLIA.**—Chinchirocoma. This herb is said to be a valuable antispasmodic and cardiac tonic.

589. **ELEPHANTOPUS TOMENTOSUS** Linné.—Elephant's Foot. *Habitat:* United States. (Herb.) Diaphoretic and expectorant; in large doses emetic. Dose: 5 to 30 gr. (0.3 to 2 Gm.).

590. **RUDBECKIA LACINIATA** Linné.—Thimble Weed. Cone Flower. This indigenous herb is used in catarrhal affections of the urinary tract. Diuretic and tonic. Dose: 15 to 60 gr. (1 to 4 Gm.).

591. **BIDENS BIPINNATA** Torrey and Gray.—Spanish Needles. An indigenous herb, popularly used as an emmenagogue. Dose: 15 to 60 gr. (1 to 4 Gm.).

592. **SENECIO AUREUS** Linné, N.F.—Life-root. Ragwort. (Herb.) Used by the Indians as a vulnerary. Emmenagogue. Dose: 30 to 60 gr. (2 to 4 Gm.), in infusion, decoction, or fluidextract.

593. **SOLIDAGO.**—Golden Rod. The herb of **Solida'go odo'ra** Aiton. (See Conspectus.) Aromatic, stimulant, carminative, and diaphoretic, in infusion. Used also to disguise the taste of other medicines.

594. LACTUCARIUM.—Lactucarium

LETTUCE-OPIUM

The concrete milk-juice of **Lactu'ca viro'sa** Linné.

Botanical Characteristics.—A biennial, rank-smelling *herb*, abounding in a milky, acrid juice. *Root* napiform; *stem* 2 to 4 feet high, erect, slender, glaucous, slightly prickly below, covered here and there with blood-red spots. *Leaves* with midrib prickly, otherwise smooth, finely toothed; radical leaves obovate, undivided, those of the stem lobed, auricled, and partly clasping. *Flower-heads* panicled, with small heart-shaped bracts; *flowers* all ligulate, perfect, light yellow.

SOURCE.—Europe; chiefly produced in Scotland, France, and Prussia.

DESCRIPTION OF DRUG.—In sections of plano-convex circular cakes, or angular pieces, of a grayish or reddish-brown color, breaking with a waxy, yellowish-white fracture; **odor** opium-like and disagreeable, characteristic; taste bitter and acrid. It is partly soluble in alcohol and ether. When triturated with water it yields a turbid solution; boiling water dissolves about 50 per cent., forming a brown infusion.

Powder.—Grayish-brown to dark brown, consisting almost entirely of irregular, angular masses, without any cellulose structure; when mounted in hydrated chloral T.S. the fragments become clear, showing a granular ground mass; from this separated rod-shaped crystals, monoclinic prisms and rosette-shaped crystal-like masses.

To powder lactucarium, the crude drug should be dried at a temperature not exceeding 70°C.

CONSTITUENTS.—Lactucin, lactucopicrin (very bitter and acrid), lactucic acid, $O_{44}H_{32}O_{21}$ (very bitter, probably an oxidation product of lactucopicrin), lactucerin (lactucone), and wax. Ash, not more than 10 per cent.

Preparation of Lactucerin, Lactucone.—Boiling alcohol extracts it in almost pure state from lactucarium, which has been deprived of resin and caoutchouc.

ACTION AND USES.—Anodyne, hypnotic, and sedative, **resembling opium in its action, but much feebler** and without the depressing after-effects. Dose: 5 to 60 gr. (0.3 to 4 Gm.).

OFFICIAL PREPARATIONS.

Tinctura Lactucarii (50 per cent.),.   Dose: 10 to 60 ♏ (0.6 to 4 mils).
Syrupus Lactucarii (10 per cent. of Tinc-
    ture),...............   ½ to 2 fl. dr. (2 to 8 mils).

595. LACTUCA SATIVA.—GARDEN LETTUCE. Popularly used as a mild anti-spasmodic to allay nervous irritability and mental worry. It yields a lactucarium during flowering, but before that period the juice is pellucid and insipid.

596. LACTUCA CANADENSIS.—WILD LETTUCE. Used as a mild soporific for children. Dose: 20 gr. (1.3 Gm.).

597. PARTHENIUM.—FEVERFEW. The herb of **Matrica'ria parthe'nium** Linné. *Habitat:* Europe; cultivated in this country. Resembles chamomile in odor and taste, in medical properties, and also in the appearance of the flowers, which differ, however, in their peculiar odor, their rounded and somewhat flattened receptacle, and the numerous large and long disk-florets which they bear.

598. COTULA.—Mayweed. Wild Chamomile. The herb of Anthe′mis cotu′la Linné. *Habitat:* Europe; naturalized in the United States. It has essentially the same properties as anthemis and chamomile, but has a disadvantage for general use in its strong, disagreeable odor. It is popularly used as a sudorific and antispasmodic, in doses of ½ to 2 dr. (2 to 8 Gm.), in infusion.

Fig. 254.—*Matricaria chamomilla*—Branch and dissected flowers.

599. MATRICARIA.—Matricaria

GERMAN CHAMOMILE

The dried flower-heads of **Matrica′ria chamomil′la** Linné.

Botanical Characteristics.—*Plant* annual; *stem* 1 to 2 feet high, much branched. *Leaves* alternate, more or less pinnate, smooth. *Heads* solitary; *ray-florets* white, pistillate, spreading, soon reflexed; *disk-florets* deep yellow, perfect; pappus none. The *flowers* have a peculiar aroma and a bitter aromatic taste.

Source.—Europe and Asia. The genus Matricaria is widely distributed; two or three species of the "wild chamomile" of this genus have been introduced into the United States.

Description of Drug.—After drying, the flower-heads are of a dull yellow or yellowish-white color, about 10 mm. (⅖ in.) broad, surrounded by a flattish, imbricated involucre; this involucre is composed of oblong scales, having a membranous, translucent margin; the receptacle is conical, internally hollow, and bears a single row of about fifteen short, toothed, reflexed ray-florets, and numerous tubular yellow disk-florets, without pappus; **disagreeably aromatic**; taste bitterish, aromatic.

Powder.—Greenish. Characteristic elements: The interesting microscopical constituent for study is found in the pollen grains with three distinct pores; seldom dispensed as powder.

Adulterations.—*Anthemis arvensis* and *A. cotula.* These have solid, chaffy receptacles.

Constituents.—Deep blue volatile oil, anthemic acid, **anthemidin,** and tannin. Ash, not more than 13 per cent.

Preparation of Anthemic Acid.—The concentrated infusion, made with water acidulated with acetic acid, is precipitated with alcohol. The alcoholic residue, after evaporation of the alcoholic solution, is treated with chloroform. The precipitate produced by alcohol contains anthemidin.

Action and Uses.—Mild **stimulant** and tonic, in large doses emetic. Dose: 15 to 60 gr. (1 to 4 Gm.) in infusion.

600. ANTHEMIS.—Anthemis, U.S.P. VIII

ROMAN CHAMOMILE. ENGLISH CHAMOMILE

The dried flower-heads of **Anthe′mis nobil′is** Linné, collected from cultivated plants.

Source.—Europe; cultivated in Germany, England (Mitcham Gardens), Surrey; introduced in United States.

Description of Drug.—There are two kinds of flower-heads, the single and the double. The latter is developed by cultivation, the disk-florets being partly or wholly converted into the white, strap-shaped, three-toothed ray-florets, forming an **almost spherical head,** dull white when dry and about 20 mm. (⅘ in.) broad; it is the kind preferred, on account of its greater aromatic properties, which reside in the rays, but as the conversion is more or less incomplete, both kinds may be found intermingled in the commercial article. It is stated, however, by some that the single flowers are more odoriferous and yield a larger proportion of volatile oil; the double flowers, being more showy, are preferred by the public. Involucre imbricate, the scales ovate-oblong, with a scarious margin; receptacle solid, conical, chaffy; odor strong, agreeable; taste aromatic and bitter.

Powder.—Straw color. Characteristic elements: Trichomes, glandular, single-celled, thick-walled; pollen and stomata present.

Constituents.—**Volatile oil** (Oleum Anthemidis, 1 per cent.), at first pale blue, becoming yellowish-brown on exposure; it is regarded as a mixture of hydro-

28

carbons with the angelic, valerianic, and tiglinic esters of butyl and amyl. Anthemis also contains a brown, bitter extractive, probably a glucoside. Ash, about 6 per cent.

ACTION AND USES.—Stimulant and tonic, in enfeebled digestion during convalescence; also carminative, and in large doses emetic. Dose: 15 to 60 gr. (1 to 4 Gm.), in infusion.

FIG. 255.—*Anthemis nobilis*—Plant and dissected flowers.

601. SANTONICA.—SANTONICA, U.S.P. VIII

LEVANT WORMSEED

The dried unexpanded flower-heads of **Artemis'ia pauciflo'ra** Weber.

BOTANICAL CHARACTERISTICS.—A low, shrubby, tomentose, aromatic plant. *Leaves* downy, pinnatifid; *flower-heads* drooping, in dense thrysoid panicles.

SOURCE.—*Artemisia pauciflora* grows on the desert plains or steppes of several parts of Russia, especially in the districts near the lower course of the Volga and Don Rivers. It is quite abundant in Persia and Turkestan, where it is known as Damanah. This Asiatic drug does not differ materially from the Russian, except that it is slightly shaggy and mixed with tomentose stalks. Of late years most of the wormseed of commerce has come from the steppes of the northern part of Turkestan, whence it finds its way to Moscow and Western Europe.

DESCRIPTION OF DRUG.—**Greenish-brown, small,** oblong-ovoid, about 2 mm. (1/8 in.) long. They consist of fifteen to eighteen imbricated scales, each having a green midrib containing oil-glands, which inclose four or five tubular florets so minute that they can scarcely be distinguished by the naked eye; odor strong, aromatic; taste bitter, aromatic, camphoraceous.

Powder.—Greenish-brown. Characteristic elements: Parenchyma cells, elongated, thin-walled; trichomes, glandular, 1 or 2 short cells or two or three pairs of cells, non-glandular, one-celled, long, slender, thin-walled; pollen mostly in masses, brown, 15 to 20μ in diam.; pores distinct.

CONSTITUENTS.—Volatile oil about 1 per cent., having a characteristic smell and taste, devoid of anthelmintic properties, which reside in the neutral principle, **santonin,** $C_{15}H_{18}O_{31}$. Santonin (Santonium, U. S.) constitutes about 2 per cent. of the drug; it occurs in colorless, rectangular, tabular crystals, which, when exposed to the light, assume a yellow hue. Soluble in 5300 parts of water, 34 of alcohol, 78 of ether and 2.5 of chloroform at 25°C. (77°F.).

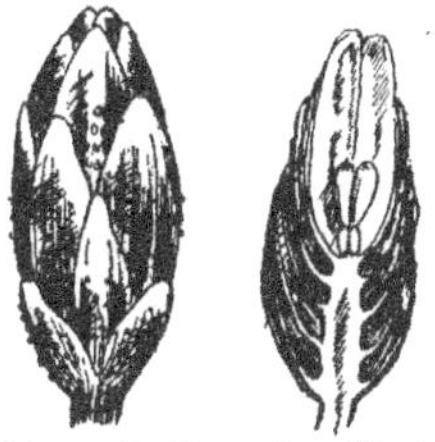

Preparation of Santonin.—Digest powdered santonica in dilute alcohol mixed with slaked lime; recover alcohol; add acetic acid in excess to residue, which separates santonin in white, shining, odorless bitter prisms, turning yellow on exposure.

FIG. 256.—Santonica—Head and longitudinal section enlarged.

This important principle is manufactured to a considerable extent in Russia, large factories at Oldberg turning out about twelve tons annually. It is well known to the natives of India, and is now imported from Germany. Much of the imported santonin is adulterated, sometimes to the extent of three-fourths of its weight, with gum and boric acid. These can easily be detected upon exposure as santonin turns yellow. The quantity of santonin in the plant diminishes as the plant grows older and the flowers expand.

Tests.—On dissolving with nitric acid and adding sulphuric acid we get a red color, and on adding Fe_2Cl_6 it changes to violet. With an alcoholic solution of KOH a pinkish-red liquid is obtained, soon becoming colorless.

On account of the fact that santonin is easily decomposed, it should be kept in amber-colored bottles, away from the sunlight, which converts it into yellow photo-santonic acid. Heating it with alkalies changes it into santoninic acid, while long boiling with baryta water changes it into santonic acid.

ACTION AND USES.—**Anthelmintic.** Dose: 15 to 60 gr. (1 to 4 Gm.), in infusion or electuary. Dose of santonin: 1/4 to 1 gr. (0.016 to 0.065 Gm.), in powder or troches. **Trochisci Santonini,** U.S. VIII., 1/2 gr. (0.03 Gm.).

602. CALENDULA.—CALENDULA, N.F.

MARIGOLD

The dried ligulate florets of **Calen'dula officina'lis** Linné.

DESCRIPTION OF DRUG.—Florets about 12 mm. (1/2 in.) long, **linear** and **strap-shaped,** delicately veined in a longitudinal direction, yellow **or orange-colored,** 3-toothed above, the short, hairy tube inclosing the remnants of a filiform style terminating in two elongated branches; odor slight and somewhat heavy; taste somewhat bitter and faintly saline.

CONSTITUENTS.—Trace of volatile oil, a bitter principle, and a peculiar gummy principle, calendulin, $C_6H_{10}O_5$, regarded by some authorities as analogous to bassorin.

Action and Uses.—It has slight stimulant and diaphoretic properties, but is used principally in the form of tincture, as a **vulnerary.** Dose: 15 to 60 gr. (1 to 4 Gm.). **Tinctura Calendulæ,** formerly official.

603. **CARTHAMUS.**—Safflower. American Saffron. The florets of **Car-tha'mus tincto'rius** Willdenow. (Official, 1820–1880.) *Habitat:* India, Levant, and Egypt; cultivated. **Orange-red;** tube long, slender, cylindrical with the two-cleft yellowish style protruding; strap divided into five narrow, lanceolate lobes; odor peculiar, aromatic; taste bitter. It contains two coloring principles, safflower-yellow, $C_{24}H_{30}O_{15}$ (24 to 30 per cent.), and a red principle, carthamin, $C_{14}H_{16}O_7$, or carthamic acid, to the latter of which its value as a dyestuff is due, and which, mixed with talc, forms rouge. Cathartic and diaphoretic in large doses of the warm infusion; in domestic practice used as a substitute for saffron to promote eruption in measles, scarlatina, etc. Dose: 8 to 15 gr. (0.5 to 1 Gm.).

604. **HELIANTHUS ANNUUS** Linné.—Our common sunflower, the seeds of which are sometimes used as a diuretic and expectorant in pulmonary and laryngeal affections. Dose of fluidextract: 1 to 2 fl. dr. (4 to 8 mils). The fixed oil expressed from them has become an article of commerce, and the growing plants themselves enjoy the reputation of purifying malarial districts.

SECTION II.—ANIMAL DRUGS

605. CANTHARIS.—Cantharides

SPANISH FLIES. BLISTER BEETLES

The beetle, **Can'tharis vesicato'ria** De Geer. (Fam. Coleoptera.) Thoroughly dried
at a temperature not exceeding 40°C. (104°F.). Should not contain more
than 10 per cent. moisture, and should contain not less than 0.6
per cent. of cantharidin.

HABITAT.—Southern and Central Europe and Northwestern Asia, feeding
on plants of the families Oleaceæ and Caprifoliaceæ.

COLLECTION.—By shaking or beating the food-plants; the insects are
then killed by heat (hot water) and rapidly dried.

DESCRIPTION.—A bronze-green beetle, with long (about 1 in. or 25 mm.)
and narrow (¼ to ⅓ in., about 7 mm.), subcylindrical body. The
vertical, rather triangular, head is constricted behind so as to form
a conspicuous neck. Odor strong and disagree-
able, caused, in the living insect, by a secreted
fluid containing uric acid, according to Maquetti.
The crushing of the dried insect yields a grayish-
brown powder containing green shining particles
(the bits of the green wing-covers and the
body-wall).

The dried insects or the powder is subject to
the attacks of several Dermestid beetles and of
several mites (*Glyciphigus*). The addition of a
little chloroform, oil of turpentine, or naphthalene balls in a tightly closed vessel will help to
keep out these cantharid-eating pests; or, if they
have established themselves in the vessels, they may be killed by
the use of carbon disulphide. (See Part III.)

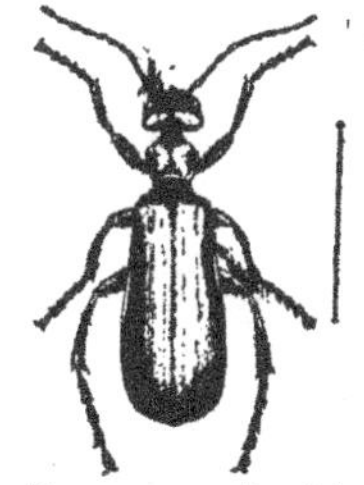

FIG. 257.—Spanish
Fly (*Cantharis vesi-
catoria* De Geer)—
(*Original*).

OTHER SPECIES.—Besides *Cantharis vesicatoria*, several other beetles of
the family Meloidæ, especially species of **Mylabris, Epicauta,** and
Macrobasis, are used to obtain vesicatory agents, and give a larger
percentage of cantharidin than the officially recognized insect.

Epicauta vittata.—The Old-fashioned Potato Beetle.* Found, often abundantly, in the United States; feeds largely on leaves of potato-plants. This insect was formerly official.

Mylabris cichorii Fab., and **M. phalerata** Pallas.—Chinese Blister Beetles. *Habitat:* Southern and Eastern Asia. *Cichorii* has its black wing-covers crossed by three broad orange-yellow bands; one band is terminal, thus rendering the apices of the wing-covers yellow.

Mylabris bifasciata.—The Two-striped Blister Beetle. *Habitat:* Northern Africa. The body is black, the wing-covers presenting two undulating narrow yellowish stripes. All these species of *Mylabris* yield about 1 per cent. of cantharidin.

ADULTERATION.—Spanish flies exhausted of their vesicating principle have been met with as substitutions. Powdered euphorbium has been spoken of as one of the adulterants, but adulteration is not common in this drug. The assay of the drug is accomplished by treating the powder with a mixture of benzine (2 vols.) and petroleum ether (1 vol.), acidulated with HCl; digesting the mixture; decanting the clear liquid, after cooling; evaporating and purifying the residue. For details, see U.S.P. IX.

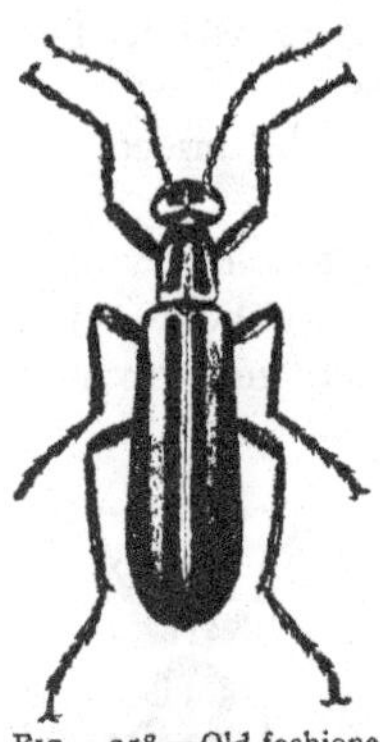

FIG. 258.—Old-fashioned Potato Beetle (*Epicauta vittata* Fab.)—(*Original*).

CONSTITUENTS.—The chief constituents are: (1) cantharidin, the active principle, a fatty crystallizable body forming shiny, colorless plates, soluble in alcohol, ether, acetic ether, glacial acetic acid, chloroform, and oils; volatilizable by heat (100°C., 212°F.) without decomposition, the vapor condensing in acicular crystals; (2) a volatile oil, giving the odor of cantharides, and said to have vesicatory properties; and (3) a green oil, the coloring principle, closely allied to chlorophyll.

Preparation of Cantharidin.—Obtained by percolating the powder with chloroform, distilling off the liquid, and purifying the resulting crystals by washing them with CS_2 to remove fat. Colorless prisms; soluble in alcohol, ether, fats, etc.

Cantharidin is associated with certain alkalies and alkaline earths in the drug, and seems to exist partly in combination with them. The principle itself has been found to combine with salifiable bases like an acid.

ACTION AND USES.—Internally cantharides acts as a powerful irritant, and has a peculiar effect on the urinary and genital organs. Large doses produce violent strangury, attended with excruciating pain and a discharge of bloody urine. The principle use of cantharides is the application, externally, of the cerate as a blistering plaster. It is seldom used as a rubefacient, but as an epispastic or vesicant it is to be preferred of all substances of this class. Its blistering action terminates in a copious secretion of serum under the cuticle. Dose: ½ gr. (0.03 Gm.).

OFFICIAL PREPARATIONS.

Ceratum Cantharidis (32 per cent.).
Collodium Cantharidatum (60 per cent.).
Tinctura Cantharidis (10 per cent.),......Dose: 1 to 5 ♏ (0.065 to 0.3 mil).

606. COCCUS.—COCHINEAL

COCHINEAL BUG. RED SCALE INSECT

The dried female insect, **Coc′cus cac′ti** Linné. (Fam. Coccidæ), enclosing the young larvæ.

HABITAT.—Mexico, Central America, and Northern South America (originally), and Spain and Algiers (introduced); feeds on various cacti, especially upon *Opuntia coccinilifera.*

COLLECTION.—Only the females (wingless) are used; they are brushed off from the food-plant, and, if alive, are killed by heat (hot water or oven). The cochineal insect is cultivated on a large scale, and large quantities are annually exported from Mexico and Peru. Humboldt estimated that 800,000 pounds of coccus (each pound representing 70,000 insects) were annually imported into Europe.

DESCRIPTION.—The females (which alone are used) are small, wingless, oval, dull purplish-brown insects, convex above, about 4 mm. (⅙ in.) long, covered, when alive, with a white cottony secretion. When the insects are dead and dry, this "cotton" rubs off, and the crushed insects yield a dark red powder; odor faint, taste slightly bitter.

VARIETIES.—These are: (1) silver, recognized by the presence of a soft, silvery white powder contained in the furrows and wrinkles; it appears to be a fatty substance as it melts on the application of heat, and the insects lose their silvery appearance. This variety is said to be the mature and fecundated insect. (2) Black cochineal, of a reddish-black color, nearly devoid of silvery powder, is supposed to be the female exhausted by propagation. (3) Granilla, an inferior kind composed of small and imperfect insects.

FIG. 259.—Cochineal Bug (*Coccus cacti*) Linné—(*Original*).

ADULTERATION.—The silvery gray variety with carbonate or sulphate of barium and lead; the black cochineal with graphite, ivory black, or manganese dioxide. "When completely incinerated, cochineal should leave not more than 5 per cent. of ash."

CONSTITUENTS.—Cochineal contains principally a red coloring matter soluble in water, alcohol, or water of ammonia.   This coloring matter is composed of carminic acid, $C_{17}H_{18}O_{10}$ (?).

Carminic acid is obtained by treating the drug first with ether to remove fat, then with alcohol.   Let alcoholic solution stand a few days, when carminic acid will deposit as a brownish-purple substance.   A vermilion-red powder (carmine), soluble in water, alcohol, and alkalies, is obtained as a combination of this acid with alumina or occasionally with oxide of tin or with albumen.   Commercial carmine is made by precipitating the decoction of cochineal with alum or cream of tartar.

MEDICAL PROPERTIES.—Cochineal has some reputation as an anodyne and antispasmodic, but it has not for many years been used as a remedial agent, its chief use being that of a coloring matter, and for this purpose it enters into the following preparation.

OFFICIAL PREPARATION.

Tinctura Cardamomi Composita (0.5 per cent.),...Dose: 1 fl. dr. (4 mils).

607. BLATTA.—COCKROACH. Periplane'ta orienta'lis Linné.   Class, Insecta; order, Orthoptera; family, Blattidæ.

HABITAT.—Asia (originally); now found in almost all parts of the world, in kitchens, laundries, and any warm, damp room.   Nocturnal in habit, feeding omnivorously on vegetable and animal products.

DESCRIPTION.—A large (1 in. long), dark brown, short-winged, broad, flat, oval insect with long, thread-like antennæ.   Wings of the female rudimentary; of the male not reaching quite to the tip of the abdomen.   Odor disagreeable.

OTHER SPECIES.—*Periplaneta americana* (American cockroach) is larger than *orientalis*, lighter brown in color, and has the wings well developed in both sexes.   Numerous in houses about the water pipes; also abundant, often in green-houses, feeding injuriously on various plants.

Ectobia germanica (German cockroach or Croton Bug), very common in New England cities; smaller than the two preceding roaches (about ½ in. long), very light (yellowish-brown) in color, with two longitudinal dark stripes upon the prothorax.

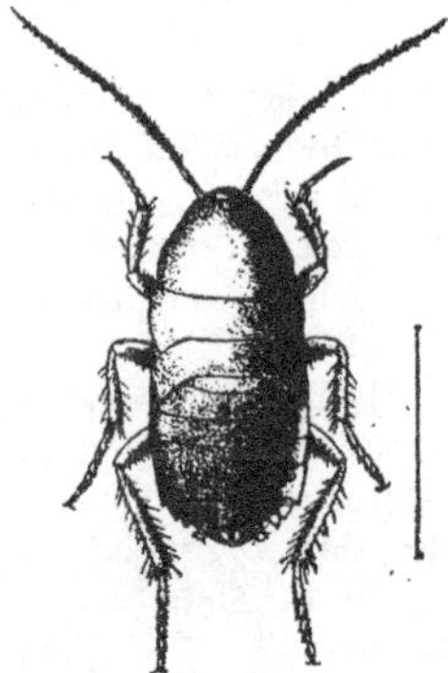

FIG. 260.—Cockroach (*Periplaneta* sp.)—(*Original*).

Blatta gigantea, found in the West Indies, attains a length of 2 inches.

CONSTITUENTS.—Fœtid oil, ammonia, trimethylamine, and a crystallizable principle, not diuretic, antihydropin.

ACTION AND USES.—Diuretic.   Dose: 5 to 10 gr. (0.3 to 0.6 Gm.), in powder or tincture.

608. HIRUDO.—LEECH. Sanguisu'ga medicina'lis Savigny.   Class, Vermes; order, Annelida; family, Hirudinea.

HABITAT.—Northern and Central Europe chiefly, but found more or less in all parts of Europe, in ponds of fresh water.

DESCRIPTION.—The body, which varies in length from 75 to 150 mm. (3 to 6 in.), is smooth and round, tapering toward both ends, and made up of about 100 soft rings or folds.   Both ends are provided with a flattened disk, the posterior being the larger, each of which is adapted to fix upon objects by suction.   The mouth has three jaws, with a double row of fine sharp teeth in each; the small anal opening is found on the under side of the last posterior wrinkle.   Color of black greenish and striped longitudinally with numerous black spots; belly somewhat lighter green.

OTHER SPECIES.—Besides *S. officinalis*, which is next to *S. medicinalis* in importance and is similar in appearance (only there are no spots, and a black line extends along each side), may be mentioned *Hirudo provincialis*, *H. obscura*, and *H. interrupta*, the species common in this country being known as *H. decora*. Leeches are said to be found in great abundance throughout India.

PRESERVATION.—The usual way of keeping leeches is to place them in clear water, in a shaded spot if possible, where the temperature will range from 10° to 20°C. (50° to 68°F.), care being taken to have a considerable quantity of charcoal, moss, and pebbles in the containing vessel.

USE.—For local blood-letting, a single leech being able to extract from 1 to 2 drachms of blood.

SPECIAL ANIMAL TISSUES AND SECRETIONS

609. SPONGIA.—SPONGE. Spon'gia officina'lis Linné. Class, Porifera; order, Ceratospongiæ.

HABITAT.—Red Sea, Mediterranean Sea, Atlantic Ocean, and other bodies of salt water, upon the rocky bottom.

COLLECTION.—The best sponges are secured by diving and cutting away their fastenings from the rocks; those of inferior quality are usually torn away with an instrument made for the purpose. The fresh sponges are exposed to the sun and washed, for the purpose of removing the animal matter with which they are filled.

DESCRIPTION.—A soft, elastic skeleton or framework of fibrous tissue surrounding the original animal matter, which, being removed, leaves a number of large and small cavities. The color is a light yellowish-brown.

VARIETIES.—The Turkey sponge is considered the best and belongs to the species *Euspongia mollisima; Euspongia zimocca*, from the coast of Greece, is harder and not so elastic. A still coarser sponge is *Euspongia equina*, collected along the north coast of Africa. The various sponges of the West Indies and Florida are different varieties of the three preceding species.

CONSTITUENTS.—A characteristic substance known as **spongin**, which yields leucin and glycocoll when treated with sulphuric acid, and when treated with KOH evolves ammonium hydrate. The ash is made up of various compounds of iodine, sodium, magnesium, calcium, etc.

USES.—Its power to absorb liquids and to expand at the same time makes sponge valuable as a surgical accessory in absorbing blood, dilating cavities, cleansing surfaces, etc., but great care should be exercised in its use, so that the same sponge may not be used more than once without being thoroughly washed in a dilute solution of carbolic acid; otherwise there is danger of contamination by infection, which is easily carried from one patient to another when the same sponge is used repeatedly. Burnt sponge is occasionally administered, on account of the iodides of sodium and potassium which it contains, in cases of goiter and scrofulous swellings.

610. CORALLIUM.—CORAL. Oculi'na virgi'nea Lamarck. Class, Polypifera; order, Hexacoralla.

HABITAT.—Atlantic Ocean and Mediterranean Sea.

DESCRIPTION.—A hard, calcareous substance produced by coral polypi. The pieces are often branched, presenting a surface more or less porous and striate, and the interior is radiate or hollow.

VARIETIES.—Besides *Oculina virginea* there are several other species, among which may be mentioned *Corallium rubrum*, the red coral.

CONSTITUENTS.—Calcium carbonate 83 per cent., animal matter 7 to 8 per cent., magnesium carbonate 3 to 4 per cent., and ferric oxide 4.25 per cent. (in the red coral).

USES.—Antacid. Used in tooth powders. Dose: 5 to 15 gr. (0.3 to 1 Gm.).

611. TESTA.—OYSTER SHELL.

SOURCE.—*O'strea virginia'na* and *O. edulis*, which excrete a calcareous bivalved covering or shell, and inhabit the shallow coast water of the Atlantic and Indian Oceans.

DESCRIPTION.—External surface rough, inner surface smooth and white, the two toothless, hinged valves made up of imbricate, foliaceous layers, presenting, when closed, an irregularly rounded, oblong, or ovate form.

CONSTITUENTS.—Largely calcium carbonate, there being only 4 per cent. or less of animal matter present and a small percentage of silica, alumina, magnesia, and calcium phosphate and sulphate.

USES.—Antacid. The shell, to be used, should first be thoroughly purified and washed in boiling water. Dose: 5 to 15 gr. (0.3 to 1 Gm.).

612. OS SEPIÆ.—CUTTLEFISH BONE.

SOURCE.—*Se'pia officina'lis* is the species from which this calcareous bone is obtained; it inhabits the Atlantic Ocean and the Mediterranean Sea.

DESCRIPTION.—A white, flattish, oval-oblong bone about 100 mm. (4 in.) in length; exterior hard and smooth, interior porous and friable; inodorous; taste somewhat saline and earthy.

CONSTITUENTS.—Mostly calcium carbonate, with from 10 to 15 per cent. of animal matter and a very small percentage of sodium chloride, calcium phosphate, and magnesia.

USES.—An antacid. Extensively employed in the manufacture of tooth powders, and used to some extent as a polishing agent.

613. CALCULI CANCRORUM.—CRABS' STONES.

SOURCE.—The stomach of the crab (*Asta'cus fluviati'lis* Fab. or *Cancer astacus* Linné), where they are formed by concretions. The crab is found in rivers throughout the North Temperate Zone.

DESCRIPTION.—The circular, plano-convex stones vary in size from 3 to 10 mm. (⅛ to ⅖ in.) in diameter, and are white and hard, changing in hot water to a rose-red; tasteless and inodorous. When treated with hydrochloric acid, they effervesce until nothing is left but a small plano-convex, cartilaginous mass.

SUBSTITUTIONS.—Artificial stones are sometimes manufactured, but can be distinguished from the true crabs' stones by treating with HCl, when, if they are artificial, they leave little or no residue.

CONSTITUENTS.—Calcium carbonate 63 per cent., calcium phosphate 17 per cent., animal matter 12 to 15 per cent., and small portions of phosphate of magnesium and sodium salts.

USES.—Antacid.

614. ICHTHYOCOLLA.—ISINGLASS. The swimming-bladder or sound of the Sturgeon, a fish found in the Black and Caspian Seas and their tributary streams. The swimming-bladders of other fish are also employed for this purpose, but the isinglass from the Russian species, *Acipenser huso, A. gulden-stadtii, A. ruthenus,* and *A. stellatus,* is considered the finest and purest. The inner layer of the swimming-bladder is separated from the outer, and after being washed is thoroughly dried. The sheets of commercial isinglass are prepared in various forms—leaf isinglass (single sheets), book isinglass (several sheets folded together), and staple isinglass. In appearance it resembles horn, is of a yellowish-white color, semi-transparent and iridescent. The substance is tough, tearing with difficulty even in the direction of the fibers, but dissolves completely in hot water, forming a transparent jelly on cooling in a solution of 24 parts of the same. *Constituents:* Gelatin (98 per cent., in the best Russian variety) and from 2 to 30 per cent. of insoluble membrane, the ash amounting to only about 0.5 per cent. Nutritive, easily digested. Emollient and protective externally.

615. AMBRA GRISEA.—AMBERGRIS.

SOURCE.—*Physe'ter macroceph'alus,* a species of whale inhabiting the Indian Ocean and the southern part of the Pacific Ocean, excretes a substance from the intestines which is found floating on the surface of the water; this is known as ambergris.

DESCRIPTION.—Waxy, grayish-brown, with streaks and dots; odor peculiar, taste slight; soluble in hot alcohol, ether, fats, and volatile oils.

CONSTITUENTS.—Ambrein (brilliant white needles precipitated from alcoholic solution) 85 per cent., a balsamic extractive, and a very small proportion of ash. On account of its high price adulterations of and substitutions for ambergris are common, but the genuine article is easily distinguished by means of its complete solubility in hot alcohol, and evaporation without evolving acrid vapor.

Preparation of Ambrien.—Obtained by crystallizing from hot alcoholic solution of ambergris; it forms white, shining, tasteless, and inodorous needles which fuse near 350°C.

Uses.—As a perfume it is highly prized. It possesses very uncertain medical properties and is very rarely administered as a remedial agent.

616. OLEUM MORRHUÆ.—Cod-liver Oil. A fixed oil obtained from the fresh livers of **Ga'dus mor'rhua** Linné, or of other species of Gadus (class, Pisces; order, Teleostei; family, Gadidæ). For tests see U.S.P. *Description:* A pale yellow thin oily liquid. Peculiar, rancid odor; bland, fishy taste. Specific gravity at 15°C. (59°F.) 0.922 to 0.927. Should be kept in dry, well-stoppered bottles. *Constituents:* Chiefly olein, palmitin, and stearin. The oil also contains dissolved in it minute quantities of the halogen elements, iodine, bromine, and chlorine, with phosphorus and sulphur. A peculiar substance named gaduin is also claimed to have been found. A crystalline substance, morrhuol, a compound body containing phosphorus, iodine, and bromine, is also said to be among the "active principles" of cod-liver oil. *Action and Uses:* A nutritive agent, generally of easy assimilation. It has long been used as a stimulant and alterative in rheumatic and strumous diseases. In pulmonary consumption it has for a long time enjoyed a great reputation. Dose: a tablespoonful (½ fl. oz.) three or four times a day.

617. CETACEUM.—Spermaceti

SPERMACETI

A peculiar concrete, fatty substance obtained from the head of the sperm whale, **Physe'ter macrocepha'lus** Linné (class, Mammalia; order, Cetacea).

Description.—A pearly-white, somewhat translucent, waxy mass, but of a somewhat granular texture, fusing at about 45°C. (113°F.). Odor faint and bland, taste mild. Insoluble in water, soluble in 50 parts of boiling alcohol; also in ether, chloroform, and carbon-disulphide. It becomes yellow and rancid on exposure to air.

Constituents.—Mainly cetin (cetyl palmitate, $C_{10}H_{33}$, $C_{16}H_{31}O_2$), with small amounts of other fatty compounds.

Uses.—Mainly as a base for cerates and ointments.

618. MEL.—Honey

HONEY

A saccharine secretion deposited in the honeycomb by the bee, **A'pis mellif'era** Linné (Fam. Apidæ).

Uses.—Mainly as a vehicle for remedial agents.

The honeycomb, from which the honey is drained, is the source of the two pharmaceutical products:

618 a. CERA FLAVA.—Yellow Wax. Beeswax. Obtained by slicing the honeycomb, draining it thoroughly, melting the residue after impurities have subsided, and allowing the melted liquid to cool. A yellowish or brownish-yellow solid, having an odor suggesting honey, and a rather agreeable taste. It melts at about 63°C. (145.4°F.).

618 b. **CERA ALBA.**—White Wax. Bleached Wax. The yellow
wax is bleached by exposing an extended surface to the light and
atmospheric influence. This is done in various ways. Bleaching
may be accomplished by chemical means, such as by the use of chlorine gas, etc. A white, shining, inodorous, insipid solid, fusing at
about 65°C. (149°F.). For Tests see U.S.P.

Uses.—As an ingredient in cerates, ointments, plasters, etc.

619. **OVUM.**—Gallinaceum, N. F. Fresh hen's egg.
SOURCE.—The egg of the common domesticated hen (probably from India
originally) is well known as an article of food throughout the country.
DESCRIPTION.—A thin, calcareous shell incloses an albuminous substance known as white of egg, which in turn incloses the vitellus or yolk.
CONSTITUENTS.—The three parts of an egg are entirely separate and distinctive in composition.
(a) *Testa Ovi*, Egg-shell.—Almost pure calcium carbonate (90 to 97 per
cent.), the remainder being made up of magnesium and calcium phosphates,
together with about equal quantities of organic matter.
(b) *Albumen Ovi*, White of Egg.—Made up mostly of a solution of albumen
and water (albumen 15 per cent., water about 85 per cent.), with slight traces of
fat and sugar, as well as KCl and NaCl, which are the chief components of the
ash. Ovi Albumen Recens, N.F. Fresh egg albumen.
(c) *Vitellus*, U.S.P. 1890.—Egg Yolk, or Yelk. Compounded of water
(about 52 per cent.), fat (30 per cent.), vitellin (16 per cent.), and inorganic
salts (1.5 per cent.), such as chloride of sodium, sulphates and phosphates of
magnesium, etc., together with coloring matter and traces of lactic acid and
sugar. Ovi Vitellum Recens, N.F. Fresh egg yolk.
ACTION AND USES.—Shell sometimes used as antacid. The white, besides
its nutriment, is valuable as an antidote when corrosive sublimate, sulphate of
copper, or other metallic poisons have been taken into the stomach. The yolk
is even more nutritious than the white, having a greater amount of digestible
solids. It is used in preparing emulsions of oils and applied as a dressing for
burns.

620. MOSCHUS.—Musk

MUSK

The dried secretion from the preputial follicles of **Mos'chus moschif'erus Linné**
(Fam. Moschidæ).

SOURCE.—Musk is obtained from a small bag or sac attached to the pre-
puce of the male Musk deer, *Mos'chus moschif'erus*, a species of horn-
less deer found in Central Asia from Thibet to China. The musk-sac
is somewhat oval and about 50 mm. (2 in.) in diameter, containing
in the mucous lining a number of delicate glands which secrete the
musk.

DESCRIPTION.—A granular substance of a brownish or reddish-black
color, having a very strong, peculiar, and penetrating odor. The
granules are irregular in size, and have a smooth, oil appearance and
a bitter taste. The color of the fresh article is considerably lighter

than that which has been dried and prepared for the market, although the commercial product is estimated to contain about 10 per cent. of moisture. The dried musk is contained in the original sac, one-half of which is smooth and the other covered with hairs arranged concentrically around two orifices. The quantity of musk in each sac amounts to about 160 grains. Not more than one-tenth of this musk is dissolved by strong alcohol, with which it forms a light yellowish-brown tincture, while as much as one-half of it can be dissolved in water, forming with it a dark brown solution having a very strong odor. Should not contain more than 15 per cent. of moisture nor 8 per cent. of ash.

VARIETIES.—Besides the Chinese or Thibetan musk, which is of the most excellent quality, there is also a Siberian musk, the quality of which is inferior. There is also an artificial musk which comes more properly under the head of adulterations. The Siberian or Russian variety is generally quite easily distinguished, the containing sac being more elongated than 'that of the Chinese variety, and the hair thinner and lighter.

ADULTERATIONS.—An artificial musk is manufactured by the Chinese and is made up chiefly of a mixture of blood and ammonia to which a small quantity of real musk is added, the whole being inclosed in a piece of the skin of the musk ox. Resin, lead, and other substances are also resorted to in preparing adulterations.

FIG. 261.—Tonquin Musk Pod. (⅓ natural size.) (Photograph.)

CONSTITUENTS.—Free ammonia, fat, albumen, an acid, wax, and gelatinous principles can be easily separated, but it has been impossible to separate the odoriferous principle. The gray ash left after burning the pure musk constitutes about 8 per cent. of the drug. The odor of musk is destroyed or greatly modified by the action of several substances, such as camphor, ergot, hydrocyanic acid, etc.

ACTION AND USES.—Antispasmodic and diffusible stimulant, together with more or less aphrodisiac action. Its powerful and lasting odor makes it valuable as a perfume, either alone or in combination with other substances. Dose: 1 to 10 gr. (0.065 to 0.6 Gm.), administered in the form of powder, pills, or enema, the powder being generally taken with milk.

OFFICIAL PREPARATION.

Tinctura Moschi (5 per cent.),..............Dose: 2 fl. dr. (8 mils).

621. FEL BOVIS.—Ox Gall

OX GALL

The fresh bile of **Bos taurus** Linné (Fam. Bovidæ).

DESCRIPTION.—The fresh bile of the ox is a brownish or dark green, viscid liquid, with a characteristic, unpleasant odor, and a nauseous, bitter taste. It is neutral or faintly alkaline. Pettenkofer's test for this liquid is as follows: Two drops in 10 mils of water, when treated, first with a drop of freshly prepared solution of one part of sugar and four parts of water, and afterward with sulphuric acid cautiously added until the precipitate first formed is redissolved, gradually acquires a brownish-red color, changing successively to carmine, purple, and violet.

PREPARATION.—**Fel Bovis Purificatum.** The method by which this medicinal preparation of the crude ox-gall is made, according to the U. S. Pharmacopœia, is as follows: Fresh ox-gall 300 mils; alcohol 100 mils. Evaporate ox-gall in tared porcelain capsule on water-bath to 100 Gm.; add to it the alcohol. When precipitation has occurred and the solution cleared, the clear liquid is decanted, the remainder filtered, and the filtrate evaporated to a pilular consistence.

Purified ox-gall is a yellowish-green, soft solid, having a peculiar odor and a sweetish, bitter taste.

Extractum Fellis Bovis U.S.P. IX.

ACTION AND USES.—The purified ox-gall only is used in medicine. It is tonic and laxative, at one time much used to increase the secretion of bile. Dose: 3 to 10 gr. (0.2 to 0.6 Gm.).

622. SANGUIS.—BLOOD.

SOURCE.—The ox (*Bos taurus* Linné) furnishes this liquid from the arterial circulation of the vascular system.

DESCRIPTION.—A red, opaque fluid, slightly heavier than water (sp. gr. 1.05), containing corpuscles in suspension, and coagulating on exposure.

CONSTITUENTS.—Chiefly water (78 per cent.), with albumen 7 per cent., salts 9 per cent., fibrin 4 per cent., and corpuscles and other constituents 13 per cent. Hæmoglobin is a peculiar coloring matter made up of globulin and hæmatin, which gives blood its red appearance.

MEDICAL PROPERTIES.—Desiccated blood has enjoyed some reputation as a nutritive or restorative, the dose being about 15 gr. (1 Gm.), but it has not been very generally adopted as an agent among therapeutists for treatment of debilitated conditions.

623. LAC. VACCINUM, Cow's milk, N.F.

SOURCE.—The mammary glands of the cow (*Bos taurus*), the well-known domestic animal.

DESCRIPTION.—A white, opaque liquid or emulsion, made up of butter and casein, and having a pleasant taste and slight odor; specific gravity about 1.030. When allowed to stand for a few hours, the oily globules rise to the surface on account of their lower specific gravity. Under the microscope these globules are seen to be separate, and each surrounded by an albuminous envelope, but when a caustic alkali is added, this envelope is destroyed, so that the globules are released and accumulate as pure butter. When exposed for a considerable time in a warm place, milk changes from sweet to sour on account of the development of an acid by chemical action between the constituents.

CONSTITUENTS.—A large percentage (about 87 per cent.) of milk is represented by water, 4 per cent. by butter, 5 per cent. by sugar and soluble salts, and only about 3.6 per cent. by casein and insoluble salts.

Butter is composed of olein (about 30 per cent.), palmitin, and stearin (68 per cent.), and about 2 per cent. of glycerides of butyric and other acids.

Casein, which is soluble in a solution of the alkalies, is a modification of albumen, and is precipitated from solution by the action of rennet or acetic acid.

Lactic acid (Acidum Lacticum, U.S.), which is developed by the action of heat, is said not to be a normal constituent of milk, but is always present in sour milk. Syrupus Calcii Lactophosphatis employs this acid. Dose: 8 mils (2 fl. dr.).

PREPARATION: LAC FERMENTATUM, N.F.

623 a. SACCHARUM LACTIS.—SUGAR OF MILK. LACTOSE. Forms about 5 per cent. of milk and is obtained from the whey by evaporation and recrystallization. A hard, somewhat gritty, slightly sweet powder, almost inodorous. Soluble in about six parts of water. For Tests see U.S.P. It has been recommended as a dietetic in wasting diseases, but in pharmacy is merely a diluent for triturations of various kinds.

ACTION AND USES.—Milk is nutritious, and its value as an article of diet is well known. In addition to this use, milk may be satisfactorily employed as a vehicle for the administration of certain remedies having an unpleasant taste.

624. OS.—BONE.

SOURCE.—The skeleton of vertebrate animals.

CONSTITUENTS.—Calcium phosphate 40 to 67 per cent., which includes a small percentage of calcium carbonate; phosphates of magnesium and other salts are also present. With the salts are also found organic substances yielding gelatine on boiling with water. The basic substance of the bony structure contains two chief constituents, namely, an organic substance, *ossein*, and the so-called bone earth inclosed in or combined with it. Ossein is generally considered identical with collagin of the connective tissue.

Preparation of Ossein: that portion of bone that is left undissolved after treatment with HCl.

USES.—For preparing bone-black, animal charcoal, and phosphates.

625. GELATINUM (U.S.).—GELATIN.

SOURCE.—Bone, cartilage, skin, tendons, and ligaments; a boiling-hot solution of these, resulting in a jelly when cooled, is dried in the air.

DESCRIPTION.—Thin, transparent sheets or porous, opaque layers or shreds, amorphous, swelling in water without dissolving, dissolving in warm water, forming a sticky liquid which solidifies on cooling. The solution is lævogyrate. Solutions of gelatin on boiling are not precipitated either by mineral acids, acetic acid, alum, lead acetate, or mineral salts in general, but precipitated by potassium ferrocyanide, tannic acid, mercuric chloride in the presence of HCl and NaCl, and by alcohol, especially when neutral salts are present. Its solution containing KCr_2O_7 yields an insoluble compound on exposure to light.

Gelatinoids.—To this group belong a number of substances occurring in bones, skins, horns, etc., having generally the property of forming a jelly with water. The organic matter in bones, usually called ossein, contains, besides albuminous substances, the two gelatinoids, collagin and gelatin, a pure mixture of which forms common glue. Chondrin resembles gelatin; it is obtained from cartilages of the ribs and non-ossifying cartilages; its aqueous solution is precipitated by alum, lead acetate, ferric salts, acetic acid, and a small quantity of mineral acid, but not precipitated by tannin or mercuric chloride. Properties: Emollient, nutritive, and protective.

626. SEVUM.—SUET

· MUTTON SUET

The internal fat of the abdomen of O'vis a'ries (Fam. Bovidæ), purified by melting and straining. Suet should be kept in well-closed vessels impervious to fat. It should not be used after it has become rancid.

DESCRIPTION.—White, unctuous, smooth solid, melting at about 48°C. (113°F.). Sevum Præparatum (U.S.) is identical with suet as above described.

CONSTITUENTS.—Stearin, palmitin, and olein, with a preponderance of the first mentioned.

USES.—Lenitive, as an external application and as a base for unctuous preparations.

627. OLEUM BUBULUM.—NEAT'S-FOOT OIL. From the fatty tissue of the feet of the ox, previously deprived of hoofs, obtained by boiling in water and skimming off the fat, which is subsequently strained and pressed. At ordinary temperatures this is a semifluid, oleaginous fat, of a peculiar odor. CONSTITUENTS.—Mainly olein, with solid fats. Used externally.

628. ADEPS.—LARD

LARD

The prepared internal fat of the abdomen of **Sus scrof'a** Linné (class, Mammalia; order, Pachydermata), purified by washing with water, melting, and straining. Lard should be kept in well-closed vessels impervious to fat, and in a cool place.

DESCRIPTION.—A white unctuous solid with faint odor and bland taste. Insoluble in water. Soluble in chloroform, carbon bisulphide and benzine. Specific gravity at 15°C. (59°F.) about 0.932.

CONSTITUENTS.—Olein, stearin, and palmitin; of the first mentioned it consists of about 50 to 60 per cent.

USES.—Emollient, and as a base for ointments and cerates.

628 a. OLEUM ADIPIS.—LARD OIL. U.S. VIII. A pale yellowish or colorless fixed oil having a slight odor and taste. It is produced by exposing lard, at a low temperature, to strong pressure.

CONSTITUENTS.—Olein, with palmitin and stearin. Used externally.

629. PEPSINUM.—PEPSIN

PEPSIN

A mixture containing a proteolytic ferment or enzyme obtained from the glandular layer of fresh stomachs of healthy pigs, and capable of digesting not less than 3000 times its own weight of freshly coagulated and disintegrated egg albumen. See details of test U.S.P. IX.

SOURCE.—Pepsin is prepared from the stomach of the ox (*Bos taurus*), the sheep (*Ovis aries*), or the hog (*Sus scrofa*), the mucous membrane being the part used. Several methods have been employed for its extraction. The ordinary methods of manufacture may be briefly stated as follows:

(1) The extraneous matter is first removed from the inner surface of the stomach by washing, and the mucous membrane scraped off

with a blunt instrument; the pulp thus obtained is placed on glass or porcelain and dried and finally reduced to a powder. This forms a rather poor quality, owing to the presence of mucus and inert matter.

(2) The finely chopped mucous coat is macerated in dilute hydrochloric acid (about 2 per cent.), and to the filtered solution common salt is added; the floating precipitate which results is carefully washed, then dried, and the dried residue mixed with sugar of milk until the strength of the article is such that 1 grain will dissolve 3000 grains of coagulated albumen, the strength directed by the United States Pharmacopœia.

(3) A scale pepsin is made by digesting the mucous lining at the temperature of about 100°C. with about 0.2 per cent. of HCl (or water acidulated with other acids to the same degree of acidity) until the membrane is completely or nearly all dissolved. The solution is neutralized by a suitable alkali and the filtered product, after reduction by evaporation at a low temperature (sometimes *in vacuo*) to a syrupy consistence, is spread on plates of glass and dried in a current of warm air, care being taken not to allow the temperature to exceed 40°C. (104°F.). The dried, transparent film is then scraped from the plates and broken into more or less fine lamellæ.

DESCRIPTION.—A yellowish-white amorphous powder or thin, pale yellowish, somewhat transparent scales, with faint odor and slight saline or acidulous taste, but no indication of decomposition; should not be hygroscopic. It invariably contains some rennin; its solutions, therefore, will coagulate milk. Incompatible with alkalies, alcohol, and heat renders it inert.

ACTION AND USES.—Pepsin has a digestive action upon the food taken into the stomach, and is employed as an artificial agent to assist digestion when there is functional derangement of the stomach. Dose: 10 gr. (0.6 Gm.).

630. PANCREATINUM.—PANCREATIN

A mixture of enzymes (Amylopsin, Trypsin, Steapsin) existing in the pancreas of warm-blooded animals capable of converting at least twenty-five times its weight of starch into sugars.

SOURCE.—Prepared from the pancreas of the hog or ox, by mixing finely chopped pancreas with half its weight of cold water and straining the liquid by pressure through cheese-cloth or flannel. To the filtrate, alcohol is added (about one volume), and the resulting precipitate collected, purified, and dried.

DESCRIPTION.—Yellowish-white amorphous powder with but slight odor and meat-like taste; slowly soluble in water, insoluble in alcohol. See U.S.P.

TEST.—Pancreatin acts best in alkaline medium (is injured by acids).　If there be added to 4 fl. oz. of tepid water contained in a suitable flask or bottle, first 5 gr. of pancreatin, 20 gr. of bicarbonate of sodium, and afterward one pint of fresh cow's milk previously heated to 38°C. (100.4°F.), and if this mixture be maintained at the same temperature for thirty minutes, the milk should be so completely peptonized that, upon adding to a small portion of it transferred to a test-tube a slight excess of nitric acid, coagulation should not occur.　This test we have found quite satisfactory as a convenient one.　An alternate method of assay is based on the property of an aqueous solution of the principle to digest (or liquefy) starch paste.　The U.S.P. IX furnishes the two tests—one indicating its power in peptonizing milk, the other its power in digesting starch.　A limit of fat is adopted as one of the standards: Two grammes of pancreatin should not yield to ether more than 0.6 Gm. of fat.

ACTION AND USES.—Used as a digestive agent, especially for "peptonizing" milk.　Dose: 10 gr. (0.6 Gm.).

RENNINUM.—Rennin, N.F.　Partially purified, milk-curdling enzyme from the calf's stomach, capable of coagulating not less than 12,500 times its weight of fresh cow's milk.　For assay see N.F

631. ADEPS LANÆ HYDROSUS.—LANOLIN

HYDROUS WOOL-FAT

The purified fat of the wool of sheep, **Ovis aries** Linné (Fam. Bovidæ), mixed with not more than 30 per cent. of water.　For Tests see U.S.P.

DESCRIPTION.—A yellowish-white unctuous mass.　Faint, peculiar odor. Insoluble in water, but miscible with twice its weight.　Melts at about 40°C. (104°F.).　**Adeps Lanæ,** U. S., is the above freed from water.

CONSTITUENTS.—Cholesterin, palmitin, olein, the first mentioned being largely represented.

USES.—As an inunction and vehicle for substances the medicinal action of which can be obtained by local application.　It is employed in several official ointments.

631 a. **HYDROCARBON FATS AND OILS.—(Petrolatum, etc.).**
　　DESCRIPTION, SOURCE, ETC.—As a most valuable addition to the list of ointment bases and oleaginous liquids there has been officially recognized: Petrolatum album (White Petrolatum); Petrolatum Liquidum (oil); Petrolatum Molle (soft Petrolatum); and Petrolatum Spissum (Hard Petrolatum). These are mixtures of the harder and softer members of the paraffin series of hydrocarbons, having different melting and congealing points, etc.　Hard paraffin consists chiefly of hydrocarbons ranging from $C_{20}H_{42}$ to $C_{30}H_{62}$; soft paraffin consists chiefly of $C_{15}H_{32}$ to $C_{20}H_{42}$; liquid consists chiefly of heptane, C_7H_{16}, and octane, C_8H_{18}.
　　USE.—As a vehicle for medicinal substances applied locally.　As such it is much less permeable through the skin than other fats.

632. **HYDRACEUM.**—A plaster mass of a blackish-brown color, occasionally used medicinally as a stimulant and antispasmodic. When warmed, it emits the odor of castor. It is an animal excretion found in Africa.

633. **CASTOREUM.**—Castor.

Source.—The preputial follicles of both sexes of *Cas'tor fiber* Linné. These follicles are not perceptible until the outer skin is removed, when they are seen to lie between the cloaca and pubic arch of the animal. This species of animals is commonly known as the beaver, and is found more or less throughout the Temperate and North Temperate Zones.

Description.—The dry, resinous, brownish contents of the fig-shaped sacs or follicles have a strong and peculiar odor, an acrid, nauseous taste, and are soluble in alcohol and ether. An aqueous decoction of castor is of a light yellowish-brown color which becomes turbid on cooling, and changes to a dark color when ferric chloride is added.

Varieties.—American or Canadian, and Russian or Siberian Castor. The Russian variety differs from the American in the size of the inclosing follicles; in the former the size varies from 2½ oz. to 8 oz. (75 to 240 Gm.) in weight, and in the latter from 1 to 4 oz. (30 to 120 Gm.). There is also a difference in the composition of the product from the different varieties, the American probably containing a larger percentage of resin.

Adulterations.—Earthy matters, as well as resin and blood, are sometimes used for this purpose, but not frequently. The product from diseased animals is also met with; this often contains as much as 50 per cent. of inert material and is of a brownish-gray color.

Constituents.—A bitter resinous substance 14 to 58 per cent., 1 to 2 per cent. of volatile oil containing carbolic acid, a small quantity of castorin (a colorless, odorless and tasteless, crystalline, non-saponifiable fat, soluble in ether and boiling alcohol), together with salicin, cholesterin, and about 3.5 per cent. ash. The resin is dark brown, slightly acid, soluble in alcohol but not in ether. The volatile oil contains the odoriferous principle and is generally colorless, having an acrid, bitter taste.

Action and Uses.—Castor enjoys some reputation as a stimulant, antispasmodic, and emmenagogue, and is employed in cases of hysteria, chorea, and epilepsy, associated with sexual disorders. On account of its disagreeable taste it is best administered in the form of a pill.

Dose.—5 to 10 gr. (0.3 to 0.6 Gm.) in the form of a pill; 1 to 4 fl. dr. (4 to 15 mils) of the tincture.

634. **CIVETTA.**—Civet.

Source.—The glandular pouch between the genitals, and anus of the male and female animals belonging to the two species Viver'ra zibe'tha Schreber, and V. **civetta** Schreber, the first of which is found in Southern Asia and the other in Africa.

Description.—The secretion, when fresh, is yellowish, becoming brown with age, soluble in hot absolute alcohol, partly soluble in ether, and insoluble in water; odor musk-like; taste acrid and nauseous.

Adulterations.—Butter or lard is not infrequently used as an adulterant of the commercial article.

Constituents.—Resinous and coloring matters are the chief components, together with volatile oil and fat.

Action and Uses.—The manufacture of perfumery is the principal use of civet, but it is also sometimes administered as a stimulant and antispasmodic in doses of 5 to 15 gr. (0.3 to 1 Gm.). As a perfume it is superior to musk, as the odors of various kinds of flowers can be successfully imitated with it.

635. ANTITOXIC SERUMS

In recent years there has grown up almost a new system of therapeutics, known as "serum therapy." This system is based upon the theory that the various infectious diseases are in most instances caused by the poisonous toxins produced by the microörganisms. As an example of the diseases thus produced, diphtheria is perhaps the most striking, because it is the most common, and the success in its treatment by this system has been universally acknowledged as phenomenal. In addition to diphtheria, we have tetanus, septicemia, glanders, cholera, etc. These diseases are now treated by hypodermic injection of the well-known animal serums containing different percentages of antitoxins. These antitoxic serums are practically produced *in the animal*—the goat, cow, or horse, for instance. The animal is gradually rendered immune to the specific microörganisms by the injection of either attenuated cultures rendered comparatively harmless by various methods.

After the animal has been rendered immune by this treatment, blood is withdrawn from it with the strictest aseptic precautions; it is then allowed to stand until the blood serum separates as a distinct layer. This blood serum, when separated and hermetically sealed in glass bulbs, etc. (containing 5 mils and upward), constitutes the remedial agent. It is needless to state that from the beginning until the very end of the process the greatest aseptic precautions are observed. The liquid before being placed in its containers is impregnated with minute quantities of such preservative material as tricresol, carbolic acid, etc. As these serums are tested only in the physiological laboratory, the pharmacist must of necessity hold the manufacturer responsible for the value of his product.

SEROTHERAPY

GENERAL INTRODUCTION

By Serotherapy is meant the injection of a serum or other like substance into the tissues of an individual for immunization or diagnostic purposes. This immunity is called *Active* if the material inoculated induces the tissues to form the immune bodies, and is called *Passive* when the immune bodies themselves (obtained from an immunized animal) are injected.

NATURAL AND ACQUIRED IMMUNITY

Immunity is the exemption to infection exhibited by an individual or species and is generally relative.

It is well known that many animals are not susceptible to numerous diseases which attack man. No one has ever succeeded in infecting the lower animals with scarlet fever, nor man with chicken cholera. Even among men, there is considerable variance of susceptibility toward certain diseases. For instance, the negro is less susceptible to yellow fever than the white man. These examples indicate that there is a resistance normally existing in the body and not brought about by having a disease. This phenomenon exhibited is termed "natural immunity."

Practically all intelligent people are aware that an attack of certain infectious diseases renders the patient relatively immune to future attacks of the same disease; also, that the injection of certain serums confers upon a person a relative, specific immunity. These, taken together, comprise "acquired immunity." Serotherapy, of course, deals entirely with acquired immunity.

BRIEF HISTORY OF SEROTHERAPY

The Chinese were the first people known to use serotherapeutic inoculation. Several centuries before serum therapy was practised in Europe the Chinese immunized themselves to small-pox by the very simple expedient of snuffing scales obtained from small-pox sores up their nose. This produced a less severe type of the disease (Variola inoculata) than that brought about by accidental infection. Small-pox was so prevalent during those ancient times that comparatively few persons escaped contracting the disease at some stage of their life. Thus, it came about that many of the people chose a convenient time from their business or other duties and inoculated themselves with the virus.

This variola inoculata, however, had the disadvantage of being contagious and also generally caused a permanent closure of the nasal passages.

Lady Mary Wortley Montague found small-pox vaccination being used in Turkey in 1718 and shortly thereafter introduced the practice into Europe. This method was used more or less until the discovery by Jenner in 1798 that the virus of cowpox was just as efficacious for vaccination as the virus of virulent small-pox. Jenner's method was to vaccinate with points prepared from vesicles of human cowpox, but the not infrequent transmission of syphilis by this means brought about the use of a vaccine prepared from the vesicles of cowpox in calves.

It was not until in 1880 that intelligent progress began in serotherapy. From 1860 the master mind of Louis Pasteur dominated the realm of microbiology and in 1880 he announced the discovery of a method of vaccination against fowl cholera, and in 1881 he published his method of vaccination against anthrax. On a farm at Pouilly le Fort sixty sheep were placed at Pasteur's disposal. Ten of these received no treatment, and twenty-five were vaccinated. Some days afterward the latter were inoculated with virulent anthrax and also twenty-five which had received no vaccine. The twenty-five non-vaccinated sheep died, the twenty-five vaccinated ones remained healthy and in the same condition as the ten control animals. This convincing experiment was followed by others.

In 1885, as the result of much animal experimentation, Pasteur related to the Academy of Science his discovery of a method of vaccination against rabies, or hydrophobia. An institute for the preparation of vaccines was built by public subscription and named the Pasteur Institute and since that date many similar establishments have been founded in different parts of the world.

As the result of the pioneer work on toxins by Roux and Yersin, Behring, in 1890, discovered the antitoxin for diphtheria; however, this serum did not come into general use as a curative until five years later. The subsequent researches on the constitutions of toxins and antitoxins by Ehrlich, Metchnikoff and others have been productive of a better understanding of the problems of immunity.

I. PROPHYLACTIC SEROTHERAPY

A. ACTIVE IMMUNITY

It has been suggested that a general principle prevails to the effect that any infection in which an attack confers strong and lasting immunity must be bacterial rather than protozoan in its etiology. On the contrary, however, it cannot be said that all bacterial diseases confer strong immunity but there are a large number of examples known. Indeed, different infections may bring about five different results in a human; for instance:

(1) It may kill the patient, or if the patient recovers (2) the disease may not only confer immunity but may actually create a predisposition for recurrence as in erysipelas and influenza; (3) where one attack confers no evident protection against a second, as gonorrhea; (4) the immunity conferred is of short duration, as cholera; (5) a lasting protection against subsequent attacks is possessed by those recovered from plague and typhoid fever.

Those serotherapeutic products classified under active immunity are called "antigens." The word "antigen" means "former of antibodies" and when an antigen is administered parenterally (that is, by the subcu-taneous, intramuscular, intraneural, intraspinal, or intravenous route, in order to reach tissues not directly accessible) its presence in the body will stimulate the production by the tissues of true antibodies. It usually requires several days to build up an immunity, but when obtained is quite permanent and may even last for years. To the class of antigens belong the vaccines, viruses, tuberculins, toxins and bacterins.

Artificial active immunity is brought about in four different ways as follows:

1. BY INOCULATION OF VIRULENT ORGANISMS OR TOXINS.—This method is principally used in experimental work. Formerly vaccination against small-pox was made with virus obtained directly from the diseased, but this practice has been long discontinued. Most of the serums used in passive immunity are obtained by this method and will be fully dis-cussed under "Prophylactic Passive Immunity." Kitt's method of vaccination against symptomatic anthrax and in immunization of rattle-snake venom was accomplished by inoculation of the bacteria or toxin directly into the blood stream.

2. BY INJECTION OF ATTENUATED ORGANISMS, VIRUS OR TOXIN.—There are several methods of attenuation used, and no single method is suitable for all organisms. Animal passage is an important one and is surprisingly variable in its results. For instance, passing the bacillus of swine erysipelas through the rabbit several times increases its virulence for the rabbit but decreases it for the swine, while passing the organism through the dove increases its virulence for swine (Pasteur). Attenua-tion may also be accomplished by exposure to air and light, as chicken cholera (Pasteur); by cultivation of the organism at high temperatures, as anthrax (Pasteur); by chemical agents, as diphtheria toxin (Behring and Roux); by desiccation, as rabies (Pasteur).

Vaccine virus and antirabic virus are the two most notably successful examples of this class.

VIRUS VACCINICUM, U.S.P.—The pustules of vaccina or cowpox removed, under aseptic conditions, from vaccinated animals of the bovine species. The vaccine pulp, after being thoroughly rubbed up in a mortar or passed through a grinder, then strained to remove coarse particles, is made into a smooth emulsion with a glycerin solution. It is usually marketed in capillary tubes or as glycerinated points.

Vaccination against small-pox is the most thoroughly tried and probably the most efficient method available for immunization against an infectious disease. There is no better substantiated fact in medicine than that vaccination protects against variola. In return for an inconvenience of a few days, protection for years is given. The length of the protection varies with individuals, but attacks of the disease long after full immunity has passed are much milder than in the unvaccinated.

Infants should be vaccinated during the first months of life (except in the presence of eczema or marasmus), at school age and every seventh year thereafter or oftener if small-pox is present in the community. Excessively sore arms, formerly due to high bacterial content of the virus at times, are now due usually to vaccination by scarification with the formation of a large crust, to the use of vaccination shields or to lack of appropriate after-care.

The skin of the upper arm at the insertion of the deltoid or other suitable site should be cleansed with soap and water or with ether. The glycerinated virus should be placed on the skin, which may be drawn taut by grasping the arm beneath, and with a needle or other sterile instrument one incision made through it an inch long parallel with the arm and penetrating to the papillary layer, but preferably not drawing blood. The lymph may be rubbed into the incision with a smooth sterile instrument. No other incision or cross-scratching should be made unless there will be no opportunity for revaccination; in such cases similar single incisions may be made on either side of the first and at least an inch away. The vaccination site should dry for a half hour before being covered with clothing. The arm must be kept clean, dressed aseptically when the vesicle appears, and, when pustulation ensues, treated antiseptically. No dressing should be allowed to remain on the arm more than two days. Vaccination is a surgical procedure and is worthy not only of the painstaking asepsis of a skilled physician at the operation, but of this intelligent after-care as well.

ANTIRABIC VACCINE.—Antirabic vaccine or antirabic virus is the virus of rabies rendered practically non-virulent for man by passage through a long series of rabbits and treated in various ways to decrease the infectivity still further. The method commonly in use in the United States is that of Pasteur as modified by the Hygienic Laboratory, Public Health Service: The spinal cords of the infected rabbits are dried over caustic potash at constant temperature for one to eight days, then cut into 0.5 cm. pieces and preserved in glycerin. For use, one of these pieces is emulsified in physiological saline solution and injected into the subcutaneous tissue of the anterior abdominal wall. Injections are continued daily for twenty-one days. Other methods of treating the virus before inoculation are dilution (Högyes), emulsification with 1 per cent. phenol (Fermi), and drying at very low temperature (Harris).

By treatment with antirabic vaccine after the bite of a rabid animal, immunity is usually established before the incubation period of the disease is concluded and rabies is thus prevented. Each dose should be emulsified as few hours as possible before the injection and it is therefore advisable that treatment should be carried out at no great distance from the place of propagation of the virus.

3. By Injection of Killed Organisms.—This class is composed almost entirely of bacterins. It is the safest method of immunization against cholera, typhoid and plague. In the Pasteur treatment of hydrophobia the first injection of the dried spinal cord probably contains the killed virus.

A bacterin is a suspension of killed bacteria in isotonic saline solution and nearly always has a preservative added such as 0.9 per cent. tri-cresol or 0.5 per cent. phenol.

An "autogenous" vaccine is one made from the bacterial strain isolated from the patient, while a "stock" vaccine is one made from laboratory cultures. Whenever practicable autogenous should be made, in some cases being absolutely necessary.

Cholera Vaccine.—Prepared from killed cholera vibrios. Cholera vaccine has been used in India as a prophylactic with in part favorable results.

Gonococcus Vaccine.—Made from the Micrococcus gonorrhœæ. Clinical experience has presented no clear evidence of the value of gonococcal vaccine in affections of the mucous surfaces. As a prophylactic against metastatic complications it may have some value. Many observers believe that these vaccines are useful in arthritis. The value of vaccines in gonococcal pelvic lesions is not clearly determined. They are of little if any value in gonococcal sepsis or in gonococcemia.

Meningococcus Vaccine.—Made from the Diplococcus intracellularis meningitidis of Weichselbaum. It has been used in a limited number of cases with apparent success for the prevention of epidemic cerebrospinal meningitis.

Pertussis Bacillus Vaccine.—Made from the bacillus of whooping-cough, isolated by Bordet and Gengou and proved the causative agent of the disease by Mallory. The evidence indicates that it is of value both for prevention and treatment, although eminent authorities state that the results have not been very satisfactory.

Plague Bacillus Vaccine.—Made from the Bacillus pestis. Vaccine has been used for the prevention of plague with results justifying its use but owing to the acute nature of the disease time is not allowed for the development of active immunity after actual infection. No practical application therefore has been made of vaccine treatment in plague.

Typhoid Vaccine.—Made from Bacillus typhosus. In some cases Bacillus paratyphosus A and Bacillus paratyphosus B are also used. Typhoid vaccine is of recognized utility in the prevention of typhoid fever. The immunity produced persists in the majority of cases from two to four years or longer. The vaccine is also of service in treatment of typhoid carriers. In such cases an autogenous vaccine is to be preferred. The same is true in the bacterial complications and sequelæ of typhoid fever, especially those that appear during convalescence or are prolonged into that stage.

The use of vaccine in the treatment of typhoid fever has given very inconclusive results. No positive evidence of harm resulting from its use has been recorded. Many clinicians, however, believe that the giving of an additional amount of the toxin of the disease may turn the balance against recovery in states characterized by marked toxemia.

4. By Injection of Bacterial and Pollen Constituents.—Buchner's plasmin, obtained by submitting microörganisms to high pressure, is only of experimental value. The products of bacterial autolysis also come under this heading. These substances, when injected, do not cause the formation of antitoxin as true toxins do, but bactericidal amboceptors and agglutinins are formed.

Pollen Extract-pollen Vaccine.—A solution of pollen protein.

Pollen extract is employed for the relief or prophylaxis of a common type of hay fever or pollinosis. Pollen extract prepared from the pollen of one plant (for instance, ragweed) is not primarily intended for use in cases due to pollen from other plants, as grasses and goldenrod, though persons subject to autumn catarrh frequently react to the pollen of more than one species. The patient's susceptibility may be tested by rubbing a small quantity of the pollen vaccine into a scratch of the skin; if the patient is sensitive to that particular pollen, an urticarial wheal results. To avoid systematic disturbance, it is recommended that no therapeutic injections be made until the reaction from this cutaneous test has subsided completely. Treatment with pollen extract has seemed to give a varying degree of relief in a number of cases. In some cases the psychic element seems to play a part, and in such instances it is difficult to determine to what degree the good results are due to suggestion. The immunity from symptoms conferred by treatment is apparently not permanent, and in most cases does not last longer than a year.

It is regarded as important that the individual dosage should be determined by testing each patient's susceptibility, as sensitiveness varies greatly and an overdose may cause disagreeable and alarming symptoms or possibly death. A method used for such test is to make a series of scratches on the patient's skin (it is important that these should be made at some distance from the scratches of the first test) and to apply to these scratches 25 per cent., 10 per cent., 1 per cent., or even weaker dilutions of the vaccine. From 5 to 10 drops of the dilution which fails to produce a definite skin reaction may be injected subcutaneously as the first dose. Injections, increasing by a few drops at first, and later by the use of a stronger dilution, may then be given at intervals of a few days or a week.

HAY FEVER VACCINE.—A non-proprietary brand complying with the standards for pollen extract.

B. PASSIVE IMMUNIZATION

The agents in passive immunity are called "antibodies" and they directly antagonize the invading bacteria or toxin. These antibodies are formed in the bodies of the larger domestic animals by active immunization and when the antibacterial and antitoxic substances have reached a high concentration the animal is bled and the serum saved. This serum is injected into other animals and these latter *immediately* become endowed with an increased resistance to the homologous infective agent. Unfortunately, this protection is of short duration as contrasted to the more permanent protection of active immunity.

1. BY INJECTION OF ANTITOXINS.—Toxins may be secreted by plants, bacteria and animals, but the typical ones are of the tetanus and diphtheria organisms.

SERUM ANTIDIPHTHERICUM, U.S.P.—A yellowish or yellowish-brown transparent or slightly turbid liquid, with sometimes a slight granular sediment; odorless, or having an odor due to the presence of an antiseptic used as a preservative. Antidiphtheric serum gradually loses its potency, the loss in one year varying between 10 and 30 per cent. The serum must come from healthy animals, must be sterile, must be free from toxins or other bacterial products, and must not contain an excessive amount of preservative (not more than 0.5 per cent. of phenol, nor 0.4 per cent. of cresol when either of these is used) and the total solids must not exceed 20 per cent. Serum of a lower potency than 250 units per mil must not be sold or dispensed. Only such Sera may be sold as have been prepared and propagated in establishments licensed by the Secretary of the Treasury of the United States. The United States law requires that each container of Serum sold by licensed establishments shall bear upon the label, in addition to the name of the Serum, the name, address and license number of the manufacturer and the date beyond which the product cannot be expected to yield its specific results. The label must also contain the laboratory number of the Serum, the name and the percentage units claimed for the contents of the container. The standard strength expressed in units of antitoxic powder shall be that established by the United States Public Health Service.

This native antitoxin serum is little used in the United States, where the methods of concentration have reached their highest development. The potency of all diphtheria antitoxin sold in interstate commerce must be stated in terms of the United States standard unit, which has been established by the United States Public Health Service and is kept at Washington under special conditions to prevent deterioration. This unit is almost exactly the same as the German unit or normal serum of Ehrlich. It is approximately the amount of antitoxin which will neutralize 100 times the minimum lethal dose of toxin for a standard 250 Gm. guinea-pig.

Diphtheria antitoxin has the power of combining with, and hence neutralizing, the toxin formed by the growth of the diphtheria bacillus. If present in the blood-stream in large amounts it not only will combine with the free toxin, but may also, by its superior attraction, dissociate some of the toxin from the tissue cells with which it has already entered into combination.

The prime object in the administration of the antitoxin, however, should be to give it early enough to nullify the toxin as fast as the latter is produced. The characteristic and usual effects of diphtheria antitoxins appear within twenty-four hours and consist of an amelioration of the general symptoms and halting of the growth of the membrane, the edges of which begin to loosen themselves from the mucous membrane. If enough antitoxin has been given in the first dose, and if this dose was given early enough, this action goes on to complete recovery. If as much as seventy-two hours elapse from the onset of the disease to the injection of the antitoxin, the remedy is not so efficient. Paralyses and cardiac complications cannot be prevented by late administration even though the primary symptoms be checked. The remedy is therefore most useful as a prophylactic. Some harm has been done as soon as clinical diphtheria is recognizable, and the antitoxin should be immediately given to prevent more injury to the organism. At the same time, no case is hopeless. The higher incidence of paralysis since antitoxin was introduced is probably due to the recovery of the severely affected patients, who formerly would have died.

The rashes, edema, and joint symptoms which sometimes follow the injection of the antitoxin (or of any foreign serum), though very distressing, have no serious effects and are to be disregarded in the presence of actual diphtheria. Cases of immediate anaphylactic shock are exceedingly rare and the possibility of their occurrence is not to be compared with the danger in withholding antitoxin. If, however, the patient has been subject to attacks of asthma when exposed to horse emanations or if hypersusceptibility to horse-serum is, for any other reason, suspected, a preliminary dose of 0.2 mil of the antitoxic serum may be given, followed in an hour by the same dose if no untoward symptoms have occurred, and doubled in successive hours until the necessary amount has been given. This protects against acute anaphylactic collapse, but not against the later manifestations of serum disease.

For the average case of diphtheria the curative dose is 10,000 units, administered subcutaneously, as soon as possible after the discovery of the disease. Every hour's delay is dangerous. For cases seen more than twenty-four hours after onset, for cases with membrane extending beyond the tonsils, or for cases with severe symptoms or seen to be of virulent, rapidly spreading type, the dose should at least be doubled.

The site of injection is usually the loose tissues of the back. In severe cases it is preferable to use the intravenous route in addition to the subcutaneous, because absorption from the subcutaneous tissue is slow. The intravenous injection should not be given more rapidly than 1 mil per minute. The serum should be free from sediment and warmed to body temperature. On account of the saving of time in this method of administration the dose need only be one-half or one-fourth as great as that necessary for subcutaneous administration in the given case. Sufficient antitoxin should be given at the initial treatment to arrest and cure the disease, but if the amount necessary has been underestimated the dose may be repeated once in six hours or increased. The prophylactic dose is 1000 units; for very young infants, 500 units. This should be repeated every two weeks if exposure continues.

The preparation consists of a solution of certain antitoxin proteins in physiologic sodium chloride solution. The antitoxin proteins are separated from the native antitetanic serum by precipitation with ammonium sulphate. The precipitate is then dissolved in saturated sodium chloride solution and the salts are removed by dialysis. After dialysis is complete, sufficient sodium chloride is added to make an 0.8 per cent. solution. Concentrated tetanus antitoxin is a transparent or slightly opalescent liquid, sometimes with a slight granular or ropy sediment; it may be more or less viscous. Its actions and uses are the same as those of tetanus antitoxin, unconcentrated.

SERUM ANTITETANICUM SICCUM, U.S.P.—This dried serum is either in the form of orange or yellowish flakes or small lumps, or a yellowish-white powder without odor. The serum is readily soluble in nine parts of distilled water, the solution being opalescent and slightly viscous. For use the serum must be dissolved in recently boiled and cooled distilled water, preferably in the original container and under the most rigid aseptic conditions. Dried antitetanic serum if kept as directed does not lose in potency, as does the liquid serum. It is sometimes used as a dusting powder or for local application to infected wounds. It must comply with the requirements for control and labeling under Serum Antitetanicum and the standard of strength, expressed in units of antitoxic power, shall be that established by the United States Health Service.

This serum is prepared from tetanus antitoxin by a method similar to that used in preparing diphtheria antitoxin, dried.

Antitoxins keep their potency longer in the dry than in the liquid state. A powdered form of this preparation may be used in wounds as an additional precaution, but this dressing should not take the place of subcutaneous injections of the liquid preparation in wounds likely to be infected with tetanus. Its actions and uses are the same as those of tetanus antitoxin, unconcentrated.

POLLANTIN, FALL.—Dunbar's Serum-Antitoxic serum from horses treated with pollen toxin derived from ragweed.

Pollantin, fall, has no pharmacologic action except the neutralization of the pollen toxin. The serum is not intended for use hypodermically. It is employed for the relief of hay fever and it seems to be effective in a proportion of cases. It may be used as a prophylactic.

One drop should be instilled by means of a pipette into the outer angle of each eye and one or two drops into one nostril, the other being closed, every morning before rising. If the first application causes sneezing or reddening of the mucous membrane of the eye, the directions are to repeat the application, even for the fourth time, if necessary.

POLLANTIN POWDER, FALL.—A powder obtained by evaporating *in vacuo* pollantin serum derived from ragweed toxin at about 45°C., and mixing it with sterilized sugar of milk.

The action and use is the same as those of the liquid.

The powder is applied to the eyes by dusting on the conjunctiva and to the nose by snuffing into one nostril, the other being closed, a piece as large as a lentil.

SERUM ANTIDIPHTHERICUM PURIFICATUM, U.S.P.—A transparent or slightly opalescent liquid with sometimes a slight granular or ropy deposit, odorless, or having an odor due to the presence of the antiseptic used as a preservative. The liquid is sometimes more or less viscous. The serum must come from healthy animals, must be sterile, must be free from toxins and other bacterial products, and must not contain an excessive amount of preservative (not more than 0.5 per cent. of phenol or 0.4 per cent. of cresol, when either of these are used), and the total solid must not exceed 20 per cent. Serum of a lower potency than 250 units per mil must not be sold or dispensed. Purified antidiphtheric serum must comply with the requirements for loss of potency, control labeling and standard for potency under Serum Antidiphthericum.

This serum is prepared from unconcentrated diphtheria antitoxin by the separation of the pseudoglobulins with which the antitoxin molecule seems to be particularly associated. The elimination of most of the other elements of the horse-serum reduces the bulk for a given unit value and probably diminishes the frequency and intensity of the urticarial rashes.

This preparation consists of a solution of certain antitoxin proteins in physiologic sodium chloride solution. These proteins are obtained by adding ammonium sulphate to the original serum, and treating the precipitate with saturated sodium chloride solution; the antitoxins pass into solution and much inert material remains in the undissolved portion of the precipitate as well as in the filtrate from the original precipitation. The saturated sodium chloride solution of the antitoxic proteins is submitted to dialysis to remove the salts and then sufficient sodium chloride is added to make an o.8 per cent. solution. It is a transparent or slightly opalescent liquid, sometimes with a slight granular or ropy sediment and may be more or less viscous. Its action and uses are the same as those of diphtheria antitoxin, unconcentrated.

SERUM ANTIDIPHTHERICUM SICCUM, U.S.P.—The dried serum occurs either in the form of orange or yellowish flakes or small lumps, or as a yellowish-white powder, without odor. The serum is readily soluble in nine parts of distilled water, the solution being opalescent and slightly viscous. For use the serum must be dissolved in recently boiled and cooled distilled water, preferably in the original container and under the most rigid aseptic conditions. Dried Antidiphtheric serum if kept as directed does not lose in potency, as does the liquid serum.

The serum is prepared from diphtheria antitoxin, concentrated or unconcentrated, by aseptic evaporation. The dried antitoxin preserves its potency much better than that in a liquid state, but is more inconvenient to use on account of the difficulty of getting it back into solution. It should be dissolved in about ten times its volume of water to restore the original bulk. Its uses are the same as those of diphtheria antitoxin, unconcentrated.

SERUM ANTITETANICUM, U.S.P.—A yellowish or yellowish-brown transparent or slightly turbid liquid with sometimes a slight granular deposit; odorless, or having an odor due to the presence of the antiseptic used as a preservative. It gradually loses its potency, the loss being greater at higher than at lower temperatures. The serum must come from healthy animals, must be sterile, must be free from toxins or other bacterial products, and must not contain an excessive amount of preservative (not more than o.5 per cent. of phenol nor o.4 per cent. of cresol, when either of these is used), and the total solids must not exceed 20 per cent. Only such sera may be sold or dispensed as have been prepared and propagated in establishments licensed by the Secretary of the Treasury of the United States. The United States law requires that each container of serum sold or dispensed by licensed establishments shall bear upon the label, in addition to the name of the serum, the name, address and license number of the manufacturer and the date beyond which the contents cannot be expected to yield its specific results. The label must also contain the laboratory number of the serum, the name and the percentage by volume of the antiseptic (if any) used and the total number of antitoxic units claimed for the contents of the container. The standard of strength, expressed in units of antitoxic powder shall be that established by the United States Public Health Service.

The serum is a fluid separated from the coagulated blood of a horse, actively immunized against tetanus toxin. Its physical properties are similar to those of diphtheria antitoxin. As in the case of the diphtheria antitoxin, most of the tetanus antitoxin used in the United States is refined (tetanus antitoxin, concentrated) by an analogous process. The United States standard unit for tetanus antitoxin is that established by the United States Public Health Service and distributed from the Hygienic Laboratory at Washington, D. C. This unit is defined as follows: "The immunity unit for measuring the strength of tetanus antitoxin shall be ten times the least quantity of an antitetanic serum (tetanus antitoxin) necessary to save the life of a 350-gram guinea-pig for ninety-six hours against the official test dose of a standard toxin furnished by the Hygienic Laboratory of the Public Health Service."

This method is also official in Belgium and Brazil (this implies that these countries use the United States Public Health Service standard toxin) and all tetanus antitoxin sold in interstate commerce in the United States must be standardized by it.

While diphtheria antitoxin is primarily curative, tetanus antitoxin is chiefly used as a prophylactic. All deep or lacerated wounds, especially those exposed to dust or dirt, and all Fourth of July and gunshot wounds, are indications for a prophylactic subcutaneous injection as soon after the injury as practicable, and a similar injection in seven days; and if suppuration of the wound ensues, the treatment should be further repeated. As the object is to keep the blood-stream saturated, these prophylactic doses may be given subcutaneously in any convenient region. The antitoxin prophylaxis is no substitute for thorough surgical opening and cleansing of the wound, but an additional precaution. In treatment of the developed disease, a procedure which is much less certain of success, the antitoxin should reach the nerve-centers as quickly as possible and in as high a concentration as possible; therefore, intraspinal injection is indicated. The blood-stream at the same time should be kept strongly antitoxic by intravenous and subcutaneous injection. For prophylactic purposes, 1500 units subcutaneously is used as a dose. In tetanus 5000 units intraspinally, which may be repeated in twenty-four hours; and 10,000 units intravenously, to be repeated subcutaneously as indicated. As a prophylactic, an injection of the serum should be given at once and the wound, after thorough antiseptic cleansing, should be dusted with dried antitoxin.

SERUM ANTITETANICUM PURIFICATUM, U.S.P.—A transparent or slightly opalescent liquid, with sometimes a slight granular or ropy deposit; odorless, or having an odor due to the presence of the antiseptic used as a preservative. The liquid is sometimes more or less viscous. The serum must come from healthy animals, must be sterile, must be free from toxins or other bacterial products, and must not contain an excessive amount of preservative (not more than 0.5 per cent. of phenol nor 0.4 per cent. of cresol, when either is used), and the total solids must not exceed 20 per cent. Purified antitetanic serum must comply with the requirements for loss of potency, control, labeling and standard for potency under Serum Antitetanicum.

2. BY INJECTION OF ANTIBACTERIAL SERUMS.—Antibacterial, bactericidal and bacteriolytic are three terms which are used in a loose, interchangeable way, although they are not strictly synonymous. A bactericidal serum is one which is able to kill bacteria; if at the same time it dissolves the organism it is bacteriolytic. In either case the serum is antibacterial. Bacteriolysis is the process in which bacteria are killed by serums and may be either with or without solution; a bacteriolysis would, then, be the substance in the serum which accomplishes the action.

Opsonins are those substances existing in the serum which are capable of rendering the bacteria susceptible to phagocytosis. Phagocytosis is the process whereby the leucocytes take up bacteria, usually destroying them. Those substances in opsonins which are thermostabile are called bacteriotropins or amboceptor and the thermo-labile part of opsonins is called complement.

Antibacterial serums are of two kinds. One, where organisms such as typhoid, paratyphoid, colon and dysentery bacilli and the vibrio of cholera are used as antigen, contains bacteriolysins, but the endotoxins released during bacteriolysis are not neutralized. The other kind, with streptococcus, staphylococcus and pneumococcus as antigen, contains endotoxins and causes the formation of neither antitoxins nor bactericidal serums, but does seem to stimulate phagocytosis.

It is believed by some authorities that the antibacterial serums are less efficacious than antitoxins because they are too specific; that is, they contain only one strain of an organism. To correct this seeming defect "polyvalent" serums are made. They are manufactured using several strains of an organism as antigen.

These serums are used principally in curative serotherapy and will be discussed under that heading.

C. MIXED ACTIVE AND PASSIVE IMMUNIZATION

This consists in the simultaneous injection of the organism (either killed or living) with its homologous immune serum.

The immune serum gives the patient immediate, temporary immunity; meanwhile, the organisms injected cause the tissues to build up a comparatively permanent, active resistance.

Serobacterins are bacterial vaccines that have been treated with an immune serum for a time, then the serum removed, the bacteria washed and suspended in isotomic saline solution. This "sensitization" of the bacteria is undoubtedly due to the union of the amboceptor of the immune serum with the bacteria and yields quick immunity, as it does away with the long period which is required for the formation of the amboceptor in the patient.

This method has been used with marked success against rinderpest and also swine erysipelas and plague, and experimentally in typhoid, cholera and plague. It has also been suggested as a general prophylactic treatment for all school children.

Tuberculosis Serum Vaccine, S.B.E.—This is a sensitized bacterial vaccine, made like tuberculin B.E., except that the pulverized bacilli are treated with a fresh antituberculous serum, which is afterward removed by washing and centrifugation.

Pneumococcus Vaccine.—Made from the Diplococcus pneumoniæ. The value of vaccination in the prophylaxis or treatment of pneumonia is very doubtful. The possibility of undermining the resistance of the patient must be considered. There is evidence that the use of vaccine alone or in conjunction with antipneumococcus serum is of advantage in the treatment of ulcus corneæ repens, an affection which is caused by the pneumococcus.

II. CURATIVE SEROTHERAPY

A. ACTIVE IMMUNITY

In certain chronic infections only the tissues immediately concerned respond by the formation of antibodies, and thus it becomes necessary to inject the dead homologous organism in order in the words of Wright, "To exploit in the interest of the infected tissues, the unexercised immunizing capacities of the uninfected tissues."

1. Injection of Killed Organisms in Small Doses.—New Tuberculin, T.R.—Tuberkelbacillin Rest, Koch—Tuberculin Residue—Tuberculin Rückstand.—This is made from living dried tubercle bacilli by thorough grinding, suspension in water and centrifuging. The supernatant fluid, containing extractives, is discarded and the sediment reground, suspended in a little water and recentrifuged. The fluid is kept this time, while the sediment is reground, suspended and centrifuged as before. This is repeated until practically no sediment remains, when all the fluid portions which have been laid aside are combined and diluted, with 20 per cent. glycerin solution to make]the final (standard) product contain the residue of 10 mg. of dried tubercle bacilli in each mil of fluid. T.R. is an uncolored slightly opalescent liquid. It is used occasionally in the treatment of tuberculosis.

New Tuberculin, B.E.—Bazillenemulsion Koch—Bacilli Emulsion.—Bacilli emulsion is practically a bacterial vaccine. It is made by suspending one part pulverized tubercle bacilli in 100 parts distilled water and 100 parts glycerin. This mixture stands one day and is then decanted from the grosser particles which have settled. One mil thus corresponds to 5 mg. of tubercle bacilli.

Tuberculin Denys, B.F.—Bouillon Filtre—Bouillon Filtrate.—This is prepared like old tuberculin without the prolonged heating and concentration; that is, it is simply a glycerin-broth culture of the tubercle bacillus, passed through a porcelain filter; it contains all the soluble products of the growth of the tubercle bacillus.

Dixon's Tubercle Bacilli Extract.—An extract of tubercle bacilli dissolved in normal saline solution.

Acne Bacillus Vaccine.—Prepared from the acne bacillus of Unna and Sabouraud. The acne bacillus is not found in all cases of acne, but in those cases in which the bacillus is found it seems to be the active pathogenic agent and the use of acne vaccine may give good results. In other cases the staphylococcus is responsible for the inflammation and the corresponding staphylococcus vaccine should be tried. If both organisms are present a mixture of the two vaccines may be indicated, but the use of a ready-prepared mixed vaccine is not rational as a routine treatment.

Colon Bacillus Vaccine.—Made from the Bacillus coli communis. The colon bacillus in a special strain is the cause of many cases of cystitis and pyelonephritis. The use of an autogenous vaccine is often highly successful in the treatment of these affections. Stock vaccines have not produced good results. If a stock vaccine must be used, it should be polyvalent, so that the special strain needed may be included if possible.

Diphtheria Bacillus Vaccine.—Made from Bacillus diphtheriæ. This vaccine has proved useful in the treatment of diphtheria bacilli carriers as a means of eliminating the bacilli which were resistant to other agents.

Friedlaender Bacillus Vaccine.—Made from the Bacillus pneumoniæ, which is found in the nasal secretion or sputum in some nasal or respiratory disorders.

Micrococcus Neoformans Vaccine.—Made from the Micrococcus neoformans, which was isolated from some malignant tumors by Doyen in 1904.

Pyocyaneus Bacillus Vaccine.—Made from the bacillus pyocyaneus.

Staphylococcus Vaccines.—Made from the staphylococcus pyogenes aureus, from staphylococcus pyogenes albus, or from the staphylococcus pyogenes citreus or from all three. It is useful in carbunculosis, furunculosis, sycosis, and certain cases of acne. An autogenous vaccine is preferable, but if this cannot be made, a stock vaccine can be used with good prospect of success. The forms of acne most likely to respond are characterized by deep-seated pustules, with considerable induration, situated on the face, chest and back. When the lesions are superficial and indolent, the acne vaccine gives good results, and when there is a mixture of active and indolent lesions, a mixture of the two vaccines is indicated.

Streptococcus Vaccine.—Made from different strains of streptococcus pyogens isolated from phlegmon, from the throat, from scarlet fever, and from erysipelas, etc. Streptococci are known to be the cause of various septic conditions and to complicate scarlet fever, and other contagious diseases. In cases of localized sepsis, a vaccine made from the organisms causing the septic condition in the particular cases is frequently useful. For this purpose an autogenous streptococcus vaccine may be useful in abscess, the septic complications of scarlet fever, such as otitis, etc. The use of vaccines in cases of chronic deforming arthritis has met with some success, but it is a mistake to rely largely on them. Stock vaccines, being less directly related to the cause of the disease, afford less prospect of success than autogenous.

Streptococcus vaccines have been suggested for the prevention of scarlet fever and for the treatment of scarlet fever, puerperal fever, acute rheumatism, ulcerative endocarditis, etc., but clinical experience affords no sufficient evidence of their value in these conditions. There is reason to believe that in conditions of general sepsis large doses of vaccines may be directly harmful.

ERYSIPELAS AND PRODIGIOSUS TOXINS (COLEY).—This preparation is practically a mixed bacterial vaccine made from strains of Streptococcus pyogenes isolated from cases of erysipelas and from bacillus prodigiosus. Its use has been advised in cases of inoperable sarcoma. This remedy is said to have produced cures in 10 per cent. of the total number of cases treated. It is worthy of trial in cases in which radium or the Roentgen ray is unsuccessful. It is given by hypodermic injection partly into the tumor or its near neighborhood and partly at a distance to secure the benefit of both local and systemic effect. A reaction consisting of chill and rise of temperature is expected to follow the injections until tolerance becomes established. Dose, 0.05 to 0.5 mil (1–8 minims).

B. PASSIVE IMMUNIZATION

1. ANTITOXIC IMMUNITY.—Rickets and Dick state 8 factors as being of importance in antitoxic therapy:

1. The concentration of the antitoxin injected.
2. Its freedom from contamination and adventitious toxins.
3. The time of its administration.
4. The quantity injected.
5. The degree of affinity between toxin and antitoxin.
6. The degree of affinity between toxin and tissue cells.
7. Amount of toxin which may be bound without fatal issue of which the vital importance of the organs involved and their recuperative powers are factors.
8. The location of the toxin in the body, *i.e.*, its accessibility for the antitoxin.

For the neutralization of the circulating toxin, a simple equivalent of antitoxin is required but where some of the toxin has combined with the tissue cells a great excess of antitoxin is required to wrest it away.

NORMAL HORSE-SERUM.—The serum of normal horse-blood obtained in a sterile manner and passed through a Berkefeld filter.

Though not a specific immunity product, normal horse-serum is classed commonly with the other serums. It is claimed to be used with success in hemorrhagic conditions to increase the coagulability of the blood.

The injection of the horse-serum is followed in certain individuals by more or less pronounced symptoms of anaphylactic shock. In its simplest form this appears as an urticarial eruption on the skin or an edematous swelling of the mucous membranes. In more severe cases there may be a fall of temperature, increased rapidity of pulse, quickened and difficult respiration, cyanosis, and occasionally convulsions. In rare cases the attack comes on with great suddenness and may terminate fatally. These cases of sudden death occur especially in asthmatics and in patients who have been sensitized by close association with horses. Ordinary serum disease manifests itself by milder but similar symptoms which appear from a few days to one or two weeks after the injection of the serum. In addition to the eruptions which are urticarial or scarlatiniform, joint pains and swelling of the joints sometimes occur. Atropine hypodermatically is a useful remedy for the severer manifestations of serum poisoning. Most cases of this poisoning have occurred after the use of antitoxic serums, but emphasis should be laid on the fact that these symptoms are not caused by antitoxin, but are due to hypersusceptibility to the proteins of horse-serum occasioned by a previous sensitization of the patient by contact with horses or by a previous injection of horse-serum.

The other antitoxic serums used in curatives serotherapy are exactly the same as previously described under "Prophylactic Antitoxic Serotherapy."

2. ANTIBACTERIAL IMMUNITY.—ANTI-ANTHRAX SERUM.—A serum prepared by immunizing horses against virulent anthrax bacilli.

Good results have generally been reported from the use of the specific serum in human anthrax. Bactericidal and bacteriotropic properties are practically absent and the virtue of the serum may possibly be ascribed to an inhibition of capsule-formation.

From 30 mils to 100 mils subcutaneously or intravenously is given. The serum should be used as early as possible and used freely, the dose being repeated several times a day in severe cases.

ANTIDYSENTERIC SERUM.—The blood-serum of horses immunized against the Shiga bacillus and other forms of the dysentery bacillus.

' A reduction in the mortality rate of bacillary dysentery from 30 to 50 per cent. through the use of some serums has been reported by some observers but not confirmed by all. It would seem that the best results may be ascribed to an antitoxic action in infections with the Shiga-Kruse type of bacillus. Infections with the Flexner, Strong, or Y strains, which are relatively poor in toxin production have not been so favorably affected, though some bactericidal action is claimed. The most favorable results are observed in the early stage of the disease. From 20 mils to 100 mils is given subcutaneously.

ANTIGONOCOCCUS SERUM.—A serum prepared by immunizing animal against the gonococcus.

Serum therapy in gonorrheal arthritis has been reported by some as successful and by others as unsuccessful. The most favorable results have been reported in the joint complications of gonorrhea. Little success has been achieved by the serum treatment of mucous membranes.

ANTIMENINGOCOCCUS SERUM.—A serum prepared by the immunization of horses with virulent cultures of the meningococcus of Weichselbaum. Greater success has attended the use of serum directed against the meningococcus than has been the case with any other antibacterial serum. There is no question as to the marked reduction in mortality. The serum must be introduced into the subdural space and its action is due probably in part to bacteriotropins, possibly to anti-endotoxins and other antibodies as well.

Average dose, 30 mils intraspinally as early as possible in the disease, and repeated as indicated. The serum should be introduced slowly by gravity after the removal of a corresponding amount of the cerebrospinal fluid. The administration of the serum should be controlled by blood-pressure readings, a drop of 10 mm. mercury during administration being the signal for withdrawal of needle.

The dried serum is sometimes used as a dusting-powder applied to the nasopharyngeal space or tonsils.

ANTIPNEUMOCOCCUS SERUM.—A serum obtained from horses immunized by injection of virulent pneumococci. The value of the serum in pneumonia as usually prepared and administered must remain on the present evidence as "not proved." It is possible, however, that early massive (from 50 to 100 mils) intravenous doses of a highly potent serum, prepared from the exact type of pneumococcus present in the case to be treated, may have a favorable influence on the general symptoms, though probably not on the local process in the lung. Investigations indicate that the pneumococcus in lobar pneumonia may be referred to one of four types in respect to its response to serum treatment. The serum used should be obtained from an animal immunized with pneumoccoci of the type corresponding to that present in the special case under treatment.

ANTISTAPHYLOCOCCUS SERUM.—A serum obtained from horses immunized by the injection of staphylococci. Well-controlled evidence of the therapeutic usefulness of antistaphylococcus serum is lacking.

ANTISTREPTOCOCCUS SERUM.—A serum obtained from horses immunized by the injection of killed or living cultures of streptococci. There is perhaps justification for the use of the serum in streptococcus infections and in scarlet fever, but it should be used early and in large doses; even then the result is doubtful. Bacteriotropins seem to be the principal antibodies present.

III. DIAGNOSTIC SEROTHERAPY

A. AGGLUTINATION REACTIONS

An agglutinogen is that constituent of bacteria which, when injected into an animal body, will stimulate the production of agglutinin. Agglutinin is that substance in the blood serum that will cause clumping of a suspension of homologous bacteria when the two are mixed in proper dilutions. The reaction is very specific.

This reaction is used for either of two purposes (1), For the identification of bacteria (agglutinogens) and, (2), For the diagnosis of certain diseases by the agglutinins formed in the blood-stream. Number 1 is seldom used as there are easier methods of identification, the difficulty being due to the fact that it requires several weeks to immunize an animal for the production of agglutinins; there are, moreover, a few agglutinating serums on the market but they are not used extensively because a considerable bacteriologic training is required for proper interpretation of results.

Under Number 2 we have the important Widal test for typhoid and paratyphoid fevers. This test is usually accomplished by pricking the patient's ear and collecting a drop of blood on a piece of paper. This is then taken up in physiologic salt solution and filtered. Then add enough of a filtered suspension of typhoid bacilli to make 50 drops. Make a control tube, using nothing but the bacterial suspension. Let both stand at 37°C. for two to four hours. If the serum contained agglutinins for bacillus typhosus there will be a fairly heavy white precipitate of agglutinated bacteria in the bottom of the serum tube but not in the control.

Boss Modification of the Widal Test: The agglutination is observed on a glass slide with the naked eye.

Bordens Modification of the Widal Test: In this test the serum of the blood is mixed with salt solution and then with a suspension of killed typhoid bacilli, so as to bring the dilution up to 1 to 50. The positive reaction is determined by noting that the clumps of bacteria sink to the bottom of the test-tube and leave a limpid, clear fluid above a small, white, flocculent, mass of agglutinated bacilli.

B. PRECIPITATION REACTION

Precipitogen, precipitin and the precipitate are necessary for this test. Bacterial precipitogen is the precipitin producing substance formed in bouillon culture of an organism and is obtained by using the filtrate of these cultures.

The precipitin is the substance formed in the animal serum by inoculating precipitogen and is analogous to agglutinin.

The precipitate is formed in a test tube as a consequence of the mixture of precipitin and precipitogen in the proper proportions.

Other substances than bacterial filtrates will act as precipitogen. Immunization with cow's milk causes a precipitin to form which causes casein of cow's milk to precipitate, but will not precipitate the casein of goat's milk or any other species. Also, the egg-white of a chicken causes a specific precipitin to form which will not precipitate any other kind of egg-white.

This precipitation reaction is important from a medicolegal standpoint. It is used to identify blood stains found in certain murder cases; also used to detect horse or other meat sold as beef.

C. COMPLEMENT-FIXATION REACTION

This test is used in diagnosis of certain diseases but principally syphilis. In 1901, Bordet and Gengou observed that when an antigen was mixed with its specific antibody in the presence of complement, the complement became "fixed" and was rendered unavailable for further reactions. As an indicator to show that the complement is bound, a hemolytic system must be used and this system is prepared as follows:

An animal, usually a rabbit, is immunized by injection of human red blood cells (erythrocytes) and its blood serum thus comes to contain an "amboceptor" which will dissolve the hemoglobin of human r.b.c. in the presence of "complement." This is called an antihuman hemolytic system, but other systems, as antichicken and antisheep, have been used with good results. The "complement" above mentioned is obtained by using fresh guinea-pig serum, and 0.2 mil of a 1 to 10 dilution of serum in isotonic salt solution is usually considered as 1 unit. One unit of amboceptor, then, is that smallest amount which, in the presence of 1 unit of complement, will produce complete hemolysis of 1 mil of a 10 per cent. suspension of homologous r.b.c. in 30 minutes at a temperature of 37°C.

The test is made by placing 2 units of complement, 2 units of antigen, and 0.1 mil of suspected serum into a test tube and incubating at 37°C. for 30 minutes in a water bath; then 2 units of amboceptor with 0.1 mil of 10 per cent. of r.b.c. are added, the tube shaken, and incubated again for 30 minutes at 37°C. A positive result is indicated by absence of hemolysis, the r.b.c. having settled to the bottom of the tube and no color above them. This, as will be readily seen, is due to the fact that the complement was fixed during the first incubation and thus made unavailable for further use in hemolysis. A negative result is indicated by hemolysis of the red blood cells.

Noguchi Modification of the Wassermann Test.—The Noguchi test for syphilis is a modification and simplification of the Wassermann test and involves the use of "amboceptor paper," a solution of "antigen" and "complement," the latter to be obtained from the blood of a guinea-pig.

The amboceptor is obtained by injecting washed human blood-corpuscles (erythrocytes) into rabbits, at intervals of from five to seven days, over a period of five or six weeks. Ten days are allowed to elapse before the last injection. The rabbits are then bled and the serum collected. Filter paper is now saturated with this serum and allowed to dry. The paper is cut into strips and set aside until wanted for use. In this form amboceptor will keep for a considerable length of time.

Amboceptor paper is standardized by measuring its specific activity. The measurement of specific activity consists in finding the amount of amboceptor necessary to cause hemolysis in 1 mil of suspended human red corpuscles, one drop of blood in 4 mils normal saline solution with 0.02 mil of fresh guinea-pig serum. This is incubated at a temperature of 37°C. for one hour. The quantity of paper necessary to cause hemolysis under these conditions is known as one unit. In the syphilis test two units are used.

Antigen.—This is made by rubbing liver and heart tissue with sand and extracting with absolute alcohol. Macerate 10 Gm. of tissue in 100 mils of alcohol for one week at 37°C., shaking the container every day. Dissolve the resulting extract in ether. Pour this solution into a large quantity of acetone. The acetone precipitates certain lipoid substances which are then collected and redissolved in methyl alcohol, in ratio of 3 per cent. This constitutes the antigen solution. For use mix 1 part of this with 9 parts, 0.9 per cent. sodium chloride solution. This dilution should not cause hemolysis in an amount of 0.4 mil and 0.4 mil should not inhibit hemolysis.

D. VACCINATION FOR DIAGNOSIS

Old Tuberculin—Tuberculin alt Koch—Concentrated Tuberculin—Crude Tuberculin.—Koch's original tuberculin is prepared from glycerin bouillon cultures of the tubercle bacillus by evaporating to one-tenth the original volume, sterilizing at 100°C. for one hour, and filtering through a Berkefeld filter. It is a clear brown syrupy liquid, with a high content of glycerin and a characteristic odor.

For diagnosis, old tuberculin may be used by hypodermic injection to show a reaction at the site of application (local), at site of suspected disease (focal) or generally (constitutionally). If positive, the tuberculin reaction merely indicates that the patient has at some time been infected with tuberculosis and not necessarily that he has clinical tuberculosis. Careful series of necropsies confirm the results of the use of tuberculin showing that perhaps 80 per cent. of adults have been infected with the tubercle bacillus, whether or not they have clinical tuberculosis requiring treatment. Moreover, in many advanced or acute cases of tuberculosis the patients do not react, so that the result of a tuberculin test is never absolute but always must be judged in the light of other findings. The occurrence of a focal reaction is good presumptive evidence of an active lesion.

For children the cutaneous test has been chiefly used. This is performed by abrading the cleansed skin of the forearm in two places 2 inches apart through a drop of undiluted old tuberculin at each site; another similar abrasion is used as a control between the two; the two drops of tuberculin are carefully wiped off after ten minutes, allowing no tuberculin to touch the control site. The reaction consists in a zone of redness, markedly larger than that at the control site. This reaction reaches its height in from twenty-four to forty-eight hours. After infancy an increasing proportion of those who react are found to be free from clinical tuberculosis. The subcutaneous test is used more frequently on adults. A two-hour temperature chart should be kept for two days preceding and two days following each injection. To an adult in good condition 0.0002 mil may be given as the initial dose, and if there is no reaction 0.001 mil and then 0.005 mil may be tried. The doses should be at least three days apart and if there is the slightest suggestion of a reaction in temperature or symptoms the dose should be repeated, not increased. Children and weak patients should receive smaller doses, but no very weak patient and none with a fever should be subjected to the danger of a subcutaneous test. A rise of temperature of 1°F. may be taken as a reaction, especially if accompanied by changes at the site of the disease. This reaction means, just as with the cutaneous test, only infection and not necessarily clinical tuberculosis and owing to the danger of large doses, patients may fail to react because, though sensitive to tuberculin, they are not sensitive to doses small enough to be used safely.

For treatment from 0.00000001 mil to 0.000001 mil may be used as the initial dose, and not more than two doses a week should be given.

LUETIN.—Luetin is an extract of the killed cultures of several strains of the Treponema pallidum, the causative agent of syphilis.

When injected into the skin, luetin provokes no reaction in normal individuals except a very small erythematous area at and around the point of injection. In certain cases of syphilitic infection, a reaction occurs consisting of papules which may become pustules. When the reaction takes the papular form, a large reddish indurated papule (usually from 7 to 10 mm. in diameter) makes its appearance in twenty-four to forty-eight hours and slowly increases for four or five days, after which the inflammatory process begins to recede. The color of the papule gradually becomes dark bluish red. The induration disappears within two weeks, as a rule.

In the pustular form, after the fourth or fifth day, the inflammatory process increases in intensity and the papules become vesicular and later purulent. The pustules rupture spontaneously and the defect caused by the escape of the pustular content becomes quickly covered by a crust that falls off within a few days. A small induration sometimes remains for a few weeks or often months, leaving a small keloid after healing.

In the torpid type of syphilis the site of injection fades to an almost invisible point within three to four days, so that it may be erroneously considered a negative reaction. After ten days or even longer, the spot suddenly begins to enlarge and goes through the same stages as seen in the pustular type.

Luetin is employed for the diagnosis of syphilis. It is of use in the examination of tertiary cases but rarely gives a positive reaction in the primary cases or in untreated secondary cases. In patients who are under treatment by mercury or salvarsan the reaction is frequently positive even in cases which fail to give a positive Wassermann reaction.

The amount of luetin to be injected for one test is 0.07 mil. The material should be properly diluted and injected into, but not under, the skin. A site should be selected on the skin of the upper arm, cleansed and sterilized and the injection made as described.

DIPHTHERIA IMMUNITY TEST (SCHICK TEST).—This test is intended to determine those persons who have not in their blood an amount of diphtheria antitoxin sufficient to render them immune. The test depends on the phenomenon that when a small amount of diphtheria toxin is injected intradermally into a person who has no free antitoxin in his blood, a circumscribed area of redness and infiltration from 1 to 2 cm. in diameter develops at the site of injection. Should the patient have free antitoxin in his blood no reaction occurs. The reaction occurs in twenty-four to forty-eight hours and is at its height in forty-eight to seventy-two hours. It remains for six to twelve days, is followed by slight scaling and leaves a brownish pigmented spot. In some persons, pseudoreaction may occur, which may be differentiated by its earlier appearance and disappearance and the fact that it is less circumscribed and is not followed by pigmentation.

The test is of special value for use in institutions and among groups of persons exposed to diphtheria, in order that it may be determined which individual should be given an immunizing dose of diphtheria antitoxin. It is also of value in the diagnosis of other conditions stimulating diphtheric infection.

Diphtheria toxin in dilute solution, such as is necessary for the tests, soon loses in potency.

DUCTLESS GLANDS

A certain number of glands in the body, some of great physiological importance, have no ducts to carry away secretion. There are two theories concerning the method by which these glands function: The theory generally accepted as correct is, briefly, that there is a secretion formed, which, when carried away by the veins or lymphatics, is responsible for the ascribed function of the gland; the other theory is to the effect, that these glands perform only excretory duties, and that when they fail to function properly, the toxic or unusual effects noticed are due to poisoning by the uneliminated substances.

Among these ductless glandular bodies, the following are the more important ones: The spleen, the adrenals, the thyroid gland, the parathyroid glands, the thymus gland, the lymphatic glands, the carotid, coccygeal, pineal and pituitary bodies.

The value of the thyroid gland, in cases of myœdema was discovered by Dr. George Murray in 1891, he administering the gland by hypodermic injections. In 1892, Drs. Hector Mackenzie, E. T. Fox, and Howetz, each working independently, showed that the gland was equally efficacious when administered by the mouth. The remedy was soon after applied to cretinism and its effects were found to be even more wonderful. Since then, preparations of other ductless glands have come more or less into practical use.

The activities of the various tissues of the body are presided over and controlled not merely by the action of the nervous system, but also by chemical substances, the result of the activity of certain organs. To these chemical substances the name of "hormones" has been given. Epinephrine is the hormone which has been most thoroughly investigated, and it is a product of the central part of the suprarenal glands. A hormone closely allied to epinephrine is derived from the pituitary body, and it causes constriction of the small arteries, except those of the kidney, which it dilates. Iodothyrin, the active principle secreted by the thyroid glands, appears to stimulate the rate of chemical exchange in the various tissues; it increases the waste of both proteins and fat.

THYROIDEUM SICCUM, U.S.P.—The thyroid glands of animals which are used for food by man, freed from connective tissue and fat, dried and powdered. One part of dried thyroid corresponds with approximately 5 parts of the fresh glands, and must contain not less than 0.17 per cent. nor more than 0.23 per cent. of iodin in thyroid combination. The average dose is 0.1 Gm.

These glands are assayed by U.S.P. method.

LIQUOR THYROIDEI is a pink turbid liquid made by macerating the fresh gland of a sheep with glycerin and phenol.

ANTITHYROID PREPARATIONS.—Rodogen is a white powder consisting of the dried milk of thyroidectomized goats, mixed with 50 per cent.

of milk sugar. In exophthalmic goiter this preparation causes a reduction of the swelling and of the pulse rate, and an increase of body weight.

ANTITHYROIDIN—SERUM ANTITHYROIDEUM.—The blood serum of sheep from which the thyroid gland has been removed at least six weeks before the blood is drawn, preserved by the addition of 0.5 per cent. of phenol.

DOSAGE.—It is administered by the mouth in doses beginning with 0.5 to 1 mil three times a day, gradually increasing the dose as necessary.

THYREOIDECTIN.—Gelatin capsules, each containing 0.33 Gm. of a powder prepared from the blood of thyroidectomized animals. Dose: one or two capsules, three times a day.

SUPRARENALUM SICCUM, U.S.P.—The suprarenal glands of animals which are used for food by man, cleaned, dried, freed from fat, and powdered, and containing not less than 0.4 per cent. nor more than 0.6 per cent. of epinephrine, the active principle of the suprarenal gland. One part of dried suprarenal represents approximately 6 parts of fresh glands, free from fat.

It is a light yellowish-brown amorphous powder, having a slight characteristic odor, partially soluble in water. Assayed by U.S.P. method. The average dose is 0.25 Gm.

PURIFIED EXTRACT OF ADRENAL GLAND.—An extract of the suprarenal gland, standardized physiologically by measuring its effect on blood pressure and so adjusted as to correspond to the effect of 4 per cent. of purified

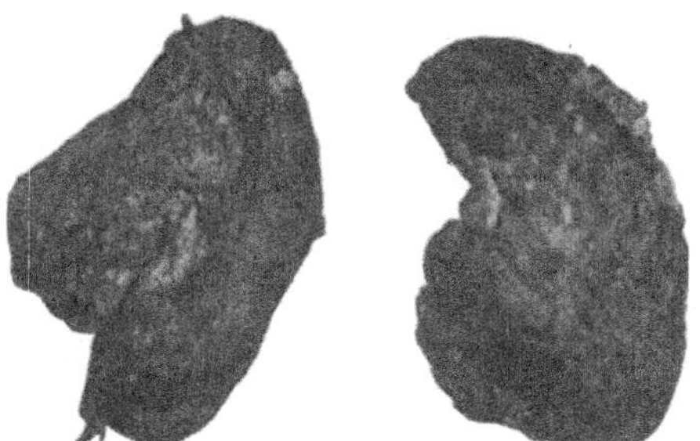

FIG. 262.—Suprarenal Capsules (⅔ natural size.) (Photograph.)

epinephrine. It has therefore approximately four times the strength of desiccated suprarenal gland U.S.P.

EPINEPHRINE is a substance, with feeble basic properties, obtained from the suprarenal gland of the sheep or other animal; also made synthetically. Its most important therapeutic actions consist in a constriction of the blood-vessels, with consequent high rise of blood pressure; a stimulation of the vagus center, with slowing of the heat, and a direct stimulant and tonic effect on the heart muscle, similar to digitalis. Large doses also cause glycosuria. Its chief use is locally in hemorrhage and in catarrhal and congestive conditions. The dose is 0.3 to 2.0 mils of a 1 to 1000 solution, every two or three hours.

SUPRARENAL SNUFF containing the dry extract with menthol and boric acid is useful in cases of hay fever.

PITUITARY GLAND.—The hyperactivity of the anterior lobe leads to gigantism; the posterior lobe contains a certain substance having marked effects upon plain muscle, especially that of the blood-vessels and the uterus.

LIQUOR HYPOPHYSIS, U.S.P.—A solution containing the water-soluble principle or principles from the fresh posterior lobe of the pituitary body of cattle.

Extract the finely minced material with slightly acidulated water, boil the solution for ten minutes and filter it. Sterilize this filtrate and preserve it in a sterile condition in glass containers.

It is standardized physiologically, using beta-iminazoyl-ethylamine hydrochloric as a standard. The average dose is 1 mil.

This preparation is beneficial in shock, pulmonary hemorrhage and intestinal paresis after abdominal operations, but is especially recommended in cases of uterine atony in postpartum and other forms of uterine hemorrhage.

Its administration by the mouth is less effective than by subcutaneous injections.

DESSICATED PITUITARY SUBSTANCE.—There are desiccated preparations of the anterior lobe, posterior lobe and of the entire gland on the market.

SECTION III.—(A) SYNOPSIS OF NATURAL ORDERS, OR FAMILIES, AND OF DRUGS, ARRANGED ACCORDING TO PART II.

ALGÆ, P. 77: Agar Agar, Chondrus, Fucus Vesiculosus, Fucus Nodosum.

FUNGI, P. 79: Ergota, Torula, Ustilago, Agaracus Albus, Agaracus Chirurgorum.

LICHENES, P. 83: Cetraria, Lacmus, Orchil, Cudbear.

LYCOPODIACEÆ, P. 87: Lycopodium.

FILICES, P. 84: Aspidium, Adiantum, Cibotium, Osmunda, Polypodium.

PINACEÆ, P. 88: Sabina, Oleum Sabinæ, Oleum Juniperi, Oleum Cadinum, Terebinthinæ, Oleum Teribinthinæ, Resina, Pix Liquida, Oleum Picis Liquidæ, Terebinthina Canadensis, Juniperis Virginiana, Pinus Strobus, Thuja, Juniperus Baccæ, Tsuga, Larix, Venice Turpentine, Pix Canadensis, Pix Burgundica, Siccinum, Oleum Succini, Damar.

GRAMINACEÆ, P. 94: Triticum, Saccharum, Zea, Amylum, Vitivera, Oleum Maydis, Avena Farina, Sago, Tapioca, Taro, Triticum Vugare, Oryza, Solanum Tuberosum, Canna, Maranta, Curcuma Leicorrhiza, Hordii Fructus, Maltum.

LILIACEÆ, P. 104: Sarsaparilla, Convallaria, Veratrum, Scilla, Colchici Cormus, Colchici Semen, Aloe, Veratrum Album, Sabadilla, Polygonatum, Chamælirium, Trillium, Asparagus, Allium, Xanthorrhiza, Erythronium.

HÆMODORACEÆ, P. 120: Aletris.

DIOSCOREACEÆ, P. 120: Dioscocorea.

IRIDACEÆ, P. 120: Crocus, Iris, Iris Florintina.

SCITAMINÆ (ZINGIBERACEÆ), P. 122: Zingiber, Cardamomum, Galenga, Zedoaria, Curcuma, Granum Paradisi.

ORCHIDACEÆ, P. 128: Cypripedium, Vanilla, Corallorrhiza, Salep.

PIPERACEÆ, P. 133: Piper, Cubeba, Oleum Cubebæ, Matico, Yerba Mansa, Piper Album, Piper Longum, Jambu Assu, Methysticum.

SALICACEÆ, P. 137: Salix, Populus, Populus Balsamifera.

MYRICACEÆ, P. 138: Myrica, Comptonia.

JUGLANDACEÆ, P. 139: Juglans.

CUPULIFERÆ, P. 139: Oleum Betulæ, Quercus Alba, Galla, Acidum Tannicum, Acidum Gallicum, Pyrogallol, Ostrya, Alnus, Fagus, Castanea.

URTICACEÆ, P. 145: Ulmus, Cannabis Indica, Humulus, Lupulinum, Ficus, Cannabis Semen, Oleum Cannabis, Morus, Urtica.

SANTALACEÆ, P. 151: Oleum Santali, Santalum Album.

LORANTHACEÆ, P. 152: Mistletoe.

ARISTOLOCHIACEÆ, P. 152: Serpentaria, Asarum.

POLYGONACEÆ, P. 154: Rheum, Rumex, Rheum Rhaponticum, Canaigre, Bistorta, Polygonium.

CHENOPODIACEÆ, P. 159: Oleum Chenopodii, Chenopodium.

PHYTOLACCACEÆ, P. 159: Phytolacca, Phytolaccæ Fructus.

CARYOPHYLLACEÆ, P. 160: Stellaria, Saponaria, Saponaria Levantica.

PORTULACACEÆ, P. 161: Portulaca.

NYMPHAEACEÆ, P. 161: Nymphæa.

RANUNCULACEÆ, P. 161: Aconitum, Cimicifuga, Hydrastis, Staphisagria, Ranunculus, Adonis Vernalis, Hepatica, Pulsatilla, Coptis, Helleborus Viridis, Helleborus Niger, Nigella, Delphinium, Actæa, Pæonia, Xanthorrhizá.

MAGNOLIACEÆ, P. 174: Magnolia, Liriodendron, Illicium, Wintera.

CALYCANTHACEÆ, P. 176: Calycanthus.

MYRISTICACEÆ, P. 176: Myristica, Oleum Myristicæ, Oleum Myristicæ Expressum, Macis.

MENISPERMACEÆ, P. 178: Calumba, Pareira, Cocculus, Menispermum.

BERBERIDACEÆ, P. 184: Podophyllum, Caulophyllum, Berberis, Jeffersonia, Berberis Radix, Berberis Cortex.

MONIMIACEÆ, P. 189: Boldus.

LAURIACEÆ, P. 189: Cinnamomum Zeylandicum, Cinnamomum Cassia, Oleum Cinnamomi, Cinnamomum Saigonicum, Camphora, Sassafras, Oleum Sassafras, Sassafras Medulla, Persea, Oleum Camphoræ, Nectandra, Sassafras Lignum, Lindera, Laurus, Oleum Lauri, Coto, Umbellularia.

PAPAVERACEÆ, P. 199: Sanguinaria, Opium, Eschscholtzia, Rhoeas, Chelidonium, Papaver, Papaveris Semen.

FUMARIACEÆ, P. 206: Corydalis.

CRUCIFERÆ, P. 206: Sinapis Alba, Sinapis Nigra, Oleum Sinapis Volatile, Armoracia, Bursa Pastoris, Oleum Sinapis Expressum.

SARRACENIACEÆ, P. 210: Sarracenia.

DROSERACEÆ, P. 210: Drosera.

CRASSULACEÆ, P. 210: Penthorum, Sedum Acre.

SAXIFRAGACEÆ, P. 210: Heuchera, Mitella Nuda, Hydrangea.

HAMAMELIDACEÆ, P. 210: Hamamelis Folia, Hamamelis Cortex, Styrax, Liquidambar.

ROSACEÆ, P. 213: Quillaja, Rubus, Cusso, Rosa Gallica, Oleum Rosæ, Prunum, Amygdala Amara, Oleum Amygdalæ Amaræ, Amygdala Dulcis, Oleum Amygdalæ Expressum, Prunus Virginiana, Spiræa, Gillenia, Cydonium, Malus, Persica, Laurocerasus, Cratægus, Rubus Idæus, Fragaria, Potentilla, Tormentilla, Choke Cherry, Geum Urbanum, Geum Rivale, Agrimonia, Rosa Centifolia, Rosa Canina.

LEGUMINOSÆ, P. 226: Acacia, Catechu, Copaiba, Oleum Copaibæ, Tamarindus, Cassia Fistula, Senna, Krameria, Hæmatoxylon, Balsamum Peruvianum, Balsamum Tolutanum, Scoparius, Tragacantha, Glycyrrhiza, Extractum Glycyrrhizæ, Kino, Santalum Rubrum, Physostigma, Erythrophlœum, Saraca, Cercis, Ceratonia, Cassia Marilandica, Baptisia, Fœnum Græcum, Melilotus, Trifolium Pratensis, Trifolium Repens, Galega, Stylosanthes, Piscidia, Abri Radix, Abri Semen, Mucuna, Dipteryx, Ararobia, Pongamia Oil, Copal.

LINACEÆ, P. 251: Linum, Oleum Lini.

ERYTHROXYLACEÆ, P. 254: Coca.

GERANIACEÆ, P. 256: Geranium, Impatiens.

ZYGOPHYLLACEÆ, P. 257: Guaiacum, Guaiaci Lignum.

RUTACEÆ, P. 260: Xanthoxylum, Buchu, Pilocarpus, Aurantii Cortex, Oleum Aurantii Corticus, Limonis Succus, Limonis Cortex, Oleum Limonis, Xanthoxyli Fructus, Ruta, Angustura, Ptelia, White Zapote, Bela, Aurantii Folia, Aurantii Flores, Oleum Bergamottæ.

SIMARUBACEÆ, P. 272: Quassia, Simaruba, Cedron, Quassia Cortex, Chaparro Amargoso, Ailanthus, Cascara Amarga.

BURSERACEÆ, P. 275: Myrrha, Elemi, Olibanum, Bdellium.

MELIACEÆ, P. 277: Naregamia Alata, Azedarach, Cocillana.

POLYGALACEÆ, P. 277: Senega, Polygala.

ANACARDIACEÆ, P. 290: Mastiche, Rhus Glabra, Anacardium, Terebinthina Chia, Rhus Toxicodendron, Rhus Aromatica, Semecarpus.

EUPHORBIACEÆ, P. 282: Oleum Tiglii, Oleum Ricini, Stillingia, Elastica, Cascarilla, Tiglium, Mercuriales, Kamala, Ricinus, Euphorbia, Euphorbia Corollata, Euphorbia Ipecacuanha, Euphorbia Pilulifera, Euphorbium, Alveloz Milk, Lacca, Curcas.

ILICINEÆ, P. 292: Ilex Opaca, Ilex Paraguayensis, Prinos.

CELASTRACEÆ, P. 292: Euonymus, Celastrus.

ACERACEÆ, P. 293: Acer Rubrum.

SAPINDACEÆ, P. 293: Guarana, Æsculus [Hippocastonum, Æsculus Glabra, Macassar Oil, Acer Rubrum.

RHAMNACEÆ, P. 295: Frangula, Rhamnus Purshiana, Rhamnus Cathartica, Ceanothus, Gouania.

AMPELIDEÆ, P. 300: Uva Passa, Ampelopsis.

TILIACEÆ, P. 301: Tilia.

MALVACEÆ, P. 301: Althæa; Cotton Derivatives: Gossypium Cortex, Gossypium Purificatum, Oleum Gossypii Seminis; Althæa Rosea, Malva.

STERCULIACEÆ, P. 305: Oleum Theobromata, Theobroma, Cola.

TERNSTRŒMIACEÆ, P. 306: Thea.

GUTTIFERÆ, P. 306: Cambogia, Mangostana.

HYPERICINEÆ, P. 308: Hypericum.

DIPTEROCARPEÆ, P. 308: Gurjun, Borneo Camphor.

FRANKENIACEÆ, P. 309: Frankenia.

CISTINEÆ, P. 309: Helianthemum.

BIXINEÆ, P. 309: Annato, Gynocardia, Oleum Gynocardiæ.

CANELLACEÆ, P. 309: Canella, Cinnamodendron.

VIOLACEÆ, P. 310: Viola Tricolor.

TURNERACEÆ, P. 310: Turnera.

PASSIFLOREÆ, P. 310: Passiflora, Carica Papaya.

CACTEÆ, P. 310: Cactus, Anhalonium.

THYMELACEÆ, P. 311: Mezereum.

PUNICACEÆ, P. 312: Granatum, Granati Fructus Cortex.

MYRTACEÆ, P. 315: Pimenta, Oleum Pimentæ, Caryophyllus, Oleum Caryophylli, Eucalyptus, Oleum Eucalypti, Oleum Cajuputi, Myrcia, Oleum Myrciæ, Caryophylli Fructus, Chekan, Jambul, Eucalyptus Rostrata.

COMBRETACEÆ, P. 323: Myrobolanus.

ONAGRARIEÆ, P. 323: Epilobium, Œnotheria Biennis.

ARALIACEÆ, P. 323: Aralia Racemosa, Aralia Hispidia, Aralis Nudicaulis, Panax.

31

UMBELLIFERÆ, P. 324: Conium, Carum, Oleum Cari, Thymol, Fœniculum, Oleum Fœniculi, Anisum, Oleum Anisi, Asafœtida, Sumbul, Eryngium, Osmorrhiza, Conii Folia, Cuminum, Apium, Cicuta, Ajowan, Petroselinum Root, Petroselinum Seed, Pimpinella, Plellandrium, Levisticum, Galbanum, Ammoniacum, Anethum, Oleum Anethi, Imperatoria, Laserpitium, Thapsia, Angelica Atropurpurea, Angelica Officinalis, Caroca.

CORNACEÆ, P. 337: Garraya, Cornus Florida.

ERICACEÆ, P. 337: Chimaphila, Oleum Gaultheriæ, Uva Ursi, Ledum, Rhododendron, Kalmia, Oxydendrum, Epigæa, Gaultheria, Arctostaphylos, Vaccinum.

MYRSENÆ, P. 342: Embelia.

PLUMBAGINACEÆ, P. 342: Anagallis.

SAPOTACEÆ, P. 342: Gutta Percha, Monesia.

EBENACEÆ, P. 343: Diospyros.

STYRACECEÆ, P. 343: Benzoinum.

OLEACEÆ, P. 345: Manna, Oleum Olivæ, Fraxinus Americana, Fraxinus Sambucifolia, Chionanthus, Ligustrum.

LOGANIACEÆ, P. 348: Gelsemium, Spigelia, Nux Vomica, Curara, Ignatia, Hoang-nan.

GENTIANEÆ, P. 355: Gentiana, Chirata, Sabbatia, Frasera, Menyanthes.

APOCYNACEÆ, P. 359: Apocynum, Strophanthus, Conessi, Alstonia Scholaris, Alstonia Constricta, Aspidosperma, Urechites, Apocynum Androsæmifolium, Oleander.

ASCLEPIADACEÆ, P. 363: Hemidesmus, Asclepias Tuberosa, Asclepias Incarnata, Asclepias Curassavica, Asclepias Cornuti, Condurango.

CONVOLVULACEÆ, P. 364: Jalapa, Scammonium, Ipomœa, False Jalaps.

POLEMONIACEÆ, P. 368: Polemonium.

HYDROPHYLLACEÆ, P. 368: Eriodictyon.

BORRAGINACEÆ, P. 369: Symphytum, Borago, Alkanna, Pulmonaria.

VERBENACEÆ, P. 369: Verbena Urticæfolia, Verbena Hastata, Lippi Mexicana, Tonga.

LABIATÆ, P. 370: Oleum Rosmarini, Scutellaria, Oleum Lavandulæ Florum, Marrubium, Salvia, Hedeoma, Oleum Hedeomæ, Oleum Thymi, Mentha Viridis, Oleum Menthæ Viridis, Mentha Piperita, Oleum Menthæ Piperitæ, Menthol, Teucrium, Rosmarinus, Lavandula, Cataria, Glechoma, Lamium, Leonurus, Betonica, Monardo Fistulosa, Monardo, Oleum Monardo, Melissa, Satureia, Hyssopus, Majorana, Origanum, Oleum Origani, Pycnanthemum, Serpyllum, Thymus, Cunila, Lycopus, Collinsonia, Ocimum, Orthosiphon, Yerba Buena.

SOLANACEÆ, P. 377: Belladonnæ Radix, Belladonnæ Folia, Scopola, Hyoscyamus, Capsicum, Stramonium, Dulcamara, Solanum Carolinense, Hyoscyamus Semen, Stramonii Semen, Tabacum, Pichi, Duboisia, Manaca, Lycopersicum Esculentum.

SCROPHULARIACEÆ, P. 388: Digitalis, Leptandra, Verbascum, Scrophularia, Chelone, Veronica Officinalis, Euphrasia.

OROBANTHACEÆ, P. 393: Epiphegus.

BIGNONIACEÆ, P. 393: Caroba, Newbouldia.

PEDALINEÆ, P. 393: Sesamum, Sesamum Leaves, Sesamum Seeds, Oleum Sesami.

PLANTAGINACEÆ, P. 394: Plantago.

RUBIACEÆ, P. 394: Cinchona Rubra, Cinchona, Ipecacuanha, Remijia, Catechu Pallidium, Galium, Mitchella, Cephanthus, Caffea, Rubia.

CAPRIFOLIACEÆ, P. 406: Viburnum Opulus, Viburnum Prunifolium, Triosteum, Sambuscus.

VALERIANACEÆ, P. 410: Valeriana, Oleum Valerianæ.

CUCURBITACEÆ, P. 413: Colocynthus, Pepo, Momordica, Luffa, Bryonia, Elaterium, Citrullus, Cucumis.

CAMPANULACEÆ, P. 417: Lobelia.

COMPOSITÆ, P. 419: Eupatorium, Grindelia, Oleum Erigerontis, Anthemis, Pyrethrum, Matricaria, Santonica, Arnica, Calendula, Lappa, Taraxacum, Lactucarium, Elephantopus, Eupatorium Purpureum, Guaca, Trilisa, Laciniaria, Solidago, Erigeron, Erigeron Canadense, Pterocaulon, Gnaphalium, Inula, Polymnia, Silphium, Parthenium, Ambrosia, Ambrosia Artemisiæfolia, Strumarium, Spinosum, Rudbeckia, Echinacea, Helianthus, Helianthella, Bidens, Helenium, Achillea, Cotula, Oleum Anthemidis, Pyrethrum Germanicum, Pyrethrum Flores, Tanacetum, Artemisia, Absinthium, Tussilago, Erechthites, Arnicæ Radix, Senecio, Lappæ Fructus, Carthamus, Carduus Benedictus, Cnicus Arvensis, Mutisia, Cichorium, Lactuca Sativa, Lactuca Canadense.

SECTION III—(B)—DRUG ASSAY PROCESSES

The Pharmacopœia gives in detail the various processes suited to the assay of the various drugs specified. By careful manipulation, using these methods, the pharmacist may reach definite results as to the quality and strength of the article under examination. The student in pharmacy should, however, be familiar with the general principles which underlie the process of drug assay. A brief outline of these principles may be here in place.

Principles of Alkaloidal Assay.—The immiscible solvents, such as chloroform, ether, benzol, amylic alcohol, etc., are employed. Any of these liquids, when shaken with water or acidulated water, will mix with the aqueous liquid only for a time. On standing for a few minutes they will separate into two distinct layers, one of these being the aqueous layer, the other the immiscible solvent (mostly ethereal in character).

If equal volumes of ether and water be shaken together and a solution of the extract of belladonna added to the mixture and a few drops of sulphuric acid, it will be found that the belladonna alkaloids will be dissolved out and will be contained in the aqueous (acidulated) layer, *not in the ethereal layer*, because it is the general property of alkaloidal salts to be soluble in water, and to be insoluble in ether. The acid having converted the atropine and the hyoscyamine of the extract into a salt (sulphate), it therefore will be taken up and retained by the aqueous layer. If to these two liquids (the ether and the acidulated solution of the alkaloidal salt) there is now added a sufficient quantity of ammonia water to neutralize the acid and make the aqueous liquid slightly alkaline, and the fluids be again mixed and allowed to stand as before, it will be found that the alkaloids (of belladonna) are no longer in the aqueous layer but in the upper (ethereal) layer. It is the general property of free alkaloids, themselves, with few exceptions, to be soluble in ether (chloroform, etc.) and to be insoluble in water. It might be stated therefore, as a general principle, that alkaloids, as a rule, are soluble in the immiscible fluids (ether, chloroform, amylic alcohol), etc., while their salts are insoluble in these fluids. Alkaloidal salts, on the other hand, are insoluble in the immiscible fluids, but are soluble in water. Advantage is taken of this property in the assay of alkaloidal drugs.

For general directions for alkaloidal assay, see Part II, No. 15 (Proximate Assays) U.S.P. IX, p. 593.

The following is the list of **assayed drugs** and preparations. Those marked with (*) are unofficial, others are found in U.S.P. or N.F.

	Percentage	Fluidextract*	Solid or Powd. Extract
Aconite Root................	0.5	0.45–0.55	1.8–2.2
Bellad. Leaf................	0.3		1.18–1.32
Bellad. Root...............	0.45	0.405–0.495	
*Scopola................	0.5	0.5	2.0
Calabar Bean (Physostig.).......	0.15		1.7–2.3
Cinchona................	5.0	4.0–5.0	
*Coca................	0.5	0.5	
Colchicum Corm.........	0.35		1.25–1.55
Colchicum Seed............	0.45	0.36–0.44	
Conium................	0.5 (N.F.)	0.45	
Guarana................	4.0	3.6–4.4	
Hydrastis................	2.5	1.8–2.2	9.0–11.0
Hyoscyamus................	0.065	0.055–0.075	0.22–0.28
Ipecac................	1.75	1.8–2.2	
Jalap (resin)................	7.0		
Nux Vomica.'................	2.5	2.37–2.63	15.2–16.8
Opium (Gum).........	9.5		19.5–20.5
Opium (Powder and deodorized).	10–10.5		
Pilocarpus................	0.6	0.55–0.65	
Stramonium................	0.25		0.9–1.1

THE ASSAYED TINCTURES

The assayed tinctures of the Pharmacopœia have the following amounts of alkaloids represented in 100 mils of tincture:

Tinctura Aconiti,	0.045–0.055	Gm. aconitine.
Tinctura Belladonnæ Foliorum,.	0.027–0.033	Gm. belladonna alkaloids.
Tinctura Cinchonæ,.....	0.8–1	Gm. alkaloids.
Tinctura Colchici Seminis,	0.036–0.044	Gm. colchicine.
Tinctura Hydrastis,...	0.36–0.44	Gm. ether Sol. alkaloids.
Tinctura Hyoscyami, ...	0.0055–0.0075	Gm. alkaloids.
Tinctura Nucis Vomicæ,.	0.237–0.263	Gm. alkaloids.
Tinctura Opii,..........	0.95–1.05	Gm. anhydrous morphine.
Tinctura Opii Deodorati,.	0.95–1.05	Gm. anhydrous morphine.
Tinctura Physostigmatis,	0.013–0.017	Gm. alkaloids.
Tinctura Stramonii, ...	0.0225–0.0275	Gm. alkaloids.

---

* Figures, representing alkaloidal strength of fluidextracts and tinctures, show the amount in 100 mils of Fl'ext.

Such drugs as Aconite, Cannabis, Digitalis, Strophanthus and Squill are assayed biologically, *q v.*

PART III

INSECTS INJURIOUS TO DRUGS

The introduction of this brief appended section on insects injurious to drugs into a text-book of materia medica, while an innovation, seems desirable to the author of the text-book on the ground of the importance of the subject. It is a fact that stored drugs are attacked by a considerable number of insects, and that a varying amount of loss from this cause is sustained by practically every druggist, wholesale and retail, in the land. If, by the acquiring of a little knowledge of the appearance and habits of these pests, and by the exertion required in a little preventive or remedial care, this loss can be lessened, the introduction of this section, which attempts to furnish the information necessary for the little knowledge and the little care, will be justified.

The necessary entomological knowledge of the pharmacist who would make some show of resistance to the insect enemies of his drugs may be limited to an acquaintanceship with these insect enemies, and a knowledge of the means of fighting them. As a basis for this acquaintanceship, however, it is necessary to glance hastily at the great class of insects in general. More numerous in species and individuals than all other animals combined, the insects are conveniently divided into several great groups or orders. All the butterflies and moths, whose wings are covered with fine scales, and who obtain their food by sucking the nectar from flowers, constitute one order; the beetles, with their horny fore-wings and their powerful jaws for biting, compose another order; the two-winged flies, of which the familiar house-fly is an example, constitute a third order; the ants, bees, and wasps, and some other highly intelligent insects are grouped together in a fourth order; the true bugs, as the chinchbug and squash-bug, with their sucking beaks, are comprised in a fifth order; the grasshoppers, crickets, cockroaches, and katydids compose a sixth order; and, finally, the gauzy-winged dragon-flies, the short-lived May-flies, and the wonderful white ants constitute a seventh order. But a simpler division of insects into two great groups is that often made, for convenience' sake, especially by the economic entomologist; namely, a division made according to mouth parts, all insects in the adult stage having mouth parts fitted for biting or mouth parts fitted for sucking. It is evident at once that the pharmacist will be especially interested in the biting insects, the ones which can attack roots and leaves, and all dry preparations. There will be little opportunity for the sucking insects to injure the pharmacist's stores. The insects may be divided according to this distinction as follows: The orders containing the beetles, the cockroaches, the dragon-flies, etc., compose the group of biting insects; the orders containing the true bugs, the butterflies and moths, and the flies, compose the group of sucking insects; while the order of the ants, bees, and wasps, and the order of mites (which are not true six-footed insects, but are closely related to them) may be said to compose a third group, in which the mouth-parts are arranged for both biting and sucking, or piercing and sucking.

But we can not thus dismiss certain of the sucking insects from our pharmaco-entomological consideration; for with wonderful adaptiveness, nature has arranged that the young of certain sucking insects shall be provided with jaws for biting. The common worm-like caterpillars, which are the larval forms, or young, of butterflies and moths, are familiar to all; most children know that the strong-jawed, foliage-eating "worm," now feeding so voraciously on the green leaves of plant or tree, will in time change into some beautiful four-winged butterfly or moth, incapable of injuring a green leaf, and taking its food only in dainty sips, by means of its sucking tubular mouth parts, from some bright flower. And most housewives know that the dreaded clothes-moth—little, brown, delicate flutterer—is, in its moth or winged stage, harmless to furs or woolens, but that the dreaded little white grub, with its sharp jaws and voracious appetite, which really does the damage, is only the young of the innocent-looking moth, and that the moth, after all, is not so innocent.

So, then, it behooves the pharmacist to keep an eye on not only those insects which all their lives are truly biting insects, but also on those insects, as the moths, which, while harmless as adults, yet in their young stages, with strong biting mouth parts, appear as ravaging caterpillars.

In setting out to fight an insect pest, the economic entomologist asks first, "What is it? Is it a beetle, or a fly, or a moth?" This question answered, he already knows much about it; whether, for example, it is a biting or a sucking insect; he knows in a general way what sort of damage it does and how it does it, and he knows, too, in a general way, what remedies are most likely to be effective in fighting it. But it is always better and usually necessary to know the exact life history of the particular pest he must fight; he must discover where and when its eggs are laid, how long it remains in the larval or grub stage, what are its times and places of feeding, and what are its favorite articles of diet. From this life history he can decide on the character of the remedy to be applied, and when and where the remedy can best be used. Therefore the pharmacist may wisely turn to his jars and boxes, his store-rooms and laboratories, and try to discover what manner and number of insects he is to array himself against.

Referring to some of the more common and destructive pests attacking stored drugs, the mites (order, Acarina) may first be noted. The mites, commonly enough represented and known in the case of the familiar flour or cheese mite, are minute, rounded-oval, eight-legged insects, with the mouth parts arranged to form a piercing beak. The body is not divided into head, thorax, and abdomen, as is the case with other insects, but all these parts are coalesced or merged into a single mass. While many mites suck the blood from animals or the juices from plants, many others feed on "dry food." Among these are the flour and cheese mites, and sugar mites with soft, smooth, whitish body (see Fig. 263), and belonging to the genera Tyroglyphus, Rhizoglyphus, and Glyciphagus. Many species of these genera of mites, besides being found in sugars, meals, and other vegetable products in the store-room, attack dried animal remains, cantharides suffering severely from the ravages of several species of *Glyciphagus* (see Fig. 264). The presence of the mites in the cantharides

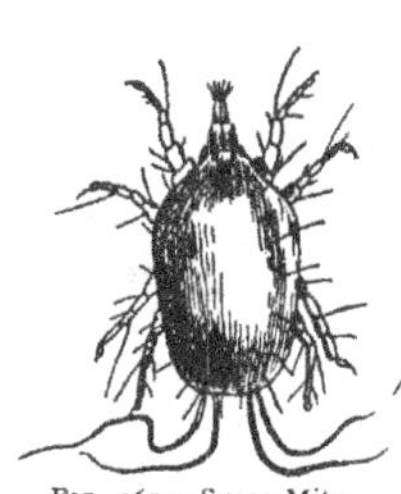

FIG. 263.—Sugar Mite.

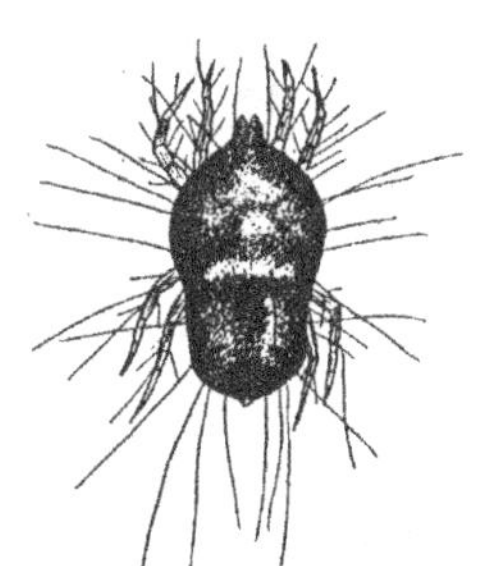

FIG. 264.—A Cantharid-eating Mite (*Glyciphagus spinipes*).—(*Fum. and Rob.*)

jars is indicated by much powder and broken bits of the beetles gathering on the bottom of the jars. In this mass of powder and fragments can be seen with the naked eye many small, moving, whitish specks, the mites. These specks, examined under the microscope, will reveal the characteristic shape and appearance of the mites.

The most abundant pest in the pharmacal store-rooms appears to be a small brown beetle, *Sitodrepa panicea*, belonging to the family Ptinidæ, a family whose members, in both larval and adult stages, feed on dead, dry vegetable and animal matter. This family comprises a number of small beetles, rarely exceeding a quarter of an inch in length, and usually brownish in color. A conspicuous and distinctive character is the hood-like prothorax, the head being so bent or drawn back under it as to be almost concealed (see *b*, Fig. 265). *Sitodrepa panicea*, the especially abundant species of this family, is from 2 to 3 mm. long, with a brown, subcylindrical body. It is almost entirely covered with many fine, short, yellowish hairs, which, on the upper surface of the body, are arranged in parallel longitudinal lines; the upper surface of the body (strictly, only the wing-covers) is finely striated (see *a*, Fig. 265). The head is almost concealed by the thorax, the front margin of the thorax reaching to the eyes. The head is also bent strongly downward. The young, or larva, of this beetle is a small white grub with three pairs of legs, and strong, dark brown jaws. The grub when lying at rest usually assumes a semicircular position (see *c*, Fig. 265). They feed voraciously on the drug, grow rapidly, and, after two or three weeks, pupate, and soon change into the perfect beetle. The beetle also feeds upon the drug by means of strong biting jaws, and the females soon lay eggs, from which another generation of larvæ, or grubs, hatch. The whole life of the insect is thus passed

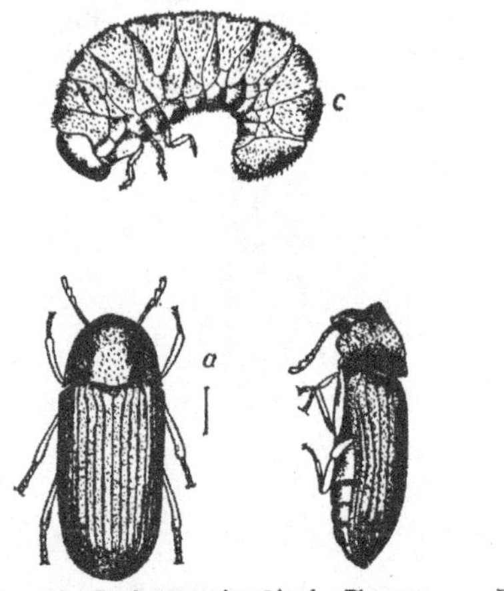

FIG. 265.—*Sitodrepa panicea* Linné. The common drug-eating insect. *a*, Dorsal view of adult beetle. *b*, Side view of adult beetle. *c*, Larva.—(*Smith.*)

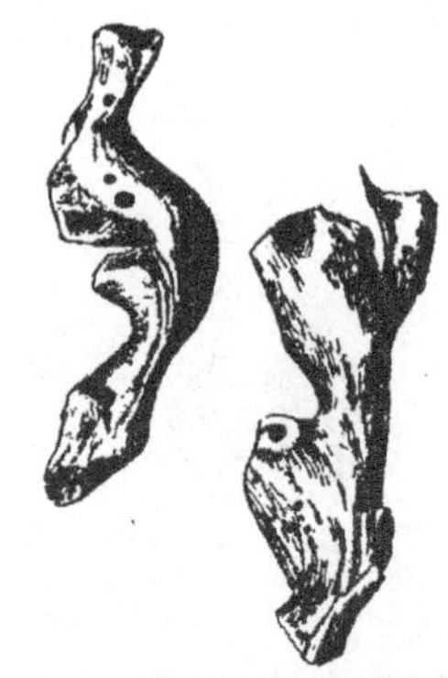

FIG. 266.—Ginger root attacked by *Sitodrepa panicea* Linné.—(*Original.*)

in the can or jar containing the drug. The presence of the pest is shown by the collecting of a considerable amount of powder on the bottom of the can or jar (if the drug is a root, stem, or leaf), and by the presence in the drug of many small holes eaten by the insects (see Fig. 266). Often the little brown beetles may be seen crawling about in the jar. If the drug is a powder, this is the easiest means of detecting their presence. *Sitodrepa panicea* is almost omnivorous in the pharmacal store-room. In the store-rooms of the department of pharmacy, University of Kansas, *Sitodrepa panicea* has been found feeding on such drugs as the following: Columbo, aconite, mustard, althæa, belladonna, poke root, ginseng, angelica, etc.

Still other species of the family Ptinidæ feed on drugs: *Lasioderma serricorne*, a small brown beetle very like *Sitodrepa panicea*, but more robust, and with the wing-covers smooth and not striated, although covered with fine hairs as in *Sitodrepa*, is not uncommon. The larva or grub is like the grub of *Sitodrepa*, and the habits are about the same. I have found *Lasioderma serricorne* attacking powdered ergot, and Prof. J. B. Smith, entomologist of Rutgers College, has found it attacking belladonna root. *Ptinus brunneus*, another species of the family, which I have found attacking musk root, powdered senna, and powdered jaborandi leaves, differs considerably in appearance from the other two members of the family just referred to. It is slightly larger, being about 4 mm. long, and it has long, slender antennæ or feelers which project forward from the head (see Fig. 267). The antennæ of *Sitrodrepa* and *Lasioderma* are usually bent back upon the body. The body of *Ptinus* is not subcylindrical, but tapers toward the head, the head itself being much narrower than the body. *Bostrichus dactilliperda*, another member of the family Ptinidæ, attacks sweet almonds.

Another family of beetles which includes several drug-attacking species

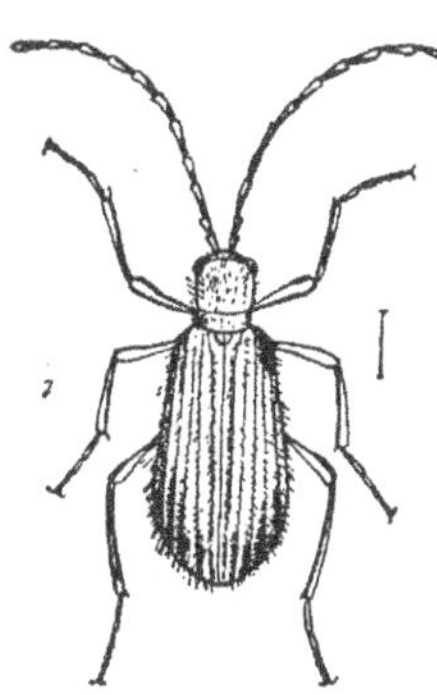

FIG. 267.—*Plinus brunneus* Duft.—(*Riley*.)

is the Dermestidæ. To this family belongs the common buffalo bug (*Anthrenus scrophulariacea*) of the house. The Dermestidæ comprise a number of beetles, mostly small, which feed on skins, furs, various dried animal substances, and, to some extent, on dried vegetable substances. *Anthrenus varius*, which I have found in jars of powdered cramp bark and fenugreek, is small, rounded-oval, with transverse black, white, and reddish-brown waved stripes (see *a*, Fig. 268). The grub differs from the larvæ of the Ptinidæ in bearing many long, bristly hairs (see *c*, Fig. 268). The adult beetle lives chiefly upon the pollen of certain plants, but the larva or grub lives indoors, and, feeding on rugs, woolen goods, collections of natural history, furs, hairs, and drugs, is a serious pest.

Another family of small beetles, the Cucujidæ, is represented among drug pests by several species of the genus *Silvanus*. The beetles belonging to this genus are about one-tenth of an inch long, light brown, flattened, and with antennæ clubbed at the tip (see Fig. 269). I have found *Silvanus surinamensis* attacking almond meal, *Silvanus advena* feeding on aconite root, and another species of *Silvanus* attacking angelica seed, quince seed, bitter-sweet, senega root, hyoscyamus, pellitory root, etc.

A large black beetle, *Tenebrio obscurus* (family Tenebrionidæ), is sometimes found attacking drugs. I have taken it in jars of parsley root. It is three-quarters of an inch long, dull black all over, with bead-like antennal joints, and with narrow, parallel, longitudinal ridges along the wing-covers. A small, shining, black beetle (genus *Paromalus*), belonging to the family Histeridæ, has been found in powdered poke root. Two species of *Ceutorynchus*, small snouted beetles or weevils, infest poppy and other seeds. Another weevil, *Calandra oryza*, imported from Europe, infests rice and ground roasted acorns.

The beetles comprise the chief drug pests, but some other orders of insects are represented by a lesser or greater number of pests.

The Lepidoptera or butterflies and moths, while possessing, in the adult stage, mouth parts adapted for sucking, have, in the young stages, strong biting-jaws. The young are the well-known caterpillars, and may be distinguished from the young or grubs of beetles by the number of legs. The larva or grub of the beetle has but three pairs of legs, and these are attached to the first three segments of the body lying just behind the head; the larva or caterpillar of a moth has, in addition to these three

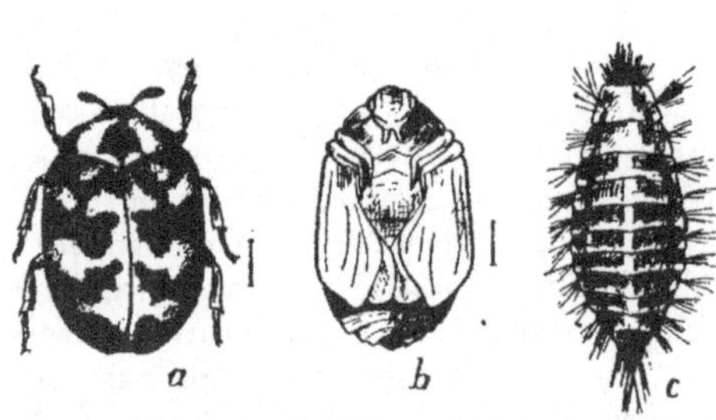

FIG. 268.—*Anthrenus varius* Fab.   *a.* Adult beetle.   *b.* Pupa.
c. Larva.—(*Original.*)

FIG. 269.—*Silvanus surinamensis* Linné.
—(*Original.*)

pairs of so-called thoracic legs, usually five more pairs of legs, four of these pairs being attached to segments in the middle region of the worm-like body, and the fifth pair being attached to the last segment of the body. The grubs of beetles sometimes have in addition to their three pairs of thoracic legs a *single* leg on the last segment of the body.

Every one knows of the clothes-moth, dread foe of the housewife, which, as a small white caterpillar, living in a cylindrical roll or case (see *d*, Fig. 270) made from the woolen cloth or fur on which it is feeding, does irreparable injury to the choicest fabrics and costliest furs. ' This moth belongs to the genus *Tinea*, of which one or more species attack drugs. Fig. 270 illustrates the life history of the moths of this genus; *c* is the larva or caterpillar; *b* is the pupa or resting stage; and *a* is the adult moth. The moth is very small and light brown in color. I have found a Tineid attacking aconite root. Another moth, known as the Angoumois grain moth (*Gelechia cerealella*), attacks, in the caterpillar stage, all kinds of stored grain. It bores holes into the grain kernels and eats out the starchy interior, leaving only a delusive hollow shell. Figure 271 shows the appearance of the infested grain kernels. The larva of *Carpocapsa amflana*, a moth of the same genus as the codlin moth, the greatest insect pest of the apple, infests the seeds of *Corylus avellana*, *Juglans regina*, and *Castanea vesca*. The larva of *Mylois ceratonia* feasts on the fruits of *Ceratonia siliqua* and *Castanea vesca*. The larva of the moth *Œcophaga olivella* inhabits the kernels of the olive, causing the dropping of the fruit and a smaller yield of oil.

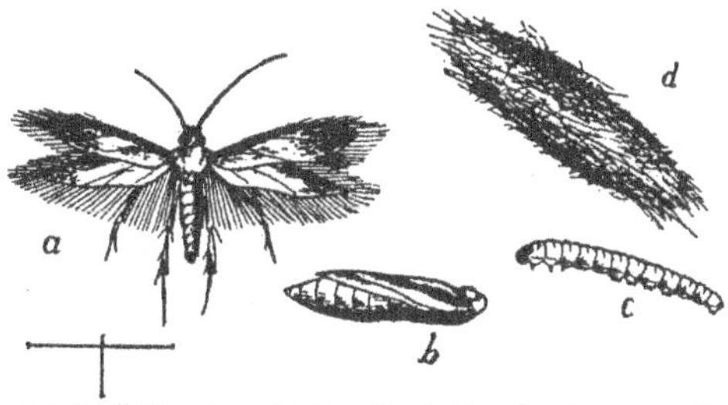

FIG. 270.—*Tinea penionella* Linné. *a.* Adult moth. *b.* Pupa. *c.* Larva. *d.* Case.—(*Original.*)

Passing now to another order of insects, the two-winged flies, we find that while the mouth parts of the adult flies are adapted for sucking or lapping, the young flies, which appear as grubs or maggots, are in many cases better prepared to partake of solid food. The olive in southern France and Italy is infested by a larva of a fly known as *Dacus oleæ;* in the kernels of fresh hazelnuts are often found the larvæ of a fly which belongs to the

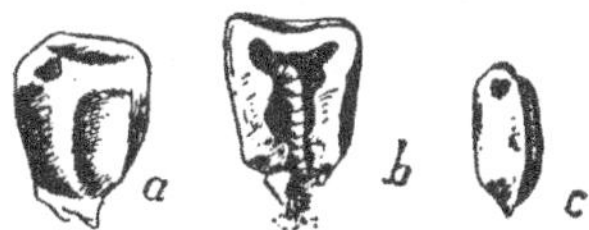

FIG. 271.—Grain kernels attacked by *Angoumois grain moth.* *a.* Infested kernel of corn. *b.* Kernel of corn cut open showing feeding larva within. *c.* Infested kernel of wheat.—(*Original.*)

same genus as that notorious wheat pest, the Hessian fly. The fly *Trypeta arnicivora* is often gathered in its youthful state with arnica flowers, and becomes developed later on, after feeding on the flowers in the pharmacist's canisters.

The book-louse insects (genus *Atropos*) have at least one representative in the list of drug pests. I have found a species (probably *divinatoria*) of this genus attacking golden seal and hyoscyamus. The insect is very small, hardly a twentieth of an inch long. When examined with a microscope it is found to be wingless, and of a general appearance as shown in Fig. 272. This insect represents the family Psocidæ, of the order Pseudoneuroptera.

The order of wingless insects *Thysanura*, which includes the "fish-moths," those active scale-covered little creatures of the household, is represented by a member of the genus *Lepisma* (probably *saccharina*) (see Fig. 273), which I have found in jars of mezereon bark and Socotrine aloes.

Finally, in jars of gall the pharmacist may find numerous little four-winged, compact-bodied "flies," which are not, however, attacking his stores, but which are only the insects which produced the galls, now issuing from them.   These little insects (see Fig. 274) are Hymenoptera, belonging to the genus *Cynips*.   The pharmacist may find other Hymenoptera (distinguished by having four clear membranous wings with almost no

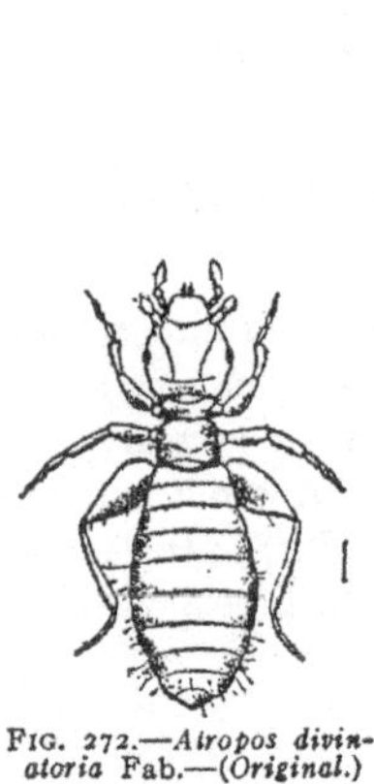

FIG. 272.—*Atropos divin-atoria* Fab.—(*Original.*)

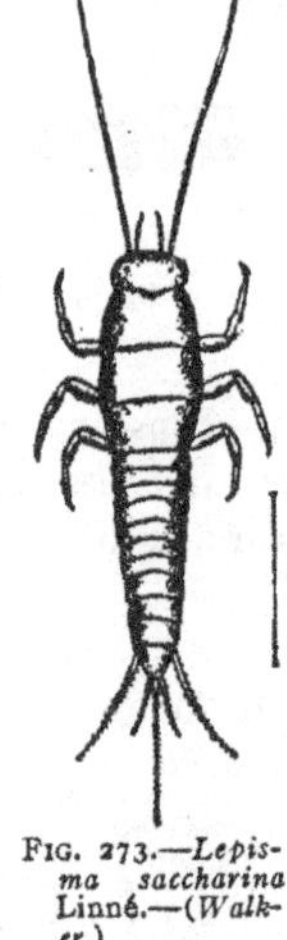

FIG. 273.—*Lepisma saccharina* Linné.—(*Walker.*)

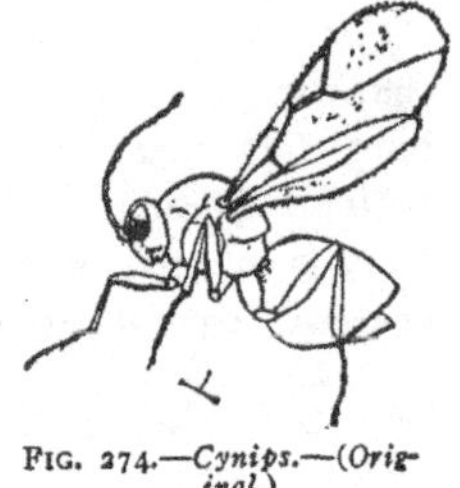

FIG. 274.—*Cynips.*—(*Original.*)

veins in them, see Fig. 274) in his jars and cans; but these insects are his benefactors.   They are parasitic on the beetles and other insect pests which are feeding on the drugs, and thus do much good.   Their eggs are laid on the body of the grub of the drug-eating beetle, and the young hymenopteron, on hatching, eats its way into the beetle-grub and lives there at the expense of its host.

REMEDIES

Coming now to the matter of remedies, a reviewing of the notes thus far presented shows that beetles are the most serious and numerous of drug pests, and that practically only insects which have biting mouth parts are injurious.   In fighting insects with biting mouth parts the common means employed by entomologists is to cover the substance attacked (usually the green foliage of plants) with a thin coating of arsenic, by means of spraying. In the nature of the case this method is out of the question in fighting drug pests, but, because the drugs are capable of being easily handled and subject to treatment in air-tight vessels, a very convenient, effective, and universally applicable method is possible, namely, treatment with vapor of bisulphide of carbon. The vapor of bisulphide of carbon is deadly to all insects in all stages, except the egg stage. The infested drug should be placed in a tight vessel (after having removed the dust and debris caused by the attacks of the insects) and a quantity of bisulphide of carbon, sufficient to charge the vessel with vapor, introduced. Any insect in the vessel will be killed. The remedy is simple, effective, and is feasible in the case of almost any drug.

Prevention of attack may be accomplished in some degree by the use of tight cases, though often the insects are introduced into the case with the drug, the drug specimens having come from an infested lot. Occasionally inspection of the jars and cans will detect the insects before they have had time to do much damage.

The ease of the detection of the presence of insects, and the ease with which the pests may be killed, makes it certainly worth the while of any druggist to devote a little time required for the effective prevention of insect damage to his stores.

PART IV
POWDERED DRUGS

CHAPTER I

A.—METHODS FOR IDENTIFICATION

Vegetable drugs frequently, perhaps in the majority of cases, reach the pharmacist in the form of powders, and it is necessary not only to identify them, but to determine their quality in this form. The old and laborious method of making powders in small quantities, by the pharmacist in his own store, has been supplanted by the specialized industry of drug milling. Thus it is that adulteration is made easier and its detection more difficult. Formerly it was considered sufficient for identification of vegetable drugs to describe gross characteristics, such as, color, odor, taste, and such other characteristics as might be brought out by hand lens; but this method is wholly inadequate, and a more detailed examination, microscopical and chemical, now is required. The enforcement of the drug and food laws will require workers skilled in microscopical technique.

Pulverization and Powdering.—Prerequisite to the microscopical study of vegetable powders is a knowledge of the processes of pulverization and drug mills, such as may be found in any well illustrated work on pharmacy, and elements of plant anatomy.

The degree of fineness of the powders is of first importance in microscopical examinations. Coarse powders can not be used and if they are too fine the fragmentary tissues and products are too small to be recognized. These degrees of fineness are represented by certain numbers. A No. 80 powder, as defined by the U.S.P. VIII, for example, is one that will pass through a sieve having 80 meshes to the inch. In the U.S.P. IX No. 80 powder is defined as "Very fine powder, has a fineness in diameter of particles less than 0.17 millimeters," and it is specified also that the larger proportion of this must not pass through a sieve of lower degree of fineness (See U.S.P. IX, Part II). To obtain the best results, microscopically, powders may vary in fineness from No. 60, a fine powder, to No. 80, a very fine powder.

During the process of pulverization the less resisting tissues, such as thin-walled parenchyma cells, which, for the most part, contain starch, proteids and crystals, are reduced rapidly to powder, while the woody and fibrous parts together with the tracheids and vessels are quite difficult to pulverize. Accordingly, frequent sifting should be resorted to during the process, so that as the broken fragments are reduced to the proper size to pass through the sieve they may be removed. The process of grinding and sifting must be continued until all the tissues have passed through the sieve. Powders in small quantities may be made by means of a mortar and pestle, and if the material is thoroughly dry the time and labor need not be great. A mortar and pestle made rough by the use of coarse carborundum powder has proved very efficient. Coarse powders in considerable quantity may be made in an ordinary small coffee mill. The process of grinding may then be continued by means of mortar and pestle and the fineness carried to any degree desired.

Color.—Vegetable powders are liable to vary greatly in color. Some of the common factors which cause this variation are light, moisture and increasing fineness. Exposure to light deadens the color, in some cases very rapidly, a light or reddish-brown soon becoming a dark or dull brown, etc. By exposure to moisture most powders grow dark in color. Increasing fineness produces varying tints and, in some instances, the quality of the color is wholly changed; for example, Spanish Licorice, in coarse powder, is yellow showing considerable portions of brown cork, while a very fine powder is almost lemon color. If the process be carried on by alternate grinding and sifting, as described above, tints from yellow to light lemon yellow will be obtained. The aging of powders, even when not exposed to light, changes them to darker tints. Powders made from plant parts, rich in oil, are likely to be dark in color and the darkening may become marked if heating is allowed to occur during the grinding. They darken rapidly on exposure to light and are likely to become rancid.

Various systems of classification by colors have been worked out for the vegetable drugs. Doctor Schneider has divided them into six groups as follows: 1, Very light; 2, yellow; 3, green; 4, gray; 5, brown; 6, very dark. Professor Henry Kraemer forms them into five main groups: 1, Greenish powders; 2, yellowish powders; 3, brownish powders; 4, reddish powders; 5, whitish powders. These groups are subdivided according to the forms of cells, nature of the cell wall and cell products. All such systems as these are more or less artificial, and although useful in many cases, have not proved wholly satisfactory in the laboratory.

Identification by Odor.—The odors from drugs are exceedingly difficult to describe, largely because we have no odor standards at command for comparing them qualitatively or quantitatively. We can understand such terms as aromatic, pungent, fragrant, agreeable, disagreeable, etc. These terms serve in a measure to indicate odor qualities.

The student is recommended to acquaint himself with such aromatic odors as cinnamon, cloves, nutmeg; with the mint family odors, such as peppermint, spearmint, pennyroyal, etc. He should acquaint himself with such odors as are furnished by the odorous fruits, of the Umbelliferæ, such as caraway, fennel, etc.; with camphoraceous odors, as eucalyptus, rosemary, and camphor; with pronounced and characteristic odors of wintergreen, sassafras, etc.; with the delicate and fragrant odors derived from the lemon, orange, orange flowers, etc. He should not omit the study of the disagreeable odors, as we find in conium, valerian, stramonium, garlic, civet, castor fiber, etc. All such odors serve as a means of comparison.

It will be seen that in order to describe an odor it becomes necessary to have some prominent characteristic odor with which to compare. The Pharmacopœia (viii) states that conium has a mouse-like odor; sumbul, a musk-like odor; lactucarium, a heavy odor; senna is described as having a tea-like odor, etc. Tarry substances that have a creosote or smoky odor are said to have an "empyreumatic odor."

Identification by Taste.—What has been said of the odor of drugs applies also to their taste. Taste is not a very distinctive property. There are some drugs that have a distinctive taste, such as gentian root, which has a *simple bitter taste;* senega, an *acrid taste;* ginger, a *pungent;* geranium, *astringent;* elm bark, *mucilaginous*, etc. Many drugs have what may be termed a mixed taste. Hence we find in descriptions such terms as: *bitter-astringent* applied to cinchona; *bitter-pungent* applied to orris root; *pungent-astringent* applied to cotton-root bark; *bitter-sweet*, applied to dulcamara; *sweetish-bitter-pungent*, applied to spigelia, etc. Many drugs are tasteless, such as lycopodium, kamala, physostigma, etc.

It is plain to be seen from the foregoing that the taste, as well as the color and odor of powders, is not distinctive enough to identify them with certainty; still, these physical properties serve in many cases as a valuable aid in their identification.

Adulterants and Their Identification.—As stated above, adulteration of drugs is made easier and the detection of adulterants is more difficult when the drugs are reduced to powders. Great skill is required in the identification of adulterants; for the art of drug adulteration is an old one and the materials employed have been selected, often ingeniously, on account of their very close resemblance to the true articles they replace. In the case of whitish powders, foreign starches, especially the common cereal starches, have been used, and not infrequently have the "scrapings" from bakeries been parched or browned to the proper degree and employed in drug and food adulteration. The endocarp of the olive, cocoanut, and walnuts; exhausted coffees; cocoa shells; and other similar substances, which are composed chiefly of stone cells, have been employed to a large extent in admixture with brownish powders. The use of wheat bran or middlings in ginger has been a common practice. Sometimes inorganic substances such as talc, chalk, clay, sand, etc., are employed. One of the most difficult means of adulteration to detect is the use of exhausted powders (the dregs left from drugs extracted by percolation). These are first dried and repowdered and mixed in various proportions with the pure article. Deteriorated drugs have been used in the same way. It goes without saying that these latter forms of adulteration can not readily be detected microscopically, but a microscopical examination in connection with careful chemical tests is of the greatest value.

A thorough knowledge of the histology of the plant part supposed to constitute the powder is necessary. And for this purpose cross and longitudinal sections, which may be prepared after soaking the dried drug materials in water, may, in many cases, be used to great advantage. By careful comparisons of sections and broken fragments, and the employment of proper reagents upon cell-products, identification is made positive. For a full account of cell-products and reagents, see Part IV, Chapters II and III.

Mounting Powders for Examination.—Powders for microscopical examination should be thoroughly mixed, so that the large and small particles will be uniformly distributed throughout the entire specimen, as before stated. In powders that have been standing for a considerable time the larger particles will be separated from the finer, so that great difficulty may be encountered in obtaining a typical mount from such a powder, unless it has been thoroughly mixed. Only a small portion of powder should be used in making a mount, the amount depending upon the size of the cover-slip to be used. When the mount is ready for examination, the particles should be spread out evenly and should not come in contact one with another so that the large ones might obscure the smaller.

Powders for examination may be mounted directly on the slide, using the proper medium, or the powder may be mixed with the mounting medium in a small test-tube, specimen tube, or homeopathic vial. If a small portion of powder be transferred to a slide, a drop of the desired mounting medium added, and the whole thoroughly mixed and covered with a cover-slip, it will furnish a mount ready for examination. However, it is frequently desirable or even necessary to use some clearing agent in order to render dark colored or opaque powders transparent. In such cases the powder should be thoroughly mixed with the reagent and left standing for twelve hours or more, when a portion may be taken up with a pipette and a drop of the mixture transferred to a slide.

Clearing Agents and Mounting Media.—For making temporary mounts of powders water is the best general medium, and should be used whenever a clearing agent is not required. In this medium delicate markings are clearly brought out, and it is especially recommended for the examination of starches. Frequently specimens are filled with air, which must be removed before a satisfactory examination can be made. For driving out air 70 per cent. or stronger alcohol should be used, but this is not a desirable medium for general use, as it evaporates rapidly and allows the specimens to dry up. However, this medium is excellent for bringing out details of structure, and may be profitably employed when a hasty examination is to be made. It can be replaced by water or other media as desired.

Equal parts of water and glycerine furnish one of the best and most useful mounting media. This mixture is especially desirable when delicate markings are not brought out in water. It acts as a clearing agent, and although the action is somewhat slow, it will render most specimens clear enough for examination. Equal parts of water, glycerine, and alcohol make a reagent to be preferred to the above in many respects, and is the most useful of the simple and cheap reagents. This mixture penetrates tissues well, acts as a clearing agent, and does not dry up. Specimens may be kept in it for days or even weeks.

In the examination of many specimens it is necessary to use a strong clearing agent, and it is frequently desirable to have one that acts rapidly. Chloral hydrate, made by dissolving five parts of chloral hydrate crystals in two parts of water, is one of the most common and useful clearing agents. Its action is rapid, but it is not a good medium for mounting in many cases, since delicate markings are not clearly brought out by it. In many specimens starch is dissolved by this reagent, and it should never be used when accurate measurements of starch grains are to be made. However, chloral-hydrate solution with iodine added is the best and most reliable agent for the detection of starch, and is especially recommended where starch occurs in small quantities or is likely to be obscured, as in chloroplasts or by proteid substances.

A clearing agent to be preferred to the above for general purposes may be made by mixing 1 part of 95 per cent. alcohol, 1 part glycerine, 1 part water, and 4 parts saturated aqueous solution of chloral hydrate. This mixture gives a reagent fairly rapid in action, and also serves well as a mounting medium. It is the most useful clearing agent and can be employed in more cases than any other.

Potassium hydrate in 2 to 10 per cent. aqueous solution is valuable as a clearing agent, and also serves well as a macerating agent. It is rapid in action, and dissolves starch. Acetic acid, 20 per cent., and hydrochloric acid, 10 to 20 per cent., may be found exceedingly useful as clearing agents in many cases. They are often valuable in removing starch from specimens where it may interfere in an examination.

In the preparation of specimens which are exceedingly difficult to clear, or in handling coarse powders where the fragments are so large that they must be broken up by macerating before mounting, javelle water and Schultz's macerating fluid will be found useful.

The action of any of the clearing agents mentioned above may be hastened or increased by the application of heat. By holding a mounted specimen over the flame of an alcohol lamp or a Bunsen burner it can be heated without injury, even to boiling, if proper care be exercised.

For more detailed directions for the use of reagents see Chapter II, on Reagents and Processes, where a complete list of them is discussed and explicit directions given for their use. This list of reagents is arranged alphabetically and is in convenient form for handy reference.

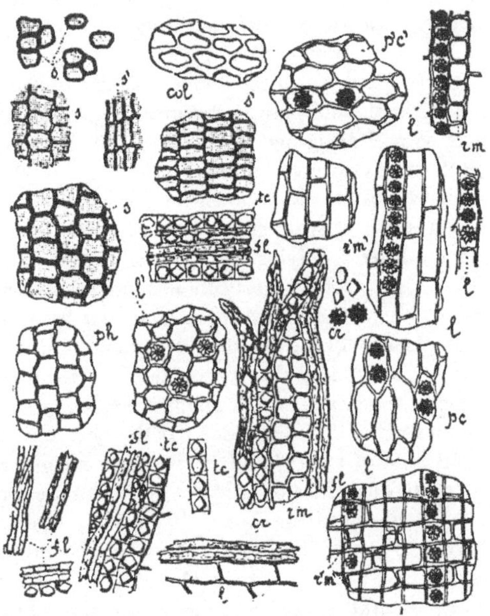

FIG. 275.—Powdered *Rhamnus Frangula*—Bark. (X 210.) *col*, Collenchyma of the cortex. *cr*, Prismatic and rosette crystals. *fi*. Bast fibers with pitted walls. *l*, *l'*, Bast in longitudinal and transverse section. *pc*, *p'c'*, Cortical parenchyma, in longitudinal and transverse section. *ph*. Phelloderm. *rm*, *r'm'*, Medullary ray in tangential and radial section. *s*, *s'*, Cork in tangential and transverse section. *tc*, Rows of crystal cells.—(*From Greenish and Collin.*)

Measurements.—The fragments of powders should be carefully measured, and the measurements used for comparison wherever it is possible to do so. Measurements should be made with an eye-piece micrometer. In preparing specimens for measurement the greatest care should be exercised in the use of reagents so that objects may not be swollen abnormally or distorted before measurements are made.

On the following pages are given a few examples to show the diagnostic characteristics of some powders which frequently, either by mistake or intentionally, are substituted one for the other.

The first example is illustrated by the barks taken from the same genus —Frangula, Fig. 275, and Cascara sagrada, Fig. 276. A comparison of the fragments composing these two powders shows them to be very similar in structure. Cascara presents one striking difference, as shown by the sclerenchymatous cells, *sc*, Fig. 276, which occur quite commonly, but occur rarely, if ever, in Frangula. In each of the specimens are bast fibers, but in Frangula the fibers have thicker walls and contain more numerous and well-defined pits than do the fibers of Cascara. Also the

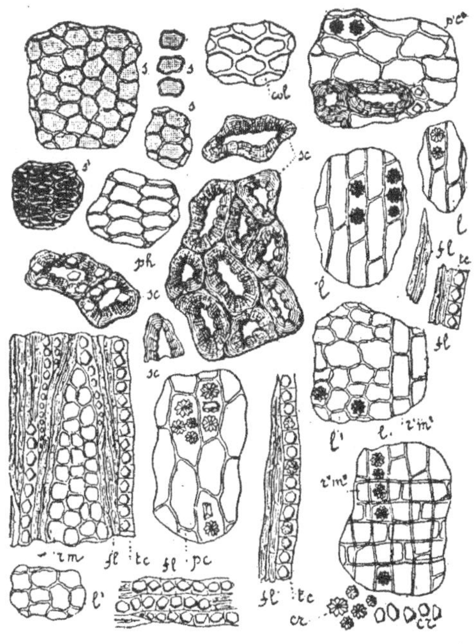

FIG. 276.—Powdered Cascara Sagrada Bark (*Rhamnus Purshiana*). (X 210.) *col*, Collenchyma of the cortex. *cr*, Prismatic and rosette crystals. *fl*, Bast fibers. *l, l'*, Bast in longitudinal and transverse section. *pc, p'c'*, Cortical parenchyma in longitudinal and transverse section. *ph*, Phelloderm. *rm*, Medullary rays, tangential section. *r'm'*, The same, transverse section. *r"m"*, The same, radial section. *s, s'*, Cork, in surface view and section. *sc*, Sclerenchymatous cells. *tc*, Rows of crystal cells.—(*From Greenish and Collin.*)

cork cells and the large parenchyma cells of the cortex show characteristics which are of diagnostic value. In Frangula the cork cells contain a deep red or purplish coloring substance, while those of Cascara have a reddish-brown coloring substance. In the large parenchyma cells of Cascara is found a substance yellowish in color which changes to orange upon the addition of potassium-hydrate solution, while in Frangula the large parenchyma cells contain a coloring substance of a much brighter yellow, which upon the addition of potassium-hydrate solution changes to a red or deep purplish color

The second example is illustrated by two roots taken from closely related species—Brazilian Ipecac, Fig. 277; Psychotria Ipecacuanha (Stokes) of the British Pharmacopœia; Cephælis Ipecacuanha (A. Richard) of the U.S.P.; and undulated Ipecac (Fig. 278), which represents species from several different genera, such as Richardsonia, Psychotria, Ionidium, etc. The starch grains from each specimen are similar in form and structure, the only difference being that the starch grains from Brazilian Ipecac, ranging in size from 4 to 15 microns, are uniformly smaller than

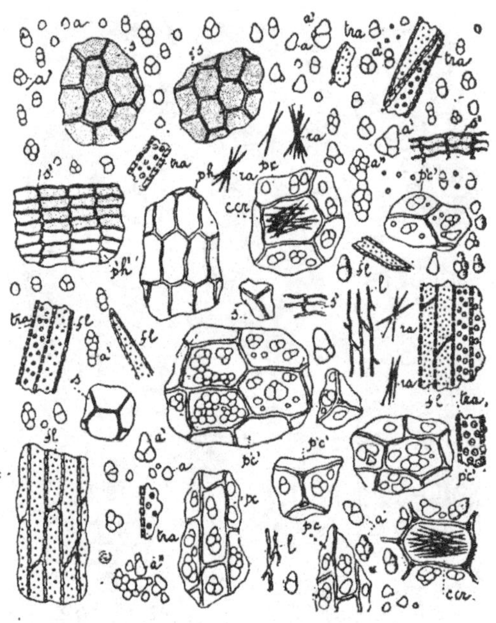

Fig. 277.—Powdered Ipecacuanha Root (*Cephælis ipecacuanha*). (X 210.) *a, a', a"*, Starch grains, simple and compound. *ccr*, Cells with calcium oxalate. *fl*, Fibrous cells. *l*, Bast. *pc, p'c'*, Cortical parenchyma in longitudinal and transverse section. *ph, p'h'*, Phelloderm in surface view and section. *ra*, Raphides. *s, s'*, Cork in surface view and profile. *tra*, Tracheids.—(*From Greenish and Collin.*)

are those of undulated Ipecac. The elements of the xylem, however, furnish a ready and reliable means of distinguishing between these two powders. The xylem of Brazilian Ipecac consists of tracheids, *tra*, Fig. 277; and of peculiar strongly pitted wood parenchyma, which somewhat resembles tracheids, *fl*, Fig. 277. Undulated Ipecac shows the presence of strongly pitted water tubes (pitted vessels), *v*, Fig. 278, and quite typical wood fibers, *fl*, Fig. 278. Brazilian Ipecac does not show water tubes, unless fragments of the stems become mixed with the roots.

As a third example, the leaves of Belladonna, Fig. 279, and Hyoscyamus, Fig. 280, furnish an excellent illustration. The epidermal cells of Belladonna are large with wavy walls and the cuticle is striated, *es*, Fig. 279; while Hyoscyamus has epidermal cells similar in every respect excepting the striated cuticle, *ei* and *es*, Fig. 280. The spongy parenchyma of Hyoscyamus contains numerous crystals of calcium oxalate, usually in the form of prisms, *cr*, *ccr*, Fig. 280, while Belladonna is without calcium oxalate excepting for crystal sand, which is contained in a few large cells of spongy parenchyma adjoining the palisade parenchyma—*c*, *cr*, Fig. 279. The presence of prismatic crystals in Hyoscyamus is the most striking diagnostic character of these two powders.

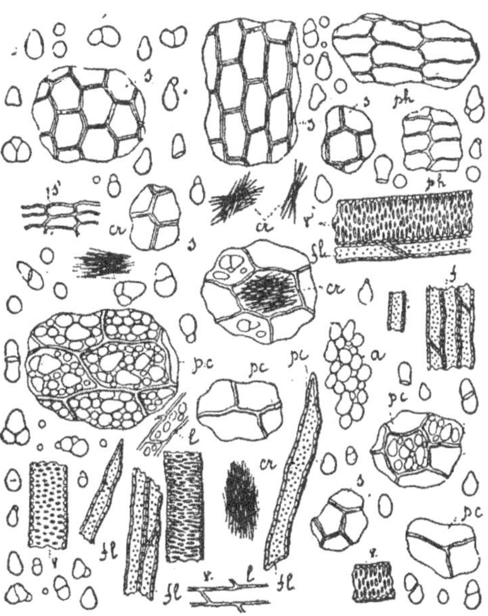

FIG. 278.—Powdered Undulated Ipecacuanha. (X 210.) *a*, Starch grains. *cr*, Acicular crystals. *ft*, Pitted wood fibers. *l*, Bast. *pc*, Cortical parenchyma. *ph*, Phelloderm. *s, s'*, Cork, in surface view and section. *v*, Pitted vessels.—(*From Greenish and Collin*.)

The trichomes furnish other valuable diagnostic characters, but they are not always reliable, since Belladonna leaves that are almost glabrous, and consequently almost devoid of trichomes, are sometimes found. Either specimen may contain both simple and glandular hairs. The simple hairs are conical and may be composed of one or more cells. In Hyoscyamus the glandular heads, which may be either bicellular or multicellular, *pg*, Fig. 280, are borne on a stalk composed of two or more cells. The glandular hairs of Belladonna are found with heads either unicellular or multicellular. The larger multicellular glands are usually borne on a stalk consisting of one or two cells, *pg*, Fig. 279, while the smaller ones

are likely to have a stalk composed of several superimposed cells.   The
unicellular glands are rounded in form and are borne on stalks of several
cells, *pg*, Fig. 279.

It should be stated that each drug has its own peculiar microscopical
elements.   Some of these, it is easy to see, are of special value in the
identification of drug powders.

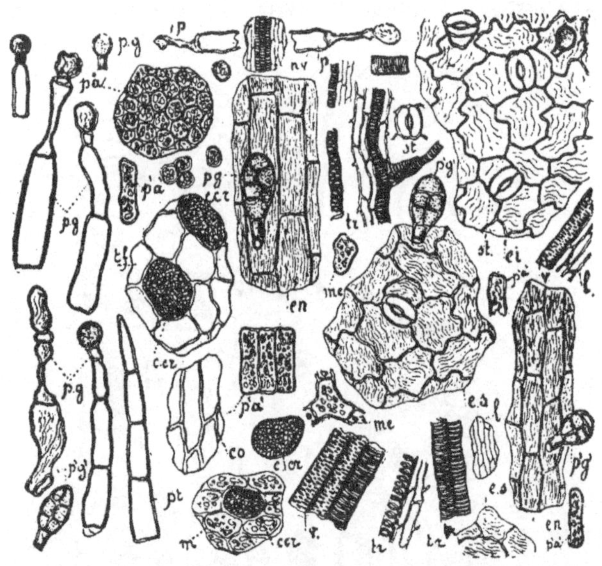

FIG. 279.—Powdered Belladonna Leaves. (X 210.)   *c, cr*, Cells with sandy crystals. *co*,
Collenchymatous cells from cortical tissues of midrib. *ei*, Epidermis of under surface. *en*, Epidermis over the veins, with striated cuticle. *es*, Epidermis of the upper surface, with striated cuticle
and occasional stomata. *l*, Bast. *me*, Branching cells of spongy parenchyma. *nv*, Fragment of
small vein. *pa*, Palisade cells, surface view. *p′a′*, Palisade cells, in longitudinal section. *pg*, Glandular hairs, long and short, with unicellular and pluricellular glands. *st*, Stomata, surrounded by
three or four cells, one of which is smaller than the others. *tf*, Cortical tissues of the midrib. *tr, v*,
Tracheids and vessels.—(*After Greenish and Collin.*)

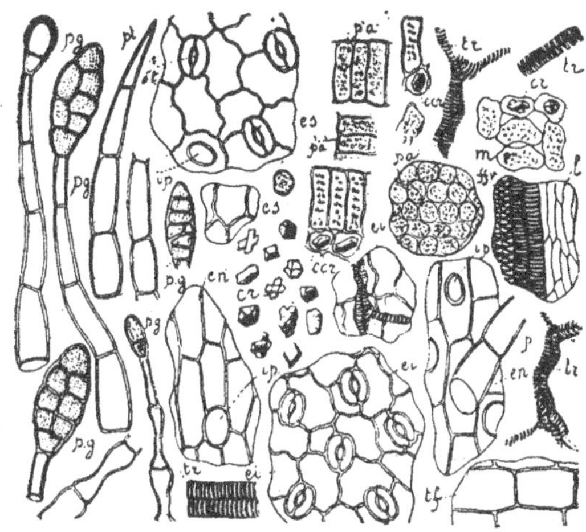

FIG. 280.—Powdered Henbane Leaves (*Hyoscyamus niger.*) (× 210.) *ccr,* Crystal cells. *cr,* Crystals of calcium oxalate. *ei,* Lower epidermis. *es,* Upper epidermis. *ffv,* Portion of fibrovascular bundle of midrib. *ip,* Scar of fallen hair. *m,* Spongy parenchyma. *pa, p'a',* Palisade cells. *pg,* Glandular hairs. *pt,* Simple hairs. *st,* Stomata. *tf,* Cortical parenchyma of midrib. *tr,* Tracheid and Vessels.—(*After Greenish and Collin.*)

GENERAL DIRECTIONS

As a general direction for the detection of adulteration or admixture it cannot be too strongly emphasized, that authentic samples of the pure drug, and of the suspected adulterant or admixture, should be carefully studied, macroscopically, and microscopically, as a preliminary process. This laboratory method supersedes all the aids in the form of representation by drawings and figures on paper.

An examination of a drug powder should never be considered complete until the sample has been compared with authentic specimens of the same drug or drugs of the same degree of fineness.

Fig. 281.—Shows Starch-granules of Ipecac. ($\times$ 750.) The cells of the bark are filled with starch. The granules are spherical, oblong, or angular, and vary much in size. The hilum is located near the center, and is often seen to be fissured. The grains are smooth, and show no concentric markings. They are often in groups of two, three, and sometimes even more grains joined together.

Fig. 282.—Shows Starch-granules of Jalap. ($\times$ 250.) The grains are very numerous in the cells; are large and have characteristic markings. They are rounded or broadly ovate, having the hilum located near the small end and surrounded by excentric lines.

Fig. 283.—Shows the Starch-grains of *Veratrum viride* ($\times$ 350), which so closely resemble those of *Veratrum album* that it would be impossible to distinguish the two by their starch-grains. Those of the former are often found in groups of twos, threes, fours, and sometimes even more. They are small, rounded, or angular, with the hilum in the center.

Fig. 284.—Represents Starch as it appears in Calumba. ($\times$ 350.) The grains are large, and in shape they are circular or oval. A few double or compound grains are found, but they do not occur frequently. The hilum is rather excentric, and is often seen to be fissured in a radial direction. The grains are smooth, and occasionally a curved line or two is to be found.

Fig. 285.—Shows Starch-grains as they appear in Galengal. ($\times$ 350.) The grains are large and mostly long ovate, but sometimes they are irregular. The hilum is located near the larger end, and is sometimes fissured. The stratification lines are plainly seen on the larger grains and but faintly, if at all, on the smaller ones.

Fig. 286.—Illustrates Starch-grains as seen in a specimen of *Iris florentina*. ($\times$ 500.) These grains are quite characteristic and very abundant. They are rather elongated, rounded or truncate at one end, and usually tapering toward the other end. Occasionally a three-lobed grain is seen. As a rule, the grains are irregular in shape. The hilum is located near the large end, and is slightly fissured. (*a*) is the most common form. A very prominent characteristic is a double line branching from the hilum and extending toward the other end.

Plate III.

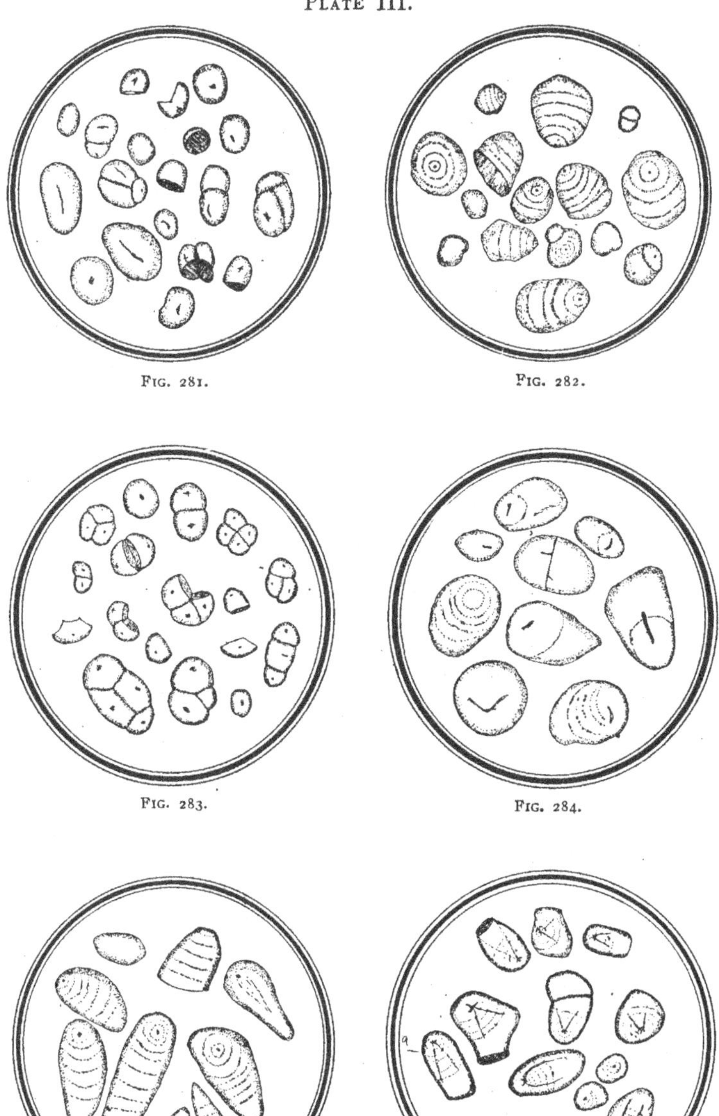

Fig. 281.

Fig. 282.

Fig. 283.

Fig. 284.

Fig. 285.

Fig. 286.

Fig. 287.—Shows Starch-grains as they appear in Caulophyllum. (× 250.) The grains are small, but quite characteristic. They are mostly gathered together in large and roundish masses, consisting of twenty-five to fifty grains. The single grains are globular, or more commonly many-sided, and without hilum or stratification lines.

Fig. 288.—Shows the Grains as they appear in *Aconitum napellus*. (× 850.) This drug is very rich in starch. The starch-grains are rather large. There are a great many compound grains composed of from two to eight granules. The single grains are round, long, and in some cases have flat faces. The hilum is located centrally, and is seen at times to be fissured slightly. The concentric markings are not discernible.

Fig. 289.—Shows Starch-grains as they appear in Geranium. (× 1200). There are specimens of Geranium in the market that contain little or no starch. This somewhat singular fact is said to be due to the season in which it is gathered. The drug usually contains starch in abundance. The grains are rather long, and appear to be thicker at one end than at the other. The hilum is located generally at the larger end, but sometimes central, and it occasionally appears at the smaller end. The stratification lines are very faintly seen at times.

Fig. 290.—Shows Starch-grains as they appear in Honduras Sarsaparilla. (× 500.) Many of the grains are seen to occur in groups of two, three, and sometimes four. The single grains are spherical or angular, with a hilum located near the center. The hilum in the larger grains is angular fissured. No concentric markings can be seen.

Fig. 291.—Shows Starch as it appears in Podophyllum. (× 550.) The grains are small and mostly single, but sometimes they are double or triple. They are spherical with a central hilum, and are seldom fissured. The hilum can hardly be seen in the smaller grains.

Fig. 292.—Shows Starch as it appears in the rhizome of Hydrastis. (× 1300.) The starch is very abundant. The grains are most commonly joined together in groups of from two to six. The grains, when single, are rounded in form. The hilum is indistinct and unfissured.

Note.—The drawings of the starches were made from authentic specimens of the crude drug of the market.

Plate IV.

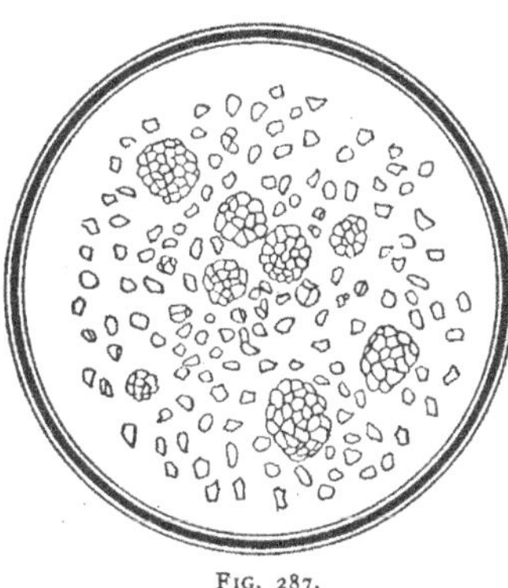

Fig. 287.

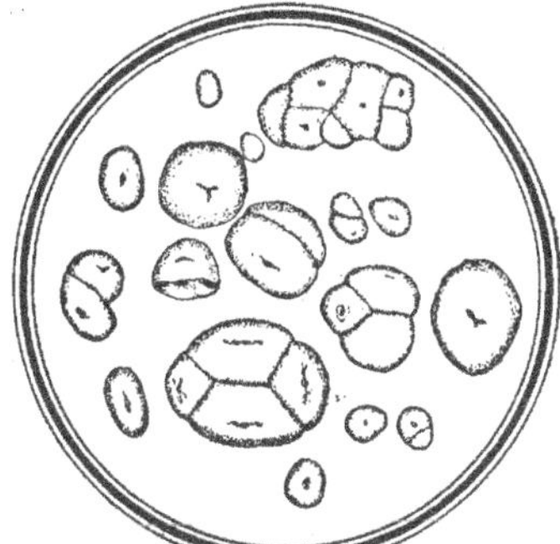

Fig. 288.

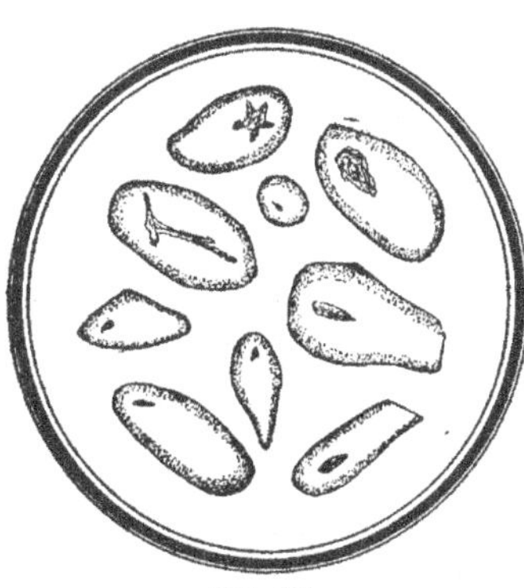

Fig. 289.

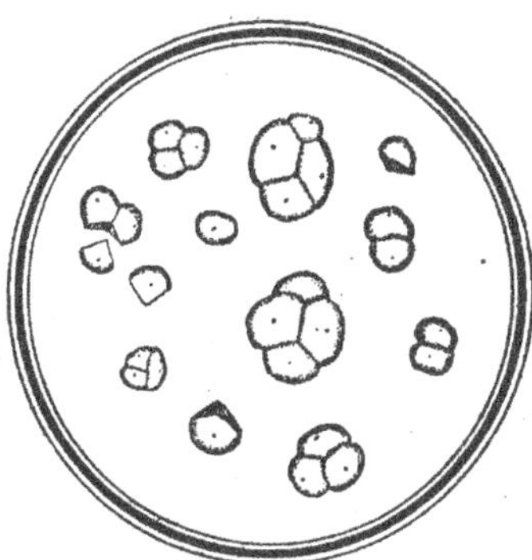

Fig. 290.

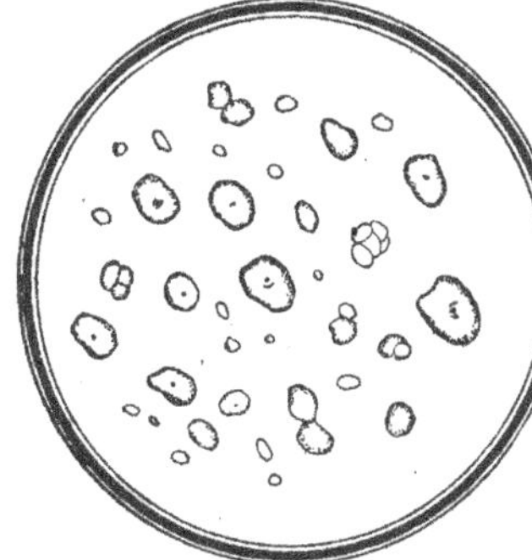

Fig. 291.

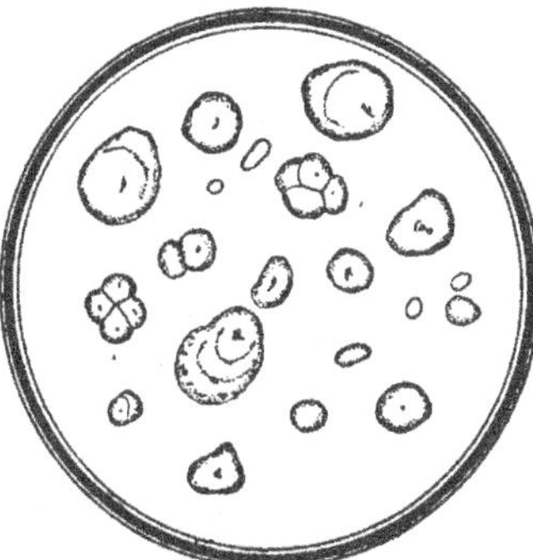

Fig. 292.

TYPES OF DRUG POWDERS

The following pages (512–519) are illustrations of some of the more important drug powders of the National Formulary and of the Pharmacopœia, designed to illustrate how characteristic elements may be selected for purposes of microscopical identification.

On pages 520–528 will be found condensed descriptions of the characteristic elements of some of the more important drug powders selected mainly to give as wide a range as possible for purposes of identification.

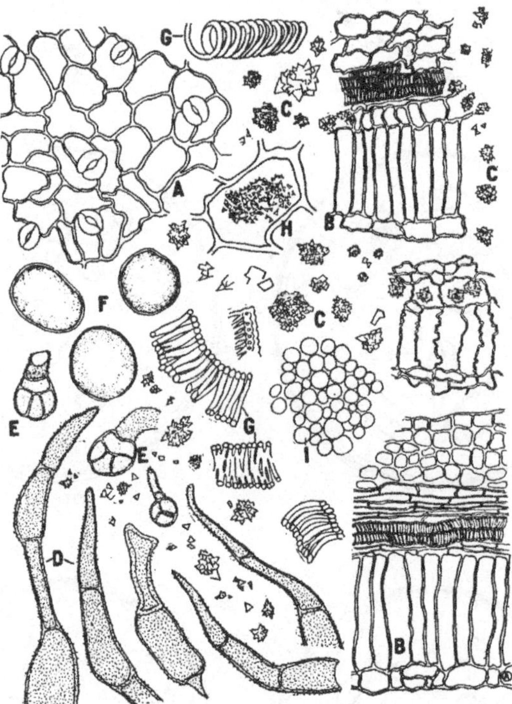

Fig. 293.—Powdered Stramonium. (× 183.) A, Epidermis, surface view. B, Transverse section of leaf. C, Aggregate crystals of calcium oxalate. D, Non-glandular trichomes. E, Glandular tricomes. F, Pollen grains. G, Water tubes. H, Mesophyll cell, containing sphenoidal crystals of calcium oxalate. I, Palisade parenchyma, surface view.

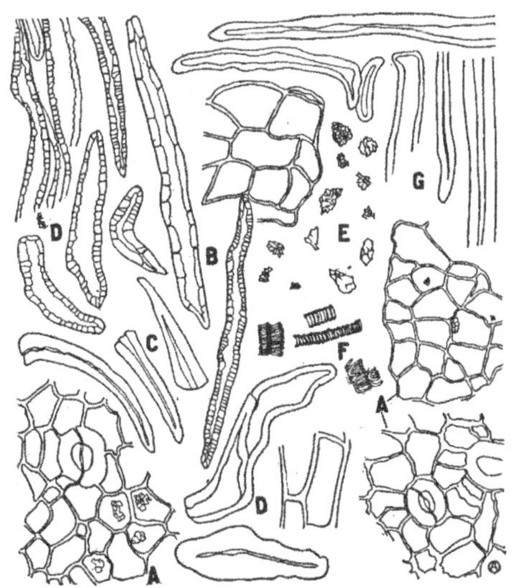

Fig. 294.—Powdered Buchu (Short). (× 183.) A, Epidermis showing crystals of carbohydrate. B, Sclerenchyma fibers. C, Trichomes. D, Sclerenchyma cell, water storage tissue near the ultimate ends of the veins. E, Aggregate crystals or calcium oxalate. F, Water tubes. G, Sclerenchyma fibers.

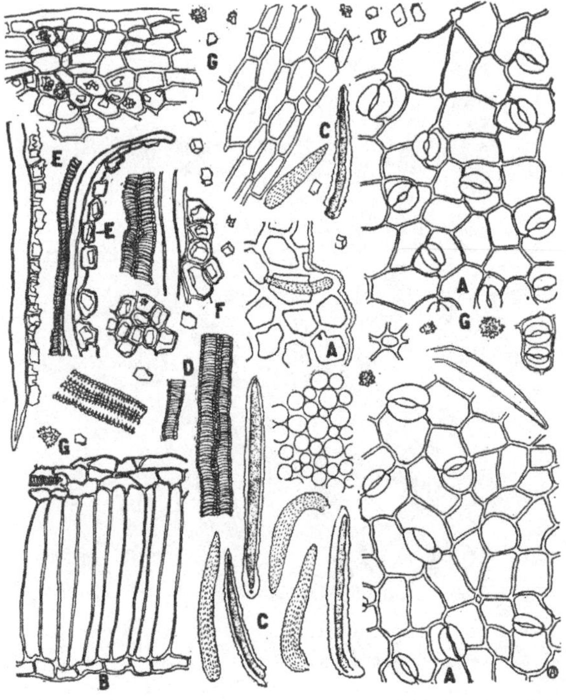

FIG. 295.—Powdered Senna (India). (× 183.)　A, Epidermis, surface view.　B, Part of transverse section of leaf.　C, Trichomes.　D, Water tubes.　E, Crystal fiber.　F, Parenchyma cells containing prisms of calcium oxalate.　G, Aggregate crystals of calcium oxalate.

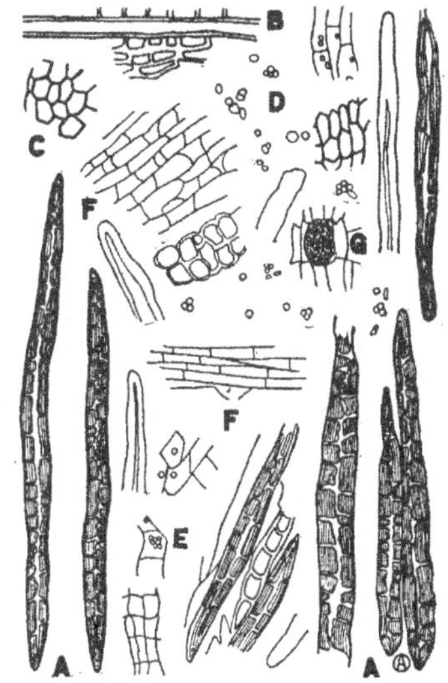

FIG. 296.—Powdered Yellow Cinchona. (× 60.) A, Bast fibers. B, Bast fibers with adjacent parenchyma cells. C, Cork, surface view. D, Starch. E, Parenchyma containing starch. F, Fragments of the phloëm. G, Parenchyma containing fine crystals of calcium oxalate.

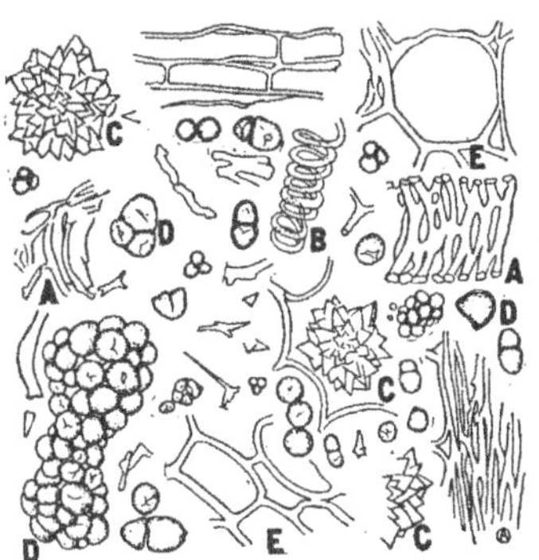

FIG. 297.—Powdered Rhubarb. (×183.) A, Fragments of reticulate water-tubes. B, Spiral water tube. C, Aggregate crystals of calcium oxalate. D, Starch. E, Parenchyma. F, Phloëm.

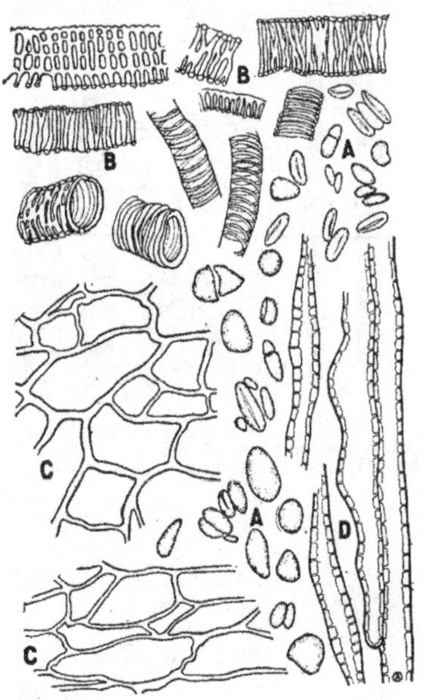

FIG. 298.—Powdered Zingiber, Jamaica Ginger. ($\times$ 183.) A, Starch. B, Water tubes. C, Parenchyma cells. D, Sclerenchyma fibers.

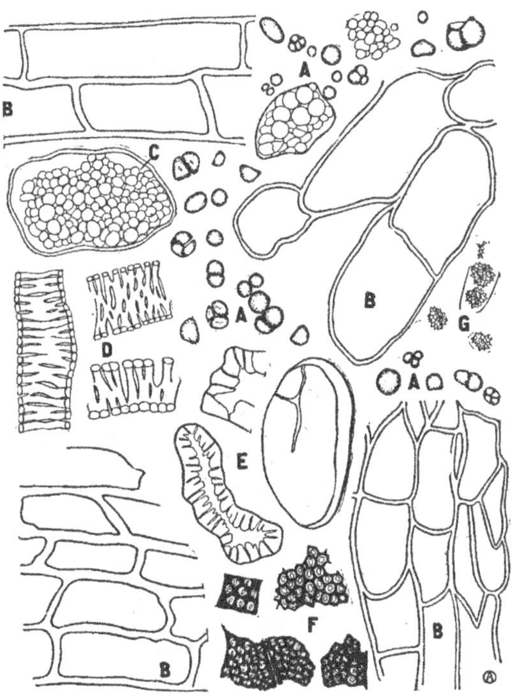

FIG. 299.—Powdered Jalapa. (× 183.) A, Starch grains. B, Parenchyma cells. C, Parenchyma cell filled with starch. D, Water tubes. E, Stone cells. F, Fragments of porous water tubes. G, Aggregate crystals of calcium oxalate.

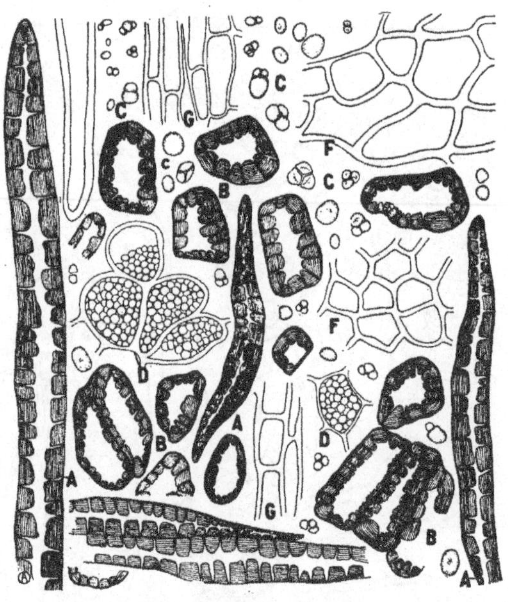

FIG. 300.—Powdered Saigon Cinnamon. (× 183.) A, Bast fibers. B, Stone cells. C, Starch. D, Parenchyma cells containing starch. F, Cork, surface view. G, Fragments of phloëm.

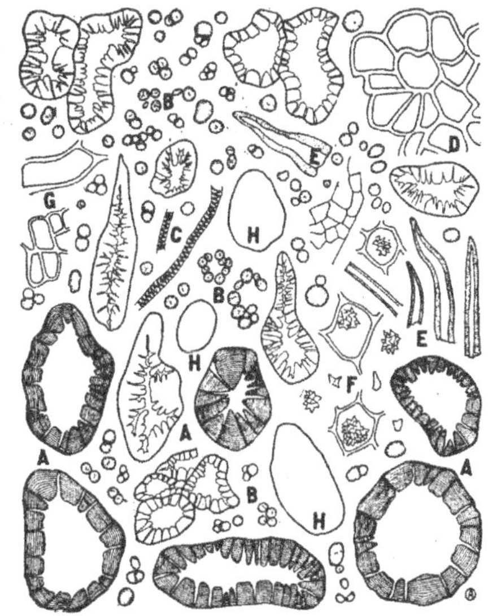

FIG. 301.—Powdered Pimenta. (× 183.) A, Stone cells.   B, Starch.   C, Spiral water tube.
D, Epidermis.   E, Trichomes.   F, Aggregate crystals of Calcium oxalate.   G, parenchyma.
H, Parenchyma cells containing red color.   See N.F., p. 329.

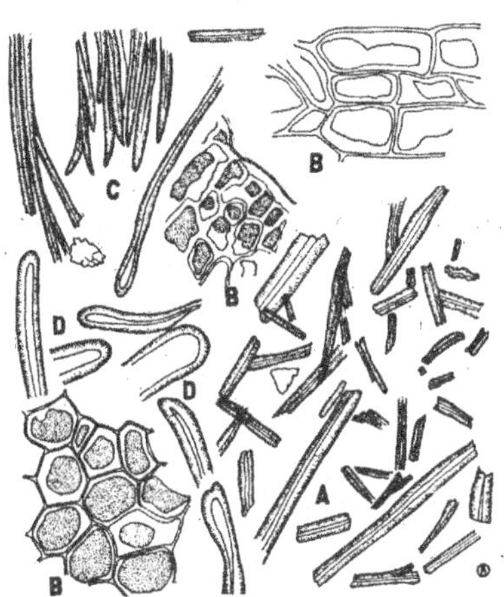

FIG. 302.—Nux Vomica. (× 183.)   A, Fragments of trichomes.   B, Thick-walled cells of the
endosperm, containing granular proteid,   C, Fragments of trichomes showing the pointed tips.   D,
Fragments of trichomes showing the rounded bases.

B—ALPHABETICAL LIST OF DRUG POWDERS WITH SYNOPSIS OF ESSENTIAL MICROSCOPICAL ELEMENTS

(For more detailed description of the microscopical elements see U.S.P.)

146.* ACONITE

Powder.—Grayish-brown; starch grains, nearly spherical, simple or 2 to 5 compound (3 to 15 μ in diam.); stone cells, large, tabular, irregular, or elongated to fibers; yellowish-brown cork fragments, few; tracheæ, spiral, reticulate or with bordered pits; parenchyma, relatively thick-walled, filled with starch.

341. ALTHÆA

Powder.—White or light yellow; starch grains, numerous (5 to 20 μ in diam.), calcium oxalate crystals in rosette aggregates (15 to 35 μ in diam.); bast fibers in groups, sometimes not strongly lignified, tracheæ scalariform or with bordered pits; numerous parenchyma fragments with large mucilage cells.

381. ANISUM

Powder.—Yellowish-brown, fragments of pericarp showing portions of yellow oil reservoirs numerous; small tracheæ accompanied by sclerenchyma fibers; endosperm cells filled with aleurone (about 6 μ in diam.), each containing a rosette crystal of calcium oxalate (about 2 μ in diam.); 1-celled hairs up to 2000 μ long, having slight projections on the surface; the endocarp is characteristic.

565. ARNICA

Powder.—Yellowish-brown, pappus consists of multicellular axis with unicellular branches; non-glandular hairs 1- to 6-celled, glandular hairs of three kinds, with unicellular stalk and unicellular head; a 4-celled stalk and unicellular head, or a 10-celled stalk of a double row of cells with a 2-celled head; pollen grains numerous, spherical (25 to 35 μ in diam.)

42. AMYLUM

Powder.—White, starch grains, polygonal, rounded or spherical (3 to 35 μ in diam.) with central cleft 3 to 5 rayed.

12. ASPIDIUM

Powder.—Greenish or brown; starch grains, numerous, oval or oblong (5 to 15 μ in diam.) in characteristic clumps; fragments of parenchyma sclerenchyma fibers and tracheids numerous; and characteristic brown fragments of the endodermis.

447. ASPIDOSPERMA

Powder.—Reddish-brown; starch grains, spherical, ovoid or plano-convex (3 to 25 μ in diam.); bast fibers, long accompanied by crystal fibers; stone cells in large groups; cork cells sometimes lignified; calcium oxalate in prisms or pyramids (8 to 30 μ long).

278. AURANTII AMARI CORTEX

Powder.—Yellowish to light brown, consists mostly of parenchyma with occasional membrane crystals of calcium oxalate (4 to 30 μ long); tracheæ, few small, spiral or with simple pores.

504. BELLADONNÆ FOLIA

Powder.—Green, consists mostly of irregular leaf fragments; calcium oxalate small in small aggregates or wedge-shaped micro-crystals; hairs, few, the non-glandular 2 to 5 cells, the glandular with 1- to 3-celled stalks and heads one to many

*Numbers refer to No. of drug in Part II.

celled; tracheæ, annular, spiral, reticulate or with bordered pits; few long thin-walled bast fibers and few pollen grains.   (See Fig. 279.)

503. BELLADONNÆ RADIX

Powder.—Light brown; starch grains, numerous, spherical, polygonal or plano-convex (3 to 30 μ in diam.) 2 to 8 or more compound; micro-crystals of calcium oxalate numerous (3 to 10 μ); cork cells, few; tracheæ, few and large, usually associated with wood fibers; long bast fibers from stem bases are often present.

274. BUCHU

Powder.—Pale green, consists mostly of parenchyma, often containing sphæro-crystals of inulin (25 to 40 μ in diam.) and numerous globules short and unicellular; aggregate crystals of calcium oxalate (15 to 25 μ in diam.); tracheids and bast fibers, few.   (See Fig. 294.)

156. CALUMBA

Powder.—Yellowish to greenish-brown; starch grains, numerous (8 to 85 μ in diam.) few 2 to 3 compound, ovoid, ellipsoidal or irregular usually with excentral hilum; stone cells, few, usually containing one or more prismatic crystals of calcium oxalate, sometimes having micro-crystals; tracheæ, few, reticulate or with bordered pits; occasionally, wood-fibers with long, oblique, slit-like pits; cork cells, yellow in regular radial rows and tangentially stretched.

112. CANNABIS

Powder.—Dark green, consists of fragments of leaves, and bracts showing yellowish lacticiferous vessels, rosette aggregates of calcium oxalate (6 to 30 μ in diam.), and fragments of fruits and stems; non-glandular hairs, unicellular, pointed, usually containing some calcium carbonate which gives a strong effervescence with dilute HCl; glandular hairs, short with 1-celled stalk, or long multicellular, the head consisting of 8 to 16 cells; palisade-like, thick-walled cells from the fruits; tissues of embryo and endosperm with numerous aleurone grains (5 to 10 μ in diam.) and oil globules.

CANTHARIS (See animal products)

Powder.—Grayish-brown, showing conspicuous shining green particles and numerous long, pointed hairs.

516. CAPSICUM

Powder.—Yellowish-brown to brownish-red, cells of epidermis of uniform size and regular arrangement; parenchyma containing numerous reddish oil globules, and chromoplasts; stone cells of endocarp with yellowish wavy, moderately thickened porous walls, those of the seed coat, yellowish, irregular, strongly thickened and much more strongly lignified than those of the endocarp.

82. CARDAMOMI SEMEN

Powder.—Greenish-brown, fragments of seed with dark brown stone cells (20 μ in diam.); polygonal in surface view; cells of endosperm and perisperm containing compound starch grains (1 to 4 μ in diam.); a few small tracheæ may be present.

385. CARUM

Powder.—Yellowish-brown; fragments of pericarp with light yellow oil ducts; tracheæ often accompanied by sclerenchyma fibers which are slightly lignified and have oblique pits; endosperm cells contain aleurone grains which usually include a rosette of calcium oxalate about 1 μ in diam

371. CARYOPHYLLUS

Powder.—Dark brown to reddish-brown; thin-walled parenchyma showing large oil reservoirs; a few small spiral tracheæ and thick-walled spindle-shaped bast fibers, rosettes of calcium oxalate (2 to 15 μ in diam.); pollen grains (15 to 25 μ in diam.).

334. CASCARA SAGRADA

Powder.—Light or dark brown; bast fibers, usually in groups accompanied by crystal fibers; thick-walled stone cells in large groups; parenchyma and medullary ray cells have numerous nearly spherical starch grains (3 to 8 μ in diam.); calcium oxalate in prisms or rosettes (8 to 20 μ in diam.); reddish-brown cork fragments. (See Fig. 276.)

133. CIMICIFUGA

Powder.—Light or dark brown; starch grains, numerous, simple or compound, spherical or polygonal (2 to 15 μ in diam.); tracheæ, mostly with bordered pits and usually associated with lignified wood fibers; yellowish-brown fragments of epidermis.

532. CINCHONA

Powder.—Reddish-brown, bast fibers, large spindle-shaped (300 to 1500 μ long) often showing lamellated walls; starch grains, simple or 2 to 5 compound, nearly spherical (3 to 12 μ in diam.); calcium oxalate in wedge-shaped micro-crystals; reddish-brown fragments of cork. (See Fig. 296.)

532. CINCHONA RUBRA

Powder.—Light brown to brown; elements similar to those of Cinchona, but starch grains are usually fewer and smaller.

169. CINNAMOMUM SAIGONICUM

Powder.—Yellowish or reddish-brown; starch grains simple or compound, ellipsoidal or polygonal (3 to 20 μ in diam.); stone cells, irregular colorless or filled with a reddish-brown amorphous substance; bast fibers having thick slightly lignified walls, single or in groups (300 to 1500 μ long); reddish-brown cork fragments. (See Fig. 300.)

167. CINNAMOMUM ZEYLANICUM

Powder.—Light or yellowish-brown; starch grains simple or compound, ellipsoidal or polygonal (3 to 20 μ in diam.); stone cells, numerous, irregular, colorless or containing reddish-brown amorphous substance; bast fibers, with thick, slightly lignified walls, spindle-shaped (300 to 1000 μ long).   Calcium oxalate raphides (5 to 10 μ long) sometimes present.

68. COLCHICI CORMUS

Powder.—Light to grayish-brown; starch grains numerous, simple or 2 to 6 compound, spherical or polygonal (3 to 30 μ in diam.); few spiral or scalariform tracheæ; few fragments of reddish-brown epidermis.

69. COLCHICI SEMEN

Powder.—Light brown; parenchyma of endosperm thick-walled with simple pits and containing aleurone grains (3 to 15 μ in diam.) and oil globules; cells of seed coat somewhat collapsed, having thin reddish-brown walls; and a few small ellipsoidal starch grains (5 to 16 μ in diam.).

544. COLOCYNTHIDIS PULPA

Powder.—Yellowish-white or buff; consists chiefly of parenchyma cells usually in fragments; tracheæ only occasional; from the seed coats few stone cells which are nearly isodiametric or irregular; few oil globules and aleurone grains.

386. CORIANDRUM

Powder.—Light brown; fragments of endosperm, filled with aleurone grains usually containing aggregates of calcium oxalate, and oil globules; sclerenchyma fibers, yellowish thick-walled, irregularly curved; few fragments of yellow oil reservoirs, and polygonal epidermis; calcium oxalate aggregates (3 to 10 μ in diam.) may be separated from the aleurone grains.

368. EUCALYPTUS

Powder.—Green; epidermis, thick-walled and strongly cuticularized; palisade, very numerous, 3 to 4 rows deep, in which occur large oil reservoirs containing yellowish contents.   Calcium oxalate of rosettes or mono-clinic prisms (15 to 30 μ in diam.), in the spongy parenchyma are vascular tissues and few slightly lignified bast fibers.

382. FŒNICULUM

Powder.—Yellowish-brown; endosperm cells filled with aleurone grains each with a rosette of calcium oxalate (about 2 μ in diam.); yellowish-brown fragments of oil reservoirs; sclerenchyma fibers few, strongly lignified; spiral or annular tracheæ, few; parenchyma cells, numerous; sometimes with thickened walls.

333. FRANGULA

Powder.—Yellowish-brown; stone cells are absent, otherwise the elements are almost identical with those of Cascara Sagrada (see page 295). Frangula gives a deeper orange color than does cascara when treated with alkalies.

105. GALLA

Powder.—Brownish-gray; starch-bearing parenchyma cells numerous; starch spherical to polygonal (11 to 35 μ in diam.); stone cells few, variable (25 to 250 μ long); tracheæ spiral or reticulate.

438. GELSEMIUM

Powder.—Dark yellow; tracheæ, spiral and with bordered pits associated with long narrow fiber-tracheids; bast fibers, long and narrow; starch grains, spherical (4 to 8 μ in diam.); calcium oxalate in monoclinic prisms (15 to 30 μ long); few very thick-walled groups of stone cells.

441. GENTIANA

Powder.—Light brown or yellowish-brown, consisting mostly of parenchyma cells varying much in size and form; tracheæ spiral, scalariform or reticulate; yellowish-brown cork.

230. GLYCYRRHIZA

Powder.—Brownish-yellow to pale yellow; starch grains oval or elliptical (3 to 25 μ in diam.); tracheæ with bordered pits associated with wood fibers, numerous; bast fibers, numerous, very long and usually in groups accompanied by crystal fibers, containing prisms of calcium oxalate (2 to 25 μ in diam.); fragments of reddish-brown cork occur in Spanish Licorice.

366. GRANATUM

Powder.—Yellowish-brown to dark brown; crystals of calcium oxalate in aggregate prisms or crystal fibers (10 to 20 μ in diam.); starch grains, spherical to polygonal, simple or compound (2 to 10 μ in diam.); cork fragments, whitish; stone cells, usually occur singly and are strongly lamellated (40 to 200 μ long).

576. GRINDELIA

Powder.—Yellowish-brown; tracheæ, annular spiral reticulate, or with bordered pits, associated with narrow wood fibers; leaf epidermis characteristic, showing large colorless multicellular glandular hairs; pollen grains, spherical spinose (about 35 μ in diam.).

329. GUARANA

Powder.—Pinkish-brown; irregular masses of parenchyma and altered starch grains; starch grains, spherical to polygonal (10 to 25 μ in diam.); few elongated, yellowish, thick-walled sclerenchyma cells, which are usually not lignified.

134. HYDRASTIS

Powder.—Yellowish-brown; starch numerous, usually simple, nearly spherical (2 to 15 μ in diam.); vascular tissues usually associated with starch-bearing parenchyma; tracheæ spiral, reticulate or with bordered pits; few thin-walled wood fibers; and occasional fragments reddish-brown cork.

509. HYOSCYAMUS

Powder.—Grayish-green; calcium oxalate crystals in 4- to 6-sided prisms (15 to 25 μ long), in spherical or rosette aggregates (about 20 μ in diam.) or in wedge-shaped micro-crystals; non-glandular hairs 2 to 10 cells long; glandular hairs with stalk 1 to 4 cells long and 1 to many celled head; stomata broadly elliptical about 30 μ long, with 3 to 4 neighboring cells; tracheæ spiral reticulate or with bordered. pits and associated with few fibers; pollen grains about 40 μ in diam., nearly smooth. (See Fig. 280.)

530. IPECACUANHA

Powder.—Light brown; starch grains, numerous, simple 2 to 6 or more compound, spherical or polygonal (2 to 18 μ in diam.); calcium oxalate raphides (15 to 40 μ long) few; tracheids numerous; occasional stone cells from stem bases (30 to 40 μ long). (See Fig. 277.)

460. JALAPA

Powder.—Light brown; starch grains, numerous, simple or 2 to 4 compound ellipsoidal to ovoid (4 to 35 μ in diam.); often swollen and somewhat altered; calcium oxalate in rosettes (10 to 40 μ in diam.); tracheæ with simple or bordered pits; laticiferous vessels containing yellowish-brown masses.   (See Fig. 299.)

264. LINUM

Powder.—Lemon yellow to light brown; the seed coat has tabular pigment cells, filled with reddish-brown insoluble substance; stone cells elongated and yellowish; oil globules numerous; aleurone grains, numerous (2 to 20 μ in diam.).

552. LOBELIA

Powder.—Dark green; cells of seed coat more or less polygonal, walls thick and yellowish; few non-glandular hairs (30 to 60 μ long); tracheæ annular, spiral or reticulate, accompanied by narrow thin-walled wood fibers; leaf epidermis with elliptical stomata about 25 μ long and with 3 to 4 neighboring cells; pollen grains nearly spherical about 25 μ in diam.

18. LYCOPODIUM

The spores are sperical tetrahedrons (25 to 40 μ in diam.) with the outer walls extended into irregular projections.

599. MATRICARIA

Powder.—Yellowish to yellowish-green; [pollen grains numerous, spinose, varying from nearly spherical to triangular (about 20 μ in diam.); glandular hairs from the corolla, and cells of the anthers are characteristic; few sclerenchyma fibers.

473. MENTHA PIPERITA

Powder.—Dark green; non-glandular hairs 1 to 8-celled; glandular hairs with stalks 1 or 3-celled and 1 to 8-celled heads; pollen grains nearly spherical, smooth (about 30 μ in diam.); tracheæ, spiral or with simple or bordered pits; thin-walled sclerenchyma fibers, few.

474. MENTHA VIRIDIS

Powder.—Similar in structure to Mentha Piperita.

365. MEZEREUM

Powder.—Light grayish-brown; numerous long bast fibers (400 to 3000 μ long) somewhat uneven and bent, non-lignified; cork cells yellowish-brown; starch grains, few, simple or 2 to 4 compound (3 to 15 μ in diam.).

154. MYRISTICA

Powder.—Dark reddish-brown; perisperm of thin-walled parenchyma cells in which are large oil reservoirs; endosperm of parenchyma filled with starch and aleurone grains; starch, simple or compound, spherical to polygonal (3 to 20 μ in diam.); few small spiral tracheæ; oil globules numerous.

294. MYRRHA

Powder.—Yellowish-brown; mounted in fixed oil shows angular fragments; when cleared and stained in chloral hydrate iodine a few spherical or irregular starch grains (10 to 35 μ in diam.) may appear; when tested with phloroglucin may show fragments of sclerenchyma fibers or stone cells.   (See Fig. 302.)

435. NUX VOMICA

Powder.—Light gray; endosperm cells thick-walled, containing oil globules and aleurone grains; numerous non-glandular, lignified hairs having pitted walls; cells of adhering fruit pulp may show few small spherical starch grains.   (See Fig. 302.)

180. OPII PULVIS

Powder.—Light brown; consists of irregular granular fragments; epidermis of poppy capsuls 4 to 5-sided or elongated, thick-walled and lignified; fragments of poppy leaves and rumex fruits.

548. PEPO

Powder.—Whitish or yellowish; outer epidermis palisade-like, cells up to 1 mm. long; stone cells variable in size and thickness of walls up to 75 μ long; parenchyma cells with peculiar reticulate markings or rather thick-walled and somewhat collapsed; cells of cotyledons, isodiametric or elongated with numerous oil globules and aleurone grains.

391a. PETROSELINUM

Powder.—Grayish-brown; fragments of pericarp of yellowish-brown cells in which are the light yellow oil reservoirs; endosperm cells having numerous aleurone grains each with a rosette of calcium oxalate (3 to 7 μ in diam.); tracheæ small, usually associated with sclerenchyma fibers.

252. PHYSOSTIGMA

Powder.—Grayish-white; ellipsoidal or reniform starch grains numerous (5 to 15 μ in diam.); cells of seed coats irregular or palisade-like; few small reticulate tracheæ.

275. PILOCARPUS

Powder.—Dark green to brownish; epidermis in surface view of 5 to 6-sided cells containing elliptical stomata about 30 μ long and usually surrounded by 4 neighboring cells; tracheæ spiral, reticulate or with bordered pits, and usually associated with wood-fibers; few bast fibers; rosettes of calcium oxalate (10 to 25 μ in diam.); fragments with large oil reservoirs; few non-glandular hairs.

89. PIPER

Powder.—Brown or grayish-brown; nearly spherical starch grains (1 to 3 μ in diam.) mostly in the angular cells of the endosperm; stone cells, varying from thick-walled isodiametric or elongated to unevenly thickened and rather thin-walled, and usually filled with reddish-brown substance; yellowish suberized oil cells; and regular epidermal cells.

161. PODOPHYLLUM

Powder.—Light brown; starch grains spherical to polygonal, simple or 2 to 6 compound (3 to 15 μ in diam.); few rosettes of calcium oxalate (about 6 to 8 μ in diam.); and occasionally raphides (about 65 μ long); tracheæ reticulate or with simple pores; numerous large thin-walled parenchyma cells; and reddish-brown cork. (See Fig. 88.)

203. PRUNUS VIRGINIANA

Powder.—Light brown; bast fibers thick-walled usually associated with crystal fibers: stone cells thick-walled, strongly lignified; nearly spherical starch grains 3 to 4 μ in diam.); calcium oxalate in prisms and rosettes (10 to 40 μ in diam.).

555. PYRETHRUM

Powder.—Light to dark brown; tracheæ reticulate or with simple pits; stone cells in groups with thick, yellowish walls; cork fragments yellowish-brown; irregular masses of inulin in thin-walled parenchyma cells or free.

287. QUASSIA

Powder.—Light yellow; tracheæ large, with simple or bordered pits, usually accompanied by thin-walled wood fibers which have oblique pits; crystal fibers with 4 to 6-sided prisms of calcium oxalate (6 to 30 μ in diam.); few nearly spherical starch grains (10 to 15 μ in diam.). In Surinam Quassia, calcium oxalate crystals are few or absent. (See Fig. 297.)

120. RHEUM

Powder.—Orange-yellow to yellowish-brown; starch grains nearly spherical simple or 2 to 4 compound (2 to 20 μ in diam.); calcium oxalate in rosettes (50 to 125 μ in diam.) tracheæ, spiral or reticulate. (See Fig. 297.)

213. ROSA GALLICA

Powder.—Reddish; upper epidermis modified to conical papillæ; lower epidermal cells rectangular, both filled with purplish-red color; loose mesophyll cells; small spiral tracheæ.

47. SABAL

Powder.—Yellowish-brown; thick-walled whitish fragments of endosperm; irregular stone cells (about 125 μ in diam.); parenchyma of the pericarp yellowish to brownish-red.

185. SANGUINARIA

Powder.—Brownish-red; starch grains, numerous, (3 to 20 μ in diam.) spherical to ovoid; simple or 2 to 3 compound; latex tissue fragments with reddish-brown masses; tracheæ with slit-like pits few.

239. SANTALUM RUBRUM

Powder.—Brownish-red; wood fibers numerous, walls thick, yellowish, up to 800 μ long; tracheæ, few with simple or bordered pits; crystal fibers with prisms of calcium oxalate (10 to 20 μ in diam.).

58. SARSAPARILLA

Powder.—Grayish-brown; starch grains, spherical to nearly tetrahedral, simple or 2 to 5 compound (3 to 25 μ in diam.); raphides of calcium oxalate (6 to 30 μ long); tracheæ scalariform, reticulate or with simple or bordered pits, often associated with thin-walled sclerenchyma fibers; cells of hypoderm and endoderm yellowish, up to 500 μ long.

170. SASSAFRAS

Powder.—Light reddish-brown; starch grains, spherical to polygonal, simple or 2 to 4 compound (3 to 20 μ in diam.); bast fibers, spindle-shaped or irregular (150 to 400 μ long and 25 μ broad) with very thick walls; numerous parenchyma cells, many containing yellowish-red masses of tannin; few brownish-red fragments of cork.

462a. SCAMMONII RADIX

Powder.—Grayish-brown; starch grains, simple or 2 to 4 compound (3 to 18 μ in diam.); calcium oxalate in prisms (10 to 45 μ long); tracheæ reticulate or with simple or bordered pits and usually associated with wood fibers; stone cells variable in form (40 to 125 μ long); few cork cells which are often lignified; fragments of phloem showing yellowish-brown resin cells.

67. SCILLA

Powder.—Light yellow; raphides of calcium oxalate (750 to 1000 μ long); parenchyma cells large, thin-walled, colorless; tracheæ spiral or reticulate; occasionally a few nearly spherical starch grains occur.

302. SENEGA

Powder.—Yellowish-gray to brown; wood fibers non-lignified (175 to 250 μ long); fragments of thin-walled parenchyma containing oil globules; tracheæ with simple or bordered pits; numerous medullary ray cells with large simple pits.

240. SENNA

Powder.—Light green (Alexandria Senna) or slightly darker green (India Senna); stomata broadly elliptical (about 20 μ in diam.); crystal fibers; calcium oxalate in rosettes (about 10 μ in diam.) or 4 to 6-sided prisms about 15 μ long); non-glandular hairs 1-celled, often curved, thick-walled and rough up to 350 μ long. In India Senna the hairs are relatively fewer. (See Fig. 295.)

118. SERPENTARIA

Powder.—Grayish-brown; starch grains spherical to plano-convex (3 to 14 μ in diam.), simple or 2 to 4 compound; tracheæ annular, spiral or reticulate; short wood fibers; small amount of cork; numerous lignified parenchyma pith cells; few non-glandular hairs from stem may be present.

188. SINAPIS ALBA

Powder.—Light yellow to brownish-yellow; parenchyma cells contain aleurone and oil; fragments of seed coats nearly colorless composed of small stone cells and large epidermal cells, the outer walls being mucilaginous; occasionally few small starch grains are present.

189. SINAPIS NIGRA

Powder.—Light brown to greenish-brown; thin-walled parenchyma of embryo contains alleurone grains and oil; fragments of seed coats composed of small yellowish stone cells with dark lumen; and large mucilaginous cells of epidermis often associated with the very large sub-epidermal cells.

439. SPIGELIA

Powder.—Grayish-brown; starch grains nearly spherical (2 to 6 μ in diam.); tracheæ and tracheids conspicuous; few long slender bast fibers; fragments of reddish-brown epidermis and brownish cork.

304. STILLINGIA

Powder.—Pinkish or reddish-brown; starch grains variable in form, mostly simple (5 to 35 μ in diam.); tracheæ with simple pits, usually associated with wood fibers; bast fibers long, narrow, thick-walled; reddish-brown cork; rosettes of calcium oxalate up to 35 μ in diam; somewhat tabular reddish-brown secretion cells.

507. STRAMONIUM

Powder.—Brownish-green; stomata elliptical about 25 μ long, usually with 3 neighboring cells; calcium oxalate in numerous rosettes (10 to 20 μ in diam.) in prisms or wedge-shape micro-crystals; non-glandular hairs with 1 to 2-celled stalks and 2 to 4-celled heads; spiral or annular tracheæ; stems have large tracheæ with annular and spiral thickening or with bordered pits usually associated with wood parenchyma and occasional wood fibers; long collenchymatous cells are often present. (See Fig. 293.)

451. STROPHANTHUS

Powder.—Grayish to dark-brown; mostly composed of thin-walled parenchyma cells; many of which are colored greenish upon addition of H_2SO_4; numerous fragments of long thin-walled hairs (relatively fewer in S. hispidus); numerous oil globules.

400. SUMBUL

Powder.—Grayish-brown; numerous large tracheæ which are mostly reticulate; long narrow collapsed fragments of phloem; few fragments of parenchyma containing starch grains (3 to 12 μ in diam.); numerous nearly colorless or yellowish to reddish-brown irregular fragments.

553. TARAXACUM

Powder.—Light brown; parenchyma cells large; thin-walled containing masses of inulin; fragments of yellowish latex vessels; reticulate tracheæ and sclerenchymatous fibers.

256. TRAGACANTHA

Powder.—Whitish; irregular fragments showing lamellated mucilaginous walls and few starch grains nearly spherical, simple or 2 to 3 compound (3 to 17 μ in diam.).

37. TRITICUM

Powder.—Yellowish; tracheæ annular, spiral or with simple pits and associated with long narrow sclerenchymatous fibers; epidermal cells rectangular strongly lignified with numerous transverse pits; numerous fragments of rectangular thin-walled parenchyma.

109. ULMUS

Powder.—Light brown; bast fibers numerous very long and slightly lignified, often associated with crystal fibers; calcium oxalate in prisms (10 to 25 μ in diam.); starch grains, mostly simple, nearly spherical (about 3 or 4 μ in diam. or up to 25 μ); fragments of large mucilage cells.

411. UVA URSI

Powder.—Olive green; epidermal cells polygonal; stomata, broadly elliptical about 25 μ long with 5 to 8 adjacent cells; tracheæ mostly spiral, often associated with sclerenchyma and crystal fibers; prisms of calcium oxalate (6 to 15 μ in diam.).

543. VALERIANA

Powder.—Light brown or grayish-brown; starch grains spherical to ploygonal, simple or 2 to 4 compound, (3 to 20 μ in diam.); tracheæ reticulate or with simple or bordered pits often accompanied by sclerenchyma fibers; fragments of epidermis with root hairs and brownish cork.

60. VERATRUM VIRIDE

Powder.—Grayish-brown to dark brown; starch grains, spherical or ellipsoidal simple or 2 to 3 compound (3 to 20 μ in diam.); calcium oxalate raphides (15 to 150 μ long); tracheæ scalariform or reticulate and usually associated with narrow sclerenchyma fibers; fragments of reddish-brown cork.

541. VIBURNUM PRUNIFOLIUM

Powder.—Dark brown; stone cells, numerous, large and thick-walled: bast fibers few, with occasional crystal fibers; calcium oxalate in prisms or rosettes (15 to 35 μ in diam.).

270. XANTHOXYLUM

Powder.—Grayish-brown; cork cells nearly colorless and lignified; parenchyma containing small starch grains, oil globules or calcium oxalate; stone cells in small groups; few bast fibers.

78. ZINGIBER

Powder.—Light yellow to brown; parenchyma cells large and thin-walled filled with starch; starch grains ovate to elliptical (15 to 60 μ long); sclerenchyma fibers long and thin-walled; tracheæ reticulate or scalariform; yellowish or brown oil and resin cells; brownish flattened cork cells, which are absent in Jamaica Ginger.   (See Fig. 298.)

BIBLIOGRAPHY

Analysis of Drugs and Medicines, by Burt E. Nelson.

Anatomischer Atlas der Pharmakognosie und Nahrungsmittlekunde, by Tschirch und Oesterle.

Anatomical Atlas of Vegetable Powders, by Greenish and Collin.

Applied and Scientific Pharmacognosy, by Henry Kraemer.

Die Wichtigsten Vegetabilischen Nahrungs und Genussmittel, by A. E. Vogl.

Einfuhrung in der Mikroskopische Analyse der Drogen pulver, by Ludwig Koch.

Elements of Applied Microscopy, by C. A. Winslow.

Food and Drugs, by Henry George Greenish.

Introduction to Pharmacognosy, by Smith Ely Jelliffe.

Lehrbuch der Pharmakognosie, by Karsten and Oltmans.

Microscopy of Vegetable Foods, by A. L. Winton.

Mikroskopische Analyse der Drogen pulver, by Ludwig Koch.

Powdered Vegetable Drugs, by Albert Schneider.

The Principal Starches used as Food, by W. Griffiths.

CHAPTER II.—REAGENTS AND PROCESSES

The different kinds of cell-walls and cell-contents may be demonstrated by the use of reagents which, in some cases, impart characteristic colors to walls and contents; in other cases act as selective solvents, dissolving some of the walls and contents, leaving others undissolved; or the reagents may produce precipitates the nature of which furnishes good evidence regarding the character of the substance which has united with the reagent to produce the precipitate.

These reagents, together with their uses, will now be given in alphabetical order.

Acetic acid dissolves most ethereal oils, while most fatty oils are insoluble in it; dissolves calcium carbonate with evolution of CO_2, while calcium oxalate is unaffected by it, and it therefore serves to distinguish between these two salts of calcium; solvent of crystals of hesperidin which have been deposited from the cell-sap of oranges, etc., when these have lain for some time in alcohol; when various lichens are treated with it, crystals of calycin in acicular form are deposited after the lichens thus treated have been powdered and dried; one per cent. solution dissolves globoids in aleurone grains, while any crystals of calcium oxalate present are unaffected by it; when pieces of potatoes, carrots, etc., are macerated in it, the separate cells become isolated. Used in the preparation of various fixatives.

Albumen.—The white of egg is used with an equal amount of glycerine and a trace of salicylate of soda for fixing microtome sections to the glass slide, the sodium salicylate acting partly as an antiseptic.

Alcannin.—This is a coloring matter, obtained from the roots of *Alcanna tinctoria.* A tincture of alcannin to be used as a reagent is prepared by Guignard as follows: 10 Gm. of alcannin are pulverized and added to 30 mils of absolute alcohol; the solution is filtered and allowed to evaporate in a drying oven; the residue is then dissolved in 5 mils of glacial acetic acid, and this solution is mixed with 50 mils of 50 per cent. alcohol. After twenty-four hours the solution is filtered and is ready for use. The solution prepared in this way is said not to be subject to precipitation on long standing. When sections are being treated with this reagent under a cover-glass, evaporation should be guarded against by the addition of drops of 50 per cent. alcohol as needed. A quicker way of preparing an alcannin solution is to extract the coloring matter from roots of Alcanna in absolute alcohol, and then to add an equal bulk of distilled water and filter the solution.

(1) Suberized and cutinized walls, when treated with a solution of alcannin for some hours, take on the color of the alcannin. (2) Alcanna tincture mixed with 1 per cent. glacial acetic or formic acid is used to fix and stain sections of elaioplasts from fresh material. (3) Where sections containing fatty oils are treated with tincture of alcannin, the oil is colored red. Sections containing ethereal oils and resins behave in the same manner, in that the ethereal oils and resins are stained red.

Alcohol.—The commercial alcohol obtained in this country is about 95 per cent. alcohol. In making alcohols from this of different strengths it answers all practical purposes to proceed as if the commercial 95 per cent. alcohol were absolute—that is, very nearly 100 per cent. Thus, if 50 per cent. alcohol is desired, 50 mils commercial alcohol and 50 mils distilled water will give sufficiently accurate results for all histological work. If absolute alcohol is desired, it may be prepared by pouring the commercial alcohol over unslaked lime, and then distilling from this over a

water-bath. Or copper sulphate may be burned until all the water of crystallization is driven off and a white powder results; then the commercial alcohol may be
poured over this, and the white powder will become blue again from the water
absorbed from the alcohol. The alcohol should then be filtered off and kept
tightly corked from the air.

(1) Used in a series of different strengths (from lower to higher) for hardening
plant tissue. (2) When sections of plant tissues containing potassium nitrate,
dulcite, asparagin, or piperine are treated under the cover-glass with strong alcohol, which is then allowed to evaporate, the substances enumerated will crystallize
out in their characteristic forms. (3) When pieces of plants containing inulin
have lain for some time in 50 or 70 per cent. alcohol, the inulin is precipitated in
the form of sphærocrystals. (4) Seventy per cent. alcohol is used for preserving
plant tissues indefinitely, but to avoid shrinkage and disintegration of the protoplasts in the case of very delicate or meristematic tissues the material should
first have been fixed and brought, by slow degree, to the 70 per cent. alcohol, as
elsewhere described. Material which has been preserved for a long time in 70
per cent. or stronger alcohol is apt to be quite brittle; when desired, the brittleness
may be removed by placing the material for an hour or so in water and then back
into the grade of alcohol from which it was taken. Alcohol is used in the preparation of various stains and reagents. The details of its use will be given under the
various formulæ where it occurs.

Alum Carmine.—A 4 per cent. aqueous solution of ammonia alum is boiled
twenty minutes with 1 per cent. of powdered carmine. Filter after it cools. (Lee).

Stain from water twelve to twenty-four hours and wash in water. No acid
alcohol is needed, since the solution does not overstain.

Alum Cochineal.—

Powdered cochineal	50 g.
Alum	5 g.
Distilled water	500 mils.

Dissolve the alum in water, add the cochineal, and boil; evaporate down to
two-thirds of the original volume, and filter. Add a few drops of carbolic acid
to prevent mold. (Stirling.).

Stain as with alum carmine.

Ammonium Molybdate.—A concentrated solution of ammonium molybdate is
made in a saturated solution of ammonium chloride. When sections containing
tannins are treated with this, a yellow precipitate is usually produced.

Ammonium Vanadate.—This is used as a test for solanin. The sections are
treated with a solution prepared by dissolving 1 part of ammonium vanadate in 1,000
parts of a mixture of 98 parts of concentrated sulphuric acid and 36 parts of water.
If solanin is present, a yellow color appears, which merges into orange, then different shades of red, and finally into violet, and then all color disappears.

Aniline Oil.—Excellent for dehydrating sections, since it will dissolve about
4 per cent. of water and may be kept dehydrated by a small piece of solid KOH
which is insoluble in it. The sections may be transferred from the aniline immediately into Canada balsam.

Aniline Sulphate.—Make a saturated aqueous solution. As a test for lignified membranes mount the sections in the solution and add a drop of sulphuric
acid, and a yellow color is given to the lignified membranes.

Balsam.—Canada balsam dissolved in xylol is, on the whole, the best medium
for making permanent mounts of sections under a cover-glass. Balsam in xylol
can be obtained ready prepared of the dealers.

Barium Chloride.—(1) This is sometimes used to distinguish calcium oxalate from calcium sulphate. When barium chloride is run under the cover-glass, calcium oxalate, if present, is left unchanged, while a fine granular layer of barium sulphate comes to incrust any crystals of calcium sulphate. (2) To determine the presence of tartaric acid, barium chloride and antimonic oxide in hydrochloric acid are run under the cover-glass, producing, with tartaric acid, rhombic crystals of antimonium-barium tartrate, whose obtuse angles measure 128°.

Beale's Carmine Solution.—Mix 0.6 Gm. with 3.75 Gm. ammonia water (10 per cent.); heat on a water-bath for several minutes; then add 60 Gm. of glycerine, 60 Gm. of water and 15 Gm. of alcohol, and filter.

Benzol.—Used in detecting caffeine, thus: Sections are heated on the slide in a drop of distilled water until bubbles arise, then the water is allowed to evaporate, and the residue is dissolved with a drop of benzol. The benzol is then allowed to evaporate and the caffeine is deposited on the edge of the drop in the form of colorless needle-crystals.

Bismarck Brown.—This is preëminently a nuclear stain. The powder is soluble with difficulty in water. It is a good plan to treat with boiling water and after a day or two to filter. Or a saturated solution may be made in 70 per cent. alcohol. Although Bismarck brown stains rapidly, it does not overstain. It may be used for staining *in toto* or for staining sections on the slide.

Bohmer's hæmatoxylin solution is prepared by mixing the two following solutions and filtering after allowing the mixture to stand for several days: (a) One part of a 3.5 per cent. alcoholic (95 per cent.) solution of hæmatoxylin and (b) three parts of a 0.4 per cent. aqueous solution of potassium alum.

Boric Acid.—Used as a mounting medium for sections containing mucilaginous membranes. The sections are cut from dry material and placed in a 10 per cent. solution of neutral lead acetate to harden the mucilaginous layers. Then the sections are stained in a solution of methyl-blue, washed in water, and mounted under a cover-glass in a 2 per cent. solution of boracic acid. The cover-glass should be sealed down with a mixture of paraffin and vaseline, which is applied with a brush while melted.

Borax-carmine.—A 4 per cent. solution of borax is made and to it is added 3 per cent. of carmine; an equal bulk of 70 per cent. alcohol is then added to this. The mixture is left standing for a day or so and then filtered. Sections should lie in the stain for about twenty-four hours, and should then be transferred without previous washing to acidulated alcohol, made by adding 4 drops of hydrochloric acid to 100 mils of alcohol. Here they should remain until they become bright and transparent. This is a useful stain for aleurone grains, for differentiating cell-contents from cell-walls when the sections are subsequently stained with methyl green, and much used also in the differentiation of the cell-contents of filamentous algæ.

Bordeaux Red.—Used in conjunction with hæmatoxylin in staining nuclear figures, particularly where Heidenhain's platinic chloride fixative has been used. The sections are placed in a weak aqueous solution of the Bordeaux until they are intensely stained; they are then rinsed and placed in a 2 to 5 per cent. solution of iron oxide-ammonium sulphate for three hours. If the sections are mounted on a slide, they should be placed upright in this solution, so that any precipitate may not gather on the slide. Then the sections are carefully washed in an abundance of water, and placed for twenty-four hours in a solution of hæmatoxylin prepared as follows: 1 Gm. of hæmatoxylin is dissolved in 10 Gm. of alcohol and 90 Gm. of water. This is allowed to stand for about four weeks, and then an equal bulk of distilled water is added. The stain is then ready for use. When the sections are taken from the hæmatoxylin, they will be found overstained; they are, therefore, rinsed and placed in a 2.5 per cent. solution of ferric-ammonium sulphate, where they remain until examination of the sections under the microscope shows them to have the proper intensity of stain. The sections are then thoroughly rinsed in water and passed through alcohol and xylol before mounting in balsam.

Borodin's Method.—To determine the nature of a precipitate Borodin treats it with a saturated solution of the same substance as the precipitate is supposed to be. Thus, if the precipitate is supposed to be asparagin, it is treated with a saturated solution of asparagin. If the precipitate dissolves by this treatment, it is then some other substance than asparagin. This method is not very reliable for substances which are very readily soluble, such as potassium nitrate. Care must be taken that the solution used for the test is entirely saturated.

Brown Discoloration of Material in Alcohol.—Some plants, such as Monotropa, are apt to become quite brown in alcohol. This can be prevented by placing the fresh material in alcohol which is acidulated by vapor of sulphuric acid in the following manner: For each 100 mils of alcohol several mils of 80 per cent. sulphuric acid are poured over ½ of a Gm. of sodium sulphite, and the vapors arising are conducted into the alcohol. This operation need require hardly more than a minute. After twenty-four hours the material should be transferred from the acid alcohol to neutral alcohol. Thereafter the material will not discolor, and will take stains very well when used for histological purposes.

Calcium Nitrate.—(1) Used to differentiate more clearly the lamellæ of starch-grains. Potato starch, for instance, is placed in a rather strong aqueous solution of methyl violet. After the grains have become deeply colored, they are treated with a weak solution of calcium nitrate, when the methyl violet becomes precipitated, particularly in the less dense lamellæ of the starch-grains. (2) Calcium oxalate is precipitated in the form of crystals when sections containing oxalic acid are treated with a solution of calcium nitrate. The calcium nitrate is thus a test for the presence of oxalic acid.

Canada balsam is used almost exclusively for mounting. Very thick balsam is very disagreeable to handle and makes unsatisfactory mounts. Very thin balsam, in drying out, allows bubbles to run under the cover. Xyol is cheaper than balsam and, consequently, the balsam on the market is likely to be too thin for immediate use. The stopper may be left out until the balsam acquires the proper consistency. Material cleared in clove oil or cedar oil may be mounted directly in xyol balsam. It is not necessary that the clearing agent should be also the solvent of the balsam.

Canarin.—This is often used as a stain for tissues which have been cleared in caustic potash. Canarin is not affected by this reagent.

Carbolic Acid (Phenol).—Used as a clearing agent. If leaves which have been hardened and bleached in alcohol are placed in 3 parts of turpentine and 1 part of carbolic acid, or in pure carbolic acid, the leaves will become so transparent that their cellular structure may be made out from one surface to the other. Pollen-grains may be made transparent in the same manner.

Carmalum, Mayer's.—Carminic acid 1 Gm., alum 10 Gm.; dissolve in 200 mils of hot distilled water; filter and add a few crystals of thymol, or 0.1 per cent. of salicylic acid, or 0.5 per cent. of sodium salicylate. This stains material well in bulk, with little danger of overstaining. If this happens, it may be corrected by washing with a 0.1 per cent. solution of hydrochloric acid. Material which has been stained in bulk with carmalum may be sectioned, and the sections may then be double-stained with some aniline stain, such as blue de Lyon. See borax-carmine for another carmine stain. Very fine double staining may be achieved by placing sections first in an aqueous solution of iodine green and then for a somewhat longer time in carmalum. By this treatment lignified membranes are stained by the iodine green, while the unlignified membranes are stained by the carmalum.

Cedar Oil.—Sections which are to be mounted in balsam may first be examined in cedar oil to determine their fitness for permanent mounts; if they are satisfactory, the cedar oil may be drained off and the balsam immediately added to the slide. Cedar oil has a clearing effect on sections which are treated with it.

Thicker cedar oil with a refractive index of about 1.515 is used as an immersion fluid for homogeneous immersion lenses.

Cedar oil is often used as an intermediary between alcohol and paraffin in paraffin-imbedding, but for plant tissues chloroform is rather to be recommended.

Celloidin.—See Steven's Plant Anatomy, pages 233–235.

Chloral Carmine.—This is useful in clearing pollen-grains and staining their nuclei at the same time. It is prepared as follows: Carmine 0.5 Gm. and 30 drops of officinal hydrochloric acid (specific gravity, 1.13 or 17° B.) are added to 30 mils of alcohol, and this is heated for about thirty minutes on the water-bath; then, after cooling, 25 Gm. of chloral hydrate are added, and the solution is filtered until clear.

Chloral Hydrate.—Dissolve 8 parts of chloral hydrate in 5 parts of water. The chloral hydrate may be taken in grams and the water in mils. This is one of the best clearing agents. Whole leaves, when boiled in this solution, clear quickly to such an extent that they may be studied by transmitted light throughout all of the cell-layers. Crystals in leaves may be plainly demonstrated in this way. This reagent is also very useful in clearing pollen-grains and embryos within the ovules.

Chloral Hydrate-iodine.—Dissolve 5 parts of chloral hydrate in 2 parts of water and add a small amount of potassium iodide-iodine. This is the best reagent for demonstrating the presence of starch in chlorophyll corpuscles and in pyrenoids, or in any situation where the starch is surrounded and obscured by other substances.

Chloroform.—Used as a solvent for paraffin in the process of imbedding in paraffin. One of the best solvents of fatty oils and of carotin. Solvent of most of the constituents of suberin.

Chloroiodide of Zinc.—Prepared by dissolving zinc in concentrated hydrochloric acid to saturation and then evaporating to the consistency of concentrated sulphuric acid, adding as much potassium iodide as can be taken up, and then crystals of iodine until no more is dissolved. Another method is to dissolve 20 parts of chloride of zinc, 6.5 parts of potassium iodine, and 1.3 parts of iodine in 10.5 parts of distilled water. Chloroiodide of zinc solutions should be kept in the dark. This reagent is one of the most generally useful in determining the character of plant membranes. By it cellulose walls are colored violet, lignified membranes a yellowish-brown, cutinized and suberized membranes from yellow to yellowish-brown. When sections containing sieve tubes are treated with chloroiodide of zinc and a rather weak solution of potassium iodide-iodine, the walls of the sieve tubes appear violet, while the pits in the sieve plates are a reddish-brown, due to the strands of protoplasm which penetrate them; the callose plates are stained a reddish-brown. Mucilaginous walls are colored violet by this reagent. Chloroiodide of zinc stains protoplasmic cell-contents from yellow to brown, and starch from purple to almost black.

Chlorophyll Solution.—A freshly-prepared strong solution of chlorophyll in alcohol is used to demonstrate suberized and cutinized membranes. When sections are kept in the chlorophyll solution for an hour or so in the dark, cutinized and suberized membranes are stained green, while lignified and cellulose membranes remain unstained. The chlorophyll solution will not keep, and should be freshly prepared whenever needed.

Chrom-acetic Acid.—Prepared by mixing 70 mils of 1 per cent. chromic acid with 5 mils of glacial acetic acid and 90 mils of water. Particularly good for fixing algæ. The algæ should remain in the fixative for twelve hours, then they should be thoroughly washed out in running water or in water which is frequently changed, and thereafter they may be preserved indefinitely in 10 per cent. glycerine, to which a bit of camphor has been added. If it is at any time desired to stain and imbed the algæ which have been fixed and preserved as above, the 10 per cent. glycerine in which the algæ are preserved may be evaporated in the drying oven until quite concentrated, and the algæ may be washed out in strong alcohol; they may then be doubled-stained in the following manner: To the strong alcohol (95 per cent.) in which the algæ are lying are added a few drops of a concentrated solution of magdala red in 95 per cent. alcohol; the algæ are quickly rinsed in alcohol and transferred to a rather dilute solution of aniline blue in 80 per cent. alcohol, where they remain for a few minutes, and are immersed for a few seconds only in a 25 per cent. hydrochloric acid-alcohol solution. The algæ are next rinsed in pure alcohol, and transferred to a 10 per cent. solution of Venetian turpentine. The turpentine is concentrated by the evaporation of the alcohol which was used as the solvent (see Turpentine) in the drying oven. Permanent mounts should be made in the concentrated turpentine.

Chromic Acid.—Solutions of 1 per cent. and 0.5 per cent. have been much used for fixing plant tissues. The material to be fixed should lie in the chromic acid for a day or more, according to the size of the pieces of material to be fixed. The material should then be thoroughly washed out in water and dehydrated by slow degrees in ascending grades of alcohol. A concentrated aqueous solution of chromic acid may be used as a macerating fluid to cause the separation of tissues into their separate cells. To this end rather thin bits of the tissue to be macerated should be placed in the chromic acid for about half a minute, and then carefully washed in water. This operation may be carried on with sections under the cover-glass. Silicious skeletons of diatoms, incrustations on the epidermis of equisetum, etc., may be prepared by allowing the material to lie in concentrated sulphuric acid until it becomes black, and then, after transferring to a 20 per cent. solution of chromic acid for some minutes, washing thoroughly in water. In the case of equisetum and the like the tissues should be scraped away from the inside down to the epidermis before treatment with the acids. Chromic acid is useful in the recognition of tannins, since sections containing tannins, when treated with a 1 per cent. solution of chromic acid, yield a brownish precipitate.

Clearing Media.—See Carbolic Acid, Cedar Oil, Chloral Hydrate, Canada Balsam, Clove Oil, Glycerine, Javelle Water (or Eau de Javelle), Origanum Oil, Turpentine, Xylol.

Clove Oil.—This is an excellent clearing medium, but it has the power of extracting certain stains, and so can not be used in all cases; it is, however, for this very reason of great advantage in the safranin-gentian violet-orange method of staining. See under this head.

Congo-red.—This stain is particularly useful in studying the growth of membranes. Old membranes are, as a rule, left unstained by it, while the newly formed membranes are colored red. In a 0.01 per cent. solution—that is, 1 part of the stain to 10,000 of water—algæ may continue to live and grow, and they are, therefore, well adapted to the study of the growth of membranes with the employment of this stain.

Copper Acetate.—Used in the determination of tannins. Small bits of the plant to be tested are placed in a saturated solution of copper acetate, where they remain for 8 or 10 days; the sections are then placed on a slide in a drop of a 0.5 per cent. solution of ferrous sulphate; after a few minutes the sections are washed in water, then in alcohol, and are finally treated with a drop of glycerine and examined under a cover-glass.

An alcoholic solution of copper acetate, to which has been added a small amount of acetic acid and glycerine, is used to demonstrate glucose in position within the cells where it occurs. The sections are laid in a mixture of the above solution, and an equal volume of sodium hydrate in alcohol, which is brought to boiling on the water-bath. Since glucose is insoluble in alcohol, the cuprous oxide which indicates the presence of glucose in this reaction is found to be deposited within the cells which contain the sugar. For other tests for sugar with a salt of copper see Fehling's Solution. See under Resin in next chapter.

Corallin.—This stain is to be dissolved in a 30 per cent. or a saturated solution of sodium carbonate. It is particularly useful in staining the callose of sieve tubes. It is best to overstain the sections and then to reduce the intensity of the color by immersing the sections in a 4 per cent. solution of sodium carbonate.

Corrosive Sublimate.—See Fixatives.

Cuprammonia.—This should be freshly prepared as needed in the following manner: Put copper filings into a bottle or flask, which is provided with a ground-glass stopper. Pour concentrated ammonia upon the filings and rock back and forth. Only sufficient ammonia should be used to cover the filings. When the solution will dissolve cotton, it is ready for use. This reagent is a solvent of cellulose. When sections are placed in it for some time and are then rinsed with ammonia and finally with distilled water, crystals of cellulose are precipitated within the cells which are stained blue with chloroiodide of zinc, and red with Congo-red. The crystals are again dissolved on the addition of cuprammonia.

Cyanin.—This stain is almost insoluble in water, and should be dissolved in 50 per cent. alcohol. This is a useful stain for fats and all ethereal oils. Sections of fresh material, or material fixed in an aqueous fixative, such as an aqueous solution of corrosive sublimate or picric acid, will be sufficiently stained when left in the cyanin solution for about half an hour. Overstaining may be reduced with glycerine. The alcoholic solution of cyanin, to which has been added an equal bulk of glycerine, is a good stain for suberized membranes, particularly after the sections have been treated with eau de Javelle, which destroys the tannins and prevents the lignified membranes from taking the stain. When sections are placed in a dilute solution of cyanin,—say 20 drops of a concentrated alcoholic solution of cyanin in 100 mils of water,—and are then washed in alcohol and placed in oil of cloves containing eosin, the lignified and suberized walls will be stained blue, while cellulose walls will be red. The sections may then be mounted in Canada balsam. When sections are placed for a quarter of an hour in a concentrated alcoholic solution of cyanin, and are then washed in alcohol and transferred for a quarter of an hour to a 5 per cent. ammoniacal solution of Congo-red, the lignified membranes will appear blue, while the unlignified membranes will appear red. After washing in alcohol, such sections may be mounted in Canada balsam.

Dahlia.—An aqueous solution of from 0.001 per cent. to 0.002 per cent. is used for staining live nuclei. The dividing nuclei of *Tradescantia virginica*, for instance, when kept in this stain for a few hours, become weakly stained. The structure of pyrenoids is well demonstrated by fixing them in equal parts of a 10 per cent. solution of potassium ferricyanide and a 55 per cent. solution of glacial acetic acid and then staining with dahlia, and finally swelling the pyrenoids somewhat in a weak solution of potassium hydrate.

Dammar Lac.—Dammar lac is dissolved in equal parts of benzol or xylol and turpentine; the solution is filtered and evaporated to the desired consistency. This is used as a mounting medium the same as Canada balsam; it has a lower refractive index than Canada balsam, and unstained tissues come out more sharply in it than in the Canada balsam.

Decalcification.—Three per cent. of nitric acid in 70 per cent. alcohol is a good decalcifying agent. The material should be left in the solution for several days. Chromic acid has a decalcifying action; a 1 per cent. to 2 per cent. solution should be used, and the material should be left in this until decalcification is found to be complete.

Decolorizing.—Material which has become brown in alcohol may be decolorized in the following solution: To each 100 mils of alcohol is added from 0.2 to 0.5 mil of concentrated sulphuric acid and as much potassium chlorate as can be transferred on the point of a knife. The material is to lie in this solution for eight to ten days, and is then to be washed a few times in pure alcohol, the material standing for some time in each change of alcohol.

Dehydration.—This is best accomplished by cutting the material into as small pieces as possible, and then placing it in 20 per cent. alcohol, and then into ascending grades of alcohol of 10 per cent. increase at intervals of about two hours. Microtome sections mounted on the slide may be transferred to strong alcohol without injury. In passing from water or aqueous stains to Canada balsam, the material should first come into strong alcohol, and then into xylol to insure complete dehydration, and to infiltrate the material with a solvent of balsam—namely, xylol. Aniline is also a good dehydrating agent. The preparations may pass directly from water into the aniline and from the aniline into the balsam. A stick of potassium hydrate placed in the aniline will keep the latter dehydrated. Potassium hydrate is not soluble in aniline. Very thin microtome sections which are found not to be injured by drying may be allowed to dry, and then may be placed in xylol and thereafter transferred to balsam.

Desilicification.—This is accomplished by hydrofluoric acid. A glass vessel is coated on the inside with melted paraffin to prevent the action of the acid on the glass. Alcohol is then poured into the vessel and the material is immersed in the alcohol; then the hydrofluoric acid is added, drop by drop. The process should be completed in a few minutes. Care must be taken not to breathe the fumes of the acid, since they attack the mucous membranes.

Diastase.—This may be prepared as follows: Germinate barley in the incubator between pieces of blotting-paper until the plumule has reached a length of about 2 mm.; then dry the barley on the water-bath and grind to a fine powder. When a diastatic solution is desired, pour over 10 Gm. of the powdered barley 1 liter of water containing 2 mils of chloroform; let stand for ten hours at about 15°C. and filter. The water filtered off will contain the diastase in solution. Add a little chloroform and preserve in a dark place. Starch-grains may be mounted in this solution under a cover-glass and kept from drying in a moist incubator, and the effect of the diastase on the starch may be studied from time to time under the microscope; or a 1 per cent. starch may be made to which about an equal amount of the diastatic solution may be added, and then at intervals samples from the mixture of starch and diastase may be tested with a solution of iodine. The starch will, after a time, be changed into dextrines and grape-sugar, and will no longer give a blue color when tested with a solution of iodine.

Digestive Fluids.—To remove from sections aleurone grains which are so numerous as to obscure the nucleus, the sections should be treated for twenty-four hours with a digestive fluid prepared by mixing 1 part of pepsin-glycerine with 1 part of pancreatin-glycerine, and 20 parts of a 0.3 per cent. solution of hydrochloric acid. Differences in the character of the protoplasmic cell-contents, and particularly in the dividing nucleus, may be demonstrated by treating sections of fixed material with a digestive fluid made by mixing 1 part of pepsin-glycerine with 3 parts of water acidified with 0.2 per cent. of chemically pure hydrochloric acid.

Double Staining.—There are certain stains which have a peculiar affinity for the lignified and suberized or cutinized membranes; others which color the cellulose membranes without affecting the modified membranes, and still others which stain all membranes, but with different degrees of intensity. The latter are known as diffuse stains, and are able to differentiate the tissues when used singly. Thus, safranin will stain the lignified membranes a cherry-red and the unlignified membranes a brownish-red. Double staining by the use of two stains, one of which has an affinity for the modified membranes and the other for the cellulose membranes, gives excellent results in differentiating the tissues. In general, the sections should first be treated with those stains which color the lignified membranes, and then, after rinsing in water or acidulated alcohol, as the case may require, the sections are to come into a stain which has a particular affinity for the cellulose walls. Fuchsin is an excellent stain for the lignified membranes, and hæmatoxylin, aniline blue, methyl blue, and Berlin blue are good stains for the cellulose membranes. Microtome sections should be left in aqueous solution of fuchsin for a quarter of an hour or longer; then they should be washed in a mixture of 1 part of concentrated alcoholic solution of picric acid and 2 parts of water. Then the sections should be placed for about an hour in one of the blue stains above named, and thereafter washed in strong alcohol, transferred to xylol, and thence mounted in Canada balsam. The lignified membranes will be stained red and the cellulose membranes blue. The cutinized and suberized membranes may be stained together with the lignified membranes in the following manner: Ammonia is added to an alcoholic solution of fuchsin until the solution attains a straw-yellow color; then the sections are placed in this and treated thereafter as described for the simple fuchsin solution.

A mixture of fuchsin and iodine green produces an excellent differentiation. One volume of a concentrated aqueous solution of fuchsin is mixed with 9 volumes of a 0.1 per cent. aqueous solution of iodine green. The sections remain in this solution for about eight minutes; then they are washed in a mixture of 100 mils of absolute alcohol, 1 mil of glacial acetic acid, and 0.1 Gm. of iodine; then transferred to xylol, and thence they are mounted in Canada balsam. The sections should be left longer in the stain if a double stain is not achieved in eight minutes. The proper time for a given material can soon be determined by experiment. In general, those sections stain best which have been fixed in a fixative containing chromic acid or corrosive sublimate. Material which has been fixed in alcohol should, just before staining, be placed for about a day in a 1 per cent. solution of chromic acid, and then washed out in water for some hours. When permanent mounts are made, staining is best carried out with sections already fixed to the slide.

The iron-hæmatoxylin method of staining imparts different intensities of gray or blue to the different tissues and cell-contents, and is one of the simplest and best differentiating stains. It is particularly useful for staining the dividing nucleus. For staining sections from which photomicrographs are to be made, it is unsurpassed. The method of procedure is as follows: The sections mounted on the slide are placed in a 4 per cent. solution of ferric ammonium sulphate for an hour or so. They are then washed in water and placed in a 0.5 per cent. aqueous solution of hæmatoxylin for an hour or more. The hæmatoxylin solution should be several weeks old. The sections are again washed in water, and placed in 1.5 per cent. solution of ferric ammonium sulphate and left there until examination with the microscope shows that the desired intensity of color is achieved. They are then washed in water, dehydrated in alcohol, cleared in xylol, and mounted in Canada balsam.

The three-color method in which safranin, gentian violet, and orange G are successively employed gives a most beautiful differentiation of the structures of the dividing nucleus, and in tissues with resting nuclei there is a sharp differentiation of the nucleus, nucleolus, and the cytoplasm, and of the cutinized, lignified, and cellulose membranes. For embryonic tissues this method of staining is unexcelled. It is not, however, so simple as the methods given above, since the time relations of the three stains employed must be accurately determined and adhered to for a given material. This stain works best with material that has been fixed in Flemming's fixative, *q.v.*, or in a fixative containing chromic acid. Sections from material which has been fixed in alcohol should be immersed for twenty-four hours in a 1 per cent. solution of chromic acid, and then washed in water for a few hours, as above suggested. The sections mounted on the slide are first immersed in a safranin solution which has been prepared by adding to a concentrated alcoholic solution of safranin an equal amount of water. Here the sections remain for twelve hours, or over night. They are then rinsed in pure alcohol, and thereafter immersed in alcohol to which is added about 0.1 per cent. of hydrochloric acid. Just as the clouds of safranin cease to come from the sections, they are rinsed in distilled water and placed in a saturated solution of gentian violet, remaining ten minutes. They are quickly rinsed in water, and while the slide is held horizontally, the sections are flooded with orange G, prepared by diluting a saturated solution of the stain with about five times its bulk of water. After about four seconds the orange G is drained off and the sections are quickly rinsed in water; then, while the slide is held slanting downward, absolute alcohol from a drop-tube is flooded over the slide, beginning at the upper edge of the rows of sections. This washes away some of the surplus gentian violet and dehydrates the sections at the same time. / The slide is now again held horizontally and the sections are covered with clove oil from a drop-tube. The process of decolorizing should be watched under the microscope. Clouds of gentian violet come off in the clove oil, and when the stain has passed from an opaque to a transparent color, the clove oil should be drained off and the slide immersed in xylol, where the sections may be further cleared without extracting more of the stain. When the sections appear clear and without any milkiness in the xylol, they may be mounted in Canada balsam under a cover-glass; or they may be first mounted in cedar oil under a cover-glass and examined under high powers to see if the differentiation has been satisfactorily attained in the finer structures. If the gentian violet should be found still too dense, the oil may be washed off in xylol, the xylol rinsed off in alcohol, and the sections again treated with the solution of orange G, alcohol, and clove oil, and thereafter brought into xylol and cedar oil for preliminary examination as before. If, on the other hand, the gentian violet is found to be too faint, the sections should be brought from the oil through xylol and alcohol to the gentian-violet bath for, say, another ten minutes, and then the process onward to the examination in cedar oil should be as at first described. The effect of the orange G is not only to impart its own color to certain of the structures, but to cause the gentian violet, which stains very intensely, to loosen its hold, to a certain extent, from the structures for which it has a special affinity, so that the absolute alcohol and the clove oil may be able to wash out the surplus stain. The critical part of the process is to allow the orange to work long enough to cause the gentian violet to loosen its hold sufficiently, but not too much. The time relations as above given produce the right result in many instances, but it will be found that they must be altered for certain subjects. In some instances the gentian-violet bath must be prolonged, and in others the orange G solution must be stronger or must be allowed to act for a longer time.

It must be remembered that sections can not be transferred from water or an aqueous stain immediately to xylol or to Canada balsam, for the reason that water does not mix with these substances. Always, in going from water to oil or a resinous solution, strong alcohol should intervene, and, of course, the same precaution must be taken in going from the oil or resinous solution to water.

Eau de Javelle.—Prepared by adding to an aqueous solution of chloride of lime a solution of potassium oxalate so long as a precipitate is formed. The solution is then filtered and diluted somewhat with water before using. Or 20 parts of a 20 per cent. solution of calcium chloride is diluted with 100 parts of water, and after this has stood for some time, a solution of 15 parts of pure potassium carbonate in 100 parts of water is added. If a film should form on the surface of this on exposure to the air, a few drops of the solution of potassium carbonate should be added and the precipitate filtered away.

Lignin is extracted from sections of woody tissues which have lain in the eau de Javelle solution for some time, and thereafter, on treating with chloroiodide of zinc, the membranes show only a cellulose reaction, staining only purple with the chloriodide of zinc.

Starch-grains included in chloroplasts may be demonstrated by first treating sections, or even whole leaves, with eau de Javelle until the chloroplasts are dissolved (this may take from one to twenty-four hours), and then treating the material with a solution of potassium iodide-iodine. The starch-grains will take on a blue or violet color. In some cases, however, the starch-grains themselves are dissolved with the eau de Javelle. In such cases, and indeed in most cases, chloral hydrate and iodine is to be preferred for demonstrating starch-inclusions in chloroplasts (see under this head).

When the forms of the cells simply are to be studied, eau de Javelle is very useful in clearing the sections by dissolving the cell-contents. If the sections become too clear in the eau de Javelle, this defect may be corrected by treating the sections with alcohol or with a solution of alum. See under Cyanin for use of eau de Javelle in differentiating cutinized and suberized membranes.

Eosin.—An aqueous solution of eosin is an excellent stain for protoplasmic cell-contents and cellulose walls. The solution should be quite dilute. For the use of eosin in double staining see under Cyanin and Gram's Method. See also in the next chapter under Aleurone Grains.

Erythrosin.—This is really an eosin, but there is some difference in the method of manufacturing. It is a more precise and a more transparent stain than eosin, and is to be preferred for nearly all staining of paraffin sections. Make a 1 per cent. solution in distilled water or in 70 per cent. alcohol. It gives good results when made up according to the general formula. Erythrosin stains rapidly, thirty seconds to two minutes being sufficient. When used in combination with other stains, erythrosin should come last.

Ether.—Used with equal parts of 95 per cent. alcohol as a solvent of collodion; solvent of ethereal and fatty oils.

Fehling's Solution.—Prepare three separate solutions: (1) 17.5 Gm. of copper sulphate in 500 mils of water; (2) 86.5 Gm. of sodium-potassium tartrate in 500 mils water; (3) 60 Gm. of caustic soda in 500 mils of water. To prepare for use mix 1 volume of each of these with 2 volumes of water. The solutions keep well separately, but the mixture becomes changed after a time, and for this reason the solutions should not be mixed until needed.

Sections may be treated with this solution on the glass slip. Two small drops of distilled water are placed on the slip with 1 drop of each of the three solutions;

then sections, not too thin, of the material which is to be tested for glucose are placed in the mixture on the slide. It is best to cut the sections without wetting the razor, and the sections should not be placed in water, but should be transferred directly to the mixture on the slide. The sections should be covered with a cover-glass and the slide carefully heated over the flame of an alcohol lamp, or a very small flame from a Bunsen burner, until bubbles arise in the solution. If glucose is present, the sections will appear reddish from very small crystals of cuprous oxide which have been reduced from the solution. If it is not desired to observe the crystals of cuprous oxide within the cells, but simply to demonstrate the presence of grape-sugar, small pieces of the tissues to be tested may be placed in a test-tube containing a few mils of the solution, which is then heated to boiling; if grape-sugar is present in considerable quantity, a copious precipitate will after a time settle to the bottom of the tube. See under Copper Acetate. This is particularly suitable for demonstrating the presence of grape-sugar in those cells which contained it in the uninjured tissues.

Ferric-ammonium Sulphate.—Used as a mordant. See under Double Staining.

Ferric Chloride.—An aqueous solution is used as a test for tannin. When sections containing tannin are placed in this solution on the slide, a color is produced which may vary from dark blue to green.

Ferricyanide of Potassium.—Used in demonstrating the structure of pyrenoids. Algæ containing pyrenoids are placed in a mixture of equal parts of a 10 per cent. solution of ferrocyanide of potassium and a 55 per cent. solution of glacial acetic acid. Then, after staining with Hofmann's violet and swelling in dilute potassium hydrate, the lamellated structure of the pyrenoids and the included hollow sphere of starch-grains can be distinguished.

Ferrocyanide of Potassium.—A mixture of this salt and hydrochloric acid diluted with 8 to 10 volumes of water is used to stain the nucleus of starch-bearing Characeæ. The starch is changed to sugar by the hydrochloric acid, and Berlin blue is produced, which is taken up by the nucleus, and, after clearing with chloral hydrate, the nucleus becomes plainly visible. See also under Berlin Blue.

Fischer's Method of Demonstrating Cilia.—The following method is highly recommended for demonstrating cilia of certain bacteria: An exceedingly small amount of the culture containing the bacteria is spread out as thinly as possible on the cover-glass. After the film has dried on the cover-glass the latter is passed through the flame of an alcohol lamp or Bunsen burner (care being taken to avoid a too excessive heat), and then a few drops of a mordant are put on the film on the cover-glass. The mordant is prepared by dissolving 2 Gm. of tannin in 20 mils of water. The cover-glass is then passed back and forth over a small flame until vapor arises from the mordant. The mordant is now washed off by means of water from a wash-bottle, and then one edge of the cover-glass is held in contact with a piece of filter paper to draw away the surplus water. Next, a concentrated aqueous solution of fuchsin is spread over the film on the cover-glass, and the cover-glass is held over a flame until the fuchsin solution begins to boil; the cover-glass is then washed off, and is allowed to dry. At any time thereafter the cover-glass, with the film side down, may be cemented to the slide with balsam. In successful preparations made by this method cilia, when present, will stand out quite sharply.

Fixatives.—All embryonic plant tissues, and those tissues, whether embryonic or mature, whose protoplasmic cell-contents are to be studied, should be first fixed and hardened preparatory to cutting thin sections from them. The object of fixing is to coagulate the protoplasmic structures in the form which they possessed during life. Subsequent dehydration and hardening prepare the material for imbedding and sectioning.

All the constituents of a fixative should penetrate the tissues quickly and at the same time, and to aid in this the material to be fixed should be cut into the smallest pieces compatible with the purpose of the study.

Absolute alcohol or 95 per cent. alcohol penetrates the tissues quickly, and for many subjects is an excellent fixative. Material fixed in alcohol does not, as a rule, yield the best results with staining media, and such material should, as already recommended, after sectioning and mounting on the slide, be placed in a 1 per cent. solution of chromic acid for about twenty-four hours, and then washed for two hours in water. Then the sections may be stained.

Flemming's Mixture.—The best all-around fixative is Flemming's mixture:

```
1 per cent. chromic acid.................................... 16 parts,
2 per cent. osmic acid...................................... 3 parts,
Glacial acetic acid ........................................ 1 part.
```

A saturated solution of corrosive sublimate in strong alcohol penetrates quickly, and tissues need to be left in it for only a few hours. The corrosive sublimate should be washed out with alcohol, to which crystals of iodine have been added, until it assumes a brown color.

A concentrated solution of picric acid in alcohol is an excellent fixative for aleurone grains in oily seeds.

It is best to thoroughly wash out all fixatives before the process of hardening in successively higher grades of alcohol is begun. This is accomplished for those fixatives which do not contain corrosive sublimate by washing in running water for about six hours, or over night. Fixatives containing corrosive sublimate should be washed in alcohol containing iodine, as above suggested.

The material to be fixed should be made to sink at once in the fixative. In alcoholic fixatives this will occur very quickly without any special means to accomplish it, but for aqueous fixatives it may be necessary to first pump the air from the tissues. This always insures a better injection of the tissues by the fixative, and is to be recommended for most material. Any sort of air-pump will answer. A bicycle-pump with the valve reversed does excellent service, and has the advantage of portability, so that it may be carried into the field where material may be fixed as gathered. In pumping out the air the material should be immersed in 0.5 per cent. chromic acid if Flemming's fixative is to be used, and after the material has been made to sink in this, the chromic acid should be immediately replaced by the intended fixative. In most aqueous fixatives, such as Flemming's, the material should be left for about forty-eight hours. In the alcoholic fixatives the material may be left from a few hours to twenty-four hours. With some material the best results are obtained by using the fixatives hot. This is true of many sporangias and zygospores, which are very difficult of penetration by the fixative. Since the osmic acid is quite volatile, special precautions must be taken in heating fixatives of which it forms a part. Good results have been obtained in the following manner: The fixative (preferably Flemming's) was poured into a tall test-tube to the height of a few centimeters. Into this was placed the material to be fixed, and the tube was tightly stoppered. Then the test-tube was immersed in a vessel of boiling water to the depth of the fixative, where it remained for about three minutes, or until the fixative began to show signs of boiling; then the test-tube was removed and plunged into cold water. After cooling, the stopper was removed and the material then sank in the fixative. Since the long test-tube extends for some distance above the boiling water, the upper part of the tube remains sufficiently cool to condense the vapors of the fixative as they are formed, and in this way any material change in the composition of the fixative is prevented. The material should remain in the cool fixative for about twenty-four hours, and then should be washed out and dehydrated in the usual manner.

The amount of the fixative employed should always be large in proportion to the material to be fixed in it.

A corrosive sublimate fixative which may be used cold or hot, as the material may require, is prepared by mixing 80 parts of a saturated aqueous solution of corrosive sublimate with 200 parts of glacial acetic acid. As with all aqueous fixatives, the air-pump should be employed when this fixative is used cold.

Fuchsin.—See under Double Staining.

Fuchsin, Acid.—Excellent for staining crystalloids. The material containing the crystalloids should be fixed in a concentrated alcoholic solution of corrosive sublimate. Then the sections should be immersed for twenty-four hours in a 0.2 per cent. solution of acid fuchsin, to which a little camphor has been added. To demonstrate crystalloids in chromatophores the sections should be treated as follows: The sections are placed in a solution of 20 per cent. acid fuchsin in 100 Gm. of aniline-water. This solution is heated somewhat while the sections remain in it from two to five minutes; they are then rinsed in a solution of 1 part of a concentrated solution of picric acid in alcohol and 2 parts of water. This solution should be warmed to about 40° C., and the sections should be rinsed in it until they cease giving off color to it. Thereafter they are dehydrated in strong alcohol, passed into xylol, and mounted in Canada balsam.

Acid fuchsin is an excellent stain for leucoplasts and chromatophores in general. The material is fixed in a concentrated alcoholic solution of corrosive sublimate in absolute alcohol, where the material remains for twenty-four hours; then the fixative is washed out in alcohol containing iodine (see under Fixatives). Sections from this material are placed in a 0.2 per cent. solution of acid fuchsin in distilled water. After remaining twenty-four hours they are taken out, washed in running water for a time, and are then examined in glycerine or are allowed to dry, after which they are mounted in Canada balsam. The sections can not be dehydrated in alcohol, because this will extract the stain from the chromatophores. The following method may also be used: The material is fixed in a solution of 5 Gm. of corrosive sublimate in 100 Gm. of absolute alcohol, which is acidulated with 10 drops of hydrochloric acid. Then the fixative is removed by placing the material in pure alcohol, which is several times replaced. Sections from this material should be stained by immersion for about twenty minutes in a solution of 2 Gm. of acid fuchsin in 200 mils of distilled water and 3 mils of aniline oil. They are then washed in a mixture of 50 mils of a saturated alcoholic solution of picric acid and 100 mils of water until color ceases to be given off from the sections. Then the picric acid is washed from the sections in pure alcohol. The sections are next placed in chloroform for ten minutes and are then ready to be mounted in Canada balsam.

When desired, sections cut from fresh material may be fixed and stained as above. Or the material may be fixed and imbedded, and after microtome sections have been cut and mounted on the slide they may be stained as above directed.

A beautiful double stain for nuclei is prepared from acid fuchsin and methyl blue as follows: The microtome sections mounted on the slide are immersed for half an hour in a 0.001 per cent. aqueous solution of acid fuchsin, then quickly washed in water, and immersed for about one minute in a 0.002 per cent. aqueous solution of methyl blue. The surplus stain is then washed off in alcohol and the preparation is allowed to dry; then the sections are immersed in olive oil from six to twenty-four hours, after which they are washed in absolute alcohol or in a mixture of absolute-alcohol and xylol until the stains are quite clear, and the preparation is ready to be mounted in Canada balsam.

Gelatine.—Motile swarm spores and the like are sometimes mounted for observation in a solution of gelatine, which renders their movements less rapid, and in this way facilitates the study of these bodies. About 1 Gm. of gelatine is dissolved in 100 mils of water; a drop of this is placed upon a slide which has been somewhat warmed, and then a drop of the fluid containing the motile bodies is added to the drop of gelatine solution and mixed with it by stirring, after which the cover-glass is put on. See also under Nutrient Media.

Gentian Violet.—A 1 or 2 per cent. solution of acetic acid, to which gentian violet is added until the solution appears of a deep violet color, is effective in instantaneously fixing and staining the nuclei of fresh tissues. Anthers and sporangia need only to be tested out with needle in this fluid or crushed under the cover-glass, when the nuclei of the pollen-grains and spores or the mother cells of these will be fixed and stained for immediate examination. See also under Double Staining and Gram's Method.

Glucose.—A mounting medium is made from glucose by mixing 140 parts of distilled water with 10 parts of camphorated alcohol, 40 parts of glucose, and 10 parts of glycerine. The water, glucose, and glycerine are first mixed, and then the alcohol is added and the mixture filtered to remove any camphor which may have been precipitated. The aniline stains are preserved particularly well in this medium.

Glycerine.—This is frequently used as a mounting medium, but since objects are apt to become very transparent in it, only those sections which have been stained should be mounted in it. Sections, such as of wood, which are not apt to shrink easily may be mounted in glycerine directly from water, but delicate tissues should first go from water into a mixture of 10 parts of water and 1 part of glycerine; this should then be allowed to concentrate by the evaporation of the water, when the sections may be mounted on the slide in a drop of pure glycerine. The cover-glass should be quite clean and the glycerine should not be allowed to run back over it. After putting on the cover-glass the surplus glycerine should be taken up with a bit of filter paper and the slide about the edge of the cover-glass should be made quite clean with a cloth moistened in water and then wiped dry with a dry cloth; then the slide may be put in position on the turntable, where a ring of Brunswick black, or of shellac to each ounce of which 20 drops of castor oil have been added, may be spun around the edge of the cover-glass. This process should be repeated several times, allowing each coat to harden before putting on the next, until a strong ring of the cement has been formed. When certain stains are used, such as hæmatoxylin, the glycerine must be entirely free from acids; but with other stains, such as the carmine stains, an acidulation with 1 per cent. of acetic acid is of advantage.

Dilute glycerine, in which sufficient chrome-alum has been dissolved to give a clear blue color, is recommended as a mounting medium for the Schizophyceæ and Florideæ, since the natural colors of these plants are retained in this medium.

Sections containing mucilaginous membranes may be mounted in a drop of pure glycerine in which the membranes will not swell, and then, by irrigating the mount with water, the process of the slow swelling of the membrane may be observed.

Glycerine-gelatine.—This is for most subjects a better mounting medium than glycerine alone. It is prepared as follows: One part by weight of the best gelatine is soaked for about two hours in 6 parts by weight of distilled water. Then 7 parts by weight of chemically pure glycerine are added, and finally, to each 100 Gm. of this mixture 1 Gm. of concentrated carbolic acid. The mixture is warmed for about fifteen minutes, and at the same time constantly sitrred until it becomes clear; then, by means of a hot-water funnel, or while kept warm in an incubator, the mixture is filtered through glass-wool or filter paper which has been washed with distilled water after being placed in the funnel.

To mount sections in glycerine-gelatine the glass slip is warmed and a small bit of the gelatine is placed upon it. If the slip is not warm enough to melt the gelatine, it should be passed back and forth above the flame of an alcohol lamp. If the sections are of a character not liable to shrink, they may be transferred directly from water to the melted gelatine; if, however, there is danger of shrinking, the sections should first be placed in a 10 per cent. solution of glycerine, which is then allowed to concentrate by evaporation of the water, and then, from the concentrated glycerine the sections may be transferred to the drop of melted glycerine-gelatine. To avoid air-bubbles the cover-glass should be put on with precaution. If several sections are being mounted under one cover-glass, and these should come to lie over each other in putting on the cover-glass, they may be properly arranged without attempting to remove the cover-glass (which usually makes the matter worse) by heating the slide until the gelatine becomes quite soft, and then drawing a hair under the cover-glass, with which the sections may be manipulated. It is sometimes a difficult matter to put just the right amount of the gelatine on the slip. To overcome this difficulty, heat the gelatine and pour it out in a thin film over a clean glass plate. When it has become cool, strip it from the glass; then cut off small squares of different size, melt them separately on glass slips, and cover with the cover-glasses of the size to be used with subsequent preparations. The film of gelatine should then be cut into wafers of the size found to exactly fill out the space under the cover-glass. These wafers should be kept from drying too much and free from dust in tightly stoppered bottles. Specially valuable for imbedding brittle objects.

Gold Chloride.—Protein crystalloids may be beautifully stained by means of a solution of gold chloride. The material is to be fixed in a 20 per cent. solution of corrosive sublimate in absolute alcohol, then the fixative is to be washed out in alcohol containing iodine, and finally in pure alcohol; and then the sections are to be placed in a 1 per cent. solution of gold chloride for about 3 hours. This process is to be carried on in the dark. Then the sections are to be placed in a 5 per cent. solution of formic acid, in which they are to remain for several hours exposed to the light. They are next to be placed in 10 per cent. glycerine, which is allowed to concentrate in a place free from dust, and from the concentrated glycerine they are to be mounted in glycerine-gelatine, under which head will be found the method of mounting. By this method of staining, the crystalloids are stained from rose to violet.

Gram's Method.—This method is specially recommended for staining bacteria, either in cover-glass preparations or in sections. The sections are stained in a mixture of 100 mils of aniline water (prepared by combining about 5 mils of aniline with 95 mils of distilled water), and 11 mils of a concentrated alcoholic solution of gentian violet, or, better, methyl violet. This is filtered, and 10 mils of absolute alcohol are added to it. The preparation is taken from the stain, rinsed in alcohol, and transferred to a solution of 2 parts of potassium iodide, and 1 part of iodine in 300 parts of distilled water, where it remains from 1 to 3 minutes. Then it is rinsed in alcohol, transferred to clove oil, and thence mounted in Canada balsam. A good double stain is obtained if the clove oil has some eosin dissolved in it.

Gunther's modification of the Gram method is as follows: The preparation is stained and passed through the potassium iodide-iodine solution as above. Then it is placed for 1 to 2 minutes in alcohol, next for 10 seconds in a 3 per cent. solution of hydrochloric acid in alcohol, then again for several minutes in pure alcohol, until no more color comes away, and then it is passed on into xylol, and finally is mounted in Canada balsam.

Greenacher's Borax Carmine.

Carmine	3 g.
Borax	4 g.
Distilled water	100 mils.

Dissolve the borax in water and add the carmine, which is quickly dissolved with the aid of gentle heat. Add 100 mils of 70 per cent. alcohol and filter (Stirling).

Gum Arabic.—The study of the spermatozoids of ferns, etc., is facilitated by adding a 10 per cent. solution of gum arabic to the drop of water containing the spermatozoids, which are then unable to move so rapidly in the thicker fluid.

Hæmatein.—Dissolve with heat 1 Gm. of hæmatein in 50 mils of 90 per cent. alcohol, and add to this a solution of 50 Gm. of alum in 1 liter of distilled water. After cooling, filter if necessary, and add a crystal of thymol to prevent the growth of fungi. The solution is ready for use at once. Sections stained in this solution should be washed in water and transferred to glycerine-gelatine for mounting, or may be dehydrated and mounted in Canada balsam. The stain may be reduced in overstained sections by allowing the preparation to stand for some time in a 1 per cent. solution of alum. A sediment is apt to settle from this solution, but this is not an indication that the stain is spoiled. The sediment can be partly prevented by adding to the solution about 2 per cent. of glacial acid, which, on the whole, increases the effectiveness of the stain. The acid should be entirely washed from the sections with water before permanent mounts are made.

Hæmatoxylin, Delafield's.—Prepared by mixing 4 mils of a saturated solution of hæmatoxylin crystals in absolute alcohol with 150 mils of a saturated solution of crystals of ammonium alum in water. After standing for a week exposed to the light, this should be filtered and mixed with 22 mils of glycerine and 25 mils of methyl alcohol. Before using this it should be allowed to stand until all precipitates have settled.

Hæmatoxylin and Eosin.—These may be combined to form a double stain by adding the above solution to a mixture of equal parts of glycerine and a saturated alcoholic or aqueous solution of eosin, until the green fluorescence of the eosin has entirely disappeared.

Hæmatoxylin and Safranin.—Sections stained in safranin and washed in water may be placed for a few minutes in a solution of 0.1 Gm. of alum in 30 mils of water to which have been added a few drops of a solution of 3.5 Gm. of hæmatoxylin in 100 Gm. of alcohol. The hæmatoxylin and alum mixture should stand a few days before using. By this treatment lignified and suberized walls are stained red and cellulose walls violet.

Hanging-drop Culture.—A hanging-drop culture is useful in the study of various microörganisms. A glass or vulcanite ring, obtained of the dealers for the purpose, is cemented to the ordinary glass slip with Canada balsam or melted paraffin. A round cover-glass, which should have a diameter at least as large as the outside diameter of the ring, should be thoroughly cleaned, and sterilized by baking in a hot-air sterilizer. Then the slide may be held with sterilized forceps, or placed on any convenient support, while a drop of nutrient substance is placed on the cover glass by means of a sterilized glass rod. By means of a sterilized platinum needle a few individuals of the organisms to be studied may be transferred from pure cultures to the drop of nutrient fluid on the slide. Then the cover-glass should be quickly inverted over the ring cemented to the glass slip, the upper surface of the ring having been previously lightly coated with vaseline, which serves to hold the cover-glass in position and, at the same time, prevents the drop from evaporating. If a single organism is desired in the hanging drop, a few of the organisms from a pure culture may be transferred by means of a sterilized platinum needle to some sterilized nutrient liquid in a test-tube; the tube should be twirled between the palms of the hands to distribute the organisms, and then a drop from the test-tube should

35

be transferred to the cover-glass. More organisms should be transferred from the pure culture to the test-tube, or more nutrient liquid should be added to that already in the test-tube, until it is found that a single individual of the organism to be studied occurs in each drop taken from the test-tube. If, in inverting the cover-glass over the ring, the drop runs to one edge or spreads out over the cover-glass so that it comes in contact with the ring, the cover-glass is to be washed off and the process repeated until the drop hangs free near the center of the cover-glass. The drop must not be so large that if the organism should sink to the lower surface of the drop it could not be brought into focus with the highest power objective to be used in studying it. The circulation in the plasmodia of myxomycetes may be conveniently studied if a bit of the substratum containing the plasmodium is moistened and placed on the cover-glass, which is then inverted over the ring as before. The preparation is then set aside in a warm, dark place until the plasmodium has grown out over the cover-glass.

It is sometimes of advantage to place a drop of water on the glass slip within the ring, so that the atmosphere within the cell will be kept quite humid. Instead of the glass or vulcanite ring, thick cardboard may be used. A piece is cut 1 inch square, if the glass slip is 3 x 1 inch, and a round or square hole is cut in the center of the cardboard somewhat smaller than the cover-glass to be used. The cardboard is sterilized in boiling water, and pressed into position on the slide. The hanging-drop culture is prepared as above described, but no medium is needed to fasten the cover-glass to the cardboard, other than the water with which the cardboard is soaked. Thereafter the cardboard should be moistened from time to time as needed. Bits of the plasmodia of myxomycetes, when suspended on the cover-glass of such a cardboard cell or of the glass or vulcanite cell, as above described, will grow out over the cover-glass, and may be studied throughout a protracted period without being disturbed. If a solid nutrient medium is required, a drop of nutrient gelatine may be placed on the cover-glass instead of the drop from the fluid nutrient medium. Under certain circumstances it is of advantage to flatten out the drop of nutrient substance, after the organisms have been planted in it, by placing over it a smaller cover-glass, which, of course, should be so small that when the larger cover-glass is inverted over the ring or piece of cardboard, the border of the smaller cover-glass will not come in contact with the inner edges of the cell.

Hardening Processes.—The hardening of tissues is accomplished by the withdrawal of water from them. This is, in most cases, best accomplished by means of successively higher grades of alcohol, as described elsewhere.

A quick method of hardening fresh tissues, and at the same time preparing them for immediate sectioning, is to freeze them by the evaporation of ether or the expansion of liquid carbonic-acid gas. This process requires the use of special apparatus, easily obtainable. For an imbedding mass, either a drop of the white of egg, or a thick solution of dextrin in a solution of carbolic acid, 1 part, water, 40 parts, may be placed about the object before freezing. If the dextrine solution is used, it would be better to pump the air from the object while immersed in the solution; then place on the object-holder, pour a small amount of the solution about it, and freeze. This method will answer very well in some cases, when it is desired to prepare a large number of sections quickly for class use, but it can by no means take the place of fixing the material in an appropriate fixative, hardening slowly in alcohol, and imbedding in paraffin or collodion.

The mucilaginous layer of certain seed coats may be hardened with a 10 per cent. solution of neutral acetate of lead. The sections are cut from dry seeds, hardened in the lead acetate, and stained with methyl blue. They are then washed in water and mounted in a 2 per cent. solution of boracic acid.

Hoyer's picro-carmine solution is made by dissolving carmine in a concentrated solution of neutral ammonium picrate. A solution of carmine and picric acid is known as Picro-carmine Solution. Carmine solutions give with cellulose, the nucleus and proteins a red color.

Hydrochloric Acid.—This reagent has such manifold application in histology that its uses are best learned in the specific cases of its application. See in the next chapter under Amylose, Berberin, Caffeine, Calcium Oxalate, Calcium Sulphate, Ethereal Oils, Elæocapsin, Magnesium Sulphate, Middle Lamella, Myrosine, Pectic Substances, Phloroglucin, Theobromine, Vanillin. See also in this chapter under Maceration.

Hydrogen Peroxide.—One part of hydrogen peroxide mixed with 20 parts of 60 per cent. alcohol will, in a few minutes, remove from sections the dark discoloration due to osmic acid which has been used as a fixative.

Infiltration.—The stony tissues of seeds, etc., which are too hard and brittle to be sectioned with a knife, and must, therefore, be ground to the requisite thinness on a stone or by means of emery powder, may be protected against breaking during this process if fairly thin sections are first cut with a fine saw and then placed in a rather thin solution of Canada balsam or copal in chloroform, which is then allowed to evaporate to the thickness of syrup; the sections are allowed to dry and are then cemented by means of a thick solution of gum arabic to a glass plate preparatory to grinding. Only a thin layer of gum arabic should be used, and this should be quite dry before the grinding is begun. The sections may be ground thin on a clean, dry Arkansas or Wichita stone. Before the section has been brought to the desired thinness, the surface should be polished by rubbing it on a piece of soft leather which has been dressed with tripoli. The stone on which the sections are ground may be cleaned of the balsam from time to time by means of a cloth dipped in xylol or turpentine. When one side has been polished, the section may be freed from the glass plate by soaking in water, and then the polished side should be cemented to the glass plate and the reverse side ground and polished as before. The sections should be examined from time to time with the microscope, so that the process of grinding may be stopped as soon as the desired transparency has been obtained. They may then be washed from the glass plate with water, and after drying should be mounted in Canada balsam.

Iodine.—The fumes from heated crystals of iodine serve well in many cases as a fixative. Small objects in drop cultures may be readily fixed by pouring over them the fumes arising from iodine heated in a test-tube. Algæ may be fixed by placing a few crystals of iodine in the bottom of a test-tube, cautiously inclining the tube slightly with the mouth downward, then placing the algæ in the test-tube near the mouth directly from the water in which they were growing, and thereafter heating the crystals so that the fumes from them pour down over the algæ. The iodine may afterward be expelled by warming the fixed material to 30° or 40° C., and the material will then need no further washing out.

Iodine has a wide application in plant histology and microchemistry. See under Aconitine, Atropine, Carotin, Cellulose, Colchicine, Gums, Gram's Method, Lipochromes, Lignin, Mucus-globules, Nicotine, Proteids, Suberin.

Iodine and Alcohol.—A good fixative for very small organisms is a solution of 3 parts of iodine in 100 parts of 70 per cent. alcohol. This, at the same time, permits the staining effect of iodine on the cell-wall and cell-contents.

Iodine and Aluminum Chloride.—Aluminum is dissolved in hydrochloric acid to saturation, and then allowed to evaporate to the consistence of syrup. Cellulose is colored a dark blue to violet color when successively acted on by this reagent and a solution of potassium iodide and iodine. It is said to act more quickly than chloroiodide of zinc and iodine and the color imparted is retained for several days.

Iodine and Glycerine.—A mixture of potassium iodide-iodine with glycerine in equal parts gives good results when the action of iodine is to be observed. The glycerine keeps the preparation from drying, and at the same time has a clearing effect.

Iodine and Potassium Iodide.—This solution is prepared by dissolving 0.5 Gm. of potassium iodide and 1 Gm. of iodine in a small amount of water, and then diluting this with 100 mils of water. The solution is left standing over any iodine which may crystallize out. This formula is commended by Arthur Meyer in his work on "Stärkekörner" as best adapted to the study of starch-grains. A rough-and-ready method of preparing an iodine solution is to dissolve a small amount of potassium iodide in distilled water and then dissolve crystals of iodine in this until a brown color is obtained. This can be diluted with water as is found necessary. A rather pale solution of iodine is sufficient to color starch blue. To stain modified cell-walls the solution needs to be stronger.

Iodine and Sulphuric Acid.—A test for cellulose. The section is first soaked in a solution of potassium iodide-iodine, and then is mounted in a drop of a mixture of 2 parts by volume of sulphuric acid and 1 part of water. Cellulose walls take on a blue color by this process.

Iodine Green.—See under Double Staining. A 2 per cent. solution of glacial acetic acid with iodine green dissolved in it serves well in the instant fixing and staining of the nuclei of fresh material.

Iodine water is prepared by adding as much iodine to distilled and sterilized water as it will dissolve (about 1 : 5,000).

Iron Acetate.—Used in the detection of tannins, which see.

Iron Hæmatoxylin.—See Double Staining.

Lactic Acid.—Dried algæ and fungi may be prepared for study with the microscope by soaking them first in water and then in concentrated lactic acid, in which they are heated until small bubbles are formed; they may then be studied in the lactic acid. A 10 per cent. solution of lactic acid is recommended for fixing bacteria. This fixative is said not to interfere in any way with the subsequent processes of staining with alcoholic solutions of aniline dyes.

Lead Acetate.—A 10 per cent. solution of neutral lead acetate is used to harden the mucilaginous layers of seed coats. For subsequent treatment see under Boracic Acid.

Lithium Carbonate.—Useful in removing from material picric acid, which has been used as a fixative. A few drops of a cold, saturated, aqueous solution of lithium carbonate are added to the alcohol, which is used to wash out the fixative.

Loeffler's Methylene Blue.—This reagent is prepared by adding 30 mils of a concentrated alcoholic solution of methylene blue to 100 mils of water containing 10 milligrams of potassium hydrate.

Maceration.—The study of the forms of cells is greatly aided by isolating the cells from each other by the process known as maceration. Various reagents may be used for this purpose. A solution of potassium chlorate in nitric acid is very commonly employed. This is known as Schulze's maceration fluid. A few pieces of potassium chlorate are put into a test-tube, where they are covered with nitric acid; not very thin longitudinal sections of the material to be macerated are put into the solution, which is then gently heated over a Bunsen burner until bubbles are violently evolved. After standing for a short time the contents of the tube are poured into a vessel containing considerable water. The sections should be transferred to a second dish of water, and then mounted in a drop of water, as needed for examination. The cells may easily be separated from each other by teasing out the section in the drop of water with two dissecting needles. The cells may be isolated from each other by this treatment, for the reason that the middle lamellæ are dissolved, and only the membranes due to secondary thickening remain. The lignin is also removed from the lignified membranes, so that these after maceration give only a cellulose reaction. All chemical manipulations involving the evolution of acid fumes as above should be carried on where the fumes may be quickly conducted out of the laboratory, since the fumes are not only irritating to the mucous membranes, but they are also injurious to delicate apparatus, such as compound microscopes.

Chromic acid may be used for maceration instead of Schulze's maceration fluid. A concentrated aqueous solution is used, and in this the sections are allowed to remain for about half a minute, when they are to be rinsed in plenty of water. They are then to be teased out in a drop of water as before. Very thick sections can not be treated by this method.

Schulze's maceration fluid is to be particularly recommended for sections containing lignified tissues, while tissues destitute of lignified membranes, or containing only a small percentage of these, may be better macerated as follows: A mixture is made of 1 part of hydrochloric acid and about 4 parts of alcohol. The sections remain in this mixture for about twenty-four hours; then they are washed in water, and mounted in a 10 per cent. solution of ammonia. A slight pressure on the coverglass will assist the cells in separating from each other. Cork-cells can be macerated to best advantage by means of a dilute solution of potassium hydrate. See also under Acetic Acid.

Magdala Red.—Make a saturated solution in 85 or 90 per cent. alcohol. Mix 1 to 3 drops of this stain with about 20 mils of 90 per cent. alcohol and let the stain act 3 to 6 hours or longer. Pour off the stain and put the material in a mixture of Venetian turpentine, 1 part, and absolute alcohol, 9 parts, and allow the turpentine to concentrate. If overstaining should occur, it may be corrected by placing the preparation on a white background in the direct sunlight. Especially useful in staining algæ.

Magnesium Sulphate and Ammonium Chloride.—A mixture of 25 volumes of a concentrated aqueous solution of magnesium sulphate, 2 volumes of a concentrated aqueous solution of ammonium chloride, and 15 volumes of water is used as a test for phosphoric acid in tissues. When sections containing salts of phosphoric acid are treated with this reagent, a crystalline precipitate of ammonium magnesium phosphate is formed.

Mercuric Chloride.—Used as a fixative in both aqueous and alcoholic solutions. An aqueous solution which has given excellent results is composed of 80 parts of a saturated aqueous solution of mercuric chloride in water and 20 parts of glacial acetic acid. See also under Fixatives.

Methyl-alcohol.—The refractive index of methyl-alcohol is 1.321, being less than that of water, which is 1.336. On account of this low refractive index methyl-alcohol is a good mounting medium for bringing out the striation in starch-grains and cell-walls.

Methyl-blue.—An excellent stain for cellulose membranes. For double staining, the sections may first be stained with safranin, then washed with alcohol and placed in a concentrated aqueous solution of methyl-blue for 15 minutes or longer. The sections are then to be washed in strong alcohol and mounted in Canada balsam. See also under Double Staining.

Methylene-blue.—A good nuclear stain. For cells filled with protein granules it is particularly good in differentiating the nucleus. Methylene-blue is useful in differentiating pectin compounds. The protoplast and lignified walls are stained a bright blue, while pectin compounds are stained a violet blue. Cells containing tannin will accumulate methylene-blue from very dilute solutions. The sections of living tissues are placed in a solution of 1 part of the stain in 500,000 parts of filtered rain-water. The cells containing tannin soon take on a distinct blue color, and, later, a deep blue precipitate is formed in them. The gelatinous sheaths of live conjugatæ may be stained by dilute aqueous solutions of methylene-blue without injury to the living organism. A 0.001 per cent. solution of methylene-blue in water will stain the living nuclei of diatoms and other simple organisms. The central body of the Cyanophyceæ may be stained by the above dilute solution if, after 24 hours' treatment, the stain is strengthened to a 0.1 per cent. solution. Methylene-blue and carmine form a good differential stain for bacteria occurring in sections of tissues.

Methylene-blue and Carbol-fuchsin.—This double staining method is used in the differentiation of Bacillus tuberculosis. The material first coughed up from the lungs by the patient on waking in the morning should be expectorated into a wide-mouthed bottle or covered jar. The person who is to make the examination should afterward pour this out into a shallow glass dish. This should be placed on a dead-black background, and one of the small, yellowish, lenticular bodies which usually occur in tuberculous sputum should be removed and placed on a cover-glass. A second cover-glass should be placed over this; then press the cover-glasses gently between the thumb and forefinger, and rub to and fro until the material is spread out in a thin film on the cover-glasses. Then slide the cover-glasses apart, and allow them to dry in the open air. When dry, hold them with a pair of forceps and pass them 3 times through the flame of the Bunsen burner or alcohol lamp. (The film should not be allowed to turn brown, else the preparation will be ruined.) Next pour over them carbol-fuchsin prepared by rubbing 1 Gm. of fuchsin with 100 mils of a 5 per cent. aqueous solution of carbolic acid, with the gradual addition of 10 mils of alcohol. Hold the cover-glasses over a flame with forceps until vapor begins to arise from the surface of the stain. Then hold away from the flame, except in intervals of gentle heating, by which they are kept warm for a minute or two. They are next washed in water and decolorized by being moved about in a 25 per cent. solution of nitric or sulphuric acid. When the previously deep-red color has changed to a greenish tint, the preparation is washed in 60 per cent. alcohol to remove the color set free by the acid. If any red color still remains, the preparation should be rinsed in water and again treated with the acid-bath. By the above process the fuchsin has been removed from everything but the tubercle bacilli. The double staining is accomplished by now pouring over the preparation a mixture of 3 parts of water with 1 part of a concentrated alcoholic solution of methylene-blue. After a few minutes the methylene-blue is washed off with water, and the preparation is allowed to dry; when dry, it may be mounted in Canada balsam. Other bacteria than the tubercle bacilli are decolorized by the acid-bath, and are subsequently stained blue by the methylene-blue.

Methyl-green.—An aqueous solution of this stain serves well to differentiate the nucleus of cells containing aleurone grains. Sections through vascular bundles which have been treated for some hours with alcohol borax-carmine, and then for a short time with methyl-green, have the protoplasmic cell-contents stained red, the lignified walls of the tracheal tubes green, and the walls of the primary phloëm portion green.

Methyl-green and Acetic Acid.—Methyl-green is dissolved in a 2 per cent. solution of acetic acid until the solution has a blue-green color. The nuclei of fresh material teased out in this become instantly fixed and stained. It is very useful for a preliminary examination of dividing nuclei.

Methyl-violet.—Starch-grains may be stained by treatment with a dark aqueous solution of methyl-violet. If the starch-grains after staining are treated with a very dilute solution of calcium nitrate, the stain becomes deposited particularly in the less dense layers of the grains. Useful as a stain for elaioplasts. See under this head in the next chapter, page 567. See also in this chapter under Staining Intra Vitam.

Millon's Reagent.—This should be prepared fresh, as needed, by dissolving mercury in an equal weight of nitric acid and then diluting this solution with an equal weight of distilled water. Proteids are colored a brick-red with this reagent. Sections to be tested are to be mounted in a drop of the reagent on a glass slip. Warming the slip hastens the reaction.

a-**Naphthol.**—When sections which are rather thick are treated on a slide with a drop of a 20 per cent. aqueous solution of *a*-naphthol and then two or three drops of concentrated sulphuric acid are added, the sections will be colored violet in a few minutes if cane-sugar, milk-sugar, glucose, lævulose, maltose, or inulin is present.

Nessler's Reagent.—Used as a test for the presence of ammonium. Prepared by dissolving 2 Gm. of potassium iodide in 5 mils of water, and then adding mercuric iodide to the solution while warm until a part remains undissolved. After cooling, 20 mils of water are added to the solution and then, after standing for a time, the solution is filtered, and to each 20 mils of it are added 30 mils of a concentrated solution of caustic potash. The solution must be filtered as often as it becomes turbid. The solution is changed to a yellow color in the presence of ammonia. However, other organic compounds may give the same reaction.

Nutrient Media.—Nutrient media must be sterilized by heat to keep them from spoiling and to make it possible to grow in them pure cultures—that is, cultures of organisms of any desired species. Sterilization may be accomplished by steaming the medium for about twenty minutes each day on three days in succession, after having poured it into test-tubes or flasks which have previously been tightly plugged with cotton rolled into the form of a stopper of the proper size and baked in an oven until the cotton is slightly scorched. The tubes and cotton plugs should be baked together. Or, if an autoclave is available in which steam can be generated under pressure, and accordingly at a higher temperature than that of boiling water at ordinary atmospheric pressure, the cotton plugs and tubes, or flasks, will not need to be baked but may be sterilized, together with the nutrient medium already poured into them, by subjecting them for fifteen minutes to a temperature of 115°C. in the autoclave. At this temperature a single sterilization suffices.

A good artificial nutrient medium for years is made by adding 0.05 per cent. of tartaric acid to a 10 per cent. solution of cane-sugar. A filtered aqueous extract of malted barley also gives good results. To prepare this, barley is germinated until the plumule just begins to protrude; the barley is then dried and ground up, and water is poured over it until there is about twice as much water by volume as of the powdered malt. The water should stand over the malt, with occasional stirring, for about an hour, when it may be filtered off and sterilized. Sterilized grape juice is also an excellent nutrient medium for yeasts. Cultures of yeasts grown in the above media may be made to produce spores in about twenty-four hours if some of the culture is transferred to the surface of sterilized bits of flowerpots which are half submerged in water and kept covered by a bell-jar.

Cohn's normal solution for the culture of bacteria is prepared as follows: Dissolve in 200 Gm. of distilled water 1 Gm. of acid potassium phosphate, 1 Gm. of magnesium sulphate, 2 Gm. of neutral ammonium tartrate, and 0.1 Gm. of calcium chloride.

An infusion of meat for the culture of bacteria is prepared by covering finely chopped lean beef with water and allowing it to stand for twenty-four hours in an ice-chest, after which it is to be filtered through a muslin bag, using pressure of the hands to make the filtration more complete. The filtrate is then cooked and again filtered, and neutralized by the gradual addition of a solution of carbonate of soda. The solution should be tested with litmus paper, and the addition of carbonate of soda should cease as soon as neutralization is accomplished. To this solution is added 0.5 per cent. of common salt. Ten Gm. of peptone may be added to a liter of the infusion.

In place of the meat infusion as prepared above, meat extract may be used in the ratio of 4 to 5 Gm. per liter of water.

Bouillon is prepared by adding 1 liter of water to 1 pound of chopped lean beef. This is cooked for half an hour, then filtered and neutralized with carbonate of soda, then again boiled for an hour to precipitate albuminoids. After a final filtering the bouillon is poured into flasks or test-tubes and sterilized.

Infusions of hay and dried fruits may also be used for nutrient media. A hay infusion for the growth of Bacillus subtilis may be prepared as follows: Chopped hay is placed in a beaker and barely covered with well water; this is kept in an incubator at a temperature of 36°C. for four hours, after which time the extract is poured off and diluted, if necessary, to a specific gravity of about 1.004. The extract is now poured into a flask which, having been closed with a cotton plug, is placed in a steam sterilizer and subjected to a gentle evolution of steam for about an hour. The flask is then placed in an incubator at 36°C. for a day or two, after which time a film produced by colonies of Bacillus subtilis will have formed over the surface of the extract. The spores of this bacterium are particularly resistant to heat, and for this reason while the spores of other bacteria are killed by the process of steaming, those of Bacillus subtilis still retain their vitality.

Solid culture media may be prepared by adding to any of the fluid culture media a sufficient amount of a gelatinous substance to keep the mixture from liquefying at the temperature of the laboratory or, if desired, at the higher temperature of an incubator. One of the most used of the solid media is prepared by adding to the peptonized infusion of meat, as above described, 10 per cent. of the best French gelatine. The gelatine may be increased up to twice this amount, as the temperature may require. One hundred grams of gelatine is allowed to soak in 1 liter of the meat infusion until the gelatine becomes swollen, and then a gentle heat is applied until the gelatine is completely dissolved. After the gelatine is dissolved the solution should again be neutralized, if necessary, with carbonate of soda. When the solution stands at a temperature of about 50°C., an egg stirred up in 100 Gm. of water is added while the mixture is stirred with a glass rod. The mixture is then kept at the boiling-point for about ten minutes. This coagulates the egg-albumen and clarifies the liquid. The clarified liquid is now filtered by means of a hot-water funnel or while kept warm in an incubator, the high temperature being necessary for the reason that the mixture would become stiff at a low temperature, and so incapable of being filtered. The medium should be distributed while warm in sterilized test-tubes or flasks, which are then stoppered with baked cotton plugs. It should then be subjected to a temperature of 100° in the steam sterilizer for 10 minutes at 4 successive intervals of 24 hours. For the reason that gelatine loses its power of solidifying at ordinary temperatures after being subjected to the temperature of boiling water for a long period, the time of each sterilization is necessarily reduced to about 10 minutes and the number of sterilizations is increased to 4; whereas with other solidifying substances, such as agar-agar, the length of each sterilization may extend to 1 hour, and the number of sterilizations need be only 2 or 3.

In pouring the filtered medium into the test-tubes care should be taken not to get any of the medium on the upper portion of the tube where the cotton plug would be likely to come in contact with it, else the plug would later be difficult of removal.

A solid nutrient medium which will remain solid at a higher temperature than the gelatine medium may be prepared from agar-agar, a substance obtained from certain gelatinous algæ, as follows: Two Gm. of the agar are broken into small pieces and soaked in cold water for 24 hours. Then the water is poured off and the swollen agar is added to 1 liter of the peptonized meat infusion. The mixture is boiled for several hours until the agar is completely dissolved. The solution is then neutralized with a solution of carbonate of soda, filtered, distributed in flasks or test-tubes, and sterilized by steaming for 1 hour at 2 or 3 successive intervals of 24 hours.

Cooked potatoes afford a solid nutrient medium which is quickly prepared and which is particularly adapted for the culture of chromogenic bacteria. Potatoes free from wounds are selected and scrubbed in water until they are perfectly clean, and the eyes and any unsound spots, if these could not be avoided, are cut out with a knife. Then the potatoes are placed for an hour in a solution of 1 part of mercuric chloride in 500 parts of water to disinfect the surface. They are next steamed for about an hour in a steam sterilizer, and after 24 hours the steaming is repeated for about half an hour. The sterilized potatoes are then placed in glass Petri dishes, are cut in halves with a sterilized table-knife, and the cut surfaces are inoculated. If the source of the inoculation is not a pure culture, an isolation of forms may be approximated by making long scratches over the surface of the potato with a sterilized platinum needle which has been in contact with the source of the inoculation. It will add to the security of the process of sterilization if each potato, before being placed in the bath of mercuric chloride, is wrapped in a piece of tissue paper, and so protected until it is cut open for inoculation.

Another method of preparing potatoes which is, on the whole, more convenient and certain, is to cut out long cylindrical plugs from sound potatoes by means of a cork-borer or any metal tube of the proper size, and then to cut the potato cylinders very obliquely in two pieces, each of which is then to be placed in the bottom of a test-tube so that the oblique surface stands uppermost. After plugging the tubes with baked cotton, the potato cylinders are subjected to a temperature of 100°C. in the steam sterilizer for one hour at three successive intervals of 24 hours. A sterilized paste made from potatoes or bread serves well for the culture of molds as well as of bacteria.

A decoction of horse-dung furnishes a good medium for the culture of mucor and various other molds. The decoction is prepared by boiling the dung in water, then filtering and sterilizing the solution. By placing the dung of different kinds of animals in a moist chamber, as, for instance, in dishes floating on water and covered with a bell-jar, characteristic fungi will after a time appear on it.

Single spore cultures of mucor may be obtained in the following manner: Glass slides are thoroughly cleaned and sterilized by baking. By means of sterilized forceps a single sporangium of mucor is picked from a spontaneous growth of this fungus on horse-dung or stale bread kept in a moist chamber. The sporangium is placed in a sterilized decoction of horse-dung contained in a sterilized watch-glass, which may be placed on an inverted tumbler in a plate of water and then covered with a bell-jar which should dip into the water and form a germ-proof moist chamber. After a few hours the sporangium will have burst open and the spores, which are now distributed through the decoction, will have swollen to several times their original diameter, and can all the more readily be discerned in subsequent manipulations. A needle which has been disinfected by heating in a flame is now dipped into the decoction and the point of it drawn along the surface of a glass slide which has been cleaned and sterilized as above directed. By this process the decoction which has adhered to the needle is drawn out in the form of a narrow streak, and if several spores of mucor are present, they will be separated from each other. A single spore may be located with a medium power of the compound microscope, and all other spores present in the streak may be wiped off with a cloth which has been sterilized by heat. Then a drop of the decoction of sterilized horse-dung should be added to the small amount containing the spore on the slide. The slide should be placed in a moist chamber where the spore will soon give rise to a mycelium visible to the naked eye, and from the mycelium numerous sporangia will be produced after a time. The slide may be taken from the moist chamber from time to time and the stages in the development of the fungus examined, but as much care as possible should be taken to prevent the contamination of the culture.

Knop's nutrient solution, which is particularly good for the culture of algæ, consists of 4 parts of calcium nitrate, 1 part of magnesium sulphate, 1 part of potassium nitrate, 1 part of potassium phosphate. These should be dissolved in sufficient water to make a 0.2 per cent. or 5 per cent. solution of the combined salts. The potassium salts should first be dissolved, then the magnesium salt, and last the salt of calcium should be added after having been dissolved by itself. By this procedure only a small amount of insoluble calcium phosphate is formed. The zoöspores of Vaucheria may be induced to form at almost any time by transferring this alga from the above solution, in which it has been growing exposed to a bright light, to pure water; or cultures in a 0.1 per cent. or 0.2 per cent. nutrient solution which have been exposed to the light need only be placed in a dark place in order to incite the production of zoöspores.

A 2 per cent. to 4 per cent. solution of cane-sugar may be used as a nutrient medium for algæ. Filaments of spirogyra may be made to conjugate by transferring them from the water in which they have been growing to a solution of cane-sugar as above, which is then placed in a well-lighted place.

The formation of zoöspores may be incited in œdogonium by transferring filaments of the alga from water at a low temperature (say at the temperature of the early morning) to a 2 per cent. or 3 per cent. solution of cane-sugar which is kept at a constant temperature of about 26°C.

Convenient flasks for the preservation of sterilized fluid nutrient media may be made from glass tubing as follows: A piece of glass tubing 0.2 inch in diameter, or larger, is held with its lower end in the flame of a blow-pipe, the tube being constantly revolved about its long axis to insure an even heating of the end of the tube until the end of the tube becomes soft and just begins to draw downward in the form of a large drop. By this time the mouth of the tube has become closed. Then quickly the tube is removed from the flame, and while the melted end of the tube is still held downward, air is blown in at the upper end of the tube by means of the mouth, so that the molten glass at the lower end of the tube is forced outward in the form of a rounded flask. After cooling so that it may be handled, the tube is held in the flame close to the bulb, and by constant turning the tube is heated equally on all sides until it becomes so soft that it may be drawn out. This process is accomplished by taking the tube from the flame and pulling on it gently so that it may be drawn out quite long and narrow. The length of the stem of the bulb should be equal to the depth of the vessel from which the nutrient medium is to be drawn into the bulb. The stem may be severed from the tube by holding it in the flame of the blow-pipe at the proper distance from the bulb, where it will soon become soft enough to be pulled off from the main tube. Then the end of the capillary neck is held in the flame until a bead is formed; in this way the flask is hermetically sealed. To fill the flask with nutrient fluid the neck is sterilized near the end by passing it through a flame, and the head is broken off with sterilized forceps. The bulb is then heated in the flame of an alcohol lamp or Bunsen burner to expand the air. The end of the neck is next quickly dipped into the nutrient fluid, which is forced up the neck into the bulb as the air in this cools. When the bulb is two-thirds full, the neck is withdrawn from the fluid and hermetically sealed in a flame. In filling the bulb the greatest care must be taken to keep the stock of nutrient medium from any source of contamination, if it has once been sterilized. Chemical flasks with narrow necks serve well for a common receptacle. These should be kept stoppered with a cotton plug, and to fill the small flasks the plugs need only to be lifted slightly while the sterilized capillary neck of the small flasks is thrust past the plug into the nutrient fluid. If the nutrient fluid is freshly prepared, and has not yet been sterilized, the small flask may be filled, sealed up in the flame, and sterilized in the steam sterilizer or in a vessel of boiling water for an hour each day on three successive days. The nutrient fluid will keep indefinitely in the little flasks, and when a drop is wanted for a drop culture, it is only necessary to sterilize the end of the capillary neck in a flame, break off the head with sterilized forceps, invert the flask, and place the palm of the hand over the bulb. The heat of the hand will expand the air over the fluid and force the latter down the neck. With a little practice just the desired amount of fluid can be forced out by the heat of the hand. The hand must not be placed on the bulb until the flask is inverted. If it is desired to make cultures within the little flasks, snip off the end of the capillary neck as before, and thrust a long platinum needle, the end of which has been in contact with the source of inoculation, down the neck into the fluid. Then withdraw the needle and hermetically seal the neck in the flame. When cultures are to be made in the flasks, these should be only one-third filled by the nutrient medium; there will then be sufficient air in the flasks for the success of the culture after the flasks have been inoculated and hermetically sealed.

Pollen grains may be made to germinate in hanging drops composed of 100 parts of well-water, 3 to 30 parts of cane-sugar, and 1.5 parts of gelatine. This should be made as needed, or it may be sterilized and kept indefinitely in the little flasks just described. The amount of cane-sugar to give the best results varies with the species of pollen, and can only be determined by experiment, but 3 parts will probably answer for most pollen-grains.

Spores of ferns may be made to germinate on pieces of flower-pot which are kept half submerged in water and are covered by a bell-jar. They should be set before a north window. They should never be exposed to the direct light of the sun, since in such a position the temperature under the bell-jar would become very great.

Orchella (Orseille).—Sections of tissues containing actinomyces may be stained to advantage by an orchella stain prepared as follows: Orchella which has been left in the open air until it is free from its ammonia is dissolved in a mixture of 20 mils of absolute alcohol, 5 mils of concentrated acetic acid, and 40 mils of distilled water, until the mixture has a dark-red appearance. Sections are left in the filtered solution for one hour. They are then washed in alcohol, stained with gentian violet, washed again in alcohol, placed for a short time in xylol, and mounted in Canada balsam. By this treatment the fungus will be double-stained red and blue.

Orseillin and Aniline Blue.—The mycelium of the Peronosporeæ may be stained blue, and the cell-walls of the plant which the fungus is parasitizing may be stained red at the same time by a combination of orseillin and aniline blue. Sections of tissues containing the parasite are bleached in Javelle water, then washed in a saturated solution of potassium hydrate in alcohol. The sections are placed for staining in acetic acid, to which have been added a few drops of an aqueous solution of orseillin BB and a drop or two of aniline blue. The solution should have a violet color. The sections may be mounted for examination in glycerine.

Osmic Acid.—The vapor of osmic acid may be used as a fixative for very small organisms. In order to accomplish this a drop of water containing the organisms need only to be inverted over a bottle containing a 2 per cent. solution of the acid. Osmic acid colors ethereal and fatty oils from brown to black, but other organic substances are also darkened by it; and as a test for oils it is not absolutely reliable. Aleurone grains in sections of Ricinus which have been freed from their oil by standing for a time in strong alcohol may be stained brown, and the crystalloid and ground substance differentiated by immersing the sections for a short time in a 1 per cent. solution of osmic acid.

Paraffin.—Paraffin of about 52° melting-point sections to good advantage at a temperature between 21° and 24°C., or 70° and 75°F. Good cells for hanging-drop cultures may be made by placing glass slides on the turntable and spinning rings on them by means of a camel's-hair brush dipped in melted paraffin.

Pepsin.—One part of pepsin-glycerine and 3 parts of water acidulated with 0.2 per cent. of chemically pure hydrochloric acid. When sections containing protoplasts are subjected to this reagent at blood temperature, certain structures of the protoplast which are insoluble in the reagent may be isolated from those which are soluble. In the dividing nucleus the kinoplasmic spindle-fibers persist after the chromosomes and nuclear plate have been dissolved by this reagent. By the action of digestive ferments on aleurone grains the ground substance is first dissolved and then the crystalloid more slowly, while the limiting membrane of the vacuole occupied by the aleurone grain persists. Digestive ferments are thus found to be excellent reagents for demonstrating the difference in constitution of the finer structures of the protoplast and protoplasmic cell-contents.

Phloroglucin.—This furnishes one of the most reliable tests for lignin. Sections are placed in alcohol containing a trace of phloroglucin, transferred to a drop of water on a slide, and covered with a cover-glass. A drop of hydrochloric acid is then applied to the edge of the cover-glass, and, and the acid comes in contact with the lignified membranes, these are colored with a bright violet red.

Phospho-molybdic Acid.—This is used as a test for proteids. Sections are treated for an hour or two with a solution of 1 Gm. of sodium-molybdium phosphate in 90 Gm. of distilled water and 5 Gm. of concentrated nitric acid. Proteid materials then take on the appearance of yellow granules.

Picric Acid.—The structures of aleurone grains are well differentiated by fixing in a concentrated alcoholic solution of picric acid and subsequent staining with eosin. The sections are to remain in the alcoholic fixative for several hours. They are then to be washed out in alcohol and stained for a few minutes in a solution of eosin in absolute alcohol. Then the sections are successively washed in absolute alcohol, transferred to oil of cloves, and mounted in Canada balsam. The ground substance is dark red, the crystalloid yellow, while the globoid remains colorless. The pyrenoids and chromatophores of algæ may be simultaneously fixed and stained by placing the algæ for an hour or longer in a concentrated solution of picric acid in 50 per cent. alcohol, to which has been added about 5 drops of a solution of 20 Gm. of acid fuchsin in 100 mils of aniline water. The aniline water is prepared by shaking up 3.5 Gm. of aniline in 96.5 Gm. of water. The algæ are then washed in alcohol, transferred to xylol, then to a thin solution of balsam in xylol, and are finally mounted in the thicker solution of Canada balsam in xylol.

Alcohol is a better solvent of picric acid than water, and accordingly it gives quicker results in washing out the acid from the fixed material than water does, but running water may be used to wash out the fixative whether the latter has been dissolved in alcohol or in water.

Picro-aniline Blue.—A double stain, which is very rapid in its action, is prepared by adding aniline blue to a saturated solution of picric acid in 50 per cent. alcohol until the solution has a blue-green color. By this treatment the unmodified cell-walls and the cell-contents are stained blue, while the lignified walls are stained by the picric acid.

Picro-nigrosin.—A solution of nigrosin in a concentrated solution of picric acid in water or 50 per cent. or 95 per cent. alcohol is a good fixative and stain for algæ and leucoplasts, and for double-staining modified and unmodified cell-walls. The solution may, in some cases, need to act for 24 hours. The strong alcoholic solution is particularly recommended for material containing chlorophyll, since this will be extracted by the strong alcohol. Nuclei and leucoplasts are stained a steel blue by the nigrosin.

Potassium Alcohol.—Used for bleaching sections. It may be prepared by mixing a concentrated aqueous solution of potassium hydrate with 90 per cent. alcohol until a sediment is formed. This is allowed to stand for 24 hours with frequent violent shaking, and then the clear liquid is poured off and is diluted for use with 2 or 3 parts of water.

Potassium Hydrate.—For general use, dissolve 5 Gm. of potassium hydrate in 95 mils of distilled water. This is used as a clearing agent for sections and small organisms. The process of clearing may require from several hours to several days. After clearing, the potash should be washed out in plenty of water, and then the preparation may be neutralized with acetic acid. This will tend to make the objects more opaque, and if too much is added, the objects may be cleared again by caustic potash or ammonia. A dilute solution of caustic potash, as above, may be used for the maceration of cork, while delicate tissues in general may be macerated by boiling for a few minutes in a 50 per cent. solution of potassium hydrate in water; the tissues should then be washed in water and teased out on a slide in a drop of water.

Ruthenium Red.—An aqueous solution is an excellent stain for pectic substances and for gums and slimes which have been derived from these. Ruthenium red is not soluble in alcohol, clove oil, or glycerine, and, therefore, preparations stained by it may be dehydrated and mounted in glycerine or balsam.

Safranin.—A saturated solution of safranin in alcohol should be made and this should be diluted with an equal bulk of water, or with an equal bulk of a saturated aqueous solution of safranin. This is an excellent general stain, and gives good differentiating effects when used singly. It is one of the few stains which are particularly adapted to the staining of pectic compounds. It also gives beautiful results in staining the cell-contents of spirogyra and other algæ. The algæ, after fixing in a fixative containing chromic acid, should lie in the alcoholic solution diluted with an equal bulk of water for 12 or 24 hours. They should be transferred to 50 per cent. alcohol, to which strong alcohol is then added, drop by drop. The color will begin to be extracted in the alcohol, and when the right intensity has been reached, the material should be transferred to dilute glycerine, where it is to remain while the glycerine slowly concentrates in a place protected against dust. Then permanent mounts may be made in glycerine or glycerine-jelly. The stain given by safranin is quite permanent. See also under Double Staining, and the directions there given for the three-color method.

Salicylate of Soda.—A clearing reagent which for small objects is not inferior to chloral hydrate is furnished by dissolving crystals of salicylate of soda in an equal weight of distilled water. With tincture of iodine added this reagent will cause starch to swell, at the same time imparting a blue color to it.

Salt.—A 4 per cent. or stronger solution of common salt, or of potassium nitrate, may be used to cause plasmolysis in living cells. This process may be all the more clearly seen by adding eosin to the salt solution.

Schulze's macerating solution is prepared by adding crystals of potassium chlorate from time to time to warm concentrated nitric acid. It is employed in the isolation of lignified cells. The material is allowed to remain in the solution for a short time or until there appears to be a disintegration of the tissues. A large excess of water is then added. The material is carefully washed, the cells teased apart and mounted in a solution of methylene-blue.

Shellac.—A thick solution of shellac in alcohol, to each ounce of which are added 20 drops of castor oil, makes an excellent sealing medium for preparations mounted in glycerine or glycerine-jelly, or in an aqueous medium.

Silver Nitrate.—A solution of silver nitrate is used to bring out more clearly the striations in bast fibers and starch-grains. Sections containing striated bast fibers are allowed to dry and are then impregnated with the silver salt. Without previous washing the sections are transferred to a 0.75 per cent. solution of common salt. They are then placed in distilled water and exposed to the light for a considerable time; thereafter they are allowed to dry and may be examined to good advantage in anise oil.

Dry starch-grains are put to soak in a 5 per cent. solution of silver nitrate. After a time they are allowed to dry superficially and are then treated with a 0.75 per cent. solution of common salt, in which they are finally exposed to the direct light of the sun to reduce the chloride of silver which has been formed within the grains. The less dense laminæ of the starch-grains will show a gray color, due to the reduced silver.

Staining Intra Vitam.—Living protoplasts may accumulate certain stains from very dilute solutions without injury to themselves. Dahlia, methyl-violet, mauvein, and methylene-blue are particularly suitable for this purpose. Solutions containing 0.001 per cent. or 0.002 per cent. of any of the first three stains have given good results in staining living nuclei, while 1 part of methylene-blue in 500,000 parts of filtered rain-water is used for staining living cells containing tannin. A large amount of these very dilute solutions should be employed in order that a sufficient amount of coloring matter may be at hand for accumulation by the living cells. Living protoplasts have the power of reducing and accumulating metallic silver from solutions of certain of the salts of silver, while dead protoplasts have not this power. The simplest method of producing this reaction is to place a few filaments of spirogyra in a liter of a mixture of 1 part of silver nitrate in 100,000 parts of water with 5 mils of lime water. The experiment will be completed in about half an hour if the temperature of the solution is raised about 30°C. By this process living protoplasts are colored black by the reduced silver, while dead protoplasts take on a yellowish or brownish color.

Sulphuric Acid.—By the action of sulphuric acid cellulose is changed to amyloid, which may be colored blue by a tincture of iodine. By the continued action of concentrated sulphuric acid cellulose becomes dissolved. Cutinized and lignified membranes remain undissolved in sulphuric acid. Silicious skeletons or incrustations may be freed of all organic matter by treating the objects with concentrated sulphuric acid until they turn black, and then with a 20 per cent. aqueous solution of chromic acid. The objects should be washed repeatedly in water before they are ready for examination with the microscope.

Tannin and Antimonium-potassium Tartrate.—These are used successfully as a mordant for methyl- and gentian-violet, fuchsin, and safranin when sections stained with these stains are to be mounted in glycerine. The sections before staining are placed in a 20 per cent. solution of powdered tannin in cold water. After washing well in distilled water, they are placed for 24 hours in a 2 per cent. solution of antimonium-potassium tartrate. After washing again in distilled water, they are transferred to the stain. From the stain the sections are washed quickly in distilled water and placed in strong alcohol, where the color is washed out until the desired degree of intensity is reached. They are now ready for mounting in glycerine, or, if desired, they may be placed in xylol and then mounted in balsam. If the sections are so deeply stained that they cannot be sufficiently washed out in alcohol, they should be placed for a time in a 2.5 per cent. solution of tannin.

Turpentine.—This may be used to dissolve paraffin from sections which have been cut from material imbedded in paraffin. See also under Carbolic Acid.

Venetian Turpentine.—To prepare a mounting medium from Venetian turpentine the product as it comes from the apothecary is diluted with an equal volume of strong alcohol, and after the mixture has become clear by long standing or by filtering after being well shaken, it is thickened somewhat on the water-bath. Objects may be mounted directly from strong alcohol into Venetian turpentine as above prepared. Objects which are found to shrink by this treatment may be transferred from strong alcohol to a mixture of 10 parts of the turpentine with 100 parts of alcohol. The alcohol is then to be withdrawn from this mixture by placing the latter, together with a dish of calcium chloride, under a bell-jar. In order to keep the mixture of turpentine and alcohol from mounting the sides of the vessel which contains it, the rim of the vessel should be coated over with hot paraffin. The turpentine hardens quite slowly, and in order quickly to fasten a cover-glass to the slide when the turpentine is being used for a permanent mount, a wire which has been heated in a flame should be quickly drawn around the edge of the cover-glass.

Xylol.—This is used as a solvent for paraffin, either in removing paraffin from sections or in preparing a dilute solution of paraffin to be used in the gradual infiltration of tissues with this substance. Used also as a solvent of Canada balsam.

Ziel's Carbol-fuchsin.—This solution is prepared by adding 15 mils of a concentrated alcoholic solution of fuchsin to 100 mils of water containing 5 Gm. of carbolic acid. •

CHAPTER III.—METHODS OF DEMONSTRATING THE CHARACTER OF CELL-WALLS AND CELL-CONTENTS

Aconitine,$C_{33}H_{43}NO_{12}$.—To demonstrate aconitine, treat sections with a solution of potassium iodide-iodine. If aconitine is present, a carmine-red color will be produced. Treatment with a mixture of equal parts of sulphuric acid and water gives a similar color. Treatment with a solution of cane-sugar and then with the dilute sulphuric acid produces a brilliant carmine-red color.

Aleurone.—The protein nature of aleurone is shown by the fact that it dissolves with a red color on the application of Millon's reagent, and assumes a yellowish or brownish color with tincture of iodine. The aleurone grains of Ricinus are best studied from material that has been fixed in a saturated alcoholic solution of picric acid, or sections from fresh material may be fixed as above, rinsed in alcohol, stained with an alcoholic solution of eosin, cleared in oil of cloves, and mounted in Canada balsam. By this process the ground substance is stained red, the crystalloid yellow, while the globoid is usually colorless. For characteristic reactions of aleurone with other reagents, see in the last chapter under Borax-carmine, Digestive Fluids, Gold Chloride, Pepsin.

Alkaloids.—Sections which are to be tested for alkaloids should be sufficiently thick to leave one cell-layer intact. In order to make the determination of the alkaloid more certain, sections for control should be soaked for a day or so in a solvent of alkaloids prepared by dissolving 1 part of tartaric acid in 20 parts of alcohol; and then the sections should be rinsed in water for a day to wash out the acid solution. Sections which have been thus treated should be mounted under the same cover-glass with untreated sections, and the reagents for testing for alkaloids applied. It is best, on the whole, to mount the sections directly in the reagent which is to be used as a test for the alkaloid, since some alkaloids might be dissolved out if the sections are first mounted in water.

With the following reagents most alkaloids are thrown down in the form of amorphous or crystalline precipitates: Potassium iodide-iodine, potassium bismuthiodide, chlor-zinc-iodide, potassium-mercuriciodide, chloride of gold, ammonium-molybdate. Phospho-molybdic acid, picric acid, chloride of platinum, or a solution of bromin in hydrochloric acid. The crystalline precipitates can be studied to best advantage by means of a polarizer attached to the microscope.

The vapor of iodine may often be used to good advantage in the detection of alkaloids in the following manner: A few grams of iodine are placed at the bottom of a small exsiccator, and a layer of sand about a centimeter deep is placed over the iodine to prevent its too rapid evaporation. Sections which have been treated with a solvent of alkaloids, and sections which have not been thus treated, are mounted on the same slide and placed on the upper part of the exsiccator, where they are to remain for several hours. The sections are then covered with a drop of paraffin oil, which will not dissolve the precipitate, and covered with a cover-glass for examination under a microscope, preferably with a polarizing attachment.

Allyl Sulphide, or Garlic Oil, $(C_3H_5)_2S$.—This may be demonstrated by treating portions of a species of Allium with a solution of palladous nitrate, which produces with allyl sulphide a kermes-brown precipitate, or the material may be treated with a 2 per cent. solution of silver nitrate, in which case sulphide of silver will be produced. The material may then be hardened in alcohol and sectioned.

Ammonia, NH₃.—The demonstration of ammonia in plant tissues is given in the preceding chapter under Nessler's Reagent.

Amygdalin.—This nitrogenous glucoside is particularly abundant in bitter almonds and in the bark, leaves, and flowers of *Prunus padus.* It may be extracted by boiling water, and from a mixture of alcohol and water it crystallizes out in the form of transparent orthorhombic crystals. From 80 per cent. alcohol it crystallizes in the form of pearly scales. It is split up into prussic acid, oil of bitter almonds, and sugar by an enzyme known as emulsin, which occurs within the plant along with the amygdalin.

Amylodextrine.—This substance occurs in those starch-grains which take on a reddish color with iodine, and it is formed by the action of diastase and acids from the amylose of those starch-grains which are colored blue with iodine. By the action of diastase on the starch of germinating seeds the amylose of the starch is converted first into amylodextrine, and this in turn into dextrine and isomaltose. The microchemical behavior of amylodextrine is given by Arthur Meyer as follows: Water at 70°C. dissolves crystals of amylodextrine slowly, while at 100° the crystals are dissolved at once. A solution of 10 Gm. of pure calcium nitrate in 14 Gm. of water dissolves crystals under the cover-glass very slowly. After some hours, if a solution of iodine is added, the calcium nitrate solution is colored brown, which indicates that the crystals of amylodextrine have at least been partially dissolved. A solution of 2 Gm. of purest potassium hydrate in 100 Gm. of water dissolves small crystals within 2 hours, while the solution of larger crystals requires a longer time. A solution of iodine, prepared as directed on page 548, colors the crystals dark brown. A 25 per cent. solution of hydrochloric acid dissolves large and small crystals immediately. When this solution is diluted with 4 parts of water, it takes on a brownish-red color with the iodine solution. When 1 drop of malt extract is added to 5 drops of a neutral solution of amylodextrine this becomes inverted within 10 minutes, so that it no longer is colored by the iodine solution. To prepare the malt extract treat 1 part of malt with 3 parts of water and filter the solution. The solution of crystals of amylodextrine by the malt extract requires several days. At a temperature of 40°C. saliva dissolves the amylodextrine crystals within 48 hours. To prepare the saliva mix human saliva with a drop of chloroform, filter, and preserve over a few drops of chloroform.

Amyloid.—This occurs as reserve material in the seeds of *Tropæolum majus, Impatiens balsamina, Pæonia officinalis,* and in many other plants. It is colored blue by dilute solution of iodine, but with a concentrated solution it is colored a brownish-orange. It is soluble in cuprammonia only after a day. Treated with a 30 per cent. solution of nitric acid it swells strongly, and finally dissolves. This is different from the amyloid produced by the action of acids and certain chlorides on cellulose.

Amylose.—Starch-grains which are colored blue by iodine—that is, most starch-grains—are, according to Meyer, composed of crystals of two kinds of amylose, named by Meyer α-amylose and β-amylose. The α-amylose has been isolated in crystalline form, but the β-amylose has not been isolated, and its microchemical behavior has only been determined by experiments with starch-grains. The microchemical behavior of the α-amylose is as follows, the reagents being prepared as directed under amylodextrine: Water at from 60° to 100°C. does not soon dissolve the crystals of this amylose. Treatment with the calcium nitrate solution for 30 minutes does not appear to affect the crystals. The solution of iodine does not color the crystals at first, but after a longer time it imparts a brownish color. The solution of hydrochloric acid dissolves the crystals at once, and the solution, diluted with four times its bulk of water, is colored deep blue with the iodine reagent; but after the solution has stood for 12 hours it is colored brownish

36

or not at all by the iodine.   The solution of potassium hydrate at ordinary temperatures affects the crystals so that they are colored blue by the iodine after the solution has been neutralized with acetic acid.   In boiling potassium hydrate the crystals are changed into viscid drops.   If the solution is now neutralized with acetic acid and diluted with four times its bulk of water, it takes on a deep blue color with the iodine reagent.

If a drop of malt extract is added to the solution formed by boiling crystals of α-amylose with the potassium hydrate solution, and exactly neutralizing with acetic acid, it is found after 5 minutes that the solution takes on a red color, due to the formation of amylodextrine by the influence of the malt extract.   Saliva and malt extract have very little effect upon α-amylose.   After treatment with these reagents for 15 days at a constant temperature of 40°C., no essential change could be detected.

β-Amylose is insoluble in cold water, but at a temperature of 70°C. it forms viscid masses or minute droplets.   The solutions of calcium nitrate, potassium hydrate, and hydrochloric acid have the same effect as water, excepting that the solution in hydrochloric acid is more complete than in water.   The solution of β-amylose acts precisely as the solution of α-amylose.   Undissolved β-amylose, however, is colored blue by the iodine solution.   The swelling of starch in hot water is probably due to the β-amylose which it contains.   Meyer considers α-amylose and β-amylose to be the same substance, but that the latter contains water of crystallization, while the former does not.

Anthochlorin.—A yellow coloring matter dissolved in the cell-sap of flowers, and differing from the yellow coloring matter xanthin occurring in chromoplasts in that it is not changed to a blue color by the action of concentrated sulphuric acid.

Anthocyanins.—These are coloring matters of flowers which impart red, violet, blue, blue-green, or green colors, the character of the color being dependent on the alkalinity or acidity of the cell-sap.   The anthocyanins are soluble in water, alcohol and ether, and are decolorized in strong alkalies.

Anthoxanthin.—This yellow coloring matter in the chromoplasts of flowers and fruits takes on a blue color with concentrated sulphuric acid.   Since the chromoplasts of flowers and fruits were first of all green, anthoxanthin is probably a derivative of chlorophyll.   Anthoxanthin is also called xanthin and xanthophyll.

Arabin.—This is the gum derived from species of Acacia and known as gum arabic.   Arabin is soluble in hot and cold water, and insoluble in alcohol and ether. The aqueous solution will mix with glycerine, but concentrated glycerine has little effect on hard gum.

Asparagin, $C_2H_3NH_2.CONH_2.COOH$.—This is a nitrogenous compound of simpler constitution that that of proteids.   It is formed within plants both analytically by the decomposition of proteid, and synthetically probably by the combination of ammonia with formic aldehyde.   Asparagin is soluble in water and in the cell-sap, and is one of the most important nitrogenous compounds which are capable of solution and circulation within the plants.   It combines with non-nitrogenous compounds to form proteids, and is apt to accumulate in those parts of plants where there is not sufficient non-nitrogenous material at hand for the formation of proteids.   The accumulation of asparagin is particularly apt to occur in plants which are grown in the dark, so that carbon assimilation does not take place. Thus, Pfeffer found that when seedlings of lupin were grown in the dark, they contained a large amount of asparagin, but when they were brought to the light, the asparagin disappeared.   He found that this was not due simply to the influence of the light, for when the seedlings were exposed to the light in an atmosphere destitute of carbon dioxide, the asparagin persisted in the seedlings.   For the ready demonstration of asparagin, tubers of dahlia may be employed. Rather thick sections are cut from a tuber while the razor is kept dry and transferred to a few drops of alcohol on a glass slide and covered with a cover-glass. On the evaporation of the alcohol crystals of asparagin in the form of rhombic plates are deposited on the cover-glass and slide. To determine whether the crystals are asparagin, they are treated with a few drops of a saturated solution of asparagin, which must be entirely saturated and of the same temperature as the preparation. If the crystals are asparagin, instead of being dissolved they will increase in size, while other substances than asparagin will dissolve in the saturated asparagin solution just as they would in water. It is characteristic of asparagin that if the crystals are heated to 100°C., they lose their water of crystallization and appear like bright droplets of oil. At 200° asparagin becomes decomposed and forms frothy brown droplets which are no longer soluble in water.

Atropine, $C_{17}H_{23}NO_3$.—In sections containing atropine a solution of potassium iodide-iodine produces a brown precipitate, while phosphomolybdic acid produces a yellow precipitate.

Bassorin.—Gum tragacanth, obtained from certain cells of the pith and medullary rays of several species of Astragalus. Swells strongly in water, but does not go into complete solution. Is not colored either by iodine or chloriodide of zinc.

Berberine, $C_{20}H_{17}NO_4 + 4\frac{1}{2}H_2O$.—This occurs in the young parenchymatous tissue, and in the older xylem portions of *Berberis vulgaris*. With potassium iodide-iodine it forms a reddish-brown precipitate which, by treatment with alcoholic potassium iodide-iodine, becomes changed into tubular or hair-like forms which have a brownish or iridescent green color. Ammonium and nitric acid impart to berberine a reddish-brown color. A solution of potassium bichromate or potassium iodide in 50 per cent. sulphuric acid produces, with berberine, an intense purplish-red color. One part of nitric acid mixed with 100 parts of water added to sections containing berberine will produce clustered acicular crystals of berberine nitrate within the berberine-bearing cells.

Betulin.—This occurs in the form of fine granules in the thinner walled cork cells of birch bark. In order that it may be studied to good advantage under the microscope, the air should be pumped from sections immersed in water, and then the sections should be examined in water under the microscope. Betulin is insoluble in water, but is soluble in alcohol. It is strongly antiseptic, and protects birch bark against the attacks of lower organisms.

Betuloretic Acid, $C_{36}H_{66}O_5$.—This is secreted by the glandular hairs on the leaves of *Betula alba*. It is obtained from the thick, pale yellow secretion by successive solution in boiling alcohol, ether, and an aqueous solution of sodium carbonate. It is colored a beautiful red by concentrated sulphuric acid.

Brucine, $C_{23}H_{26}N_2O_4 + 4H_2O$.—Brucine occurs in the seeds of various species of strychnos. Ammonium vanadate in sulphuric acid gives with brucine a yellowish-red color. When sections containing brucine are treated with a mixture of nitric and hydrochloric acids, the cell-contents are colored a reddish-orange, which merges into yellow.

Caffeine, $C_8H_{10}N_4O_2 + H_2O$.—When sections containing caffeine (theine, methyl-theobromine, trimethyl-xanthin) are treated with a drop of concentrated hydrochloric acid, and then after a minute with a drop of a 3 per cent. gold chloride solution, somewhat slender, yellowish, silken crystals of a double chloride of gold and caffeine begin to be formed on the evaporation of the reagent. However, theobromine forms quite similar crystals when treated as above. Another method for the detection of caffeine is to place sections in a few drops of water and heat to boiling; then to allow the water to evaporate slowly and to treat the residue with a drop of benzol. On the evaporation of the benzol, caffeine appears in the form of fine needle-crystals.

Calcium.—When the ash of plants is treated with sulphuric acid, this unites with the calcium present to form crystals of gypsum. If calcium sulphate is already present in the ash, its characteristic crystals may be detected when an aqueous solution of the ash is allowed to dry slowly. If calcium is present in sections, it may be deposited in the form of crystals of calcium oxalate if the sections are treated with a solution of ammonium oxalate.

Calcium Carbonate, $CaCO_3$.—This rarely occurs in the crystalline form within the cells. It may, however, be found imbedded in, or incrusted on, the cell-walls. Calcium carbonate dissolves with effervescence when treated with dilute acetic acid. When treated with concentrated hydrochloric acid, it dissolves with the evolution of carbon dioxide gas. The ingrowths from the walls of certain cells of the leaves of *Ficus elastica*, known as cystoliths, are thickly incrusted with calcium carbonate, and afford excellent material for the demonstration of this salt within plant tissues.

Calcium Phosphate, $Ca_3(PO_4)_2$.—This salt of calcium occurs usually, if not always, in solution in the cell-sap. It may be deposited in the form of sphærocrystals when plant tissues containing it are kept for a long time in strong alcohol. When treated with sulphuric acid, the sphærocrystals are dissolved and crystals of calcium sulphate are formed in their stead. When sections containing calcium phosphate are heated on a slide in a drop of ammonium molybdate acidulated with nitric acid, a yellow precipitate is produced. This reaction may be hindered by the presence of certain organic compounds, such as potassium tartrate, in which case the sections should be treated with a mixture of 25 volumes of a concentrated aqueous solution of magnesium sulphate with 2 volumes of a concentrated aqueous solution of ammonium chloride and 15 volumes of water. In this case a crystalline precipitate of ammonio-magnesium phosphate is formed.

Calcium Oxalate, CaC_2O_4.—Crystals of calcium oxalate occur so commonly in plants that it is safe to assume that any crystals observed in fresh tissues are of this substance until the contrary is demonstrated. The crystals may occur singly in the cells, in which case their definite crystalline form can be made out, or in the form of agglomerated star-shaped clusters of crystals, or in bundles of parallel needle-shaped crystals, or they may occur very numerously in cells in the form of very minute crystals. The crystals are insoluble in water and acetic acid, but dissolve without effervescence in hydrochloric acid. When they are treated with sulphuric acid, crystals of calcium sulphate are formed in their place. Calcium oxalate appears to be an excretion formed by the union of salts of calcium, which have been absorbed from the soil, with oxalic acid which is formed by the plant.

Calcium Sulphate, $CaSO_4$.—Minute crystals of calcium sulphate occur in many desmids. They are insoluble in concentrated sulphuric acid. A solution of barium chloride dissolves them with the formation of barium sulphate.

Callose.—Callose occurs in sieve tubes, where it may close up the sieve pores. It also occurs commonly in cystoliths, and in the membranes of pollen-grains and various fungi. Callose is insoluble in water, alcohol, and cuprammonia, but it is readily soluble in cold sulphuric acid, calcium chloride, and concentrated chloride of zinc. It is insoluble in cold alkaline carbonates, but swells up without dissolving in ammonium. Corallin, aniline blue, and a mixture of soluble blue and vesuvin, or of vesuvin and orseillin, are suitable stains for callose. The corallin should be dissolved in a saturated solution of sodium carbonate. After remaining in this solution for a time, the sections should be examined in glycerine. If the sections are over-stained, the intensity of the stain may be reduced in a 4 per cent. solution of sodium carbonate. The aniline blue should be used in dilute aqueous solutions, in which the sections are to remain for about half an hour. Over-staining may be remedied by washing out in glycerine.

Calycin, $C_{18}H_{12}O_5$.—This occurs in the tissues of many lichens. Its presence may be demonstrated by moistening some of the powdered lichen with glacial acetic acid, and when the preparation dries, the long, doubly refractive crystals of calycin are deposited. When a section of lichen containing calycin is treated on the slide with a few drops of chloroform and a drop of sodium hydrate, that portion of the section which contains calycin assumes a color varying from brick-red to blue-red.

Cane-sugar (Sucrose), $C_{12}H_{22}O_{11}$.—This is of common occurrence in plant tissues. At $15°C$. it is soluble in ⅓ part of water. It is difficultly soluble in alcohol. When boiled with Fehling's solution, it does not at first precipitate cuprous oxide, but on longer boiling it becomes converted into glucose and lævulose, which are capable of reducing Fehling's solution. If rather thick sections containing cane-sugar (the sugar-beet affords good material) are placed for a short time in a concentrated solution of cupric sulphate, and then quickly rinsed in water, transferred to a solution of 1 part of potassium hydrate in 1 part of water, and heated to boiling, the cells containing the sugar take on a sky-blue color. A blue color is also produced by Fehling's solution when sections containing cane-sugar are heated in a drop of the solution on a slide until bubbles arise.

Carotin, $C_{26}H_{38}$.—Carotin occurs in the orange and red chromatophores of many flowers and fruits; it seems also to be an essential part of chlorophyll; it occurs in crystalline form in the roots of carrots, which have a yellow color in consequence. To demonstrate the presence of carotin in chloroplasts place pieces of fresh leaves in a 20 per cent. solution of potassium hydroxide in 40 per cent. alcohol, and leave them thus in a tightly closed vessel for several days. When the chlorophyll has been extracted from the leaves, they should be washed in distilled water and sections from them should be mounted in glycerine. Yellowish and red crystals will then be found in the cells which formerly contained chlorophyll. Carotin is insoluble in water and with difficulty in alcohol, but is readily soluble in petroleum ether, benzol, and benzine. When freshly dried leaves or roots of carrots are powdered and treated with one of these solvents, and the solution is allowed to dry or is treated with alcohol, carotin crystallizes out in the form of reddish or yellowish crystals. With a solution of iodine carotin is colored greenish or bluish; with concentrated sulphuric acid it is colored from violet to indigo blue.

Cellulose, $C_6H_{10}O_5$.—Cellulose is one of the most important constituents of cell-walls; the first-formed walls are nearly always of cellulose, together with certain pectic compounds. Modified cell-walls—namely, those which have become cutinized or lignified—have arisen by the chemical modification of cellulose, or by the infiltration of new material between the cellulose molecules, or by both of these processes. Cellulose is characterized by being soluble in sulphuric acid and cuprammonia; by being colored from violet to blue by sulphuric acid and iodine, chloroiodide of zinc, chloroiodide of calcium, iodine and aluminum chloride, iodine and phosphoric acid. See under these heads in the chapter on Reagents.

Chitin.—The walls of many fungi consist of chitin instead of cellulose. This may be demonstrated by cutting the pileus of an agaracus into small pieces, which are then to be treated successively with dilute potassium hydrate, dilute sulphuric acid heated to boiling, alcohol, and finally ether. When this process is completed, a white substance remains which becomes hard and horny on drying, and which is insoluble to all reagents except concentrated acids, and in all other respects possesses the characteristics of chitin.

Chlorophyll.—Chlorophyll may be extracted from the chloroplasts by means of strong alcohol. When this extract is shaken up with benzol and a few drops of water, and allowed to stand for a short time, the benzol which rises to the top will contain two pigments, amorp bous chlorophyll-green and carotin; while the lower stratum of alcohol will cont ain a crystallizable chlorophyll-green and xanthophyll. The amorphous and the crystallizable chlorophyll-green differ in the character of their spectra and in t heir solubility in different reagents. The amorphous form is soluble in alcohol, petroleum ether, carbon bisulphide, and benzine; while the crystallizable is soluble only in the alcohol.

Colchicine, $C_{22}H_{25}NO_6$.—This occurs in a few rows of cells immediately sur-rounding the vascular bundles of the corm of *Colchicum autumnale*. Treated with a mixture of 1 part of sulphuric acid and 3 parts of water colchicine is colored yellow, and this color is changed to a brownish-violet by the addition of a crystal of potassium nitrate. Iodine stains it brown, and potassic-mercuric iodide and hydrochloric acid produce with it a yellow precipitate.

Corydalin, $C_{18}H_{19}NO_4$.—This is an alkaloid which is found in the idioblasts of the Fumariaceæ. When corydalin is present, ammonia produces a dark gray color, picric acid a yellow, and potassium iodide-iodine a deep reddish-brown color.

Crocin (Saffron-yellow), $C_{44}H_{70}O_{28}$.—This is a glucoside occurring in the stigmas of *Crocus sativus*. When concentrated sulphuric acid is added to crocin, a deep blue color is produced which passes into violet, cherry-red, and then brown. Nitric acid also produces a blue color which passes into brown.

Curarin.—This occurs in the parenchyma and bast of several species of Strychnos. Concentrated nitric acid produces with it a blood-red, and dilute or concentrated sulphuric acid a carmine-red, color.

Curcumin, $C_{14}H_{44}O_4$.—Curcumin occurs, dissolved in an ethereal oil, in certain cells of the ground parenchyma of the rhizome of *Curcuma longa*. ·It crystallizes in the form of yellow needles which have a bluish tint by reflected light. Lead acetate forms a brick-red precipitate with curcumin, and sulphuric acid gives it a crimson color.

· **Cutin.**—Cutin is a substance which is nearly related to suberin (which see), but is not identical with it. None of the acids derived from cutin is identical with any derived from suberin. However, the micro-chemical reactions of suberin and cutin are the same. They are insoluble in concentrated sulphuric acid and cuprammonia and are colored from yellow to brown with the iodine reagents. When heated with concentrated potassium hydrate, they form yellowish droplets and granular masses. When heated in nitric acid and potassium hydrate, they form droplets which melt between 30° and 40° C., and which are soluble in boiling alcohol, ether, benzol, chloroform, and dilute potassium hydrate. Both suberized and cutinized walls resist concentrated chromic acid at ordinary temperatures. Chemical analysis shows that cutin is composed of compound esters and fatty acids, and when heated to 300° in glycerine, it behaves as a fatty body. For staining cutinized walls, see under Cyanin, Alcannin, Chlorophyll Solution, Double Staining.

Cytisin, $C_{20}H_{27}N_3O$.—This alkaloid occurs in the seeds of *Cytisus laburnum* and of other species of Cytisus. It occurs in less abundance in other parts of the plant, such as the petals and peripheral tissues of the stem. Potassium iodide-iodine produces with it a reddish-brown, granular precipitate which is soluble in sodium hyposulphite. Chloride of iron gives an orange-red color with cytisin. With phosphomolybdic acid a light yellow precipitate is produced, and picric acid when added to thin sections containing cytisin produces crystal groups of a reddish-yellow color.

Datiscin, $C_{21}H_{22}O_{12}$.—This glucoside is found in the cell-walls of the wood and bark of *Datisca cannabina*. Lime and baryta waters produce with it a yellow solution which loses its color on the addition of acetic or dilute hydrochloric acid. In the presence of datiscin, acetate of lead and chloride of zinc produce a yellow, oxides of copper a greenish, and chloride of iron a dark bluish-green, precipitate.

Dextrine, $C_{12}H_{20}O_{10}$.—This is one of the intermediate products between starch and maltose (see Amylose). It is easily soluble in water, and from its aqueous solution it may be precipitated by strong alcohol. It is not colored by iodine, and does not reduce salts of copper.

Dextrose (Glucose, Grape-sugar).—See under Glucose.

Diastase.—To demonstrate the presence of diastase in sections they are laid for a time in a dark brown solution of guaiacum in absolute alcohol. When the sections are completely infiltrated with this solution the alcohol is allowed to evaporate, and then the sections are placed in a rather dilute solution of hydrogen peroxide. By this treatment cells containing diastase are colored a beautiful blue. See also under Diastase Solution in the preceding chapter.

Dulcite, $C_6H_{14}O_6$.—Dulcite may be demonstrated in sections from one-year-old stems of *Euonymus japonicus*. The sections are placed on a slide in a few drops of alcohol, and covered with a cover-glass. After the alcohol has slowly evaporated from under the cover-glass, crystals of dulcite will be deposited in the form of long, branched prisms or needles radiating from a common center. They are distinguished from crystals of potassium nitrate by dissolving in diphenylamine and sulphuric acid without coloration, and by being insoluble in a concentrated solution of dulcite.

Elaioplasts.—They are rounded or irregularly polygonal, more or less granular bodies, consisting of a protoplasmic stroma and inclosed oil, which occur closely applied to the nucleus in the epidermal cells of many monocotyledonous and some dicotyledonous plants. In old cells the elaioplasts have the appearance of a sponge saturated with oil. The oil in the elaioplast of *Ornithogalum umbellatum* may be almost instantly dissolved by means of alcohol. These elaioplasts may be fixed and stained at the same time by treating sections containing them with a dilute solution of alcannin in 1 per cent. acetic or formic acid. The acid fixes the protoplasmic stroma, while the alcannin stains the oil a beautiful red. The fixing and staining process should be complete in 5 minutes. If desired, the sections may be double-stained by transferring them from the alcannin to a solution of iodine-green and glycerine, after which they may be mounted in glycerine-jelly. The sections may also be stained in a mixture of a dilute solution of alcannin and a solution of iodine-green in 50 per cent. alcohol and 1 per cent. acetic acid.

Emulsin.—This is a glucoside-splitting ferment which occurs in certain cells of the almond and of the bundle-sheath of the leaves of *Prunus laurocerasus*. When sections are treated with Millon's reagent, the cells containing emulsin take on an orange-red color, while the surrounding cells are colored a pale rose-red. A solution of copper sulphate and caustic potash produces a violet color in the emulsin-bearing cells.

Ethereal Oils.—Ethereal oils are distinguished from fatty oils in that they may be distilled from plants along with vapor of water, and are soluble in glacial acetic acid, and an aqueous solution of chloral hydrate. At 130°C. all ethereal oils may be driven from sections, while the fatty oils remain behind. Ethereal oils are only slightly soluble in water, but they impart their smell strongly to it. They are easily soluble in ether, chloroform, and fatty oils. The spot produced on paper by ethereal oils soon disappears. They agree with the fatty oils in being browned or blackened by osmic acid, and in being stained by alcannin and cyanin.

Eugenol, $C_6H_3.OH.OCH_3.C_3H_5$.—Eugenol occurs in clove and pimento oil. When sections containing either of these oils are treated with a concentrated solution of potassium hydrate, long columnar or needle-shaped crystals of potassium caryophyllate are produced. When sections of cloves are used, they often become covered by the forming crystals.

Fats and Fatty Oils.—These are insoluble in cold and hot water, and, with the exception of castor oil, hardly soluble in alcohol, but readily soluble in ether, chloroform, benzol, ethereal oils, acetone, and wood spirit. They make a spot on paper which does not disappear, as in the case of ethereal oils. Most fats and fatty oils are colored brown or black by 1 per cent. osmic acid. When a drop of fat or fatty oil is placed on a glass slide in a drop of a mixture of equal parts of concentrated potassium hydrate and ammonium, the oil becomes saponified, and may assume a form like a bunch of grapes, or it may be partly or wholly changed into clusters of soap crystals. Vapor of hydrochloric acid has been used to distinguish between ethereal and fatty oils. A large and a small glass ring, such as are used for hanging-drop cultures, are cemented to a glass slide, the small one being shallower than the large one, and placed within it concentrically. Hydrochloric acid is placed into the space between the rings, and the sections to be tested are placed on a cover-glass in a drop of glycerine containing a large amount of sugar.

The cover-glass is then inverted and placed on the larger ring. After the vapor of hydrochloric acid has had time to act, any ethereal oil present in the sections will take on the form of bright yellow drops which finally disappear. Fatty oils do not form yellow drops by this treatment. A solution of alcannin colors the fats red, but several hours may be required to accomplish this. A solution of cyanin in 50 per cent. alcohol is also a good stain for fats. The sections will not need to lie in this stain longer than half an hour. If sections are overstained, they may be washed out in glycerine or a concentrated solution of potassium hydrate.

Frangulin, $C_{20}H_{20}O_{10}$.—This glucoside occurs in the cortex of species of Rhamnus. It forms yellow crystalline masses which are insoluble in water, but soluble in alkalies, which produce with it a cherry-red color. Concentrated sulphuric acid produces with frangulin an emerald-green, which changes into purple, and finally the frangulin dissolves with a dark red color. Water will precipitate it from this solution.

Fungus Cellulose.—The membranes of very few fungi give the reactions of cellulose. The membranes of most fungi are insoluble in cuprammonia, and are colored from yellow to brown by chloroiodide of zinc, sulphuric acid and iodine. Neither do they react in the same manner as suberized and lignified membranes. They are, therefore, considered to be a distinct substance, which is termed fungus cellulose. See also under Chitin.

Gelatinous Sheaths.—The homogeneous gelatinous sheaths which cover the entire outside of certain algæ—notably, species of Spirogyra and Zygnema—may be demonstrated by the use of certain stains and other substances, such as India ink, which may become deposited in the sheaths. Aqueous solutions of vesuvin, methyl-violet, and methylene-blue will stain both the cell-walls and gelatinous sheaths, but the latter with less intensity. Chloroiodide of zinc will stain the walls without affecting the sheaths. Turnbull's blue may be deposited in the gelatinous sheaths in the following manner: A small number of zygnema filaments, for instance, may be tied together with a thread and placed for about 2 minutes in a 2 per cent. solution of ferrous lactate, then quickly washed in water, and transferred to a 0.2 per cent. solution of ferricyanide of potassium. A small amount of Turnbull's blue will then be deposited in the gelatinous sheaths. This process should be repeated several times, until the deposit of Turnbull's blue is sufficiently dense to cause the sheaths to stand out quite sharply. By this method very instructive double stains may be achieved with algæ which have been growing in a dilute solution of congo-red (see under this head in the preceding chapter), which stains the cell-walls, but not the gelatinous sheaths. See also in the preceding chapter under India ink.

Globoids.—The globoids found in aleurone grains consist of a double phosphate of calcium and magnesium, which is insoluble in alcohol and dilute potassium hydrate but is soluble in dilute mineral acids and in acetic, oxalic, and tartaric acids. In an ammoniacal solution of ammonium phosphate the globoids are replaced by groups of crystals of ammonium-magnesium phosphate. Treated with ammonium oxalate, they become replaced by crystals of calcium oxalate. The globoids may be isolated to a certain extent by extracting the oil from sections of endosperm containing them by means of alcohol or alcohol and ether, and then dissolving the ground substance and crystalloid by means of 1 per cent. potassium hydrate. If crystals of calcium oxalate are present along with the globoids, they may be distinguished by means of the polarizer, since they are doubly refractive, while the globoids are not.

Glucose, $C_6H_{12}O_6$.—This occurs in sweet fruits and in the leaves and other members of plants, being one of the most common forms in which carbohydrates circulate within the plant. The warty crystals of glucose which are deposited from aqueous and alcoholic solutions at low temperatures melt at 86°, and become free from water at 110°C. At from 30° to 35°C. glucose crystallizes from concentrated solutions in water, ethyl- and methyl-alcohol in the form of hard crusts, which melt at 146°C. The presence of glucose may be easily demonstrated in the fruit of the pear, for instance, and in the leaves of balsamina, or other rather translucent leaves which have been cut from the parent plant and kept fresh under a bell-jar for several days. Pieces of the flesh of a ripe pear may be put into a test-tube with Fehling's solution and brought to a boil, when a reddish precipitate of cuprous oxide will be thrown down. This reaction is characteristic of dextrose, maltose, lactose, lævulose, and many glucosides. In this instance, however, we are dealing with dextrose. This reaction may also be carried out on the microscopic slide. Sections from the pear three or four cell-layers thick should be placed on the slide in a few drops of the solution, the cover-glass should then be put on, and the solution heated until bubbles begin to arise. The microscope will then reveal the granular precipitate of cuprous oxide within the cells. Portions of the leaf of the balsamina may be treated on the slide as directed for the sections from the pear. See under Fehling's Solution in the preceding chapter.

Glycogen, $C_6H_{10}O_6$.—This is a colorless, amorphous, highly refractive substance occurring quite commonly in the cells of fungi. It is soluble in water, but within the cells it may be stained a reddish-brown by means of iodine.

Gums.—These are amorphous, transparent substances which dissolve in water more or less completely and form a sticky solution. They may be precipitated from their aqueous solutions by alcohol. Those gums which dissolve in water completely, such as the gum of the cherry, apricot, peach, and gum arabic, are called true gums, while those which contain cellulose and are not completely soluble in water are known as mixed gums. Gum tragacanth is an example. One of the most striking characteristics of gums which may be used in their identification is their great capacity to swell in water. To follow the process of swelling with the microscope, sections should be cut from dry material with a razor which may be wetted with alcohol, but not with water. The sections should be placed on a slide in a drop of strong alcohol, the cover-glass should be put on, and a drop of water placed on the slide so that it just touches the edge of the cover-glass. As the water mixes with the alcohol and comes in contact with the section a slow swelling of the gum will begin, which may be followed very accurately through the microscope. For directions for staining and making permanent preparations of sections containing mucilages and gums see under Boric Acid in the preceding chapter.

Hemicelluloses.—These are reserve materials which are deposited as additions to the cell-walls in the endosperm of seeds and in wood parenchyma and wood-fibers. By means of enzymes they may be converted into gums and sugars, in which forms they may be transported to those parts where growth is taking place. The hemicellulose or reserve cellulose in the endosperm of the date seed acts like ordinary cellulose in being colored blue by chloroiodide of zinc and in dissolving in cuprammonia. The reserve cellulose in the endosperm of the seeds of *Lupinus luteus* is not dissolved in cuprammonia, and does not assume a blue color when treated with chloroiodide of zinc.

Hesperidin, $C_{21}H_{26}O_{12}$.—This glucoside occurs dissolved in the cell-sap of many plants. It may be precipitated from its solution in the cell-sap by means of alcohol. The precipitate is in the form of crystals, which are colorless or slightly yellow, and are doubly refractive, so that they may be studied to good advantage by means of the polarizer. Hesperidin is also precipitated on the drying up of the cell-sap. The crystals of hesperidin are insoluble in cold and boiling water, alcohol, ether, benzine, and dilute acids, but they are soluble in solutions of caustic potash and soda, and in ammonia, yielding a yellowish color to the solvent. Hesperidin may readily be obtained for study in the unripe fruit of the orange and in the epidermal cells of *Capsella bursa-pastoris*. Hesperidin may become deposited in the form of spærocrystals, when the tissues containing it have lain for some time in strong alcohol or glycerine, acting in this respect similarly to inulin. The constituent acicular crystals of the hesperidin sphærites can be more easily distinguished than those of inulin, and when the hesperidin sphærites are treated with a drop of α-naphtol, and then with two or three drops of concentrated sulphuric acid, they dissolve with a yellow color, while, with like treatment, inulin sphærites dissolve with a violet color.

Indican.—The glucoside indican is a substance of the consistency of syrup, and of a yellowish or brownish color. It is found in *Isatis tinctoria, Phajus grandifolius* and in other indigo-bearing plants. When tissues containing indican are exposed to the air, they may take on a blue color due to the conversion of the indican to indigotin, which may be precipitated in alcohol in the form of small, tabular, bluish crystals. To demonstrate the presence of indican, tissues containing it should be placed under a bell-jar and over a dish of absolute alcohol. After standing exposed to the vapor of alcohol for 24 hours, the tissues will be colored blue by the indigo blue which will have been forced from the indican. A piece of moistened filter-paper should be placed under the bell-jar to keep the tissues from drying.

Inulin, $C_{12}H_{20}O_{10}$.—Inulin is a carbohydrate which occurs dissolved in the cell-sap of many plants, particularly among the Compositæ. It may be deposited from its solution in the cell-sap by means of alcohol. To study the sphærocrystals of inulin, pieces of dandelion or dahlia roots should be placed in 50 per cent. alcohol for a week or more, and then thin sections should be prepared and examined in a drop of the alcohol under the microscope. The sections should not be placed in water, since the crystals of inulin are soluble in it. The sphærites will appear applied to the walls of the cells. When the alcohol is replaced by water which is then heated over a flame, the sphærites will dissolve. If sections containing inulin sphærites are treated with a 20 per cent. solution of α-naphthol, and then 2 or 3 drops of concentrated sulphuric acid are added, the sphærites will be seen to dissolve with a violet color. Inulin does not reduce Fehling's solution.

Leucin, $C_6H_{10}NH_2\cdot COOH$.—Leucin belongs to the amido-compounds. It has been found in etiolated leaves of *Paspalum elegans* and *Dahlia veriabilis*, and associated with asparagin in seedlings of various Leguminosæ, particularly in those of lupinus. Leucin crystallizes in thin plates, which are lighter than water and have the appearance of mother-of-pearl. If sections containing leucin are carefully heated on a slide under a cover-glass to a temperature of 170°C., the cover-glass will become covered with minute, scale-like crystals, which are doubly refractive and may be studied to advantage by means of the polarizer. The crystals of leucin may also be obtained if sections are treated with alcohol under a cover-glass, and the alcohol is then allowed to evaporate slowly.

Leucoplasts.—For methods of fixing and staining leucoplasts, see in the last chapter under Acid Fuchsin, Gold Chloride, and Picronigrosin.

Lignified Membranes.—Lignified membranes are distinguished from cellulose membranes by their insolubility in cuprammonia and by being colored from yellow to brown by iodine or chloroiodide of zinc. One of the most reliable tests for lignified membranes will be found in the last chapter under Phloroglucin. Aniline sulphate is also a good test for lignified membranes. The sections are first mounted in a drop of a concentrated solution of aniline sulphate, and then this is replaced by a drop of concentrated sulphuric acid. By this treatment lignified membranes are stained a golden yellow. See also under Double Staining in the last chapter.

Lipochromes.—These are yellow and red pigments which are for the most part dissolved in fatty substances within the cells, and which are colored blue by sulphuric or nitric acid, and green by potassium iodide-iodine.

Magnesium.—To demonstrate the presence of magnesium within plant tissues, sections are placed on the slide in a drop of sodium phosphate or sodium-ammonium phosphate, and a little ammonium is added. In the presence of magnesium, crystals of ammonio-magnesium phosphate are then formed, which have a coffin-lid form. When the ash of tissues containing magnesium is treated as above, the crystals are apt to form in x- or *-shaped groups.

Maltose.—Maltose is a sugar which is produced from starch by the action of diastase. Maltose reduces Fehling's solution, but only about two-thirds as much as grape-sugar (dextrose, glucose).

Morphine, $C_{17}H_{19}NO_3$.—When the latex containing morphine is treated with potassium iodide-iodine, a reddish-brown precipitate is produced, with potassium-bismuth iodide a reddish-orange, and with potassio-mercuric iodide a yellowish-white precipitate, while phospho-molybdic acid produces a yellow precipitate. A solution of 5 drops of methylal in 1 mil of concentrated sulphuric acid gives a violet color to latex containing morphine.

Mucilages (see also under Gums).—Mucilage contained in sections of plant tissues may be differentiated by staining with methylene-blue. The sections may be mounted on the glass slip in a drop of a solution of 0.4 Gm. of methylene-blue in 100 mils of equal parts of water, glycerine, and 95 per cent. alcohol. The mucilage will be seen under the microscope to be stained more deeply blue than the cell-walls or other cell contents. If the sections are taken from fresh materials, the razor should be moistened with alcohol. Dry materials should be soaked in water to soften for cutting. If it is found that the mucilage dissolves too much in the water, the mucilage may be hardened and the tissues softened at the same time in the lead acetate solution described under Gums.

Mustard Oil, C_3H_5CNS.—Seeds and the vegetable organs of the Cruciferæ, Resedaceæ, Capparidaceæ, Tropæolaceæ, and Lemnanthaceæ contain peculiar nitrogenous glucosides which become decomposed into sulphur-bearing substances, long known as mustard oils, by means of the enzyme myrosin.

Myrosin.—Myrosin is an enzyme occurring in certain specialized cells in the seeds and other parts of many Cruciferæ.   The cells containing myrosin are stained a deep red by Millon's reagent, while the surrounding cells may be stained a pale rose color.   When sections containing myrosin are heated in a concentrated solution of hydrochloric acid which contains a drop of a 10 per cent. aqueous solution of orcin in each mil, a violet color is produced in the cells containing the myrosin. Myrosin produces allylic mustard oil from potassium myronate, a glucoside occurring in the parenchyma cells which are associated with those containing myrosin.

Narceine, $C_{23}H_{29}NO_9$.—This is an alkaloid occurring in the latex of *Papaver somniferum.*   When a yellow color follows the addition of methylal to the latex, the presence of narceïne is indicated.

Narcotine, $C_{22}H_{23}NO_7$.—Sodium selenate produces an orange-red color with the latex of *Papaver somniferum*, indicating the presence of narcotine.

Nicotine, $C_{10}H_{14}N_2$.—When sections containing nicotine are treated with potassio-mercuric chloride, a yellowish-white precipitate is produced.   Phospho-molybdic acid gives, with nicotine, an abundant yellow precipitate.   In the presence of nicotine mercuric chloride produces a white, and platinum chloride a yellow, precipitate, while potassium iodide-iodine causes first a carmine-red color and finally a reddish-brown precipitate, which gradually bleaches out.

Nitrates.—When nitrates are present in a solution, a drop of barium chloride added to a drop of the solution will produce a precipitate of octahedral crystals of barium nitrate.   See also under Diphenylamine in the preceding chapter.

Nucleus.—The nucleus can best be demonstrated in tissues which have been fixed according to the directions given under Fixatives in the last chapter.   Also under Iodine-green and Acetic Acid, and Methyl-green and Acetic Acid, are given directions for instantly fixing and staining nuclei.   The three-color method of staining detailed on page 537 gives the best results for the dividing nucleus.

Oils.—Ethereal and fatty oils have already been discussed under separate heads, where the methods for distinguishing the two will be found.   See also in the preceding chapter under Alcannin, Cyanin, and Osmic Acid.

Oxalic Acid.—When calcium nitrate is added to sections containing oxalic acid, crystals of calcium oxalate are formed.   With uranyl acetate crystals of uranium oxalate are formed in tissues containing oxalic acid.   The crystals are rhombic, of rectangular form, and when large, appear of a yellow color, and, being doubly refractive, they may be studied to advantage with the polarizer.

Paragalactan.—This occurs as a thickening of the cell-walls in the cotyledons of *Lupinus luteus.*   When it is heated with nitric acid, mucic acid is formed, and when heated with dilute sulphuric acid, galactose, $C_6H_{12}O_6$, and a pentaglucose are formed.   When heated with phloroglucin and hydrochloric acid, a cherry-red color is produced.   Paragalactan is not dissolved by cuprammonia, and is stained slightly or not at all by chloroiodide of zinc.

Paramylum.—Paramylum grains are flattened, cylindrical, stratified bodies occurring in the bodies of the Euglenæ and in the cysts of *Leptophrys vorax.*   The paramylum grains are hardly affected by water, alcohol, ether, nitric acid, or concentrated chromic acid; and while they are hardly soluble in 5 per cent. potassium hydrate, they are easily soluble in a 6 per cent. solution.   They may also be dissolved in concentrated sulphuric acid.   They are not stained by iodine, chloro-iodide of zinc, or by any of the organic coloring matters.

Pectic Compounds.—The pectic substances (pectin, pectose, and pectic acids) are widely distributed in the membranes of plants.   Pectose occurs associated with cellulose in the membranes of embryonic tissues, where it is distributed throughout the entire thickness of the membrane. Pectose also occurs in most lignified, suberized, and cutinized membranes. The middle portion of cell-walls—the so-called middle lamella—consists, for the most part of calcium pectate. When thin sections of plant tissues are treated for several hours with a mixture of 1 part of hydrochloric acid and 4 parts of alcohol, the calcium pectate becomes changed, so that pectic acid is liberated and calcium chloride is formed. The pectic acid is insoluble in water, but is soluble in a 10 per cent. solution of ammonia, so that after rinsing the sections in water and treating with the ammonia solution, the cells may be separated from each other by a slight pressure on the cover-glass. When the sections are placed for a considerable time in cold alkaline solutions, a double pectate is formed which swells in cold water and finally dissolves in it. After the calcium pectate of the middle lamella has been removed, the pectose which permeates the cell-wall still remains, but by treatment with cuprammonia it may be removed from sections which have already been acted on by dilute hydrochloric acid. The pectic substances may be stained only in neutral or slightly acid solutions. For this reason it is a good plan to place sections for a short time in a 3 per cent. solution of acetic acid, and then to wash them in water before transferring then to the staining solutions. Safranin, methylene-blue, bleu de nuit, and ruthenium-red are excellent stains for pectic substances. Safranin stains the protoplasts and the lignified, suberized, and cutinized cell-membranes a cherry-red, while the pectic compounds are stained orange-yellow. Methylene-blue and bleu de nuit stain the protoplasts and the lignified membranes blue, and the pectic substances a violet color. See also in the last chapter under Ruthenium-red.

Pezizin.—Pezizin is an orange-red coloring matter which occurs in solution within the paraphyses of *Peziza aurania* and *P. convexula*. It is soluble in alcohol and ether, and is not altered by alkalies and organic acids. It dissolves without color in hydrochloric acid and is colored bright green by nitric acid.

Phloridzin, $C_{21}H_{24}O_{10}$.—A glucoside occurring in the leaves and in the cortex of the roots and stems of the Pomaceæ. When tissues of *Pirus malus* containing phloridzin are treated with ferric chloride, a dark brown solution is formed, while treatment with ferrous sulphate causes a yellowish-brown precipitate. The tissues of the pear, cherry, and plum are apt to contain large amounts of tannins which produce a green color with salts of iron, and so mask the phloridzin reaction.

Phloroglucin, $C_6H_3(OH)_3$.—This occurs in solution in the cell-sap. To demonstrate its presence treat previously dried sections with a solution prepared by dissolving 0.005 Gm. of vanillin in 0.5 Gm. of alcohol, and adding 0.5 Gm. of water and 3 Gm. of concentrated hydrochloric acid. When phloroglucin is present, this solution produces a light red color.

Phosphoric Acid, H_3PO_4.—This can be best demonstrated in the ash. The ash is dissolved in hydrochloric acid and the solution is evaporated to dryness; then the residue is treated with ammonium molybdate, which, if phosphoric acid is present, produces a precipitate of ammonium phospho-molybdate, the crystals of which have a greenish-yellow color under the microscope. If the presence of phosphoric acid is to be sought for in fresh tissues, sections should be heated in a drop of ammonium molybdate on the glass slide. This method also produces a precipitate of crystals of ammonium phospho-molybdate in the presence of phosphoric acid. If ammonium tartrate is present in the tissues, ammonium molybdate does not serve so well as a test for phosphoric acid. In such a case a solution should be used, consisting of 25 volumes of a concentrated aqueous solution of magnesium sulphate, 2 volumes of a concentrated aqueous solution of ammonium chloride, and 15 volumes of water. With phosphoric acid this solution produces a precipitate of ammonio-magnesium phosphate the crystals of which are frequently formed in x- and *-shaped clusters.

Phycoerythrin.—The red coloring matter in the Florideæ or red algæ. It is soluble in fresh water, leaving chlorophyll behind in the plastids, while in ether the chlorophyll is extracted and the phycoerythin is left.

Phycocyanin.—The blue coloring matter in the blue-green algæ. It is soluble in cold water, glycerine and alkalies, giving a blue solution with red fluorescence. It is insoluble in alcohol and ether.

Phycophæin.—The brown coloring matter of the brown algæ. It is soluble in fresh water and more readily in hot water, leaving chlorophyll and carotin behind in the plastids. It is insoluble in strong alcohol, ether, etc.

Piperine, $C_{17}H_{19}NO_3$.—Piperine is an alkaloid occurring in the fruit of the Piperaceæ. Very thin sections may be rubbed out somewhat under a cover-glass to press out the ethereal oil, which will then evaporate and leave the piperine to crystallize in the form of minute short needles. A section becomes of a deep red color when treated with concentrated sulphuric acid, while with nitric acid an orange color is produced. When sections are moistened with sodium molybdate, and then treated with concentrated sulphuric acid, they take on a blue color. Piperine is easily soluble in acetic acid.

Proteids (Albuminoid Substances).—Proteids are stained from yellow to brown by a dark solution of potassium iodide-iodine. The dilute solution of iodine recommended for starch should not be used, for proteids are stained less readily than starch. Millon's reagent (see under this head in the preceding chapter) colors proteids a brick-red color. If the solution is old and has lost its efficiency, a few drops of a solution of potassium nitrate will probably restore it. Concentrated nitric acid colors proteids yellow, and the addition of ammonium produces a still deeper yellow. When sections lie for an hour or two in a solution of 1 Gm. of sodium phospho-molybdate in 90 Gm. of distilled water and 5 Gm. of nitric acid, which has been filtered after standing for several days, the proteid substances appear in the form of yellowish granules. A concentrated solution of nickel sulphate colors proteid granules yellow or blue. When rather thin sections are placed in a concentrated solution of copper sulphate for about half an hour, and then are placed in water for about an hour, and then are transferred to a concentrated solution of potassium hydrate, proteids are colored red or violet, which becomes deeper when the solution in which the sections are lying is heated somewhat. The pepsin-glycerine and pancreatin-glycerine ferments prepared by Dr. G. Grübler in Leipzig are solvents of proteids. Sections are treated for an hour at a temperature of 40°C. with a mixture of 1 part of pepsin-glycerine and 3 parts of water, to which is added 0.2 per cent. of chemically pure hydrochloric acid. Pancreatin-glycerine is employed in the same manner as the pepsin-glycerine.

Protein Crystalloids.—Under Aleurone in the preceding chapter are given methods of differentiating crystalloids in aleurone grains. Protein crystalloids also occur in the cytoplasm, nucleus, and chromatophores, and in all of these cases the crystalloids have essentially the same nature, but they may vary considerably in form. For staining crystalloids, see in the preceding chapter under Acid Fuchsin.

Protoplasm.—The protoplasm of the cell can be studied to advantage by means of the microscope only after being killed and fixed by fixative reagents, *q.v.* The different constituents of the protoplasm can then be differentiated by means of stains or by means of digestive ferments, such as pepsin and pancreatin. Iron hæmatoxylin, or a combination of fuchsin and iodine green, or of safranin, gentian violet, and orange G, as recommended under Double Staining in the preceding chapter, are specially to be recommended for differentiating the different parts of the protoplasm. For staining the leucoplasts and chromatophores in general, see Fuchsin, Acid.

Protoplasmic Connections.—The protoplasmic connections between the plates of sieve tubes may be strongly stained by acid fuchsin and aniline water (see page 542). More delicate protoplasmic connections require the use of a swelling agent for their demonstration. Sections of fresh material may be fixed with a solution of 0.05 Gm. of iodine and 0.2 Gm. potassium iodide in 15 Gm. of water, and then the iodine should be replaced by chloroiodide of zinc, which should be allowed to act for about 12 hours. At the end of this time the membranes traversed by the protoplasmic connections will be swollen to greater or less extent, so that the chloroiodide of zinc may be washed out in water and the sections stained by acid fuchsin and aniline water, as already suggested. Sulphuric acid may be used instead of the chloroiodide of zinc as the swelling agent. For demonstration purposes sections through the endosperm of the Gramineæ, or tangential sections through the green bark of *Rhamnus frangula*, may be used. Sections are placed on a cover-glass in a drop of sulphuric acid. After a few seconds the acid is washed away by immersing the cover-glass and moving the sections about in a dish filled with water. The sections remain in the water for only a short time, and are then to be stained in an aqueous solution of aniline blue, washed in water, and mounted for examination in dilute glycerine. Or the sections may be stained in a saturated solution of picric acid in 50 per cent. alcohol, to which aniline blue is added until the solution has a blue-green color.

Pyrenoids.—The pyrenoids may be simultaneously fixed and stained by placing the material in a concentrated solution of picric acid in 55 per cent. alcohol, to a watch-glass of which is added about 5 drops of the acid fuchsin and aniline water solution described on page 542. The material should remain for about 2 hours in a watch-glass of this solution. It should then be washed for a quarter of an hour in alcohol and mounted for examination in dilute glycerine. If permanent mounts are desired, the material should be placed in a watch-glass of dilute glycerine, which should then be allowed to concentrate in a place free from dust. The material should finally be mounted in glycerine-jelly. The material may be mounted in Canada balsam by transferring it from the alcohol in which it was washed to successively stronger solutions of balsam in xylol until the ordinary solution used for mounting is reached.

See under Dahlia in the previous chapter for other methods of treating pyrenoids.

Reserve Cellulose.—Those hemicellulose thickenings of cell-walls in seeds, etc., which are essentially reserve food materials, and are made soluble by diastatic ferments and employed as food material in the germination of seeds, are known as reserve cellulose. Sections taken from a sprouted date seed and treated with potassium hydrate and stained with alizarine show the inner layers of the cell-walls which have been acted on by the diastase unstained, while the outer layers which have not yet been affected by the diastase are stained an intense violet. If Congo-red is used instead of the alizarine, the intact layers are scarcely stained, while the layers which have come under the influence of the diastase are stained a dark red. See under Hemicellulose.

Resin.—When sections containing resin are treated for some time with a tincture of alcannin, the resin assumes a red color. When sections from tissues which have lain for about a week in a concentrated aqueous solution of copper acetate are examined under the microscope, the resin will be seen to be colored an emerald-green.

Ruberythric Acid, $C_{26}H_{28}O_{14}$.—This occurs in the roots of *Rubia tinctorium*, and is the chief constituent of the madder dye obtained from the roots of this plant. It gives a yellow color to the cell-sap of the young roots; the cell-walls of old roots, however, have absorbed it and are colored by it. It is colored a purple-red by potassium hydrate, and an orange color by acids. In dry roots it takes on the form of red flakes, and in the injured cells of fresh material it assumes the same form. It may be extracted by alcohol from its yellow solution in uninjured tissues, but in the red flake form it is not dissolved by alcohol.

Rutin, $C_{42}H_{50}O_{25}$.—This glucoside is widely distributed in plants. It crystal-lizes from an aqueous solution in the form of minute light yellow crystals. The yellow color of straw is, in part, due to it. When treated with ammonia or lime-water, rutin forms a deep yellow solution, which turns to brown on exposure to the air.

Salicin, $C_{13}H_{18}O_7$.—Salicin is a glucoside which occurs in particular abundance in the cortex of many poplars and willows. It may be dissolved by water, but more readily by boiling water, by aqueous solutions of alkalies, and by acetic acid. It is insoluble in ether. It crystallizes in the form of needles, scales, or thin plates. It is colored by concentrated sulphuric acid, and, on the addition of a little water, a red powder is thrown down in the sulphuric acid solution.

Santalin.—Santalin is the coloring matter of the red sandal-wood, *Pterocarpus santalinus*. Santalin is insoluble in water, but is soluble in ether with a yellow color and with 80 per cent. alcohol it gives a blood-red solution. Stronger alcohols give the same result. It is also soluble in acetic acid and in aqueous alkaline solutions.

Saponin, $C_{19}H_{30}O_{10}$.—This glucoside occurs in solution in the cell-sap. When treated with a mixture of equal parts of alcohol and sulphuric acid, a yellow color is produced which soon changes to red, and later to violet. If it is then treated with a concentrated solution of chloride of iron, a brown or bluish-brown precipi-tate is formed, the intensity of the bluish color increasing with the amount of saponin present.

Seminose.—Seminose is one of the products resulting from the hydrolysis of hemicellulose by sulphuric acid. It is dextrorotary, reduces Fehling's solution, and is fermentable.

Silica, SiO_2.—Silica occurs in the skeletons of diatoms, and as incrustations over the epidermis of the Equisetaceæ and Gramineæ. It also sometimes occurs in masses in the interior cells. It may be isolated from the organic matter with which it is associated by burning over a flame bits of epidermis incrusted by it, or diatoms, which are placed in a drop of concentrated sulphuric acid on a piece of platinum foil. By this treatment the organic matter will be destroyed, and the silica will remain behind as a pure white ash. The silica may also be obtained pure by plac-ing bits of tissues incrusted by it in a drop of concentrated sulphuric acid, and then after a time adding 20 per cent. chromic acid, and following this with additions of still stronger chromic acid until a considerable strength has been reached, and, finally, washing in water and alcohol. Silica is distinguished by being insoluble in all the acids excepting hydrofluoric acid. Silicious skeletons may be removed from diatoms by placing the latter in hydrofluoric acid which is contained in a platinum vessel. The vessel should be kept on a water-bath, and the diatoms should remain in the acid for 24 hours. At the end of this time the acid should be thoroughly washed out from the diatoms. On examination with the microscope, the diatoms will then be found to have lost their silicious skeletons. In some instances a thin exterior membrane which is stained brown by iodine is to be ob-served; but in other instances this membrane has been a too insignificant part of the skeleton to retain its identity after the removal of the silica.

Sinapine, $C_{16}H_{22}NO_5$.—This is an alkaloid occurring in the seeds of the white mustard. When sections of these seeds are placed in a concentrated solution of potassium hydrate, they assume a yellow color, which changes to orange on warming. This reaction loses some of its value, however, from the fact that a glucoside called sinalbine also occurs in the seeds of the white mustard and turns yellow on the application of potassium hydrate.

Solanin, $C_{42}H_{75}NO_{15}$.—This glucoside occurs in the tissues of *Solanum tuberosum*. To demonstrate its presence, sections should be placed in a mixture of 1 part of ammonium vanadate and 1,000 parts of a mixture of 98 parts of sulphuric acid with 36 parts of water. This produces with solanin a yellow color, which changes successively into orange, purple-red, brown, red-orange, carmine-red, raspberry-red, and blue-violet. The color then passes into a grayish-blue and disappears. With concentrated sulphuric acid solanin assumes a yellow color, which changes to red, and then violet, and then passes into gray and disappears.

Spergulin, $C_6H_7O_2$.—This occurs in the seed coats of species of Spergula. It is soluble in alcohol with a blue fluorescence, in ether, and in concentrated sulphuric acid with a deep blue color. It is insoluble in chloroform, benzine, and in the fatty and ethereal oils. When caustic potash is added to an alcoholic solution of spergulin, an emerald-green fluorescence is produced.

Starch, $C_6H_{10}O_6$.—A solution of potassium iodide stains starch from pale violet to purple, depending on the strength of the iodine solution. Chloroiodide of zinc stains starch-grains purple, and at the same time swells them. A solution of chloral hydrate and iodine dissolves the protoplasmic cell-contents and stains included starch-grains purple. This reagent is particularly adapted to demonstrate the presence of starch in chloroplasts or amyloplasts. The bleaching effect of the chloral hydrate is so great that starch may be demonstrated in whole leaves by the chloral hydrate and iodine reagent. For the further treatment of starch with reagents, see in the preceding chapter under Eau de Javelle, Calcium Nitrate, Diastase, Methyl-violet, Silver Nitrate.

Strychnine, $C_{21}H_{22}N_2O_2$.—When sections containing strychnine are treated with a solution of 1 Gm. of ammonium vanadate in 100 mils of sulphuric acid, they quickly take on a violet-red color, which after a time changes to brown. If sections of the seeds of *Strychnos nux-vomica* are treated with sulphuric acid containing an excess of ceric sulphate, the walls of the cells are colored a bluish-violet. The sections must have been previously treated with petroleum ether and absolute alcohol to remove the fatty oils, grape-sugar, and brucine. The reagent should be applied immediately before the observation is to be made. If sections are treated with concentrated sulphuric acid, and crystals of potassium bichromate are then added, a violet color is produced.

Suberin and Suberized Walls.—Suberized walls are stained green when treated for about an hour in the dark by a freshly-prepared strong solution of chlorophyll. A cold concentrated solution of potassium hydrate colors suberized walls yellow. When the potassium hydrate is heated yellow drops and granular masses are formed. When suberized walls are heated in a solution of potassium chlorate in nitric acid, they become changed into droplets which melt between 30° and 40° C., and which are soluble in hot chloroform, alcohol, ether, benzol, or dilute potassium hydrate. At ordinary temperatures concentrated chromic acid solutions have little effect on suberized walls. A solution of potassium iodide-iodine and chloroiodide of zinc colors suberized membranes from yellow to brown. After long treatment with a dilute solution of potassium hydrate, suberized membranes may be stained violet with chloroiodide of zinc. Alcannin stains suberized walls red. Under the polariscope suberized walls are seen to be doubly refractive. They lose this property on heating and regain it on cooling. It may be deduced from this that the constituents of the walls are in part, at least, in crystals which are melted by heat, but reappear on cooling. See under Methyl-blue and Double Staining.

37

Syringin.—Syringin is a glucoside occurring in the cortex, and to a certain extent in the xylem and medullary rays of *Syringa vulgaris*. It is especially abundant in early spring. Sections containing syringin, when treated with concentrated sulphuric acid, acquire a dark blue color, which changes to violet. Nitric acid dissolves syringin with a blood-red color. Syringin crystallizes from aqueous solutions in the form of colorless, needle-like crystals which are grouped in the form of a star. The crystals are dissolved with difficulty in cold water, but readily in boiling water or in alcohol.

Tannins.—Various substances occurring in plants having an astringent taste, and turning blue or green with salts of iron, are termed tannins or tannic acid. Oak-galls furnish excellent material to illustrate the demonstration of tannin. When sections are treated with an aqueous solution of ferric chloride, they take on a deep blue color, due to the presence of tannin. Aqueous solutions of ferrous sulphate give the same result. If the reaction is watched under the microscope, it is noticed that at first a deep blue precipitate is formed, which soon dissolves and imparts its color to the surrounding fluid. When sections are placed in a 10 per cent. aqueous solution of potassium bichromate, a flocculent reddish-brown precipitate is formed in the tannin-bearing cells. When sections are placed in a concentrated solution of ammonium molybdate in concentrated ammonium chloride, the same character of precipitate is produced as when potassium bichromate is used. Lead acetate produces a white precipitate with tannins. The following method may be employed; Sections are placed in a 7 per cent. solution of copper acetate for a week or longer, and are then placed on a slide in a drop of a 0.5 per cent. solution of ferrous sulphate. After a few minutes, and before the cell-walls begin to turn brown, the sections are washed in water and transferred to a watch-glass of alcohol to drive out air-bubbles and extract chlorophyll, if any is present. The sections are then mounted in glycerine for examination under the microscope. By this treatment an insoluble brown precipitate is produced in the presence of tannins. The sections may be transferred from the glycerine to glycerine-jelly if permanent mounts are desired. If the sections are taken from the alcohol in which they were placed to remove the chlorophyll, etc., and placed in a solution of iron acetate, a blue or a green color will be produced, according to the kind of tannin present. If it is desired to fix the cell-contents while testing for tannins, the sections should be placed in a concentrated alcoholic solution of iron acetate instead of in the aqueous solution, as above. When living tissues are placed in a solution of 1 part of methylene-blue in 500 parts of distilled water, those cells which contain tannins take on a blue color, and later a deep blue precipitate is formed in these cells. Cells containing phloroglucin act in the same way to this reagent as those containing tannins.

Theobromine, Dimethyl-xanthin, $C_7H_8N_4O_2$.—This alkaloid occurs in the cocoa-bean. Its presence may be demonstrated by the use of hydrochloric acid and chloride of gold, as directed under Caffeine. The reactions for caffeine and theobromine are sometimes difficult to distinguish. When sections containing theobromine are heated in distilled water on the slide to the boiling-point, and the sections are allowed to dry slowly, and a drop of benzol is added to the residue, crystals of theobromine appear in the form of a fine powder on the evaporation of the benzol; whereas, when sections containing caffeine are treated in like manner, the crystals containing caffeine take on the form of needles.

Tyrosin, $C_6H_4OH.CH_2.CHNH_2.COOH$.—Tyrosin may be demonstrated in abundance in the tubers of the dahlia. When sections are mounted under a cover-glass in glycerine for several days, needle-shaped crystals of tyrosin are deposited in radiating groups. In an abundance of glycerine the crystals are not deposited, for the reason that the tyrosin becomes too much diffused through the glycerine. The crystals appear brownish by transmitted, and white by reflected light. When a portion of a dahlia tuber is placed in a dish of about the same size as itself, and covered for about two-thirds of its length with alcohol, an abundance of tyrosin crystals will collect at the exposed cut surface. The crystals of tyrosin are colored a deep red by means of Millon's reagent, and when nitric acid is poured over them, and then evaporated, a yellow residue is left. ·

Vanillin, $C_6H_3.OH.OCH_3.CHO.$—Vanillin occurs abundantly in solution in the pods of vanilla, or in the dry pods it occurs in an amorphous condition. It is often found in a crystalline condition on the surface of dried pods. It is soluble in alcohol and ether, and to a certain extent in hot water, but it is soluble with difficulty in cold water. When sections containing vanillin are wetted with a 4 per cent. solution of orcin, and then treated with concentrated sulphuric acid, they take on a deep carmine-red color; and when they are treated in like manner with a solution of phloroglucin in place of the orcin, a brick-red color is produced. It seems probable that vanillin is always present in lignified walls, judging from the colors which these assume with certain aromatic compounds.

Veratrine, $C_{37}H_{53}NO_{11}.$—Veratrine occurs in the tissues of *Veratrum album.* When sections are placed in a mixture of 1 drop of concentrated sulphuric acid and 2 drops of water on a glass slide, and examined under a microscope, it is to be seen that the walls or cell-contents containing veratrine assume a yellow color, which soon changes to an orange-red, and finally to a dusky violet.

Wax.—Wax frequently occurs in plants as a crust-like, or granular, or rod-like layer over the cuticle. It consists of fats and free fatty acids, together with other substances. Wax is insoluble in water, but it will melt and form droplets in water at 100°C. It is hardly soluble in cold alcohol, but will quickly dissolve in boiling alcohol. When sections containing wax are heated in a solution of alcannin in 50 per cent. alcohol, the wax runs together in droplets, which become stained red by the alcannin. Wax is not wetted by water, and sections are best mounted for study in cold alcohol, which will dissolve the wax but little, if at all.

Wound Gum.—The wounded surfaces of deciduous trees become protected by the formation of wound gum from starch contained in the live cells. Sections taken through the wounded surfaces of such plants several days after the wound has been inflicted show brownish granules of wound gum in the medullary rays, tracheal tubes, and wood-cells. The wound gum may be found lying free in the cytoplasm or surrounding starch-grains which have contributed to the formation of the gum. Wound gum is not soluble in warm water, but may be dissolved in hot nitric acid or in eau de Javelle after several hours. It is not soluble in sulphuric acid, potassium hydrate, alcohol, or ether, but it may be dissolved in alcohol after treatment for a few minutes with a solution of potassium chlorate in dilute hydrochloric acid. It may be stained with a solution of fuchsin, iodine green, safranin, or methyl-green. It is stained red by phloroglucin and hydrochloric acid.

Xanthin, $C_5H_4N_4O_2.$—Xanthin occurs in an amorphous condition or in the form of granules in yellow chromoplasts. It differs from carotin in being more soluble in alcohol, and in being deposited in amorphous and resin-like masses on the evaporation of its solvent. It is insoluble in water, and but little soluble in ether and benzine. It becomes green and then blue when treated with sulphuric acid, and with potassium iodide-iodine it is colored green.

INDEX

Made in the USA
Monee, IL
07 July 2026

56550020R00364